Edited by
Roger J. Mortimer, David R. Rosseinsky, and
Paul M. S. Monk

Electrochromic Materials and Devices

Related Titles

Ogale, S.B., Venkatesan, T.V., Blamire, M. (eds.)

Functional Metal Oxides

New Science and Novel Applications

2013
Print ISBN: 978-3-527-33179-6; also available in electronic formats

Tiwari, A., Mishra, A., Mishra, Y.K., Kobayashi, H., Turner, A.P.

Intelligent Nanomaterials

2012
Print ISBN: 978-0-470-93879-9; also available in electronic formats

Pacchioni, G., Valeri, S. (eds.)

Oxide Ultrathin Films

Science and Technology

2012
Print ISBN: 978-3-527-33016-4; also available in electronic formats

Decher, G., Schlenoff, J.B. (eds.)

Multilayer Thin Films

Sequential Assembly of Nanocomposite Materials
Second Edition

2012
Print ISBN: 978-3-527-31648-9; also available in electronic formats

Knoll, W., Advincula, R.C. (eds.)

Functional Polymer Films

2 Volume Set

2011
Print ISBN: 978-3-527-32190-2; also available in electronic formats

Urban, M.W. (ed.)

Handbook of Stimuli-Responsive Materials

2011
Print ISBN: 978-3-527-32700-3; also available in electronic formats

Ariga, K. (ed.)

Organized Organic Ultrathin Films

Fundamentals and Applications

2012
Print ISBN: 978-3-527-32733-1; also available in electronic formats

Edited by
Roger J. Mortimer, David R. Rosseinsky, and Paul M. S. Monk

Electrochromic Materials and Devices

WILEY-VCH
Verlag GmbH & Co. KGaA

The Editors

Prof. Dr. Roger J. Mortimer †
Loughborough University
Department of Chemistry
Loughborough
LE11 3TU Leicestershire
UK

Dr. David R. Rosseinsky
University of Exeter
School of Physics
EX4 4QL Exeter
UK

Rev Dr. Paul M. S. Monk
St. Barnabas' Vicarage
1 Arundel Street
OL4 1NL Clarksfield, Oldham
UK

Cover
Sprayed films of electrochromic polymers developed at the University of Florida and Georgia Institute of Technology with the John Reynolds Research Group. Artistic concept and photography by Aubrey Dyer, Keith Johnson, and Justin Kerszulis.

Library of Congress Card No.: applied for

British Library Cataloguing-in-Publication Data
A catalogue record for this book is available from the British Library.

Bibliographic information published by the Deutsche Nationalbibliothek
The Deutsche Nationalbibliothek lists this publication in the Deutsche Nationalbibliografie; detailed bibliographic data are available on the Internet at <http://dnb.d-nb.de>.

Print ISBN: 978-3-527-33610-4
ePDF ISBN: 978-3-527-67988-1
ePub ISBN: 978-3-527-67987-4
Mobi ISBN: 978-3-527-67986-7
oBook ISBN: 978-3-527-67985-0

Typesetting SPi Global, Chennai, India
Printing and Binding Printed and bound by CPI Group (UK) Ltd, Croydon, CR0 4YY

Printed on acid-free paper

C9783527336104_180624

In memoriam

1956 – 2015

We record with deep sadness the untimely death of our co-editor Roger J. Mortimer,
a fine academic colleague and friend, and initiator of this book.
We dedicate this book to his memory.

Contents

Preface

Electrochromic materials have the property of a change, evocation or bleaching, of colour as effected either by an electron-transfer (redox) process or by a sufficient electrochemical potential. Although materials are usually said to be electrochromic when light is modulated by reflectance or absorbance in the visible region of the electromagnetic spectrum – colour changes perceptible to the human eye – interest in electrochromic devices (ECDs) for multispectral energy modulation, to include the infrared and microwave regions, has extended the working definition [1].

While the topic of electrochromism has a history dating back to the nineteenth century, only in the last quarter of the twentieth has its study gained a real impetus. So, applications have hitherto been limited, apart from the astonishing success of the 250 million Gentex anti-dazzle car mirrors that have been sold since 1987 and the adjustable-darkening windows of the Boeing Dreamliner aircraft. The ultimate goal of contemporary studies is the provision of large-scale electrochromic 'smart' windows/glazing for buildings at modest expenditure which, applied widely in the United States, would save billions of dollars in air-conditioning costs. In tropical and equatorial climes, savings would be proportionally greater: Singapore, for example spends one-quarter of its Gross Domestic Product (GDP) on air conditioning, a *sine qua non* for tolerable living conditions there. Importantly, note that only weak anti-thermal protection is provided by colour alone, the electrochrome itself getting heated. Hence, transparent metal oxides that can be electrochemically or otherwise reduced to form shiny metallic reflectors are of fundamental importance, though this process is widely under-emphasised. Numerous other applications have been contemplated, and for some, prototype devices have been developed. Applications include electrochromic strips as battery state-of-charge indicators, electrochromic sunglasses, reusable price labels, protective eyewear, controllable aircraft canopies, glare-reduction systems for offices, devices for frozen-food monitoring, camouflage materials, chameleonic fabrics, spacecraft thermal control, an optical iris for a camera lens and (non-emissive) controllable light-reflective or light-transmissive display devices for optical information and storage.

This edited book follows our earlier research monographs [2, 3] now with invited contributions from the main experts across the globe. Part One concerns electrochromic materials and processing and covers metal oxides, Prussian blue, viologens, conjugated conducting polymers, transition metal coordination complexes and polymers, organic near-infrared materials and metal hydrides. Part Two concerns nanostructured electrochromic materials and device fabrication and covers nanostructures in electrochromic materials, advances in polymer electrolytes for ECD applications, gyroid-structured electrodes for electrochromic (and supercapacitor) applications, layer-by-layer assembly of electrochromic materials and plasmonic electrochromism of metal oxide nanocrystals. Part Three describes the applications of electrochromic materials and covers solution-phase ECDs and systems, electrochromic smart windows and fabric electrochromic displays. Part Four covers device case studies, environmental impact issues and elaborations and includes a case study of an electrochromic foil, life cycle analysis (LCA) of electrochromic windows, a case study of the installation, operation, monitoring and user experience of electrochromic glazing in a UK office and photoelectrochromic devices. The book closes with an Appendix, where electrochromic materials and device performance parameters are defined, to include some cautions about their comparisons between different research laboratories.

References

1. Rauh, R.D. (1999) Electrochromic windows: an overview. *Electrochim. Acta*, **44**, 3165–3176.
2. Monk, P.M.S., Mortimer, R.J., and Rosseinsky, D.R. (1995) *Electrochromism – Fundamentals and Applications*, Wiley-VCH Verlag GmbH, Weinheim.
3. Monk, P.M.S., Mortimer, R.J., and Rosseinsky, D.R. (2007) *Electrochromism and Electrochromic Devices*, Cambridge University Press, Cambridge.

Acknowledgements

We are indebted to numerous colleagues and correspondents with whom we have collaborated over many years in electrochromics research, in particular Professor Hassan Kellawi of Damascus University, who introduced DRR to Electrochromism, which then led to RJM and PMSM becoming interested in this research field. Special thanks are due to Professor Paulo R. Bueno of São Paulo State University (UNESP), whose papers reviewed in Chapter 3 helped clarify our general summary of that research area.

From the staff of Wiley-VCH, we wish to thank Dr Bente Flier (Associate Commissioning Editor), who first commissioned the book, Dr Martin Preuss (Senior Commissioning Editor), Lesley Belfit (Project Editor, STMS Books) and the production team at Wiley-VCH, for their professionalism, care and help.

Professor Roger J. Mortimer
Loughborough, UK
Dr. David R. Rosseinsky
Exeter, UK
Rev. Dr. Paul M. S. Monk
Oldham, UK

List of Contributors

Harlan J. Byker
Pleotint LLC
6722 18th Avenue
Jenison, MI 49428
USA

Hsin-Wei Chen
National Taiwan University
Department of Chemical Engineering
No. 1, Sec. 4, Roosevelt Road
10617 Taipei
Taiwan

Susana I. Córdoba de Torresi
Universidade de São Paulo
Instituto de Química
C.P. 26077
05513-970 São Paulo, SP
Brazil

Aubrey L. Dyer
Clayton State University
Department of Natural Sciences
College of Arts and Sciences
2000 Clayton State Blvd.
Morrow, GA 30260
USA

and

Center for Organic Photonics and Electronics
Georgia Institute of Technology
School of Chemistry and Biochemistry
School of Materials Science and Engineering
901 Atlantic Dr. NW
Atlanta, GA 30332
USA

Alice Lee-Sie Eh
Nanyang Technological University
School of Materials Science and Engineering
50 Nanyang Avenue
Singapore 639798
Singapore

Claes-Göran Granqvist
Uppsala University
Department of Engineering
Sciences
The Ångström Laboratory
PO Box 534, SE-75121
Uppsala
Sweden

Matthias Harsch
LCS Life Cycle Simulation GmbH
Aspacher Strasse 9
D-71522 Backnang
Germany

Kuo-Chuan Ho
National Taiwan University
Department of Chemical
Engineering
No. 1, Sec. 4, Roosevelt Road
10617 Taipei
Taiwan

Chih-Yu Hsu
Electronic Functional Materials
Group, Polymer Materials Unit
National Institute for Materials
Science (NIMS)
1-1 Namiki
305-0044 Tsukuba
Japan

Chih-Wei Hu
National Taiwan University
Department of Chemical
Engineering
No. 1, Sec. 4, Roosevelt Road
10617 Taipei
Taiwan

and

National Institute of Advanced
Industrial Science and
Technology (AIST)
Anagahora 2266-98
Shimoshidami, Moriyama-ku
Nagoya 463-8560
Japan

Fritz Huguenin
Universidade de São Paulo
Faculdade De Filosofia
Ciências e Letras de Ribeirão
Preto
Av. Bandeirantes 3900
14040-901 Ribeirão Preto, SP
Brazil

Michael A. Invernale
University of Connecticut
Department of Chemistry and
the Polymer Program
97 North Eagleville Road U-2136
Storrs, CT 06269-3136
USA

Bjørn Petter Jelle
SINTEF Building and Infrastructure
Department of Materials and Structures
Høgskoleringen 7A
NO-7465 Trondheim
Norway

and

Norwegian University of Science and Technology (NTNU)
Department of Civil and Transport Engineering
Høgskoleringen 7B
NO-7491 Trondheim
Norway

Keith E. Johnson
Center for Organic Photonics and Electronics
Georgia Institute of Technology
School of Chemistry and Biochemistry
School of Materials Science and Engineering
901 Atlantic Dr. NW
Atlanta, GA 30332
USA

Pooi See Lee
Nanyang Technological University
College of Engineering
School of Materials Science and Engineering
50 Nanyang Avenue
639798
Singapore

Anna Llordes
The University of Texas at Austin
McKetta Department of Chemical Engineering
Austin, TX 78712
USA

and

The Molecular Foundry
Lawrence Berkeley National Laboratory
Berkeley, CA 94720
USA

Sebastien D. Lounis
The University of Texas at Austin
McKetta Department of Chemical Engineering
Austin, TX 78712
USA

and

The Molecular Foundry
Lawrence Berkeley National Laboratory
Berkeley, CA 94720
USA

and

The University of California
Graduate Group in Applied Science & Technology
Berkeley, CA 94720
USA

Xuehong Lu
Nanyang Technological University
College of Engineering
School of Materials Science and Engineering
50 Nanyang Avenue
Singapore 639798
Singapore

John Mardaljevic
Loughborough University
School of Civil and Building Engineering
Loughborough
Leicestershire LE11 3TU
UK

Jose R. Martins Neto
Universidade de São Paulo
Instituto de Química
C.P. 26077
05513-970 São Paulo, SP
Brazil

Delia J. Milliron
The University of Texas at Austin
McKetta Department of Chemical Engineering
Austin, TX 78712
USA

Paul M. S. Monk
St. Barnabas' Vicarage
1 Arundel Street
Clarksfield
Oldham OL4 1NL
UK

Roger J. Mortimer[†]
Loughborough University
Department of Chemistry
Loughborough
Leicestershire LE11 3TU
UK

Anna M. Österholm
Center for Organic Photonics and Electronics
Georgia Institute of Technology
School of Chemistry and Biochemistry
School of Materials Science and Engineering
901 Atlantic Dr. NW
Atlanta, GA 30332
USA

Michael T. Otley
University of Connecticut
Department of Chemistry and the Polymer Program
97 North Eagleville Road U-2136
Storrs, CT 06269-3136
USA

Birgit Painter
De Montfort University
Institute of Energy and Sustainable Development
The Gateway, Leicester
LE1 9BH
UK

Uwe Posset
Center for Applied Electrochemistry
Fraunhofer-Institut für Silicatforschung ISC
Neunerplatz 2
D-97082 Würzburg
Germany

John R. Reynolds
Center for Organic Photonics and Electronics
Georgia Institute of Technology
School of Chemistry and Biochemistry
School of Materials Science and Engineering
901 Atlantic Dr. NW
Atlanta, GA 30332
USA

David R. Rosseinsky
University of Exeter
School of Physics
Exeter EX4 4QL
UK

Evan L. Runnerstrom
The University of Texas at Austin
McKetta Department of Chemical Engineering
Austin, TX 78712
USA

and

The Molecular Foundry
Lawrence Berkeley National Laboratory
Berkeley, CA 94720
USA

and

Department of Materials Science and Engineering
Berkeley, CA 94720
USA

Maik R.J. Scherer
Adolphe Merkle Institute
Chemin des Verdiers
1700 Fribourg
Switzerland

D. Eric Shen
Center for Organic Photonics and Electronics
Georgia Institute of Technology
School of Chemistry and Biochemistry
School of Materials Science and Engineering
901 Atlantic Dr. NW
Atlanta, GA 30332
USA

Ullrich Steiner
Adolphe Merkle Institute
Chemin des Verdiers
1700 Fribourg
Switzerland

Gregory A. Sotzing
University of Connecticut
Department of Chemistry and the Polymer Program
97 North Eagleville Road U-2136
Storrs, CT 06269-3136
USA

Thomas S. Varley
University College London
Department of Chemistry
Gower Street
WC1H 0AJ London
UK

Marcio Vidotti
Universidade Federal de Paraná
Departamento de Química
CP 19081
81531-990 Curitiba
Brazil

Xinhua Wan
Peking University
Beijing National Laboratory for Molecular Sciences
Key Laboratory of Polymer Chemistry and Physics of Ministry of Education
College of Chemistry and Molecular Engineering
No.5 Summer Palace Road, Haidian District
Beijing 100871
China

Ruth Kelly Waskett
Institute of Energy and Sustainable Development
De Montfort University
Leicester
The Gateway LE1 9BH
UK

Shanxin Xiong
Xi'an University of Science and Technology
College of Chemistry and Chemical Engineering
58 Yanta Road
710054 Xi'an
P R China

Bin Yao
Peking University
Beijing National Laboratory for Molecular Sciences
Key Laboratory of Polymer Chemistry and Physics of Ministry of Education
College of Chemistry and Molecular Engineering
No.5 Summer Palace Road, Haidian District
Beijing 100871
China

Kazuki Yoshimura
Structural Materials Research Institute
National Institute of Advanced Industrial Science and Technology (AIST)
Anagahora, Shimoshidami, Moriyama-ku
Nagoya
Aichi 463-8560
Japan

Jie Zhang
Peking University
Beijing National Laboratory for Molecular Sciences
Key Laboratory of Polymer Chemistry and Physics of Ministry of Education
College of Chemistry and Molecular Engineering
No.5 Summer Palace Road, Haidian District
100871 Beijing
China

Yu-Wu Zhong
Beijing National Laboratory for Molecular Sciences
CAS Key Laboratory of Photochemistry
Chinese Academy of Sciences, Institute of Chemistry
2 Bei Yi Jie, Zhong Guan Cun
Beijing 100190
China

Part I
Electrochromic Materials and Processing

Electrochromic Materials and Devices, First Edition.
Edited by Roger J. Mortimer, David R. Rosseinsky and Paul M. S. Monk.

1
Electrochromic Metal Oxides: An Introduction to Materials and Devices

Claes-Göran Granqvist

1.1
Introduction

Electrochromic materials are able to change their properties under the action of an electrical voltage or current. They can be integrated in devices that modulate their transmittance, reflectance, absorptance or emittance. Electrochromism is known to exist in many types of materials. This chapter considers electrochromic metal oxides and devices based on these.

Figure 1.1 shows a generic electrochromic device comprising five superimposed layers on a single transparent substrate or positioned between two transparent substrates [1]. Its variable optical transmittance ensues from the electrochromic films, which change their optical absorption when ions are inserted or extracted via a centrally positioned electrolyte. The ion transport is easiest for small ions, and protons (H^+) or lithium ions (Li^+) are used in most electrochromic devices. Transparent liquid electrolytes as well as ion-containing thin oxide films were employed in early studies on electrochromics [2], but polymer electrolytes became of interest subsequently [3, 4] and paralleled the developments in electrical battery technology.

The ions are moved in the electrochromic device when an electrical field is applied between two transparent electrical conductors, as illustrated in Figure 1.1. The required voltage is only of the order of 1 V DC, so powering is, in general, easy and can be achieved by photovoltaics [5]. In small devices, the voltage can be applied directly to the transparent conductors but large devices – such as 'smart' windows for buildings – require 'bus bars', that is, a metallic frame partly or fully around the circumference of the transparent conducting thin film in order to achieve a uniform current distribution and thereby sufficiently fast and uniform colouring and bleaching. The transparent substrates are often of flat glass, but polymers such as polyethylene terephthalate (PET) or polycarbonate can also be used. The permeation of gas and humidity through foils may or may not be an issue for devices; barrier layers can be applied if needed [6].

An electrochromic device contains three principally different kinds of layered materials: The electrolyte is a pure *ion conductor* and separates the two

Electrochromic Materials and Devices, First Edition.
Edited by Roger J. Mortimer, David R. Rosseinsky and Paul M. S. Monk.

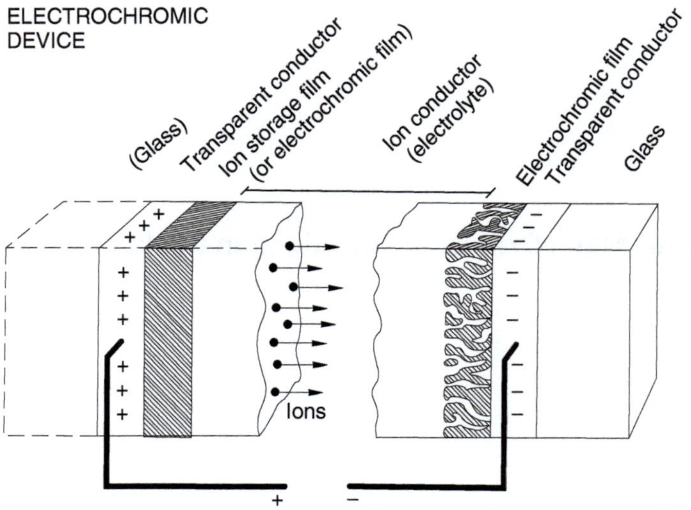

Figure 1.1 Generic five-layer electrochromic device design. Arrows indicate movement of ions in an applied electric field. From Ref. [1].

electrochromic films (or separates one electrochromic film from an optically passive ion storage film). The electrochromic films conduct both ions and electrons and hence belong to the class of *mixed conductors.* The transparent conductors, finally, are pure *electron conductors.* Optical absorption occurs when electrons move into the electrochromic film(s) from the transparent conductors along with charge-balancing ions entering from the electrolyte. This very simplified explanation of the operating principles for an electrochromic device emphasises that it can be described as an 'electrical thin-film battery' with a charging state that translates to a degree of optical absorption. This analogy has been pointed out a number of times but has only rarely been taken full advantage of for electrochromics.

Electrochromic devices have a number of characteristic properties that are of much interest for applications. Thus, they exhibit *open circuit memory,* just as electrical batteries do, and can maintain their optical properties and electrical charge for extended periods of time without drawing energy (depending on the quality of electrical insulation of the electrolyte). The *optical absorption can be tuned* and set at any level between states with minimum and maximum absorption. The *optical changes are slow* and have typical time constants from seconds to tens of minutes, depending on physical dimensions, which means that the optical changes can occur on a timescale comparable with the eyes' ability to light-adapt. Furthermore, the optical properties are based on processes on an atomic scale, so electrochromic windows can be *without visible haze*; this latter property has been documented in detailed spectrally resolved measurements of scattered light [7]. By combining two different electrochromic films in one device, one can adjust the optical transmittance and reach better *colour neutrality* than with a single electrochromic film. Finally, the *electrolyte can be functionalised*, provided it is a solid and adhesive bulk-like polymer, so that the smart window combines its optical performance with spall shielding, burglar protection, acoustic damping, near-infrared damping and perhaps even more features.

This chapter is organised as follows: Section 1.2 serves as a background and gives some notes on early work on electrochromic materials and devices. Section 1.3 provides an in-depth discussion of EC materials and covers optical and electronic effects and, specifically, charge transfer in tungsten oxide. It also treats ionic effects with foci on the inherent nanoporosity in electrochromic oxides and on possibilities to augment the porosity by choosing appropriate thin-film deposition parameters. A number of concrete examples on the importance of the deposition conditions are reported, and Section 1.3 ends with a discussion of the electrochromic properties of tungsten–nickel oxide films across the full compositional range. Section 1.4 surveys properties of transparent conducting electrode materials as well as transparent electrolytes. Section 1.5 gives a background to electrochromic devices, specifically delineating a number of hurdles for practical device manufacturing as well as principles for some large-area devices. Conclusions are given in Section 1.6.

1.2 Some Notes on History and Early Applications

Electrochromism in thin films of metal oxides seems to have been discovered several times through independent work. A vivid description of electrically induced colour changes in thin films of tungsten oxide immersed in sulfuric acid was given in an internal document at the Balzers AG in Liechtenstein in 1953 (cited in a book on inorganic electrochromic materials [1]). Later work on W oxide films by Deb at the American Cyanamid Corporation during the 1960s yielded analogous results, which were reported in two seminal papers in 1969 and 1973 [8, 9]; these publications are widely seen as the starting point for research and development of electrochromic devices. Deb's early work was discussed in some detail much later [2, 10]. Another very important electrochromic oxide is Ni oxide, whose usefulness became clear in the mid-1980s [11, 12]. Parallel developments took place in the Soviet Union, and a paper from 1974 [13] quotes 'USSR Author's Certificates' and patents by Malyuk *et al.* dating back to 1963; this work dealt with Nb oxide films.

Early research on electrochromic materials and devices in the United States, Soviet Union, Japan and Europe was motivated by potential applications in *information displays*, and there were strong research efforts during the first half of the 1970s at several large companies. Generally speaking, these efforts became of less relevance towards the end of the 1970s, and liquid-crystal-based constructions then started to dominate the market for small displays. Electrochromic-based variable-transmission glass was of interest in the context of cathode ray tubes for some time [14]. At present, there is a strong resurgence in the interest in electrochromic-based display-oriented devices (such as 'electronic paper'), and much research and development is devoted to organics-based full-colour electrochromic displays with excellent viewing properties, cheap printable electrochromic 'labels' and, very recently, to 'active' authentication devices [15].

Electrochromic-based *rear-view mirrors* for cars and trucks provide another applications area, and it appears that research and development goes back to the late 1970s [16]. However, the market for 'active' rear-view mirrors was largely taken by a device design that is not identical to the one discussed here [17].

Oxide-based electrochromics came into the limelight during the first half of the 1980s, when it became widely accepted that this technology can be of great significance for *energy-efficient fenestration* [18, 19]. The term 'smart' window (alternatively 'intelligent' or 'switching' window) was coined in 1984/1985 [20, 21] and got immediate attention from researchers as well as from media and the general public.

Electrochromic *eyewear* has attracted interest on and off for decades. Photochromic glass or plastic has been widely used in sunglasses and goggles, but photochromic devices have clear limitations: their coloration relies on ultraviolet solar irradiation, so they are not of much use inside buildings and vehicles, and the coloration and bleaching dynamics are undesirably slow especially for bleaching at low temperatures. Electrochromic-based eyewear does not have these drawbacks, which motivates a lingering interest. Studies have been reported for sun goggles [22] and helmet visors [23].

A different type of application regards electrochromic-controlled thermal emittance for *thermal control of spacecraft*. Their exposure to cold space or solar irradiation can yield temperature differences between −50 and +100 °C, which may lead to unreliability in electronic components, detectors and so on. Mechanical shutters can be used for emittance control but tend to be impractical. Investigations on electrochromic-based thermal management have been conducted at least since the early 1990s and remain active today [24–26]. Related devices can be employed for military camouflage in the infrared [27].

The aforementioned applications are not the only ones of interest for electrochromic devices, but the possibility of achieving colour-changing systems based on inorganic oxides or organic materials seems to serve as a catalyst for human inventiveness, and recently mentioned potential uses for electrochromics include fingerprint enhancement in authentication devices and forensics [28] as well as fashion spandex (Lycra — a polyurethane–polyurea copolymer) [29].

1.3
Overview of Electrochromic Oxides

Electrochromic materials and associated devices have been researched continuously ever since the discovery of electrochromism. Looking specifically at electrochromic metal oxides, the literature up to 2007 has been covered in some detail in several prior publications [1, 30–33], and it was noted that some 50–100 scientific and technical papers were published per year. Section 1.3.1 gives a rather superficial presentation of work on electrochromic oxide films presented from 2007. From the sheer number of publications, one may conclude that the research field is thriving and that a large amount of work is devoted to materials based on oxides of W and Ni, as has been the case for the last two decades. Section 1.3.2 discusses optical and electronic properties in electrochromic oxides from a general perspective, and Section 1.3.3 treats these properties in detail for the most widely used electrochromic material, that is, W oxide. Section 1.3.4 discusses ion conduction in electrochromic oxides, again with focus on W oxide. The composition of the electrochromic oxides and their deposition parameters are very important, and some examples from recent studies are given in Section 1.3.5. Finally, Section 1.3.6 reports data on electrochromic properties for films in the full compositional range of W–Ni oxide.

It should be pointed out that much can be inferred about the electrochromic oxides also from studies on alternative applications such as gas sensors and gasochromics, electrical batteries, photocatalysts and so on [34], so the actual knowledge basis for the electrochromic oxides is much larger than the one outlined in this chapter.

1.3.1 Recent Work on Electrochromic Oxide Thin Films

There are two principally different kinds of electrochromic oxides: 'cathodic' ones colouring under ion insertion and 'anodic' ones colouring under ion extraction. Figure 1.2 indicates metals capable of forming oxides of these two categories and also points out that oxides based on vanadium are intermediate [1]. This scheme serves as a natural starting point for a survey of the recent scientific literature on electrochromic oxides. In the following, we first treat the cathodic oxides, then the anodic oxides and finally the intermediate one.

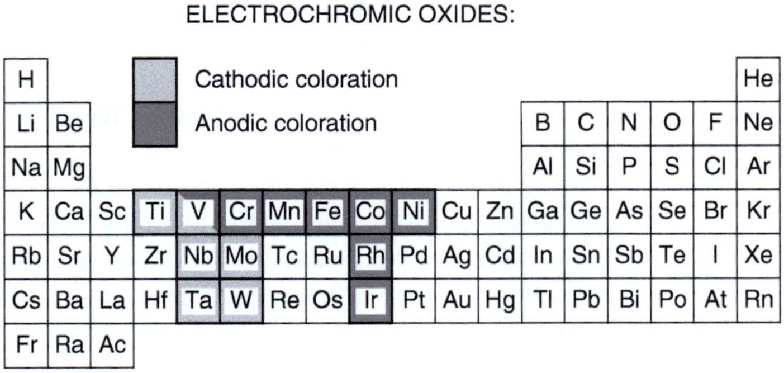

Figure 1.2 Periodic system of the elements (except the lanthanides and actinides). Differently shaded boxes indicate transition metals with oxides capable of giving cathodic and anodic electrochromism. From Ref. [1].

Tungsten oxide remains the most widely studied electrochromic oxide, and films of this *cathodically colouring* material have been prepared by a huge number of techniques, including traditional thin-film preparation with physical and chemical vapour deposition, a plethora of chemical methods, electrochemical methods and others. Surveys of these techniques can be found in a number of books and review papers [35–40]. Nanoparticles have sometimes been used as intermediate steps for the films, and substrate templating has been employed to improve the electrochromic performance.

Regarding physical vapour deposition, data have been reported on W oxide films prepared by thermal evaporation, sputtering and pulsed laser deposition. Other work has used chemical vapour deposition and related spray pyrolysis, and a large number of investigations have been based on chemical routes for making films. Furthermore, electrodeposition, anodisation and electrophoretic deposition have been employed. A useful review has been published on properties, synthesis and applications of nanostructured W oxide [41].

Mixed oxides based on tungsten can exhibit properties that are superior, in one way or another, to those of the pure oxide. One well-studied option is W–Ti oxide, where the addition of titanium leads to significantly enhanced durability under electrochemical cycling, as has been known for many years [42, 43]. Recent work on electrochromism of W–Ti oxide has been reported for films made by sputtering, spray pyrolysis, chemical techniques, electrodeposition and anodisation. A key result is that addition of Ti stabilises a highly disordered structure [44, 45], as also reported in other work on W–Ti oxide that was not specifically on electrochromism [46]. Additional research on mixed oxides has been carried out on films of $W-M_W$ oxide with M_W being Li, C, N, V, Ni, Nb, Mo, Ru, Sn and Ta. Furthermore, investigations have been reported on W oxide containing coinage metal nanoparticles capable of giving plasmon-induced optical absorption of visible light, specifically for nanoparticles of Ag, Pt and Au; using, instead, nanoparticles of indium-tin oxide (ITO) makes it possible to have plasmon absorption at infrared wavelengths. Finally, a number of hybrid nanomaterials have been studied, such as WO_3–PEDOT:PSS, where the second component is poly(3,4-ethylenedioxythiophene):poly(styrenesulfonate).

Electrochromic Mo oxide has many similarities to W oxide, and studies have been performed on films prepared by evaporation, chemical vapour deposition, wet chemical techniques and electrodeposition. Mixed oxides have been researched for $Mo-M_{Mo}$ oxide with M_{Mo} being C, Ti, V, Nb and Ce.

Concerning Ti oxide, electrochromic properties have been reported for films made by sputtering, chemical vapour deposition and spray pyrolysis, various wet chemical techniques, doctor blading and anodisation. This oxide has also been used as anchoring agent for organic chromophores (such as viologens, see Chapter 3) and for inorganic compounds (such as 'Prussian Blue', see Chapter 2). Work has been published for films of $Ti-M_{Ti}$ with M_{Ti} being V and Zr. Finally, studies have been reported on electrochromism of Nb oxide and Nb–Mo oxide.

Regarding *anodically colouring* electrochromic oxides, we first note work on Ir oxide prepared by sputter deposition and by sol–gel technology, and

Ir–Sn oxide prepared by evaporation and Ir–Ta oxide by sputtering. However, iridium oxide is too expensive for most applications, even if diluted with less costly Sn or Ta. Ni oxide is a good alternative whose electrochromic activity was discovered many years ago, and Ni-oxide-based films are included in several of today's electrochromic devices. Recent work has been reported on Ni oxide films prepared by thermal evaporation, sputter deposition, chemical vapour deposition and spray pyrolysis, various chemical techniques and electrodeposition. Furthermore, Ni oxide pigments have been deposited from water dispersion. Mixed binary oxides of Ni–M_{Ni} have been studied for M_{Ni} being Li, C, N, F, Al, Ti, V, Mn, Co, Cu and W. One reason why additions to Ni oxide are of interest is that the luminous transmittance can be enhanced [47]. Further improvements are possible with ternary or more complex oxides with M_{Ni} being (Li,W) [48], (Li,Al) [49], (Li,Zr) [50] and LiPON [51]. Films of the latter materials were produced recently by sputter deposition and are of particular interest for practical applications in future electrochromic-based fenestration. Hybrid films have been made of Ni oxide and one of several polymers or graphene oxide.

V-pentoxide-based films with *intermediate* electrochromic properties have been prepared by vacuum evaporation, sputter deposition, spray pyrolysis, chemical techniques, electrodeposition and inkjet printing. Mixed oxides based on V–M_V have been investigated with M_V being C, Na, Ti, Mo, Ag, Ta and W. Mixed V–Ti oxides can serve as good counter electrodes in electrochromic devices [52, 53].

1.3.2
Optical and Electronic Effects

Most oxide-based electrochromic devices employ two electrochromic films, and it is clearly advantageous to combine one cathodic oxide (e.g. based on W, Mo, Ti or Nb) and another anodic oxide (e.g. based on Ni or Ir). Applying a voltage in order to transport ions and electrons between the two electrochromic films in one direction makes both of these films colour, and transporting ions and electrons in the other direction makes both of them bleach. By combining cathodic and anodic oxides, one can accomplish a rather neutral visual appearance. The most commonly used oxides are based on tungsten and nickel, which exhibit cathodic and anodic electrochromism, respectively, according to the highly schematic reactions [1]

$$[WO_3 + H^+ + e^-]_{bleached} \leftrightarrow [HWO_3]_{coloured} \tag{1.1}$$

and

$$[Ni(OH)_2]_{bleached} \leftrightarrow [NiOOH + H^+ + e^-]_{coloured} \tag{1.2}$$

for the case of proton insertion/extraction. Electrons are denoted e^-.

We now consider the origin of the electrochromic properties in these oxides and first look at their crystalline nature. The pertinent structures fall into three categories: (defect) perovskites, rutiles and layer/block configurations.

These structures can be treated within a unified framework with 'ubiquitous' MO_6 octahedra (where M denotes metal) connected via joint corners and/or joint edges [1, 54]. Edge-sharing is associated with some degree of octahedral deformation. Only two oxides are problematical within this description: the first one is vanadium pentoxide (V_2O_5), whose crystal structure can be constructed from heavily distorted VO_6 octahedra or, alternatively, from square pyramidal VO_5 units [55, 56]; and the second example is hydrous nickel oxide – which is the actual electrochromic material rather than pure NiO [57, 58] – which is thought to contain layers of edge-sharing NiO_6 octahedra.

Octahedral coordination is essential for the electronic properties of electrochromic oxides [1, 54]. It was shown in detail many years ago [59] that oxygen *2p* bands are separated from metal *d* levels, and octahedral symmetry leads to splitting of these latter levels into bands with the conventional designations e_g and t_{2g}. Figure 1.3 illustrates three cases of importance for the electrochromic oxides. Left-hand panel, for H_xWO_3, indicates that the O2*p* band is separated from the split *d* band by an energy gap. Pure WO_3 has a full O2*p* band and an empty *d* band, and the band gap is wide enough to render films of this material transparent. Inserting small ions and charge-balancing electrons – according to the aforementioned 'battery' model – leads to a partial filling of the *d* band along with optical absorption as discussed later. The middle panel in Figure 1.3 is adequate for anodically colouring electrochromic oxides. The pure oxides have unoccupied t_{2g} states, and insertion of ions and electrons may fill these states to the top of the band, so the material exhibits a gap between the e_g and t_{2g} levels. The material then becomes transparent, assuming that the band gap is large enough. Finally, the right-hand part of Figure 1.3 indicates that V_2O_5 – with both cathodic and anodic features in its electrochromism – has a principally

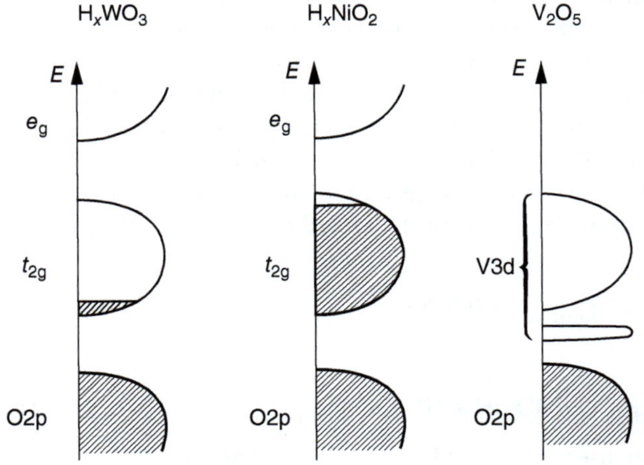

Figure 1.3 Schematic band structures for different types of electrochromic oxides, as discussed in the main text. Shaded regions signify filled states and *E* denotes energy. From Ref. [1].

different electronic structure. The deviation from octahedral coordination is significant enough that the *d* band displays a narrow split-off part in the band gap. Insertion of ions and electrons into V_2O_5 may fill this narrow band so that the optical band gap is widened. These features of the band structure can account for electrochromism of V_2O_5 [60], at least in principle, as well as band gap widening during photo-injection of hydrogen into this material [61].

The detailed mechanism for optical absorption in the electrochromic oxides is often poorly understood. The exception may be W oxide, which is considered in the following section, whose properties were reviewed some years ago [33]. However, optical absorption in electrochromic oxides is generally believed to be connected with charge transfer, and polaron absorption accounts for at least most of the significant features [33, 62–65]. A simplified, but closely related, model for the absorption considers intervalence charge transfer transitions [66].

1.3.3
Charge Transfer Absorption in Tungsten Oxide

Electrons inserted together with ions are localised on metal ions and, for the case of tungsten oxide, change some of the W^{6+} sites to W^{5+}. Charge transfer between sites *i* and *j* can be expressed, schematically, as

$$W_i^{5+} + W_j^{6+} + \text{photon} \rightarrow W_i^{6+} + W_j^{5+} \tag{1.3}$$

More specifically, the electrons are thought to enter localised states positioned 0.1–0.2 eV below the conduction band and displace the atoms surrounding them so that they form a potential well; strong electron–phonon interaction then leads to the creation of small polarons with a size of 0.5–0.6 nm [33].

Figure 1.4, taken from the work of Berggren *et al.* [65], reports data on the optical absorption coefficient of sputter-deposited W oxide films electrochemically intercalated with Li^+ ions to a number of different levels x (defined as the number of Li^+ ions per W atom). It is found that Li^+ intercalation yields a broad and asymmetric peak at an energy of ~1.3 eV. For large intercalation levels, this peak is shifted slightly towards higher energies.

Figure 1.5 shows data from a comparison of the spectral absorption coefficient at two intercalation levels with Bryksin's theory of polaron absorption [63], which is based on intraband transitions between localised energy levels in a Gaussian density of states [65]. Theory and experiments are found to agree well both for $x = 0.04$ and $x = 0.36$, as seen from Figure 1.5(a) and (b), respectively. A comparison with the more recent theory by He [64] does show equally good correspondence, especially not at low intercalation. Successful comparison between Bryksin's theory and experimental optical data was presented several years ago also in other work on W oxide films [67], and further data on polaron absorption in the same material have been given recently [68].

The simple model for charge exchange, outlined earlier, is credible only as long as transitions can take place from a state occupied by an electron to another state capable of receiving the electron. If the ion and electron insertion levels are large,

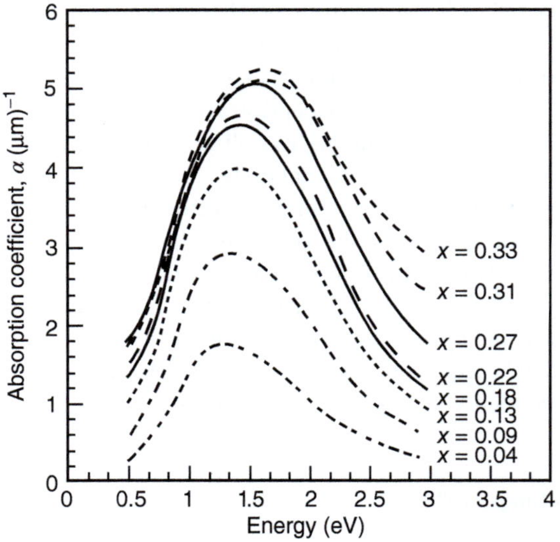

Figure 1.4 Spectral absorption coefficient for W oxide films with different intercalation levels, where x denotes the Li^+/W ratio. From Ref. [65].

this assumption is no longer the case and then 'site saturation' [69] becomes important, as considered next.

A detailed investigation of the optical properties at different amounts of lithium insertion into sputter-deposited W oxide films was reported recently by Berggren and Niklasson [70] and Berggren *et al.* [71], and data are available also for hydrogen-containing material [72]. Figure 1.6a reports spectral optical absorption coefficient α for different levels of Li^+ intercalation into a W oxide film. A broad absorption band evolves at an energy of ~1.5 eV for low values of x, as also observed earlier, while the absorption peak shifts towards higher energies for large amounts of Li. The strong increase of α above 3.5 eV is associated with the fundamental band gap of W oxide. It is interesting to subtract the absorption due to the unintercalated material, and the corresponding absorption coefficients, denoted α^+, are given in Figure 1.6b which indicates that the data have unambiguous peak structures. The data could be modelled with three Gaussian peaks, whereas modelling with only two peaks was unsuccessful. Two of the peaks were at the positions shown in Figure 1.6b, and the third peak could be located at intermediate energies. Figure 1.7 indicates the integrated strengths of these peaks.

The origin of the Gaussian peaks can be reconciled with 'site saturation' by considering three types of sites, namely W^{4+}, W^{5+} and W^{6+}. Beginning with empty states ($x = 0$), most of the states will be singly occupied at the start of the intercalation, and electron transitions between empty and singly occupied states will be prevalent. As more single states are filled, the probability that doubly occupied states also will be formed will be higher. Analytical expressions for the number of possible electronic transitions can be given for $W^{6+} \leftrightarrow W^{5+}$, $W^{5+} \leftrightarrow W^{4+}$ and

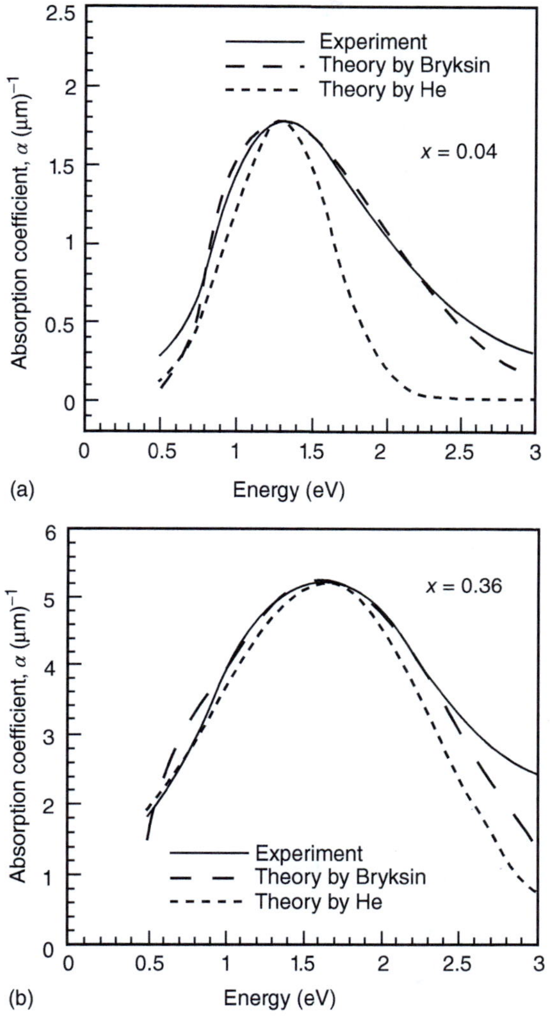

Figure 1.5 Spectral absorption as measured (cf. Figure 1.4) and calculated from two theories of polaron absorption at two intercalation levels, where x denotes Li^+/W ratio. From Ref. [65].

$W^{6+} \leftrightarrow W^{4+}$ by $2x(2-x)^3$, $2x^3(2-x)$ and $x^2(2-x)^2$, respectively [71]. Corresponding curves are given in Figure 1.8, and it is clear that there is good similarity with the integrated peak structure in Figure 1.7, which hence indicates that 'site saturation' takes place. The intensities of the curves do not coincide, which is not surprising since the absorption strength per transition is likely to be different for the three cases. Considering a practical electrochromic device, long-term cycling durability demands that x is kept low, perhaps not exceeding 0.3–0.35, and then the $W^{5+} \leftrightarrow W^{6+}$ transitions are clearly dominating.

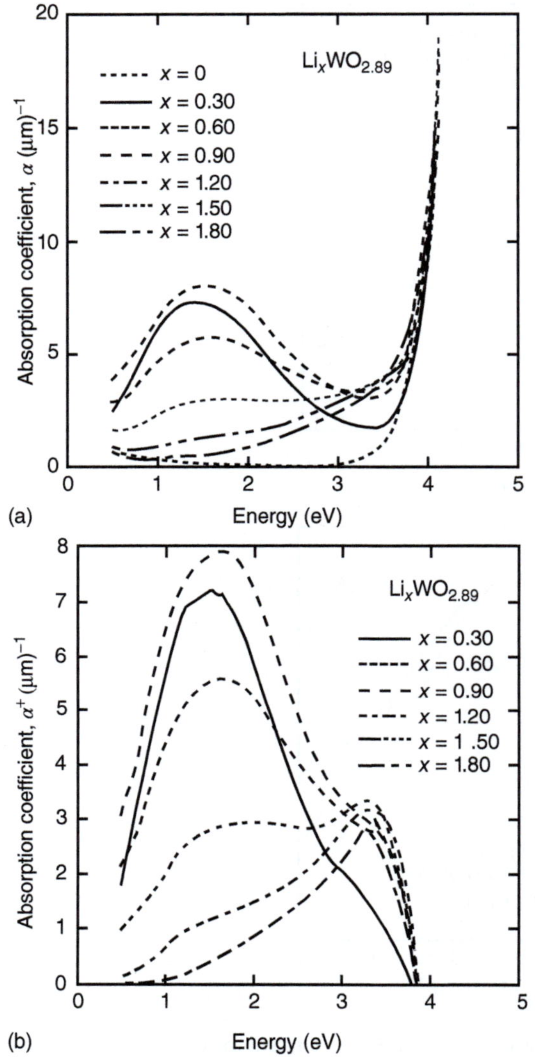

Figure 1.6 Panel (a) shows spectral absorption coefficient α for a slightly sub-stoichiometric W oxide film intercalated to the shown Li^+/W ratios x, and panel (b) shows corresponding absorption α^+ when the absorption of the unintercalated film in panel (a) has been subtracted. From Ref. [71].

1.3.4 Ionic Effects

Most electrochromic oxides consist of octahedral units in various arrangements, as discussed in detail in Section 1.3.2. These oxides are appropriate both because of their electronic features and because the spaces between the octahedral units

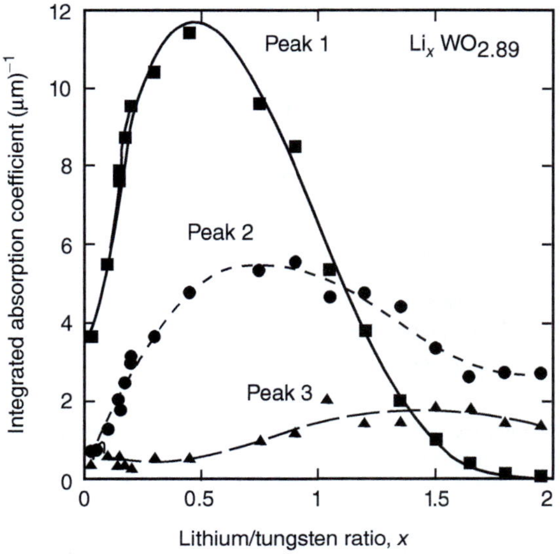

Figure 1.7 Integrated absorption coefficient as a function of Li^+/W ratio x for three Gaussian peaks representing the data in Figure 1.6b. From Ref. [71].

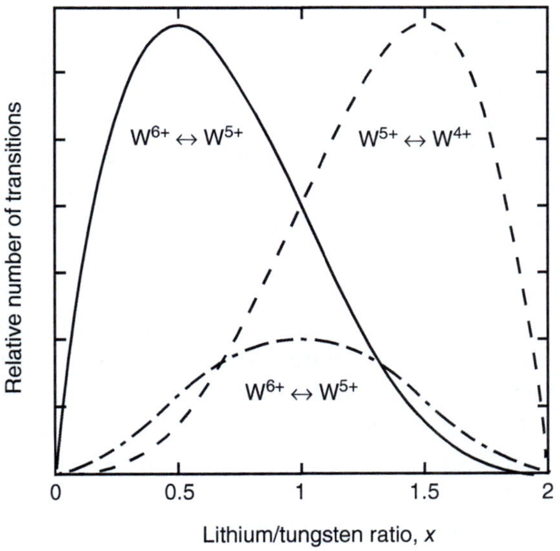

Figure 1.8 Relative number of transitions of the types $W^{6+} \leftrightarrow W^{5+}$, $W^{5+} \leftrightarrow W^{4+}$ and $W^{6+} \leftrightarrow W^{4+}$ as a function of Li^+/W ratio x. From Ref. [71].

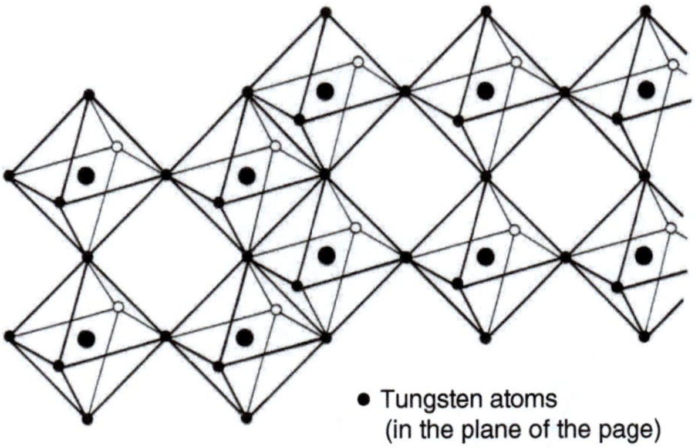

Figure 1.9 Schematic image of corner-sharing and edge-sharing octahedra in slightly sub-stoichiometric crystalline W oxide. From Ref. [1].

are sufficiently large to allow facile transport of small ions. Clusters of octahedra are able to form disordered and more or less loosely packed aggregates with large porosity, and it follows that nanostructures enter at two or more length scales. The following discussion is again focused on W oxide, which has been investigated in great detail. It is not obvious that the anodic electrochromic oxides can be understood on the same premises however, and grain boundaries may then be of large significance, but the situation remains unclear.

Figure 1.9 illustrates nanostructural features of W oxide and shows WO_6 octahedra, each with six oxygen atoms surrounding a tungsten atom. Stoichiometric WO_3 has a structure wherein each octahedron shares corners with neighbouring octahedra. WO_3 and similar transition-metal-based oxides easily form sub-stoichiometric oxides, which include a certain amount of edge-sharing octahedra. The three-dimensional structure comprised by the octahedra yields a three-dimensional 'tunnels' structure conducive for ion transport.

The crystalline structure in Figure 1.9 is simplified and refers to a cubic structure (cf. Figure 1.10a), and a tetragonal structure usually prevails in WO_3 at normal temperature and pressure. The tetragonal structure is more favourable for ion transport than the cubic one since the separations among the octahedral units are larger, as shown in Figure 1.10b. Hexagonal structures, indicated in Figure 1.10c, are easily formed in thin films [1, 73, 74] of W oxide, and the structure is then even better suited for the transport of ions.

Figure 1.11 indicates structural data based on modelling of X-ray scattering from films made by evaporation onto substrates at different temperatures, from room temperature to 300 °C [75]. Cluster-type structures are apparent and include hexagonal-type units, which grow in size and interconnect at high substrate temperatures.

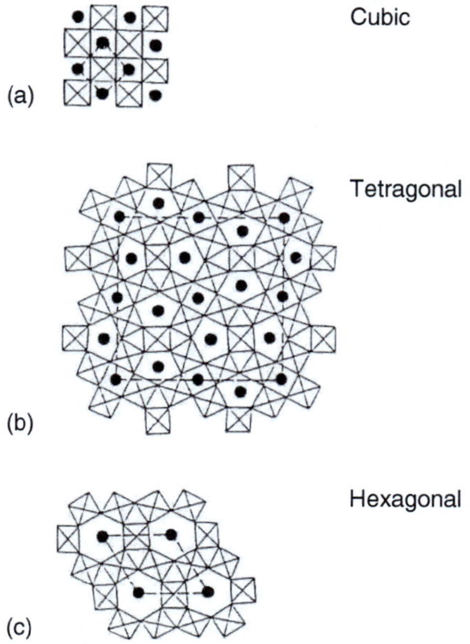

Figure 1.10 W oxide with cubic (a), tetragonal (b) and hexagonal (c) structure. Dots signify sites for ion insertion in spaces between the WO_6 octahedra. Dashed lines show extents of the unit cells. From Ref. [1].

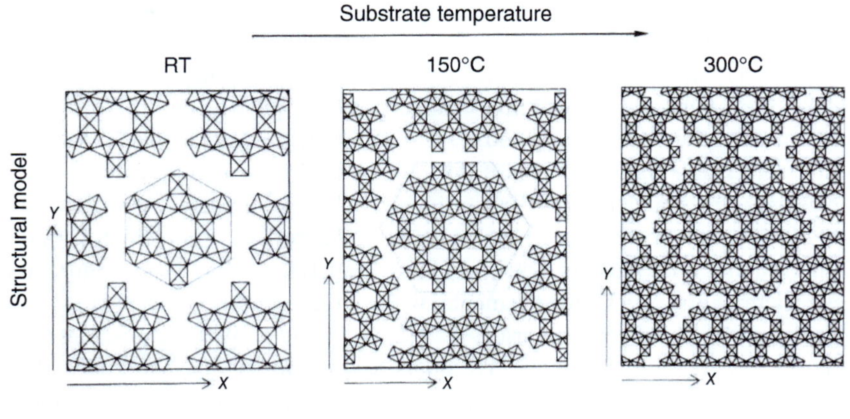

Figure 1.11 Structural models for WO_6 octahedra in W oxide films prepared by evaporation onto substrates at the shown temperatures (RT denotes room temperature). Arrows in the x and y directions are 2 nm in length. After Ref. [75].

Larger nanostructures than those created by the aggregation of octahedral units can occur as a consequence of the limited mobility of the deposition species, and this mechanism is important for thin films prepared by most techniques. Regarding sputter deposition, the main features are captured in a 'zone diagram', commonly referred to as a 'Thornton diagram' [76]. It is shown in Figure 1.12. More elaborated or specialised versions of this diagram have been presented in subsequent work. It is evident that low substrate temperature and high pressure in the sputter plasma lead to nanoporous features appropriate for

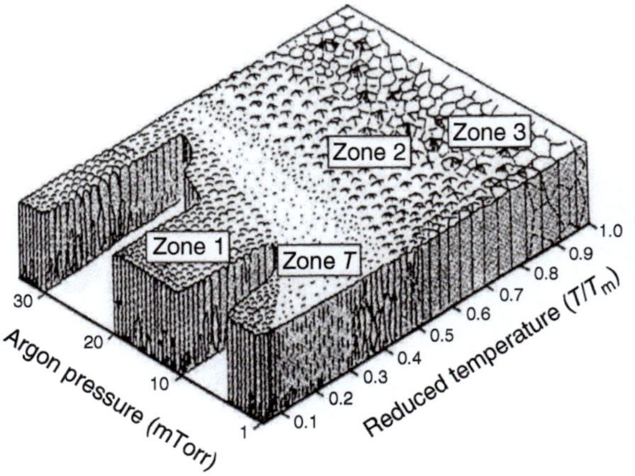

Figure 1.12 Schematic nanostructures in sputter-deposited thin films formed at different pressures in the sputter plasma and at different substrate temperatures. The melting temperature of the material is denoted T_m. From Ref. [76].

ion transport across the film thickness and hence for electrochromic devices. A direct correlation between sputter gas pressure and electrochromic performance was reported recently [77]. Oblique angle deposition can promote porosity still further, as specifically shown for W-oxide-based films [78–80]. Annealing (cf. Figure 1.11) and ion irradiation [81] are other ways to influence film density.

W oxide films can involve molecular deposition species, and it is of interest to consider the lowest energy structures of $(WO_3)_q$ clusters, which may form the deformable building blocks for octahedra-based nanostructures. Such clusters were evaluated from first-principle calculations in recent work by Sai *et al.* [82]. Data for $2 \leq q \leq 12$ are depicted in Figure 1.13, from which it can be inferred that small clusters (with q equal to 3 and 4) have ring-like configurations with alternating W–O arrangements, while larger clusters (with $q \geq 8$) can be described as symmetric spherical-like cages. Trimeric W_3O_9 molecules form during evaporation [83] and correspond to $q = 3$ in the cluster model; these structures are characterised by a hexagonal configuration. Aqueous solutions, useful for liquid-phase film deposition, can have a preponderance of $(W_6O_{19})^{2-}$ ions, known as 'Lindqvist anions' [84].

1.3.5
On the Importance of Thin-Film Deposition Parameters

Irrespective of the technique for making EC thin films, the detailed deposition conditions usually play a decisive role. However, each technique is unique to a considerable extent and little can be said in general apart from the guidelines inferred from the previous section.

We first look at films made by physical vapour deposition, which is notable for its possibilities to accomplish process control and reproducibility. A clear

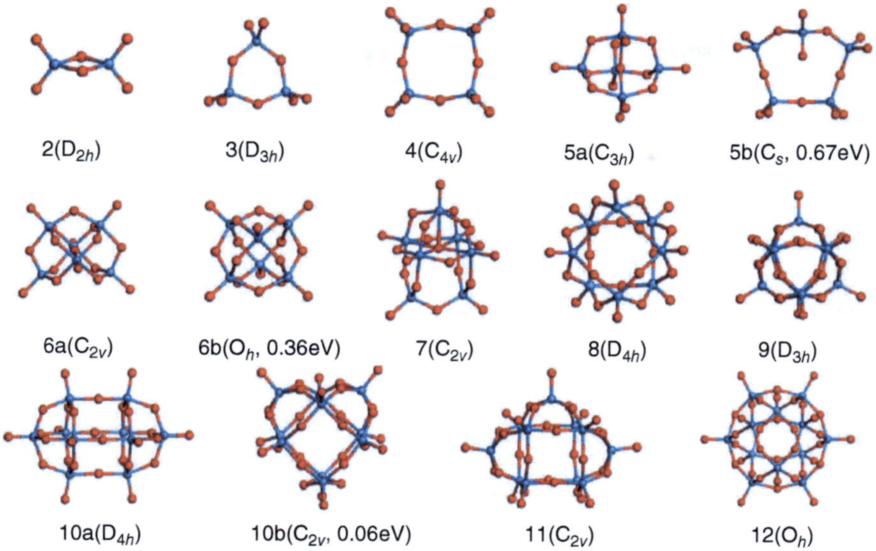

Figure 1.13 Lowest-energy structures of $(WO_3)_q$ clusters with $2 \leq q \leq 12$ and metastable isomers (denoted 5b, 6b and 10b). Crystallographic symmetry designations are indicated at the various clusters. From Ref. [82].

illustration of the applicability of the Thornton diagram in Figure 1.12 for electrochromic films was found in recent work on titanium-oxide-based films made by sputtering at various pressures p onto substrates at different temperatures τ_s [85, 86]. Figure 1.14, reproduced from work by Sorar *et al.* [85], shows cyclic voltammograms indicating current density under charge insertion and extraction as a function of applied voltage. Clearly, $p > 10$ mTorr and $\tau_s < 100\,°C$ are required to accomplish a structure porous enough for facile ion transport.

The composition of the sputter gas is very important, in addition to its pressure, as elucidated in Figure 1.15 for water vapour added to $Ar + O_2$ during sputter deposition of Ni oxide films; the results were presented recently by Green *et al.* [87]. Addition of water vapour to a pressure of $\sim 7 \times 10^{-2}$ Pa increased the current density during ion insertion and extraction and enlarged the optical modulation by allowing a darker state. Other work on sputtering of Ni oxide in the presence of water vapour has been reported elsewhere [88, 89].

The properties of the substrate for the electrochromic film may be of large significance. One example of this dependence is reported in Figure 1.16a, taken from work by Yuan *et al.* [90], which shows optical transmittance in bleached and coloured states for a Ni oxide film prepared by electrodeposition onto a substrate with and without a layer of self-assembled polystyrene nanospheres. Clearly, the sphere templating yields a lower coloured-state transmittance. The visual appearance of films on untreated and templated substrates is shown in Figure 1.16b.

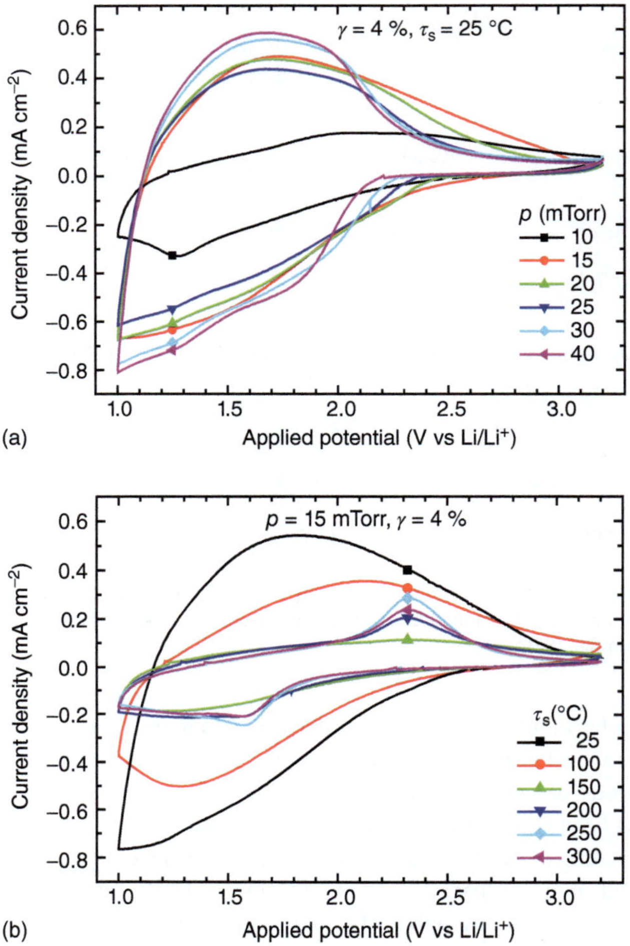

Figure 1.14 Cyclic voltammograms for sputter-deposited TiO_2-based film in a Li^+-conducting electrolyte. Film deposition took place at the pressure p and oxygen/argon ratio γ onto substrates at temperature τ_s. Data in panels (a) and (b) refer to the effects of varying p and τ_s, respectively. From Ref. [85].

The deposition rate is a very important parameter, which often more or less determines the cost for industrial coating production. However, this parameter is not commonly reported in scientific papers. Among high-rate techniques, it is particularly interesting to consider reactive-gas-flow sputtering, which was studied recently by Oka *et al.* [91] to make W oxide films. Figure 1.17 shows deposition rate, oxygen/tungsten stoichiometry and density for films made by sputtering at two different oxygen flows but under otherwise identical conditions. The deposition rate was found to be as high as ~4 nm s^{-1}, which is considerably higher than for typical reactive DC magnetron sputtering. The density depended on the oxygen gas flow and could be as low as half of the bulk value for WO_3 (7.16 g cm^{-3}).

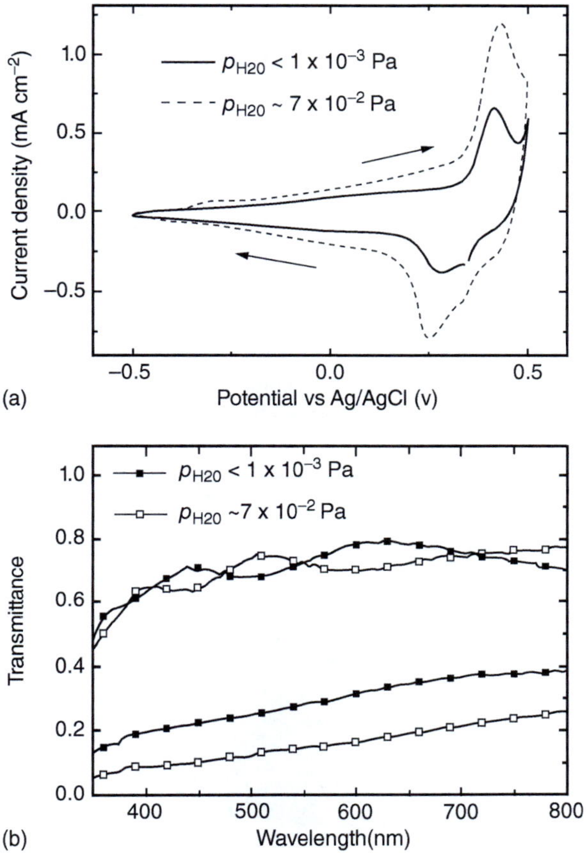

Figure 1.15 Current density during ion insertion/extraction (a) and spectral transmittance at maximum and minimum intercalation (b) for sputter-deposited Ni oxide films prepared with the shown partial pressures of H_2O in the sputter gas. From Ref. [87].

Hence, the film structure is highly porous and suited for electrochromic devices. Not surprisingly, investigations of the electrochromic performance showed good results. Another contemporary high-rate deposition technique, appropriate for W oxide, is high-power impulse magnetron sputtering (known as HiPIMS) [92].

1.3.6 Electrochromism in Films of Mixed Oxide: The W–Ni-Oxide System

Electrochromism has been investigated in many binary, ternary and so on, oxides, as surveyed in Section 1.3.1. Nevertheless, there is a general lack of comprehensive investigations across the full compositional range between the component oxides. An exception is the W–Ni oxide system, denoted Ni_xW_{1-x} oxide, which was studied by Green *et al.* [93–98] and clearly combines cathodically and anodically colouring oxides. Tungsten-rich sputter-deposited films were found to

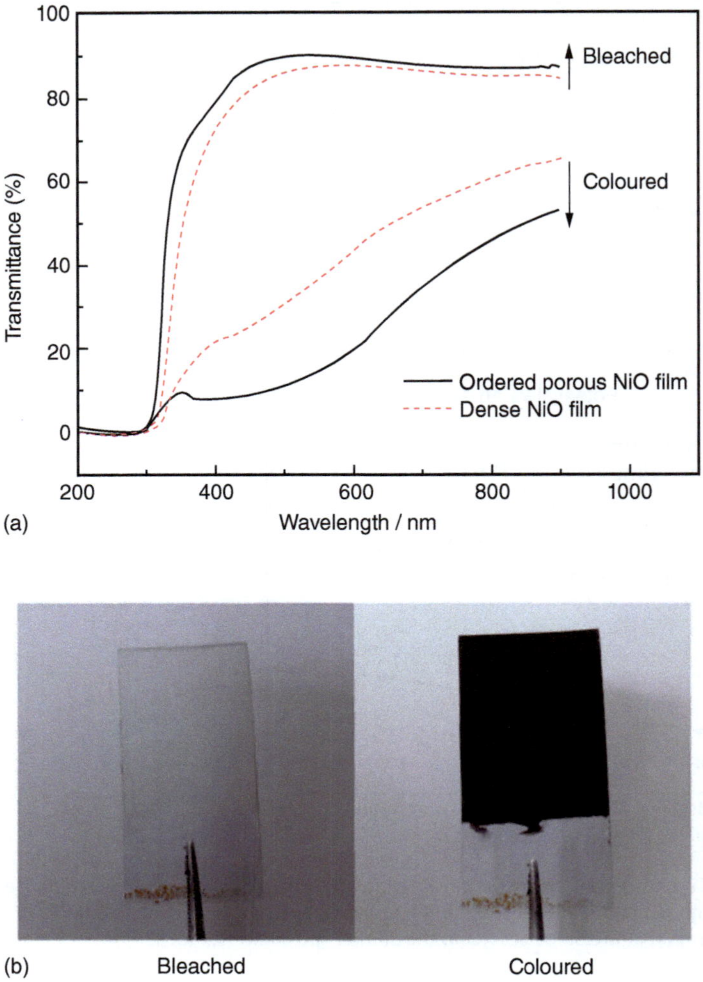

Figure 1.16 Spectral transmittance (a) and visual appearance (b) of an electroplated Ni oxide film on a substrate with and without a template layer consisting of polystyrene nanospheres. The film is in its fully coloured and bleached states. From Ref. [90].

consist of a mixture of amorphous WO_3 and nanocrystalline $NiWO_4$, with equal amounts of W and Ni the structure was dominated by $NiWO_4$, and nickel-rich films were made up of nanocrystalline NiO and $NiWO_4$.

Figure 1.18 shows optical absorption coefficients at a mid-luminous wavelength of 550 nm for fully coloured and bleached films. The electrochromism is seen to be much stronger in W oxide than in Ni oxide. For W-rich films, the absorption coefficient drops as Ni fraction x is increased, except at the composition $x \approx 0.12$ where a pronounced peak can be seen in the absorption coefficient. For Ni-rich films, the absorption coefficient rises as x approaches unity. Films in the compositional range $0.3 < x < 0.7$ do not display electrochromism. Another important parameter

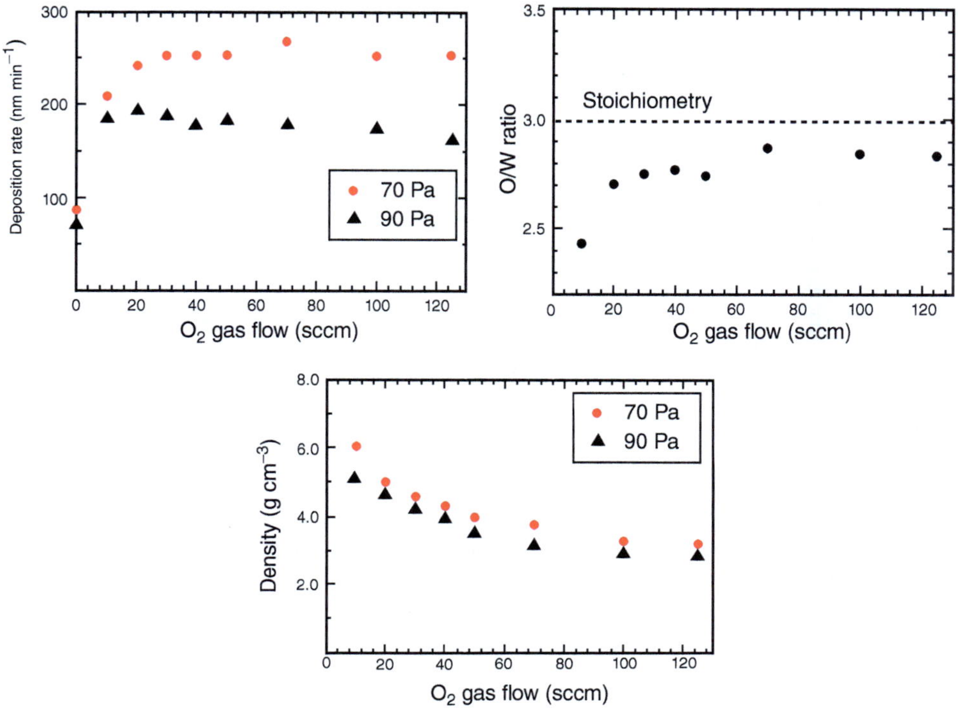

Figure 1.17 Deposition rate, oxygen/tungsten stoichiometry and density of thin films of electrochromic W oxide prepared by reactive-gas-flow sputtering at different oxygen gas flows. The total gas pressure was 70 Pa, and for two of the data sets also 90 Pa. From Ref. [91].

for electrochromic films is coloration efficiency η, defined by $\eta = \mathrm{OD}/\Delta Q$ where optical density (OD) is absorption coefficient multiplied by film thickness and ΔQ is inserted/extracted charge density. Figure 1.19 shows data on coloration efficiency and allows easy comparison with Figure 1.18. Obviously, η increases slightly for increasing values of x, except for compositions around $x \approx 0.5$ where η is approximately zero.

The strong optical absorption at $x \approx 0.12$ is interesting for electrochromic device applications, and Figure 1.20 reports spectral coloration efficiency for films of pure W oxide and for Ni_xW_{1-x} oxide films with two values of x. It is evident that good performance of films with $x \approx 0.12$ is found in the whole luminous wavelength range, that is, for 400–700 nm.

1.4 Transparent Electrical Conductors and Electrolytes

An electrochromic device does not only include electrochromic thin films but also incorporates transparent electrical conductors and an electrolyte, as seen

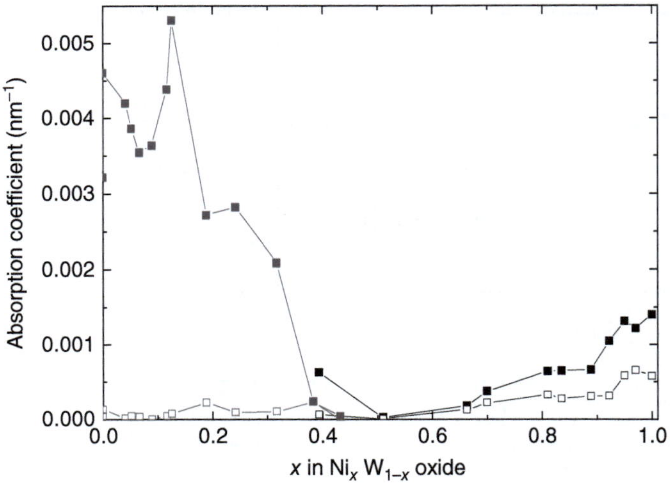

Figure 1.18 Mid-luminous absorption coefficient as a function of composition for electrochromic thin films of W–Ni oxide. Filled and open symbols denote fully coloured and bleached states, respectively. From Ref. [95].

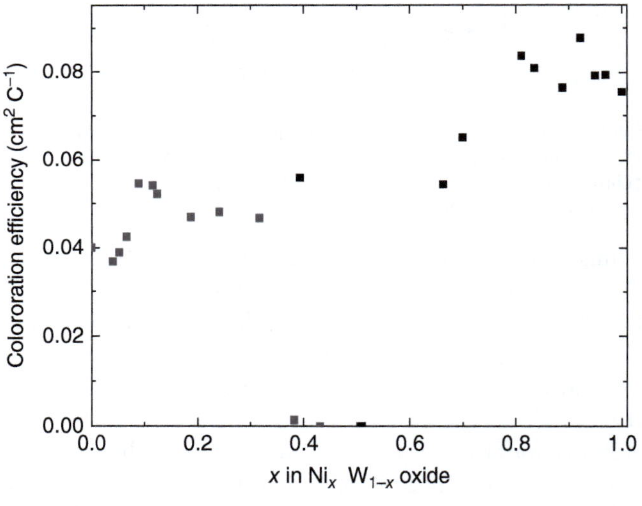

Figure 1.19 Mid-luminous coloration efficiency as a function of composition for electrochromic thin films of W–Ni oxide. From Ref. [95].

in Figure 1.1. The transparent electrical conductors may be the most costly part in the device – especially if they are based on an indium-containing oxide – and clearly deserve attention. They are of critical importance not only in electrochromics but also for thin-film solar cells, light emitting devices and so on. There are several recent reviews covering the field of transparent conductors [32, 99–101], and therefore only a bird's-eye view is given here, though with some

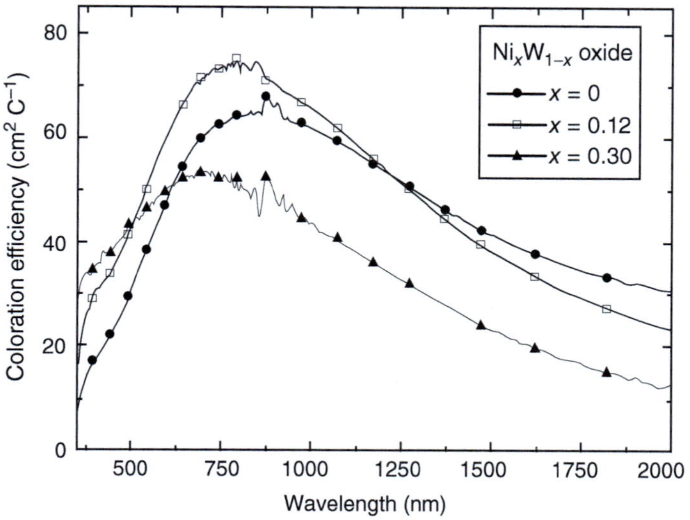

Figure 1.20 Spectral coloration efficiency for W–Ni oxide films with the shown compositions. From Ref. [94].

attention to very recent work and to transparent conductors suitable for flexible substrates. There are several types of transparent conductors with specific pros and cons; semiconductor-based films are treated in Section 1.4.1, metal-based films in Section 1.4.2 and nanowire-based and other alternatives in Section 1.4.3. Electrolytes for electrochromic devices are surveyed in Section 1.4.4 with foci on thin films and polymer layers.

1.4.1 Transparent Electrical Conductors: Oxide Films

Thin films of heavily doped wide-bandgap conducting oxides are commonly used in electrochromic devices. These materials include In_2O_3:Sn (ITO), In_2O_3:Zn, ZnO:Al, ZnO:Ga, ZnO:In, ZnO:Si, ZnO:B, SnO_2:F (FTO), SnO_2:Sb and TiO_2:Nb; their doping levels are typically a few atomic percent. Several of the oxides can combine a resistivity as low as $\sim 1\times10^{-4}\,\Omega\,cm$ with excellent luminous transmittance and durability. Films of ITO, ZnO:Al and ZnO:Ga deposited by reactive DC magnetron sputtering onto glass and PET typically have a resistivity of $\sim 2\times10^{-4}$ and $\sim 4\times10^{-4}\,\Omega\,cm$, respectively. High-quality FTO films are normally made by spray pyrolysis in conjunction with float glass production, and high temperatures are necessary also for TiO_2:Nb films. All of these oxides are transparent across most of the solar spectrum, as seen in Figure 1.21 for FTO-coated glass [102]. The oxide-based transparent conductors are very well understood theoretically [103–105], which means that detailed and accurate simulations of optical properties can be made for multilayer configurations such as electrochromic devices.

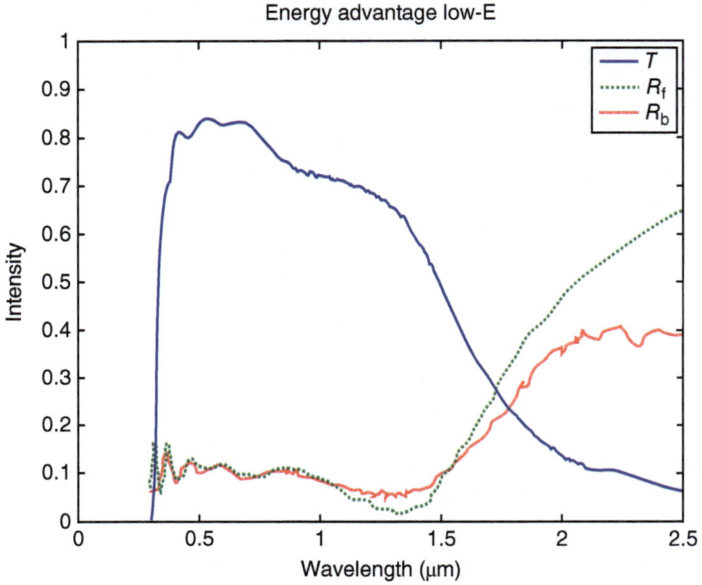

Figure 1.21 Spectral transmittance T and reflectance from the front (coated) side (R_f) and the back side (R_b) for commercial FTO-coated glass. The luminous transmittance is 83% and the solar energy transmittance is 71%. From Ref. [102].

Deposition onto flexible substrates introduces risks for cracking and delamination, and thereby lowered electrical conductivity, if the bending radius is smaller than a few centimetres [106, 107]. In general, the fracture behaviour is dependent on film thickness and bending direction. If needed, several techniques can be used to diminish the propensity for cracking.

ITO films are widely used but expensive, even if indium is not one of the rarer elements in the earth's crust [108], and its deposition requires careful process control. ZnO:Al and ZnO:Ga films can reach similar optical and electrical properties but tend to demand even more stringent process supervision, and good FTO and TiO_2:Nb films must be deposited onto hot glass. Clearly, each type of transparent conducting oxide films presents particular challenges, and there is no 'best' option for electrochromic-based and other devices.

Health aspects of indium-based films require attention, and their preparation can lead to pulmonary disorders [109]. This affliction is named 'indium lung' and obviously is a concern for large-scale production of ITO films and particles.

1.4.2 Transparent Electrical Conductors: Metal-Based Films

Metal-based transparent conductors can show excellent properties and avoid the electronic activation barrier that may exist between, for example, W oxide and ITO [110]. Coinage metals (Cu, Ag and Au) have electrical conductivities around 2 orders of magnitude higher than for the best transparent conducting oxide, which means that they can be thin enough that the luminous absorption is only some per cent. Metal films are stretchable to a much larger degree than oxide-based films.

Metal films to be used in transparent conductors must be extremely thin, which means that their properties are strongly affected by the peculiarities of thin-film growth. When metal is deposited onto a substrate of glass or polymer, the condensate goes through a number of more or less distinct growth stages [111]: tiny metallic nuclei form initially; they grow and create increasingly irregular 'islands'; these 'islands' interconnect and produce a contiguous meandering (percolating) network at a thickness that corresponds to 'large-scale coalescence'; the network is then transformed to a film comprising holes; and finally, a well-defined metallic film may be created. The relevant films have thicknesses of the order of 5–10 nm [112].

Interface reflectance limits the luminous transmittance of the coinage metal film to ~50%, but this transmittance can be radically enhanced if the metal film is positioned between high-refractive-index transparent layers anti-reflecting the metal. This effect is shown in Figure 1.22 for an 8.0-nm-thick Au film and for such a film between 55-nm-thick TiO_2 layers, from the work by Lansåker *et al.* [113]. The anti-reflection increases the transmittance to ~80%. This figure also indicates that computations based on thin-film optics can model the optical properties accurately except in the infrared. The ultrathin Au films were durable enough to be used in electrochromic devices [113]. Many other combinations of materials can be employed as alternatives to the one illustrated here, and transparent conductors based on oxide/metal/oxide are of intense interest not only for energy-related applications but also for transparent and flexible electronics and so on.

Silver-based coatings are used on a massive scale in today's fenestration, and it is possible to devise multilayer structures with a luminous transmittance of 80% and minimum throughput of solar energy (known as a 'solar-control' coating) [114, 115]. These coatings may combine two or more metal films in an advanced construction with more than 10 individual layers. Figure 1.23 shows an example of a commercial coating of this type on glass [102]. The figure allows direct comparison with data for the semiconductor-based coating in Figure 1.21.

1.4.3
Transparent Electrical Conductors: Nanowire-Based Coatings and Other Alternatives

Numerous alternatives to transparent conductors based on thin films of oxides and metals are currently investigated. Carbon-based materials are one of these [116]. We first consider graphite, which consists of loosely bound layers of carbon atoms on a honeycomb lattice. Such layers can create nanoparticles, for example, 60-atom units called 'fullerene molecules' as well as *carbon nanotubes* [117]. Single-wall carbon nanotubes can be several centimetres in length and consist of one layer of carbon atoms rolled into a seamless cylinder with a radius up to a few nanometres. There are two types of such carbon nanotubes, with metallic and semiconducting properties; the two fractions can be separated by several

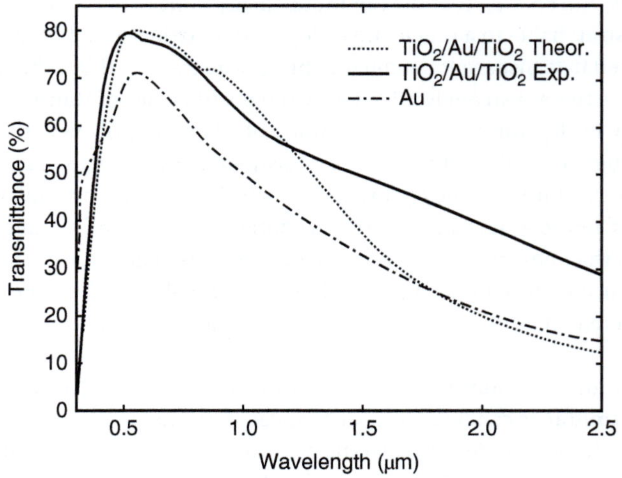

Figure 1.22 Experimental and computed spectral transmittance for Au and $TiO_2/Au/TiO_2$ films. Film thicknesses were 8 nm for Au and 55 nm for TiO_2. From Ref. [113].

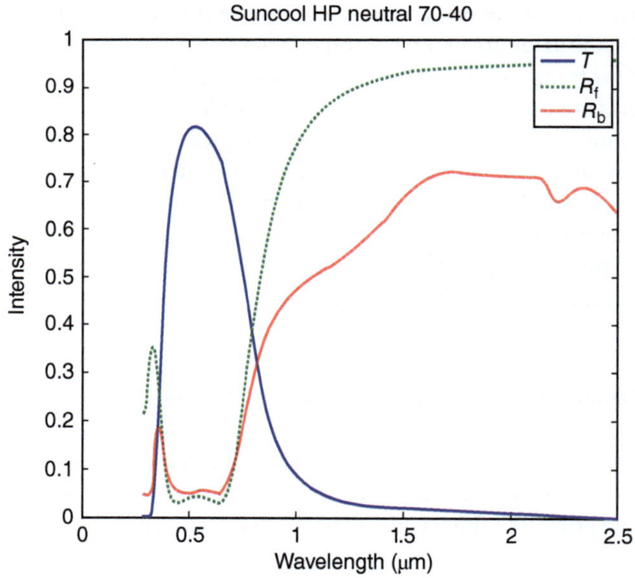

Figure 1.23 Spectral transmittance T and reflectance from the front (coated) side (R_f) and the back side (R_b) for a commercial glass with an Ag-based coating. The luminous transmittance is 80% and the solar energy transmittance is 41%. From Ref. [102].

techniques. Carbon nanotube meshes can be used as transparent conductors and attached to substrates by roll-to-roll coating [118].

Another carbon-based alternative is *graphene*, which consists of atomically thin layers of sp^2-hybridised carbon atoms arranged on a honeycomb lattice [119]. Graphene layers are usually prepared via mechanical or chemical exfoliation of graphite into individual sheets [120] or by chemical vapour deposition [121]; doping can be used to enhance the properties. Roll-to-roll production of graphene coatings have been demonstrated, and up to 30-in.-wide ribbons yielded a sheet resistance as low as ~30 Ω along with ~90% optical transmittance for films comprising four graphene layers [122]. Hot-press lamination onto PET and ink-jet printing are other techniques that can be used for large-area manufacturing.

Meshes of *metal-based nanowires* are another possibility. They can be prepared cheaply by wet-chemical techniques and applied to substrates by 'simple' techniques such as electrospraying and brush painting. A vast amount of research has been performed recently, mainly on silver-based nanowires. The electrical connection between the nanowires can be improved by different techniques such as 'nanowelding' and application of electrical fields. Recent work on metal 'nanotrough' networks demonstrated particularly good optical and electrical properties [123]. It should be noted that nanowire assemblies can have good electrical properties, but the physical dimensions of the individual nanowires make it hard to avoid some diffuse light scattering, as studied recently for silver nanowires [124]. This haze tends to limit applications in electrochromic devices for see-through applications.

Some *organic materials* can be used as transparent conductors, and a good example is poly(3,4-ethylenedioxythiophene) (PEDOT) (see Chapter 5). This material cannot quite compete with the other alternatives with regard to performance but is nevertheless of considerable interest since preparation is possible by printing at very low cost [125].

A number of *hybrid transparent conductors* have attracted attention during recent years; they combine at least two of the options discussed earlier. One may include the oxide/metal/oxide coatings in this class, provided that the oxide is a transparent conductor in its own right. Other alternatives of this rapidly growing group of transparent conductors include silver nanowires combined with ITO films or nanoparticles or with carbon nanotubes, graphene and so on. Furthermore, *macroscopic metal meshes, grids and arrays* can be of interest, although they may lead to unwanted optical scattering.

1.4.4
Electrolytes: Some Examples

Three types of electrolytes are of importance for electrochromic devices: thin solid films, polymer layers and ionic liquids. Electrolytes are generally proprietary in devices, and little is known about the ones that are used in practice.

Concerning thin solid films, those of porous oxide can be made H^+ or Li^+ conducting by co-deposition, by chemical or electrochemical post-treatment or by

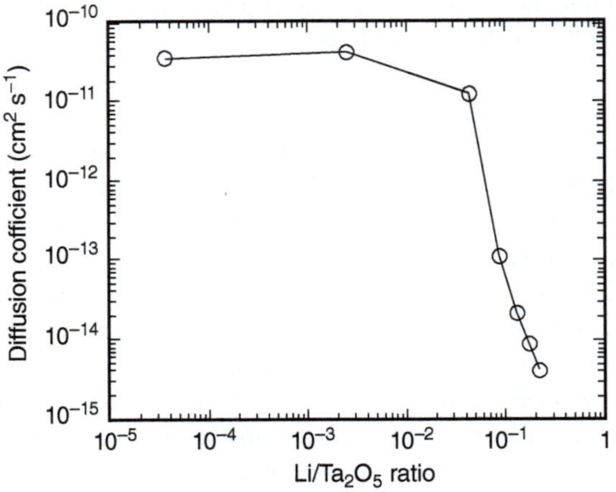

Figure 1.24 Diffusion coefficient for Ta oxide films with different amounts of Li^+. After Ref. [126].

exposure to humidity. A particularly well-studied example is tantalum pentoxide, whose applicability in electrochromic devices has been investigated in detail. For example, proton-conducting films have been prepared by sputter deposition of Ta in the presence of O_2 and H_2O and have shown an ion conductivity of $\sim 10^{-9}$ S cm^{-1} for as-deposited films; heat treatment led to a rapid decrease in their ion conductivity. Electrochemical lithium incorporation is another possibility and can yield amorphous-like films with Li^+ diffusion constants of 10^{-10} to 10^{-11} cm^2 s^{-1}, at least for Li/Ta_2O_5 ratios below 0.01, as evidenced in Figure 1.24, which reproduces data from Frenning *et al.* [126]. More complex thin-film electrolytes have been investigated in other work, such as LiPON [127] and $LiBO_2-Li_2SO_4$ (known as LiBSO) [128].

Polymers and ionic liquids can be used in electrochromic devices that are more rugged than those possible with thin-film ion conductors. There are almost innumerable options. *In situ* polymerisation may be required for sealing purposes, and research with this object has been reported for polymer electrolytes [129] and ionic-liquid-based gel electrolytes [130]. Polymer electrolytes are discussed further in Chapter 10.

1.5 Towards Devices

Electrochromic device technology has been under development for many years, as noted in Section 1.1.2. There may be several reasons why this development has taken so many years, one being that electrochromic operation is 'difficult' and combines electrochemistry with optical technology and often also with thin-film science in order to produce devices that should work for decades under harsh conditions while being observed by persons used to benchmarking against today's high-quality optical products (such as windows). Section 1.5.1 below considers a number of hurdles that must be overcome in order to have functional EC 'smart' windows. Section 1.5.2 then surveys some practical oxide-based devices. More in-depth discussions of device performance are presented elsewhere in this book.

1.5.1 Six Hurdles for Device Manufacturing

The first hurdle is that the electrochromic and the counter electrode films, shown in Figure 1.1, must have well-developed *nanoporosity* over their full areas, which may require that thin films are produced under non-standard conditions. These aspects were treated briefly in Section 1.3.5.

Secondly, the *transparent conductors* connecting electrochromic and counter electrode films must combine excellent electrical conductivity and high optical transparency, which is demanding particularly for polymer substrates. Transparent conductors of several different types were discussed in Sections 1.4.1 – 1.4.3.

The third difficulty is easily understood by considering the electrochromic device as a thin-film battery, which highlights that *charge insertion/extraction and charge balancing* are essential. In practical manufacturing, these processes must be accomplished by highly controllable and industrially viable techniques. Ozone exposure is one of several possible techniques [49, 131, 132].

Fourth, the *electrolyte* in the centre of the electrochromic device must combine good ion conductivity with very low electronic conductivity and high durability under solar irradiation. In some devices, it must also serve as an adhesive and reliably join two parts of the device.

Long-term *cycling durability* is of obvious interest for most electrochromic devices and is the fifth item. It hinges on good strategies for voltage and current control, which points at analogies between electrochromics and electrical batteries. These strategies, and their implementations via electronic drive circuitry design, are another area where proprietary considerations are dominating. However, it is evident that simple switching of voltage levels between two set values is far from an optimised strategy.

The sixth and final item is *large-scale manufacturability,* which is an obvious key to cost-effective electrochromic devices and hence to their acceptance on the consumer market. Clearly, it is necessary to avoid time-consuming production steps such as protracted thin-film deposition, long post-treatment times, elaborate steps for electrochemical charge insertion and extraction and so on. Life-cycle assessment is an important ingredient for judging the suitability of a specific manufacturing technology [133, 134]; these aspects are discussed in Chapter 18.

1.5.2
Practical Constructions of Electrochromic Devices

An uncritical reading of the scientific and technical literature may give an impression that many types of oxide-based electrochromic devices are available, but few of these are ready for practical use. However, there are some full-scale products and prototypes, especially for 'smart window' applications [115, 135–137], and those currently (2014) delivered to early adopters and other customers are introduced next. This presentation also allows a view of practical device designs and some insights into their pros and cons. The following literature references are to scientific journals, and some of these papers are rather old and probably more or less outdated. Thin films of tungsten oxide are a significant part of the electrochromic functionality for all of these devices.

We first consider a *five-layer 'monolithic' design* based on a single glass pane and developed by Sage Electrochromics Inc. in the United States [138] and Saint-Gobain Recherche in France [139]. Details are obscure, but it is clear that the electrolyte is a thin solid film. Such an arrangement makes it hard to avoid some leakage current between the electrochromic films through structural imperfections, and repeated electrical 'refresh' pulses may be needed to maintain the device in its dark state and transmittance changes can develop unevenly over large surfaces. The design may also lead to visible 'pin holes'. A related construction was developed by the US company View (earlier known as Soladigm Inc.). Somewhat analogous work was performed at the Optical Coating Laboratory Inc. (OCLI) in the United States during the 1990s [140, 141].

A laminated design embodying two double-layer-coated glass panes, joined by a polymer electrolyte, has been developed by Flabeg/EControl Glass GmbH in Germany. Some data are available for this device [142]. The electrolyte is injected in fluid form in a millimetre-wide gap between the glass panes via vacuum filling [143].

Another laminated device has been presented by Gesimat GmbH in Germany [144]. Here, the electrolyte is based on poly vinyl butyral (PVB), which is a standard material for glass lamination, and the transparent electrical conductors are produced by low-cost spray pyrolysis during float-glass production. The complementary electrochromic material is a film of iron(III) hexacyanoferrate(II) ('ferric ferrocyanide' or 'Prussian Blue'; see Chapter 2) prepared by electrodeposition.

A principally different *polymer-foil-based laminated design* has been developed by ChromoGenics AB in Sweden. It employs flexible PET foil substrates and allows roll-to-roll web coating [33, 145], that is, production by low-cost technology [146]. Another distinguishing feature is that the foil-based electrochromic device allows 'free-form design', which implies that product definition can be made at a later stage of device manufacturing. One PET foil is coated with a transparent and electrically conducting thin film and W oxide, another PET foil is coated with a transparent conductor and Ni oxide and the electrochromic-film-coated surfaces of the two foils are joined by an electrolyte via continuous lamination with an adhesive polymer. This type of electrochromic foil is discussed in more detail in Chapter 17.

1.6 Conclusions

This chapter has presented the conceptual and materials-oriented basis of oxide-based electrochromics with special attention to thin films of tungsten oxide and nickel oxide. The connection to the electronic properties and to nanostructural features was emphasised, and the importance of the thin-film manufacturing technology was stressed. Transparent electrical conductors and electrolytes were discussed too as were some general manufacturing aspects.

The electrochromic properties of a number of transition metal oxides have been known for decades and their application in devices have been attempted for many years. Promises of applications in 'smart windows' and elsewhere have been made repeatedly by many, but few products have emerged some products have swiftly been retracted. Nevertheless, research and development have progressed at a steady pace, and oxide-based electrochromic technology may finally (2015) be ready for market introduction on a massive scale.

Acknowledgement

Financial support during the writing of this chapter was received from the European Research Council under the European Community's Seventh Framework Program (FP7/2007–2013)/ERC Grant Agreement No. 267234 (GRINDOOR).

References

1. Granqvist, C.G. (1995) *Handbook of Inorganic Electrochromic Oxides*, Elsevier, Amsterdam.
2. Deb, S.K. (1995) Reminiscences on the discovery of electrochromic phenomena in transition metal oxides. *Sol. Energy Mater. Sol. Cells*, **39**, 191–201.
3. Andersson, A.M., Granqvist, C.G., and Stevens, J.R. (1989) Electrochromic Li_xWO_3/polymer laminate/$Li_yV_2O_5$ device: toward an all-solid-state smart window. *Appl. Opt.*, **28**, 3295–3302.
4. Passerini, S., Scrosati, B., Gorenstein, A., Andersson, A.M., and Granqvist, C.G. (1989) An electrochromic window based on $Li_xWO_3/(PEO)_8LiClO_4/NiO$. *J. Electrochem. Soc.*, **136**, 3394–3395.
5. Lampert, C.M. (2003) Large-area smart glass and integrated photovoltaics. *Sol. Energy Mater. Sol. Cells*, **76**, 489–499.
6. Fahlteich, J., Fahland, M., Schönberger, W., and Schiller, N. (2009) Permeation barrier properties of thin oxide films on flexible polymer substrates. *Thin Solid Films*, **517**, 3075–3080.
7. Rönnow, D., Kullman, L., and Granqvist, C.G. (1996) Spectroscopic light scattering from electrochromic tungsten-oxide-based films. *J. Appl. Phys.*, **80**, 423–430.
8. Deb, S.K. (1969) A novel electrophotographic system. *Appl. Opt. Suppl.*, **3**, 192–195.
9. Deb, S.K. (1973) Optical and photoelectric properties and colour centres in thin films of tungsten oxide. *Philos. Mag.*, **27**, 801–822.
10. Deb, S.K. (1992) Opportunities and challenges of electrochromic phenomena in transition metal oxides. *Sol. Energy Mater. Sol. Cells*, **25**, 327–338.
11. Svensson, J.S.E.M. and Granqvist, C.G. (1986) Electrochromic hydrated nickel oxide coatings for energy efficient

windows: optical properties and coloration mechanism. *Appl. Phys. Lett.*, **49**, 1566–1568.

12. Estrada, W., Andersson, A.M., and Granqvist, C.G. (1988) Electrochromic nickel-oxide-based coatings made by reactive dc magnetron sputtering: preparation and optical properties. *J. Appl. Phys.*, **64**, 3678–3683.
13. Palatnik, L.S., Malyuk, Y.I., and Belozerov, V.V. (1974) An X-ray diffraction study of the mechanism of reversible electrochemical dielectric ↔ semiconductor transformations in Nb_2O_5. *Dokl. Akad. Nauk SSSR*, **215**, 1182–1185 [English translation: *Dokl. Chem. Technol.*, **215** (5), 68–71].
14. de Vries, G.C. (1999) Electrochromic variable transmission glass for picture tubes. *Electrochim. Acta*, **44**, 3185–3193.
15. Baloukas, B., Lamarre, J.-M., and Martinu, L. (2011) Active metameric security devices using an electrochromic material. *Appl. Opt.*, **50**, C41–C49.
16. Baucke, F.G.K. (1991) Electrochromic applications. *Mater. Sci. Eng., B*, **10**, 285–292.
17. Byker, H.J. (1994) in *Proceedings of the Symposium on Electrochromic Materials II* (eds K.-C. Ho and D.A. MacArthur), The Electrochemical Society, Pennington, NJ, pp. 3–13.
18. Lampert, C.M. (1984) Electrochromic materials and devices for energy efficient windows. *Sol. Energy Mater.*, **11**, 1–27.
19. Svensson, J.S.E.M. and Granqvist, C.G. (1984) Electrochromic tungsten oxide films for energy efficient windows. *Sol. Energy Mater.*, **11**, 29–34.
20. Svensson, J.S.E.M. and Granqvist, C.G. (1985) Electrochromic coatings for "smart windows". *Sol. Energy Mater.*, **12**, 391–402.
21. Svensson, J.S.E.M. and Granqvist, C.G. (1985) Electrochromic coatings for smart windows: crystalline and amorphous WO_3 films. *Thin Solid Films*, **126**, 31–36.
22. Ma, C., Taya, M., and Xu, C. (2008) Smart sunglasses based on electrochromic polymers. *Polym. Eng. Sci.*, **48**, 2224–2228.
23. Buyan, M., Brühwiler, P.A., Azens, A., Gustavsson, G., Karmhag, R., and Granqvist, C.G. (2006) Facial warming and tinted helmet visors. *Int. J. Ind. Ergon.*, **36**, 11–16.
24. Demiryont, H. and Moorehead, D. (2009) Electrochromic emissivity modulator for spacecraft thermal management. *Sol. Energy Mater. Sol. Cells*, **93**, 2075–2078.
25. Metts, J.G., Nabity, J.A., and Klaus, D.M. (2011) Theoretical performance analysis of electrochromic radiators for space suit thermal control. *Adv. Space Res.*, **47**, 1256–1264.
26. Teissier, A., Dudon, J.-P., Aubert, P.-H., Vidal, F., Remaury, S., Crouzet, J., and Chevrot, C. (2012) Feasibility of conducting semi-IPN with variable electro-emissivity: a promising way for spacecraft thermal control. *Sol. Energy Mater. Sol. Cells*, **99**, 116–122.
27. Sauvet, K., Sauques, L., and Rougier, A. (2009) IR electrochromic WO_3 thin films: from optimization to devices. *Sol. Energy Mater. Sol. Cells*, **93**, 2045–2049.
28. Brown, R.M. and Hillman, A.R. (2012) Electrochromic enhancement of latent fingerprints by poly(3,4-ethylenedioxythiophene). *Phys. Chem. Chem. Phys.*, **14**, 8653–8661.
29. Invernale, M.A., Ding, Y., and Sotzing, G.A. (2010) All-organic electrochromic spandex. *ACS Appl. Mater. Interfaces*, **2**, 296–300.
30. Granqvist, C.G. (2000) Electrochromic tungsten oxide films: review of progress 1993–1998. *Sol. Energy Mater. Sol. Cells*, **60**, 201–262.
31. Granqvist, C.G., Avendaño, E., and Azens, A. (2003) Electrochromic coatings and devices: survey of some recent advances. *Thin Solid Films*, **442**, 201–211.
32. Granqvist, C.G. (2007) Transparent conductors as solar energy materials: a panoramic review. *Sol. Energy Mater. Sol. Cells*, **91**, 1529–1598.

33. Niklasson, G.A. and Granqvist, C.G. (2007) Electrochromics for smart windows: thin films of tungsten oxide and nickel oxide, and devices based on these. *J. Mater. Chem.*, **17**, 127–156.
34. Granqvist, C.G., Azens, A., Heszler, P., Kish, L.B., and Österlund, L. (2007) Nanomaterials for benign indoor environments: electrochromics for "smart windows", sensors for air quality, and photo-catalysis for air cleaning. *Sol. Energy Mater. Sol. Cells*, **91**, 355–365.
35. Pulker, H. (1999) *Coatings on Glass*, 2nd edn, Elsevier, Amsterdam.
36. Gläser, H.J. (2000) *Large Area Glass Coating*, von Ardenne Anlagentechnik, Dresden.
37. Ohring, M. (2002) *The Materials Science of Thin Films: Deposition and Structure*, 2nd edn, Academic Press, New York.
38. Mattox, D.M. (2010) *Handbook of Physical Vapor Deposition (PVD) Processing*, 2nd edn, Elsevier/William Andrew, Norwich, NY.
39. Granqvist, C.G. (2012) Preparation of thin films and nanostructured coatings for clean tech applications: a primer. *Sol. Energy Mater. Sol. Cells*, **99**, 166–175.
40. Piegari, A. and Flory, F. (2013) *Optical Thin Films and Coatings: From Materials to Applications*, Woodhead, Cambridge.
41. Zheng, H., Ou, J.Z., Strano, M.S., Kaner, R.B., Mitchell, A., and Kalantar-zadeh, K. (2011) Nanostructured tungsten oxide: properties, synthesis, and applications. *Adv. Funct. Mater.*, **21**, 2175–2196.
42. Hashimoto, S. and Matsuoka, H. (1991) Lifetime of electrochromism of amorphous WO_3-TiO_2 thin films. *J. Electrochem. Soc.*, **138**, 2403–2408.
43. Göttsche, J., Hinsch, A., and Wittwer, V. (1993) Electrochromic mixed WO_3-TiO_2 thin films produced by sputtering and the sol–gel technique: a comparison. *Sol. Energy Mater. Sol. Cells*, **31**, 415–428.
44. Lin, F., Cheng, J., Engtrakul, C., Dillon, A.C., Nordlund, D., Moore, R.G., Weng, T.-C., Williams, S.K.R., and Richards, R.M. (2012) *In situ* crystallization of high performing WO_3-based electrochromic materials and the importance for durability and switching kinetics. *J. Mater. Chem.*, **22**, 16817–16823.
45. Ramana, C.V., Baghmar, G., Rubio, E.J., and Hernandez, M.J. (2013) Optical constants of amorphous, transparent titanium-doped tungsten oxide thin films. *ACS Appl. Mater. Interfaces*, **5**, 4659–4666.
46. Manciu, F.S., Yun, Y., Durrer, W.G., Howard, J., Schmidt, U., and Ramana, C.V. (2012) Comparative microscopic and spectroscopic analysis of temperature-dependent growth of WO_3 and $W_{0.95}Ti_{0.05}O_3$ thin films. *J. Mater. Sci.*, **47**, 6593–6600.
47. Avendaño, E., Azens, A., Niklasson, G.A., and Granqvist, C.G. (2004) Electrochromism in nickel oxide films containing Mg, Al, Si, V, Zr, Nb, Ag, or Ta. *Sol. Energy Mater. Sol. Cells*, **84**, 337–350.
48. Gillaspie, D., Norman, A., Tracy, C.E., Pitts, J.R., Lee, S.-H., and Dillon, A. (2010) Nanocomposite counter electrode materials for electrochromic windows. *J. Electrochem. Soc.*, **157**, H328–H331.
49. Lin, F., Nordlund, D., Weng, T.-C., Moore, R.G., Gillaspie, D.T., Dillon, A.C., Richards, R.M., and Engtrakul, C. (2013) Hole doping in Al-containing nickel oxide materials to improve electrochromic performance. *ACS Appl. Mater. Interfaces*, **5**, 301–309.
50. Lin, F., Nordlund, D., Weng, T.-C., Sokaras, D., Jones, K.M., Reed, R.B., Gillaspie, D.T., Weir, D.G., Moore, R.G., Dillon, A.C., Richards, R.M., and Engtrakul, C. (2013) Origin of electrochromism in high-performing nanocomposite nickel oxide. *ACS Appl. Mater. Interfaces*, **5**, 3643–3649.
51. Cha, I.Y., Park, S.H., Lim, J.W., Yoo, S.J., and Sung, Y.-E. (2013) The activation process through a bimodal transmittance state for improving electrochromic performance of nickel oxide thin film. *Sol. Energy Mater. Sol. Cells*, **108**, 22–26.
52. Burdis, M.S., Siddle, J.R., Batchelor, R.A., and Gallego, J.M. (1998)

$V_{0.50}Ti_{0.50}O_x$ thin films as counter-electrodes for electrochromic devices. *Sol. Energy Mater. Sol. Cells*, **54**, 93–98.
53. Marcel, C., Brigouleix, C., Vincent, A., Plessis, D., Nouhaud, G., Hamon, Y., Sabary, F., and Campet, G. (2003) in *Electrochromic Materials and Applications* (eds A. Rougier, D. Rauh, and G.A. Nazri), The Electrochemical Society, Pennington, NJ, pp. 218–230.
54. Granqvist, C.G. (1994) Electrochromic oxides: a unified view. *Solid State Ion.*, **70/71**, 678–685.
55. Enjalbert, R. and Galy, J. (1986) A refinement of the structure of V_2O_5. *Acta Crystallogr., Sect. C*, **42**, 1467–1469.
56. Haber, J., Witko, M., and Tokarz, R. (1997) Vanadium pentoxide I: structures and properties. *Appl. Catal., A*, **157**, 3–22.
57. Avendaño, E., Azens, A., Niklasson, G.A., and Granqvist, C.G. (2005) Proton diffusion and electrochromism in hydrated NiO_y and $Ni_{1-x}V_xO_y$ thin films. *J. Electrochem. Soc.*, **152**, F203–F205.
58. Avendaño, E., Rensmo, H., Azens, A., Sandell, A., Azevedo, G.d.M., Siegbahn, H., Niklasson, G.A., and Granqvist, C.G. (2009) Coloration mechanism in proton-intercalated hydrated NiO_y and $Ni_{1-x}V_xO_y$ thin films. *J. Electrochem. Soc.*, **156**, P132–P138.
59. Goodenough, J.B. (1971) in *Progress in Solid State Chemistry*, vol. **5** (ed H. Reiss), Pergamon, Oxford, pp. 145–399.
60. Talledo, A. and Granqvist, C.G. (1995) Electrochromic vanadium-pentoxide-based films: structural, electrochemical, and optical properties. *J. Appl. Phys.*, **77**, 4655–4666.
61. Gavrilyuk, A., Tritthart, U., and Gey, W. (2011) Photoinjection of hydrogen and the nature of a giant shift of the fundamental absorption edge in highly disordered V_2O_5 films. *Phys. Chem. Chem. Phys.*, **13**, 9490–9497.
62. Schirmer, O.F., Wittwer, V., Baur, G., and Brandt, G. (1977) Dependence of WO_3 electrochromic absorption on crystallinity. *J. Electrochem. Soc.*, **124**, 749–753.
63. Bryksin, V.V. (1982) Optical intra-band absorption in disordered systems with strong electron–phonon interaction. *Fiz. Tverd. Tela*, **24**, 1110–1117 [English translation: *Sov. Phys. Solid State*, **24**, 627–631].
64. He, T. (1995) Optical absorption of free small polarons at high temperatures. *Phys. Rev. B*, **51**, 16689–16694.
65. Berggren, L., Azens, A., and Niklasson, C.G. (2001) Polaron absorption in amorphous tungsten oxide films. *J. Appl. Phys.*, **90**, 1860–1863.
66. Faughnan, B.W., Crandall, R.S., and Heyman, P.M. (1975) Electrochromism in WO_3 amorphous films. *RCA Rev.*, **36**, 177–197.
67. Gabrusenoks, J.V., Cikmach, P.D., Lusis, A.R., Kleperis, J.J., and Ramans, G.M. (1984) Electrochromic colour centres in amorphous tungsten trioxide thin films. *Solid State Ion.*, **14**, 25–30.
68. Saenger, M.F., Höing, T., Hofmann, T., and Schubert, M. (2008) Polaron transitions in charge intercalated amorphous tungsten oxide thin films. *Phys. Status Solidi A*, **205**, 914–917.
69. Denesuk, M. and Uhlmann, D.R. (1996) Site-saturation model for the optical efficiency of tungsten-oxide-based devices. *J. Electrochem. Soc.*, **143**, L186–L188.
70. Berggren, L. and Niklasson, G.A. (2006) Optical charge transfer absorption in lithium-intercalated tungsten oxide thin films. *Appl. Phys. Lett.*, **88**, 081906/1–081906/3.
71. Berggren, L., Jonsson, J.C., and Niklasson, G.A. (2007) Optical absorption in lithiated tungsten oxide thin films: experiments and theory. *J. Appl. Phys.*, **102**, 083538/1–083538/7.
72. Yamada, Y., Tajima, K., Bao, S., Okada, M., and Yoshimura, K. (2009) Optical charge transfer absorption in proton injected tungsten oxide thin films analyzed with spectroscopic ellipsometry. *Solid State Ion.*, **180**, 659–661.
73. Balaji, S., Djaoued, Y., Albert, A.-S., Ferguson, R.Z., and Brüning, R. (2009) Hexagonal tungsten oxide based electrochromic devices: spectroscopic evidence for the Li ion occupancy

of four-coordinated square windows. *Chem. Mater.*, **21**, 1381–1389.

74. Kharade, R.R., Patil, K.R., Patil, P.S., and Bhosale, P.N. (2012) Novel microwave assisted sol–gel synthesis (MW-SGS) and electrochromic performance of petal like *h*-WO_3 thin films. *Mater. Res. Bull.*, **47**, 1787–1793.
75. Nanba, T. and Yasui, I. (1989) X-ray diffraction study of microstructure of amorphous tungsten trioxide films prepared by electron beam vacuum evaporation. *J. Solid State Chem.*, **83**, 304–315.
76. Thornton, J.A. (1977) High-rate thin film growth. *Annu. Rev. Mater. Sci.*, **7**, 239–260.
77. Sun, X., Liu, Z., and Cao, H. (2010) Effects of film density on electrochromic tungsten oxide thin films deposited by reactive dc-pulsed magnetron sputtering. *J. Alloys Compd. (Suppl.)*, **504S**, S418–S421.
78. Le Bellac, D., Azens, A., and Granqvist, C.G. (1995) Angular selective transmittance through electrochromic tungsten oxide films made by oblique angle sputtering. *Appl. Phys. Lett.*, **66**, 1715–1716.
79. Beydaghyan, G., Renaud, J.-L.M., Bader, G., and Ashrit, P.V. (2008) Enhanced electrochromic properties of heat treated nanostructured tungsten trioxide thin films. *J. Mater. Res.*, **23**, 274–280.
80. Gil-Rostra, J., Cano, M., Pedrosa, J.M., Ferrer, J.F., García-García, F., Yubero, F., and González-Elipe, A.R. (2011) Electrochromic behavior of $W_xSi_yO_z$ thin films prepared by reactive magnetron sputtering at normal and glancing angles. *ACS Appl. Mater. Interfaces*, **4**, 628–638.
81. Nagata, S., Fujita, H., Inouye, A., Yamamoto, S., Tsuchiya, B., and Shikama, T. (2010) Ion irradiation effects on the optical properties of tungsten oxide films. *Nucl. Instrum. Methods Phys. Res., Sect. B*, **268**, 3151–3154.
82. Sai, L., Tang, L., Huang, X., Chen, G., Zhao, J., and Wang, J. (2012) Lowest-energy structures of $(WO_3)_n$ $(2 \leq n \leq 12)$ clusters from first-principles global search. *Chem. Phys. Lett.*, **544**, 7–12.
83. Maleknia, S., Brodbelt, J., and Pope, K. (1991) Characterization of the reactive and dissociative behavior of transition metal oxide cluster ions in the gas phase. *J. Am. Soc. Mass Spectrom.*, **2**, 212–219.
84. Lang, Z.-L., Guan, W., Yan, L.-K., Wen, S.-Z., Su, Z.-M., and Hao, L.-Z. (2012) The self-assembly mechanism of the Lindqvist anion $[W_6O_{19}]^{2-}$ in aqueous solution: a density functional theory study. *Dalton Trans.*, **41**, 11361–11368.
85. Sorar, I., Pehlivan, E., Niklasson, G.A., and Granqvist, C.G. (2013) Electrochromism of DC magnetron sputtered TiO_2 thin films: role of deposition parameters. *Sol. Energy Mater. Sol. Cells*, **115**, 172–180.
86. Sorar, I., Pehlivan, E., Niklasson, G.A., and Granqvist, C.G. (2014) Electrochromism of DC magnetron-sputtered TiO_2: role of film thickness. *Appl. Surf. Sci.*, **318**, 24–27.
87. Green, S.V., Watanabe, M., Oka, N., Niklasson, G.A., Granqvist, C.G., and Shigesato, Y. (2012) Electrochromic properties of nickel oxide based thin films sputter deposited in the presence of water vapor. *Thin Solid Films*, **520**, 3839–3842.
88. Ueta, H., Abe, Y., Kato, K., Kawamura, M., Sasaki, K., and Itoh, H. (2009) Ni oxyhydroxide thin films prepared by reactive sputtering using $O_2 + H_2O$ mixed gas. *Jpn. J. Appl. Phys.*, **48**, 015501/1–015501/4.
89. Abe, Y., Ueta, H., Obata, T., Kawamura, M., Sasaki, K., and Itoh, H. (2010) Effects of sputtering gas pressure on electrochromic properties of Ni oxyhydroxide thin films prepared by reactive sputtering in H_2O atmosphere. *Jpn. J. Appl. Phys.*, **49**, 115802/1–115802/4.
90. Yuan, Y.F., Xia, X.H., Wu, J.B., Chen, Y.B., Yang, J.L., and Guo, S.Y. (2011) Enhanced electrochromic properties of ordered porous nickel oxide thin film prepared by self-assembled colloidal crystal template-assisted electrodeposition. *Electrochim. Acta*, **56**, 1208–1212.

91. Oka, N., Watanabe, M., Sugie, K., Iwabuchi, Y., Kotsubo, H., and Shigesato, Y. (2013) Reactive-gas-flow sputter deposition of amorphous WO_3 films for electrochromic devices. *Thin Solid Films*, **532**, 1–6.
92. Hemberg, A., Dauchot, J.-P., Snyders, R., and Konstantinidis, S. (2012) Evaporation-assisted high-power impulse magnetron sputtering: the deposition of tungsten oxide as a case study. *J. Vac. Sci. Technol., A*, **30**, 040604/1–040604/4.
93. Green, S.V., Kuzmin, A., Purans, J., Granqvist, C.G., and Niklasson, G.A. (2011) Structure and composition of sputter-deposited nickel–tungsten oxide films. *Thin Solid Films*, **519**, 2062–2066.
94. Green, S.V., Pehlivan, E., Granqvist, C.G., and Niklasson, G.A. (2012) Electrochromism in sputter deposited nickel-oxide films. *Sol. Energy Mater. Sol. Cells*, **99**, 339–344.
95. Green, S.V. (2012) *Electrochromic Nickel–Tungsten Oxides: Optical, Electrochemical and Structural Characterization of Sputter-Deposited Thin Films in the Whole Composition Range*, Digital Comprehensive Summaries of Uppsala Dissertations from the Faculty of Science and Technology, vol. **963**, Acta Universitatis Upsaliensis, Uppsala, ISBN 978-91-554-8444-6.
96. Green, S.V., Granqvist, C.G., and Niklasson, G.A. (2014) Structure and optical properties of electrochromic tungsten-containing nickel oxide films. *Sol. Energy Mater. Sol. Cells*, **126**, 248–259.
97. Valyukh, I., Green, S.V., Granqvist, C.G., Niklasson, G.A., Valyukh, S., and Arwin, H. (2011) Optical properties of thin films of mixed Ni–W oxide made by reactive DC magnetron sputtering. *Thin Solid Films*, **519**, 2914–2918.
98. Valyukh, I., Green, S.V., Granqvist, C.G., Gunnarsson, K., Arwin, H., and Niklasson, G.A. (2012) Ellipsometrically determined optical properties of nickel-containing tungsten oxide thin films: nanostructure inferred from effective medium theory. *J. Appl. Phys.*, **112**, 044308/1–044308/6.
99. Ellmer, K., Klein, A., and Rech, B. (eds) (2008) *Transparent Conductive Zinc Oxide: Basics and Applications*, Springer, Berlin.
100. Ginley, D.S., Hosono, H., and Paine, D.C. (eds) (2010) *Handbook of Transparent Conductors*, Springer Science+Business Media, New York.
101. Ellmer, K. (2012) Past achievements and future challenges in the development of optically transparent electrodes. *Nat. Photonics*, **6**, 809–817.
102. WINDOW International Glazing Database, Lawrence Berkeley National Laboratory, Berkeley, CA, *http://windows.lbl.gov/software/window/window.html* (accessed 19 February 2015).
103. Hamberg, I. and Granqvist, C.G. (1986) Evaporated Sn-doped In_2O_3 films: basic optical properties and applications to energy-efficient windows. *J. Appl. Phys.*, **60**, R123–R159.
104. Jin, Z.-C. and Granqvist, C.G. (1988) Optical properties of sputter-deposited ZnO:Al thin films. *J. Appl. Phys.*, **64**, 5117–5131.
105. Stjerna, B., Olsson, E., and Granqvist, C.G. (1994) Optical and electrical properties of radio frequency sputtered tin oxide films doped with oxygen vacancies, F, Sb, or Mo. *J. Appl. Phys.*, **76**, 3707–3817.
106. Cairns, D.R., Witte, R.P. II,, Sparacin, D.K., Sachsman, S.M., Paine, D.C., Crawford, G.P., and Newton, R.R. (2000) Strain-dependent electrical resistance of tin-doped indium oxide on polymer substrates. *Appl. Phys. Lett.*, **76**, 1425–1427.
107. Kim, E.-H., Yang, C.-W., and Park, J.-W. (2011) The crystallinity and mechanical properties of indium tin oxide coatings on polymer substrates. *J. Appl. Phys.*, **109**, 043511/1–043511/8.
108. Schwartz-Schampera, U. and Herzig, P.M. (2002) *Indium: Geology, Mineralogy, and Economics*, Springer, Berlin.
109. Cummings, K.J., Nakano, M., Omae, K., Takeuchi, K., Chonan, T., Xiao, Y.-L., Harley, R.A., Roggli, V.L., Hebisawa, A., Tallaksen, R.J., Trapnell, B.C., Day, G.A., Saito, R., Stanton, M.L.,

Suarthana, E., and Kreiss, K. (2012) Indium lung disease. *Chest*, **141**, 1512–1521.

110. Ostermann, R. and Smarsly, B. (2009) Does mesoporosity enhance thin film properties? A question of electrode material for electrochromism of WO_3. *Nanoscale*, **1**, 266–270.

111. Smith, G.B., Niklasson, G.A., Svensson, J.S.E.M., and Granqvist, C.G. (1986) Noble-metal-based transparent infrared reflectors: experiments and theoretical analyses for very thin gold films. *J. Appl. Phys.*, **59**, 571–581.

112. Hövel, M., Gompf, B., and Dressel, M. (2010) Dielectric properties of ultrathin films around the percolation threshold. *Phys. Rev. B*, **81**, 035402/1–035402/8.

113. Lansåker, P.C., Backholm, J., Niklasson, G.A., and Granqvist, C.G. (2009) TiO_2/Au/TiO_2 multilayer thin films: novel metal-based transparent conductors for electrochromic devices. *Thin Solid Films*, **518**, 1225–1229.

114. Smith, G.B. and Granqvist, C.G. (2010) *Green Nanotechnology: Solutions for Sustainability and Energy in the Built Environment*, CRC Press, Boca Raton, FL.

115. Jelle, B.P. (2013) Solar radiation glazing factors for window panes, glass structures and electrochromic windows in buildings: measurement and calculation. *Sol. Energy Mater. Sol. Cells*, **116**, 291–323.

116. Roth, S. and Park, H.J. (2010) Nanocarbonic transparent conductive films. *Chem. Soc. Rev.*, **39**, 2477–2483.

117. Niu, C. (2011) Carbon nanotube transparent conducting films. *MRS Bull.*, **36**, 766–773.

118. Feng, C., Liu, K., Wu, J.-S., Liu, L., Cheng, J.-S., Zhang, Y., Sun, Y., Li, Q., Fan, S., and Jiang, K. (2010) Flexible, stretchable, transparent conducting films made from superaligned carbon nanotubes. *Adv. Funct. Mater.*, **20**, 885–891.

119. Novoselov, K.S., Fal'ko, V.I., Colombo, L., Gellert, P.R., Schwab, M.G., and Kim, K. (2012) A roadmap for graphene. *Nature*, **490**, 192–200.

120. De, S., King, P.J., Lotya, M., O'Neill, A., Doherty, E.M., Hernandez, Y., Duesberg, G.S., and Coleman, J.N. (2010) Flexible, transparent, conducting films of randomly stacked graphene from surfactant-stabilized, oxide-free graphene dispersions. *Small*, **6**, 458–464.

121. De Arco, L.G., Zhang, Y., Schlenker, C.W., Ryu, K., Thompson, M.E., and Zhou, C. (2010) Continuous, highly flexible, and transparent graphene films by chemical vapor deposition for organic photovoltaics. *ACS Nano*, **4**, 2865–2873.

122. Bae, S., Kim, H., Lee, Y., Xu, X., Park, J.-S., Zheng, Y., Balakrishnan, J., Lei, T., Kim, H.R., Song, Y.I., Kim, Y.-J., Kim, K.S., Özyilmaz, B., Ahn, J.-H., Hong, B.H., and Iijima, S. (2010) Roll-to-roll production of 30-inch graphene films for transparent electrodes. *Nat. Nanotechnol.*, **5**, 574–578.

123. Wu, H., Kong, D., Ruan, Z., Hsu, P.-C., Wang, S., Yu, Z., Carney, T.J., Hu, L., Fan, S., and Cui, Y. (2013) A transparent electrode based on a metal nanotrough network. *Nat. Nanotechnol.*, **8**, 421–425.

124. Preston, C., Xu, Y., Han, X., Munday, J.N., and Hu, L. (2013) Optical haze of transparent and conductive silver nanowire films. *Nano Res.*, **6**, 461–468.

125. Elschner, A. and Lövenich, W. (2011) Solution-deposited PEDOT for transparent conductive applications. *MRS Bull.*, **36**, 794–798.

126. Frenning, G., Engelmark, F., Niklasson, G.A., and Strømme, M. (2001) Li conduction in sputtered amorphous Ta_2O_5. *J. Electrochem. Soc.*, **148**, A418–A421.

127. Yoo, S.J., Lim, J.W., and Sung, Y.-E. (2006) Improved electrochromic devices with an inorganic solid electrolyte protective layer. *Sol. Energy Mater. Sol. Cells*, **90**, 477–484.

128. Yang, H., Wang, C., Diao, X., Wang, H., Wang, T., and Zhu, K. (2008) A new all-thin-film electrochromic device using LiBSO as the ion conducting layer. *J. Phys. D Appl. Phys.*, **41**, 115301/1–115301/5.

129. Soutar, A.M., Rosseinsky, D.R., Freeman, W., Zhang, X., How, X., Jiang, H., Zeng, X., and Miso, X. (2012) Electrochromic cell with UV-curable

electrolyte polymer for cohesion and strength. *Sol. Energy Mater. Sol. Cells*, **100**, 268–270.

130. Sydam, R., Deepa, M., and Srivastava, A.K. (2012) Electrochromic device response controlled by an *in situ* polymerized ionic liquid based gel electrolyte. *RSC Adv.*, **2**, 9011–9021.
131. Azens, A., Kullman, L., and Granqvist, C.G. (2003) Ozone coloration of Ni and Cr oxide films. *Sol. Energy Mater. Sol. Cells*, **76**, 147–153.
132. Aydogdu, G.H., Ruzmetov, D., and Ramanathan, S. (2010) Metastable oxygen incorporation into thin film NiO by low temperature active oxidation: influence of hole conduction. *J. Appl. Phys.*, **108**, 113702/1–113702/6.
133. Papaefthimiou, S., Syrrakou, E., and Yianoulis, P. (2009) An alternative approach for the energy and environmental rating of advanced glazing: an electrochromic window case study. *Energy Build.*, **41**, 17–26.
134. Posset, U., Harsch, M., Rougier, A., Herbig, B., Schottner, G., and Sextl, G. (2012) Environmental assessment of electrically controlled variable light transmittance devices. *RSC Adv.*, **2**, 5990–5996.
135. Baetens, R., Jelle, B.P., and Gustavsen, A. (2010) Properties, requirements and possibilities of smart windows for dynamic daylight and solar energy control in buildings: a state-of-the-art review. *Sol. Energy Mater. Sol. Cells*, **94**, 87–105.
136. Granqvist, C.G. (2012) Oxide electrochromics: an introduction to devices and materials. *Sol. Energy Mater. Sol. Cells*, **99**, 1–13.
137. Jelle, B.P., Hynd, A., Gustavsen, A., Arasteh, D., Goudey, H., and Hart, R. (2012) Fenestration today and tomorrow: a state-of-the-art review and future research opportunities. *Sol. Energy Mater. Sol. Cells*, **96**, 1–28.
138. Sbar, N., Badding, M., Budziak, R., Cortez, K., Laby, L., Michalski, L., Ngo, T., Schulz, S., and Urbanik, K. (1999) Progress toward durable, cost effective electrochromic window glazings. *Sol. Energy Mater. Sol. Cells*, **56**, 321–341.
139. Béteille, F., Boire, P., and Giron, J.-C. (1999) Highly durable all-solid-state electrochromic glazings. *Proc. Soc. Photo-Opt. Instrum. Eng.*, **3788**, 70–74.
140. Mathew, J.G.H., Sapers, S.P., Cumbo, M.J., O'Brien, N.A., Sargent, R.B., Raksha, V.P., Lahaderne, R.B., and Hichwa, B.P. (1997) Large area electrochromics for architectural applications. *J. Non-Cryst. Solids*, **218**, 342–346.
141. O'Brien, N.A., Gordon, J., Mathew, H., and Hichwa, B.P. (1999) Electrochromic coatings: applications and manufacturing issues. *Thin Solid Films*, **345**, 312–318.
142. Zinzi, M. (2006) Office worker preferences of electrochromic windows: a pilot study. *Build. Environ.*, **41**, 1262–1273.
143. Xu, C., Ma, C., Kong, X., and Taya, M. (2009) Vacuum filling process for electrolyte in enhancing electrochromic polymer window assembly. *Polym. Adv. Technol.*, **20**, 178–182.
144. Kraft, A. and Rottman, M. (2009) Properties, performance and current status of the laminated electrochromic glass of Gesimat. *Sol. Energy Mater. Sol. Cells*, **93**, 2088–2092.
145. Azens, A., Gustavsson, G., Karmhag, R., and Granqvist, C.G. (2003) Electrochromic devices on polyester foil. *Solid State Ion.*, **165**, 1–5.
146. Bishop, C.A. (2011) *Vacuum Deposition onto Webs, Films, and Foils*, 2nd edn, William Andrew, Waltham, MA.

2
Electrochromic Materials Based on Prussian Blue and Other Metal Metallohexacyanates

David R. Rosseinsky and Roger J. Mortimer

2.1
The Electrochromism of Prussian Blue

2.1.1
Introduction

The widely used pigment [1] Prussian blue (PB) was fortuitously invented by J. J. Diesbach in Berlin in 1704 [2]. The chromic centre ('chromophore') of 'PB' is $Fe^{III}[Fe^{II}(CN)_6]$, or 'ferric ferrocyanide' colloquially. The Roman superscripts, the oxidation states, just approximate the actual charges. Under 'white' illumination, Blue arises from the electron-transferring optical transition

$$Fe^{III}[Fe^{II}(CN)_6] + h\nu \rightarrow Fe^{II}[Fe^{III}(CN)_6] \quad (2.1)$$

the product – an excited state – rapidly decaying back to the original.

The driven electron transfer at a film of PB on a transparent electrode that comprises its electrochromism from blue (PB) to clear (labelled 'ES', as explained as follows):

$$Fe^{III}[Fe^{II}(CN)_6] + e^- \rightarrow Fe^{II}[Fe^{II}(CN)_6], \text{ that is, } PB + e^- \rightarrow ES \quad (2.2)$$

at a sufficiently negative potential applied to the substrate conductor. Typically in aqueous chloride-containing electrolyte, the switch occurs between 0.6 and −0.2 V versus Ag/AgCl. The product in Eq. (2.2) is the colourless form, which in bulk is called 'Everitt's Salt', ES. On reversing the applied potential, on current flow the process re-colourises the film to PB:

$$Fe^{II}[Fe^{II}(CN)_6] \rightarrow Fe^{III}[Fe^{II}(CN)_6] + e^- \quad (2.3)$$

Much of the earlier work is encompassed in several monographs [3–5]. An outline of the chemistry follows.

Electrochromic Materials and Devices, First Edition.
Edited by Roger J. Mortimer, David R. Rosseinsky and Paul M. S. Monk.

2.1.2
Electrodeposited PB Film and Comparisons with Bulk PB

PB film shows a maximum absorbance ranging over 650–720 nm. The range arises because the precise chemical composition does vary even in carefully reproduced preparations: the high insolubility of the substance imposes a marked composition sensitivity to precise preparative conditions and ions (cations usually) incorporated in the lattice [6] and also on the extent of crystallinity or amorphism. Furthermore, the maximum may shift with thickness of the PB film [7]. In this chapter, acid-containing aqueous KCl solution is assumed to be the electrolyte in contact with the electrochrome, unless otherwise stated. Other electrolytes may involve different mechanistic analyses, but if aqueous, usually not dissimilar to KCl. In balancing the 3+ of Fe^{III} and the anionic 4−, the required cation has earlier been deemed 'supernumerary', implicitly unrelated to the electrochromic responses of Eqs. (2.2) and (2.3), but revised views now hold as outlined next. Such a 'supernumerary' could be $1/3(Fe^{3+})$, but K^+ or another M^+ ion could serve if present.

Lattice energy calculations [8], for bulk PB precipitated from component solutions, show the K^+-containing species to be thermodynamically more stable than the $1/3(Fe^{3+})$-containing PB though the title as published [8] is undoubtedly incorrect regarding PB films as follows. For historically incorrect reasons, bulk PB of composition $(Fe^{3+})_{1/3}$ $(Fe^{3+}[Fe^{II}(CN)_6]^{4-})$ comprising visible dispersed solid was termed 'Insoluble PB' (hereinafter InsolPB). This substance contrasted with the K^+-containing colloid – in early days not visible as such – called 'Soluble PB' (hereinafter SolPB). It is important to note that 'Soluble' and 'Insoluble' regarding PB clearly lack their ordinary chemical usages, which is to be borne in mind in reading much of the last decade's literature. The calculated lattice energy [8] accords with experiment: in bulk preparations from mixing Fe^{3+} and $[Fe^{II}(CN)_6]^{4-}$ containing solutions, if K^+ is present, SolPB invariably results. In a masterly understatement [9], 'The iron hexacyanoferrate compound ... known as Prussian blue is widely studied, although its chemistry is still not totally understood'.

Electrochromic PB film is often deposited on metals or transparent glass-based conductors such as indium-tin oxide (ITO) coated on glass [4, 5], by electroreduction from aqueous solutions of soluble salts of Fe^{3+} and $[Fe^{III}(CN)_6]^{3-}$ ions, basically represented as

$$Fe^{3+}[Fe^{III}(CN)_6]^{3-} + e^- \rightarrow Fe^{3+}[Fe^{II}(CN)_6]^{4-} \tag{2.4}$$

In reality, solute is adsorbed on the electrode [10] preceding the complete reduction, Eq. (2.5),

$$3Fe^{3+}([Fe^{III}(CN)_6]^{3-}) + Fe^{3+} + 3e^- \rightarrow (Fe^{3+})_4([Fe^{II}(CN)_6]^{4-})_3 \tag{2.5}$$

Analyses [8, 10] always show that the form in which film PB is initially electrodeposited corresponds closely in composition and structure [11] with InsolPB, Eq. (2.5), but it is claimed that *vacancies* have an important (actually dominant) role in PB-film electrochromism [12–14], as explained next. The new conclusions contradict long-held views. Thus, a part-transformation of initial InsolPB into

SolPB on PB/ES cycling was inferred [10] from both the relative bulk stabilities and the K^+ retention in PB/ES cycling, and copious such inferences (summarised in Refs. [3–5]) have been supported, for example, by XPS [15, 16]. The new conclusions have been drawn from notable chemical physics that has developed in recent years [17, 18] (regarding which many further references are summarised [9, 12–14, 19]). This includes XRD of PB film scraped from electrodes, spectroscopy and, most importantly, Electrochemical Quartz MicroBalance (EQMB) observations during the PB $\leftrightarrow$ ES voltammetric cycling [17].

From EQMB, all the original Fe^{3+} of the InsolPB composition is shown to be retained (as Fe^{2+}) in the reduction to ES. So, on re-oxidation to PB, the InsolPB composition recurs, that is, no Fe ions are 'supernumeraries'. As was found by Ludi's group in bulk crystals [11], about 25% of the ferrocyanide sites in film PB are in fact vacancies [9] (up to about 31% [19]). Ludi and co-workers [11] assigned O bonds (from H_2O or OH^-) to the Fe^{III} otherwise lacking a ligand at the vacancy. From film XRD [9], it was likewise inferred that into every vacancy, four Fe^{III} ions protrude from the lattice, each with an OH^- ligand replacing the lacking –NC (ferrocyanide) link [9]. For charge balance, completely electro-reduced $(Fe^{II})_4[Fe^{II}(CN)_6]_3$ acquires four K^+ in the vacancy:

$$\underset{\text{blue}}{(Fe^{III})_4[Fe^{II})(CN)_6]_3} + 4K^+ + 4e^- \rightarrow \underset{\text{colourless}}{K_4(Fe^{II})_4[Fe^{II}(CN)_6]_3} \quad (2.6)$$

The EQMB study [17] shows that, at part of the way through the reduction, Eq. (2.6), there is a transformation ('changeover') where only some 23% of the Fe^{III} of PB is found to be reduced, this reduction immediately preceding both a mass incursion of K^+ together with H_2O into the vacancy and the electrochromic colour change, Eq. (2.6). This 23% is deemed [9, 12] to be the $Fe^{III}–OH^-$ species in the vacancies, hence nicely approximating the vacancy population [20, 21].

K^+ can be retained in subsequent PB/ES cycles, accompanied by Cl^- or OH^- to preserve electroneutrality in the ES $\rightarrow$ PB step. Regarding pH, solutions of pH = 2 in the electrochromic cell [23, 24] allow much swifter PB/ES transformations than do higher pH solutions: possibly the more mobile H^+ rather than K^+ is the electroneutrality-preserving cation acquired or ejected in PB/ES switches [24] (cf. further views [9, 19]).

The new compositions established for PB transformations are extremely important for the field. However, these penetrating results, mainly from French, Spanish and Brazilian laboratories, are often unclearly expounded. The following examples exemplify: 'The spectroscopic band detected at 380 nm records the electronic charge transfer from iron(II) ions into high-spin sites to iron(III) ions into low-spin sites' [20]. In Figure 2.1, the arrows indicate the preferential $Fe^{3+}(NC)_5OH^-$ clusters or entities where electronic charge is preferentially occupied prior to the changeover process taking place [9].

Spin does need attention here. The useful ligand field theory shows that in free Fe^{II} or Fe^{III} ions, electrons will occupy five 3d orbitals of equal energy. In hexacoordinate compounds, two orbitals fall on the coordinate axes and suffer repulsion by ligands, and depending on the 'ligand field' magnitude, which

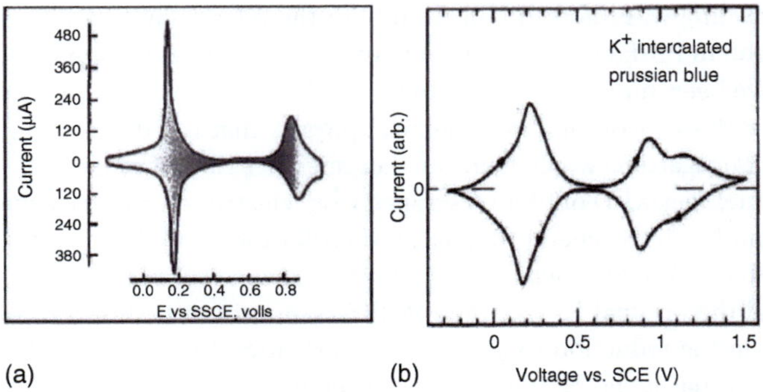

Figure 2.1 Cyclic voltammograms (see monographs [3–5]) in KCl solution of PB deposited (a) some few nanometres thick and (b) a thicker sample with KCl incorporated in PB, described in the main text. In (a), a single oxidative peak encompasses the steps of blue through (a visible) green to yellow PY, while in (b) there are two distinct peaks, one for blue/green and the second for green/yellow [3–5]. Reproduced with the permission of The Electrochemical Society [10].

correlates with the strength of the bonding, these e_g orbitals are raised in energy cf. the three interaxial t_{2g} orbitals escaping much of that repulsion. Fe–CN bonds are strong (hence, high ligand fields) while Fe–NC (and $FeOH_2$] bonds are weak (→low ligand fields]. Thus, in $Fe^{II}(CN)_6$ and $Fe^{III}(CN)_6$, the e_g orbitals are raised to such a level that the electrons overcome Pauli repulsion to pair up in the lower-energy t_{2g} orbitals, giving 'low-spin ferrocyanide'. The e_g orbitals are unoccupied. But $Fe^{III}(OH_2)_6$, for example, has electrons distributed over t_{2g} and e_g orbitals because of the weak ligand fields here. Thus, the optical transition conferring the blue coloration involves

$$Fe^{3+}(t_{2g}^3e_g^2)\ [Fe^{II}(t_{2g}^6e_g^0)(CN)_6] + h\nu \rightarrow {}^*Fe^{2+}[{}^*Fe^{III}(CN)_6] \tag{2.7}$$

where the conjectured spin states of the (asterisked) excited species are $Fe^{2+}(t_{2g}^4e_g^2)\ [Fe^{III}(t_{2g}^5e_g^0)(CN)_6]$, viz. a high-spin Fe^{II} and a low-spin Fe^{III}.

Assignment [20] of the aforementioned small 380 nm absorbance peak mentioned requires $Fe^{3+}(t_{2g}^5e_g^0)\ [Fe^{II}(t_{2g}^4e_g^2)(CN)_6]$ in the ground-state wave function, which appears somewhat exotic. (Were these configurations merely trace amounts of thermally populated species, then their already raised energies before optical charge transfer would lead to an absorbance lower in energy than the main 'blue' band, not higher at 380 nm, leaving an unresolved issue.)

It has been inferred [17] that H_2 is generated during the PB → ES reduction, which had not been seen in the multiple cycles noted: the reported 10^5 cycle [23–25] attests the robustness and integrity of the PB/ES system.

A useful PB-with-WO_3 complementary system [26] with the electrochromes roughly deposited and then coated with ammonium acetate gives efficient switching and colour intensity.

Other aspects [9], for example, the inference of some $Fe^{II}(CN)_5OH^-$ occurrence, are beyond the scope of this chapter, also the magnetic results [17, 27] which are suggested [9] to give rise to non-electrochemical contributions to observed currents. Wave-mechanical treatments seem unproductive in materials as variable in structure as PB and its other oxidation states [27], treating all such states as having the SolPB structure providing just such an example. The inference of semi-metallic conductivity in ES seems especially questionable; electrochemical impedance spectroscopy shows ionic conductivity to be the more likely [20, 21]. So far only the PB ↔ ES conversions have been considered. But PB has been oxidised all the way to the yellow $Fe^{3+}[Fe^{III}(CN)_6{}^{3-}]$, which we call Prussian yellow or PY (Figure 2.1) [10].

Cyclic voltammograms in the right conditions (e.g. slow enough scans) show an intermediate peak that corresponds to the pigment Berlin Green ('BG'). The thicker sample needs a wider potential range for its electrochromism, possibly from the modest conductivity [28] of the iron hexacyanoferrates across the range, the higher resistances requiring greater potential changes for bulk transformations to PY of the thicker material.

Colloid-derived PB films [29] show large voltammetric peaks for BG formation (and extended potential ranges as just alluded to). Analyses lacking, BG could contain a 50% : 50% anionic ratio, in a formula such as $(Fe^{3+})_n$ $[Fe^{II}(CN)_6]^{4-}$ $[Fe^{III}(CN)_6]^{3-}$, requiring $n = 7/3$ for that ratio. The final PY is intrinsically $Fe^{3+}[Fe^{III}(CN)_6]^{3-}$. However, PY films have not been analysed. With 10^5 PB/ES cycles [22, 23, 25], the −0.2 to +1.4 V scan with PY as the positive limit has not been so tested for endurance and may not be similarly reproducible if the extreme product PY film resembles the rather soluble [10] orange–yellow $Fe^{3+}([Fe^{III}(CN)_6]^{3-})$ complex occurring in aqueous solution. But hitherto untested non-aqueous or polymeric electrolytes should obviate such dissolution. In considering possible retention of the InsolPB iron-ion framework, positive-electrode repulsion would, however, seem to prevent retention by PY of more Fe^{2+} or extra Fe^{3+} to $Fe^{3+}([Fe^{III}(CN)_6]^{3-})$ during the ES ↔ PY cycling. But shedding K^+, or acquiring Cl^- , in the no-doubt persistent vacancies as seen in reductively cycled PB, could allow retention of that InsolPB-type Fe framework throughout ES ↔ PY cycling, as could be demonstrated by EQMB study [20].

Sometimes, the fully oxidised solid, often over-simplified as $Fe^{3+}([Fe^{III}(CN)_6]^{3-})$, in films as thick as 500 nm [29], is referred to as 'Berlin Green', but the observation of green could be attributed to the outer, visible layers being deprived of the final electron loss, forming inner orange–yellow PY, again owing to the modest conductivities of the species [28].

The aforementioned structural results contrast with the amorphous nature [30] found for different preparations of PB films (as also with higher current electrodeposition), emphasising the preparation-dependent nature of different PBs: any particular asserted structure or composition could be specific to that research.

2.1.3
PB Prepared from Direct Cell Reaction, with No Applied Potential

Ho [31] first suggested making PB by the spontaneous cell reaction with silver as second sacrificial electrode, short-circuited to the PB-deposition electrode, in a modification of Eq. (2.5):

$$3Fe^{3+}[Fe^{III}(CN)_6]^{3-} + Fe^{3+} + 3Ag \rightarrow (Fe^{3+})_4([Fe^{II}(CN)_6]^{4-})_3 + 3Ag^+ \quad (2.8)$$

a process with both benefits and drawbacks. The need for a driven electrochemical cell is obviated. But silver is expensive, and, for an even PB deposit, an equal area of Ag should oppose the depositing PB film. Furthermore, while but a few Ag^+ will penetrate PB, the long-term effects of Ag^+ ions in contact with PB are deleterious (PB incorporates Ag^+ spontaneously and deleteriously), and the formed PB film must be rapidly removed from the solution and washed.

A modification [32] of this process that obviates solid Ag is to use silver paint on the required electrode to be dipped into the reactant solution, around the area where PB is to be deposited, which cuts down on making a short-circuited two-electrode cell as just described, but would need fine Ag-paint tracks across the PB-deposition surface again for uniform deposition. Then also, all the exposed Ag paint has to be coated with insulant, to preserve PB against deleterious Ag^+. However, printing techniques could be devised if PB were on a flexible electrode.

2.1.4
Layer-by-Layer Deposition of PB

By alternate immersions in the PB-forming solutions containing Fe^{3+} and then $[Fe(CN)_6]^{4-}$, controlled layers of PB are deposited [33]. An identical so-called 'multiple-sequential deposition' method has been employed by Tieke and co-workers [34–36] for several systems.

2.1.5
PB on Graphene

A novel conductive substrate, graphene, has recently been tested [37]. A 1 cm^2 transparent graphene film of CVD origin was used to electrodeposit PB from a solution of 0.1 M KCl and HCl containing 5 mM $FeCl_3$ and $K_3Fe(CN)_6$. Several thicknesses of film (about 100 nm up to 1.2 µm needed for an 80% transmittance change) were examined. The times for coloration/decoloration were up to about 38 s (about twice that of a comparable PB/ITO system), that is, satisfactorily fast. The flexibility of graphene and large-scale versions appear attractive.

2.1.6
Alternative Preparations of PB: PB from Colloid and Similar Origins

SolPB has been deposited by electrostatic forces from the colloidal state directly onto positive Pt electrodes, and 185-nm-thick deposits have been produced [29], which were voltammetrically cycled between −0.4 and +1.2 V. As noted regarding Figure 2.1, only thin-film voltammetry shows the fully oxidised form to be yellow, as expected from the orange–yellow appearance of the $Fe^{3+}[Fe^{III}(CN)_6]^{3-}$ solutions used in electroreductive PB preparations.

PB has been obtained from other colloids. In a study [38, 39] (curiously interpreted in view of the well-known PB sol properties), acid aqueous $Na_4[Fe^{II}(CN)_6]$ and $Fe^{III}(NO_3)_3$ in stoichiometric amounts deposited InsolPB. After being stirred for 3 days in solution with Na salt, 'a transparent blue solution' ensued [38], no doubt colloidal Na-containing 'soluble' PB. After being dried by evaporation, the powder was re-suspended in water. A blue solid was insecurely attached from this liquid to an ITO-coated electrode by spin coating. Then multiple PB/ES cycles in a KPF_6/acetonitrile solution produced a stably attached eletrochromic film, XRD indicating several hundred nanometre thicknesses. 10^4 cycles showed no deterioration of colour intensity. Many similar claims of using truly 'soluble' PB have been made by these and associated authors [40].

Also claimed in patents [41], 'nanoparticle' PB applied as wet powder to conductive electrodes led to reliable devices, with an average PB particle size of 10–200 nm resulting from mechanical crushing and a claimed PB coating thickness of 50–1000 nm. This work follows earlier PB/polymer thin-film constructions [42, 43].

2.1.7 Alternative Electrolytes Including Polymeric for PB Electrochromism

Aqueous solutions are not necessarily the best for electrochromism. One [44], involving *inter alia* PB and poly(3,4-ethylenedioxythiophene) (PEDOT) electrochromes, uses flexible, transparent and self-supporting electrolyte films comprising poly(trimethylene carbonate)/poly(ethylene oxide) (pTMC/PEO) inter-penetrating networks containing $LiClO_4$, which allows a number of switches, but needs 10 min to attain full coloration. This is an occasional problem with polymer electrolytes, widely studied by this group. The Gesimat organisation, efficiently targeting commercial prospects [45], uses, amongst others, the polymer PVB (polyvinyl butyral) with Li salt as electrolyte, which works for numerous cycles of a PB-with-WO_3 cell. A comprehensive review of commercially promising systems [46] also includes polymer electrolytes, favouring WO_3 as electrochrome sometimes opposed with (the reliable) PB. Commercial claims regarding reliability, however, are said not always to be met [46].

A study was conducted [47] of a cell comprising a polymer electrolyte with PB and the opposing electrode a novel WO_3 made from acetic peroxytungstate dropped onto heated ITO-coated substrate. A 4 mol% $LiClO_4$-doped sol–gel material of 3-methacryl-oxypropyl-trimethoxy-silane (MEMO), together with zirconium *n*-propoxide, forms the electrolyte. Efficiency prevails: monomeric electrolyte is wiped onto the two electrochrome surfaces, which when combined are exposed to UV illumination to effect polymerisation without damage to electrochromes. The 100th and 200th PB/ES voltammetry cycles were virtually identical, so stability appears assured.

PB and WO_3 electrodes (often with KCl electrolyte) are widely used complementarily, and a number of alternatives for PB are listed in Ref. [5], notably plasma-polymerised ytterbium *bis*(phthalocyanine) and further combinations including a three-electrode array.

2.2 Metal Metallohexacyanates akin to Prussian Blue

Of the many prospectively electrochromic compounds amongst PB analogues, Ruthenium purple promises the most marked coloration changes, though redox colorations are not lacking among the rest. In a study [48] of PB analogues and their solid solutions, a number of redox colour changes were seen, which may promise electrochromism, not however sought in that study [48].

2.2.1 Ruthenium Purple RP

The coloured species $K^+Fe^{3+}[Ru^{II}(CN)_6]$ is established from bulk precipitate in the form of colloid [49], with powder XRD showing a cubic structure like PB [50]. As for PB, deposited films ensue [51–53] from electroreduction of solutions of Fe^{3+} and $[Ru^{III}(CN)_6]^{4-}$ ions or use of a sacrificial Cu anode (though the possible hazard to stability of Ag^+ with PB, comparable here, needs be noted). The absorbance maximum lies at about 550 nm. Reversible reduction to $(K^+)_2Fe^{2+}[Ru^{II}(CN)_6]$ decolourises the film, but unlike PB, oxidation evokes no further colour changes. The layer-by-layer deposition method introduced for PB [33] has been utilised by Jain *et al.* [54] to produce high-contrast coloration in RP.

2.2.2 Vanadium Hexacyanoferrate

While $(VO_2{}^+)_3[Fe^{III}(CN)_6]$ is conceivably the fully oxidised form of this substance [55, 56] and a possible electrochrome candidate, both metal ions being capable of one-electron reduction (vanadium to VO^{2+}, via $VO_2^+ + 2H^+ + e^- \rightarrow VO^{2+} + H_2O$), interactive electrochromism has not been found, possibly owing to variable oxide complexity of the vanadium ions.

2.2.3 Nickel Hexacyanoferrate

Nickel hexacyanoferrate from Ni-metal electrooxidation [57] in $[Fe^{III}(CN)_6]^{3-}$ solution, or the electroreduction [58] of Ni^{2+} and $[Fe^{III}(CN)_6]^{3-}$ solution, shows only a weak yellow to colourless electrochromism [59], which could just represent ferricyanide transitioning into ferrocyanide. However, a more novel Fe anion $[Fe^{II}(CN)_4(bipy)]^{2-}$ results [60] from adding as ligands both CN^- and 2,2′-bipyridine; the complex is itself bright red and electro-reducible to colourless. With Ni^{2+}, linked to Fe via the 2,2′-bipy, a film on an electrode can be cycled between red and transparent, apparently vividly. It is further speculated [61] that the complexes $[Ru^{II}(CN)_4(bipy)]^{2-}$ and $[Os^{II}(CN)_4(bipy)]^{2-}$ could further be prospective orange/transparent and green/transparent electrochromes, respectively.

2.3 Copper Hexacyanoferrate

The preparation here is more complicated than usual [62–65]. Thin Cu metal is electroplated from $Cu(NO_3)_2$ and $KClO_4$ solution onto glassy carbon or ITO-coated electrodes (which is pre-oxidised in this solution by cycling between 0.4 and 0.5 V). Cu was deposited then from 0.3 to −0.5 V. Cycling between −0.5 and +0.5 V was repeated until a reproducible voltammetry ensued. Then the Cu electrodeposited at −0.5 V in its preparative solution received an injected aliquot of $K_4[Fe^{II}(CN)_6]$ solution, depositing an immediate red-brown colloid, while a recyclable $Cu_2^{II}[Fe^{III}(CN)_6]$ film ensued on the electrode, capable of at least 1000 cycles of red-brown (reduced) to yellow (when oxidised). The latter is attributable to the internal 420 nm $Fe^{III}(CN)_6$ CT. In the reduced state, the Fe^{II} to Cu transition gives a broad band with $\lambda_{\varepsilon_{max}} = 490$ nm and extinction coefficient $\varepsilon = 2 \times 10^3\ dm^3\ mol^{-1}\ cm^{-1}$.

The elaborate preparation obviates the instability of films electrodeposited directly from Cu^{2+} and $K_3[Fe^{III}(CN)_6]$ solutions, underlining the physicochemical complexity of such films.

2.3.1 Palladium Hexacyanoferrate

Potential cycling [66] of a solution containing $PdCl_2$ and $K_3Fe(CN)_6$ provides an electrochromic film, but Pd^{II} is inactive, mixed $Fe^{III}(CN)_6$–$Fe^{II}(CN)_6$ interactions being believed to effect coloration, light yellow in the oxidised state and yellow-green in the reduced state [67].

2.3.2 Indium Hexacyanoferrate and Gallium Hexacyanoferrate

Solutions of $InCl_3$ and $K_3Fe(CN)_6$ on electrochemical reduction by electrodes [68–71] provide weakly absorbing electrochromic films of In^{III} ferrocyanide, white on reduction and yellow on oxidation. As a film on the metal, gallium ferricyanide ensues from Ga metal oxidation in a ferricyanide/KCl solution [72], but hitherto not tested as an electrochrome.

2.3.3 Miscellaneous PB Analogues as Hexacyanoferrates

Several further PB analogues, some only as possible electrochromes, are tabulated as follows.

Electrochrome	Coloration ('Rev' = Reversibly)	References
Cadmium hexacyanoferrate	Rev white to colourless on reduction	[59, 73]
Chromium hexacyanoferrate	Rev blue to pale blue–grey on reduction	[59, 74]
Cobalt hexacyanoferrate	Rev green–brown to dark green on reduction	[59, 75]
Manganese hexacyanoferrate	Pale yellow to colourless on reduction	[59, 76]
Molybdenum hexacyanoferrate	Pink to red on reduction	[59, 77]
Platinum hexacyanoferrate	Pale blue to colourless on reduction	[59, 78]
Rhenium hexacyanoferrate	Pale yellow to colourless on reduction	[59]
Rhodium hexacyanoferrate	Pale yellow to colourless on reduction	[59]
Titanium hexacyanoferrate	Rev brown to pale yellow on reduction	[59, 79]
Osmium hexacyanoferrate	Possible electrochrome	[80]
Osmium(IV) hexacyanoruthenate	Possible electrochrome	[81]
Ruthenium oxide–hexacyanoruthenate	Possible electrochrome	[82]
Samarium hexacyanoferrate	Possible electrochrome	[83]
Silver hexacyanoferrate	Possible electrochrome	[84]
Uranium hexacyanoferrate	Possible electrochrome	[85]
Zirconium hexacyanoferrate	Possible electrochrome	[86]

2.3.4
Mixed-Metal and Mixed-Ligand PB Analogues Listed

Ruthenium oxide–hexacyanoferrate with ruthenium hexacyanoferrate [87].
Silver-'cross-linked' nickel hexacyanoferrate [88] (reversibly yellow to white on reduction [59]).
Copper heptacyanonitrosyl-ferrate [89].
Iron(III) carbonylpentacyanoferrate [84].
Iron(III) pentacyanonitroferrate [84].

References

1. Kirby, J. and Saunders, D. (2004) Fading and colour change of Prussian blue: methods of manufacture and the influence of extenders. *Nat. Gallery Tech. Bull.*, **25**, 73–99.
2. Bartoll, J. (2008) The early use of Prussian blue in paintings, Art 2008. Proceedings of the 9th International Conference on Non-Destructive Testing of Art, Jerusalem, Israel, May 25–30, 2008.
3. Granqvist, C.-G. (1995) *Handbook of Inorganic Electrochromic Materials*, Elsevier, Amsterdam, pp. 431–497.
4. Monk, P.M.S., Mortimer, R.J., and Rosseinsky, D.R. (1995) *Electrochromism – Fundamentals and Applications*, VCH Publishers, Weinheim, pp. 101–119.
5. Monk, P.M.S., Mortimer, R.J., and Rosseinsky, D.R. (2007) *Electrochromism and Electrochromic Devices*, Cambridge University Press, Cambridge, pp. 282–302.

6. Rosseinsky, D.R., Lim, H., Jiang, H., and Chai, J.W. (2003) Optical charge-transfer in iron(III)hexacyanoferrate(II): electro-intercalated cations induce lattice-energy-dependent ground-state energies. *Inorg. Chem.*, **42**, 6015–6023.
7. Kellawi, H., Mortimer, R.J. and Rosseinsky, D.R. (2003) Preliminary experiments. Unpublished results, University of Exeter.
8. Rosseinsky, D.R., Glasser, L., and Jenkins, H.D.B. (2004) Thermodynamic clarification of the curious ferric/potassium ion exchange accompanying electrochromic redox reactions of Prussian blue, iron(III) hexacyanoferrate(II). *J. Am. Chem. Soc.*, **126**, 10473–10477.
9. Bueno, P.R., Giménez-Romero, D., Ferreira, F.F., and Setti, G.O. (2010) Electrochemical capacitance spectroscopy and capacitive relaxation of the changeover process in iron hexacyanoferrate molecular compound. *Electrochim. Acta*, **55**, 6147–6155.
10. Rosseinsky, D.R. and Glidle, A. (2003) EDX, spectroscopy, and composition studies of electrochromic iron(III) hexacyanoferrate(II) deposition. *J. Electrochem. Soc.*, **150**, C641–C645.
11. Buser, H.J., Schwarzenbach, D., Petter, W., and Ludi, A. (1977) The crystal structure of Prussian blue: $Fe_4[Fe(CN)_6]_3.xH_2O$. *Inorg. Chem.*, **16**, 2704–2710.
12. Bueno, P.R., Ferreira, F.F., Giménez-Romero, D., Setti, G.O., Faria, R.C., Gabrielli, C., Perrot, H., Garcia-Jareño, J.J., and Vicente, F. (2008) Synchrotron structural characterization of electrochemically synthesized hexacyanoferrates containing K^+: a revisited analysis of electrochemical redox. *J. Phys. Chem. C*, **112**, 13264–1327.
13. Bueno, P.R., Giménez-Romero, D., Ferreira, F.F., Setti, G.O., Garcia- Jareño, J.J., Agrisuelas, J., and Vicente, F. (2009) Electrochromic switching mechanism of iron hexacyanoferrates molecular compounds: the role of $Fe^{2+}(CN)_6$ vacancies. *J. Phys. Chem. C*, **113**, 9916–9920.
14. Emrich, R.J., Traynor, L., Gambogi, W., and Buhks, E. (1987) Surface analysis of electrochromic displays of iron hexacyanoferrate films by x-ray photoelectron spectroscopy. *J. Vac. Sci. Technol., A: Vac. Surf. Films*, **5**, 1307–1310.
15. Giménez-Romero, D., Bueno, P.R., Gabrielli, C., Garcia-Jareño, J.J., Perrot, H., and Vicente, F. (2006) Kinetic aspects of ion exchange in $K_hFe_k[Fe(CN)_6].mH_2O$ compounds: a combined electrical and mass transfer functions approach. *J. Phys. Chem. B*, **110**, 19352–19363.
16. García-Jareño, J.J., Sanmatías, A., Vicente, F., Gabrielli, C., Keddam, M., and Perrot, H. (2000) Study of Prussian blue (PB) films by ac-electrogravimetry: influence of PB morphology on ions movement. *Electrochim. Acta*, **45**, 3765–3776.
17. Bueno, P.R., Giménez-Romero, D., Gabrielli, C., Garcıa-Jareño, J.J., Perrot, H., and Vicente, F. (2006) Changeover during in situ compositional modulation of hexacyanoferrate (Prussian blue) material. *J. Am. Chem. Soc.*, **128**, 17146–17152.
18. Giménez-Romero, D., Bueno, P.R., Garcıa-Jareño, J.J., Gabrielli, C., Perrot, H., and Vicente, F. (2006) Mechanism for interplay between electron and ionic fluxes in $K_hFe_k[Fe(CN)_6]_l$ mH_2O compounds. *J. Phys. Chem. B*, **110**, 2715–2722.
19. Ferreira, F.F., Bueno, P.R., Setti, G.O., Giménez-Romero, D., Garcia-Jareño, J.J., and Vicente, F. (2008) Resonant x-ray diffraction as a tool to calculate mixed valence ratios: application to Prussian blue materials. *Appl. Phys. Lett.*, **92**, 264103.
20. Agrisuelas, J., Bueno, P.R., Ferreira, F.F., Gabrielli, C., García-Jareño, J.J., Giménez-Romero, D., Perrot, H., and Vicente, F. (2009) Electronic perspective on the electrochemistry of Prussian blue films. *J. Electrochem. Soc.*, **156**, P74–P80.
21. Wojdeł, J.C., Moreira, I.de.P.R., Bromley, S.T., and Ilias, F. (2009) Prediction of half-metallic conductivity in Prussian blue derivatives. *J. Mater. Chem.*, **19**, 2032–2036.
22. Stilwell, D.E., Park, K.H., and Miles, M.H. (1992) Electrochemical studies of

the factors influencing the cycle stability of Prussian blue films. *J. Appl. Electrochem.*, **22**, 325–331.

23. Rodrigues, L.F.F.T.T.G. and Sequeira, C.A.C. (2008) Electrochromic studies of Prussian blue intercalated compounds. *Mater. Sci. Forum*, **587-588**, 288–296.
24. Garcia- Jareño, J.J., Navarro-Laboulais, J., and Vicente, F. (1996) Charge transport in Prussian blue films deposited on ITO electrodes. *Electrochim. Acta*, **41**, 835–841.
25. Roig, A., Navarro, J., Garcia-Jareño, J.J., and Vicente, F. (1994) Voltammetric study of the stability of deposited Prussian blue films against successive potential cycling. *Electrochim. Acta*, **39**, 437–442.
26. Jiao, Z., Wang, J., Ke, L., Liu, X., Demir, H.V., Yang, M.F., and Sun, X.W. (2012) Electrochromic properties of nanostructured tungsten trioxide (hydrate) films and their applications in a complementary electrochromic device. *Electrochim. Acta*, **63**, 153–160.
27. Middlemiss, D.S. and Wilson, C.C. (2008) Ferromagnetism and spin transitions in Prussian blue: a solid-state hybrid functional study. *Phys. Rev. B*, **77**, 155129–155142.
28. Rosseinsky, D.R. and Tonge, J.S. (1987) Electron-transfer rates by dielectric relaxometry and the direct-current conductivities of solid homonuclear and heteronuclear mixed-valence metal cyanometallates and of the methylene blue-iron dithiolate adduct. *J. Chem. Soc., Faraday Trans. I*, **83**, 245–255.
29. Cisternas, R., Muñoz, E., Henríquez, R., Córdova, R., Kahlert, H., Hasse, U., and Scholz, F. (2011) Irreversible electrostatic deposition of Prussian blue from colloidal solutions. *J. Solid State Electrochem.*, **15**, 2461–2468.
30. Demiri, S., Najdoski, M., and Velevska, J. (2011) A simple chemical method for deposition of electrochromic Prussian blue thin films. *Mater. Res. Bull.*, **46**, 2484–2488.
31. Ho, K.-C. (1994) On the deposition of Prussian blue by the sacrificial anode method. *Proc. Electrochem. Soc.*, **94-2**, 170–184.
32. Barton, R.T., Kellawi, H., Marken, F., Mortimer, R.J., and Rosseinsky, D.R. (2012) Simplest Prussian-blue deposition from ferric ferricyanide solution by a reducing Ag spot put onto an ITO Substrate. *J. Solid State Electrochem.*, **16**, 3723–3724.
33. Millward, R.C., Madden, C.E., Sutherland, I., Mortimer, R.J., Fletcher, S., and Marken, F. (2001) Directed assembly of multi-layers – the case of Prussian blue. *Chem. Commun.*, 1994–1995.
34. Pyrasch, M. and Tieke, B. (2001) Electro- and photoresponsive films of Prussian blue prepared upon multiple sequential adsorption. *Langmuir*, **17**, 7706–7709.
35. Pyrasch, M., Toutianoush, A., Jin, W., Schnepf, J., and Tieke, B. (2003) Self-assembled films of Prussian blue and analogues: optical and electrochemical properties and application as ion-sieving membranes. *Chem. Mater.*, **15**, 245–254.
36. Jin, W., Toutianoush, A., Pyrasch, M., Schnepf, J., Gottschalk, H., Rammensee, W., and Tieke, B. (2003) Self-assembled films of Prussian blue and analogues: structure and morphology, elemental composition, film growth, and nanosieving of ions. *J. Phys. Chem. B*, **107**, 12062–12070.
37. Ko, J.H., Yeo, S., Park, J.H., Choi, J., Noh, C., and Son, S.U. (2012) Graphene-based electrochromic systems: the case of Prussian blue nanoparticles on transparent graphene film. *Chem. Commun.*, **48**, 3884–3886.
38. Omura, A., Tanaka, H., Kurihara, M., Sakamoto, M., and Kawamoto, T. (2009) Electrochemical control of the elution property of Prussian blue nanoparticle thin films: mechanism and applications. *Phys. Chem. Chem. Phys.*, **11**, 10500–10505.
39. Hara, S., Shiozaki, H., Omura, A., Tanaka, H., Kawamoto, T., Tokumoto, M., Yamada, M., Gotoh, A., Kurihara, M., and Sakamoto, M. (2008) Color-switchable glass and display devices fabricated by liquid processes with electrochromic nanoparticle "ink". *Appl. Phys Express*, **1**, 104002-1–104002-3.
40. Gotoh, A., Uchida, H., Ishizaki, M., Satoh, T., Kaga, S., Okamoto, S., Ohta,

M., Sakamoto, M., Kawamoto, T., Tanaka, H., Tokumoto, M., Hara, S., Shiozaki, H., Yamada, M., Miyake, M., and Kurihara, M. (2007) Simple synthesis of three primary colour nanoparticle inks of Prussian blue and its analogues. *Nanotechnology*, **18**, 345609(6 pp).

41. Jang, J.H., Lee, S. and Shin, C.S. (2007) Preparation of Prussian blue coating film for electrochromic device. US Patent 8221829 B2, European Patent 1991636 B1.
42. DeLongchamp, D.M. and Hammond, P.T. (2004) High-contrast electrochromism and controllable dissolution of assembled Prussian blue/polymer nanocomposites. *Adv. Funct. Mater.*, **14**, 224–232.
43. DeLongchamp, D.M. and Hammond, P.T. (2004) Multiple-color electrochromism from layer-by-layer-assembled polyaniline/Prussian blue nanocomposite thin films. *Chem. Mater.*, **16**, 4799–4805.
44. Barbosa, P.C., Rodrigues, L.C., Silva, M.M., Smith, M.J., Parola, A.J., Pina, F., and Pinheiro, C. (2010) Solid-state electrochromic devices using pTMC/PEO blends as polymer electrolytes. *Electrochim. Acta*, **55**, 1495–1502.
45. Kraft, A. and Rottmann, M. (2009) Properties, performance and current status of the laminated electrochromic glass of Gesimat. *Sol. Energy Mater. Sol. Cells*, **93**, 2088–2092, *http://www.gesimat.de/*. (accessed 18 November 2014).
46. Baetens, R., Petter Jelle, B., and Gustavsen, A. (2010) Properties, requirements and possibilities of smart windows for dynamic daylight and solar energy control in buildings: a state-of-the-art review. *Sol. Energy Mater. Sol. Cells*, **94**, 87–105.
47. Soutar, A.M., Rosseinsky, D.R., Freeman, W., Zhang, X., How, X., Jiang, H., Zeng, X., and Miao, X. (2012) Electrochromic cell with UV-curable electrolyte polymer for cohesion and strength. *Sol. Energy Mater. Sol. Cells*, **100**, 268–270.
48. Widmann, A., Kahlert, H., Petrovic-Prelevic, I., Wulff, H., Yakhmi, J.V., Bagkar, N., and Scholz, F. (2002) Structure, insertion electrochemistry, and magnetic properties of a new type of substitutional solid solutions of copper, nickel, and iron hexacyanoferrates/hexacyanocobaltates. *Inorg. Chem.*, **41**, 5706–5715.
49. Robin, M.B. (1962) The colour and electronic configurations of Prussian blue. *Inorg. Chem.*, **1**, 337–342.
50. Wilde, R.E., Ghosh, S.N., and Marshall, B.J. (1970) The Prussian blues. *Inorg. Chem.*, **9**, 2512–2516.
51. Rajan, K.P. and Neff, V.D. (1982) Electrochromism in the mixed-valence hexacyanides. 2. Kinetics of the reduction of ruthenium purple and Prussian blue. *J. Phys. Chem.*, **86**, 4361–4368.
52. Itaya, K., Ataka, T., and Toshima, S. (1982) Electrochemical preparation of a Prussian blue analog: iron-ruthenium cyanide. *J. Am. Chem. Soc.*, **104**, 3751–3752.
53. Abe, T., Toda, G., Tajiri, A., and Taneko, M. (2001) Electrochemistry of ferric ruthenocyanide (Ruthenium Purple), and its electrocatalysis for proton reduction. *J. Electroanal. Chem.*, **510**, 35–42.
54. Jain, V., Sahoo, R., Jinschek, J., Montazami, R., Yochum, H.M., Beyer, F.L., Kumar, A., and Heflin, J.R. (2008) High-contrast solid state electrochromic devices based on ruthenium purple nanocomposites fabricated in layer-by-layer assembly. *Chem. Commun.*, 3663–3665.
55. Carpenter, M.K., Conell, R.S., and Simko, S.J. (1990) Electrochemistry and electrochromism of vanadium hexacyanoferrate. *Inorg. Chem.*, **29**, 845–850.
56. Dong, S.J. and Li, F.B. (1987) Researches on chemically modified electrodes: Part XVI. Electron diffusion coefficient in vanadium hexacyanoferrate film. *J. Electroanal. Chem.*, **217**, 49–63.
57. Bocarsly, A.B. and Sinha, S. (1982) Chemically-derivatized nickel surfaces: synthesis of a new class of stable electrode interfaces. *J. Electroanal. Chem.*, **137**, 157–162.
58. Joseph, J., Gomathi, H., and Prabhakara Rao, G. (1991) Electrochemical characteristics of thin films of nickel hexacyanoferrate formed on carbon substrates. *Electrochim. Acta*, **36**, 1537–1541.

59. Dillingham, J.L. (1999) Investigation of bipyridilium and Prussian blue systems for their potential application in electrochromic devices, Chapter 5, A survey of the transition metal hexacyanoferrates. PhD thesis, Loughborough University.
60. Sinha, S., Humphrey, B.D., Fu, E., and Bocarsly, A.B. (1984) The coordination chemistry of chemically derivatized nickel surfaces – generation of an electrochromic interface. *J. Electroanal. Chem.*, **162**, 351–357.
61. Monk, P.M.S., Mortimer, R.J., and Rosseinsky, D.R. (2007) *Electrochromism and Electrochromic Devices*, Cambridge University Press, Cambridge, p. 294.
62. Siperko, L.M. and Kuwana, T. (1983) Electrochemical and spectroscopic studies of metal hexacyanometalate films. 1. Cupric hexacyanoferrate. *J. Electrochem. Soc.*, **130**, 396–402.
63. Siperko, L.M. and Kuwana, T. (1986) Electrochemical and spectroscopic studies of metal hexacyanoferrate films. 2. Cupric hexacyanoferrate and Prussian blue layered films. *J. Electrochem. Soc.*, **133**, 2439–2440.
64. Siperko, L.M. and Kuwana, T. (1987) Electrochemical and spectroscopic studies of metal hexacyanometalate films. 3. Equilibrium and kinetic studies of cupric hexacyanoferrate. *Electrochim. Acta*, **32**, 765–771.
65. Siperko, L.M. and Kuwana, T. (1987) Studies of layered thin-films of Prussian-blue-type compounds. *J. Vac. Sci. Technol., A*, **5**, 1303–1306.
66. Jiang, M. and Zhao, Z.F. (1990) A novel stable electrochromic thin-film: a Prussian blue analogue based on palladium hexacyanoferrate. *J. Electroanal. Chem.*, **292**, 281–287.
67. Lezna, R.O., Romagnoli, R., de Tacconi, N.R., and Rajeshwar, K. (2003) Spectroelectrochemistry of palladium hexacyanoferrate films on platinum substrates. *J. Electroanal. Chem.*, **544**, 101–106.
68. Kulesza, P.J. and Faszynska, M. (1988) Indium(III) hexacyanoferrate as a novel polynuclear mixed-valent inorganic material for preparation of thin zeolitic films on conducting substrates. *J. Electroanal. Chem.*, **252**, 461–466.
69. Kulesza, P.J. and Faszynska, M. (1989) Indium(III)-hexacyanoferrate(III,II) as an inorganic material analogous to redox polymers for modification of electrode surfaces. *Electrochim. Acta*, **34**, 1749–1753.
70. Dong, S.J. and Jin, Z. (1989) Electrochemistry of indium hexacyanoferrate film modified electrodes. *Electrochim. Acta*, **34**, 963–968.
71. Jin, Z. and Dong, S.J. (1990) Spectroelectrochemical studies of indium hexacyanoferrate film modified electrodes. *Electrochim. Acta*, **35**, 1057–1060.
72. Eftekhari, A. (2004) Electrochemical behavior of gallium hexacyanoferrate film directly modified electrode in a cool environment. *J. Electrochem. Soc.*, **151**, E297–E301.
73. Luangdilok, C.H., Arent, D.J., Bocarsly, A.B., and Wood, R. (1992) Investigation of the structure reactivity relationship in the $Pt/M_xCdFe(CN)_6$ modified electrode system. *Langmuir*, **8**, 650–657.
74. Jiang, M., Zhou, X., and Zhao, Z. (1990) A new zeolitic thin-film based on chromium hexacyanoferrate on conducting substrates. *J. Electroanal. Chem.*, **287**, 389–394.
75. Joseph, J., Gomathi, H., and Prabhakar Rao, G. (1991) Electrodes modified with cobalt hexacyanoferrate. *J. Electroanal. Chem.*, **304**, 263–269.
76. Bharathi, S., Joseph, J., Jeyakumar, D., and Prabhakara Rao, G. (1991) Modified electrodes with mixed metal hexacyanoferrates. *J. Electroanal. Chem.*, **304**, 341–345.
77. Dong, S. and Jin, Z. (1988) Molybdenum hexacyanoferrate film modified electrodes. *J. Electroanal. Chem.*, **256**, 193–198.
78. Liu, S., Li, H.L., Jiang, M., and Li, P. (1997) Platinum hexacyanoferrate: a novel Prussian blue analogue with stable electroactive properties. *J. Electroanal. Chem.*, **426**, 27–30.
79. Jiang, M., Zhou, X., and Zhao, Z. (1990) Preparation and characterization of mixed-valent titanium hexacyanoferrate film modified glassy carbon electrode. *J. Electroanal. Chem.*, **292**, 289–296.

80. Chen, S.-M. and Liao, C.-J. (2004) Preparation and characterization of osmium hexacyanoferrate films and their electrocatalytic properties. *Electrochim. Acta*, **50**, 115–125.

81. Cox, J.A. and Das, B.K. (1987) Characteristics of a glassy carbon electrode modified in a mixture of osmium tetroxide and hexacyanoruthenate. *J. Electroanal. Chem.*, **233**, 87–98.

82. Kulesza, P.J. (1987) A polynuclear mixed-valent ruthenium oxide/cyanoruthenate composite that yields thin coatings on a glassy carbon electrode with high catalytic activity toward methanol oxidation. *J. Electroanal. Chem.*, **220**, 295–309.

83. Wu, P., Lu, S., and Cai, C. (2004) Electrochemical preparation and characterization of a samarium hexacyanoferrate modified electrode. *J. Electroanal. Chem.*, **569**, 143–150.

84. Itaya, K., Uchida, I., and Neff, V.D. (1986) Electrochemistry of polynuclear transition metal cyanides: Prussian blue and its analogs. *Acc. Chem. Res.*, **19**, 162–168.

85. Jiang, M., Wang, M., and Zhou, X. (1992) Facile attachment of uranium hexacyanoferrate to carbon electrode by reductive electro-deposition. *Chem. Lett.*, 1709–1712.

86. Liu, S.-Q., Chen, Y., and Chen, H.-Y. (2001) Studies of spectroscopy and cyclic voltammetry on a zirconium hexacyanoferrate modified electrode. *J. Electroanal. Chem.*, **502**, 197–203.

87. Chen, S.-M., Lu, M.-F., and Lin, K.-C. (2005) Preparation and characterization of ruthenium oxide/hexacyanoferrate and ruthenium hexacyanoferrate mixed films and their electrocatalytic properties. *J. Electroanal. Chem.*, **579**, 163–174.

88. Kulesza, P.J., Jedral, T., and Galus, Z. (1989) A new development in polynuclear inorganic films: Silver(I)/"crosslinked" nickel(II) hexacyanoferrate(III, II) microstructures. *Electrochim. Acta*, **34**, 851–853.

89. Gao, Z., Zhang, Y., Tian, M., and Zhao, Z. (1993) Electrochemical study of copper-heptacyanonitrosylferrate film modified electrodes: preparation, properties and applications. *J. Electroanal. Chem.*, **358**, 161–176.

3
Electrochromic Materials and Devices Based on Viologens

Paul M. S. Monk, David R. Rosseinsky, and Roger J. Mortimer

3.1
Introduction, Naming and Previous Studies

Bipyridilium species are electrochromic. They are formed by diquaternising 4,4′-bipyridyl to form 1,1′-disubstituted-4,4′-bipyridilium salts (**I**). The positive charge shown localised on N is better viewed as being delocalised over the respective rings. There is generally little delocalisation between the two quaternised pyridine rings owing to a lack of π-overlap, which is a consequence of a significant dihedral angle between the two rings.

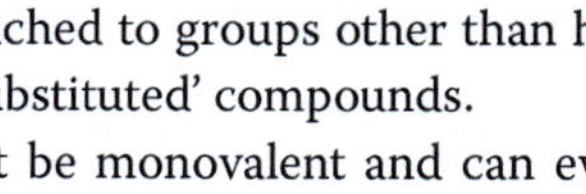

I

The compounds are formally named 1,1′-di-substituent-4,4′-bipyridilium if the two substituents attached to the nitrogen are the same and 1-substituent-1′-substituent′-4,4′-bipyridilium when they differ. When the carbon atoms of the bipyridyl core are attached to groups other than hydrogen, these species are usually named 'nuclear-substituted' compounds.

The anion X^- need not be monovalent and can even be attached to a polymer backbone. The molecules are zwitterionic (i.e. they bear both positive and negative charge concentrations at different molecular regions or sites) when the substituents at nitrogen bear a negative charge [1, 2].

A convenient abbreviation for any bipyridyl unit regardless of its redox state is 'bipm', with its charge indicated. In this notation, the dication is $bipm^{2+}$ and so on.

These compounds have several trivial, non-IUPAC names. By far the most common is 'viologen' following Michaelis [3, 4], who noted the intense violet colour formed when 1,1′-dimethyl-4,4′-bipyridilium undergoes a one-electron reduction to form a radical cation (Eq. (3.1)). In this nomenclature, the simplest viologen (1,1′-dimethyl-4,4′-bipyridilium) is styled 'methyl viologen' (MV). Another extensively used name is 'paraquat' (PQ) after the ICI brand name for methyl viologen, which ICI developed for herbicidal use. In this latter style, MV^{2+}

Electrochromic Materials and Devices, First Edition.
Edited by Roger J. Mortimer, David R. Rosseinsky and Paul M. S. Monk.

is named 'PQ' and bipyridilium species with substituents other than the dimethyl are called '*substituent* paraquat'.

There are several reviews of this field. The most recent of any length is the chapter 'The viologens' in *Electrochromism and Electrochromic Devices* by Monk, Mortimer and Rosseinsky (2007) [5]. The most substantial is still the 1998 book *The Viologens: Physicochemical Properties, Synthesis, and Applications of the Salts of 4,4′-Bipyridine* by Monk [6]. Other works are becoming dated, but some still incorporate valuable bibliographic data, including the chapter 'Bipyridilium systems' in *Electrochromism: Fundamentals and Applications* by Monk *et al.* (1995); [7]; 'The Bipyridines' (1984), by Summers [8], deals at length with syntheses and properties of 4,4′-bipyridine; and Summers' book *The Bipyridinium Herbicides* (1980) [9] comprises copious detail. Although dated, 'The Electrochemistry of the Viologens' (1981) by Bird and Kuhn [10] is particularly relevant to this chapter. 'Formation, properties and reactions of cation radicals in solution' (1976) by Bard *et al.* [11] has a good section on bipyridilium radical cations. The review 'Chemistry of viologens' (1991) by Sliwa*et al.* [12] also alludes to electrochromism.

3.2
Redox Chemistry of Bipyridilium Electrochromes

There are three common bipyridilium redox states: a dication ($bipm^{2+}$), a radical cation ($bipm^{+\cdot}$) and a di-reduced neutral compound ($bipm^{0}$). The dicationic salt is the most stable of the three and is the species purchased or synthesised. It is colourless when pure unless charge-transfer-type interactions with the counter anion or other charge donors in solution generate optical bands. Such absorbances are generally feeble for anions such as chloride or sulfate but are stronger for charge-transfer interactive anions such as bromide or iodide [13]; thus, $PQ^{2+}\ 2Br^{-}$ is rich yellow in colour, and $PQ^{2+}\ 2I^{-}$ is brilliant scarlet.

Reductive electron transfer to the dication forms a radical cation, Eq. (3.1):

$$\underset{\text{(colourless)}}{bipm^{2+}} + e^{-} \rightarrow \underset{\text{(intense colour)}}{bipm^{+\cdot}} \tag{3.1}$$

Bipyridilium radical cations are amongst the most stable of all organic radicals and can be prepared as air-stable solid salts [14, 15].

In solution, the colour of the radical will persist almost indefinitely in the absence of oxidising agents such as periodate or ferricyanide [16]; however, its reaction with molecular oxygen is particularly rapid [17]. The stability of the radical cation is attributable to the delocalisation of the radical electron throughout the π-framework of the entire two-ring bipyridyl nucleus – not merely the individual pyridine rings – because the ion becomes planar on reduction, thereby enabling π-overlap and hence conjugation between the two rings. Furthermore, ESR measurements demonstrate how the 1 and 1′ substituents also bear some of the charge, particularly if aryl or otherwise are capable of additional conjugation.

The potential needed to effect the reduction reaction in Eq. (3.1) depends on many factors including the solvent, ionic strength, choice of electrode substrate but, more importantly, on the choice of the two substituents at nitrogen and on further substituents if the electrochrome has nuclear (i.e. bipyridyl) substitution. For example, Hünig and coworkers [18–20] prepared a wide range of bipyridilium species and obtained for each a polarographic value of $E_{1/2}$. They correlated each $E_{1/2}$ with the results of theoretical calculations and with parameters from linear free-energy relationships (LFERs) from within the corpus of physical organic chemistry such as σ and σ^*. They found clear links between the molecular properties that determine variations such as electron densities and electronic shifts and λ_{max}.

Electrochromism occurs in bipyridilium species because, in contrast to the bipyridilium dications, radical cations display an intense colour. This colour derives from an optical charge transfer between the (formally) +1 and 0 valent nitrogens. This is a simplified view of the phenomenon: in fact, because of the delocalisation referred to earlier, the source of the colour is better viewed as an intramolecular photo-effected electronic excitation.

The colours of radical cations depend on the substituents on nitrogen. As noted earlier, simple alkyl groups such as methyl promote a blue-violet colour, whereas aryl groups can generally impart a variety of colours to the radical cation, the exact colour depending on the choice of substituent(s). Indeed, manipulation of the substituents attached to N, N', and the bipyridyl 'nucleus' can generally modify the appropriate molecular orbital energy levels and thereby, in principle, tailor the colour as desired. Values of λ_{max} from electronic spectra readily correlate with σ and σ^* from physical organic chemistry, as noted earlier [18–20].

The molar absorptivity ε for the methyl viologen radical cation is large: for example, in water, $\varepsilon = 13\,700\ dm^3\ mol^{-1}\ cm^{-1}$ when extrapolated to zero concentration [21]. ε is usually somewhat solvent dependent [22]. A few values of wavelength maxima and ε are listed in Table 3.1. The data refer to monomeric radical-cation species unless stated otherwise (Table 3.1).

The colours of viologen radical cations species change with concentration. For viologens with short alkyl chains, the radical cation is monomeric in solution and has a blue colour becoming blue purple when the solution is more concentrated. There is a tendency towards crimson as the length of the alkyl chain increases. This change is largely due to an increasing incidence of dimerised radical cation: the monomer of alkyl-substituted radical cations is blue and the respective dimer is red [24].

By comparison, aryl-substituted viologens generally form a solid green or dark-red deposit. The viologen radical cation within the solid is neutral and therefore generally referred to as a 'salt'. Also, dication solubility and radical-cation stability (in thin-film form) are both greatly improved by using aryl substituents. These physicochemical considerations explain why ICI chose aryl-substituted viologens for their ECD, particularly the p-cyanophenyl variant (cyanophenyl paraquat (CPQ, **II**)) since the electrochromic colour of the heptyl viologen radical cation was deemed insufficiently intense: the molar absorptivity (and therefore the CR,

Table 3.1 Optical data for some bipyridilium radical cations.

R	Anion	Solvent	λ_{max} (nm)	ε ($dm^3\ mol^{-1}\ cm^{-1}$)	References
Methyl	Cl^-	H_2O	605	13,700	[22]
Methyl	I^-	H_2O-MeCN	605[a)]	10,060	[23, 24]
Methyl	Cl^-	H_2O	606	13,700	[21]
Methyl	Cl^-	MeCN	607	13,900	[22]
Methyl	Cl^-	MeOH	609	13,800	[22]
Methyl	Cl^-	EtOH	611	13,800	[22]
Methyl	Cl^-	H_2O	604	16,900	[25]
Ethyl	ClO_4^-	DMF	603	12,200	[26]
Heptyl	Br^-	H_2O	545[b),c)]	26,000	[27]
Octyl	Br^-	H_2O	543[c)]	28,900	[28]
Benzyl	Cl^-	H_2O	604	17,200	[29]
p-CN-Ph	BF_4^-	PC	674	83,300	[30]
p-CN-Ph	Cl^-	H_2O	535[b),c)]	–	[31]

a) Estimated from reported spectra.
b) Solid film on an optically transparent electrode, OTE.
c) Solution-phase radical-cation dimer.

as defined in Appendix) of aryl-substituted viologens is always far greater than that of alkyl-substituted viologens (see Table 3.1). Furthermore, the green radical cation of CPQ is more stable than the other aryl-viologen radical cations [30, 31].

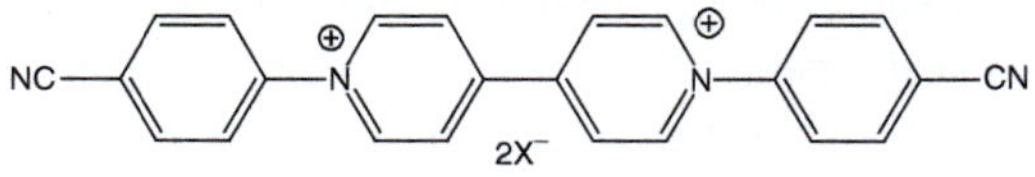

II

In contrast with $bipm^{2+}$ or $bipm^{+\cdot}$, comparatively little is known about the third redox form of the bipyridilium series, the di-reduced or the so-called 'di-hydro' compounds which form by one-electron reduction of the respective radical cation (Eq. (3.2)).

$$\underset{\text{intense colour}}{bipm^{+\bullet} + e^-} \rightarrow \underset{\text{weak colour}}{bipm^0} \tag{3.2}$$

This product may also be formed by direct two-electron reduction of the dication:

$$bipm^{2+} + 2e^- \rightarrow bipm^0 \tag{3.3}$$

Di-reduced compounds are often termed 'bi-radicals' [32] because of their extreme reactivity, but magnetic susceptibility measurements show that they are diamagnetic [33] in the solid state, indicating that the two spins are paired. In fact, di-reduced $bipm^0$ compounds are simply reactive di-amines [34]. The intensity of the colour exhibited by $bipm^0$ species is often low since no obvious optical charge transfer or internal transition corresponding to visible wavelengths is accessible.

3.3 Physicochemical Considerations for Including Bipyridilium Species in ECDs

We now discuss the electrochromism of viologen species according to electrochrome type.

3.3.1 Type-1 Viologen Electrochromes

Most viologen species can operate as a type-1 electrochrome. In the context of device construction, a straightforward monomeric viologen is dissolved in a non-aqueous solvent such as γ-butyrolactone, MeCN or PC. The solvent must be dried most rigorously to obviate the many electrochemical complications that arise in aqueous solvents (see Section 3.3.2).

The write–erase efficiency for ECDs using a type-1 electrochrome is likely to be high during long observation periods, but will decrease (possibly badly) as the observation period becomes shorter. This statement merely reflects the requirement for the coloured product of electron transfer to diffuse back towards the electrode within a finite timescale. Stated differently, type-1 electrochromes generally have a poor response time if a good contrast ratio *CR* is required. As a direct consequence, many type-1 electrochromes are constructed for applications able to accommodate slower response times such as anti-glare mirrors and long-term data displays. This criterion does not preclude faster coloration times, for example, when forming the coloured state of an electrochrome.

Virtually all the viologen-based ECDs currently on the market operate as type-1 devices. Examples include the world's best-selling type-1 ECD, the Gentex mirror (described in Section 3.4 and Chapter 14) in which the primary electrochrome is a viologen (which may be zwitterionic to ensure that the radical-cation product of reduction is charge-neutral). Ho *et al.* [35, 36] has modelled the electrochemical behaviour and cycle life of this kind of electrochromic device, particularly one with heptyl viologen as the primary electrochrome and tetramethylphenylenediamine (TMPD) as the secondary. The cycle lives are high: figures are not available for the Gentex mirror, but an ECD utilising benzyl viologen (as the BF_4^- salt) in PC has a cycle life of greater than 4×10^4 cycles [37].

3.3.2 Type-2 Viologen Electrochromes

Type-2 electrochromes are in solution when colourless but are solid in their coloured forms. Viologen species act as type-2 electrochromes if the *N*-substituent is aryl or the alkyl chain is long. The slow response times of type-1 electrochromes can be avoided by dissolving the viologen in a water-containing electrolyte. To become properly type-2, the viologen requires an aryl or long-chain alkyl substituent at nitrogen, because the coloured radical-cation product of reduction reaction in Eq. (3.1) has a considerably lowered solubility than with, say, R = methyl in formula I of Section 3.1. Such ECDs do not have the solubility–diffusion problem of type-1 devices and are also faster at bleaching.

The formation and removal of colour in a type-2 device are generally complicated and require significant discussion, as follows.

Firstly, the solid will be an ionic salt albeit with some slight covalency. The process of forming a solid viologen radical-cation at an electrode (reaction (3.1)) is usually termed 'electrodeposition.' Strictly, that term could imply that the solid product is the *immediate* product of electron transfer, when the formation of viologen radical-cation salts is generally regarded as a three-step process:

1) Radical cation forms at the electrode according to (Eq. (3.1)) followed by
2) The acquisition of an anion X^- in solution and thence
3) Precipitation of the salt from solution:

$$\text{bipm}^{+\bullet}(\text{aq}) + \text{X}^-(\text{aq}) \rightarrow [\text{bipm}^{+\bullet}\text{X}^-](\text{s}) \tag{3.4}$$

Equation (3.4) represents the chemical step of a so-called 'EC'-type process since the product of electron transfer ('*E*') undergoes a chemical reaction. Such an overall EC reaction is strictly 'electro-*precipitation*' rather than 'electrodeposition'. (If electro-precipitation occurs by two steps that are in effect instantaneously sequential, 'electrodeposition' is then an adequate description.)

3.3.2.1 The Effect of the Bipyridilium-N Substituent

van Dam and Ponjeé (of Texas Instruments) [38] examined the effect that variations in the length of the alkyl chain have on the film-forming properties of the radical cation as the bromide salt (Table 3.2). Redox potentials have been added to this table from Ref. [10]. The data show the effect of increasing the length of the alkyl chain: the pentyl chain produces the first consistently insoluble viologen radical-cation salt, but the heptyl is the first salt for which the solubility product is sufficiently low for realistic device usage (Table 3.2).

Table 3.2 shows that stable solid films only form if the viologen has an effective chain length in excess of four CH_2 units. The radical-cation salt of CPQ is far less soluble in water than is $HV^{+\bullet}$ Br^-, yet the dicationic salt is extremely soluble. In water, the solubility product K_{sp} of $HV^{+\bullet}$ Br^- is [39] 3.9×10^{-7} mol^2 dm^{-6}.

3.3.2.2 The Effect of Micellar Viologen Species

Association of bipyridilium species to form π-dimers is well documented for the viologen radical cation [23, 24], but is not so well attested for the dication; Refs. [14, 39–41] are some of the few citations of note. A particular problem with aqueous solutions of viologen can be the formation of micelles of dication, particularly if the substituents are aryl groups or alkyl chains (heptyl or longer) that are hydrophobic. Barclay *et al.* [39] (of IBM) quotes a critical micelle concentration *cmc* for the HV^{2+} dication of 10^{-2} mol dm^{-3} in aqueous bromide solution. The analogy between such viologens and quaternary ammonium cationic surfactants in this regard is strong [42].

Table 3.2 Symmetrical viologens: the effect of varying the alkyl chain length on radical-cation film stability [10, 38].

Substituent of *R*	Effective length (units of CH_2)	Solid bromide salt film on Pt	Colour	$E_1^{\ominus}$ (mV)
Methyl	1	No	Blue	−688 (Cl^-)
Ethyl	2	No	Blue	−691 (Cl^-)
Propyl	3	No	Blue	−690 (Br^-)
Butyl	4	No	Blue	−686 (Br^-)
Pentyl	5	Yes	Purple	−686 (Br^-)
Hexyl	6	Yes	Purple	−710 (Br^-)
Heptyl	7	Yes	Mauve	−600 (Br^-)
Octyl	8	Yes	Crimson	−705 (Br^-)
iso-pentyl	4	Yes	Purple	−696 (Br^-)
Benzyl	4–5	Yes	Mauve	−573 (Cl^-)
$CH_3(Cl)CH_2OCH_2-$	4	No	–	—
$CH_3-CH{=}CH-CH_2-$	4	No	–	—
$H-CH{=}CH-(CH_2)_3-$	4–5	No	–	—
$NCM-C_3H_6-$	4–5	No	–	−362[a] (Cl^-)

$E_1^{\ominus}$ values are quoted against the SCE and refer to viologen salts with the parenthesised anion.
a) Polarographic $E_{1/2}$ value.

Electrochemistry below the *cmc* is straightforward enough. Reduction is considerably more complicated when the concentration of the dication is higher than the *cmc*, as follows. Reduction occurs preferentially on the outside of the micelle as it approaches a cathodically biased electrode, to form $bipm^{+\cdot}$, while the inside of the micelle remains fully oxidised dication [14, 43]. A similar explanation has been advanced to explain the complicated cyclic voltammetric behaviour observed during the formation of solid radical-cation salt of CPQ [14, 44], and Heyrovský has postulated the existence of solution-phase mixed-valence species of methyl viologen in water [45–47].

Although such mixed-valence viologens are not involved often and comprise very small amounts of material, they are capable of greatly complicating the speciation and electrochemistry of solutions that contain them.

In an effort to mimic the properties of bipyridilium species encapsulated within a micelle environment, Kaifer and Bard [40] investigated the electrochemistry of methyl viologen in the presence of various surfactants (anionic, cationic or non-ionic), showing how the properties of methyl viologen were largely unaffected by the presence of surfactant when below the *cmc*, but above it, the properties of the methyl viologen were markedly different, for example, the ESR spectrum of $MV^{+\cdot}$ lost all hyperfine coupling due to rapid inter-radical spin swapping.

Engelman and Evans [48, 49] also investigated the electrochemical reduction of MV^{2+} in the presence of micellar anions. In the presence of anionic surfactant, the position of the monomer–dimer equilibrium was displaced significantly in favour of the monomeric form when above the *cmc*, whereas cationic and non-ionic surfactants did not affect the equilibrium either way. Cyclic voltammetry also differed above and below the *cmc*.

Given the complexity of the overall electro-precipitation mechanism, and the different specifications in solution during each of the steps during electroprecipitation, it is again worth remarking that comparison of results from different authors is difficult since each step is defined by the number and nature of each anion in solution, and different experimental conditions have been employed by each author. In fact, the anion is occasionally quite unspecified.

3.3.2.3 The Effect of Film Morphology

Barna [50] (of Texas Instruments) addressed the morphology of $HV^{+\bullet}$ films. Electrodeposited films are largely amorphous but partially crystalline. The proportion of crystalline material increases with time; this acquisition of crystallinity is probably associated with many additional sharp peaks observed during cyclic voltammetry of heptyl viologen films [10, 38, 51, 52]. Goddard *et al.* [53] also observed these time-dependent changes within deposits of $HV^{+\bullet}$ on an OTE, as studied using UV–vis spectroscopy via a novel potential cycling technique (now rendered obsolete by use of diode-array technology).

Bewick [54, 55] used diode-array spectroscopy to investigate HV^{2+} dibromide and an array of asymmetric bipyridilium salts (i.e. with different substituents attached to N and N'). He concluded the initial product of mono-reduction was a solid salt of $HV^{+\bullet}$ radical cation in which some unreduced HV^{2+} dication was incorporated [55]. He then explained the subsequent ageing effects and previously inexplicable additional CV peaks in terms of this composite form of solid deposit. A similar explanation has been advanced for the complicated cyclic voltammetric behaviour observed during the formation of solid $CPQ^{+\bullet}$ radical-cation salt [14, 44].

3.3.2.4 The Effect of the Counter Anion

The counter anion in the viologen salt usually has a crucial effect on ECD performance. Different counter ions yield solid radical-cation products of electrodeposition having a wide range of morphologies, solubilities and chemical stabilities [30]. ICI used the SO_4^{2-} salt of CPQ^{2+} in their prototype ECDs [56] because other anions yielded a radical salt that was too unstable. For example, $CPQ^{+\bullet}$ is oxidised chemically by the nitrate ion via a rapid but complicated mechanism [30]. Studies of counter-ion effects may be performed by using cyclic voltammetry (e.g. Ref. [51]) or by observing the time dependence of an ESR trace, which demonstrates the $bipm^{+\bullet}$ concentration [30].

The properties of heptyl viologen radical-cation films also depend on the anion as shown by van Dam and Ponjeé [38]. Jasinski [51] (Texas Instruments) found that the optimum anion in water was dihydrogen phosphate. Anions found useful for ECDs were acetate, dihydrogen phosphate, fluoride, formate and sulfate. Bromide, chloride, perchlorate and tetrafluoroborate also proved satisfactory (as also concluded by van Dam and Ponjeé [38]). The dication salts of heptyl viologen are water insoluble when the anion is bicarbonate (at pH 5.5),

Table 3.3 Reduction potentials[a)] of heptyl viologen: the effect of supporting electrolyte anion and of electrode substrate.

Anion[a)]	Epc(1)/V on Ag	Epc(2)/V on Ag	Epc(1)/V on Pt	Epc(2)/V on Pt	Epc(1)/V on Ag	Epc(2)/V on Ag
Bromide (0.3)	−0.698	−1.008	−0.708	(<−0.818)	−0.708	−0.978
$H_2PO_4^-$ (2.0)	−0.668	−1.048	−0.668	(<−0.818)	−0.668	—
Formate (0.4)	−0.848	−0.928	—	—	−0.828	−0.948
HCO_3^- (1.0)	−0.768	−0.958	—	—	−0.778	−0.948
Acetate (0.5)	−0.828	−0.928	—	—	—	—
Fluoride (1.0)[b)]	−0.818[c)]	−0.878[c)]	—	−0.868	—	−0.808
Sulfate (0.3)	−0.818[c)]	−0.928[c)]	−0.848[b)]	−0.898	−0.798	−0.918

a) The concentration in parenthesis is expressed in mol dm^{-3}. All reduction potentials were determined at pH 5.5.
b) Millimolar viologen dication employed for measurement.
c) No colour formed.
The values of peak potential E_{pc} are cited against the SCE.
Table reproduced from Ref. [51].

thiocyanate, tetrahydroborate, hexafluorophosphate, tetrafluoroantimonate or tetrafluoroarsenate. Like $CPQ^{+\bullet}$, $HV^{+\bullet}$ is also oxidised by the nitrate ion [51] and presumably by a related mechanism.

Jasinski studied the effect of electrode substrate on heptyl viologen's reduction potential [51]. His results for aqueous HV^{2+} on various metals and with a variety of anions are given in Table 3.3. Many other redox potentials for mono-reduction of bipyridilium salts are quoted in the reviews by Monk [6] and by Bird and Kuhn [10].

The speciation chemistry of aqueous viologen is generally complicated. For example, the anion in a viologen-containing solution often participates in charge-transfer (CT)-type interactions with the viologen dication. In fact, some studies suggest that the CT complex must dissociate prior to reductive electron transfer [57]. Reduction is therefore a two-step process: ion-pair dissociation → reduction, although electron transfer could occur via a special type of S_N2 reaction in which the 'nucleophile' is the electron and the leaving group is an anion [57].

The rates at which the CT complexes of methyl viologen dissociate vary: the complex with iodide dissociates at a rate of 8.7×10^5 s^{-1}, while that with chloride dissociates faster with $k = 26.3 \times 10^5$ s^{-1}. The anion may thus also influence the speed of electrochromic coloration, as discussed more fully as follows [57].

3.3.2.5 The Use of Electron Mediators and the Formation of Electro-Inactive Oils

Sometimes a film of $HV^{+\bullet}$ salt will not electro-erase. The reasons are many and complicated. The first way to prevent non-erasure is to add an auxiliary redox couple (i.e. an electron mediator) to the dication-containing electrolyte solution. The most commonly used mediators include cerous ion [58, 59], ferrocyanide [56, 60, 61], ferrous ion [56] and hydroquinones [62]. (Ferrocene in acetonitrile has also been used in a type-1 device [63, 64].)

The mediator works in this way. During electro-coloration, $bipm^{2+}$ ion is reduced to form coloured $bipm^{+\bullet}$ but during re-oxidation at a positive potential it is the mediator (e.g. ferrocyanide, $[Fe^{II}(CN)_6]^{4-}$) that is oxidised at the electrode. The oxidised form of the mediator (in this example, ferricyanide $[Fe^{III}(CN)_6]^{3-}$) enables *chemical* oxidation of the radical-cation film, to re-form the dication. Such oxidation is always extremely rapid [16]. Mediators are generally added to solutions of heptyl viologen to facilitate electro-oxidation of the radical cation. For aryl viologens in aqueous solution, a mediator is *always* necessary to ensure complete colour removal on re-oxidation [6]. As ferrocyanide is known to form a charge-transfer complex with methyl viologen dication [65, 66] and also with the dications of CPQ [6, 13, 67] and HV [60], it will be the free-anion equilibrium fraction of the species that can be assumed to act.

Despite the use of mediators such as ferrocyanide, unsightly yellow–brown stains still persist on the electrode, even with the HV^{2+} and CPQ^{2+} systems [56, 61]. Some workers find that adding a complexing sugar such as β-cyclodextrin to the voltammetry solution will impede the formation of yellow–brown stains [60], probably by encapsulating the dication within the cavity of its cyclodextrin structure in a form of guest–host relationship. Because complexed dications cannot juxtapose in this way, association of $bipm^{2+}$ cations in solution [39] is largely prevented and so alignment of $bipm^{+\bullet}$ species within the solid deposit becomes impossible. However, ICI found that such 'oiling' still occurred 'ultimately' especially for aryl viologens despite the use of complexing sugars (Mr J. G. Allen, ICI Ltd., personal communication, 1987).

Other attempts to stop the ageing phenomenon have used different, modified bipyridilium compounds [68–70]. For example, Bruinink *et al.* [68] prepared the compound (**III**) in which the two pyridinium rings are separated by methylene linkages; to a similar end, Barna and Fish [69] prepared asymmetric bipyridilium salts with $R_1 \neq R_2$ (cf. **I**). These new structures, for example a compound with $R_1 = C_7H_{15}$ and $R_2 = C_{18}H_{37}$, do apparently inhibit the crystallisation process. With similar reasoning, Barltrop and Jackson [70] prepared asymmetric viologens and the diquaternised 3,8-phenanthroline salt (**IV**) and a series of species in which substituents are directly bonded to carbon in the pyridine rings – the so-called nuclear-substituted bipyridyls. Again, films with superior write–erase properties were formed.

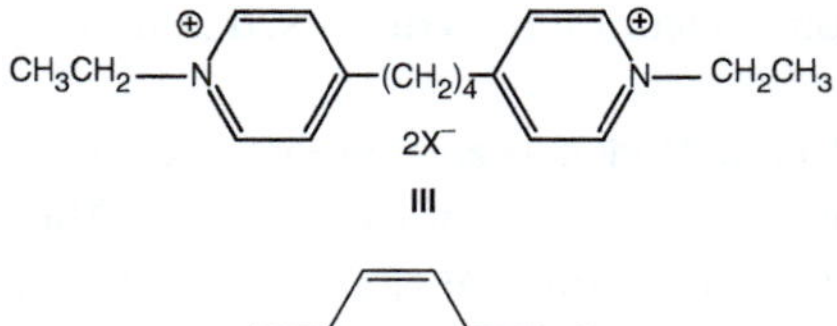

III

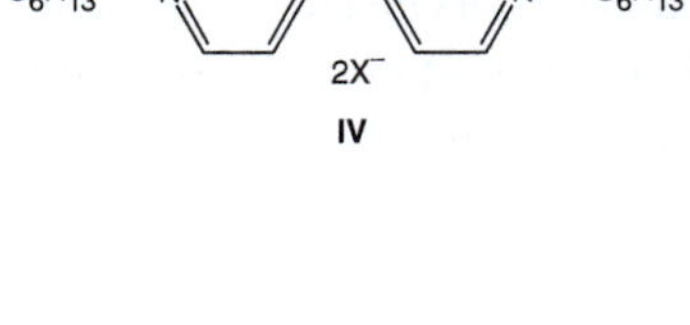

IV

3.3.2.6 The Effect of Dimerised Radical Cations

There is a further complication in the speciation of aqueous viologen species prior to the transfer of an electron.

Fresh films of $HV^{+\bullet}$ are amorphous, visually even [51] and smooth [71], yet the films appear patchy as ageing process occurs. The change is rapid, occurring soon after deposition (<10 s) [53] and probably involves ordering (crystallisation) of the radical moieties. Re-oxidation of the film to bleach the colour is rapid for fresh $HV^{+\bullet}$ films, but patchy films that show that these signs of ageing are more difficult to oxidise, that is, they require a higher potential or a longer re-oxidation time.

Secondly, as stated earlier, after prolonged cycling between the coloured and bleached states, bipyridilium ECD devices form an unsightly yellow–brown stain on the electrode. Some evidence now suggests that this stain is a form of crystalline radical-cation salt [30] containing spin-paired radical-cation dimer or intervalence species comprising both $bipm^{2+}$ and $bipm^{+\bullet}$. For example, SERS study [72] of a disulfide-containing dimeric viologen adsorbed on rough silver implies the presence of a 'liquid-like' environment at the electrode surface following reduction to form dimerised radical $(bipm)_2^{2+}$.

As discussed earlier, almost certainly some form of reordering of the radical species ('recrystallisation') occurs within the electro-precipitated viologen deposit soon after it forms. In order to understand these processes, thin films of $HV^{+\bullet}$ salt have been studied by many techniques including UV–vis spectroelectrochemistry [55, 73–75], ESR [76], Raman spectroscopy [61, 77, 78], photoacoustic spectroscopy [79, 80], photothermal spectroscopy [81] and the electrochemical quartz-crystal microbalance (EQCM) [71].

Gołden and Przyłuski [82] looked at phase changes within radical-cation salt deposits of $HV^{+\bullet}$, and Scharifker and Wehrmann [83] investigated solid deposits of both $HV^{+\bullet}$ and benzyl viologen radical cation. The separate groups assigned the ageing phenomenon, in part at least, to the dimerisation of radical cation in solution prior to precipitation.

Belinko [32] investigated the write–erase efficiency of $HV^{+\bullet}$ films by cyclic voltammetry, making the lower scanning limit progressively more negative to deliberately generate $bipm^0$. He found a clear correlation between device failure and the production of di-reduced bipyridilium ($bipm^0$) as a minor electrode product. In common with Michaelis in 1933 [3, 4], Belinko discerned how electrochemical oxidation of $bipm^0$ is slow in aqueous solution, so formation of HV^0 (at largely negative potentials) should be avoided because viologen species can undergo comproportionation, as follows.

After $bipm^0$ forms, it may react with $bipm^{2+}$ from the solution according to Eq. (3.5):

$$bipm^{2+} + bipm^0 \rightarrow (bipm^{+\bullet})_2 \rightarrow 2\,bipm^{+\bullet} \tag{3.5}$$

where the immediate product of Eq. (3.5) is the radical-cation dimer. *Solution*-phase dimer will dissociate soon after formation to form monomeric radical cations [84], but in the solid state, deposits of 'bipm$^{+\bullet}$' often exhibit often spectroscopic IR bands attributable to the spin-paired $(bipm^{+\bullet})_2$ dimer long after formation [85]. The spin-paired dimer is effectively 'locked' into solid deposits of viologen radical cation.

At best, the radical-cation dimer is electrochemically quasi-reversible, that is, its oxidation at an electrode is slow within a realistic timescale [3, 4, 84, 86], hence in that way ECDs containing traces of dimer soon fail. That comproportionation occurs in the *solid* state has been confirmed conclusively for CPQ (**II**), where the CPQ^{2+} precursor was initially in aqueous solution [30].

The review by Monk [6] demonstrates how widely comproportionation processes occur in viologen redox chemistry. Its fast rate constant and moderate equilibrium constant make it virtually certain that comproportionation processes will occur whenever $bipm^0$ is formed at an electrode. In fact, invoking the participation of Eq. (3.5) can greatly simplify Belinko's otherwise complicated mechanistic observations. (In this context, see also the way comproportionation can simplify mechanistic observations in Ref. [87].)

Engelman and Evans have also published [88] studies of potentiostatically deposited MV^0, EtV^0, BzV^0 and HV^0; each was formed at a glassy-carbon rotated ring–disc electrode (RRDE) from solutions containing the respective dications. The reductions represent a concerted two-electron process. To summarise their findings, deposition is initiated by nucleation of supersaturated $bipm^0$ close to the electrode; the rate of deposition decreases as the bulk of the deposit increases, that is, as the surface of the disc becomes 'blocked'. Comproportionation of $bipm^{2+}(aq)$ from the solution with $bipm^0(s)$ on the disc becomes increasingly important with time, so the total amount of $bipm^0$ on the disc decreases until the amount reaches a steady state.

The mechanism of comproportionation differs when ferrocyanide is involved. This result may be important because this ion is a popular choice for electron mediator [89].

Benzyl viologen has also been investigated extensively in this context because it too forms an insoluble film of radical-cation salt following one-electron reduction [53, 74, 83, 90].

3.3.3
Type-3 Viologen Electrochromes

A different way of improving the write–erase efficiency of bipyridilium-based ECDs is either to retard the rate at which the radical-cation product of electron transfer diffuses away from the electrode and into the solution bulk or to stop that diffusion altogether. The former approach is best achieved by immobilising the viologen species within a semi-solid electrolyte. The latter approach is best achieved either by immobilising the viologen, for example, by tethering the dication to the surface of an electrode, thereby forming a chemically modified ('derivatised') electrode (Section 3.3.3).

3.3.3.1 Immobilising Viologen Electrochromes

The first way of creating a type-3 electrochrome is to embed the bipyridilium salt within a matrix of viscous or nearly solid electrolyte. The matrix cannot be entirely rigid because the electrochrome needs to move through it. The rates of coloration and bleaching of such devices are generally rather sluggish because ionic diffusion through the matrix is itself intrinsically slow.

Several workers have fabricated ECDs by incorporating a viologen in the water-based electrolyte poly(2-acrylamido-2-methylpropanesulfonic acid) 'poly(AMPS)'. Sammells and Pujaru [58, 59] were probably the first when they suspended heptyl viologen (HV) in polyAMPS, while Calvert *et al.* [91] used methyl viologen. Both groups reported excellent long-term write–erase efficiencies and a good electrochromic memory. Different carboxylate polymers are also available for ECD usage [92].

Viologen species readily form ionic complexes in which the electrochrome is held electrostatically inside the columnar structure of a liquid crystal. Kominami and coworkers [93] prepared the first liquid-crystalline viologen complex having electrochromic properties; Asaftei *et al.* [94] prepared complexes that were both thermochromic and electrochromic. The latter study analysed several viologens. With methyl viologen, for example, the thermochromic transition was yellow to blue–green on heating above 75 °C with an absorption maximum at 600 nm. The electrochromic response was unremarkable. Also, germane is Kato's [95] study of viologen-based redox-active ionic liquid crystals.

Viologen monomers have been incorporated into a matrix of styrene that was photopolymerised by UV irradiation [96, 97] and semi-crystalline succinonitrile [98].

Another means to a similar end is to employ a normally liquid solvent containing a gelling agent such as silica [99], which is just as effective in immobilising viologens. In much the same way, Kubo [100] prepared a 'paper-like' ECD in which the viologen was dissolved in a glycol-based electrolyte.

3.3.3.2 Derivatised Electrodes

The second way of increasing the write–erase efficiency is by derivatising an electrode surface with a viologen. This method is generally superior and is more truly type 3.

Wrighton [101, 102] was one of the first to derivatise an electrode with viologen-based electrochromes. His first choice of substituent consisted of a short alkyl chain terminating in the trimethoxysilyl group, which can bond to the oxide lattice on the surface of an optically transparent electrode such as ITO on glass. With chemical tethering of this type, Bookbinder and Wrighton [101] attached the viologen (**V**) and benzyl viologen [103] species to electrode surfaces.

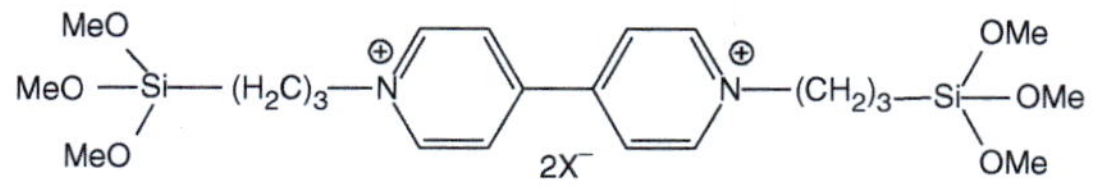

V

Wrighton also diquaternised a bipyridilium nucleus with a short alkyl chain terminating in pyrrole (which was bonded to the alkyl chain attached to the nitrogen [102]) – see (**VI**); anodic polymerisation of the pyrrole formed an adherent film of the linked poly(pyrrole) to derivatise the electrode surface [102], thereby attaching the bipyridilium units.

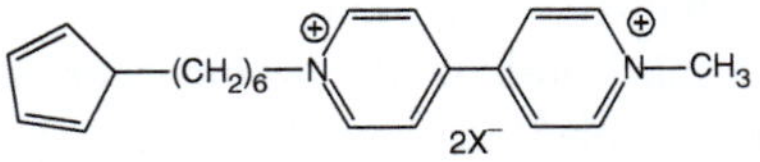

VI

An identical analogue has been prepared with thiophene as the polymerisable heterocycle [104]. The electroactivity of the poly(thiophene) backbone in this latter polymer degraded rapidly after only a few doping/de-doping cycles, but the electroactivity of the viologen moiety remained high.

Itaya [105] used polymeric electrolytes, but with an electrochromic salt in which the bipyridilium electrochrome was bonded electrostatically to a poly(styrene sulfonate) electrolyte. A bipyridilium salt of poly(*p*- or *m*-xylyl)-4,4′-bipyridilium bromide (**VII**, shown here as the *p*-form) was employed in this manner: the interaction is coulombic between the cationic bipyridilium nucleus and the sulfonyl group. The electrode was prepared by dipping the conducting substrate into solutions of electrochrome-containing polymer which, after drying, is insoluble in aqueous solution. Polymeric bipyridilium salts have also been prepared by Berlin *et al.* [106], Factor and Heisolm [107], Lieder [108], Sato and Tamamura [109] and Willman and Murray [110].

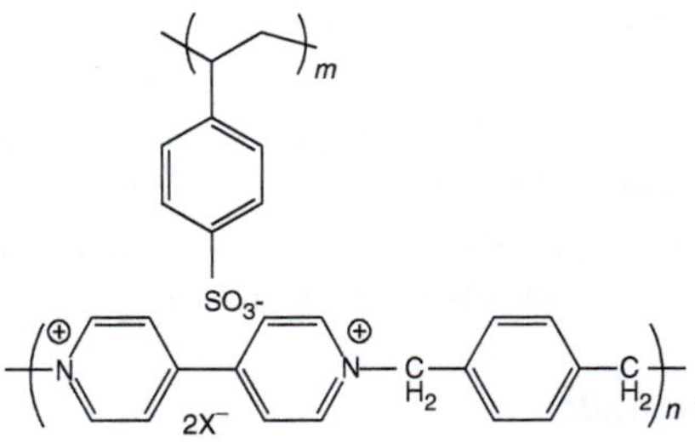

VII

ECDs can be constructed by means of a Langmuir–Blodgett type of structure that accumulates following layer-by-layer deposition. For example, Moon [111] fabricated a device in which the viologen–perylenediimide hybrid (**VIII**) alternated with layers of poly(3,4-ethylenedioxythiophene)–poly(styrenesulfonate) (PEDOT:PSS). This ECD exhibited a sharp colour change between deep red and dark blue as the applied potential was switched from +2.0 to −2.0 V.

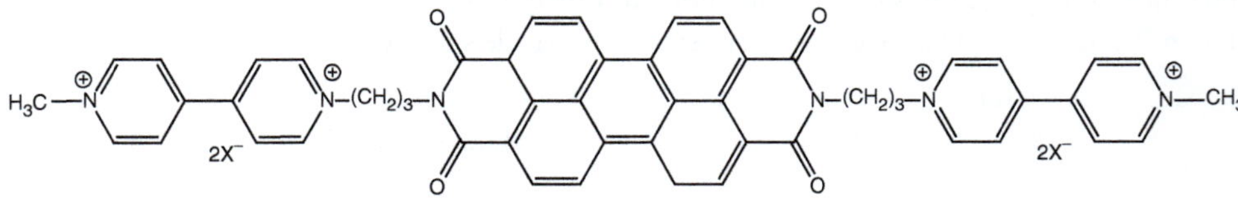

VIII

Other viologen ECDs have been fabricated by LB methods such as Heflin's poly(silsesquioxane) nanoparticles, which use viologen bridges [112].

Other workers have preferred to make mixed organic–inorganic systems. Fitzmaurice *et al.* were probably the first to use a viologen adsorbed onto layers of transition metal oxide [113]. Fitzmaurice's work was followed up by NTera of Eire, who devised a so-called NanoChromics™ device with the viologen (**IX**) bonded via a strong chemisorptive interaction to the surface of metal oxide. NTera chose nanostructured titanium dioxide (in the anatase allotrope) as their oxide owing to the ease of making it as a thin film of high surface area. The amount of (**IX**) adsorbed was therefore high, leading to a good CR. The electrochromism of (**IX**) is discussed at more length in Section 3.4.

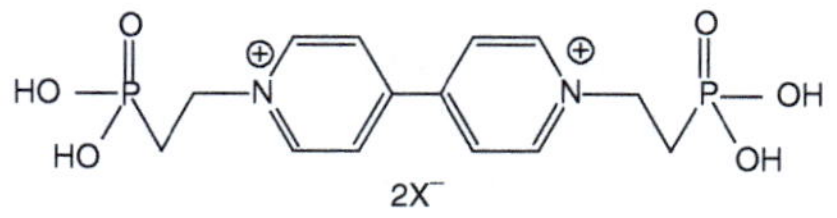

IX

Peter and coworkers [114] adsorbed a similar viologen on anatase and studied its electrochromism while the viologen film was immersed in γ-butyrolactone.

Corr *et al.* [115] (of NTera) also studied the electrochromic properties of an analogue of (**V**), in which the phosphonate substituent was replaced with a simple alkyl chain to make a 'paper-like' device. Viologens with similar phosphorus-containing substituents have also been prepared and adsorbed on TiO_2, but the bipyridyl core is nuclear substituted [116].

By contrast, Sun and Wang [117] used a viologen adsorbed on a thin film of zinc oxide (a 'nanowire array') grown on ITO glass. Their ECD had switching times of 170 and 142 ms for coloration and bleaching, respectively; the coloration efficiency was 196 $cm^2\,C^{-1}$.

Hoshino [118] prepared an ECD in which a viologen electrochrome (1,1′-didodecyl-4,4′-bipyridinium) was dispersed within a composite of ITO nanoparticles. The reductive electrochromic colour change was off-white to violet. The response time was 500 ms for coloration and 380 ms for bleaching, and the coloration efficiency was 140 $cm^2\,C^{-1}$. They report a long-term stability of >500 test cycles and an open-circuit memory of >24 h.

Finally, Diamond [119] prepared a similar-looking viologen, (**X**) in which the phosphonyl chain cannot actually bond to an oxide surface. By remaining liquid, the layer of viologen acts also as its own viscous electrolyte. The coloration efficiency is low at 10.7 $cm^2\,C^{-1}$, and the response time is minutes in duration.

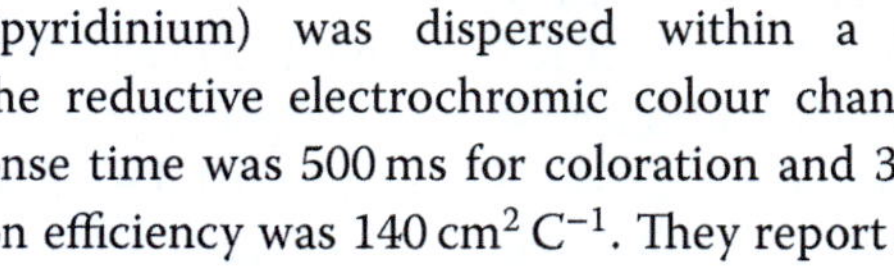

X

3.4 Exemplar Bipyridilium ECDs

The exemplar devices are described in more or less the order of their appearance in patents or as marketable products.

3.4.1 The Philips Device

The Philips device The first ECD using bipyridilium salts was reported in 1973 by Schoot *et al.* [120] (of Philips in the Netherlands), although Philips seem to have started work on electrochromic devices at some time during the mid-1960s. Their first patent was Dutch and submitted in 1970 [62] for heptyl viologen (HV = 1,1′-diheptyl-4,4′-bipyridilium, as the dibromide salt). The HV^{2+} dication is soluble in water but forms an insoluble film of crimson-coloured radical-cation salt that adheres strongly to an electrode surface following a one-electron reduction. The Philips ECD had a CR of 20:1, an erase time of 10–50 ms [120] and cycle life of more than 10^5 cycles. Philips chose heptyl viologen for their ECD rather than a viologen with a shorter chain, because reduction of the HV^{2+} dication formed a durable film on the electrode, whereas shorter alkyl chains yield somewhat soluble radical-cation salts. The Philips device was never marketed.

3.4.2 The ICI Device

The ICI device In 1971, ICI patented the use of an aryl-substituted viologen 1,1′-*bis*(*p*-cyanophenyl)-4,4′-bipyridilium (CPQ, (**II**)) [56]. This electrochrome electroprecipitates according to Eq. (3.4) to form a green electrochrome. ICI preferred CPQ to the simpler, cheaper HV owing to its greater extinction coefficient (and hence a device of higher coloration efficiency η), and therefore its faster response time. It is also more resistant to aerial oxidation.

Figure 3.1 shows a schematic of ICI's simple ECD cell. The conducting layer of the ITO (of fairly high resistance, ~80 Ω per square) acts as the working electrode that displays the colour. A strip of insulating cellulose acetate is placed near opposing edges of the base, and a stripe of conducting silver paint is applied to its upper surface to facilitate an ohmic contact with the electrode surface. The electrolyte layer is gelled with agar (5%) to improve its stability, containing the electrochrome in a concentration of 10^{-3} mol dm^{-3} in sulphuric acid or potassium chloride (either of concentration 0.1 mol dm^{-3}). The layer is applied over the platinum wire counter electrode, which is positioned over the insulatory layer. The device is completed by encapsulating the electrolyte layer, so a sheet of plain non-conducting glass covers the device. Electrical connection is made to the counter electrode and the exposed end of the silver paint (Figure 3.1).

A potential of −0.2 V (relative to a small internal silver/silver chloride electrode) is applied to the silver paint to effect electrochromic coloration (cf. Eq. 3.1), to form

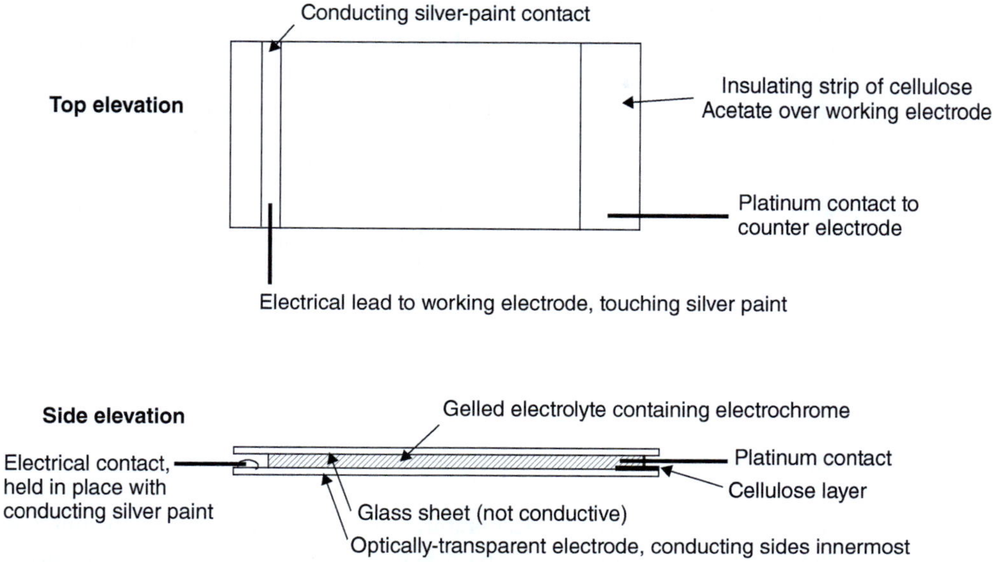

Figure 3.1 Schematic representation of an ECD operating by a type-2 electro-coloration mechanism, with colourless CPQ^{2+} in solution being electro-reduced to form a coloured film of radical-cation salt. Figure redrawn after an original in Ref. [56].

a thin, even layer of insoluble, green (**II**) radical-cation salt:

$$CPQ^{2+}(\text{soln.}) + e^- \rightarrow CPQ^{+\cdot}(\text{s}) \tag{3.6}$$

It is best to prevent the formation of a further reduction product, the pale-red species CPQ^0 (oxidation of which is electrochemically slow), so the reducing potential should not exceed −0.4 V.

The intense olive-green colour of the $CPQ^{+\cdot}$ radical is stable on open circuit with the colour persisting for many tens of hours. Reversing the polarity and applying a potential of +1.0 V (measured vs. silver/silver chloride electrode) oxidatively removes the electro-generated colour. The bleaching time is about 1 min.

The Pt counter electrode in Figure 3.1 is pre-coated with solid $CPQ^{+\cdot}$ and undergoes the reverse of Eq. (3.6) during coloration on the ITO. Then for bleaching at the ITO (i.e. the reverse of reaction (3.6)), the reaction (3.6) takes place at the Pt counter electrode in a confined, invisible volume. This pre-coating procedure represents an ingenious resolution of the often problematic choice of counter electrode. (In demonstration devices, often electrolysis of solvent is allowed to take place at the counter electrode, which in progressively destroying solvent will of course not serve in long-term use.)

A less-anodic potential of +0.4 V (vs. AgCl−Ag) can be used if the electrolyte is gelled and also contains sodium ferrocyanide (0.1 mol dm^{-3}) as an electron mediator to facilitate electro-bleaching.

The ICI ECD was first marketed as a data-display device in the early 1970s after extensive and successful field trials, but liquid-crystal displays (LCDs) entered the market for displays at about the same time. LCDs had faster response times τ and rapidly captured an unassailable market share. The slow kinetic response of the ICI type-2 cell was the result of gelling the electrochrome-containing electrolyte with agar. Removing the agar allows for considerable improvements in device response times (to seconds or tenths of a second), but the electrochromic image is generally streaky and uneven. A yellow-brown 'oil' eventually stains the electrode surface after many write–erase cycles. Yasuda *et al.* [60] (of the Sony Corporation) added encapsulating sugars such as β-cyclodextrin to aqueous heptyl viologen to circumvent the problem of 'oil' formation; but ICI believed that this molecular encapsulant failed to improve the long-term write–erase efficiency of the different viologen CPQ (J.G. Allen, ICI Ltd., personal communication, 1987). The origin of oiling is a result of dimer formation by the radicals.

3.4.3 The IBM Device

The IBM device In 1970, IBM laboratories [121, 122] manufactured an impressive device using heptyl viologen on a 64 × 64 pixel device. The 1-in. square silicon chip had eight levels of grey tone capable of detailed images. These devices were not exploited further owing to competition from LCDs.

3.4.4 The Gentex Device

The Gentex device To date, the best-selling electrochromic mirror is the Gentex night-vision system [123, 124] (NVS®), of which many millions have been sold if not hundreds of millions – indeed, probably 90–95% of all self-darkening mirror sales [125]. The Gentex device is described in greater detail in Chapter 14.

At heart, each NVS® mirror incorporates a front electrode of ITO-coated glass and a metallic rear electrode having a highly reflective surface. A sub-millimetre gap separates these two parallel electrodes and thus forms the basis of the cell. (In a similar device containing heptyl viologen and TMPD in PC, the cell would only function when the gap was narrower than 0.28 mm [36].)

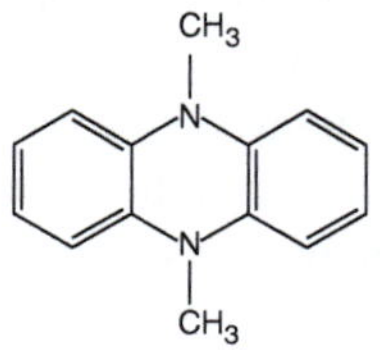

XI

The Gentex mirror uses two solution-phase electrochromes: a viologen species and 9,10-dimethylphenazine (DMP, **XI**) [123, 124]. The solution containing both electrochromes is injected into the cavity between the electrodes.

The primary electrochrome is the substituted viologen species 'bipm', undoubtedly in solution as the $bipm^{2+}$ dication. Reductive coloration occurs at the cathode, Eq. (3.1). The electrolyte in the Gentex mirror is free of additional swamping electrolyte, so mass transport occurs by ohmic migration when the mirror is switched on as the positive charge of the uncoloured precursor propels the viologen towards the cathode.

The other, secondary, electrochrome is molecular DMP (**XI**), which is initially present in its neutral, reduced form. Electro-oxidation of (**XI**) at an anode evokes a second, different, colour to the viologen, Eq. (3.7):

$$(\mathbf{XI})(\text{soln.}) \rightarrow (\mathbf{XI})^{+\cdot}(\text{soln.}) + e^- \tag{3.7}$$

Electro-oxidation of $(\mathbf{XI})^0$ depletes the solution of uncharged species, so mass transport of (**XI**) ensues by diffusion alone.

The terminology of electro-bleaching here is borrowed from the dyestuffs industry. The Greek descriptor *leucos* ('white') is used to describe the colourless form of a redox dye, so charged, oxidised (**XI**) is green-blue but colourless following reduction to form the neutral radical – the so-called *leuco*-DMP.

In operation, the colour in a commercially available NVS® mirror is an intense blue-green. The colour-forming reduction process, $\text{bipm}^{2+} + e^- \rightarrow \text{bipm}^{+\cdot}$, and the complementary oxidation reaction, $(\mathbf{XI}) \rightarrow (\mathbf{XI})^{+\cdot} + e^-$, occur in tandem in a dual electro-coloration process. Both coloured species then diffuse away from the respective electrodes and meet in the intervening solution where their mutual reaction ('radical annihilation') ensues, Eq. (3.8)

$$(\mathbf{XI})^{+\cdot}(\text{soln.}) + \text{bipm}^{+\cdot}(\text{soln.}) \rightarrow (\mathbf{XI})(\text{soln.}) + \text{bipm}^{2+}(\text{soln.}) \tag{3.8}$$

This reaction regenerates both the original uncoloured species. These reactions are depicted schematically in Figure 3.2.

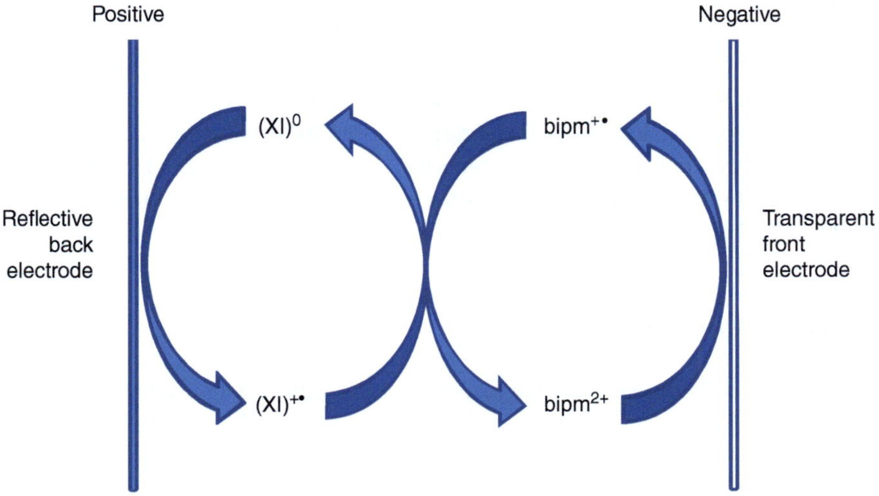

Figure 3.2 Schematic representation of the redox cycles occurring within the Gentex Night-Vision System NVS®. Coloration occurs electrochemically at both electrodes; bleaching occurs chemically mid-way between the two parallel electrodes by a process of radical annihilation.

The radical annihilation in Eq. (3.8) represents a divergence from one of the benefits of electrochromism since the 'memory effect' is lost. In consequence, maintaining the colour requires the passage of a continuous (albeit minute) current to replenish the coloured electrochromes lost by this annihilation process. Conversely, the solution-phase reaction (3.8) obviates any need to electro-bleach the Gentex NVS® mirror since colour fades spontaneously following the cessation of current, when the device is switched 'off'. For this reason, the Gentex NVS® is sometimes described as a *self-erasing* mirror. US law requires the 'failure mode', on loss of current, to be the clear condition, with which these mirrors comply.

3.4.5 The NTERA Device

The NTERA device This device was originally designed as a spin-off from the Grätzel cell and was devised in 1997 by Fitzmaurice and co-workers at the Dublin-based NTERA Ltd. They fabricated nanostructured electrodes by spreading a concentrated colloidal suspension over a conducting substrate and firing the resulting gel film at 450 °C [126]. Such electrodes have been widely investigated for use in dye-sensitised photoelectrochemical cells [126, 127]. The rough surface of the porous titanium dioxide film consists of a network of interconnected semi-conducting or conducting metal oxide nanocrystals. Because the oxide crystals are so small, such films have an extraordinarily high internal surface area.

The ratio between the internal surface area and the electrode's smooth geometrical area (its 'roughness factor') approaches 1000 for a film that is only 4 μm thick [128]. This means that a high number of electrochromic viologen molecules can be contained in a relatively small area, leading to a high coloration efficiency η. Furthermore, as they are surface-confined, the viologen molecules do not need to diffuse to the electrode surface, which leads to shorter switching times. Nanostructured titanium dioxide in its anatase form can be deposited as a thin film of high surface area. Viologens are readily adsorbed strongly on its surface owing to their electron deficiency. Such systems have long been investigated in the context of dye-sensitised solar-cell research, for example, in Grätzel's photoelectrochemical cell [127, 128].

3.4.6 The NanoChromics Cell

The NanoChromics cell. Fitzmaurice *et al.* [129] later developed a 'next-generation display technology' called NanoChromics.™ [115, 129–131] with manufacturing facilities in Ireland and Taiwan.

NTera described their ECD as a 'paper quality' electrochromic display, that is, an ECD of very high definition. An assembled NanoChromics™ electrochromic device uses two metal oxide films – one at the negative electrode and, unusually, the other at the positive electrode. In a typical construction [131, 132], the negative F-doped tin-oxide conducting glass electrode (the cathode on coloration) is coated with a wide band-gap titanium dioxide film of 4 μm thick, followed by a monolayer of self-assembled, chemisorbed phosphonated viologen molecules. The positive F-doped tin-oxide conducting glass counter electrode (the anode on coloration) carries a film of heavily doped antimony tin oxide (SnO_2:Sb) of 3 μm thick, followed by a monolayer of self-assembled, chemisorbed phosphonated phenothiazine molecules. The TiO_2 film is further modified with an adsorbed monolayer of viologen (**IX**), *bis*(2-phosphonoethyl)-4,4′-bipyridilium dichloride. The electrolyte was γ-butyrolactone containing $LiClO_4$ (0.2 mol dm^{-3}) and ferrocene (0.05 mol dm^{-3}) [130, 131]. In trials, their device had a coloration efficiency η of 170 $cm^2\,C^{-1}$ at 608 nm [131] and was said to be stable over 10^4 'standard' test cycles.

The counter electrode has a high capacitance which assists charge storage during coloration. The ECD is sealed with a thermoplastic gasket and a UV-curable epoxy resin. Application of a potential of 1.2 V reduces the dicationic viologen to its blue radical cation and oxidises the phenothiazine from its weak yellow colour to red. The overall colour change is therefore from virtually colourless to a blue-red purple.

Placing a diffuse reflector between the electrodes, for example, a layer of an ion-permeable nanostructured solid film of titanium dioxide, gives on coloration the visual effect of ink on pure white paper. Without the intermediate TiO_2 layer, the display is transparent while retaining readability. Different colours can be achieved in ECDs depending on the nature of the substituent(s) on the viologen molecule [130]. In such devices, many thousands of switches are possible before there is significant degradation of performance. Some open-circuit memory persists, with the colour remaining for more than 10 min after the voltage is switched off, but readily regenerated. Electrodes can be micro-patterned for display applications.

Fitzmaurice described his display as 'ultra-fast', although the criterion behind his claim is wholly unclear, since the switching time is 1 s for a change in absorbance of only 0.60 [130]. But this switching time is certainly faster than that of most of the other viologen-based devices, since the anchored viologen electrochrome avoids the diffusion delay before electron transfer. Fitzmaurice notes that charge compensation within the viologen layer is also fast because many counter-ions are also adsorbed on the TiO_2 layer.

If the counter electrode is covered with a secondary electrochrome such as a phenothiazine, the value of η increases to about 270 $cm^2\,C^{-1}$ and the response time is decreased to 250 ms [132]. Published spectra suggest an OD change of about 0.55, again at 608 nm.

NTera demonstrated a NanoChromics™ display operating in a converted *iPod* [133], and their website [129] provides 'consumer product reference designs' for digital clocks and an eight-digit calculator.

NTera also stated that their flexible display prototype can, in principle, be applied to all the product types: dimmable window laminates; applications in toys and games; flexible electronic paper displays and displays; windows and mirrors, giving rise to products such as 'smart card' displays (here 'smart' means 'clever' rather than the traditional English, 'neat, tidy, ordered'). NTera noted how signs using NanoChromics™ display technology are ideal for sports-player substitution boards. They claim that the current LED boards can become bleached out and difficult to read in bright daylight in sports stadia and that NanoChromics™ display signs are perfect for this application as they are easy to read in bright daylight and at all angles.

3.4.7
The Grätzel Device

The Grätzel device Grätzel *et al.* [134, 135] prepared devices with a series of viologens having aryl as well as alkyl substituents. In each case, the anchored group attaching the viologen to the titania was benzoate, salicylate or phosphonate (as in (**IX**)). Electrochromic devices constructed included shutters and displays. Their shutter device had the geometry, OTE|TiO_2-poly(viologen)|glutaronitrile-$LiN(SO_2CF_3)_2$|Prussian blue|OTE, and exhibited an optical density change of about 2, and the colour changes on reduction were transparent to blue, or yellowish to green, and (at higher potentials) to red–brown. They report switching times in the range of 1–3 s. Higher optical density changes are possible if the switching times are slower [134, 135]. Grätzel also obtained a variety of cell geometries for ECDs operating in a reflective sense. The viologens in such devices were generally oligomers rather than polymers.

In a similar way, Boehlen *et al.* [136] prepared a salt of 2,2′-bipyridine (**XII**) but calling it a 'viologen' and generated a pink colour on reduction (which could indicate that a proportion of the viologen exists as radical-cation dimer).

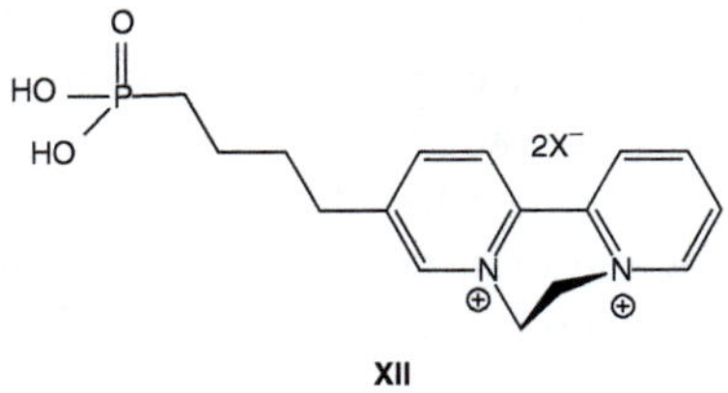

XII

Edwards *et al.* [137–141] prepared many similar systems for devices, with viologen electrochromes adsorbed on titania and named them 'electric paint'. They generally employed the viologen (**IX**), to produce images of amazing clarity with a typical response time of about 0.5 s.

3.5
Elaborations

The majority of the new developments reported here aim to enhance the rate of coloration in bipyridilium-based ECDs.

3.5.1 The Use of Pulsed Potentials

Pulses of current have been shown to enhance the rate at which electrochromic colour is formed, relative to coloration with a continuous potential [142]. It has been called a 'hammer drill' effect. The procedure relies on the solution-phase redox reaction between $bipm^{2+}$ (from the bulk solution) and $bipm^0$ electro-generated during the current pulse. The reaction is comproportionation, Eq. (3.5), so a sufficiently cathodic potential must be applied at the working electrode.

The amounts of $bipm^{2+}$ and $bipm^0$ at the electrode and in the region around the electrode depleted of $bipm^{2+}$ will govern the rate of comproportionation and hence the rate of product colour formation. Thus, for a given concentration of $bipm^{2+}$ and $bipm^0$ in such a region, the most intense colour will ensue when the two species are in equal concentration. It is envisaged that the pulse procedure possibly favours this equality.

3.5.2 Electropolychromism

Bipyridilium salts may typically possess three colours, one for each oxidation state in (**I**) although the dication in solution is essentially colourless. Viologen electrochromes comprising *n*-bipyridilium units may thus, in principle, exhibit $2n + 1$ colours. Several such approaches have employed a number of bipyridilium units connected with either alkyl linkages [143, 144] or benzylic moieties [145]. A further useful combination, the juxtaposition of bipyridilium and Prussian blue electrochromes, allows the fabrication of a five-colour ECD [146].

3.5.3 Viologen Electrochemiluminescence

'Electrochemiluminescence' is the phenomenon whereby the product of an electron-transfer process is luminescent. The literature on systems is vast, and even that concerning methyl viologen and ruthenium *tris*(2,2′-bipyridyl), $Ru(bipy)_3{}^{2+}$ is too great to cite in full [147–149].

Santa-Nokki *et al.* [150] adapted Fitzmaurice's cell to fabricate an ECD that doubles as a dye-sensitized solar cell. The sensitiser of choice was ruthenium *tris*(2,2′-bipyridyl). Applying a potential of 1.0 V to an ECD varied the transmittance between 39% (bleached) and 16% (coloured), with a response time of less than a second. The device therefore acts as both an electrochromic display and a solar cell.

Materials in which an intrinsic photoluminescence is modulated by redox processes are known as 'electrofluorochromic'. The application to viable electrochromic devices is obvious, provided the fluorescence quantum yield is high for at least one of the redox states. The idea is not new: Wrighton patented devices relying on these ideas in 1990 [151].

Viologen-based displays have been fabricated in which the colours generated are due to emission rather than absorption, thereby producing a display with a significantly higher brilliance [152]. For example, Beneduci *et al.* [153] reduced a thienoviologen electrochrome to yield a fast and reversible electrofluorochromic response in both columnar and smectic phases of liquid crystals.

Kobayashi [154] demonstrated a device capable of both electro-coloration and emission. The active material was a composite containing a luminescent complex of europium(III) and heptyl viologen. The photoluminescence of the Eu(III) complex was controlled by the electrochemical coloration of HV^{2+} *via* an intermolecular energy transfer mechanism. The fluorescent europium complex emits light in a straightforward enough manner, hence the device acts as a fluorescent display. Electroreduction of the viologen yields the blue radical $HV^{+\bullet}$ which, by an intermolecular process, reduces the europium complex and thereby quenches the emission completely. The emission is therefore modulated simply by applying a reductive potential. When acting as a passive (non-emissive) device, the electrogenerated colour of the viologen was sufficiently intense to hide the colour of the europium complex. The contrast and switch rate were improved by incorporating a complementary electrochrome, in their case, Prussian blue.

Park *et al.* [155] devised a similar system with the electrochrome loaded into the perfluorinated polymer Nafion®.

3.5.4
Viologens Incorporated within Paper

Viologen electrochromes have been incorporated within paper, to effect electrochromic writing. These include methyl viologen [156–159], heptyl viologen [159] and the asymmetric system, methyl–benzyl paraquat (**XIII**) [157]. The adsorption of methyl viologen onto the carbohydrate structures of paper follows Langmuir adsorption isotherms that imply chemisorptive behaviour [156].

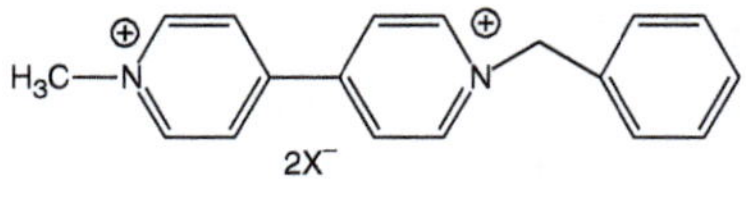

XIII

While MV in paper is electrochromic [156, 158], its response time is prohibitively slow. The speed is faster if the paper is layered with the polyelectrolyte poly(AMPS), presumably because it provides an additional source of suitable counter ions. With MV^{2+}, the speed of response depends critically on the paper's relative moisture. To summarise the results, provided the paper retains a modicum of humidity, the solution-phase electrochemistry of viologens can be reproduced as though in a standard electrochemical cell.

Alternatively, incorporation within Nafion® has been shown to produce good results. Several viologen electrochromes have been incorporated into Nafion® as a host matrix [160, 161] in which the viologen cation is immobilised by electrostatic interactions. Coloration is faster than bleaching. Mortimer fabricated an ECD evincing five-colour electropolychromism, comprising a bi-layer structure with a film of Prussian blue beneath a film of Nafion® containing a viologen [146, 161].

However, commercial utilisation of the processes just outlined seems at present somewhat questionable, as colour printing in, say, newsprint is now commonplace.

References

1. Kamagawa, H. and Suzuki, T. (1985) Organic solid photochromism *via* a photoreduction mechanism: photochromism of viologen crystals. *J. Chem. Soc., Chem. Commun.*, 525–526.
2. Sariciftci, N.S., Mehring, M., and Neugebauer, N. (1991) *In situ* studies on the structural mechanism of zwitter-viologen system during electrochemical charge-transfer reactions. *Synth. Met.*, **41–43**, 2971–2974.
3. Michaelis, L. (1935) Semiquinones, the intermediate steps of reversible organic oxidation–reduction. *Chem. Rev.*, **16**, 243–286.
4. Michaelis, L. and Hill, E.S. (1933) The viologen indicators. *J. Gen. Physiol.*, **16**, 859–873.
5. Monk, P.M.S., Mortimer, R.J., and Rosseinsky, D.R. (2007) *Electrochromism and Electrochromic Devices*, Cambridge University Press, Cambridge.
6. Monk, P.M.S. (1998) *The Viologens: Physicochemical Properties, Synthesis and Applications of the Salts of 4,4′-Bipyridine*, John Wiley & Sons, Ltd, Chichester.
7. Monk, P.M.S., Mortimer, R.J., and Rosseinsky, D.R. (1995) *Electrochromism: Fundamentals and Applications*, VCH Publishers, Weinheim.
8. Summers, L.A. (1984) The bipyridines. *Adv. Heterocycl. Chem.*, **35**, 281–394.
9. Summers, L.A. (1980) *The Bipyridinium Herbicides*, Academic Press, London.
10. Bird, C.L. and Kuhn, A.T. (1981) The electrochemistry of the viologens. *Chem. Soc. Rev.*, **10**, 49–82.
11. Bard, A.J., Ledwith, A., and Shine, H.J. (1976) Formation, properties and reactions of cation radicals in solution. *Adv. Phys. Org. Chem.*, **13**, 155–278.
12. Sliwa, W., Bachowska, B., and Zelichowicz, N. (1991) Chemistry of viologens. *Heterocycles*, **32**, 2241–2273.
13. Rosseinsky, D.R. and Monk, P.M.S. (1993) Comproportionation in propylene carbonate of substituted bipyridiliums. *J. Chem. Soc., Faraday Trans.*, **89**, 219–222.
14. Rosseinsky, D.R. and Monk, P.M.S. (1994) Solid-state conductivities of CPQ [1,1′-bis(*p*-cyanophenyl)-4,4′-bipyridilium] salts, redox-state mixtures and a new intervalence adduct. *J. Chem. Soc., Faraday Trans.*, **90**, 1127–1131.
15. Emmert, B. and Varenkamp, O. (1923) Über chinhydronartige. Verbindungen der *N,N′*-dialkyl-[dihydro-γ,γ′-dipyridyle]. *Berichte*, **56**, 490–501.
16. Levey, G. and Emmertson, T.W. (1983) Methyl viologen radical reactions with several oxidizing reagents. *J. Phys. Chem.*, **87**, 829–832.
17. Leest, R.E.V.D. (1973) The coulometric determination of oxygen with the electrochemically generated viologen radical-cation. *J. Electroanal. Chem.*, **43**, 251–255.
18. Hünig, S. and Schenk, W. (1979) Einfluß von N-substitutenten in 4,4′-bipyridylen auf das redox-verhalten, die radikalstabilität und die elektronenspektren. *Liebigs Ann. Chem.*, 1523–1533.
19. Čársky, P., Hünig, S., Scheutzow, D., and Zahradník, R. (1969) Theoretical study of redox equilibria. *Tetrahedron*, **25**, 4781–4796.
20. Hünig, S. and Groß, J. (1968) Reversible redoxsysteme vom Weitz-typ: eine polarographische studie. *Tetrahedron Lett.*, **21**, 2599–2604.
21. Thorneley, R.N.F. (1974) A convenient electrochemical preparation of reduced methyl viologen and kinetic

study of the reaction with oxygen using an anaerobic stopped-flow apparatus. *Biochim. Biophys. Acta*, **333**, 487–496.
22. Watanabe, T. and Hondo, K. (1982) Measurement of the extinction coefficient of the methyl viologen cation radical and the efficiency of its formation by semiconductor photocatalysis. *J. Phys. Chem.*, **86**, 2617–2619.
23. Kosower, E. and Cotter, J.L. (1964) Stable free radicals, II: the reduction of 1-methyl-4-cyanopyridinium ion to methylviologen radical cation. *J. Am. Chem. Soc.*, **86**, 5524–5527.
24. Schwarz, W. Jr, (1962) Stable free radicals, II: the reduction of 1-methyl-4-cyanopyridinium ion to methylviologen radical cation. PhD thesis. University of Wisconsin.
25. Stargardt, J.F. and Hawkridge, F.M. (1983) Computer decomposition of the ultraviolet-visible adsorption spectrum of the methyl viologen cation radical and its dimer in solution. *Anal. Chim. Acta*, **146**, 1–8.
26. Imabayashi, S.-I., Kitamura, N., Tazuke, S., and Tokuda, K. (1988) The role of intramolecular association in the electrochemical reduction of viologen dimers and trimers. *J. Electroanal. Chem.*, **243**, 143–160.
27. Anonymous (1986) *New Electron.*, **7**, 66.
28. Monk, P.M.S., Hodgkinson, N.M., and Ramzan, S.K. (1999) Spin pairing ('dimerisation') of the viologen radical cation: kinetics and equilibria. *Dyes Pigm.*, **43**, 207–217.
29. Müller, F. and Mayhew, S.G. (1982) Dimerisation of the radical cation of Benzyl Viologen in aqueous solution. *Biochem. Soc. Trans.*, **10**, 176–177.
30. Compton, R.G., Waller, A.M., Monk, P.M.S., and Rosseinsky, D.R. (1990) Electron paramagnetic resonance spectroscopy of electrodeposited species from solutions of 1,1′-*bis*(*p*-cyanophenyl)-4,4′-bipyridilium (cyanophenyl paraquat, CPQ). *J. Chem. Soc., Faraday Trans.*, **86**, 2583–2586.
31. Mori, H. and Mizuguchi, J. (1987) Green electrochromism in the system *p*-cyanophenyl viologen and potassium ferrocyanide. *Jpn. J. Appl. Phys.*, **26**, 1356–1360.
32. Belinko, K. (1976) Electrochemical studies of the viologen system for display applications. *Appl. Phys. Lett.*, **29**, 363–365.
33. Müller, E. and Bruhn, K.A. (1953) Über das merichinoide, *N,N*′-dibenzyl-γ,γ′ dipyridinium-subechlorid. *Chem. Ber.*, **86**, 1122–1132.
34. Carey, J.E., Cairns, J.E., and Colchester, J.E. (1969) Reduction of 1,1′-dimethyl-4,4′-bipyridilium dichloride to 1,1′-dimethyl-1,1′-dihydro-4,4′-bipyridilyl. *J. Chem. Soc., Chem. Commun.*, 1290–1291.
35. Ho, K.-C., Fang, J.G., Hsu, Y.-C., and Yu, F.-C. (2003) A study on the electro-optical properties of HV and TMPD with their application in electrochromic devices. *Proc. Electrochem. Soc.*, **2003–2017**, 266–278.
36. Ho, K.-C., Fang, Y.-W., Hsu, Y.-C., and Chen, L.-C. (2003) The influences of operating voltage and cell gap on the performance of a solution-phase electrochromic device containing HV and TMPD. *Solid State Ion.*, **165**, 279–287.
37. Byker, H.J. (1990) Single-compartment, self-erasing, solution-phase electrochromic devices, solutions for use therein and uses thereof. US Patent 4,902,108, Gentex Corporation.
38. van Dam, H.T. and Ponjeé, J.J. (1974) Electrochemically generated colored films of insoluble viologen radical compounds. *J. Electrochem. Soc.*, **121**, 1555–1558.
39. Barclay, D.J., Dowden, A.C., Lowe, A.C., and Wood, J.C. (1983) Viologen-based electrochromic light scattering display. *Appl. Phys. Lett.*, **42**, 911–913.
40. Kaifer, A.E. and Bard, A.J. (1985) Micellar effects on the reductive electrochemistry of methylviologen. *J. Phys. Chem. B*, **89**, 4876–4880.
41. Rosseinsky, D.R. and Monk, P.M.S. (1992) Electrochromic cyanophenylparaquat (CPQ: 1,1′-bis-cyanophenyl-4,4′-bipyridilium) studied voltammetrically, spectroelectrochemically and by ESR. *Sol. Energy Mater. Sol. Cells*, **25**, 201–210.

42. Shaw, D.J. (1980) *Introduction to Colloid and Surface Chemistry*, 3rd edn, Butterworths, London, p. 74.
43. Hoshino, K. and Saji, T. (1987) Electrochemical formation of thin film of viologen by disruption of micelles. *Chem. Lett.*, 1439–1442.
44. Rosseinsky, D.R., Monk, P.M.S., and Hann, R.A. (1990) Anion-dependent aqueous electrodeposition of electrochromic 1,1′-*bis*-cyanophenyl-4,4′-bipyridilium (cyanophenylparaquat) radical cation by cyclic voltammetry and spectroelectrochemical studies. *Electrochim. Acta*, **35**, 1113–1123.
45. Heyrovský, M. (1983) Catalytic and photocatalytic reduction of water by the reduced forms of methylviologen. *J. Chem. Soc., Chem. Commun.*, 1137–1138.
46. Heyrovský, M. (1986) Effect of light upon electroreduction of 4,4′-bipyridyl and methyl viologen in aqueous solutions. *J. Chem. Soc., Faraday Trans. 1*, **82**, 585–596.
47. Heyrovský, M. (1987) The electroreduction of methyl viologen. *J. Chem. Soc., Chem. Commun.*, 1856–1857.
48. Engelman, E.E. and Evans, D.H. (1992) Explicit finite-difference digital simulation of the effects of rate-controlled product adsorption or deposition in double-potential-step chronocoulometry. *J. Electroanal. Chem.*, **331**, 739–749.
49. Engelman, E.E. and Evans, D.H. (1994) Treatment of the electrodeposition of alkyl sulfate salts of viologen radical cations as an equilibrium process governed by solubility product. *Anal. Chem.*, **66**, 1530–1534.
50. Barna, G.G. (1980) The morphology of viologen films on transparent oxide electrodes. *J. Electrochem. Soc.*, **127**, 1317–1319.
51. Jasinski, R.J. (1977) The electrochemistry of some *n*-heptyl viologen salt solutions. *J. Electrochem. Soc.*, **124**, 637–641.
52. Jasinski, R. (1979) On the cathodic growth of *n*-heptylviologen radical cation films. *J. Electrochem. Soc.*, **126**, 167–170.
53. Goddard, N.J., Jackson, A.C., and Thomas, M.G. (1983) Spectroelectrochemical studies of some viologens used in electrochromic display devices. *J. Electroanal. Chem.*, **159**, 323–335.
54. Bewick, A., Lowe, A.C., and Wederell, C.W. (1983) Recrystallisation processes in viologen-based electrochromic deposits: voltammetry coupled with rapid time-resolved spectroscopy. *Electrochim. Acta*, **28**, 1899–1902.
55. Bewick, A., Cunningham, D.W., and Lowe, A.C. (1987) Electrochemical and spectroscopic characterisation of structural reorganisation in *N,N'*-dipyridinium cation radical deposits. *Makromol. Chem. Macromol. Symp.*, **8**, 355–360.
56. Kenworthy, J.G. (1973) Variable light transmission device. British Patent 1,314,049, ICI Ltd.
57. Monk, P.M.S. and Hodgkinson, N.M. (1998) Charge-transfer complexes of the viologens: effects of complexation on the rate of electron transfer to methyl viologen. *Electrochim. Acta*, **43**, 245–255.
58. Sammells, A.F. (1987) Semi-conductor/Solid Electrolyte Junctions for Optical Information Storage, US Government Reports and Announcements Index, **87**, abstract No. 703,869, as cited in *Chem. Abstr.* **107**, 86,064*m*.
59. Sammells, A.F. and Pujare, N.U. (1986) Electrochromic effects on heptylviologen incorporated within a solid polymer electrolyte cell. *J. Electrochem. Soc.*, **133**, 1270–1271.
60. Yasuda, A., Mori, H., Takehana, Y., and Ohkoshi, A. (1984) Electrochromic properties of *n*-heptyl viologen–ferrocyanate system. *J. Appl. Electrochem.*, **14**, 323–328.
61. Yasuda, A., Kondo, H., Itabashi, M., and Seto, J. (1986) Structure changes of viologen + β-cyclodextrin inclusion complex corresponding to the redox state of viologen. *J. Electroanal. Chem.*, **210**, 265–275.
62. Philips Ltd (1971) Image display apparatus. British Patent 1,302,000.
63. Yasuda, A. and Mori, H. (1985) Electrochromic display devices. Jpn. Kokai Tokkyo Koho JP 60,198,521, as cited

in *Chem. Abstr.* **104**, P99,571n. Sony Corp.

64. Chen, P.-Y., Chen, C.-S., and Yeh, T.-H. (2014) Organic multiviologen electrochromic cells for a color electronic display application. *J. Appl. Polym. Sci.*, **131**, 40485–40545.
65. Nakahara, A. and Wang, J.H. (1963) Charge-transfer complexes of methylviologen. *J. Phys. Chem.*, **67**, 496–498.
66. Murthy, A.S.N. and Bhardwaj, A.P. (1982) Electronic absorption spectroscopic studies on charge-transfer interactions in a biologically important molecule: *N,N′*-dimethyl-4,4′-bipyridylium chloride (paraquat or methyl viologen) as an electron acceptor. *Spectrochim. Acta*, **38A**, 207–212.
67. Kramarenko, S.F., Krainov, I.P., Pretsenko, E.G., and Vargalyuk, B.F. (1985) Electrochemical chromism of 1,1′-diaryl-4,4′-bipyridilium perchlorates. *Ukr. Khim. Zh.*, **51**, 501–504, (in Russian), as cited in *Chem. Abstr.*, **103**, 112, 232.
68. Bruinink, J., Ponjeé, J.J., and Kregting, C.G.A. (1977) Modified viologens with improved electrochemical properties for display applications. *J. Electrochem. Soc.*, **124**, 1854–1858.
69. Barna, G.G. and Fish, J.G. (1981) An improved electrochromic display using an symmetric viologen. *J. Electrochem. Soc.*, **128**, 1290–1292.
70. Barltrop, J.A. and Jackson, A.C. (1984) The synthesis and electrochemical study of new electrochromic viologen-based materials. *J. Chem. Soc., Perkin Trans.* 2, 367.
71. Ostrom, G.S. and Buttry, D.A. (1988) Quartz crystal microbalance studies of deposition and dissolution mechanisms of electrochromic films of diheptylviologen bromide. *J. Electroanal. Chem.*, **256**, 411–431.
72. Tang, X.Y., Schneider, T., and Buttry, D.A. (1994) A vibrational spectroscopic study of the structure of electroactive self-assembled monolayers of viologen derivatives. *Langmuir*, **10**, 2235–2240.
73. Beden, B., Enea, O., Hahn, F., and Lamy, C. (1984) Investigation of the adsorption of methyl viologen on a platinum electrode by voltammetry coupled with 'in situ' UV-visible reflectance spectroscopy. *J. Electroanal. Chem.*, **170**, 357–361.
74. Crouigneau, P., Enea, O., and Beden, B. (1987) 'In situ' investigation by simultaneous voltammetry and UV-visible reflectance spectroscopy of some viologen radicals absorbed on a platinum electrode. *J. Electroanal. Chem.*, **218**, 307–317.
75. Reichman, B., Fan, F.-R.F., and Bard, A.J. (1980) Semiconductor electrodes, XXV: the p-GaAs/heptyl viologen system: photoelectrochemical cells and photoelectrochromic cells. *J. Electrochem. Soc.*, **127**, 333–338.
76. Crouigneau, P., Enea, O., and Lamy, C. (1986) A comparative electron spin resonance study of adsorbed cation-radicals generated 'in situ' by electrochemical and photoelechemical reduction of some viologen derivatives. *Nouv. J. Chem.*, **10**, 539–543.
77. Lu, T. and Cotton, T.M. (1987) *In situ* Raman spectra of the three redox forms of heptylviologen at platinum and silver electrodes: counterion effects. *J. Phys. Chem.*, **91**, 5978–5985.
78. Osawa, M. and Suëtaka, W. (1989) Electrochemical reduction of heptyl viologen at platinum studied by time-resolved resonance Raman spectroscopy. *J. Electroanal. Chem.*, **270**, 261–272.
79. Sawada, T. and Bard, A.J. (1982/3) Laser-photoelectric-photoacoustic observation of the electrode surface. *J. Photoacoust.*, **1**, 317–327.
80. Malpas, R.E. and Bard, A.J. (1980) *In situ* monitoring of electrochromic systems by piezoelectric detector photoacoustic spectroscopy of electrodes. *Anal. Chem.*, **52**, 109–112.
81. Brilmyer, G.H. and Bard, A.J. (1980) Application of photothermal spectroscopy to in-situ studies of films on metals and electrodes. *Anal. Chem.*, **52**, 685–691.
82. Gołden, A. and Przyłuski, J. (1985) Studies of electrochemical properties of *N*-heptylviologen bromide films. *Electrochim. Acta*, **30**, 1231–1235.

83. Scharifker, B. and Wehrmann, C. (1985) Phase formation phenomena during electrodeposition of benzyl and heptyl viologen bromides. *J. Electroanal. Chem.*, **185**, 93–108.
84. Monk, P.M.S., Fairweather, R.D., Duffy, J.A., and Ingram, M.D. (1992) Evidence for the product of viologen comproportionation being a spin-paired radical cation dimer. *J. Chem. Soc., Perkin Trans. 2*, 2039–2041.
85. Poizat, O., Sourisseau, C., and Corset, J. (1986) Vibrational and electronic study of the methyl viologen radical cation MV^+ in the solid state. *J. Mol. Struct.*, **143**, 203–206.
86. Rosseinsky, D.R. and Monk, P.M.S. (1990) Kinetics of the comproportionation of the bipyridilium salt *p*-cyanophenyl paraquat in propylene carbonate studied by rotating ring-disc electrodes. *J. Chem. Soc., Faraday Trans.*, **86**, 3597–3601.
87. Monk, P.M.S. (1998) Comment on: 'Dimer formation of viologen derivatives and their electrochromic properties'. *Dyes Pigm.*, **39**, 125–128.
88. Engelman, E.E. and Evans, D.H. (1992) Investigation of the nature of electrodeposited neutral viologens formed by reduction of the dications. *J. Electroanal. Chem.*, **349**, 141–158.
89. Monk, P.M.S. (1997) The effect of ferrocyanide on the performance of heptyl viologen-based electrochromic display devices. *J. Electroanal. Chem.*, **432**, 175–179.
90. Rosseinsky, D.R., Slocombe, J.D., Soutar, A., Monk, P.M.S., and Glidle, A. (1989) Simple diffuse reflectance monitoring of emerging surface-attached species. *J. Electroanal. Chem.*, **259**, 233–239.
91. Calvert, J.M., Manuccia, T.J., and Nowak, R.J. (1986) A polymeric solid-state electrochromic cell. *J. Electrochem. Soc.*, **133**, 951–953.
92. Zakirov, M.I., Shandryuk, G.A., Bondarenko, G.N., Nodova, E.L., Kryl'skii, D.V., and Talroze, R.V. (2012) Formation of electrochromic systems based on noncovalently associated polymer–viologen complexes. *Polym. Sci. Ser. B*, **54**, 50–60.
93. Tabushi, I., Yamamura, K., and Kominami, K. (1986) Electric stimulus-response behavior of liquid-crystalline viologen. *J. Am. Chem. Soc.*, **108**, 6409–6410.
94. Asaftei, S., Ciobanu, M., Lepadatu, A.M., Song, E., and Beginn, U. (2012) Thermotropic ionic liquid crystals by molecular assembly and ion pairing of 4,4'-bipyridinium derivatives and tris(dodecyloxy)benzenesulfonates in a non-polar solvent. *J. Mater. Chem.*, **22**, 14426–14437.
95. Tanabe, K., Yasuda, T., Yoshio, M., and Kato, T. (2007) Viologen-based redox-active ionic liquid crystals forming columnar phases. *Org. Lett.*, **9**, 4271–4274.
96. Gao, L., Zheng, G., Zhou, Y., Jiang, J., Ding, G., and Wang, Y. (2014) Preparation, characterization of styrene-functional hyperbranched viologen, and its application in optical writing and electrical erasing device. *Soft Mater.*, **12**, 149–158.
97. Cao, L.-C., Mou, M., and Wang, Y. (2009) Hyperbranched and viologen-functionalized polyglycerols: preparation, photo- and electrochromic performance. *J. Mater. Chem.*, **19**, 3412–3418.
98. Hu, C.-W., Lee, K.-M., Chen, K.-C., Chang, L.-C., Shen, K.-Y., Lai, S.-C., Kuo, T.-H., Hsu, C.-Y., Huang, L.-M., Vittal, R., and Ho, K.-C. (2012) High contrast all-solid-state electrochromic device with 2,2,6,6-tetramethyl-1-piperidinyloxy (TEMPO), heptyl viologen, and succinonitrile. *Sol. Energy Mater. Sol. Cells*, **99**, 135–140.
99. Byker, H.J. (2001) Electrochromics and polymers. *Electrochim. Acta*, **46**, 2015–2022.
100. Sanehira, Y., Uchida, S., Kubo, T., and Segawa, H. (2008) A distinguished retentive memory using polyethylene glycol electrolyte solvent for viologen modified titania electrochromic device. *Electrochemistry*, **76**, 150–153.
101. Bookbinder, D.C. and Wrighton, M.S. (1983) Electrochromic polymers covalently anchored to electrode surfaces:

optical and electrochromic properties of a viologen-based polymer. *J. Electrochem. Soc.*, **130**, 1081–1087.

102. Dominey, R.N., Lewis, T.J., and Wrighton, M.S. (1983) Synthesis and characterization of a benzylviologen surface-derivatizing reagent: *N,N'*-bis[*p*-(trimethoxysilyl) benzyl]-4,4'-bipyridilium dichloride. *J. Phys. Chem.*, **87**, 5345–5354.
103. Shu, C.-F. and Wrighton, M.S. (1988) Synthesis and charge-transport properties of polymers derived from the oxidation of 1-hydro-1'-(6-(pyrrol-1-yl) hexyl)-4,4'-bipyridinium bis(hexafluorophosphate) and demonstration of a pH-sensitive microelectrochemical transistor derived from the redox properties of a conventional redox center. *J. Phys. Chem.*, **92**, 5221–5229.
104. Ko, H.C., Park, S.-A., Paik, W.-K., and Lee, H. (2002) Electrochemistry and electrochromism of the polythiophene derivative with viologen pendant. *Synth. Met.*, **132**, 15–20.
105. Akahoshi, H., Toshima, S., and Itaya, K. (1981) Electrochemical and spectroelectrochemical properties of polyviologen complex modified electrodes. *J. Phys. Chem.*, **85**, 818–822.
106. Berlin, A.A., Zherebtsova, L.V., and Rabazobovskii, Y.F. (1964) Polymers with a conjugated system, XXXVII: synthesis of polymers with charged heteroatoms in the macromolecular chain (onium polymerization). *Polym. Sci. (USSR)*, **6**, 67–74.
107. Factor, A. and Heisolm, G.E. (1971) Polyviologens – a novel class of cationic polyelectrolyte redox polymers. *Polym. Lett.*, **9**, 289–295.
108. Lieder, M. and Schlapfer, C.W. (1997) Synthesis and electrochemical properties of new viologen polymers. *J. Appl. Electrochem.*, **27**, 235–239.
109. Sato, H. and Tamamura, T. (1979) Polymer effect in electrochromic behavior of oligomeric viologens. *J. Appl. Polym. Sci.*, **24**, 2075–2085.
110. Willman, K.W. and Murray, R.W. (1982) Viologen homopolymer, polymer mixture and polymer bilayer films on electrodes: electropolymerization, electrolysis, spectroelectrochemistry, trace analysis and photoreduction. *J. Electroanal. Chem.*, **133**, 211–231.
111. Kim, S.-H., Shim, N., Lee, H., and Moon, B. (2012) Synthesis of a perylenediimide-viologen dyad (PDI–2V) and its electrochromism in a layer-by-layer self-assembled multilayer film with PEDOT:PSS. *J. Mater. Chem.*, **22**, 13558–13563.
112. Jain, V., Khiterer, M., Montazami, R., Yochum, H.M., Shea, K.J., and Heflin, J.R. (2009) High-contrast solid-state electrochromic devices of viologen-bridged polysilsesquioxane nanoparticles fabricated by layer-by-layer assembly. *ACS Appl. Mater. Interfaces*, **1**, 83–89.
113. Marguerettaz, X., O' Neill, R., and Fitzmaurice, D. (1994) Heterodyads: electron transfer at a semiconductor electrode-liquid electrolyte interface modified by an adsorbed spacer–acceptor complex. *J. Am. Chem. Soc.*, **116**, 2629–2630.
114. Garcia-Canadas, J.L., Peter, L.M., and Wijayantha, K.G.U. (2003) Characterisation of electrochromic viologen-modified nanocrystalline TiO_2 films by frequency-resolved optical transmission spectroscopy. *Electrochem. Commun.*, **5**, 199–202.
115. Corr, D., Bach, U., Fay, D., Kinsella, M., McAtamney, C., O'Reilly, F., Rao, S.N., and Stobie, N. (2003) Coloured electrochromic 'paper-quality' displays based on modified mesoporous electrodes. *Solid State Ion.*, **165**, 315–321.
116. Bar, G., Larina, N., Grinis, L., Lokshin, V., Gvishi, R., Kiryuschev, I., Zaban, A., and Khodorkovsky, V. (2012) RGB organic electrochromic cells. *Sol. Energy Mater. Sol. Cells*, **99**, 123–128.
117. Sun, X.W. and Wang, J.X. (2008) Fast switching electrochromic display using a viologen-modified ZnO nanowire array electrode. *Nano Lett.*, **8**, 1884–1889.
118. Nakajima, R., Yamada, Y., Komatsu, T., Murashiro, K., Saji, T., and Hoshino, K. (2012) Electrochromic properties of ITO nanoparticles/viologen composite film electrodes. *RSC Adv.*, **2**, 4377–4381.

119. Kavanagh, A., Fraser, K.J., Byrne, R., and Diamond, D. (2013) An electrochromic ionic liquid: design, characterization, and performance in a solid-state platform. *Appl. Mater. Interfaces*, **5**, 55–62.
120. Schoot, C.J., Ponjeé, J.J., van Dam, H.T., van Doorn, R.A., and Bolwijn, P.J. (1973) New electrochromic memory device. *Appl. Phys. Lett.*, **23**, 64–65.
121. Barclay, D.J., Bird, C.L., Kirkman, D.H., Martin, D.H., and Moth, F.T. (1980) An integrated electrochromic data display. *SID Digest*, 124–125.
122. Barclay, D.J. and Martin, D.H. (1984) in *Technology of Chemicals and Materials for the Electronics Industry* (ed. E.R. Howells), Ellis Horwood, Chichester, pp. 266–276.
123. Byker, H.J., Gentex Corporation (1990) Single-compartment, self-erasing, solution-phase electrochromic devices, solutions for use therein and uses thereof. US Patent 4,902,108.
124. Gentex Corporation *http://www.gentex.com/automotive/automotive-products* (accessed 25 February 2014).
125. Byker, H.J. (1994) in *Proceeding of the Symposium on Electrochromic Materials II* (eds K.-C. Ho and D.A. MacArthur), The Electrochemical Society Inc., pp. 3–13.
126. Nazeeruddin, M.K. and Grätzel, M. (2004) in *Comprehensive Coordination Chemistry II: From Biology to Nanotechnology* (eds J.A. McCleverty and T.J. Meyer), Elsevier, Oxford, pp. 719–758.
127. O'Regan, B. and Grätzel, M. (1991) A low-cost, high-efficiency solar-cell based on dye-sensitized colloidal TiO_2 films. *Nature (London)*, **353**, 737–740.
128. Uppasala Universitet *http://uu.diva-portal.org/smash/record.jsf?searchId=1&pid=diva2:67917* (accessed 21 April 2014).
129. YouTube *http//www.youtube.com/watch?v=mnP7puQvA-c* (accessed 24 February 2014) or The Sol-Gel Gateway *http://www.solgel.com/articles/Mar02/ntera.asp* (accessed 24 February 2014).
130. Cinnsealach, R., Boschloo, G., Nagaraja Rao, S., and Fitzmaurice, D. (1999) Coloured electrochromic windows based on nanostructured TiO_2 films modified by adsorbed redox chromophores. *Sol. Energy Mater. Sol. Cells*, **57**, 107–125.
131. Cinnsealach, R., Boschloo, G., Nagaraja Rao, S., and Fitzmaurice, D. (1998) Electrochromic windows based on viologen-modified nanostructured TiO_2 films. *Sol. Energy Mater. Sol. Cells*, **55**, 215–223.
132. Cummins, D., Boschloo, G., Ryan, M., Corr, D., Rao, S.N., and Fitzmaurice, D. (2000) Ultrafast electrochromic windows based on redox-chromophore modified nanostructured semiconducting and conducting films. *J. Phys. Chem. B*, **104**, 11449–11459.
133. Graham-Rowe, D. (2005) Mirror Trick Leads Chase for Electronic Paper. New Scientist (Feb. 26 2005), p. 27.
134. Grätzel, M. (2001) Ultrafast colour displays. *Nature*, **409**, 575–576.
135. Campus, F., Bonhôte, P., Grätzel, M., Heinen, S., and Walder, L. (1999) Electrochromic devices based on surface-modified nanocrystalline TiO_2 thin-film electrodes. *Sol. Energy Mater. Sol. Cells*, **56**, 281–297.
136. Boehlen, R., Felderhoff, M., Michalek, R., and Walder, L. (1998) A new 2,2'-bipyridinium salt with pink electrochromism for the modification of nanocrystalline TiO_2-electrodes. *Chem. Lett.*, **27**, 815–816.
137. Edwards, M.O.M., Andersson, M., Gruszecki, T., Pettersson, H., Thunman, R., Thuraisingham, G., Vestling, L., and Hagfeldt, A. (2004) Charge–discharge kinetics of electric-paint displays. *J. Electroanal. Chem.*, **565**, 175–184.
138. Edwards, M.O.M., Boschloo, G., Gruszecki, T., Pettersson, H., Sohlberg, R., and Hagfeldt, A. (2001) 'Electric-paint displays' with carbon counter electrodes. *Electrochim. Acta*, **46**, 2187–2193.
139. Edwards, M.O.M., Gruszecki, T., Pettersson, H., Thuraisingham, G., and Hagfeldt, A. (2002) A semi-empirical model for the charging and discharging

of electric-paint displays. *Electrochem. Commun.*, **4**, 963–967.

140. Edwards, M.O.M. and Hagfeldt, A. (2003) Switch-speed considerations for viologen–metal oxide displays. *Proc. Electrochem. Soc.*, **2003–2017**, 305–309.

141. Pettersson, H., Gruszecki, T., Johansson, L.-H., Edwards, M.O.M., Hagfeldt, A., and Matuszczyk, T. (2004) Direct-driven electrochromic displays based on nanocrystalline electrodes. *Displays*, **25**, 223–230.

142. Monk, P.M.S., Fairweather, R.D., Ingram, M.D., and Duffy, J.A. (1993) Pulsed electrolysis enhancement of electrochromism in viologen systems: influence of comproportionation reactions. *J. Electroanal. Chem.*, **359**, 301–306.

143. Porat, Z., Tricot, Y.-M., Rubinstein, I., and Zinger, B. (1991) New multi-charged viologen derivatives, 1: electrochemical behaviour in Nafion films. *J. Electroanal. Chem.*, **315**, 217–223.

144. Porat, Z., Tricot, Y.-M., Rubinstein, I., and Zinger, B. (1991) New multi-charged viologen derivatives, 2: unusual electrochemical behaviour in solution. *J. Electroanal. Chem.*, **315**, 225–243.

145. Rosseinsky, D.R. and Monk, P.M.S. (1994) Studies of tetra-(bipyridilium) salts as possible polyelectrochromic materials. *J. Appl. Electrochem.*, **24**, 1213–1221.

146. Mortimer, R.J. (1991) Five color electrochromicity using Prussian blue and Nafion/methyl viologen layered films. *J. Electrochem. Soc.*, **138**, 633–634.

147. Nanni, E.J. Jr.,, Angelis, C.T., Dickson, J., and Sawyer, D.T. (1981) Oxygen activation by radical coupling between superoxide ion and reduced methyl viologen. *J. Am. Chem. Soc.*, **103**, 4268–4270.

148. Kim, C.S., Lee, S., Tinker, L.L., Bernhard, S., and Loo, Y.-L. (2009) Cobaltocene-doped viologen as functional components in organic electronics. *Chem. Mater.*, **21**, 4583–4588.

149. Clark, C.D., Debad, J.D., and Yomemoto, E.H. (1997) Effect of oxygen on linked $Ru(bpy)_3{}^{2+}$–viologen species and methylviologen: a reinterpretation of the electrogenerated chemiluminescence. *J. Am. Chem. Soc.*, **119**, 10525–10531.

150. Santa-Nokki, H., Kallioinen, J., and Korppi-Tommola, J. (2007) A dye-sensitized solar cell driven electrochromic device. *Photochem. Photobiol. Sci.*, **6**, 63–66.

151. Leventis, N. and Wrighton, M.S. (1990) Electrochromic, electroluminescent and electrochemiluminescent displays. US Patent 5189549 A, filed Feb. 26, 1990.

152. Audebert, P. and Miomandre, F. (2013) Electrofluorochromism: from molecular systems to set-up and display. *Chem. Sci.*, **4**, 575–584.

153. Beneduci, A., Cospito, S., La Deda, M., Veltri, L., and Chidichimo, G. (2014) Electrofluorochromism in π-conjugated ionic liquid crystals. *Nat. Commun.*, **5**, Article number: 3105.

154. Nakamura, K., Kanazawa, K., and Kobayashi, K. (2011) Electrochemically controllable emission and coloration by using europium(III) complex and viologen derivatives. *Chem. Commun.*, **47**, 10064–10066.

155. Park, H., Park, Y., Bae, E., and Choi, W. (2009) Photoactive component-loaded Nafion film as a platform of hydrogen generation: alternative utilization of a classical sensitizing system. *J. Photochem. Photobiol., A*, **203**, 112–118.

156. Monk, P.M.S., Delage, F., and Costa Vieira, S.M. (2001) Electrochromic paper: utility of electrochromes incorporated in paper. *Electrochim. Acta*, **46**, 2195–2202.

157. Rosseinsky, D.R. and Monk, J.L. (1989) Thin layer electrochemistry in a paper matrix: electrochomography of Prussian blue and two bipyridilium systems. *J. Electroanal. Chem.*, **270**, 473–478.

158. Monk, P.M.S., Turner, C., and Akhtar, S.P. (1999) Electrochemical behaviour of methyl viologen in a matrix of paper. *Electrochim. Acta*, **44**, 4817–4826.

159. Mortimer, R.J. and Warren, C.P. (1999) Cyclic voltammetric studies of Prussian blue and viologens within a paper matrix for electrochromic printing applications. *J. Electroanal. Chem.*, **460**, 263–266.

160. John, S.A. and Ramaraj, R. (1995) Role of acidity on the electrochemistry of Prussian blue at plain and Nafion film-coated electrodes. *Proc. Ind. Acad. Sci.*, **107**, 371–383.

161. Mortimer, R.J. and Dillingham, J.L. (1997) Electrochromic 1,1′-dialkyl-4,4′-bipyridilium-incorporated Nafion electrodes. *J. Electrochem. Soc.*, **144**, 1549–1553.

4
Electrochromic Devices Based on Metal Hexacyanometallate/Viologen Pairings

Kuo-Chuan Ho, Chih-Wei Hu, and Thomas S. Varley

4.1
Introduction

Electrochromic devices (ECDs) are examples of electrochemical cells, that is, devices that perform electrochemical reactions when sufficient electrical potential is applied. Typically, ECDs have a two-electrode configuration, where at least one electrode is optically transparent and the electrodes are separated by a layer of supporting electrolyte. The physical properties of a device, such as its colour and switching time, are predominantly dictated by the electrochromic materials used in construction. It is common to create an ECD with materials that are complementary with respect to their coloration redox chemistry, that is, where one material is oxidised at an electrode's surface the second material is reduced at the opposite electrode's surface.

Prussian blue (PB, iron(III) hexacyanoferrate(II)) as a material was first synthesised in 1704 and primarily used as a bright blue colour pigment [1]. Only in 1978 were its electrochromic properties first recognised by Neff [2]. It is now renowned for its electrochromic properties (covered in Chapter 2) [3, 4]. PB belongs to a class of materials called metal hexacyanometallates, $M'_x[M''(CN)_6]_y$, where M′ and M″ are transition metals and in the case of PB both are Fe. PB as an electrochromic material is typically deployed as a thin film that possesses multi-electrochromism through its four independent redox states: PB itself is blue (Fe^{II}/Fe^{III}) and can be reduced to colourless Everitt's salt (Fe^{II}/Fe^{II}), oxidised to Berlin green (possibly $[Fe^{III}/Fe^{III}]_2[Fe^{II}/Fe^{III}]$) and subsequently further oxidised to Prussian yellow (Fe^{III}/Fe^{III}) [5–7]. Therefore, when in its colourless state, PB is an anodically colouring material, which has previously been coupled with cathodically colouring electrochromic materials such as WO_3 [8–23], conducting polymers [24–32] and viologens [33], to form complementary ECDs with high transmittance changes. M′ and M″ can be other transition metals (e.g. M′ = Fe and M″ = Ru is ruthenium purple), which exhibit different redox chemistry and associated chromophores depending on the metals used [34].

In contrast, viologens (1,1′-disubstituted-4,4′-bipyridinium salts) are classic organic electrochromic materials that have three redox states; these are the

Electrochromic Materials and Devices, First Edition.
Edited by Roger J. Mortimer, David R. Rosseinsky and Paul M. S. Monk.

dication (V^{2+}), radical-cation ($V^{\bullet+}$) and di-reduced (V^0) states [35–39]. The reduction of dissolved methyl viologen (MV) dication to the radical cation shows an obvious colour change from transparency to deep blue with high reversibility [40]. This redox transformation is reversible and can be cycled many times without suffering from significant side reactions. However, further reduction to the di-reduced state is much less reversible as this species can undergo several complex reactions with other species present in solution (e.g. comproportionation reactions with the radical-cation species). The colour of the viologen radical cation is dependent on the nitrogen-atom substituents. For example, the radical cation for *n*-heptyl viologen (HV) is blue-purple, whilst the same redox state for cyanophenyl viologen is green [41]. Coupling the anodically colouring viologen with the cathodically colouring metal hexacyanometallates and by tailoring the substituents attached to the viologen's bipyridinium core and the transition metals used in metal hexacyanometallates, a wide variety of ECDs can be fabricated. A review of the viologens as electrochromic materials is provided in Chapter 3.

In this chapter, complementary ECDs based on the classical combination of PB and viologen, an inorganic/organic ECD, are reviewed. Depending on the formation of both materials, film-type and hybrid-type devices can be fabricated. The PB thin film can be obtained by electrochemical deposition [35–39] or chemical synthesis [42–48], while viologens exist in solution or as solid thin films.

4.1.1
Overview of Prussian Blue and Viologen Electrochromic Devices

Table 4.1 summarises ECDs based on solid-with-solution (or 'hybrid') PB and viologen pairings reported in the literature. Hybrid ECDs have one electrode modified with a PB thin film, whilst the other electrode is left as a bare transparent

Table 4.1 A list of Prussian blue with viologen hybrid-type ECDs reported in the literature.

Viologen[a)]	Electrolyte[b)]	Colour[c)]	V_{cell} (V)	%ΔT (T_d/T_b) or ΔA at λ_{max}[d)]	τ_d/τ_b (s)[e)]	η (cm^2 C^{-1})	References
HV	1.0 M $KCl_{(aq.)}$/IPA	C–B	+1.0/0	59 (20/79)	8/100	N/A	[49]
MV	0.5 M $KCl_{(aq.)}$	C–B	+1.5/−1.0	$\Delta A \approx 1.8$	10/10	170	[50]
MV	0.5 M $KCl_{(aq.)}$	C–P	+1.5/−0.5	$\Delta A \approx 0.42$	4/10	N/A	[51]
IL_1	N/A	LY–DB	+2.0/−2.0	30.5 (35/65.5)	2/3	725	[52]

a) HV: heptyl viologen, MV: methyl viologen and IL_1: 1,1′-*bis*[4-(5,6-dimethyl-1H-benzimidazole-1-yl)butyl]-4,4′-bipyridinium dibromide.
b) IPA: isopropyl alcohol.
c) C: colourless, B: blue, P: purple, LY: light yellow and DB: deep blue.
d) %ΔT: percentage change in transmittance, T_d: transmittance at darken state, T_b: transmittance at bleach state and ΔA: absorbance change.
e) τ_d: response time for coloration and τ_b: response time for bleaching measured at the wavelength which the maximum transmittance or absorbance change occurs.

Table 4.2 A list of Prussian blue with viologen solid–solid ECDs (thin-film type) reported in the literature.

Viologen[a)]	Electrolyte[b)]	Colours[c)]	V_{cell} (V)	%ΔT (T_d/T_b) or ΔA at λ_{max}	τ_d/τ_b (s)	η (cm² C⁻¹)	References
MV	Nafion®	Y–G–B–C–P	+1.0/–0.9	N/A	N/A	N/A	[53]
MV and HV	Nafion®	B–G–Y	+1.0/–0.2	N/A	N/A	N/A	[54]
BV	PVPI/K_2SO_4	C–B	+0.8/–0.9	$\Delta A \approx 1.3$	3.0/3.0	634	[55]
TiO_2/PV	GTN-$LiN(SO_2CF_3)_2$	W–B	+1.5/0	N/A	7/N.A	N/A	[56]
TiO_2/PhV	0.1 M ACN	W–B	+2.0/0	68(11/79)	1.0/1.0	270	[57]
PBV	0.1 M $LiBF_4$/SN	C–B	+0.7/–1.0	65(8/73)	2.0/9.4	163	[58]
Viologen/ PEDOT	IL_2	LB–DB	+1.5/–1.5	$\Delta A \approx 0.8$	N/A	187	[59]
TiO_2/PhV	0.1 M $LiClO_4$	W–B	+2.0/–3.0	64.7(0.1/64.8)	6.2/10.3	912	[60]

a) MV: methyl viologen, HV: heptyl viologen, BV: benzyl viologen, PV: poly(viologen), PBV: poly(butyl viologen), PEDOT: poly(3,4-ethylenedioxythiophene) and PhV: viologens modified with phosphate.
b) PVPI: poly(vinylpyrrolidone), GTN: glutaronitrile, CH_3CN: acetonitrile, SN: succinonitrile and IL_2: 1-butyl-1-methylpyrrolidiniumtrifluoromethanesulfonate.
c) Y: yellow, G: green, B: blue, C: colourless, P: purple, W: white, LB: light blue and DB: deep blue. Comments made in Table 4.1 also apply here.

conducting substrate, in direct contact with a solution containing the dissolved viologen species. Table 4.2 summarises solid–solid-type ECDs consisting of PB and viologen thin films. The construction of a viologen thin film is achieved by several different methods. The most common techniques employed are (1) electropolymerisation of the viologens via cyclic voltammetry or constant current; (2) incorporation of the viologens into a gel-type electrolyte, such as Nafion®, to fabricate a stand-alone thin film; (3) electrostatic attraction, whereby viologens serve as counter ions to conducting-ionic polymers and (4) viologens are functionalised with anchoring groups that are adsorbed or covalently attached to electrode materials such as titanium oxide (TiO_2). In the following sections, the construction and properties of these devices are discussed.

4.2 Hybrid (Solid-with-Solution) Electrochromic Devices

4.2.1 Prussian Blue and Heptyl Viologen Solid-with-Solution-Type ECD

Lin *et al.* [49] reported a hybrid-type complementary ECD based on a PB thin film and a $HV(BF_4)_2$ solution. An equal mixture (v/v) of deionised water and isopropanol solution containing potassium chloride (0.1 M) was used as the electrolyte solution. A PB-$HV(BF_4)_2$-based ECD (abbreviated to PHECD) formulated as ITO-glass|PB|$HV(BF_4)_{2(aq./IPA)}$,$KCl_{(aq./IPA)}$|ITO-glass (where ITO is Indium-Tin Oxide) can be switched reversibly between blue and colourless states. The time required to darken the device was found to be dependent on the applied potential and the charge capacity of the PB film. A blue film was formed evenly on the clean ITO-glass surface by deposition of the radical-cation salt, $HV^{\bullet+}$, during darkening of the PHECD. The device exhibited slow discoloration due to the re-oxidation of the $HV^{\bullet+}$ species being under diffusion control.

4.2.1.1 Preparation and Characterisation of PB Thin Film and $HV(BF_4)_2$

The PB films were prepared by electrochemical deposition method from a solution containing 10 mM $FeCl_3$, 10 mM $K_3Fe(CN)_6$, 0.1 M KCl and 0.1 M HCl [3, 25]. A constant cathodic current density of 20 μA cm^{-2} was applied to the ITO-glass substrate for 5 min. In order to prepare the HV^{2+}-containing solution, $HV(BF_4)_2$ powder was obtained according to the following equation:

$$HVBr_{2(aq.)} + 2NaBF_{4(aq.)} \rightarrow HV(BF_4)_{2(s)} + 2NaBr_{(aq.)} \qquad (4.1)$$

Saturated $HVBr_2$ and $NaBF_4$ aqueous solutions each 100 ml were mixed for 10 min, and the precipitate was filtered out and recrystallised using ethanol. After ethanol was evaporated by heating the solution to 70 °C, a light-yellow $HV(BF_4)_2$ powder was obtained. The $HV(BF_4)_2$ solution, used to fabricate the ECD, was prepared as follows: $HV(BF_4)_2$ powder was dissolved in an equal-volume mixture of deionised water and isopropanol solution with 1.0 M KCl to form a solution of 0.05 M $HV(BF_4)_2$, adjusted to pH 2.5 by adding HCl. Finally, the solution was filtered to remove any remaining precipitant.

4.2.1.2 Redox Behaviours and Visible Spectra of the PB Film and $HV(BF_4)_2$ Solution

The as-deposited PB thin film is blue and can be reduced to form colourless Everitt's salt in a K^+-containing environment (Eq. (4.2)). The colour of the PB film was reversibly changed from a colourless state at 0.0 V (vs. Ag/AgCl) to a blue state at +0.5 V. The redox reactions of the PB thin film and dissolved $HV(BF_4)_2$ can be expressed as follows:

$$\underset{\text{(blue)}}{KFe^{III}[Fe^{II}(CN)_6]_{(s)}} + K^+ + e^- \rightleftharpoons \underset{\text{(colourless)}}{K_2Fe^{II}[Fe^{II}(CN)_6]_{(s)}} \qquad (4.2)$$

$$HV^{2+}_{aq} + e^- \rightleftharpoons HV^{\cdot+}_{(s)}$$
$$\text{(di-cations, colourless) (radical-cation, blue)} \qquad (4.3)$$

The absorbance in the visible wavelength for both the PB film and the $HV(BF_4)_2$ solution under different applied potentials are shown in Figure 4.1a,b, respectively. From Figure 4.1a, an absorbance peak at around 690 nm can be observed under an applied potential of +0.5 V, and the absorbance peak decreases to near zero when the applied potential drops to 0 V. The absorbance of 0.01 M $HV(BF_4)_2$ increases as the reductive applied potential becomes more negative, as shown in Figure 4.1b. Three distinguishable absorbance peaks can be observed at 605, 664 and 735 nm.

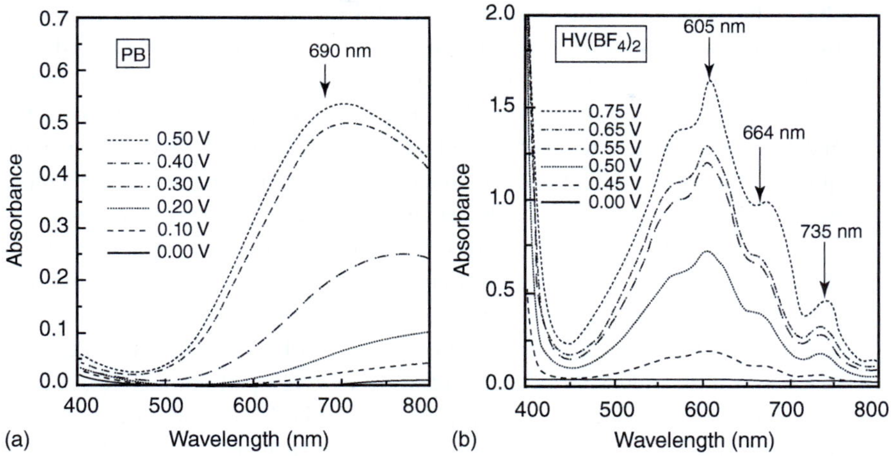

Figure 4.1 (a) Absorbance spectra of the PB thin film measured at different oxidising potentials. (b) Absorbance spectra of 0.01 M $HV(BF_4)_2$ measured at different reducing potentials (ITO-glass subtracted). (From Ref. [49].)

4.2.1.3 **Operating Parameters and Properties of PHECD**

An appropriate operating potential of this, or any, ECD needs to be determined when considering long-term operation. At extreme potentials, undesired side reactions can occur, decreasing the long-term stability of an ECD. Figure 4.2

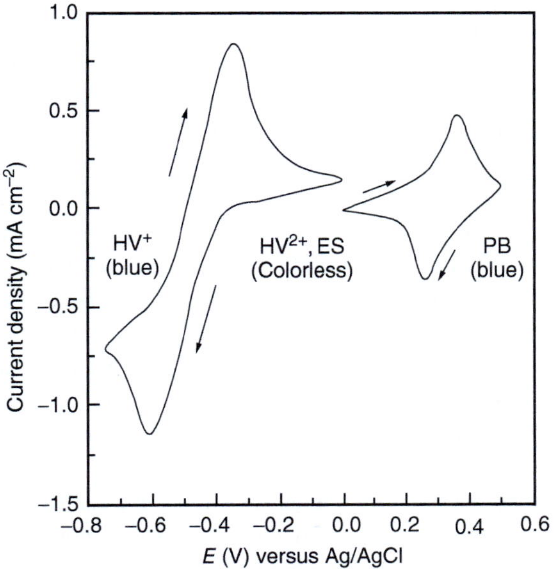

Figure 4.2 Three-electrode cyclic voltammogram for PB and $HV(BF_4)_2$ in isopropanol and H_2O with 1.0 M KCl (pH 2.5), $v = 50\,mV\,s^{-1}$ (From Ref. [49].)

shows the cyclic voltammograms for a PB film and the $HV(BF_4)_2$ solution, when the ITO-glass working electrode is coated with a thin film of Everitt's salt and a clean ITO-glass counter electrode is in contact with the solution. A positive applied potential to this setup oxidises the Everitt's salt to PB and reduces the HV^{2+} in the solution to $HV^{\bullet+}$, producing a dark-blue-coloured film. Consequently, if a potential difference of about 1.0 V is applied, the device will colour without the occurrence of any side reactions. The device returns to its colourless state when a potential difference of 0 V is applied. For a complementary ECD, the reactions at each electrode must be balanced, that is, device performance is greatly affected by the charge capacity ratio of two electrodes. Theoretically, the transmittance window and the coloration efficiency of a complementary ECD will reach maximum values if the charge capacity ratio is equal to unity [23]. Unlike thin-film ECDs, in which the charge capacities for both electrodes can be predetermined, the charge capacity of PHECD (and similar hybrid solid-with-solution devices) is time dependent with the viologen being under diffusion control. Subsequently, the charge capacity is increased with increasing time whilst a potential is applied. For such devices, in order to keep the charge capacity ratio equal to unity during darkening, a limit on the application time for the darkening potential is required.

When 1.0 V was applied to darken and 0.0 V to bleach, the PHECD produced highly reversible, colourless-to-blue electrochromism. The transmittance window at 609 nm was 59%.

4.2.1.4 **Analogous Devices**

Mortimer and Varley [50] constructed a device similar to that of Lin *et al.* [49] instead replacing HV with the solution-phase MV. Key differences in the design and performance of the device are that Mortimer and Varley used a solely aqueous solvent-electrolyte system (potassium chloride, 0.5 M) and the colour of the MV is a deep purple (a result of the radical-cation presence as both a monomer and a dimer) [61]. The final colour of this device would differ from that of Lin *et al.*, which would be quantifiable through colour space analysis [62]. The structure of a PB-$MV(Cl)_2$ ECD is (ITO glass|PB |PB-$MV(Cl)_{2(aq.)}$, $KCl_{(aq.)}$| ITO-glass). The device switching time was reported as about 10 s for both coloration and bleaching. The total change in transmission and coloration efficiency between on and off states was found to be a function of viologen concentration. The respective values of 83.0% and 170 $cm^2\,C^{-1}$ were found for 5 mM MV, which became 93.1% and 86 $cm^2\,C^{-1}$ when operating at 10 mM.

Finally, it is worth mentioning that in 1989 Rosseinsky and Monk published a paper using the same electrochromic pairing of PB and MV, but within a paper matrix. This paper demonstrates not only the possibility of electrochromic printing but also the construction of a reflective-type ECD [63]. This work was furthered by Mortimer and Warren in 1999 [54].

4.2.2
PB Thin Film and Viologen in Ionic Liquid–Based ECD

Sydam *et al.* [52] synthesised the novel viologen 1,1′-*bis*-[4-(5,6-dimethyl-1H-benzimidazole-1-yl)-butyl]-4,4′-bipyridinium dibromide (IBV) salt and dissolved it in the ionic liquid (1-ethyl-3-methylimidazolium dicyanamide). The liquid functioned as both solvent and electrolyte with a relatively high ionic conductivity ($0.25\ mS\ cm^{-1}$). Additionally, the liquid has a thermal decomposition temperature >200 °C, has a voltage window of +4 V and is colourless. All of these features are desirable properties for a model ECD. A PB thin film was used to create an ECD with the ionic liquid containing IBV. The device has reversible transitions between transparent and deep blue, with good optical contrast (30.5% at 605 nm), high coloration efficiency of $725\ cm^2\ C^{-1}$ at 605 nm and moderate switching times of 2–3 s.

4.3
All-Solid Electrochromic Devices

4.3.1
Prussian Blue and Poly(butyl viologen) Thin-Film ECD

Kuo *et al.* [58] fabricated an all-solid-state ECD using PB and poly(butyl viologen) (PBV) as the electrochromic materials with a solid-state succinonitrile (SN)-based electrolyte (ITO-glass|PBV |$LiBF_4$ in SN| PB|ITO-glass). PBV can be deposited onto an electrode by electropolymerisation of its respective viologen monomer [64]. PBV has three redox states similar to its parent viologen counterpart, the dication PBV^{2+} that is colourless and the most stable, the purple radical-cation $PBV^{\bullet+}$ and the di-reduced species PBV^0, which is yellowish-brown. Succinonitrile is a *plastic crystal* that is a solid in ambient conditions with a melting temperature of 58 °C. In its plastic-crystalline phase, SN is highly polar, allowing various salts to be dissolved producing solid electrolytes with high ionic conductivities over a wide temperature range [65]. The use of solid-state electrolytes instead of conventional liquid electrolytes in electrochemical systems improves the operational safety through reducing vapour pressure and eliminating electrolyte leakage and improves device manufacture.

4.3.1.1 Preparation of Poly(butyl viologen) Thin Film

PBV thin films were electropolymerised from *bis*(4-cyano-1-pyridino)butane dibromide ($BVBr_2$). The deposition solution was composed of $BVBr_2$ (20 mM), KH_2PO_4 (10 mM), K_2HPO_4 (90 mM) and $K_4Fe(CN)_6$ (100 mM). The viologen monomer was reduced to the radical-cation state at −0.75 V. The electropolymerisation then proceeds by the attack of $PBV^{\bullet+}$ onto a CN group of another $BVBr_2$ molecule. The chemical structure of polymerised PBV is shown in Figure 4.3.

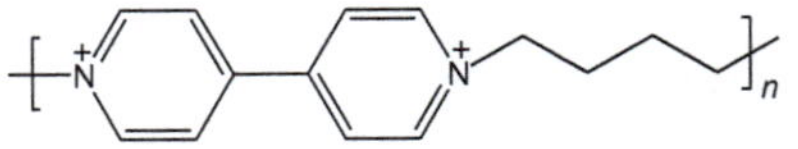

Figure 4.3 The chemical structure of polymerised PBV thin film (excluding counter ions).

4.3.1.2 **Electrochemical and Optical Properties of Poly(butyl viologen) Thin Films**

There are three pairs of redox peaks observed during potential cycling of PBV films. For the reduction process, peaks are found at about −0.70 V ($PBV^{2+} \rightarrow PBV^{\bullet+}$, colourless to purple), about −1.2 V ($PBV^{\bullet+} \rightarrow PBV^{0}$, purple to yellowish-brown) and a third peak not associated with any electrochromism at +0.20 V. The equilibrium absorbance spectra in the visible region for the PBV thin film are shown in Figure 4.4. The absorbance peak becomes more prominent as the potential is decreased from 0.0 to −0.8 V (measured at intervals of 0.1 V). The absorbance spectra showed no change between 0 and −0.4 V. However, above this and up to −0.8 V, the absorbance increased, with the wavelength of maximum absorbance found at about 550 nm, a wavelength the human eye is especially susceptible to. When the applied potential was more negative than −0.8 V, the intensity of the absorbance peak centred at 550 nm decreased and almost vanished at −1.0 V (Figure 4.4 insert). This was due to the onset of the second reduction process, generating the yellowish-brown PBV^{0} species. As a

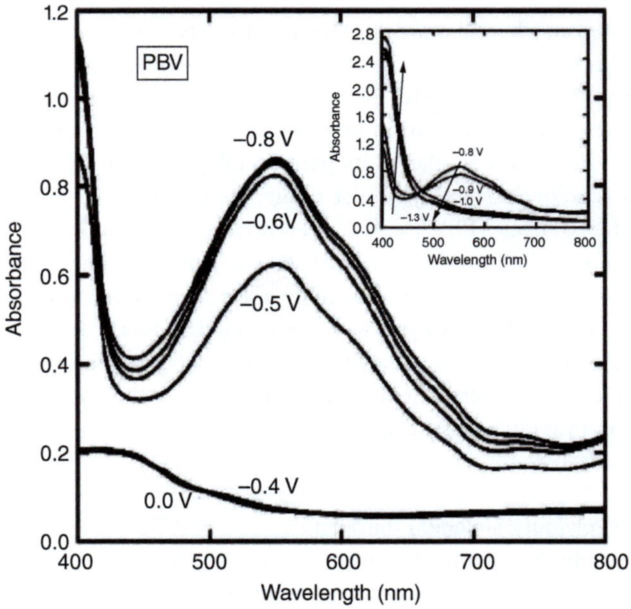

Figure 4.4 Equilibrium absorbance spectra of PBV thin film at different potential stepping from 0 to −0.8 V with 0.1 V decrement, measured in 0.5 M KCl aqueous solution. The inset shows the spectra of continuing the applied potential from −0.8 to −1.3 V. (From Ref. [58].)

result, the operating potential for complementary ECDs employing this material is limited to the potentials of first redox couple ($PBV^{2+}/PBV^{\bullet+}$).

4.3.1.3 Electrochromic Performance of PBV-PB ECD

The redox reactions that occur in an ECD consisting of PB and PBV in the solid-state electrolyte, SN, containing $LiBF_4$ are illustrated by Eqs. (4.4) and (4.5):

$$([PBV]^{2+}[BF_4]_2^-) + e^- \rightleftharpoons ([PBV]^{\bullet+}\,[BF_4]^-) + [BF_4]^- \tag{4.4}$$

$$Li_2Fe^{II}[Fe^{II}(CN)_6] \rightleftharpoons LiFe^{III}[Fe^{II}(CN)_6] + Li^+ + e^- \tag{4.5}$$

Thus, the overall device (electrochemical cell) reaction of the complementary ECD is

$$\begin{aligned}&([PBV]^{2+}[BF_4]_2^-) + Li_2Fe^{II}[Fe^{II}(CN)_6] \rightleftharpoons\\&([PBV]^{\bullet+}\,[BF_4]^-) + LiFe^{III}[Fe^{II}(CN)_6] + Li^+ + [BF_4]^-\end{aligned} \tag{4.6}$$

A two-stage reaction of the ECD was observed from the cyclic voltammogram: the first between −1.0 and −0.2 V and the second between −0.2 and +0.7 V. PBV accounted for most of the observed current in the first stage, with PB being the most dominant during the second stage. These observations were verified by the reported absorbance spectra and electrochemically, whereby a reversible cyclic voltammogram was shown for the PBV-PB ECD between −1.0 and 0.7 V. When the voltage reached about −0.2 V, PB absorbance features appear (a peak at about 690 nm). As the applied voltage was increased in the negative direction from −0.2 to −1.3 V, a second peak at about 550 nm appears due to contributions from the PBV.

The coloration efficiency of the device was calculated as 183 $cm^2\,C^{-1}$. Switching times for the bleaching and darkening processes were reported to be 9.4 and 2.0 s, respectively. Consecutive operation for 4000 cycles of the device resulted in transmittance attenuation decreasing from 65% to 56%, which is considered to demonstrate good electrochemical cycling stability.

4.3.2 Prussian Blue and Viologen Anchored TiO_2-Based ECD

Bonhôte *et al.* [56] and Campus *et al.* [57] proposed an ECD based on PB functionalised viologens that were covalently attached to a TiO_2-coated electrode. TiO_2 electrodes consist of a nanocrystalline TiO_2 thin film on top of a transparent conductive substrate (TCS). The viologens were synthesised with benzoate, salicylate or phosphonate functional groups. An illustration of the TiO_2 and viologen electrode is shown in Figure 4.5. Filters and displays were constructed whereby the filter devices ranged from 12 to 100 cm^2 and were typically setup as TCS|TiO_2-poly-viologen|glutaronitrile|$LiN(SO_2CF_3)_2$|spacer|PB|TCS. These devices demonstrate a high optical density change, up to 2, within a short switching time of 1–3 s.

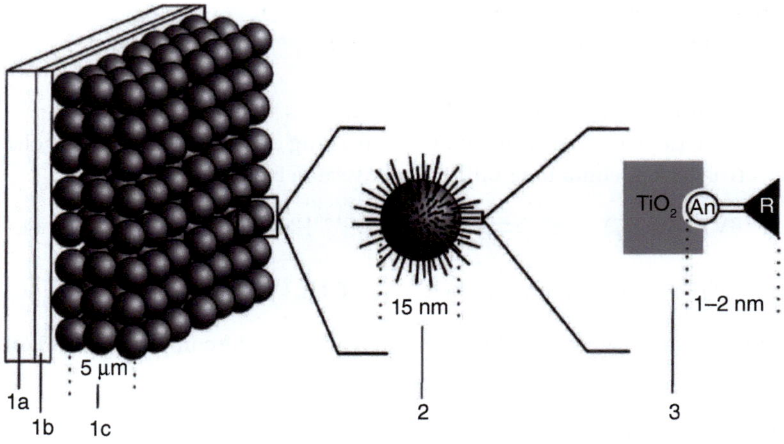

Figure 4.5 Exploded view of a thin-film nanocrystalline TiO_2 electrode modified with an electrochromophore, R: 1(a) glass, 1(b) F-doped SnO_2 layer, 1(c) nanocrystalline TiO_2 film (5 μm, roughness factor: 100 μm^{-1}); (2) TiO_2 nanoparticle (about 20 nm) and (3) electrochromophore, R (about 1–2 nm) adsorbed via an anchoring group (An) on a TiO_2 coordination site. The nanoparticles are schematically drawn as closely packed spheres. However, in reality they represent an accidental, porous structure. (From Ref. [56].)

The design of these devices was modified in 2011 by Rong *et al.* [60], who reported an ECD constructed of a PB thin film coated on antimony tin oxide (ATO) nanocomposites and viologen molecules anchored onto TiO_2 nanoparticles. A polymer gel electrolyte was used between the PB and viologen layers as the ionic transport layer. The constructed 2.5 cm × 2.5 cm ECD exhibited a high contrast, 64.8%, at 600 nm between on and off states, with very low transmittance of 0.1%, at 600 nm, when coloured. The ECD also had comparatively fast switching times of 0.6 and 0.7 s, for coloration and bleaching, respectively, and a high coloration efficiency of 912 $cm^2\,C^{-1}$ at 600 nm. The configuration of the device is shown in Figure 4.6. As yet no one has attempted to mimic this setup, but replacing the ATO nanoparticles with TiO_2. The enhanced surface area should lead to a greater optical density change (two to four) but also offer a more cost-effective substrate.

4.3.3
Polypyrrole-Prussian Blue Composite Film and Benzylviologen Polymer–Based Thin-Film-Type ECD

Leventis and Chung [55] reported a complementary ECD based on polypyrrole-Prussian blue (PP-PB) composite film, which was employed as the anodically colouring material, and a viologen polymer derivative, pBPQ (Figure 4.7), was used as the cathodically colouring material. A viscous aqueous solution of poly(vinylpyrrolidone) (PVP)/K_2SO_4 was used as the electrolyte.

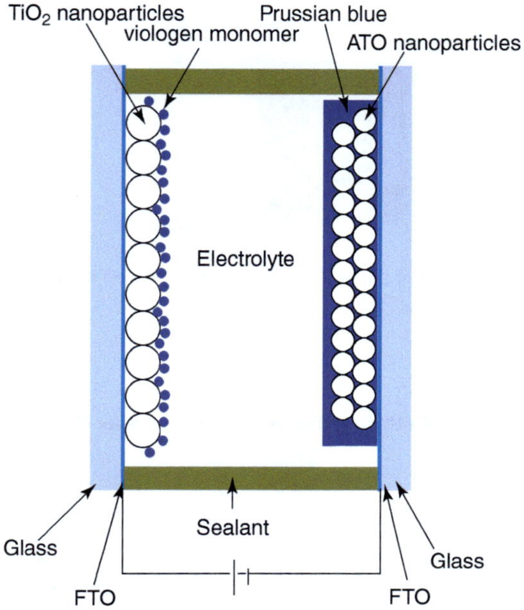

Figure 4.6 Construction of an ECD based on viologen anchored on TiO_2 nanoparticles as cathodic electrode and PB/ATO-nanocomposites as anodic electrode. (From Ref. [60].)

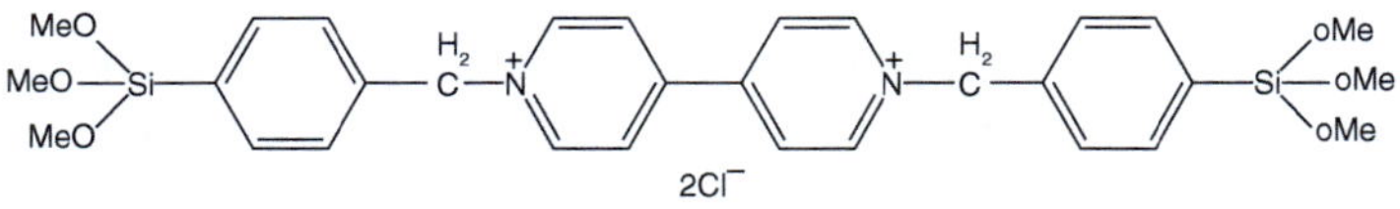

Figure 4.7 Monomer structure of viologen polymer derivative, BPQ dichloride.

4.3.3.1 **Preparation of PP-PB Thin-Film**

First, polypyrrole was electrodeposited from a solution of pyrrole (0.2 M) in $NaClO_4/CH_3CN$ solution (1.0 M). PB was then subsequently electrodeposited onto the polypyrrole-coated electrode from an aqueous K_2SO_4 solution (0.5 M) containing $K_3[Fe(CN)_6]$ (5 mM) and $FeCl_3$ (5 mM), by cycling the applied potential between 0.60 and 0.35 V at 50 mV s^{-1}, until the desired coverage was obtained (achieved by monitoring the amount of charge passed).

4.3.3.2 **Performance of the PP-PB Thin-Film and pBPQ-Based Electrochromic Device**

Two devices were constructed, one of transmissive design and the other reflective (only one transparent electrode). The devices were $4 \times 4\,in.^2$ and able to switch from colourless ($Abs._{400-800} \approx 0.15-0.25$) to blue ($Abs._{650} \approx 1.35$) in about 3–5 s, upon the application of 0.8 V across the two electrodes (Figure 4.8).

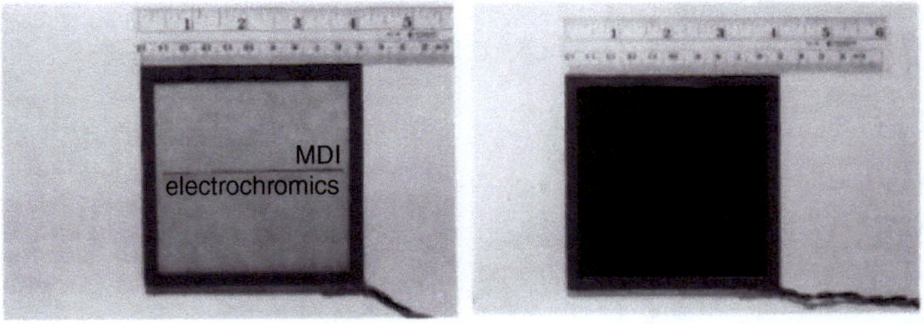

Figure 4.8 The ECD on top of a white sheet of paper with the letters typed on it. (a) $E_{App.} = -0.9$ V and (b) $E_{App.} = +0.8$ V. (From Ref. [55].)

4.3.4 PB Thin-Film and Viologen-Doped Poly(3,4-ethylenedioxythiopene) Polymer–Based ECD

Bhandari *et al.* [59] reported an electrochromic zwitterion viologen, *N,N′*-*bis*(3-sulfonatopropyl)-4-4′-bipyridinium, for doping in poly(3,4-ethylenedioxythiophene) (PEDOT) films during electropolymerisation at +1.2 and +1.8 V, respectively. The PEDOT-viologen thin film formed at +1.8 V showed higher coloration efficiency, larger electrochemical charge capacity and better redox activity in comparison to its analogue grown at +1.2 V.

A PEDOT-viologen film grown at +1.8 V was coupled with PB to create an ECD. The ionic liquid, 1-butyl-1-methylpyrrolidinium trifluoromethanesulfonate was used as the supporting electrolyte solution. Flexible electrodes were transparent polyethylene terephthalate sheets coated with a transparent conductive layer of PEDOT-polystyrene sulfonate, which created a flexible ECD showing reversible coloration between pale and dark purple as depicted in Figure 4.9.

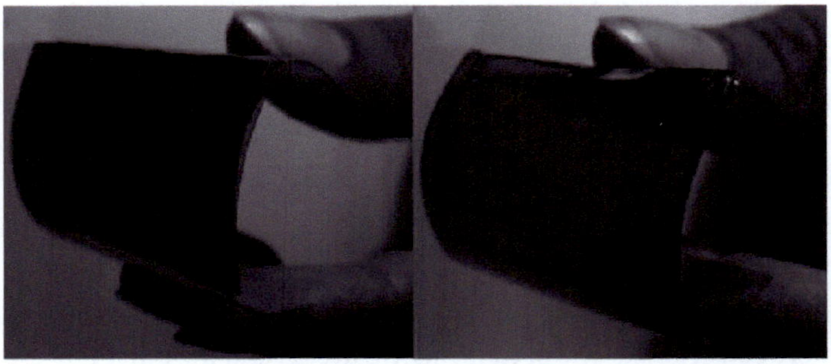

(a) (b)

Figure 4.9 Photographs of the flexible PEDOT-viologen (1.8) |IL gel| PB device in (a) coloured ($E_{App.} = -1.5$ V) and (b) bleached ($E_{App.} = +1.5$ V) state. (From Ref. [59].)

4.3.5
Other Solid-State Viologens

In addition to those mentioned in the previous subsection, other methods of designing viologens to be used as solid-state or thin-film systems have been reported, but not yet utilised in ECDs coupled with PB. In this section, we briefly mention some of these and their associated electrochromic properties.

Methods similar to those developed by Bonhôte *et al.* [56] and Campus *et al.* [57] have been proposed for tethering viologens onto the surface of particles, with particular focus on utilising TiO_2 and exhibiting switching times down to 250 ms. [66–69]. Ryu *et al.* [70] attached viologens to polymeric microspheres that had a chlorine functionalised surface, to permit attachment of the viologen species. They went on to employ their technology in an ECD consisting of only viologen species [71]. Similar material derivatisation using viologens has been employed on carbon nanotubes [72, 73] and by direct surface modification [74], the latter of which reports switching speeds as fast as 5 ms. These are just a few examples of published research on the covalent attachment of viologen species to electrodes in order to improve properties such as switching times and energy efficiency (ensuring that the materials exhibit the memory effect). Although these publications do not describe pairing with PB, they focus on improving on one-half of this important electrochromic material pairing.

Dominey *et al.* [75] prepared the electrochromic dibenzyl viologen polymer *N,N′-bis*[p-(trimethoxysilyl)benzyl]-4,4′-bipyridium dichloride, which upon reduction forms an insoluble thin-film polymer via cross-linking of the methoxysilyl groups. The thin film can undergo a one-electron reduction, resulting in a colourless-to-blue-violet transition. Later, Leventis and Chung [76] prepared a novel bilayer electrochromic electrode using WO_3 and dibenzyl viologen that exhibited a colourless-to-blue transition upon reduction in several stages.

In 1991 Mortimer [53] showed that a single electrode could exhibit five different colours, through the use of two different electrochromes, PB and MV. This was achieved by electrodepositing the PB, onto which a layer of Nafion® was drop-cast. The negatively charged perfluorinated ionomer allowed the formation of a trapped, concentrated thin layer of the positively charged viologen molecules in close proximity to the electrode's surface. This technique was originally pioneered by Anson *et al.* [77]. This method of creating a trapped thin layer of viologen material close to the electrode surface is yet to be utilised in a complete ECD; its elegance means a solid-with-solution hybrid device can be easily transformed into an all-solid-type device via the appropriate pretreatment of the viologen associated ITO-glass electrode with Nafion®. Similar 'dual electrochrome' electrodes have been fabricated using a layer-by-layer approach that was developed by Decher and Hong [78] and subsequently by other research groups who took into account various viologen systems (polybutylviologen [79, 80], polyhexylviologen [81] and polyoctylviologen [82]). This straightforward method involves exposing a surface to alternating materials of opposite charge, thus producing a multilayer. Good electrochromic properties have been reported that include switching times of 1–4 s and transmittance changes of about 82%. [81] As yet, these dual electrochromes or layer-by-layer viologen electrodes have not been paired with PB to produce an ECD, neither as a single electrode composed of alternating layers of PB and viologen species, both exciting prospective systems.

The majority of the work presented in this chapter focuses on PB and the viologens separately (with the exception of the previous paragraph). However, in 2001 Kostina *et al.* [83] reacted ethyl viologen (EV) with metal hexacyanometallates to create the ion pair $EV_{1.5}$ $K[M(CN)_6]{\cdot}12H_2O$ (where M = Fe or Ru). Despite performing extensive characterisation of this new compound, its electrochromic properties were not investigated, an omission considering it consists of two well-known and established electrochromic materials.

4.4
Other Metal Hexacyanometallate-Viologen-Based ECDs

The majority of this chapter has focused on the various ways by which scientists have coupled viologens with PB thin films, with the majority of the research being incorporated into the viologen aspect of this pairing. However, there have been several publications that focus on changing the transition metals within the metal hexacyanometallate to create new ECDs.

Mortimer and Varley [51] reported the first colour reinforcing ECD that paired MV with ruthenium purple, iron(III) hexacyanoruthenate(II). At −0.5 V, the device was colourless but became purple by applying +1.5 V. The device exhibited a complementary wavelength of 565 nm, almost identical to the wavelength of maximum absorbance for the individual components. The device was constructed as a solid-with-solution system and exhibited coloration and bleaching switching times of 4 and 10 s, respectively. Devices were constructed using two different viologen concentrations and are shown in their coloured (on) and bleached (off) state in Figure 4.10.

In a follow-on publication [84], Mortimer and Varley again showed the construction of a ruthenium purple/viologen device, this time utilizing di-*n*-HV, which created a hybrid solid-with-solution/solid–solid device due to the decreased solubility of the di-*n*-heptyl radical cation. This device exhibited a complementary wavelength of 555 nm, with switching times for coloration and bleaching of 33 and 3 s, respectively.

As for the viologens, research away from devices and focusing solely on improving the material's properties is prevalent. A good example is that of Leventis and Chung [85], who improved the electrochemical stability and electrode adherence of PB by co-depositing PB with the conducting polymer polyaniline, the latter itself an electrochromic material (an overview of electrochromic conjugated conducting polymers is contained in Chapter 5). It is thought that the performance

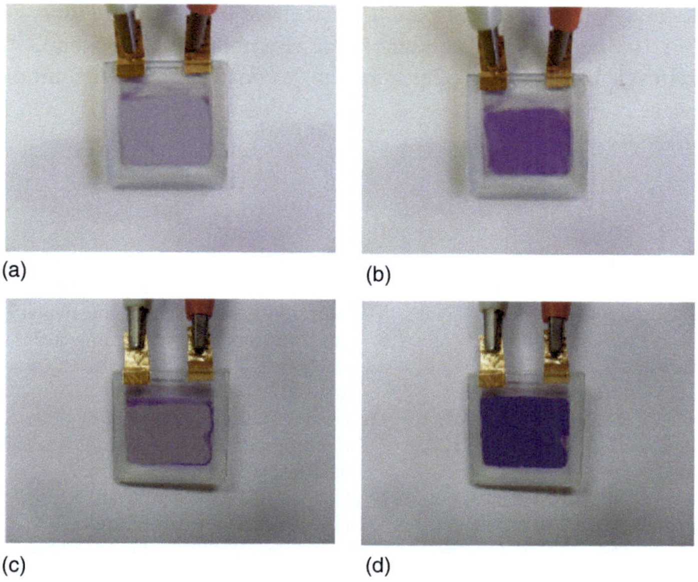

Figure 4.10 Photographs of a ruthenium purple/methyl viologen ((a, b) 5 mM, (b, d) 10 mM) ECD in their colourless (off, (a, c)) and coloured (on, (b, d)) states. Both devices have the dimensions 25 × 25 × 2.3 mm. (From Ref. [51].)

enhancement occurs by utilising the fast switching speed of the polyaniline and the intense colour of PB.

4.5 Prospects for Metal Hexacyanometallate-Viologen-Based ECDs

For PB materials: PB thin film can be obtained from nano-PB particles [86–88] or mesoporous particles [89, 90]. The electrochromic properties for nanoarchitectured PB film are improved from tradition thin-film preparations, thus are more suitable for use in ECDs [91–93]. Furthermore, the effect on the colour made by replacing one, or both, of the iron atoms within PB with another transition metals (e.g. ruthenium, resulting in a purple material) allows new colours to be formed. Coupling this adaptability with the expansive array of colours already available for viologens (via changing the functional groups attached to the pyridinium nitrogen) provides opportunities for the discovery of new pairings of between these two materials groups.

For viologens: viologen molecules functionalised with conducting polymeric monomers should create multicolour thin films with fast optical switching time attributable to the solid-state nature of such polymeric materials [94–97]. Additionally, new methods of attaching viologen, and PB-type, electrochromes to electrodes will improve existing device properties.

For electrolytes: gel and solid-transparent electrolytes [98–102] and ionic liquids [103–106] are potential candidates for assembling PB-viologen ECDs for longer cycle lifetimes and offer safer device construction over solution-type devices.

References

1. Diesbach, H. (1704) *Handbuch der Anorganischen Chemie*, Frankfurt am Main, Deutsche Chemische Gesellschaft, 1930, Vol. 59, Teil B (cited in Gmelin).
2. Neff, V.D. (1978) Electrochemical oxidation and reduction of thin films of Prussian Blue. *J. Electrochem. Soc.*, **128**, 886.
3. Itaya, K., Shibayama, K., Akahoshi, H., and Toshima, S. (1982) Prussian-blue-modified electrodes – An application for a stable electrochromic display device. *J. Appl. Phys.*, **53**, 804–805.
4. Mortimer, R.J. (1997) Electrochromic materials. *Chem. Soc. Rev.*, **26**, 147–156.
5. Greenberg, C.B. (1994) Optically switchable thin-films – A review. *Thin Solid Films*, **251**, 81–93.
6. Guo, Y.Z., Guadalupe, A.R., Resto, O., Fonseca, L.F., and Weisz, S.Z. (1999) Chemically derived Prussian blue sol-gel composite thin films. *Chem. Mater.*, **11**, 135–140.
7. Vittal, R., Jayalakshmi, M., Gomathi, H., and Rao, G.P. (1999) Surfactant promoted enhancement in electrochemical and electrochromic properties of films of Prussian blue and its analogs. *J. Electrochem. Soc.*, **146**, 786–793.
8. Honda, K., Fujita, M., Ishida, H., Yamamoto, R., and Ohgaki, K. (1988) Solid-state electrochromic devices composed of Prussian blue, WO_3, and poly(ethylene oxide)-polysiloxane hybrid-type ionic conducting membrane. *J. Electrochem. Soc.*, **135**, 3151–3154.
9. Oyama, N., Ohsaka, T., Menda, M., and Ohno, H. (1989) Electrochemical and electrochromic properties of all solid-state WO_3-Prussian blue based electrochromic devices. *Denki Kagaku*, **57**, 1172–1177.
10. Habib, M.A., Maheswari, S.P., and Carpenter, M.K. (1991) A tungsten-trioxide Prussian blue complementary electrochromic cell with a polymer electrolyte. *J. Appl. Electrochem.*, **21**, 203–207.
11. Granqvist, C.G. (1992) Electrochromism and smart window design. *Solid State Ionics*, **53**, 479–489.
12. Habib, M.A. and Maheswari, S.P. (1992) Effect of temperature on a complementary WO_3-Prussian blue electrochromic system. *J. Electrochem. Soc.*, **139**, 2155–2157.
13. Habib, M.A. and Maheswari, S.P. (1993) Electrochromic characteristics of a complementary tungsten trioxide Prussian blue cell. *J. Appl. Electrochem.*, **23**, 44–50.
14. Jelle, B.P. and Hagen, G. (1993) Transmission spectra of an electrochromic window based on polyaniline, Prussian blue and tungsten-oxide. *J. Electrochem. Soc.*, **140**, 3560–3564.
15. Ho, K.C., Rukavina, T.G., and Greenberg, C.B. (1994) Tungsten-oxide Prussian blue electrochromic system based on a proton-conducting polymer electrolyte. *J. Electrochem. Soc.*, **141**, 2061–2067.
16. Inaba, H., Iwaku, M., Nakase, K., Yasukawa, H., Seo, I., and Oyama, N. (1995) Electrochromic display device of tungsten trioxide and Prussian-blue films using polymer gel electrolyte of methacrylate. *Electrochim. Acta*, **40**, 227–232.
17. Su, L.Y., Fang, J.H., Liang, B.J., and Lu, Z.H. (1997) All-solid-state electrochromic display device of Prussian blue WO_3 particulate film. *Jpn. J. Appl. Phys., Part 2*, **36**, L684–L686.
18. Su, L.Y., Fang, J.H., Xiao, Z.D., and Lu, Z.H. (1997) An all-solid-state electrochromic display device of Prussian

blue and WO_3 particulate film with a PMMA gel electrolyte. *Thin Solid Films*, **306**, 133–136.
19. Su, L.Y., Hong, Q.Y., and Lu, Z.H. (1998) All solid-state electrochromic window of Prussian blue and electrodeposited WO_3 film with PMMA gel electrolyte. *J. Mater. Chem.*, **8**, 85–88.
20. Su, L.Y., Wang, H., and Lu, Z.H. (1998) All-solid-state electrochromic window of Prussian blue and electrodeposited WO_3 film with poly(ethylene oxide) gel electrolyte. *Mater. Chem. Phys.*, **56**, 266–270.
21. Su, L.Y., Wang, H., and Lu, Z.H. (1998) All solid-state electrochromic smart window of electrodeposited WO_3 and Prussian blue film with PVC gel electrolyte. *Supramol. Sci.*, **5**, 657–659.
22. Su, L.Y., Xiao, Z.D., and Lu, Z.H. (1998) All solid-state electrochromic window of electrodeposited WO_3 and Prussian blue film with PVC gel electrolyte. *Thin Solid Films*, **320**, 285–289.
23. Ho, K.C. (1999) Cycling and at-rest stabilities of a complementary electrochromic device based on tungsten oxide and Prussian blue thin films. *Electrochim. Acta*, **44**, 3227–3235.
24. Jelle, B.P., Hagen, G., and Nodland, S. (1993) Transmission spectra of an electrochromic window consisting of polyaniline, Prussian blue and tungsten-oxide. *Electrochim. Acta*, **38**, 1497–1500.
25. Tung, T.S. and Ho, K.C. (2006) Cycling and at-rest stabilities of a complementary electrochromic device containing poly(3,4-ethylenedioxythiophene) and Prussian blue. *Sol. Energy Mater. Sol. Cells*, **90**, 521–537.
26. Deepa, M., Awadhia, A., Bhandari, S., and Agrawal, S.L. (2008) Electrochromic performance of a poly(3,4-ethylenedioxythiophene) – Prussian blue device encompassing a free standing proton electrolyte film. *Electrochim. Acta*, **53**, 7266–7275.
27. Deepa, M., Awadhia, A., and Bhandari, S. (2009) Electrochemistry of poly(3,4-ethylenedioxythiophene)-polyaniline/Prussian blue electrochromic devices containing an ionic liquid based gel electrolyte film. *Phys. Chem. Chem. Phys.*, **11**, 5674–5685.
28. Agnihotry, S.A. (1996) Electrochromic devices: present and forthcoming technology. *Bull. Electrochem.*, **12**, 707–712.
29. Kang, J.-H., Paek, S.-M., Bin Choy, Y., Hwang, S.-J., and Choy, J.-H. (2007) Enhanced contrast of electrochromic full cell systems with nanocrystalline PEDOT-Prussian blue. *J. Nanosci. Nanotechnol.*, **7**, 4131–4134.
30. Takahashi, Y., Oyaizu, K., Honda, K., and Nishide, H. (2007) Low-energy driven electrochromic devices using radical polymer as transparent counter electroactive material. *J. Photopolym. Sci. Technol.*, **20**, 29–34.
31. Chen, K.-C., Hsu, C.-Y., Hu, C.-W., and Ho, K.-C. (2011) A complementary electrochromic device based on Prussian blue and poly(ProDOT-Et_2) with high contrast and high coloration efficiency. *Sol. Energy Mater. Sol. Cells*, **95**, 2238–2245.
32. Kalagi, S.S., Mali, S.S., Dalavi, D.S., Inamdar, A.I., Im, H., and Patil, P.S. (2011) Limitations of dual and complementary inorganic-organic electrochromic device for smart window application and its colorimetric analysis. *Synth. Met.*, **161**, 1105–1112.
33. Mortimer, R.J. (1999) Organic electrochromic materials. *Electrochim. Acta*, **44**, 2971–2981.
34. de Tacconi, N.R., Rajeshwar, K., and Lexna, R.O. (2003) Metal hexacyanoferrates: electrosynthesis, in situ characterization, and applications. *Chem. Mater.*, **15**, 3046–3062.
35. Kawata, T., Yamamoto, M., Yamana, M., Tajima, M., and Nakano, T. (1975) Characteristics of viologen derivatives for electrochromic display. *Jpn. J. Appl. Phys., Part 1*, **14**, 725–726.
36. Bruinink, J., Kregting, C.G.A., and Ponjee, J.J. (1977) Modified viologens with improved electrochemical properties for display applications. *J. Electrochem. Soc.*, **124**, 1854–1858.
37. Fletcher, S., Duff, L., and Barradas, R.G. (1979) Nucleation and charge-transfer kinetics at the viologen-SnO_2 interface

in electrochromic device applications. *J. Electroanal. Chem.*, **100**, 759–770.

38. Yasuda, A., Mori, H., Takehana, Y., Ohkoshi, A., and Kamiya, N. (1984) Electrochromic properties of the normal-heptyl viologen ferrocyanide system. *J. Appl. Electrochem.*, **14**, 323–327.
39. Monk, P.M.S. (1997) The effect of ferrocyanide on the performance of heptyl viologen-based electrochromic display devices. *J. Electroanal. Chem.*, **432**, 175–179.
40. Bird, C.L. and Kuhn, A.T. (1981) Electrochemistry of the viologens. *Chem. Soc. Rev.*, **10**, 49–82.
41. Mizuguchi, J. and Karfunkel, H. (1993) Semiempirical calculations on the optical-absorption of methylviologen and p-cyanophenylviologen in different oxidation states. *Ber. Bunsenges. Phys. Chem.*, **97**, 1466.
42. Hara, S., Tanaka, H., Kawamoto, T., Tokumoto, M., Yamada, M., Gotoh, A., Uchida, H., Kurihara, M., and Sakamoto, M. (2007) Electrochromic thin film of Prussian blue nanoparticles fabricated using wet process. *Jpn. J. Appl. Phys., Part 2*, **46**, L945–L947.
43. Song, Y.-Y., Jia, W.-Z., Li, Y., Xia, X.-H., Wang, Q.-J., Zhao, J.-W., and Yan, Y.-D. (2007) Synthesis and patterning of Prussian blue nanostructures on silicon wafer via galvanic displacement reaction. *Adv. Funct. Mater.*, **17**, 2808–2814.
44. Hong, S.-F. and Chen, L.-C. (2008) Stability-enhanced indium hexacyanoferrate electrodes: morphological characterization, in situ EQCM analysis in nonaqueous electrolytes and application to a WO_3 electrochromic device. *Electrochim. Acta*, **53**, 5306–5314.
45. Omura, A., Shiozaki, H., Hara, S., Kawamoto, T., Gotoh, A., Kurihara, M., Sakamoto, M., and Tanaka, H. (2008) Electrochromic thin film of water-dispersible Prussian-blue nanoparticles. *IEICE Trans. Electron.*, **E91C**, 1887–1888.
46. Shan, Y., Yang, G., Gong, J., Zhang, X., Zhu, L., and Qu, L. (2008) Prussian blue nanoparticles potentiostatically electrodeposited on indium tin oxide/chitosan nanofibers electrode and their electrocatalysis towards hydrogen peroxide. *Electrochim. Acta*, **53**, 7751–7755.
47. Shiozaki, H., Kawamoto, T., Tanaka, H., Hara, S., Tokumoto, M., Gotoh, A., Satoh, T., Ishizaki, M., Kurihara, M., and Sakamoto, M. (2008) Electrochromic thin film fabricated using a water-dispersible ink of Prussian blue nanoparticles. *Jpn. J. Appl. Phys., Part 1*, **47**, 1242–1244.
48. Ko, J.H., Yeo, S., Park, J.H., Choi, J., Noh, C., and Son, S.U. (2012) Graphene-based electrochromic systems: the case of Prussian blue nanoparticles on transparent graphene film. *Chem. Commun.*, **48**, 3884–3886.
49. Lin, C.F., Hsu, C.Y., Lo, H.C., Lin, C.L., Chen, L.C., and Ho, K.C. (2011) A complementary electrochromic system based on a Prussian blue thin film and a heptyl viologen solution. *Sol. Energy Mater. Sol. Cells*, **95**, 3074–3080.
50. Mortimer, R.J. and Varley, T.S. (2012) In situ spectroelectrochemistry and color measurement of a complementary electrochromic device based on surface-confined Prussian blue and aqueous solution-phase methyl viologen. *Sol. Energy Mater. Sol. Cells*, **99**, 213–220.
51. Mortimer, R.J. and Varley, T.S. (2011) Novel color-reinforcing electrochromic device based on surface-confined ruthenium purple and solution-phase methyl viologen. *Chem. Mater.*, **23**, 4077–4082.
52. Sydam, R., Deepa, M., and Joshi, A.G. (2013) A novel 1,1′-bis 4-(5,6-dimethyl-1H-benzimidazole-1-yl)butyl-4,4′-bipyridinium dibromide (viologen) for a high contrast electrochromic device. *Org. Electron.*, **14**, 1027–1036.
53. Mortimer, R.J. (1991) 5 color electrochromicity using Prussian blue and Nafion/methyl viologen layered films. *J. Electrochem. Soc.*, **138**, 633–634.
54. Mortimer, R.J. and Warren, C.P. (1999) Cyclic voltammetric studies of Prussian blue and viologens within a paper matrix for electrochromic printing applications. *J. Electroanal. Chem.*, **460**, 263–266.

55. Leventis, N. and Chung, Y.C. (1992) New complementary electrochromic system based on polypyrrole Prussian blue composite, a benzylviologen polymer, and poly(vinylpyrrolidone) potassium-sulfate aqueous-electrolyte. *Chem. Mater.*, **4**, 1415–1422.
56. Campus, F., Bonhôte, P., Gratzel, M., Heinen, S., and Walder, L. (1999) Electrochromic devices based on surface-modified nanocrystalline TiO_2 thin-film electrodes. *Sol. Energy Mater. Sol. Cells*, **56**, 281–297.
57. Bonhote, P., Gogniat, E., Campus, F., Walder, L., and Grätzel, M. (1999) Nanocrystalline electrochromic displays. *Displays*, **20**, 137–144.
58. Kuo, T.H., Hsu, C.Y., Lee, K.M., and Ho, K.C. (2009) All-solid-state electrochromic device based on poly(butyl viologen), Prussian blue, and succinonitrile. *Sol. Energy Mater. Sol. Cells*, **93**, 1755–1760.
59. Bhandari, S., Deepa, M., Pahal, S., Joshi, A.G., Srivastava, A.K., and Kant, R. (2010) A dual electrochrome of poly-(3,4-ethylenedioxythiophene) doped by N,N′-bis(3-sulfonatopropyl)-4-4′-bipyridinium-redox chemistry and electrochromism in flexible devices. *ChemSusChem*, **3**, 97–105.
60. Rong, Y., Kim, S., Su, F.Y., Myers, D., and Taya, M. (2011) New effective process to fabricate fast switching and high contrast electrochromic device based on viologen and Prussian blue/antimony tin oxide nanocomposites with dark colored state. *Electrochim. Acta*, **56**, 6230–6236.
61. Kosower, E.M. and Cotter, J.L. (1964) Stable free radicals. II. The reduction of 1-methyl-4cyanopyridinium ion to methylviologen cation radical. *J. Am. Chem. Soc.*, **86**, 5524–5527.
62. Mortimer, R.J. and Varley, T.S. (2011) Quantification of color stimuli through the calculation of CIE chromaticity coordinates and luminance data for application to in situ colorimetry studies of electrochromic materials. *Displays*, **32**, 35–44.
63. Rosseinsky, D.R. and Monk, J.L. (1989) Thin layer electrochemistry in a paper matrix: electrochromography of Prussian blue and two bipyridilium systems. *J. Electroanal. Chem. Interfacial Electrochem.*, **270**, 473–478.
64. Hsu, Y.C. and Ho, K.C. (2005) Anionic effect on the intercalation and spectral properties of poly(butyl viologen) films. *J. New Mater. Electrochem. Syst.*, **8**, 49–57.
65. Alarco, P.-J., Abu-Lebdeh, Y., Abouimrane, A., and Armand, M. (2004) The plastic-crystalline phase of succinonitrile as a universal matrix for solid-state ionic conductors. *Nat. Mater.*, **3**, 476–481.
66. Cinnsealach, R., Boschloo, S., Rao, S.N., and Fitxmaurice, D. (1998) Electrochromic windows based on viologen nanostructured TiO2 films. *Sol. Energy Mater. Sol. Cells*, **55**, 215–223.
67. Cinnsealach, R., Boschloo, S., Rao, S.N., and Fitzmaurice, D. (1999) Colored electrochromic windows based on nanostrucutred TiO_2 films modified by adsorbed redox chromophores. *Sol. Energy Mater. Sol. Cells*, **57**, 107–125.
68. Cummins, D., Boschloo, G., Ryan, M., Corr, D., Rao, S.N., and Fitzmaurice, D. (2000) Ultrafast electrochromic windows based on redox-chromophore modified nanostructured semiconducting films. *J. Phys. Chem. B*, **104**, 11449–11459.
69. Fitzmaurice, D., Rao, R.N., Cinnsealach, R. and Enright, B. (2003) Electrochromic system. US Patent 6,605,239 B2, Aug. 12, 2003.
70. Ryu, J.-H., Shin, D.-O., and Suh, K.-D. (2005) Preparation of a reflective-type electrochromic device based on monodisperse, micrometer-size-range polymeric microspheres and viologen pendants. *J. Polym. Sci., Part A: Polym. Chem.*, **43**, 6562.
71. Ryu, J.-H., Lee, J.-H., Han, S.-J., and Suh, K.-D. (2008) Influence of viologen length on the response time of the reflective electrochromic display prepared by monodisperse viologen-modified polymeric microspheres. *Colloids Surf., A*, **315**, 31–37.
72. Constantin, V.-A., Cao, L., Sadaf, S., and Walder, L. (2012) Oligoviologen/SWCNT nano-composites:

preparation and characterization. *Phys. Status Solidi B*, **249**, 2395–2398.

73. Alvaro, M., Aprile, C., Atienzar, P., and Garcia, H. (2005) Preparation and photochemistry of single wall carbon nanotubes having covalently anchored viologen units. *J. Phys. Chem. B*, **109**, 7692–7697.
74. Bookbinder, D.C. and Wrighton, M.S. (1983) Electrochromic polymers covalently anchored to electrode surfaces. Optical and electrochemical properties of a viologen-based polymer. *J. Electrochem. Soc.*, **130**, 1080–1087.
75. Dominey, R.N., Lewis, T.J., and Wrighton, M.S. (1983) Synthesis and characterization of a benzylviologen surface-derivatizing reagent. N,*N'*-*bis*[*p*-(trimethoxysilyl)benzyl]-4,4'-bipyridinium dichloride. *J. Phys. Chem.*, **87**, 5345.
76. Leventis, N. and Chung, Y.C. (1993) Preparation and characterization of tungsten trioxide/dibenzyl viologen polymer bilayer electrochromic films. *J. Mater. Chem.*, **3**, 833.
77. Shigehara, K., Tsuchida, E., and Anson, F.C. (1984) Comparison of the electrochemical responses of methylviologen and poly(xylylviologen) incorporated in Nafion coatings on graphite electrodes. *J. Electroanal. Chem.*, **175**, 291–298.
78. Decher, G. and Hong, J.D. (1991) Buildup of ultrathin multilayer films by a -assembly process: II. Consecutive adsorption of anionic and cationic bipolar amphiphiles and polyelectrolytes on charged surfaces. *Ber. Bunsen Ges. Phys. Chem.*, **95**, 1430–1434.
79. Stepp, J. and Schlenoff, J.B. (1997) Electrochromism and electrocatalysis in viologen polyelectrolyte multilayers. *J. Electrochem. Soc.*, **144**, L155–L157.
80. Schlenoff, J.B., Laurent, D., Ly, H., and Stepp, J. (1998) Redox-active polyelectrolyte multilayers. *Adv. Mater.*, **10**, 347–349.
81. DeLongchamp, D.M., Kastantin, M., and Hammond, P.T. (2003) High-contrast electrochromism from layer-by-layer polymer films. *Chem. Mater.*, **15**, 1575–1586.
82. Wang, X., Huang, H.-X., Liu, A.-R., Liu, B., Wakayama, T., Nakamura, C., Miyake, J., and Qian, D.-J. (2006) Layer-by-layer assembly of single-walled carbon nanotubes-poly(viologen) derivative multilayers and their electrochemical properties. *Carbon*, **44**, 2115–2121.
83. Kostina, S.A., Ilyukhin, A.B., Lokshin, B.V., and Kotov, V.Y. (2001) Ion pairs in the crystal structure of potassium ethyl viologen hexacyanometallates(II). *Mendeleev Commun.*, **11**, 12–13.
84. Mortimer, R.J. and Varley, T.S. (2013) Electrochromic devices based on surface-confined Prussian blue or Ruthenium purple and aqueous solution-phase di-*n*-heptyl viologen. *Sol. Energy Mater. Sol. Cells*, **109**, 275–279.
85. Leventis, N. and Chung, Y.C. (1990) Polyaniline-Prussian blue novel composite material for electrochromic applications. *J. Electrochem. Soc.*, **137**, 3321–3322.
86. Gotoh, A., Uchida, H., Ishizaki, M., Satoh, T., Kaga, S., Okamoto, S., Ohta, M., Sakamoto, M., Kawamoto, T., Tanaka, H., Tokumoto, M., Hara, S., Shiozaki, H., Yamada, M., Miyake, M., and Kurihara, M. (2007) Simple synthesis of three primary color nanoparticle inks of Prussian blue and its analogues. *Nanotechnology*, **18**, 345609.
87. Hara, S., Shiozaki, H., Omura, A., Tanaka, H., Kawamoto, T., Tokumoto, M., Yamada, M., Gotoh, A., Kurihara, M., and Sakamoto, M. (2008) Color-switchable glass and display devices fabricated by liquid processes with electrochromic nanoparticle "ink". *Appl. Phys. Express*, **1**, 104002.
88. Omura, A., Tanaka, H., Kurihara, M., Sakamoto, M., and Kawamoto, T. (2009) Electrochemical control of the elution property of Prussian blue nanoparticle thin films: mechanism and applications. *Phys. Chem. Chem. Phys.*, **11**, 10500–10505.
89. Ariga, K., Yamauchi, Y., Rydzek, G., Ji, Q.M., Yonamine, Y., Wu, K.C.W., and

Hill, J.P. (2014) Layer-by-layer nanoarchitectonics: invention, innovation, and evolution. *Chem. Lett.*, **43**, 36–68.

90. Wang, J.M., Sun, X.W., and Jiao, Z.H. (2010) Application of nanostructures in electrochromic materials and devices: recent progress. *Materials*, **3**, 5029–5053.

91. Tokarev, A., Agulhon, P., Long, J., Quignard, F., Robitzer, M., Ferreira, R.A.S., Carlos, L.D., Larionova, J., Guerin, C., and Guari, Y. (2012) Synthesis and study of Prussian blue type nanoparticles in an alginate matrix. *J. Mater. Chem.*, **22**, 20232–20242.

92. Hu, M., Furukawa, S., Ohtani, R., Sukegawa, H., Nemoto, Y., Reboul, J., Kitagawa, S., and Yamauchi, Y. (2012) Synthesis of Prussian blue nanoparticles with a hollow interior by controlled chemical etching. *Angew. Chem. Int. Ed.*, **51**, 984–988.

93. Cheng, K.C., Chen, F.R., and Kai, J.J. (2007) Electrochromic property of nano-composite Prussian blue based thin film. *Electrochim. Acta*, **52**, 3330–3335.

94. Yamaguchi, I. and Makishi, S. (2013) Synthesis and chemical properties of electrochromic -conjugated polyphenylenes with pendant viologen-TCNQ salts. *J. Appl. Polym. Sci.*, **129**, 397–403.

95. Gadgil, B., Damlin, P., Aaritalo, T., Kankare, J., and Kvarnstrom, C. (2013) Electrosynthesis and characterization of viologen cross linked thiophene copolymer. *Electrochim. Acta*, **97**, 378–385.

96. Zakirov, M.I., Shandryuk, G.A., Bondarenko, G.N., Nodova, E.L., Kryl'skii, D.V., and Talroze, R.V. (2012) Formation of electrochromic systems based on noncovalently associated polymer-viologen complexes. *Polym. Sci. Ser. B*, **54**, 50–60.

97. Krompiec, M., Grudzka, I., Filapek, M., Skórka, Ł., Krompiec, S., Łapkowski, M., Kania, M., and Danikiewicz, W. (2011) An electrochromic diquat-quaterthiophene alternating copolymer: a polythiophene with a viologen-like moiety in the main chain. *Electrochim. Acta*, **56**, 8108–8114.

98. Rodriguez, J., Navarrete, E., Dalchiele, E.A., Sanchez, L., Ramos-Barrado, J.R., and Martin, F. (2013) Polyvinylpyrrolidone-$LiClO_4$ solid polymer electrolyte and its application in transparent thin film supercapacitors. *J. Power Sources*, **237**, 270–276.

99. Raphael, E., Avellaneda, C.O., Aegerter, M.A., Silva, M.M., and Pawlicka, A. (2012) Agar-based gel electrolyte for electrochromic device application. *Mol. Cryst. Liq. Cryst.*, **554**, 264–272.

100. Ha, H.J., Kwon, Y.H., Kim, J.Y., and Lee, S.Y. (2011) A self-standing, UV-cured polymer networks-reinforced plastic crystal composite electrolyte for a lithium-ion battery. *Electrochim. Acta*, **57**, 40–45.

101. Kaplan, A., Korin, E., Soifer, L., and Bettelheim, A. (2012) Ion-conductive and transparent resorcinol-formaldehyde hydrogels for electrochemical and solar applications. *Electrochem. Solid-State Lett.*, **15**, F1–F4.

102. Howlett, P.C., Ponzio, F., Fang, J., Lin, T., Jin, L.Y., Iranipour, N., and Efthimiadis, J. (2013) Thin and flexible solid-state organic ionic plastic crystal-polymer nanofibre composite electrolytes for device applications. *Phys. Chem. Chem. Phys.*, **15**, 13784–13789.

103. Ogoshi, T., Onodera, T., Yamagishi, T., Nakamoto, Y., Kagata, A., Matsumi, N., and Aoi, K. (2011) Transparent ionic liquid-phenol resin hybrids with high ionic conductivity. *Polym. J.*, **43**, 421–424.

104. Brennan, L.J., Barwich, S.T., Satti, A., Faure, A., and Gun'ko, Y.K. (2013) Graphene-ionic liquid electrolytes for dye sensitised solar cells. *J. Mater. Chem. A*, **1**, 8379–8384.

105. Sharma, N., Lakhman, R.K., Zhou, Y.X., and Kasi, R.M. (2013) Physical gels of BMIM BF_4 by N-tert-butylacrylamide/ethylene oxide based triblock copolymer self-assembly:

synthesis, thermomechanical, and conducting properties. *J. Appl. Polym. Sci.*, **128**, 3982–3992.

106. Lee, S.Y., Yasuda, T., and Watanabe, M. (2010) Fabrication of protic ionic liquid/sulfonated polyimide composite membranes for non-humidified fuel cells. *J. Power Sources*, **195**, 5909–5914.

5
Conjugated Electrochromic Polymers: Structure-Driven Colour and Processing Control

Aubrey L. Dyer, Anna M. Österholm, D. Eric Shen, Keith E. Johnson, and John R. Reynolds

5.1
Introduction and Background

The field of electrochromics spans a large and diverse class of materials, as is demonstrated by the breadth of those covered in this book. Here, conjugated polymers have the advantage that key properties can usually be readily tuned through synthetic methods. Being organic materials, they allow wide possibilities of altering and controlling redox, electronic, optical, mechanical, and processing properties. The extensive and successful effort over the past several decades in utilising synthetic methods to alter electrochromic properties is demonstrated by the number of reviews in the area [1–6]. This chapter covers the structure–property relationships that control electrochemical properties (redox potentials), optical/colorimetric properties (electronic absorptions in the various redox states) and solubility/processability. Only polymers that are thin electrode-supported films are considered here, rather than polymers in devices, because the largest influencing factors – device architecture and components (electrodes, electrolyte, sealant, etc.) – are highly variable from one research group to another. Similarly, precise quantifications of electrochromic properties (switching speed, coloration efficiency, lifetime, etc.) are highly dependent on film thickness, processing method, electrolyte system, electrochemical cell geometry, and many other factors. Thus, discussions are limited to those properties most influenced by repeat unit structure alone: oxidation potentials and colour properties, along with solubility and processability.

5.1.1
Source of Electrochromism in Conjugated Polymers

Conjugated electrochromic polymers (ECPs) are conducting polymers able to conduct charge via their alternating single and double bonds along the polymer backbone, resulting in unique electronic properties. This widely extended conjugated macromolecular structure, allowing for delocalisation of charges induced on oxidation or reduction, distinguishes them from other molecular

Electrochromic Materials and Devices, First Edition.
Edited by Roger J. Mortimer, David R. Rosseinsky and Paul M. S. Monk.

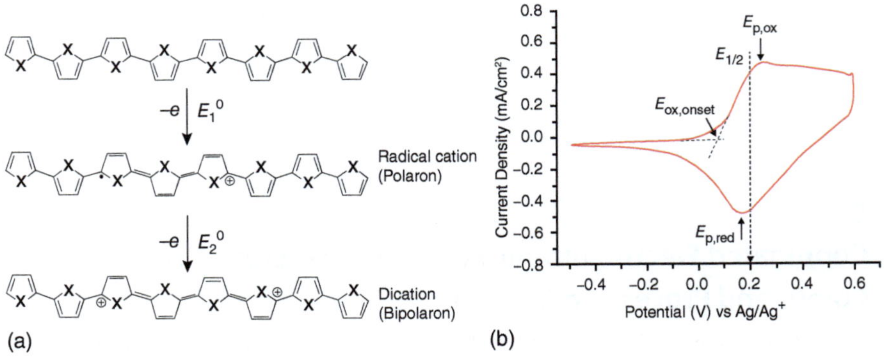

Figure 5.1 (a) Representative polyheterocycle repeat unit (octamer shown for clarity) structure for neutral (top), radical cation (centre) and dication (bottom) redox states. (b) Cyclic voltammogram, for example, polymer (ECP-Red) showing approximate locations of oxidation onset potential ($E_{onset,\ ox}$), peak oxidation potential ($E_{p,\ ox}$), peak potential for reduction to the neutral state ($E_{p,\ red}$) and half-wave potential ($E_{1/2}$). Adapted from Ref. [7].

conjugated systems (e.g. viologens, porphyrins, phthalocyanines) discussed in this book. Figure 5.1 depicts the process that occurs on oxidation of a conjugated polymer (for clarity, a generic five-membered heterocyclic ring is provided where X represents a heteroatom), with Figure 5.1a showing a representative structure and Figure 5.1b an example cyclic voltammogram. For a large majority of conjugated polymers in this chapter, the oxidation is of an electrode-supported polymer film, prepared as discussed in Section 5.3. The two-electron oxidation proceeds with the first electron removal producing a radical cation (a 'polaron', in semiconductor terminology), and a dication ('bipolaron') resulting from removal of a second electron. However, only rarely can two separate redox peaks be observed: most cyclic voltammograms are broad and ill-defined as shown in Figure 5.1b. The values of importance are the onset of oxidation ($E_{onset,\ ox}$) and the potential at peak current for oxidation ($E_{p,\ ox}$), the latter often not clearly discernible, especially at higher scan rates.

Electrochromic optical changes from such electron processes are as shown in Figure 5.2 for ECP-Red which is a copolymer of an acyclic branched alkoxy thiophene (3,4-OEtHx-Th) with a dimethoxy thiophene (3,4-OMe-Th). The electronic structure of these conjugated polymers, via the high level of π-orbital overlap, uniquely results in the evolution of accessible electronic energy bands. The highest occupied molecular orbital (HOMO) forms the valence band, and the lowest unoccupied molecular orbital (LUMO) provides the conduction band, in semiconductor terminology. The optical 'bandgap energy', or E_g, comprises an electronic transition from the top of the valence band to the bottom of the conduction band – see the left panel of Figure 5.2a. The bandgap is determined by the onset of absorption from the low-energy side of the spectrum as shown for the example polymer in Figure 5.2b. The polymer neutral state has a single, broad peak in the visible region centred at 525 nm, giving it a red colour as

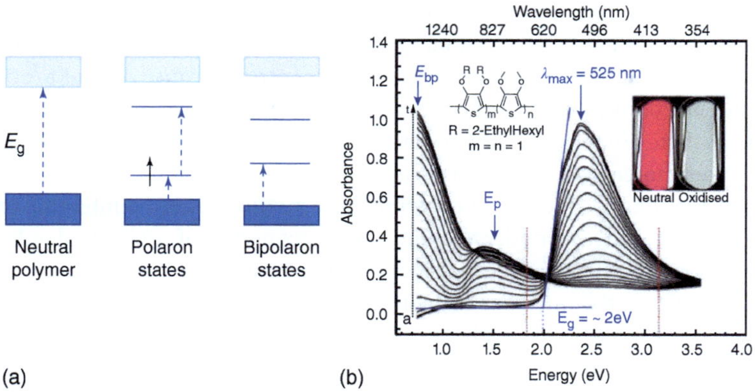

Figure 5.2 (a) Allowed electronic transitions for neutral, radical-cation (polaron) and dication (bipolaron) states in conjugated polymers. (b) Example spectroelectrochemistry for ECP-Red showing absorptions as progression from neutral to fully oxidised state. Inset photographs show the colour of ECP-Red in the neutral (left) and fully oxidised (right) state as a thin film on a transparent electrode. Adapted from Ref. [7].

shown in the inset photograph. During oxidation to the radical cation (Polaron States in Figure 5.2a centre panel), mid-gap states are formed, resulting in lower energy electronic transitions seen as absorptions emerging at longer wavelengths and growing in intensity with successively higher potentials. At intermediate oxidation states, the spectra will exhibit contributions from both the neutral material and the oxidized form. As the polymer is further oxidised, the dication is formed and the only allowed electronic transition is from the top of the valence band to a mid-gap state with the bandgap transition no longer allowed (Bipolaron States in Figure 5.2a right panel). This is seen as a broad absorption extending from the near-infrared (NIR) with little absorbance in the visible region, and the polymer becomes a highly transmissive light blue. This process is quite reversible with the coloured state regenerated by applying a voltage low enough to reduce the polymer back to the neutral state. This chapter covers the oxidative process, also called p-doping. There is a very small subset of polymers [2, 5, 6] that exhibit reductive electrochromic behaviour via n-doping: for details, see the cited references.

Electronic transitions involving valence-band and conduction-band energies and those of mid-gap states therefore determine material colour, adjustments of which contribute to their versatility. Clearly, these are not the only factors that dictate redox and optical properties. Other factors include polymer backbone planarity (and hence conjugation length), varying the amount of electron-rich and -poor building blocks incorporated, integration of fused heterocycles that reduce the polymer bond-length alternation and tuning properties through copolymerisation of differing monomer units. Other strategies have been to utilise blends, laminates and composites with various conjugated polymers or other electrochromic materials, to tune colour properties. More details can be found in many reviews of the field [2, 5, 6, 8].

The most common synthetic strategies involve the use of electron-rich polyheterocycles as building blocks in homopolymers and copolymers designed for electrochromic applications. A recent strategy to control polymer optical properties is the use of alternating donor and acceptor units – see Section 5.1.1.2. Additionally, there are two methods commonly used to control the extent of conjugation – steric interactions and fused ring systems: – see limited representative examples in Sections 5.2 and 5.3. In covering these methods, it is the intention of this chapter to introduce an understanding of how properties of conjugated polymers applicable to electrochromics are manifested and altered.

It should be noted that in order to avoid conflict with formal electrochemical terminology (which has a tangled history), 'oxidation potential' here is the plain English term meaning just the positive potential applied to the substrate in order to effect the oxidation being referred to.

5.1.1.1 Common Polyheterocycles

Even though linear polyacetylene is the simplest structure, it fails as an electrochrome due to its high oxidation potential and lack of solubility. Likewise, high oxidation potentials explain why poly(*p*-phenylene) and poly(*p*-phenylene vinylene), and their derivatives, are less viable in electrochromics while succeeding as organic transistors, photovoltaics and light-emitting devices. To overcome the high oxidation potential observed in the conjugated polymers previously mentioned, electron-rich conjugated five-membered ring polyheterocycles are commonly used and can include thiophenes, pyrroles, selenophenes and furans. These systems are highly synthetically flexible in regard to altering repeat unit structure to elicit desired properties, as shown schematically in Figure 5.3. The identity of the heteroatom (depicted in red in Figure 5.3) strongly influences both the oxidation potential (via the HOMO energy level) and the bandgap (via the HOMO/LUMO energy level difference). Substituting the heteroatom alters the polymer oxidation potential, bandgap, and colour properties. An example is the differences between poly(3,4-ethylenedioxythiophene) (PEDOT, E_g of 1.6 eV, allowing a blue to transmissive switch), poly(3,4-ethylenedioxyselenophene)

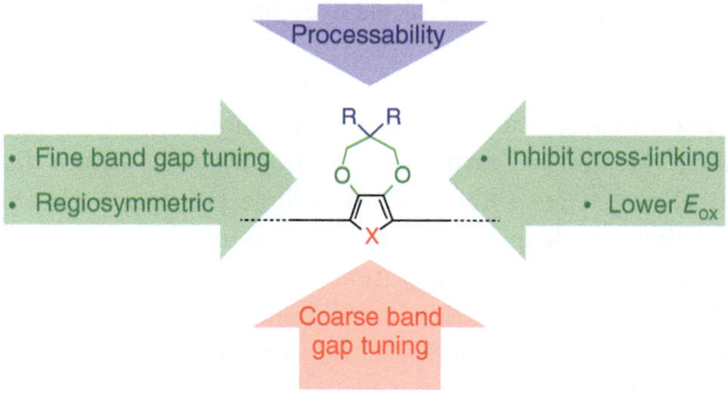

Figure 5.3 How colour and oxidation potentials are tuned using simple heterocycles as an example.

(PEDOS, E_g of 1.4 eV, allowing a deep blue to transmissive switch), and poly(3,4-ethylenedioxypyrrole) (PEDOP, E_g of 2.0 eV, allowing a red to transmissive switch), the polymer oxidation potential of PEDOP being lower than that of PEDOS, which is lower than that of PEDOT.

An additional synthetic route to altering the electrochromic properties of common polyheterocycles is through substitution at the 3- and/or 4-position of the heterocycles (shown in green in Figure 5.3), the identity and placement of these substituents being widely variable. For example, one can choose to utilise electron-withdrawing substituents such as fluorine to lower the HOMO energy level, (raising the oxidation potential) or electron-donating substituents, such as methyl groups, to raise the HOMO energy level (lowering the oxidation potential). For example, substitution at the 3-position of thiophene with fluoromethyl groups raises the oxidation potential and shifts the neutral-state absorption to higher energies compared to the 3-methyl-substituted thiophene [9]. An added benefit of substitution at the 3- and 4-position of the heterocycle is that it inhibits polymerisation or cross-linking at these reactive positions when oxidative polymerisation conditions are employed, which can result in loss of processability and induce conjugation breaks.

The selection of substituents at these positions can furthermore influence steric interactions (Section 5.1.1.3) and solubility (Section 5.3). The 3,4-substitution can be acyclic, such as is shown for the example ECP-Red polymer given in Figure 5.2, or rendered cyclic using bridging alkyl units as shown in the schematic in Figure 5.3. The bridging alkyl units are typically ethylene, propylene or butylene. For poly(3,4-alkylenedioxythiophene) derivatives, a comparison of the influence of ring size has shown that as the ring size increases, so does the optical contrast [6, 10]. One of the most common substitution patterns is that of ethylenedioxy units, such as with PEDOT – one of the most successful commercially available conjugated polymers – as will be discussed in later sections. The propylene bridging unit allows for synthetic flexibility in imparting processability through addition of solubilising substituents at the 2-position of the ring as shown by the R-groups in blue in Figure 5.3, further discussed in Section 5.3. Additionally, interchain separation can result from introducing branched and bulky units, enhancing optical contrast as well [6, 10].

While substitutions at the 3- and 4-position are most common, an additional option for substitution, and subsequent influence on optoelectronic properties, involves the electron-rich pyrrole heterocycle via *N*-functionalisation [11]. Pyrroles have lower oxidation potentials than their thiophene, selenophene, and furan analogues due to their high electron density, which results in a higher HOMO. With the addition of electron-donating substituents at the 3,4-position, such as with the poly(3,4-alkylenedioxypyrrole)s or PXDOPs, the electron density is further increased and the oxidation potential is further lowered [12–14]. Functionalisation at the nitrogen can allow for further tuning of electron density in addition to inducing twisting along the polymer backbone via steric interactions as is shown in Section 5.2.2. This added torsional strain results in a lowering of π-orbital overlap along the polymer chain, increasing the bandgap. How much torsional strain, and energy level adjustment occurs depends highly on the nature of the substituent and can be quite dramatic. For example, the unsubstituted poly(3,4-propylenedioxypyrrole) (PProDOP) has a bandgap of 2.2 eV and is orange in the neutral state, while the *N*-alkyl-substituted PProDOP has a bandgap ranging from 3.0 eV for the *N*-methyl PProDOP to 3.4 eV for the *N*-propyl and *N*-octyl PProDOP with the neutral-state absorption now in the ultraviolet region of the spectrum. This results in a colourless neutral state, and the polymers are anodically colouring while maintaining a low oxidation potential [2, 5, 11].

The most straightforward approach to control optoelectronic properties is through homopolymerisation of an electron-rich heterocycle, tuning optical and electronic properties through substituents as outlined earlier; specific polymer structures will be covered later. Material properties can be further controlled by copolymerisation with a variety of electron-rich monomer units through a combination of electronic and steric effects. Such electron-rich monomers chosen are often based on thiophenes, pyrroles, and carbazoles. However, the use of electron-poor co-monomers provides yet another method to tune optoelectronic properties.

5.1.1.2 Donor–Acceptor Approach – The Push–Pull of Electrons

Another approach to optoelectronic property control is via copolymerisation of electron-rich units with those that are electron-poor. The electron-rich units (heterocycles such as thiophene or pyrrole) can be considered electron donors; the electron-poor units (such as cyanovinylene (CNV) or heterocycles such as benzothiadiazole or benzobisthiadiazole) are then electron acceptors, as is shown schematically in Figure 5.4. This approach was first aimed at developing systems that can be both electrochemically oxidised and reduced within a stable potential range [15, 16]. Electron-poor units in the polymer backbone lower the LUMO, thereby lowering the bottom of the conduction band. The polymer bandgap is lowered, and the polymer absorption is red-shifted as a result. One commonly chosen acceptor unit is an sp^2 nitrogen such as in pyridine. Copolymerising the donor thiophene with acceptors of increasing nitrogen content (i.e. proceeding from pyridine to pyridazine to tetrazine) further red-shifts the optical absorption in the neutral state [2, 17, 18]. Rather than using pyridine-based acceptors, further bandgap lowering can be achieved by introducing fused heterocycles like the aforementioned pyridopyrazine, 2,1,3-benzothiadiazole (BTD) or benzobisthiadiazole [2].

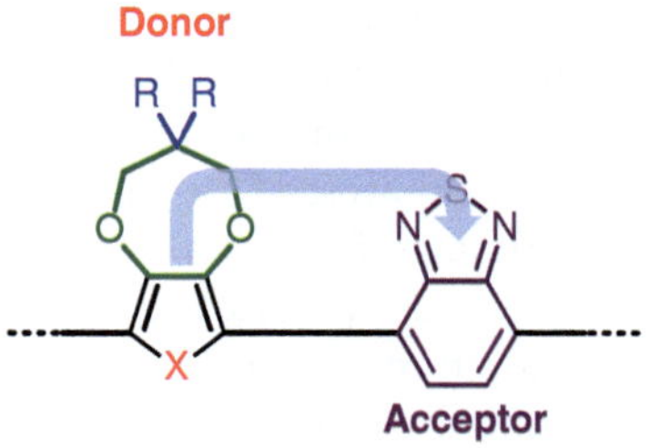

Figure 5.4 Donor–acceptor concept using donor heterocycles and BTD acceptor as an example.

On incorporating these donor and acceptor units in an alternating copolymerisation, not only is the bandgap lowered, but also an absorption profile containing two distinct absorption peaks, rather than a single broad absorption, is observed. These two absorption peaks are theoretically predicted to result from the formation of low-lying unoccupied energy bands within the bandgap of the conjugated polymer in its neutral state [19, 20]. Such a new unoccupied energy band lies below the conduction band, allowing a new electronic transition from the top of the valence band to this new mid-gap band in addition to the usual bandgap transition from the top of the valence band to the conduction band. It is thought that the position of this mid-gap band is determined by the electron-poor acceptor unit. The absorbance spectrum of this donor–acceptor polymer in the neutral state has two absorptions: the true bandgap transition occurring at shorter wavelengths, with that from the valence band to the mid-gap band occurring at longer wavelengths.

Through judicious selection of the donor and acceptor units (electron-donating strength and electron-accepting strength) and relative concentration, the location, width and separation of these two peaks, relative to one another, can be controlled with impressive results. This is demonstrated by the photograph in Figure 5.5: a series of solutions of conjugated oligomers differing by the choice of donor and acceptor units, and their placement relative to one another along the oligomer backbone, is shown [21]. Two specific examples are highlighted for an oligomer containing benzothiadiazole, flanked by two thiophenes on either side and capped with phenylene units, which is a pink-red in solution (PTT–BTD–TTP), and a benzobisthiadiazole flanked by two ethylenedioxythiophenes, capped with alkyl groups, which is green in colour (EE–BBT–EE).

While these are conjugated oligomers, not conjugated polymers, the concept is demonstrated and can be routinely extended to polymeric systems with closely similar outcomes. The result of varying donor/acceptor strength (e.g.

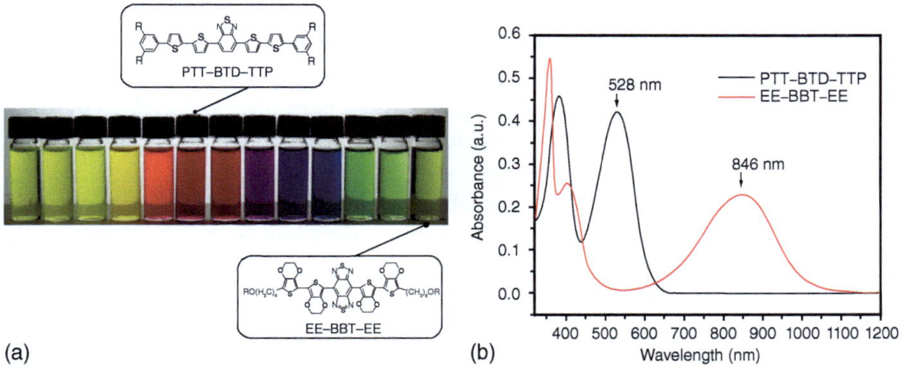

Figure 5.5 Demonstration of D–A concept using oligomers. (a) Evolution of colours achieved by varying donor and acceptor strengths. (b) Solution spectra of two donor–acceptor oligomers showing evolution of spectra with variation in donor and acceptor units. Adapted from Ref. [21].

electron-rich/-poor character, number of units, placement) can be clearly seen in the solution spectra of the two example oligomers discussed in Figure 5.5b. The oligomer PTT–BTD–TTP has two relatively narrow absorptions; one centred at ~400 nm and the other at 528 nm. On use of the stronger donor (more electron-rich) EDOT (3,4-ethylenedioxythiophene) in place of thiophene, and a stronger acceptor (more electron-poor) benzo[1,2-c,3,4-c′]bis[1,2,5]thiadiazole (BBT) in place of benzodithiophene (BDT) for EE–BBT–EE, the high-energy absorption peak is in a similar position, but the lower energy peak is broadened and substantially red-shifted to 846 nm. This concept has been utilised with relative reliability to generate a large spectrum of colours of conjugated polymers, from vibrant blues and greens, to broadly absorbing inky blues and blacks [2]. While the resulting spectral properties can be tailored using this molecular-engineering approach, secondary interactions, such as between neighbouring units, have to be taken into consideration; they cannot always be predicted.

5.1.1.3 Steric Interactions

Yet another approach to controlling optoelectronic properties in conjugated polymers is to take advantage of steric interactions within a polymer chain. Referring back to the ECP-Red example, judicious choice of substituents on the heterocycle can either induce interactions, resulting in torsional strain or twisting of the main chain, or reduce interactions, relieving strain and causing a relaxation of the main chain. By incorporation of bulky substituents, adjacent units can twist out of plane from one another. If the twist is sufficient, π-orbital overlap is affected and the extent of conjugation can be decreased or broken all together. This causes a blue-shift of the neutral-state absorption, resulting in dramatic shifts in colour, as can be seen for the ECP-Orange polymer in Figure 5.6 with bulky bis(ethylhexyloxy) substituents. Alternatively, the substitution pattern can be chosen such that there is little to no interaction between adjacent units, causing planarisation of the backbone and extension of conjugation, red-shifting the absorption, as can be seen for the ECP-Red in Figure 5.6 where there is random incorporation of dimethoxythiophene units, relieving strain. Not only does 'bulk' have to be considered, but other secondary effects, such as sulfur–oxygen electronic interactions as is observed in dioxythiophenes or hydrogen bonding as has been observed in dioxypyrroles, come into play. These types of interactions can result in a stronger driving force for planarity and extended conjugation.

An added interaction to consider when designing these organic systems is regiochemistry and influence on steric distortion of the conjugated chain. For example, in mono-substituted thiophenes (at the 3-position), the length of the alkyl chain does not have a strong influence on the electrochromic properties, but can allow for increased solubility using alkyl chains of sufficient length. However, when the regioregularity is controlled, there can be large changes in electrochromic properties. The regiorandom poly(3-alkylthiophene) has an orange-red neutral state, switching to blue on oxidation [22, 23]. When regioregularity is designed into the synthesis, as is shown in Figure 5.7, the substitution pattern can have strong effects on the steric result [24]. For those alkylthiophenes that have a head-to-head

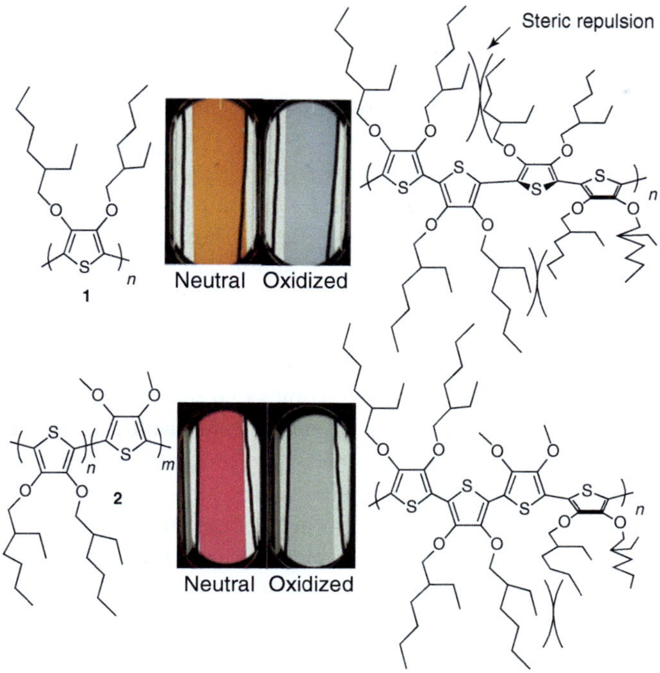

Figure 5.6 Repeat unit structures for ECP-Orange (top, 1) and ECP-Red (bottom, 2). Structures to the right illustrate the steric repulsions occurring between side chains and resulting backbone twisting. Photographs show the polymer films in their neutral (left) and oxidised (right) states. Reproduced from Ref. [7].

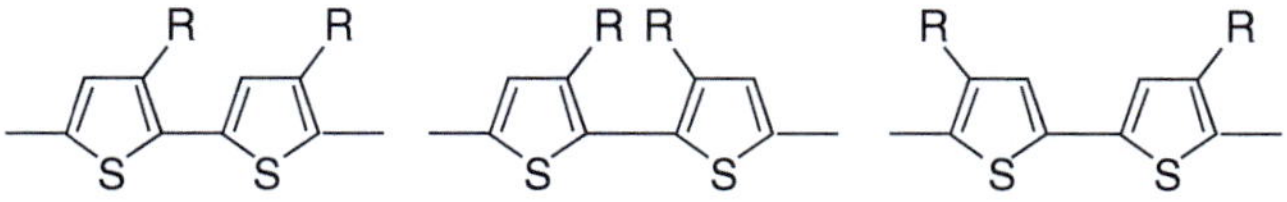

Figure 5.7 Structures of thiophene dimers substituted at the 3-position showing variations in regioregularity possible with head-to-tail (HT), head-to-head (HH) and tail-to-tail (TT) arrangement.

(HH) or tail-to-tail (TT) arrangement, the steric interactions between adjacent units can shorten the conjugation length as was previously shown in the literature for electropolymerised methyl-substituted bithiophenes [22, 25]. The TT example switches from orange to blue-purple on oxidation, while the HH example is yellow when neutral and blue-purple on oxidation. When regioregularity is designed, as with the head-to-tail (HT) arrangement, steric interactions are lessened and the polymer exhibits red to blue-purple switching. Further examples containing longer alkyl chains have shown similar results [22].

Another approach to relieving steric congestion is through the use of unsubstituted conjugated spacers. These are typically placed between two heterocycles that would, if placed next to one another, interact and increase torsional strain. The conjugated spacer relieves that strain, allowing relaxation of the structure and extension of the conjugation length. One example includes thienylenevinylene copolymers where the neutral-state colour is a deep blue with a lowering of the bandgap to 1.5–1.7 eV [26–29]. Further, by using the same approach, a copolymer of two EDOTs for every single vinylene spacer was reported to have a lower bandgap (1.4–1.5 eV) compared to the EDOT polymer (1.6 eV) with a deep purple neutral state [30, 31]. Thus, sterics have many influences on the optical properties with large or subtle variations in polymer bandgap.

5.1.1.4 **Fused Aromatics**

Yet another method to lower the energy gap in conjugated polymers is through the use of building blocks that reduce the extent of bond-length alternation by stabilising the quinoidal resonance geometry, or promote planarisation. These building blocks are fused aromatic systems with high resonance energies [32]. This was first achieved in poly(isothianaphthene) homopolymers that yielded the first reported cathodically colouring conjugated polymers, switching from a blue neutral state to a highly transmissive green-yellow on oxidation with a 1 eV bandgap [33–35]. Various groups then took advantage of the high resonance energies and ability to increase the backbone planarity, narrowing the energy gap via homopolymerisation of a variety of fused ring heterocycles or copolymers containing those as building blocks. Examples include thieno[3,4-b]thiophenes [36, 37], thieno[3,4-b]furans [38, 39], cyclopentadithiophenes [40, 41], *N*-substituted dithienopyrroles [42] and diphenyl-substituted dithienothiophenes [43]. There resulted a large variety of neutral-state colours with bandgaps that range from 0.85 eV for the thieno[3,4-b]thiophene to 2.18 eV for the diphenyl-substituted dithienothiophene, in systems covered in more detail in Section 5.2.1.3.

While the interactions discussed can be designed into the polymer repeat unit to affect the neutral-state colour quite reliably, such control is more complicated when considering the oxidised state colours. In fact, while a molecular design approach has successfully resulted in the completion of a neutral-state colour palette in ECPs, the major aim still remains out of reach with regard to controlling both the neutral state and oxidised state optical properties simultaneously [44]. While multicoloured systems can switch from a coloured state when neutral to another coloured state when oxidised, many applications require polymers that switch from highly absorptive in the one oxidisation state (e.g. neutral) to colourless and transmissive in the other state (e.g. fully oxidised), such as for display and window applications. The following section discusses representative examples of polymers that switch between coloured and transmissive states (anodically colouring or cathodically colouring), those that switch between multiple coloured states, and those that take advantage of long-wavelength optical changes. This is followed by Section 5.3, which details how solubility and processability are designed in these materials.

5.2 Representative Systems

There is a large body of work towards developing conjugated polymers for electrochromics that cannot be effectively covered here. This chapter is a starting point for understanding how chemists utilise synthetic design principles to tailor optical and electrochemical properties. One can delve deeper into the literature for further understanding of systems not covered here. The names of the polymers used are as they were originally represented in the literature.

5.2.1 Coloured-to-Transmissive Polymers

The large majority of electrochromes are those that switch between an absorptive coloured state and a highly transmissive, near-colourless state. Many applications, such as those for electrochromic windows, glasses, mirrors, and displays, require the device to alternate between absorptive and transmissive or reflective; hence, at least one transmissive state is required. Some researchers suggest that polymers exhibiting multiple coloured states can be useful in vibrantly coloured displays or that multiple colours demonstrated by a single material can be practical. However, it remains that most often the displays/devices are desired to switch to an 'off' state (i.e. not coloured) and besides, multiple colours in a single display can be achieved via colour mixing as in the emissive (light-emitting displays, liquid crystal displays) and static-print (newspapers, magazines, etc.) displays that currently dominate the market. In this section, examples of those polymers that switch between at least one coloured state and a non-coloured (or near-colourless) state are presented.

As with the other electrochromic materials, coloration can be achieved by cathodic colouring (colouring on reduction/switching to transmissive on oxidation) or anodic colouring (colouring on oxidation/switching to transmissive on reduction). As most conjugated polymers are neutral in their native state and lack stable reductive electrochemistry, the cycling is between a neutral state and an oxidised state with re-neutralisation on reduction (see Figure 5.1b for an example cyclic voltammogram). On oxidation, those that are cathodically colouring are absorptive in the visible region in their reduced state and become colourless on oxidation; those that are anodically colouring are colourless when neutral and become absorptive in the visible region on oxidation. Many of the polymers reported in the literature are cathodically colouring, as in this subsection.

In order of progression of the colours in the neutral state, polymers start from those that absorb at the high-energy end of the spectrum (yellow) and proceed across the spectrum to orange/red, blue/purple and finishing with those that are green (requiring two absorptions) and black (those requiring a broad absorption). To demonstrate the large number of hues that are available in cathodically colouring electrochromic polymers, Figure 5.8 shows the absorption spectra (in the neutral state) of a number of dioxythiophene-based conjugated polymers,

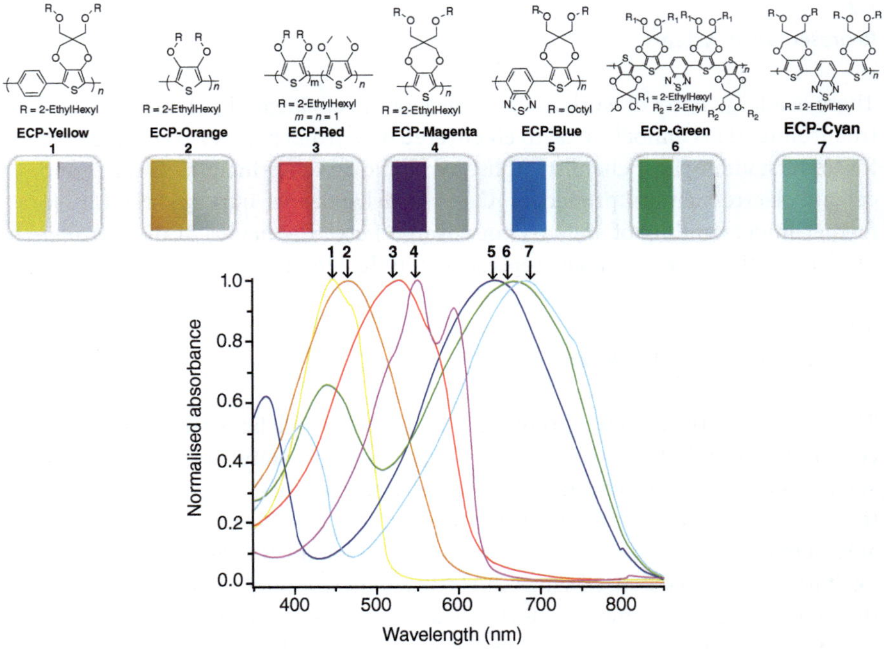

Figure 5.8 Repeat unit structures (top) for a series of cathodically colouring conjugated polymers. Photographs (centre) of the polymer films (on ITO/glass) in their neutral (left) and oxidised (right) states. Absorption spectra of the polymer films in their fully neutralised states. Reproduced from Ref. [44].

with photographs of the polymer films in their neutral (left) and oxidised (right) states, and repeat unit structures.

5.2.1.1 Yellow

To date, there have been a few yellow-to-transmissive switching conjugated polymers, owing to the difficulty in achieving a polymer with sufficiently high bandgap to shift the neutral-state absorption to the blue region of the spectrum, yet have the polaron and bipolaron absorptions exclusively in the NIR, having also a low oxidation potential. There are several examples of neutral-state yellow polymers, such as those based on EDOTs, thiophenes, and 3,4-propylenedioxythiophenes (ProDOTs) combined with pyrenes and naphthalenes or PProDOT substituted at the 1,3-position with t-butyl groups [45–48]. However, each of these examples switches from yellow to either blue or green oxidised states. Even with this difficulty, yellow-to-transmissive switching polymers remain highly important, as they are one of the key components in achieving the full palette of subtractive primaries of red–yellow–blue or cyan–magenta–yellow.

An early system that comes close to displaying this yellow-to-transmissive switch is a conjugated oligomer with sufficient solubility that films could be spray-cast from organic solvents, then UV-cross-linked through acrylate functionality to achieve insoluble films [49]. As shown in Figure 5.9, the yellow films were oxidised to a dark blue cation-radical state and to a transmissive neutral grey dication state.

Three years later, the first fully conjugated polymer was reported (Figures 5.10 and 5.8), which could both exhibit both a yellow neutral state, and switch to a highly transmissive, near-colourless oxidised state, as in the inset photographs [50]. The polymer is formed via copolymerisation of a ProDOT unit with a phenylene unit. Upon reporting of this yellow cathodically colouring polymer, the way has been paved for full colour displays by colour mixing via blends, stacked layers, or pixelated devices as this polymer allows completion of the subtractive primary colour palette. Notably, this polymer is highly soluble in common organic solvents (tetrahydrofuran (THF), toluene, etc.) and can be processed from solution as discussed in Section 5.3. More recent efforts have extended the available yellow-to-transmissive family of polymers to include ProDOT copolymerised with a variety of arylene units such as fluorene, carbazole, bis(ProDOT)-phenylene and a dimethoxy-substituted phenylene [51].

5.2.1.2 **Orange and Red**

As for yellow-to-transmissive polymers, the difficulty in achieving an orange-to-transmissive switch is due to the relatively high bandgap necessary for the neutral state. While achieving an orange neutral state on its own is not difficult, doing so with a transmissive oxidised state together with an accompanying stable switching potential is where the difficulties lie. Only two examples of conjugated electroactive polymers that achieve a stable orange-to-transmissive switch as thin films are

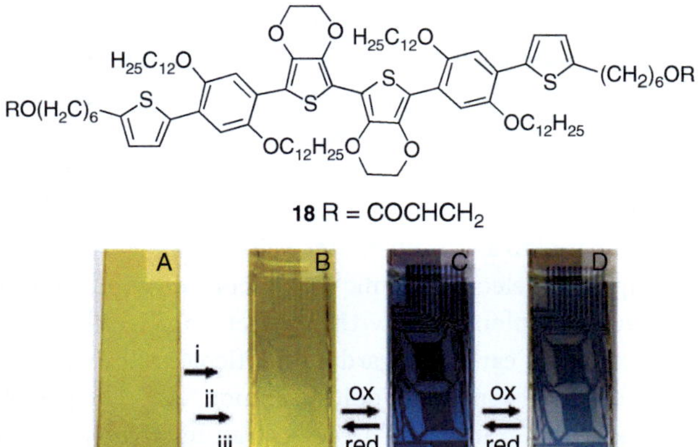

18 R = $COCHCH_2$

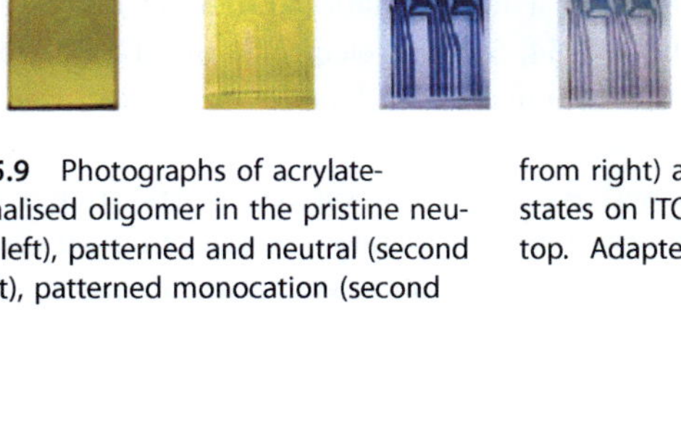

Figure 5.9 Photographs of acrylate-functionalised oligomer in the pristine neutral (far left), patterned and neutral (second from left), patterned monocation (second from right) and patterned dication (far right), states on ITO/glass. Oligomer structure at top. Adapted from Ref. [49].

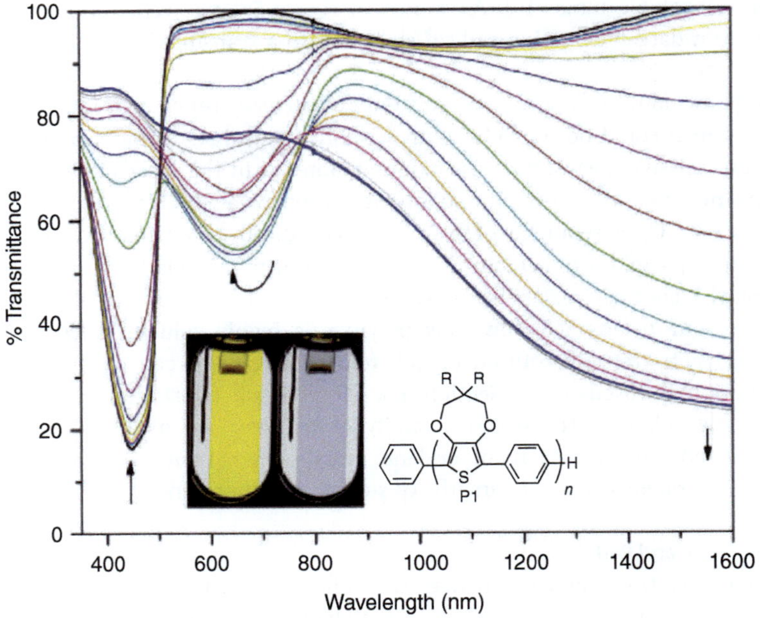

Figure 5.10 Transmittance spectroelectrochemistry for a yellow-to-transmissive switching conjugated polymer. Inset shows photographs of the polymer film in the neutral (left) and oxidised (right) states along with the repeat unit structure. Reproduced from Ref. [50].

reported. The first is 3,4-propylenedioxypyrrole (ProDOP) – an electrochemically polymerisable material shown in Figure 5.11. This polymer exhibits a bandgap of 2.2 eV and a high transmittance difference at λ_{max} (~70%) on oxidation to a transmissive state, exhibiting also a brown intermediate state [13, 14, 53]. The second orange-to-transmissive example is a solution-processable thiophene homopolymer – ECP-Orange [7]. This earlier-noted polymer (see Figure 5.6) has sufficient torsional strain, owing to the bulky branched 3,4-substitution, to blue-shift the neutral-state absorption compared to an unsubstituted thiophene.

There are several examples of electrochromic polymers with red neutral states (polythiophene is one example), but few that switch to a highly transmissive oxidised state as explained earlier. Regarding multicoloured ECPs (see Section 5.2.3), red-to-blue electrochromism is quite common. Another pyrrole analogue – PEDOP – is an early example of a red-to-transmissive switching ECP that is electrochemically polymerisable. Like PProDOP, it exhibits a high valence band energy, resulting in a low oxidation potential and a relatively wide bandgap of ~2.0 eV [12–14, 54, 55]. On electrochemical oxidation, a red neutral state switches to a highly transmissive blue-grey fully oxidised state as is shown in Figure 5.11 [13]. A solution-processable red polymer is ECP-Red (Figure 5.6), as already discussed. As mentioned, the red neutral state is achieved by releasing steric strain induced by the pendant chains in ECP-Orange by random incorporation of relatively small dimethoxythiophene units. The ease in

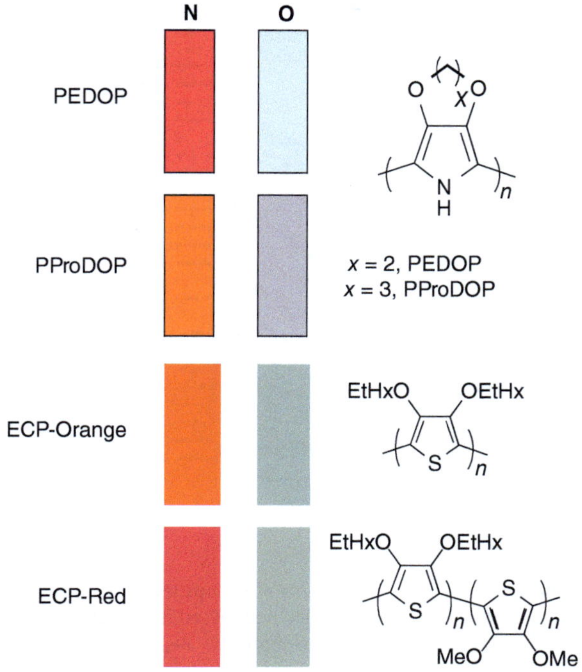

Figure 5.11 Neutral state colours (left) and oxidised state colours (right) for the conjugated electrochromic polymers shown. Repeat unit structures are to the right. Adapted from Refs. [7] and [52].

backbone twisting is sufficient to red-shift the neutral-state absorption, allowing for a red-to-transmissive switching polymer with a bandgap of 2.0 eV [7].

5.2.1.3 Blue and Purple

Thiophene-Based Homopolymers The most well-known, and commercially successful, conjugated polymer is PEDOT. First developed by researchers at Bayer AG in the late 1980s, it was found to possess a highly conductive, transparent and stable oxidised state, owing to its low bandgap, making PEDOT a breakthrough material for the field of intrinsically conductive polymers [5, 56–60]. Since the initial development, this polymer is used in wide ranging applications including antistatic coatings, transparent conductive coatings and electrodes, electronic components and displays, and is commercially available in multi-ton quantities under various trade names.

The polymer can be prepared *in situ* by oxidative polymerisation (electrochemically or chemically) of EDOT to form an insoluble film. The water-processable dispersion of the oxidised PEDOT with the polyelectrolyte poly(styrene sulfonate) (PSS), to form the complex PEDOT:PSS, has led to the marked success of this conjugated polymer. Detailed in Section 5.3, films of PEDOT:PSS can be

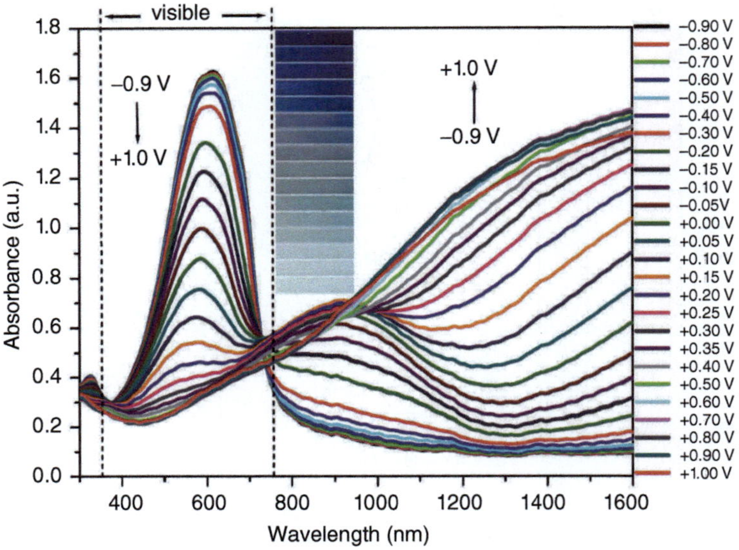

Figure 5.12 Spectroelectrochemistry of PEDOT at various oxidising voltages (provided in legend) between −0.9 and +1.0 V versus Ag/Ag^{+}. Inset shows photographs of PEDOT films at various levels of oxidation from fully neutralised (top) through to fully oxidised (bottom). Reproduced from Ref. [62].

prepared via spin coating and doctor blade coating [61]. Neutral films of PEDOT exhibit a bandgap of 1.6 eV, switching from a dark blue to a highly transmissive oxidised state at low potentials – see Figure 5.12. The absorption begins in the NIR and peaks around 600 nm. On oxidation, there is a small amount of residual absorbance in the visible region due to the tailing from the NIR of the polaron/bipolaron transitions resulting in a highly transmissive, albeit light blue, oxidised state.

Efforts at producing organic-soluble/processable neutral states of PEDOT have focused on the introduction of solubilising side chains on the alkylenedioxy bridge [10, 27, 63–66]. Comparing the electrochromic performance of these polymers to the unsubstituted PEDOT shows a shift in the bandgap to higher energies (1.75–1.8 eV depending on the bulkiness of the side chains) resulting in a blue/purple neutral state [63]. However, similar to the ECP-Orange/ECP-Red example discussed in the previous section, the solubilising side chains in the regiorandom PEDOTs can induce significant repulsion between adjacent units, reducing the conjugation length in the neutral state, yielding a purple colour compared to the blue hue for the unsubstituted PEDOT. In addition to the shift in neutral-state colour, the addition of side chains influences the switching speed and contrast of these PEDOT derivatives. For example, the tetradecyl-substituted PEDOT switched between the fully neutralised state and the fully oxidised state in 5 s compared to 10 s for the unsubstituted PEDOT (both in 0.1 M tetrabutylammonium perchlorate (TBAP)/acetonitrile (ACN)) [10, 63, 64]. Also, slightly improved bleached-state transmissivities were observed with highly transmissive faint green to light grey oxidised states, showing a transmittance difference at λ_{max} ranging from 44% to 63%, attributed to the bulkiness of the side chains causing a higher interchain separation. This allows higher levels of doping as more charge balancing ions are able to diffuse into the polymer film.

A study (thiophene-based homopolymers) of the influence of the size of the bridging ring on the electrochromic properties, compared ethylene with propylene (PProDOT) and butylene (PBuDOT) [10, 67]. As noted previously, increasing ring size increases contrast [6, 10]. The propylene bridging unit can allow further increases in contrast as the tetrahedral substitution pattern at the central carbon provides for two substituents to be present orthogonal to the main chain, increasing interchain separation and enhancing optical contrast as well [6, 10]. Hence, the electrochemically polymerised dibenzyl-substituted ProDOT derivative exhibited a high transmittance difference at λ_{max} of 89% with switching times as fast as 0.4 s [68], as seen in Figure 5.13a. Here the X-ray crystal structure of ProDOT-Bz_2 shows the benzyl groups nearly orthogonal to the thiophene ring, and Figure 5.13b shows the change in transmittance (monitored at a single wavelength) when the polymer film is switched between the fully neutral and fully oxidised states, illustrating the rapid switch and high contrast.

Further research on disubstituted ProDOTs also showed high optical contrasts with dimethyl and diethyl derivatives showing 78% and 75% Δ%T, respectively [69, 70]. Increasing the substituent chain length to butyl, producing PProDOT-Bu_2, aids in solubilising the polymer for solution casting. However, for this polymer, the electrochemically synthesised films exhibited a higher transmittance difference at λ_{max} (transmittance difference Δ%T of 75%) than those prepared by solution casting the chemically polymerised polymer (Δ%T of 45%) [71]. Soluble and processable PProDOTs were subsequently developed with the utilisation of sufficiently long side chains (*n*-hexyl, 2-ethylhexyloxy and *n*-octadecyloxy) with the 2-ethylhexyloxy derivative providing a high transmittance difference at λ_{max} of 80% and switching rate of 0.6 s (at 95% of full switching). Similar to the dibenzyl derivative, the branched side chains appear to increase inter-chain separation, opening up the polymer film morphology to allow charge-balancing ions more rapid access through the film [72, 73]. The ProDOT-based polymers exhibit dark blue to purple neutral states (depending on substitution and film preparation method) and highly transmissive, near-colourless oxidised states. Solubility and processing of these polymers are discussed in Section 5.3.

Selenophene-Based Homopolymers A selenium-based analogue of EDOT (EDOS) was reported in 2001 to produce a polymer film with a dark blue neutral state and transmissive oxidised state when prepared electrochemically or chemically [74]. As selenophene is more electron-rich than thiophene, EDOS has a lower bandgap (1.4 eV) and lower oxidation potential than EDOT. In electrochemical studies of PEDOS, conductivities up to 30 S cm^{-1} were found in pressed pellets [75]. The electrochromic properties of PEDOS have shown it to be a promising material with a highly transmissive oxidised state, a high transmittance difference at λ_{max} of 55%, and bleaching and colouring times measured as 0.4 and 0.6 s,

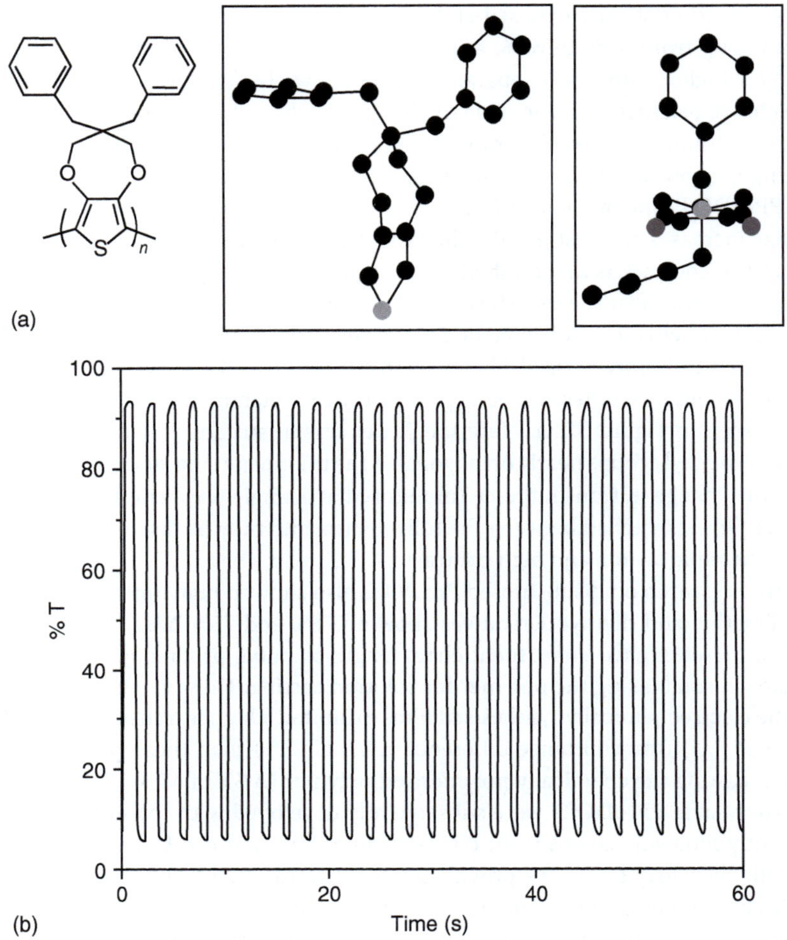

Figure 5.13 (a) Repeat unit structure for PProDOT-Bz_2 (left) and X-ray crystal structure (right) for the ProDOT-Bz_2 monomer showing the benzyl groups orthogonal to, and on the opposite sides of, the thiophene ring. (b) Optical switching of a PProDOT-Bz_2 film on a transparent electrode monitored at 632 nm when switched between fully neutralised (−1.0 V) and fully oxidised (+1.0 V) states (vs Ag/Ag^+). The Δ%T is 89% and a switch speed of 0.4 s to reach 95% of the full contrast. Reproduced from Ref. [68].

respectively [76, 77]. On substitution with a hexyl chain, PEDOS-C6, though not soluble, reached a Δ%T of 89% and a switch speed of 0.7 s (at 95% of the full contrast). The spectroelectrochemical series and photographs for this polymer in Figure 5.14 show that the polymer achieves a highly transmissive bleached state with little absorption in the visible region, which is nearly colourless.

Similar to alkylenedioxythiophene polymers, work towards examining the electrochromic properties of selenophene derivatives increasing the size of alkylene bridge was performed. The derivative PProDOS-$(decyl)_2$ shows a blue neutral state and highly transmissive oxidised state. Unlike the ProDOT

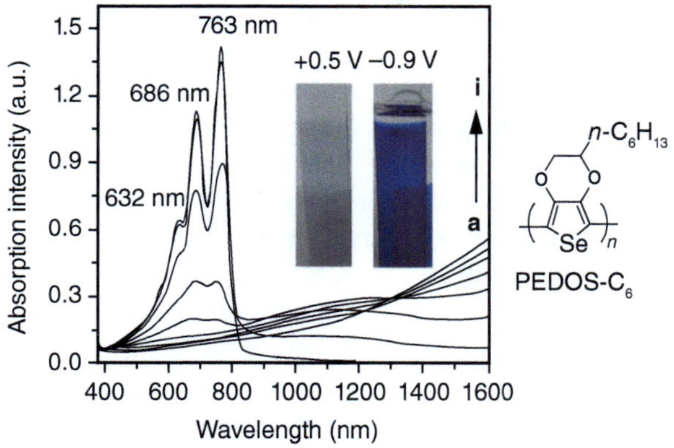

Figure 5.14 Spectroelectrochemistry for the hexyl derivative of PEDOS (PEDOS-C6) when switched between fully neutralised −0.9 V and fully oxidised +0.5 V versus Ag/AgCl. Inset photographs show the fully neutralised (right) and fully oxidised states (left). Reproduced from Ref. [77].

derivatives, the contrast for this polymer is lower than PEDOS-C6, but similar to the unsubstituted parent polymer, PEDOS with a Δ%T of 56% [78].

Donor–Acceptor Polymers As already noted, one of the first known structures using the donor–acceptor approach towards electrochromic polymers was the electropolymerisation of the bis(thienyl)cyanovinylene donor–acceptor–donor precursor [79]. The electron-withdrawing CNV unit was utilised to further lower the bandgap of the bis(thienyl)vinylene polymer. Poly(*trans*-1,2-bis(thienyl)vinylene) is deep purple in the neutral state, switching to transmissive on oxidation while the bis(thienyl)cyanovinylene polymer is blue-to-transmissive switching. This approach was further extended to include EDOT, EDOP and thiophene units as the donors with CNV as the acceptor. In this work, the effect of the electron-rich thiophene-based donor on the resulting polymer bandgap was studied. While the bandgaps varied from 1.1 to 1.6 eV, the polymers all exhibited similar electrochromic properties, switching from a neutral blue state to a light blue transparent oxidised state [80]. Moving on from electropolymerisation of macromonomer precursors, this concept was extended to a family of solution-processable analogues utilising dialkyl ProDOT and dialkoxybenzene as the donor units [81, 82]. These polymers exhibited bandgaps ranging from 1.8 to 2.3 eV, and absorption maxima in the 610–620 nm range, switching from blue or purple neutral state to transmissive light blue grey on oxidation. While the colour property alterations were subtle, the voltages required to cause oxidation increased (0 to +0.5 V vs Fc/Fc^+) compared to that for PEDOT (−0.4 V vs Fc/Fc^+) [10]. This allows for an air-stable neutral state as the polymer valence-band energy levels are deeper than the air oxidation threshold (5.2–5.3 eV).

Expanding on the donor–acceptor approach to achieve improvements over PEDOT (deeper blue neutral state, higher contrast, colourless oxidised state, faster switch speed and solution processability), several researchers have explored the use of fused ring acceptors in combination with EDOT and ProDOT units of varying ratios. One approach utilised an alternating copolymer of dialkyl-ProDOT and BTD, yielding a soluble blue-to-transmissive polymer with a Δ%T at λ_{max} of 52% and sub-second switching speeds [83]. At about the same time, other researchers explored the donor–acceptor–donor approach using either EDOT or dialkyl-ProDOT as the donor and benzotriazole as the acceptor to provide blue-to-transmissive switching polymers. Benzotriazole is a weaker acceptor than BTD, yielding a slight red-shift from PEDOT and a deeper blue neutral state, as can be seen in Figure 5.15. An advantage of the benzotriazole acceptor is the additional site for functionalisation with solubilising groups at the central nitrogen, further improving solubility. The bisEDOT-benzotriazole polymer has a bandgap identical to PEDOT (1.6 eV) with a higher Δ%T at λ_{max} of 53% and faster switch speed of 1.1 s. Following on that work, solution-processable polymers based on bisProDOT-benzotriazole and a bisEDOT-benzotriazole incorporating a dialkoxythiophene unit have been reported [84, 85]. Other donor–acceptor systems explored include those with *meta*-phenyl-substituted quinoxaline units [86] of substituted benzimidazoles [87] as the acceptor with thiophene-based donor polymers produced with blue to blue/purple

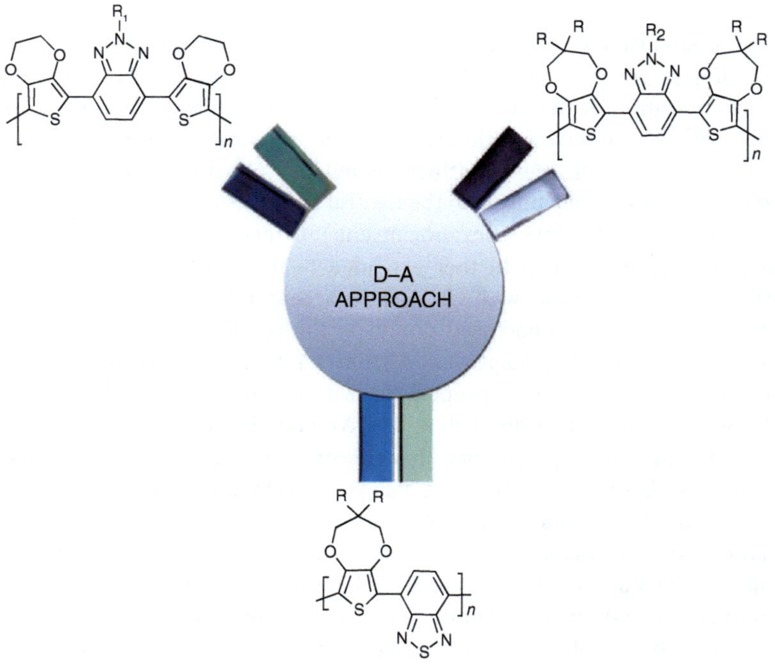

Figure 5.15 Photographs of donor–acceptor and donor–acceptor–donor polymer films in their neutral (left) and oxidised (right) states along with repeat unit structures. Polymers include EDOT-benzotriazole-EDOT (top left), ProDOT-benzotriazole-ProDOT (top right) and ProDOT-benzothiadiazole (bottom). Reproduced from Ref. [6].

neutral states and highly transmissive oxidised states with contrasts comparable to PEDOT.

Fused Ring Systems As already noted, the use of fused ring systems also allows a lower polymer bandgap. This has allowed shifting the colour of soluble thiophene homopolymers (e.g. poly(3-hexylthiophene)), which are red in the neutral state, to lower energies yielding blue electrochromic polymers. One such example is the cyclopentadithiophene (CPDT) polymer, poly(4,4-dioctyl-cyclopenta-dithiophene), which is a soluble polymer switching from blue to highly transmissive with a 60% Δ%T at λ_{max} [88]. The fused CPDT ring has a lower bandgap than bithiophene as the effective conjugation length is extended due to the ring fusion that planarises the conjugated backbone. The substitution of the alkyl chains at the quaternary carbon bridge also allows for higher doping levels, and hence higher contrast, because interchain interactions are reduced as in the disubstituted ProDOTs mentioned previously [89].

For blue-to-transmissive switching polymers comprising fused rings, homopolymers of thieno[3,4-b]thiophene [36, 37, 90], thieno[3,4-b]furan [38, 39] and thieno-[3,2-b]thiophene were employed, all switching from various hues of blue to highly transmissive oxidised states [91].

5.2.1.4 Cyan/Green

A green/cyan-to-transmissive switching electrochromic polymer has been a valuable recent target, as combinations of green with red and blue and combinations of cyan with magenta and yellow would complete both the additive and subtractive primaries, respectively. A green neutral-state polyheterocycle proved difficult to synthesize as the polymer needs to absorb two wavelengths simultaneously – one in the red and the other in the blue spectrum region, with light transmission around 550 nm. Early work achieved a neutral-state green polymer through use of an electropolymerised donor–acceptor macromonomer containing two thiophene units flanking a dithienothienylpyrazine (DTP). It had a distinct green neutral state, fast switching times and a good cycling stability. However, it showed a low Δ%T at λ_{max} (12–23%) due to the yellow-brown oxidised state. This is attributed to the appended thiophene units that contributed to the high-energy absorption, as they are not in conjugation with the main chain, leaving an incomplete bleaching of these units and a persistent hue in the oxidised state [92]. Prior to this, it was demonstrated that poly(bis-EDOT pyridopyrazines) exhibited strong green neutral colours, but could not bleach upon oxidation, so were discounted for practical electrochromism [93, 94]. Varying combinations of donor and acceptor moieties were thus explored to provide a neutral-state green, but the two absorption bands would need to be depleted simultaneously to yield a highly transmissive oxidised state.

The first reported green-to-transmissive switching electrochromic polymer was a previously reported electrochemically polymerised bisEDOT-BTD polymer [95–98]. It was found to switch rapidly from a saturated green neutral state to a transmissive light blue oxidised state (see Figure 5.16). Extensive efforts followed

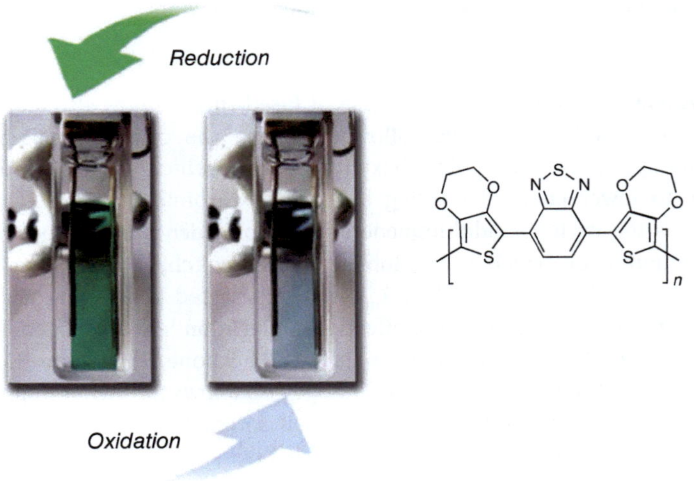

Figure 5.16 Photographs of bis(EDOT)-BTD polymer films on ITO/glass in the neutral (left) and oxidised (right) states. Repeat unit structure on the right. Reproduced from Refs. [6] and [99].

in establishing the role that the donor and acceptor units and relative ratios had on the placement of these two absorption bands [83, 84, 92, 100–102]. It was found was that the two absorption bands could be tuned to yield polymers with hues that ranged from green to cyan to blue (as discussed earlier). As shown in Figure 5.17a, increasing the ratio of donor (ProDOT) to acceptor (BTD) heterocycles shifts the short wavelength peak to lower energies [101]. The colour of the neutral polymer can then be tuned from blue (a 1 : 1 ratio of donor:acceptor) through to various shades of green based on the position and intensity of the high energy peak. In addition to varying the ratio of donor to BTD, the nature of the donor (unsubstituted thiophene, acyclic thiophene or ProDOT), and introduction of vinylene spacers, was explored with a large family of green-to-transmissive solution-processable polymers prepared as shown in Figure 5.17b [100, 103]. These polymers exhibited transmittance contrasts (measured as Δ%T) up to 53%, measured at the long-wavelength absorption maxima and maintained the high contrast even at switch speeds as fast as 1 s.

Thus, it became evident that a strong donor–acceptor interaction can constrain the two absorption peaks in the visible region of the spectra while simultaneously keeping an absorption minimum in the green light region (~550 nm). Continued efforts in exploring various donors and acceptors included substituting the sulfur in BTD with selenium in 2,1,3-benzoselenadiazole (BSD) [84, 104]. The resulting polymers also exhibited varied hues of green in the neutral state, depending on the donor strength. The bisProDOT-BSD analogue exhibited a red-shifted absorption, resulting in a saturated green neutral colour, when compared to the turquoise bisProDOT-BTD. The soluble polymer demonstrated similar contrast and switching speed values on the order of 42% (again, measured at the longer wavelength peak) and sub-second switching.

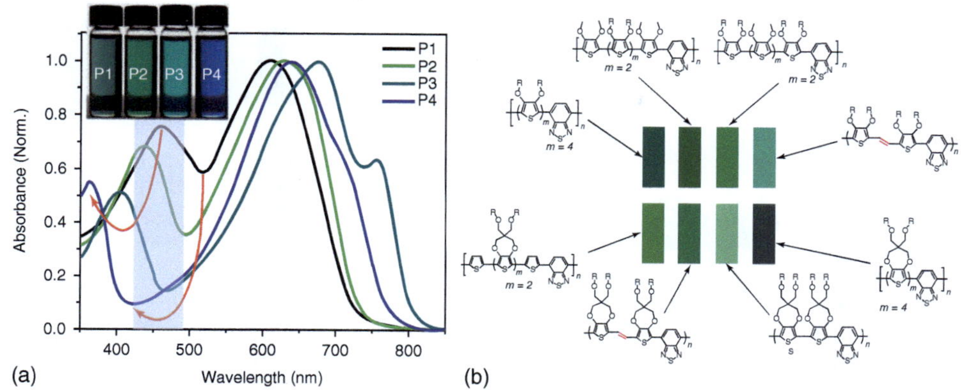

Figure 5.17 (a) Solution spectra for a series of soluble D–A polymers based on ProDOT donors and BTD acceptors with the ratio of D–A varied. (P1 = 6 : 1, P2 = 4 : 1, P3 = 2 : 1 and P4 = 1 : 1). Inset photograph of polymer solutions. (b) Photographs of polymer films in their neutral state. Polymer repeat unit structures provided. Reproduced from Refs. [83] and [103].

Another donor–acceptor derivative yielding polymers with saturated green states of varying hues can be seen in Figure 5.18. This family of bisEDOT-quinoxaline polymers differed not on the donor–acceptor strength previously demonstrated, but on the nature of the functionality of the quinoxaline acceptor [99]. The electropolymerised macromonomer containing thiophene pendant groups exhibited a saturated green neutral state, but a coloured oxidised state. This is attributed to the shorter conjugation length thiophene-based chromophore appended on each acceptor, as with DTP (above). On substitution with phenyl-based units on the quinoxaline acceptor, the resulting polymers exhibited more transmissive oxidised states, as shown in Figure 5.18 [102, 105]. The bulkier tert-butyl phenyl polymer exhibited higher contrasts on switching (43% at 711 nm) than the phenyl-substituted polymer (29% at 732 nm). As thought earlier, diffusion of counter ions within the polymer film could be facilitated by the bulkier pendant group, allowing for a higher level of oxidation, and depletion of residual colour, in the doped state. Addition of two alkoxy solubilising chains at the pendant phenyl allowed for chemical polymerisation of bisEDOT-quinoxaline, affording spray-processable polymer films with similar contrasts and colours in both coloured and bleached states.

Additional attempts towards green-to-transmissive switching polymers tested a variety of donor and acceptor units, like CPDT donors combined with BTD acceptors [88], thiophene donors and thienopyrazine acceptors [106], EDOT donors and ferrocene-functionalised quinoxaline acceptors, [107] and pyrrole donors with quinoxaline acceptors [108], to name a few.

5.2.1.5 **Black**

Similar extensive effort has gone towards achieving black electrochromic polymers, with limited success. A saturated black-to-colourless polymer would find

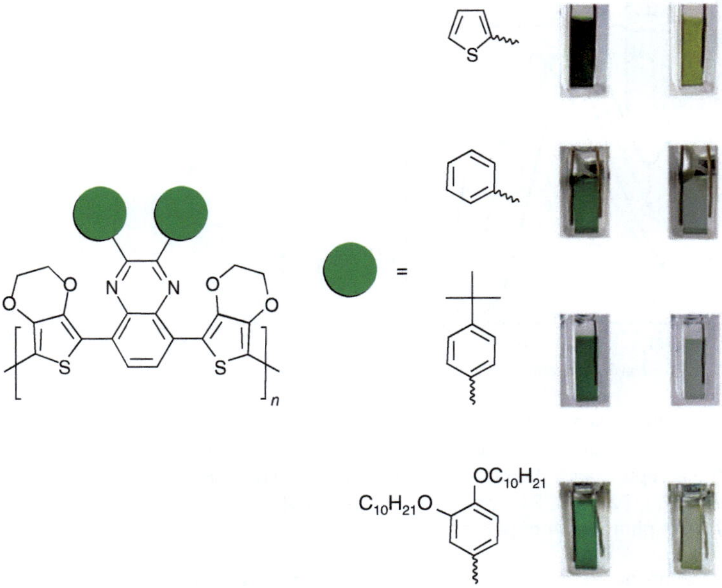

Figure 5.18 Structures of bisEDOT-quinoxaline-based polymers and photographs of polymer films in their neutral (left) and oxidised (right) states. Reproduced from Ref.[6].

many applications ranging from smart windows to information displays and e-readers. However, the difficulty lies in developing a single material that can absorb evenly across the entire visible spectrum, then switch to colourless with film thicknesses that can provide a high level of saturation when coloured, but none when bleached. An ideal case would not exhibit intermediate colours, but achromatic degrees of grey at transitional voltages [2, 44]. Blending or colour mixing has been tried to achieve such a film or device [109–111], making use of a combination of chromophores. It was the successes in developing the green-to-transmissive electrochromic polymers that laid the groundwork for black electrochromic polymers [101] as demonstrated in Figure 5.19a, where the solution spectra for three green electrochromic polymers (based on copolymers of BTD with varying amounts of ProDOT) are overlaid with that of the magenta polymer, PProDOT-$(CH_2OEtHx)_2$ (Section 5.2.1.3). The green/cyan polymers provide the low- and high-energy spectral absorbances while the magenta polymer fills the trough in absorbance centred at 550 nm. In addition, as each of these polymers switches to highly transmissive oxidised states individually, their combination would ideally achieve the same. The first example of a black-to-transmissive switching electrochromic polymer utilised this concept via copolymerisation of a bis(ProDOT)-BTD trimer with the ProDOT monomer (labelled as ECP-black 1) as shown in Figure 5.19b [101]. This first example of a black-to-transmissive switching electrochromic polymer was also solution processable from common organic solvents (toluene, THF, etc.). The polymer was spray-cast onto indium–tin oxide (ITO)/glass electrodes and exhibited a

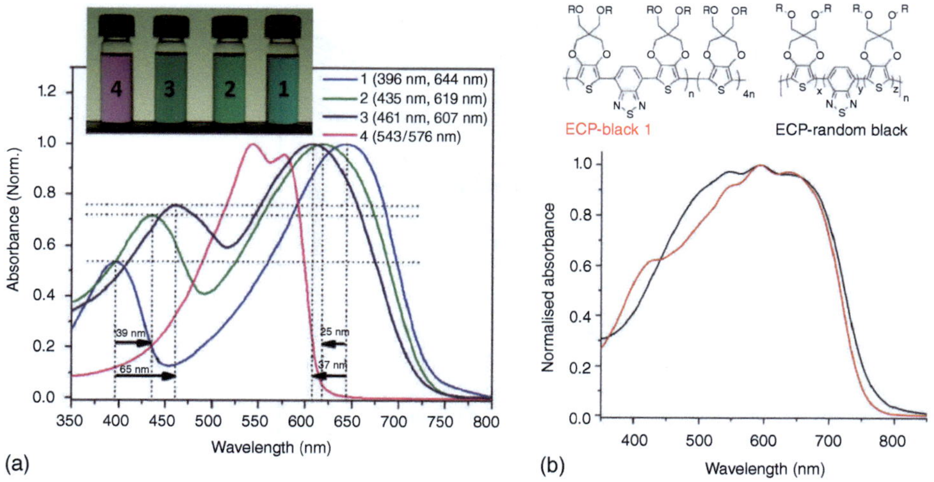

Figure 5.19 (a) Solution spectra and photographs of D–A–D polymers (1–3) based on ProDOT and BTD in varying ratios and PProDOT-(CH_2OEtHx)$_2$ (4). (b) Film spectra in the neutral state for ECP-black 1 and ECP-Random black with polymer structures. Adapted from Refs. [101] and [112].

transmittance change (Δ%T) of 50% at 592 nm with a relative luminance (Δ%Y) change up to 52%, on oxidation.

Further improvement by utilising a polymerisation route of single-ring heterocycles allowed for a higher degree of 'randomness' in incorporating the donor and acceptor units [112]. This random incorporation of ProDOT and BTD units gave a broadening of the absorbance, especially on the high-energy end of the spectrum (Figure 5.19b), and a more straightforward synthesis, as the preparation of the ProDOT-BTD-ProDOT trimer is avoided. The spray-processed polymer films (Figure 5.20) exhibit broad absorption across the visible region, with drop-off at the extremes, switching to a highly transmissive oxidised state; insets show photographs of a polymer film on ITO.

Similar concepts of copolymerisation of blue/green polymer chromophores with those that would broaden the visible region absorption had varying degrees of success: see three of these examples in Figure 5.21. One random copolymerisation of a quinoxaline derivative, a benzotriazole derivative and thiophene to yield a polymer (top left) exhibits a black neutral state, with the idea that D–A copolymers of quinoxaline and thiophene are green in their neutral state and benzotriazole copolymerised with thiophenes is red [114]. The resulting polymer films achieved a 23% change in transmittance at 600 nm with a transmissive, yet grey oxidised state. Another black neutral-state polymer is, again, a random copolymer of ProDOT, BSD, and benzotriazole (bottom left) to yield a polymer with a relatively colour neutral absorbing state ($L^* = 14.3$, and a^*, b^* values less than 1), switching to transmissive grey on oxidation with a 15% change in transmittance at 522 nm [115]. The final example shown here

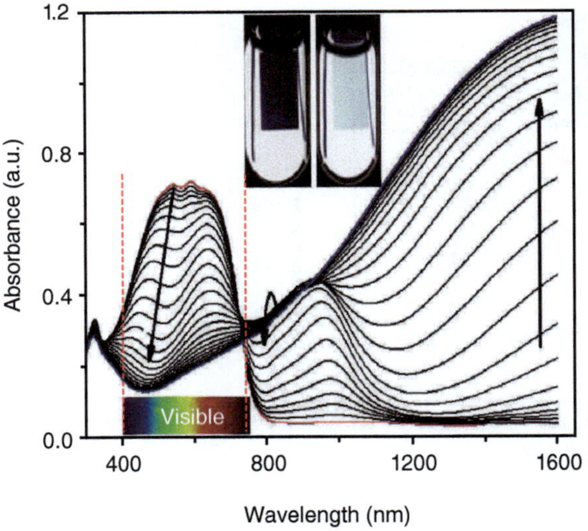

Figure 5.20 Spectroelectrochemistry of ECP-Random Black and inset photographs showing neutral (left) and oxidised (right) states. Reproduced Ref. [112].

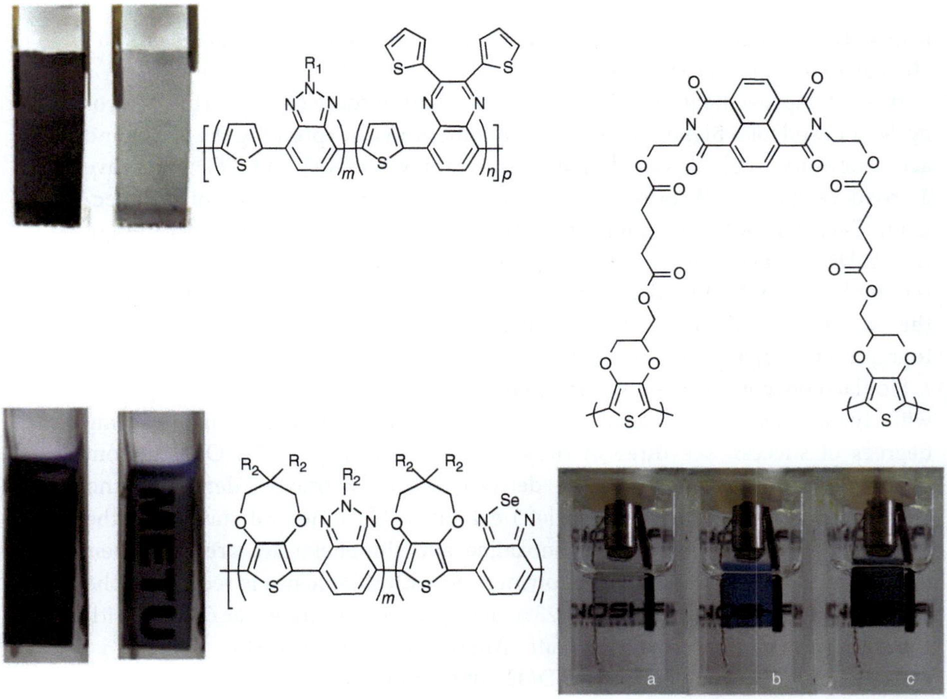

Figure 5.21 Photographs of polymer films on ITO in their neutral and oxidised states with polymer structures. Adapted from Refs. [6] and [113].

uses two EDOT monomers bridged by a naphthalenediimide unit (right) to achieve an electropolymerised film switching between highly transmissive and dark blue/black states (with an intermediate blue) and a change in transmittance at 600 nm of approximately 70% [113]. Other efforts have revolved around colour combinations, and/or appending small molecule chromophores on a conjugated polymer backbone, plus the multilayer and polymer blending efforts (as mentioned earlier), with varying successes [116, 117].

5.2.2
Anodically Colouring

Anodically colouring systems show several advantages, while showing several challenges. As these materials are high bandgap (typically >3.0 eV), their neutral-state absorption occurs in the UV, and they appear nearly colourless. However, a high HOMO level is needed to maintain a low oxidation potential, and control over the colour attained in the oxidised state is difficult to achieve with multiple colours and poor saturation often the result. But several examples show effective switching from colourless to a variety of coloured states, and there are those that exclusively switch outside the visible region (UV to NIR electrochromism).

Unusually, polycarbazoles and polyfluorenes have been utilised as electrochromes with specific examples showing anodic colouring, like the *N*-substituted polycarbazoles, such as the *N*-butyl- and *N*-dodecyl-substituted carbazoles that are organic soluble and the *N*-oligoether and *N*-alkylsulfonate examples that are water soluble [118–121]. The unsubstituted polycarbazole exhibits a yellow transmissive neutral state, switching to green on oxidation [122]. The *N*-alkyl substitution results in a blue-shift with those polymers exhibiting colourless neutral states, green partially oxidised and blue fully oxidised states, while the water-soluble analogues switch from colourless to dark green. Similar to polycarbazoles, polyfluorenes are not commonly used as electrochromes but are employed as blue emitters in polymer light-emitting devices and as hole-transport materials [123]. Limited examples of anodically colouring analogue of both polymers exist. One such example is poly(9-fluorenecarboxylic acid) prepared via electropolymerisation under highly specific conditions to achieve a clear-to-brown-orange switching polymer film [124]. Another is a dispirofluorene-indenofluorene polymer prepared via electropolymerisation to achieve a clear-to-purple switch on oxidation [125].

A class of anodically colouring polymers includes the large set based on triarylamines [126]. The electron-rich triarylamine can be easily oxidised at the nitrogen centre, and when the *para*-position on the phenyl is blocked, a stable radical cation forms with a distinct colour change. The bulky propeller shape of the triphenylamine moiety allows for a polymer with thermal stability, amorphous character, and solubility in aprotic solvents with the ability to form films on casting from solution without the need to rely on added functional groups such as long alkyl chains. Triphenylamine polymers are also unique compared to the fully conjugated systems discussed so far. They are considered as mixed-valence materials with one or more redox centres to allow multiple redox stages. This affords several colours on oxidation, as in Figure 5.22 where triphenylamine-based polymers with increasing complexity, and triarylamine centres, are shown along with photographs of the polymer films in their colourless neutral states and coloured oxidised states. More triphenylamine centres increase the number of coloured states. The examples shown here are polyamides with triphenylamine redox centres, all blocked at the *para*-position to curb coupling during oxidation. Of several aryl linkages explored, those based on a flexible ether gave the best anodic colouring electrochromic properties, as shown here. The polymer containing one triphenylamine centre switches from colourless to green on oxidation to the radical cation. The polymer containing two centres has an added blue state on oxidation of the second redox centre. The polymer containing three centres switches from colourless to yellow, followed by dark purple and dark blue, and the polymer containing four redox centres has a colourless to yellow, to purple, to light blue, to dark blue switch [126–128]. Another interesting example incorporates the triarylamine, spirobifluorene and carbazole moieties in one polymer as shown in Figure 5.23. Electropolymerisation occurs first through the triarylamine functionalities at low oxidation potentials with further cross-linking occurring through the carbazole functionality at higher oxidation potentials to afford thin polymer films that switch from transparent through lime green to blue-green when fully oxidised [129].

As mentioned previously, not only is a high bandgap (to attain a transmissive neutral state) desired, but also a high HOMO to allow for stable and repeated oxidative cycling, for example, those based on *N*-substituted ProDOP [11]. Figure 5.24 shows the geometric structure of an *N*-H ProDOP dimer where the torsion angle between pyrrole rings is 0°. The resulting *N*-H ProDOP polymer has a bandgap of 2.2 eV, yielding the aforementioned orange neutral state [5]. However, substitution at the nitrogen, as shown in the bottom image in Figure 5.24, increases steric hindrance, which decreases planarity and decreases the effective conjugation length. The example provided here, a ProDOP dimer with an *N*-oligoether substituent, has a torsion angle of over 70°, resulting in a decrease in π-orbital overlap and a lowering of conjugation in the resulting polymer. This ultimately results in a shifting of the λ_{max} into the UV as the polymer has a bandgap of 3.4 eV and a colourless neutral state. This route has successfully been utilised to achieve a number of anodically colouring polymers, with a variety of *N*-substituents [14, 130–132]. These include *N*-methyl, *N*-propyl, *N*-octyl, *N*-glycol, and *N*-propane sulfonate PProDOPs with bandgaps that range from 3.0 eV for the short chain alkyl and propane sulfonate-derivatised polymers to as high as 3.4 eV for the aforementioned *N*-glycol and those containing longer alkyl chains. While the bandgap transition occurs in the UV region, the relatively broad polaron absorption occurs in the visible and the bipolaron occurs in the NIR with some tailing into the visible. This results in polymers that have oxidised states with low colour saturation. Many of the *N*-substituted PProDOPs have pale blue or grey blue oxidised states that, while not vibrant or saturated as desired for window and display applications, are complementary to many of the saturated

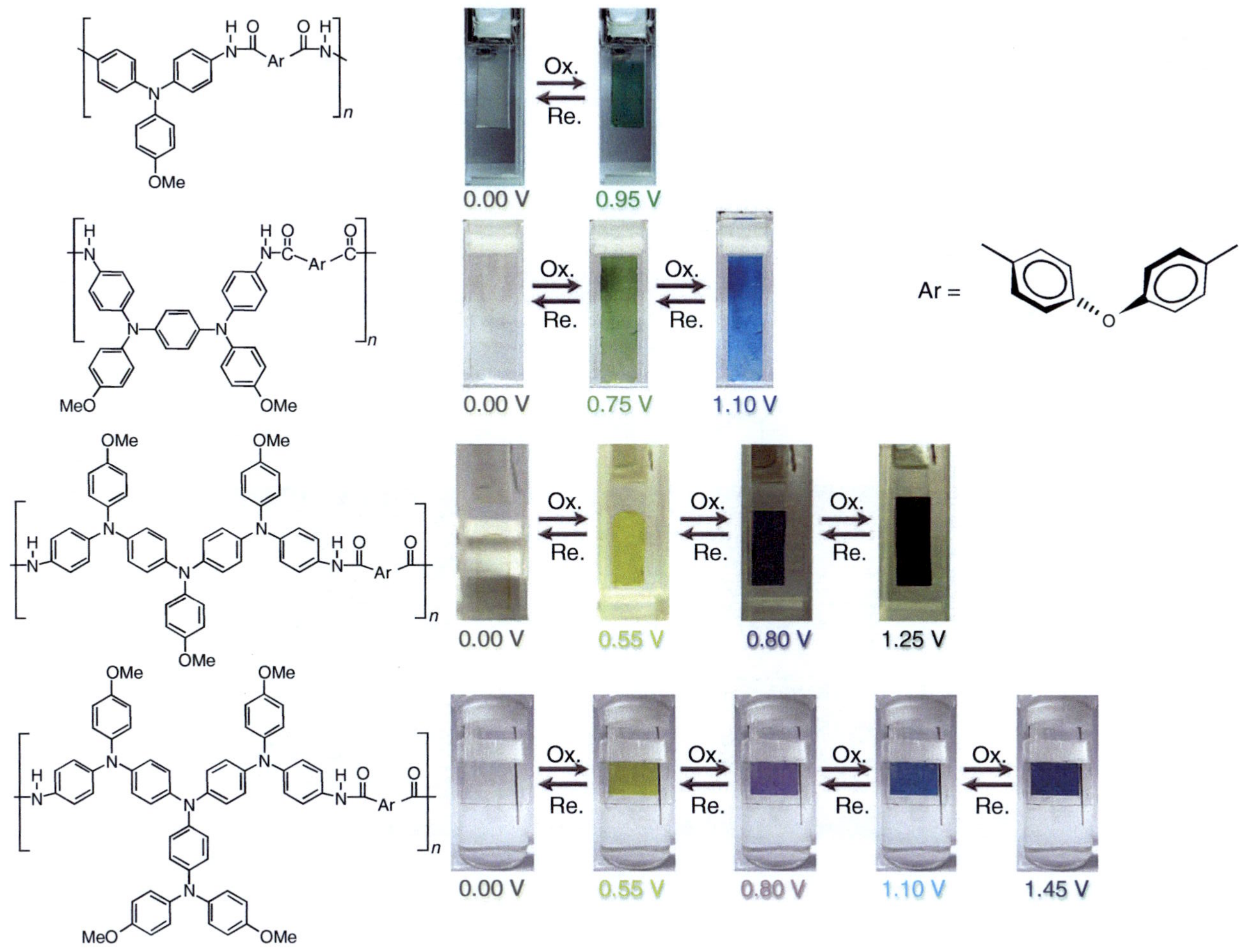

Figure 5.22 A series of triphenylamine-containing polyamide polymers along with photographs of films on transparent electrodes in the neutral state (far left) and states at increasing oxidising potentials with voltages (vs Ag/AgCl) indicated. Reproduced from Ref. [126].

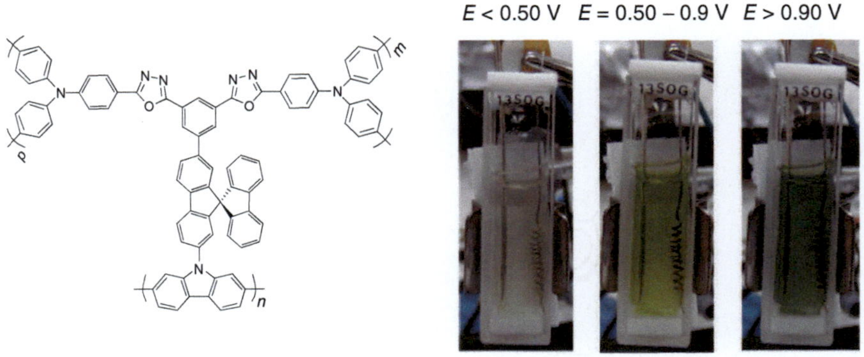

Figure 5.23 Structure of a starburst polymer with tris(oxadiazole)-phenylene core, bridged by a spirofluorene unit with terminal triphenylamine and carbazole units through which electropolymerisation occurs. Photographs (right) of the polymer film in the neutral (far left) and oxidised states (two right) with voltages indicated versus Fc/Fc^{+}. Reproduced from Ref. [129].

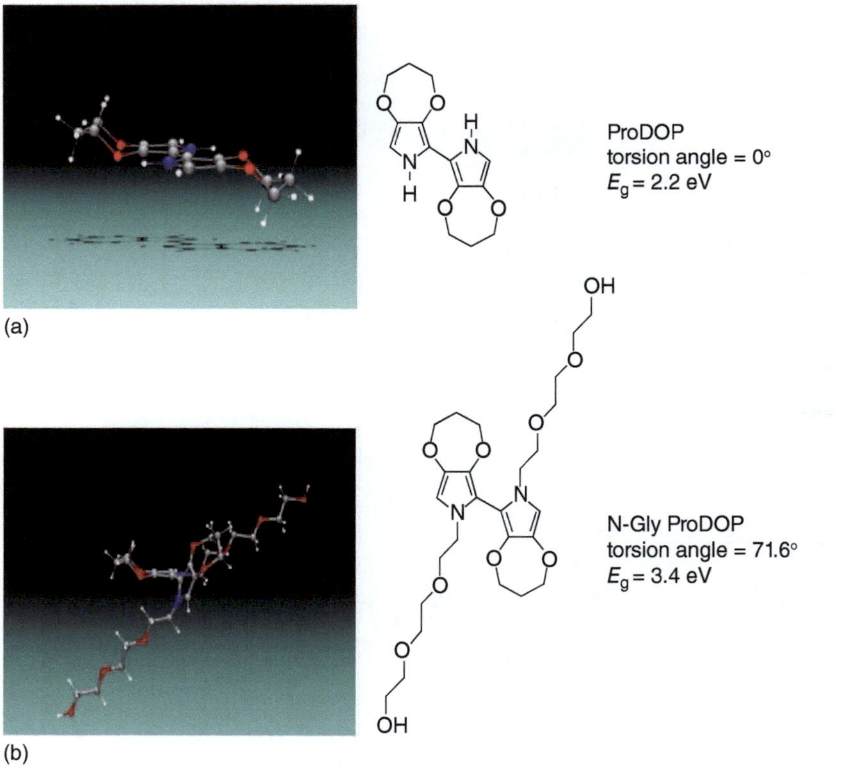

Figure 5.24 Geometric representation for N–H ProDOP dimer (a) and N-Gly ProDOP dimer (b). The value of the angle θ indicates the degree of twisting between the two pyrrole rings. Reproduced with permission from Ref. [129].

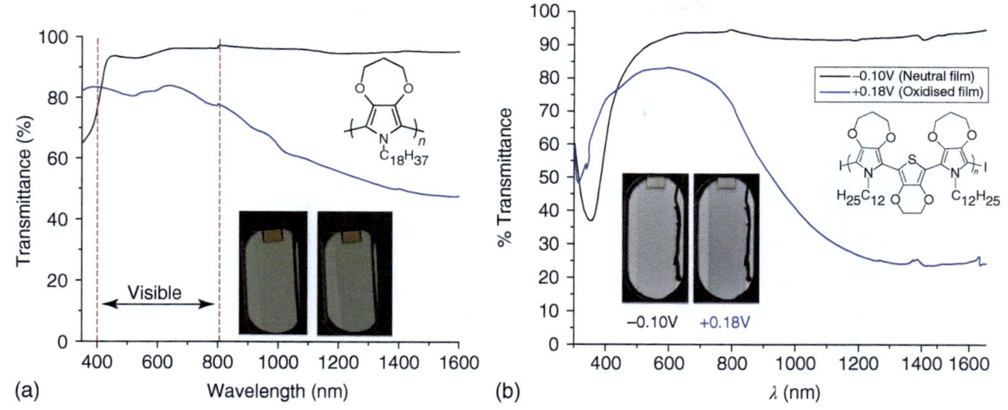

Figure 5.25 Spectroelectrochemistry for the neutral (black line) and oxidised (blue line) states of PProDOP-N-C18 (a) and P(ProDOP-N-C12-EDOT-ProDOP-N-C12) (b) copolymer. Inset photographs show the neutral (left) and oxidised (right) states. Reproduced from Refs. [134] and [135].

cathodically colouring polymers previously mentioned and are utilised as the counter electrode material in dual polymer devices [133].

In some examples, the polaron absorption in the visible can be rather weak, with barely noticeable colour, and the bipolaron absorption shifted entirely into the NIR such that the polymer exhibits a minimal colour in both the neutral and oxidised states. This is the case for the *N*-octadecyl-substituted PProDOP shown in Figure 5.25a, also called minimally colour changing polymer or MCCP, and is solution-processable from organic solvents due to the long alkyl chain [134]. Another polymer that exhibits a colourless to colourless switch is the bis(*N*-dodecyl ProDOP)-EDOT copolymer shown in Figure 5.25b [135]. For both polymers, it is the high torsional strain that results in a neutral-state absorption in the UV region of the spectrum. It is thought that on oxidation, the strain relaxes and the polymer planarises, shifting the bipolaron absorption into the NIR. As with the previous anodically colouring *N*-substituted PProDOP examples, these polymers have found utilisation as transparent counter electrode materials in dual polymer electrochromic devices (ECDs) with higher contrasts and the ability to fabricate devices without loss of colour purity [134, 136, 137].

5.2.3 Inducing Multicoloured States in ECPs

It is evident from the results and photographs shown in Figures 5.5 and 5.8 that conjugated oligomers and polymers can attain a full palette of colours with essentially any hue accessible in the neutral state by using specific repeat unit structures and by blending separate polymers for colour mixing. The anodically and cathodically colouring polymer systems illustrated demonstrate how appropriately designed polymers can be made to switch from the vibrant coloured

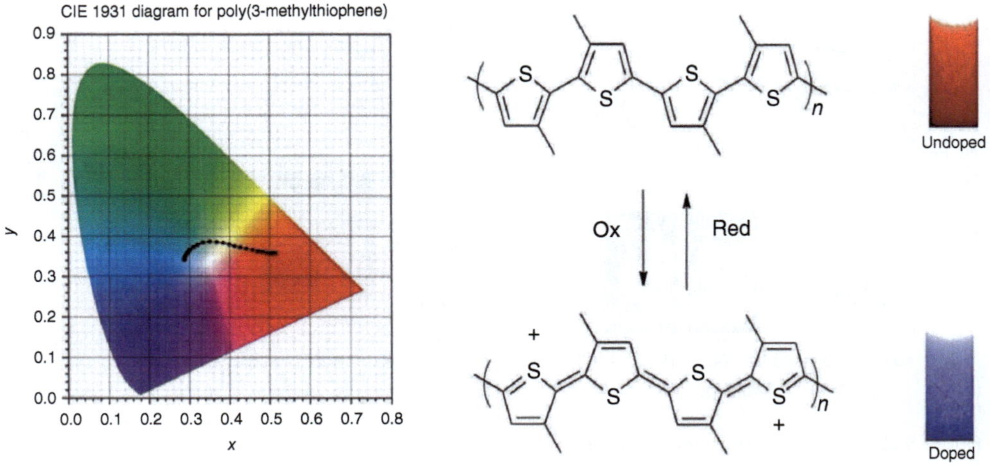

Figure 5.26 CIE 1931 colour track illustrating the change in hue along with neutral/doped state structures and colours, for poly(3-methylthiophene) during oxidation.

state to a highly transmissive form. This is because the optical absorption of the neutral polymer located in the visible region of the spectrum is transferred sufficiently far into the NIR to be outside of the wavelength limit of human eye sensitivity. In many instances though, the low-energy carrier transitions are not moved into the NIR but appear instead in the visible spectrum. With this, the polymer can express two, or more, distinct colours in the extreme neutral and various doped states, as illustrated in Figure 5.26 for poly(3-methyl thiophene) (P3MT) [138]. This is in contrast to the continuous loss in colour saturation of the anodically and cathodically colouring materials and is best termed multicoloured electrochromism. Illustrated by the CIE 1931 colour track in Figure 5.26, P3MT expresses a vibrant red-orange colour in its neutral form. As it is oxidised, and the intensity of the interband transition is depleted, an intermediate polaron state is induced in the visible, ultimately transitioning to a long-wavelength absorption that tails into the visible, absorbing a strong portion of the red spectrum causing the polymer to appear blue to the human eye. This red-to-blue electrochromism is common for many intermediate-gap conjugated polymers and is, in fact, more the norm for polymers not specifically designed to be especially anodically or cathodically colouring. The charge carrier bands also tend to be especially broad in character relative to the neutral π-to-π^* transition leading to colours having a low saturation and a pastel-like appearance. In early work examining the colorimetric properties of a series of electrochromic polymers, Thompson *et al.* [52] used XY colour coordinates measured with a colorimeter and Photoshop™ to create colour swatches that illustrated the electrochromic switching properties of a broad family of polymers. This is shown in Figure 5.27 for the electrochemically prepared PBEDOT-*N*MeCz and PBEDOT-BP, along with the soluble polymer linear poly(3,4-ethylenedioxythiophene)-2,5-didodecyloxybenzene

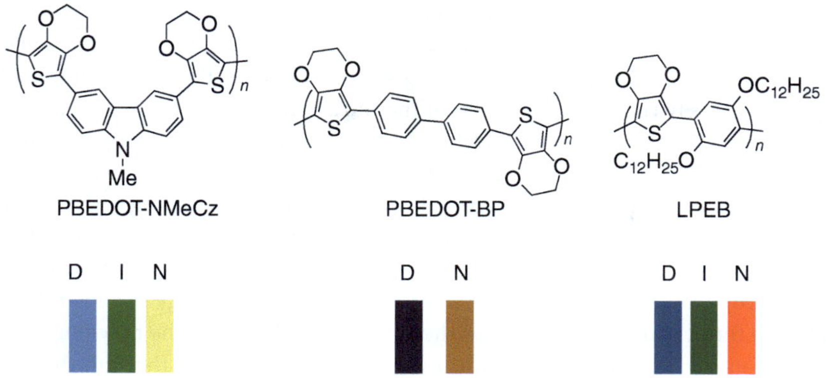

Figure 5.27 Repeat unit structures for the ECPs PBEDOT-NMeCz, PBEDOT-BP and LPEB showing colorimetrically determined colour swatches for neutral (N), intermediate (I) and fully doped (D) forms.

(LPEB). Examining the repeat unit structures, it can be seen that PBEDOT-BP is a fully conjugated polymer while PBEDOT-*N*MeCz has broken conjugation due to the bonding pattern at the 3,6-positions of the carbazole unit. The EC properties of the PBEDOT-BP are similar in concept to those of P3MT, transitioning from a neutral orange [30] form to a distinctly purple-doped state. PBEDOT-*N*MeCz, on the other hand, shows three distinct colour states due to the presence of the neutral form, along with discrete cation-radical and dication forms causing the polymer to attain yellow, green, and blue forms that can be stably accessed. Intermediate colours can also be found in a number of other fully conjugated polymers [128, 139–141] as evident by the colours accessed by LPEB. Here, an orange-red form similar to P3MT and PBEDOT-BP is seen for the neutral material. Oxidation leads to depletion of the interband transition, while a polaron state, having a relatively discrete peak, at 708 nm grows in. At an intermediate oxidising potential, the remaining interband absorption at 502 nm, coupled with this polaron absorption, leads to a transmission window at 570 nm giving the film a distinct green appearance. Further oxidation leads to complete bleaching of the inter-band transition, growth of long-wavelength transitions in the NIR that tail into the visible yielding the final blue oxidised state as seen for many of these other multicoloured electrochromic polymers.

In a fundamental sense, it is difficult to finely control the colour states in these multicoloured ECPs *a priori* as design of a material where one can predict the spectra for the neutral and the doped polaron and bipolaron states is complicated. In essence, multicoloured ECPs tend to give colour states that are unique for each polymer system, but it is difficult to simultaneously tune both the neutral and doped states in the structures to attain desired colours. At the same time, there are some logical structural handles that the chemist can work with in designing these material types as illustrated in the next five sections.

5.2.3.1 **Polyaniline: A Model ECP with Multiple Redox States**

The *a priori* establishment of a set of accessible redox states along a polymer backbone is a straightforward way of creating a multicolour ECP. This is exemplified by what is likely the oldest, and one of the most well-known, of the conducting and electrochromic polymers; PANI [5, 142, 143]. PANI has been studied extensively as an ECP due to its ease of synthesis via both chemical and electrochemical polymerisation routes, yielding highly electroactive films. Beginning with the leucoemeraldine form, which is in essence poly(phenyl amine), a yellow-toned film can be formed at low potential between −0.2 and 0.0 V, as shown in Figure 5.28 [144–146]. Partial oxidation to the emeraldine form, particularly the emeraldine salt (ES) form, which can be seen as the cation radical form of the polymer, leads to a green colour of the polymer near 0.4 V. Further oxidation through the second redox process between 0.6 and 0.9 V leads to the blue and ultimately violet pernigraniline form of the polymer.

PANI can be derivatised at both the nitrogen atom via *N*-alkylation [147, 148] and the phenylene ring [149, 150] of the main chain where the derivatives provide similar yellow to green to blue electrochromism as seen for the parent polymer. Self-doped PANIs [151] and template-synthesised PANI with polyelectrolyte dopants such as poly(2-acrylamido-2-methyl-1-propanesulfonic acid) (PAMPSA) [152] provide distinctly different physical and electrochemical properties, including water solubility and aqueous electrolyte compatibility, but only minor effects on the electrochromism.

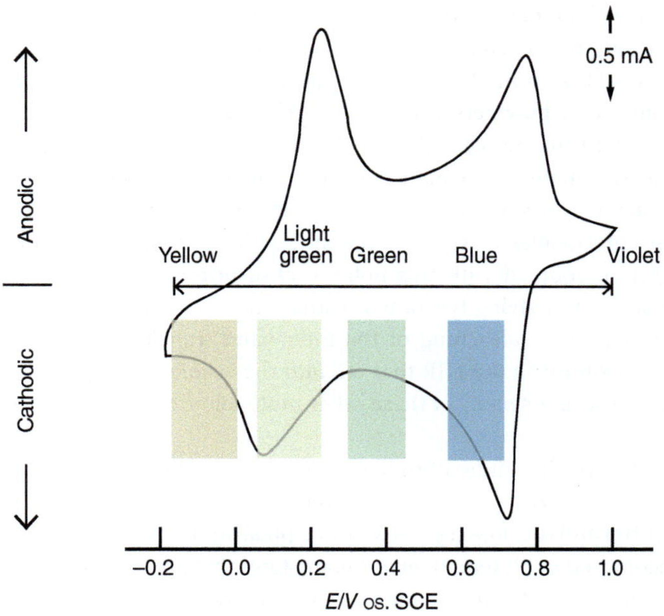

Figure 5.28 Cyclic voltammetry and accessible colour states for polyaniline cycled in 1 M aqueous HCl demonstrating multi-colour nature with various redox states. Adapted from Ref. [144].

5.2.3.2 Colour Control via Copolymerisation

Copolymerisation of multiple monomers into a single polymer chain is a well-known, and often exploited, method for controlling the mechanical and physical properties of macromolecular systems. Similar concepts have been utilised in conjugated and electrochromic polymers, as illustrated by conversion of the homopolymer ECP-orange to the random copolymer ECP-red discussed earlier [7]. As one example for developing a family of multicolour ECPs using copolymerisation, BEDOT-*N*MeCz was reacted with the dimer of EDOT (BiEDOT) under oxidative polymerisation conditions [153]. Given the close match of the oxidation potentials of the two monomers, it is expected that the copolymer composition will approximately mirror the feed composition. This provides a copolymer having the nominal repeat unit structure illustrated in Figure 5.29 where the conjugation breaking 3,6-linked carbazoles are separated by varying lengths of EDOT moieties as a function of their composition in the polymerisation feed. As shown by the photograph, copolymers with neutral purple, red, and orange colours were accessible as the spectral properties ranged from the deep blue 100% BiEDOT polymer to the yellow 100% BEDOT-*N*MeCz polymer. It should be noted that each of these copolymers then expresses a set of colours as they can be converted into the intermediate cation-radical and subsequently dication forms as was illustrated for the BEDOT-*N*MeCz homopolymer in Figure 5.27. As such, with the compositionally

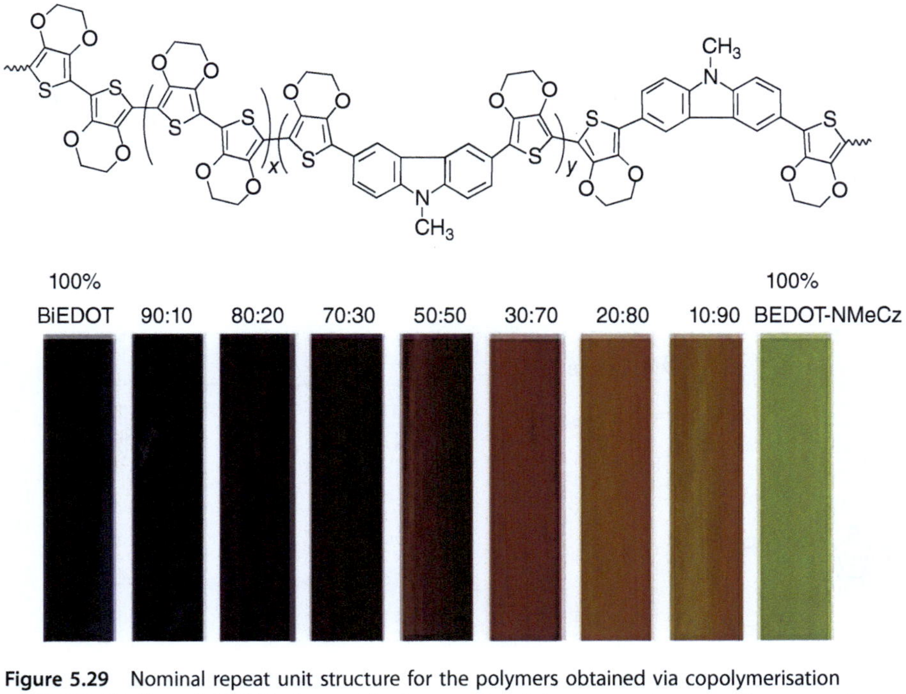

Figure 5.29 Nominal repeat unit structure for the polymers obtained via copolymerisation of BEDOT-NMeCz and BiEDOT, along with photographs of the homopolymers and a range of copolymers in their charge neutral forms. Adapted from Ref. [153].

continuous tuning of the colour of the neutral form, a broad colour palette was accessible.

5.2.3.3 Appended Electrochromes on ECPs

Another route to multicolour electrochromic polymers involves systems where the conjugated backbone provides one set of spectral changes and covalently attached chromophores provide a second response. As the potential at which these redox processes occur is controlled by the nature of the separate electrochromes, colour control can be quite specific.

An example of covalent attachment of molecular electrochromes to ECPs is provided in the work of Toppare and co-authors [154], who appended azobenzene (Az), coumarine (Co) and fluorescein (Flo) derivatives onto electropolymerisable terthiophene (PTT) scaffolds. As shown in Figure 5.30, distinct colour differences are observed as a function of the appended electrochrome. PTTCo shows the extreme states expected in a polythiophene, again the red-orange ($a^* = 43, b^* = 48$) to blue ($a^* = -12$, $b^* = -19$) transition, with subtle earth-tone brown and grey intermediate states. PTTFlo, on the other hand, switches from a yellow-orange ($a^* = 7$, $b^* = 46$) neutral form to a pale yellow-green ($a^* = -22$, $b^* = 42$) oxidised form with a more distinct yellow ($a^* = -4$, $b^* = 60$) intermediate. Examination of the spectroelectrochemical series for these polymers shows a unique narrowing of the π-to-π^* transition in PTTFlo with oxidation, lowering the absorption of light

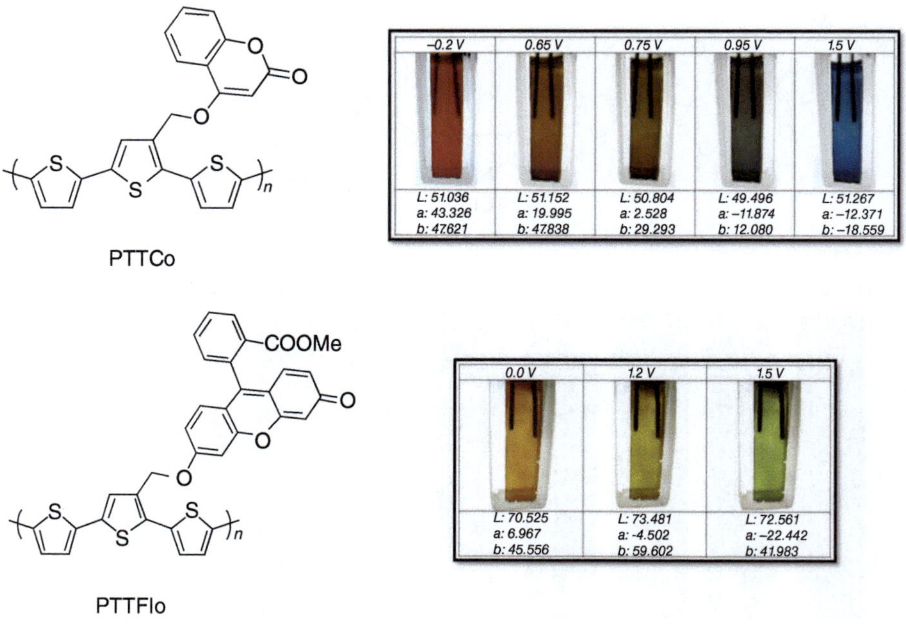

Figure 5.30 Repeat unit structures for coumarine (PTTCo) and fluorescein (Flo) functionalised poly(ter-thiophenes) along with photographic and colorimetric depictions of their accessible colour states. Adapted from Ref. [154].

in the 500–800 nm region of the visible spectrum and causing the perception of the yellow colour.

5.2.3.4 **Surface-Confined Polymerisation**

In addition to the direct film formation methods of electrochemical polymerisation, along with solution processing of soluble polymers as will be discussed for the formation of ECP films, Skene and co-authors [155] have developed a unique electrode confined chemical polymerisation condensing amine and aldehyde functional groups to prepare polymeric imines (the so-called Sken-imines), which provide multicolour electrochromic properties. This is nicely illustrated by the polymerisation of 2,5-diaminothiophene with 4,4-triphenylamine dialdehyde to form a fully conjugated polyazomethine having the repeat unit structure shown in Figure 5.31. Solution spray coating of mixtures of these monomer solutions followed by heating on the substrate at 120 °C provided polymeric films in 30 min. As shown in Figure 5.31, the polyazomethine formed is bright red in its as-prepared state due to the highly conjugated nature of the backbone as illustrated by the *a***b** coordinates and photograph I. Electrochemical oxidation of this polymer at 1.1 V yielded the oxidised form of the material with a deep blue colour (photograph II), interestingly bringing the red-blue colour change so often seen in multicolour systems. Cyclic voltammetry showed this material to switch reproducibly between neutral and oxidised forms. Neutralisation of form II via electrochemical reduction leads to a transmissive sky blue form III when −0.7 V was applied. While this deep blue-to-transmissive blue electrochromism could be accessed repeatedly, the original deep red state was not re-accessible even at large negative potentials. Application of an especially negative potential (−1.9 V) leads to the yellow form IV, which could subsequently be reconverted to forms II and III upon oxidation. This is an excellent example of how multi-vibrant coloured states can be induced in electrochromic polymers, albeit without *a priori* control of colour. Due to the high level of flexibility possible in diaminoaryl

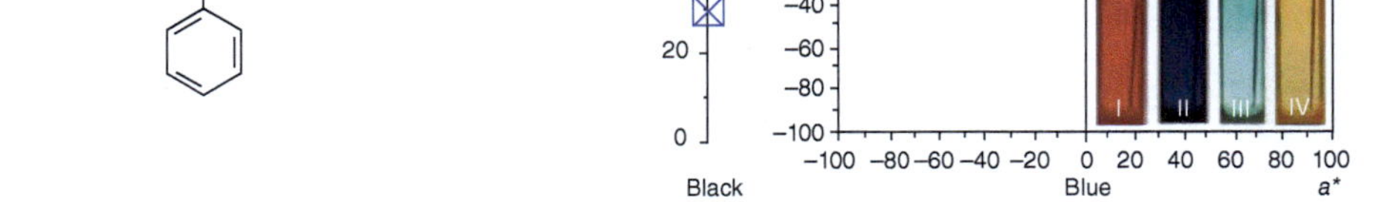

Figure 5.31 Repeat unit structure for the polyazomethine prepared from condensation of 2,5-diaminothiophene with 4,4-triphenylamine dialdehyde, along with photographs and *a***b** colour representation for various charged states of the polymer. Adapted from Ref. [155].

and dialdehyde-functionalised compounds, there is much promise in these polyazomethines as ECPs as new systems are investigated.

5.2.3.5 Combining Redox States – Oxidation and Reduction in a Single Material

Up to this point, the EC conversions that have been discussed involve oxidising a neutral conjugated polymer in one colour state to the p-type doped state, which provides another colour or possibly a high level of transmissivity. The ability to incorporate not only electron-rich donor units into the backbone but also electron-poor acceptor units provides another opportunity for electrochromism as reduced, n-type doped states can be made accessible. By designing polymer systems with a combination of donor and acceptor units with accessible oxidations and reductions, multicolour ECPs can be generated due to the multiple redox states available. This concept is illustrated by the work of DuBois *et al.* [93, 94, 156] via the electropolymerisation of bisEDOT pyridines and pyridopyrazines shown in Figure 5.32 where the terminal EDOT units provide BiEDOT linkages along the backbone separated by the heterocycle acceptor units.

These polymers are multicolour electrochromics having three or more distinct states. Poly(bisEDOT pyridine) has a red neutral state, an oxidised blue-purple state and a reduced sky blue state. In addition, due to the basicity of the pyridine unit, protonation of the neutral polymer leads to a separate navy blue state. Poly(bisEDOT pyridopyrazine-biphenyl) is also multi-chromic having a neutral lime green state, an oxidised light grey state and a reduced burgundy red state. The pyridopyrazine heterocycle is a sufficiently strong electron acceptor that it can stabilise a further reduced (dianionic) dark grey state. In a further example, Camurlu and Karagoren [157] have explored the use of Naphthaleneimides as reducible species appended onto p-dopable 2,5-dithienylpyrrole (SNS) backbones [158–161] using copolymerisation with EDOT as a method of providing neutral-state colour control as illustrated earlier. All of the polymers exhibited ambipolar redox behaviour with reversible n- and p-type doping observed. The homopolymer displayed yellow-blue colour changes upon doping, while the copolymers displayed multi-chromic behaviour. While a wide variety of colours can be made available with electro-reducible species, n-type systems tend to be more reactive to ambient conditions providing less switching stability than their p-type counterparts – certainly an area worth further exploration.

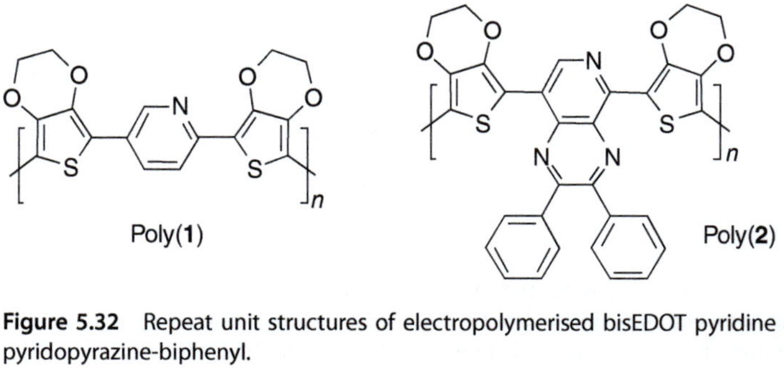

Figure 5.32 Repeat unit structures of electropolymerised bisEDOT pyridine and pyridopyrazine-biphenyl.

5.2.3.6 Composite Formation with Electrochrome Dopants

In one final approach to multicolour ECPs, the ability to combine the conjugated polymer with electroactive and electrochromic dopant systems allows multi-coloured states to be introduced based on the separate EC properties of the two components in the system. This is illustrated by the work of Song *et al.* [162], who used 2,2′-azino-bis(3-ethylbenzthiazoline-6-sulfonic acid) (ABTS) as a molecular electrochrome in polypyrrole. Palmore and co-authors [163] capitalised on this by preparing a polymeric ABTS derivative that served as the dopant ion for electrochemical polymerisation of EDOT. The resultant PEDOT/PABTS (poly(2,2′-azino-bis(3-ethylbenzthiazoline-6-sulfonic acid))) composite material thus contains two distinct electrochromes having the structures shown in Figure 5.33. Initially, both PEDOT and PABTS are relatively transmissive and colour neutral when held at 0 V. At this potential, PEDOT is in its oxidised form charge-balanced by the pendant sulfonate ion of the charge-neutral PABTS. Reduction of this composite at −0.8 V leads to introduction of a vibrant blue colour (see photographs in Figure 5.33), as the PEDOT is neutralised and a cation from the electrolyte penetrates into the composite film to charge-balance the sulfonate. This reversible blue-to-colour-neutral electrochromism is then complimentary with subsequent oxidation +0.8 V causing the ABTS to oxidise to the cation radical form. At this point, while the doped PEDOT remains transmissive, the PABTS converts into a vibrant green. One important benefit of these conjugated polymer/electrochromic dopant ion composites is that both electrochromes are embedded and entrapped in the material and do not diffuse out with switching. As such, the electrochromic switching is quite stable.

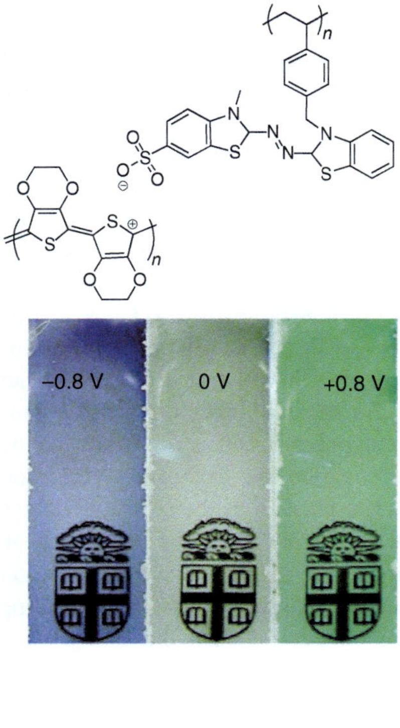

Figure 5.33 PEDOT/PABTS composite polymer repeat unit structures, along with three colour states made possible by utilisation of three electrochromes. Adapted from Ref. [163].

5.3 Processability of Electrochromic Polymers

ECPs possess two characteristics that advantageously impact their processability. First, owing to the ease with which synthetic modifications can be made to the structure of the polymer through the addition of polar or non-polar substituents, a wide range of solubilities in various solvents can be achieved. Second, polymers readily form robust and high-quality thin films on a variety of substrates. In this section, we describe a number of synthetic strategies that can be utilised to make ECPs processable from both organic and aqueous solutions.

5.3.1 Electrochemical Polymerisation

Despite the fact that electrochemical deposition of ECPs is unlikely to be the method of choice for industrial production, it is still an important technique for fundamental studies on conjugated monomers and polymers. It is also the deposition method of choice for many ECPs that are not soluble in their polymeric form, such as P3MT and PEDOT. Electrochemical deposition is typically performed in a three-electrode electrochemical cell containing monomer and electrolyte dissolved in an unreactive solvent. Monomers that can be successfully electropolymerised are those with oxidation potentials that are accessible in the electrolyte-solvent system, and that form radical-ion intermediates with stabilities suitable for further coupling reactions [5, 164]. The exact mechanism is not fully understood and of continuing debate, but there are several reviews that discuss the various theories in great detail [164, 165].

It is understood that the electropolymerisation process is highly dependent on conditions (electrochemical method, electrolyte, solvent, temperature, electrode material, etc.) and monomer identity. During electrochemical polymerisation, the electrolyte is incorporated in the film, resulting in an as-deposited film morphology that is templated to incorporate the counter ions and solvent molecules necessary for subsequent switching of the film. The various polymerisation conditions significantly influence the structure and morphology of the film, and these, in turn, influence the electrochemical and electrochromic properties. Of the parameters listed, the polymerisation method has a large influence with examples of these techniques being cyclic voltammetry, potentiostatic, and galvanostatic polymerisation. The cyclic voltammetry method can provide a large amount of information including monomer oxidation potential, rate of polymer growth/deposition, and polymer redox potentials during polymerisation. An example cyclic voltammogram for electropolymerisation of polypyrrole is shown in Figure 5.34. As can be seen in the repeated voltage cycling, the monomer oxidation potential (E_{pa}^{Mon}) and resulting polymer peak oxidation potential (E_{pa}^{Pol}) can be determined. The first cycle shows oxidation of the monomer. This is followed by a series of coupling reactions, resulting in deposition of oxidised polymer. During the reverse scan the as-deposited

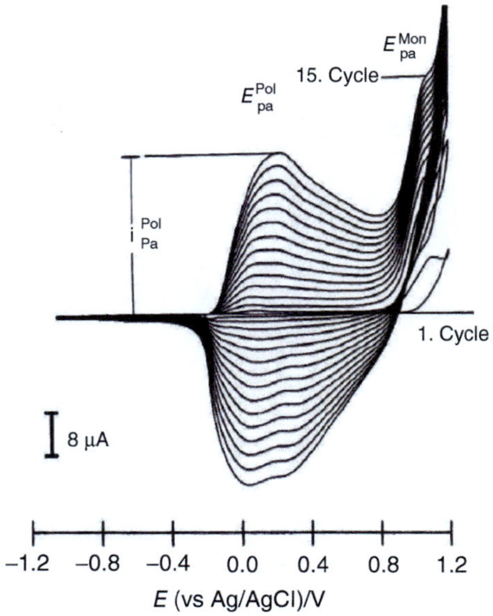

Figure 5.34 Typical cyclic voltammetry polymerisation/deposition of polypyrrole in 0.1 M $TBAPF_6$/acetonitrile (+1% water) at a scan rate of $100\,mV\,s^{-1}$ and at 298 K. Reproduced from Ref. [164].

oxidised polymer is reduced to its neutral state. Subsequent oxidative cycles cause further oxidation and coupling reactions to occur with polymer growth and deposition monitored by the increase in polymer oxidation current density. The driving force for polymerisation is different for each method (cyclic voltammetry, constant potential or constant current), and thus, the film quality and morphology will be highly dependent on the deposition method employed.

The choice of solvent affects the solubility of oligomers created during the initial stages of polymerisation, which can ultimately affect the chain length of the polymer as the material precipitates on the electrode surface. This, in turn, affects spectral properties such as λ_{max}, spectral bandwidth, and electrochromic contrast, as well as physical properties such as film roughness [166]. The distribution of chain lengths and morphology also impacts the doping level, which is known to influence contrast ratio and coloration efficiency [167, 168]. It has been shown that the choice of electrolyte used during polymerisation, in comparison to other parameters, has a moderate effect on electrochemical and electrochromic properties of the film [166, 169, 170]. For example, an evaluation of the polymerisation solvent and electrolyte showed these parameters had a strong influence on the resulting polymer film morphology during electropolymerisation of PEDOT films as is shown in Figure 5.35 for the scanning electron microscope images for a series of films. It was found that the smoother morphology accessible by use of propylene carbonate, rather than acetonitrile, provided improved electrochromic properties such as higher transmittance contrast and higher coloration efficiency.

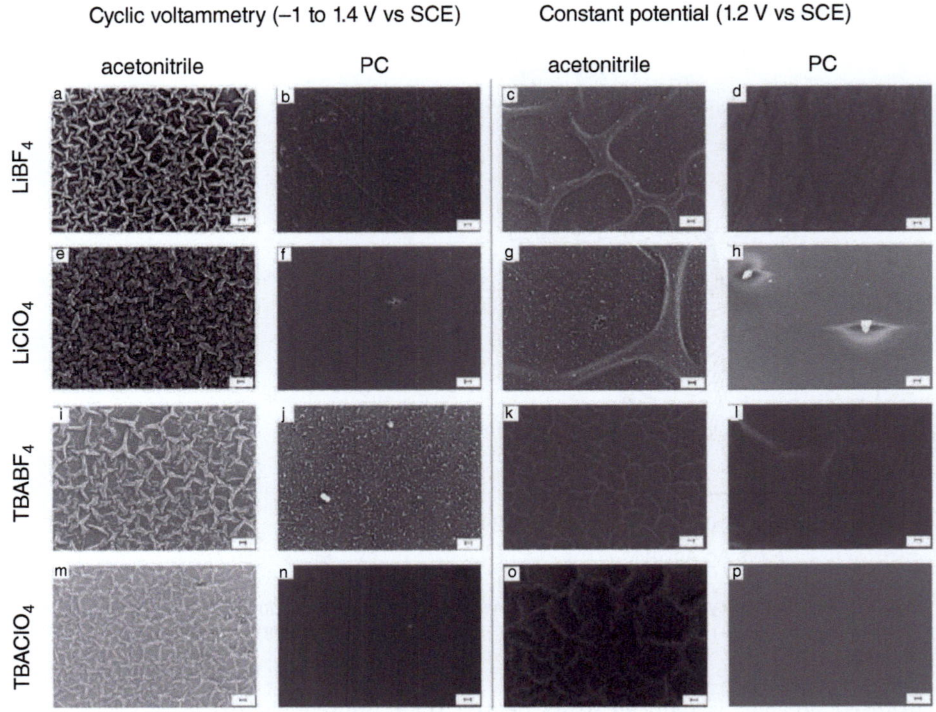

Figure 5.35 Scanning electron microscope images for a series of PEDOT polymer films on ITO/glass electropolymerized using cyclic voltammetry (a, b, e, f, i, j, m, n) and using potentiostatic polymerization (c, d, g, h, k, l, o, p) in acetonitrile (left image in each pair) and PC (right imagine in each pair). The supporting electrolytes are $LiBF_4$ (a–d), $TBABF_4$ (i–l), and $TBAClO_4$ (m–p). Magnification ~1000, scale bar 10 μm, working distance is 3 mm. Reproduced from Ref. [166].

A major advantage of electrochemical polymerisation is that only a small amount of monomer is required (concentrations typically in the millimolar range), and there is no need for isolation and additional purification of the product as there is for chemically polymerised materials given that the polymer film is prepared directly on the electrode surface. A caveat however is that much of the monomer is not incorporated into the electropolymerised film and is therefore wasted. Another advantage is the accuracy and precision of electropolymerisation; even and relatively homogeneous polymer deposition can be achieved on intricately patterned electrodes as the deposition is limited to just the conducting portions of the substrate. Electrochemical polymerisation typically results in well-adhering films on a wide range of electrode substrates.

Some of the challenges related to electrochemically polymerised ECP films are limitations related to substrate material and substrate size. Many applications require the use of optically transparent electrodes and traditionally, ITO coated glass or plastic has been the substrate of choice. However, PEDOT:PSS has also become a popular electrode material for developing all organic ECDs due to the high conductivity and high level of transparency attainable in many formulations [59, 171]. Both ITO and its organic alternatives have lower conductivity and are more prone to defects than metal electrodes, and this invokes challenges for electrochemical deposition of ECPs, especially on larger surfaces. When using large area electrodes there will be a drop in electric field across the electrode which is likely to result in uneven films. This issue is even more pronounced when using electrodes with higher resistance than ITO, such as PEDOT:PSS. As such, special care must be taken to ensure that proper electrical contacts are made to avoid significant drops in potential across the electrode [172]. For these reasons, electrochemical polymerisation is better suited to fabricating small area films.

Through appropriate choice of starting conjugated monomer (or set of monomers), virtually any colour polymer can be realised using electrochemical polymerisation. There has been a significant body of work published and reviewed on the electrochromism of derivatives of thiophenes, pyrroles, anilines and different donor–acceptor materials that can be electrochemically polymerised [2, 5, 6, 173]. Electropolymerisation offers a simple route towards tuning the coloration of a film through copolymerisation of different monomers and varying their feed ratio or use of monomers with incorporated small molecule chromophores, as was discussed in Section 5.2.3 [153, 174].

Recent advances in electrochemically polymerised systems have focused on conquering some of the shortcomings in using electropolymerisation. For example, one approach for reducing the amount of precursor waste is to spray-coat the monomer or monomer blends followed by a surface-confined electrochemical polymerisation where all of the monomer is converted into its corresponding polymer. This also allows for tuning of the resulting film colour through the use of monomer blends in differing ratios [175–178]. It was also demonstrated that this method can be used to fabricate an electrochromic polymer by *in situ* polymerisation as shown in Figure 5.36. A precursor electrolyte

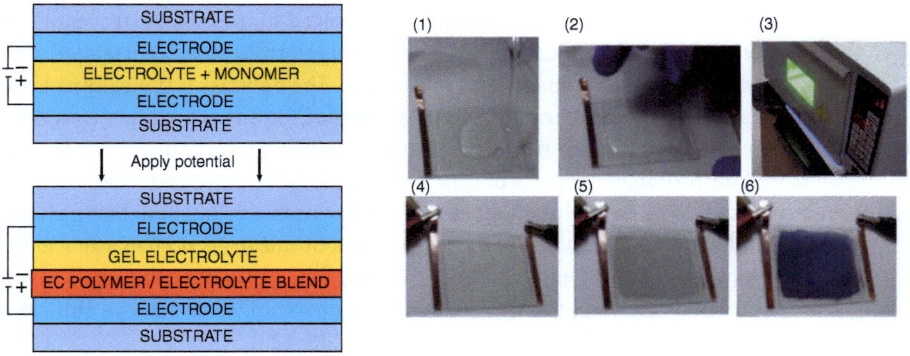

Figure 5.36 Electrochromic devices fabricated by first dissolving the monomer of the ECP in a precursor electrolyte and sandwiching between two substrates (1) and (2), followed by curing the electrolyte (3) and (4), and applying a potential to polymerise the monomer and achieve the ECP in its fully oxidised (5) and fully neutralised state (6). Reproduced from Ref. [175].

containing the conjugated monomer is prepared and sandwiched between two transparent electrodes. Photo or thermal cross-linking of the electrolyte is then induced, followed by application of a potential bias that induces polymerisation of the monomer. While it is unlikely that electropolymerisation is a technique that will be able to fabricate large area films, on a smaller scale, these recent advances make this technique suitable for fabricating electrochemically polymerised films on patterned substrates.

5.3.2 Functionalisation of ECPs for Achieving Organic Solubility

Functionalisation of the polymer repeat unit with solubilising groups is the most commonly employed strategy to develop soluble electrochromic polymers. The insolubility of most unfunctionalised conjugated polymers derives from the rigidity of the backbone, lack of conformational freedom in solution and strong intermolecular interactions that induce π-stacking between polymer chains. To overcome these factors, solubilising groups often take the form of long or branched alkyl or alkoxy chains, which serve to introduce more conformational freedom and sterically hinder aggregation. The synthetic routes towards functionalised systems are well-developed and understood, and the synthesis in many cases is straightforward and inexpensive. Furthermore, side chains can be introduced or tuned to promote or inhibit solubility with only minor effects on the colour properties of the material. Side chain modification can also be used to achieve more pronounced colour properties if so desired, as discussed in Section 5.1. The addition of functional groups not only improves solubility but also often has the added benefit of improving the electrochromic switching behaviour.

Functionalisation is typically carried out during monomer synthesis. Chemical polymerisation is then performed in solution, and the presence of solubilising groups allows the growing chain to remain dissolved up to molecular weights of at least tens of kilodaltons. Larger molecular weights are typically correlated with improved film-formation quality and robustness. The polymer is typically isolated from the reaction by precipitation, filtering, washing (often via Soxhlet extraction) and then drying to remove residual solvent. Many electrochromic polymers can be stored for prolonged periods of time in the dark, to be dissolved when ready to be processed. While attaching longer and greater numbers of solubilising chains aids in inducing solubility [102], most examples of linear alkyl chains in the literature have between 6 and 18 carbons [42, 88, 134, 179]. Typically, chains of at least six carbons are necessary to induce enough solubility to achieve sufficient molecular weights to cast high-quality films [73, 86, 132].

The first thiophene-based soluble polymers were reported almost simultaneously by Elsenbaumer and co-workers [180, 181], as well as by Heeger and Wudl and co-workers [182]. In these instances, solubility was obtained through alkyl functionalisation of the thiophene unit in the 3-position. The solubility of the chemically polymerised PThs was investigated as a function of the alkyl chain length, and it was found that butyl chains were sufficient to obtain molecular weights of 2.5 kDa [180]. These polymers were found to be soluble in a variety of organic solvents, including THF, toluene and anisole, to name a few.

The effects of attaching functional groups on the electrochromic properties are diverse, but fairly well understood. In many cases, it is desired to attach functional groups without drastically changing the colour of the material. Because PEDOTs and PProDOTs can be functionalised on the alkylenedioxy ring separated from the conjugated backbone, these are good examples of this. The first solution-processable coloured-to-transmissive polymer electrochrome was a purple-to-colourless PEDOT derivative that was tetradecyl-substituted (PEDOT-C14) on the ethylene bridge [63]. PEDOT-C14 was synthesised through oxidative polymerisation with an M_n of 8.2 kDa. While unsubstituted PEDOT is insoluble in organic solvents, PEDOT-C14 is soluble in chloroform and THF. The C14-chain, located on an sp^3 hybridised carbon, protrudes out of the plane of the conjugated backbone, hindering π-stacking, while the extended chain length provides solubility through adding more degrees of conformational freedom. Finally, polymerisation proceeds in a regiorandom fashion, leading to an atactic polymer, which also disturbs the packing of the material. Presence of the alkyl chains gives rise to steric-induced changes in the intra- and intermolecular interactions of the polymer compared to the unsubstituted polymer. In spite of the steric bulk of the C14 chains, there is only a slight change in the colorimetric properties between the two with a difference in λ_{max} around 20 nm. This shifts the spectra from a blue colour in PEDOT to an purple colour in PEDOT-C14. While early work focused on systems such as alkylated-EDOTs, these compounds are rarely employed nowadays as their polymerisation leads to a regiorandom polymer. As shown in Section 5.1, solubilising groups are typically placed in locations that maintain the regioregularity or regiosymmetry of the polymer [183].

Capitalising on the high contrast found in the alkylated PEDOT derivatives, regiosymmetric polymers based on PProDOTs where functionalisation at the 2-position of the propylene bridge only minimally alters the electrochromic properties were developed. To study the effects of functionalising with various soluble groups, PProDOTs were disubstituted with butyl, hexyl, ethylhexyl, ethylhexyloxy and octadecyl chains, as shown in Figure 5.37 [73]. With the exception of the butyl functionalised compound, the polymers attained M_n molecular weights >33 kDa and were readily soluble in THF, $CHCl_3$ and toluene. Examining the spectra of the films, in spite of the differences in steric bulk and chain length of the substituents, the λ_{max} were all within 30 nm of each other, and their electrochemical half-wave potentials closely ranged from 0.07 to 0.20 V. However, the switching speeds, as well as the contrast, differed between the branched and linear derivatives. The linear polymers displayed switching times of seconds and transmittance differences of 51% at λ_{max}. In comparison, the branched polymers afforded switching times on the order of tens of milliseconds, with full transmittance differences at λ_{max} of 80%. Optical microscope and atomic force microscopy (AFM) images, shown in Figure 5.38, comparing 2-methylbutyl and hexyl substituted PProDOT illustrate the underlying differences between films

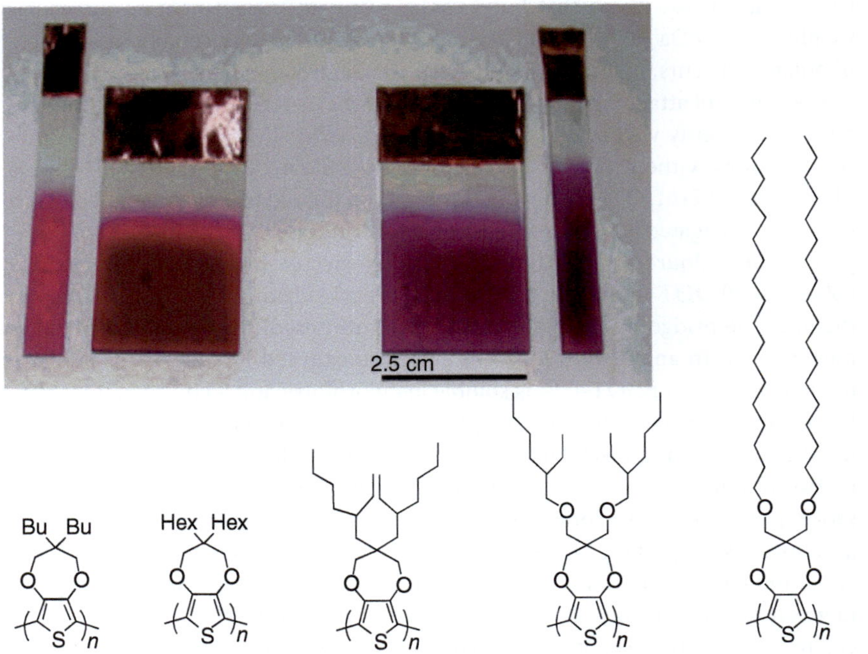

PProDOT(Bu)$_2$ PProDOT(Hx)$_2$ PProDOT(EtHx)$_2$ PProDOT(CH$_2$OEtHx)$_2$ PProDOT(C18)$_2$

Figure 5.37 Repeat unit structures for several ProDOTs functionalised with various solubilising groups, and photographs of spray-cast films of PProDOT-(CH$_2$OEtHx)$_2$ (left) and PProDOT-(C18)$_2$ (right). Reproduced from Ref. [73].

formed from branched and linear derivatives [184]. Films from hexyl-substituted PProDOTs possessed a relatively smoother morphology, while films from methylbutyl-substituted PProDOTs were far rougher. This study also showed that the evolution of the colour of the two films as a function of the applied voltage upon oxidation and reduction was different. This is believed to be due to the more porous morphologies arising from the branched systems that allow electrolyte ions to readily migrate through the bulk of the film, while smooth morphologies from linear systems form denser films that can possibly impede ion diffusion. The negligible colour change observed for PProDOTs functionalised at the 2-position is due to the substituents protruding out of the plane of the main chain and not interfering with the conjugation. In contrast, by attaching substituents that affect the conjugated backbone, either sterically or electronically, there are large changes in the coloration, as discussed in Section 5.1.

5.3.3 Aqueous Processability and Compatibility

The ability to dissolve conjugated electrochromic polymers in aqueous solutions or alcohols can allow lower cost and more environmentally benign processing

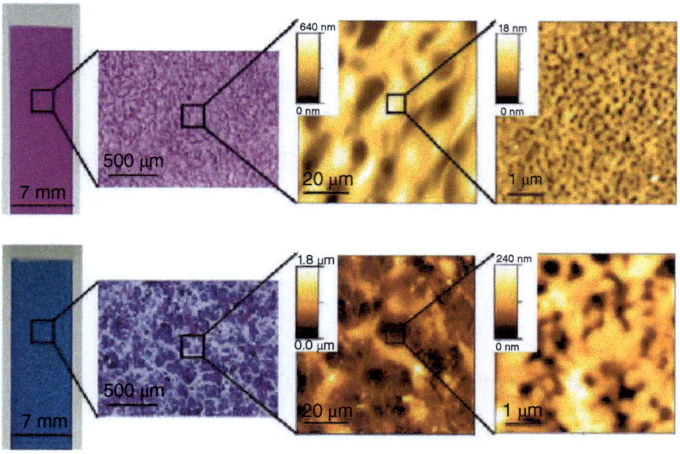

Figure 5.38 Photographs (left), optical microscope images (centre) and AFM images (two far right) illustrating the differences in the film morphology between PProDOTs with linear (top) and branched (bottom) chains [73]. The top images are for dihexyl-substituted PProDOT and the bottom images are for the branched, methylbutyl-substituted PProDOT spray-cast films. Reproduced from Ref. [184].

methods to be adopted. However, with the exception of PEDOT:PSS and PANI, examples of such ECPs and devices have been few [185–187]. This is due to not only the challenges related to obtaining such solubility but also the challenges in high-quality film formation from polar solvents with low vapour pressure. In the next section, the three main routes to achieve water processable and/or compatible ECPs are detailed and examples of their deposition and performance in devices are highlighted. With these approaches, it should be apparent that, while such materials are not currently widespread and challenges persist, the capability to synthesise and process such materials has been developed and could be applicable for a variety of systems.

5.3.3.1 **Use of Charged Polymers**

One successful strategy towards achieving water- or alcohol-soluble materials is by polymerising the electrochromic polymer in the presence of a water-soluble polyelectrolyte host, giving rise to a dispersion. Next, we illustrate the fundamental principles of this technique using PEDOT:PSS as the main example, one of the first and most successful examples of a water-dispersible conjugated electrochromic polymer. While PEDOT itself is an insoluble material that was developed in the late 1980s at Bayer [57, 58, 188], polymerising in the presence of water-soluble PSS gives rise to a polyelectrolyte complex that is dispersible in water and can readily be cast into thin films [189–191]. In addition to electrochromics, PEDOT:PSS has been used for a variety of applications, ranging from electrode materials to antistatic coatings for photographic film. Dispersions of PEDOT:PSS are commercially available with a range of molecular weights and molecular weight dispersities, as well as in varying weight ratios

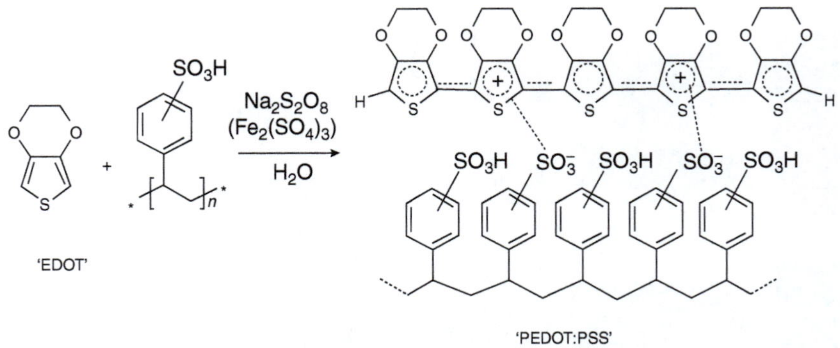

Figure 5.39 Synthesis and structure of PEDOT:PSS.

of PEDOT to PSS giving rise to materials with varying conductivities and viscosities.

The reaction to form PEDOT:PSS is shown in Figure 5.39 and begins with the polymerisation of EDOT in water using a combination of peroxydisulfates with Fe(III) in the presence of PSS^- [59] resulting in a colloidal solution where PEDOT is doped with one charge per 3–4 repeat units. While PEDOT is normally insoluble in water, the presence of excess PSS (typically 1–3 equiv.) as the host polyelectrolyte complexes with the PEDOT polycation, giving rise to a dispersion that is directly suitable for aqueous processing [190, 191].

In spite of the challenges related to processing from water, PEDOT:PSS has been successfully deposited using a variety of techniques and incorporated into displays. While many of the techniques developed are in other areas of organic electronics, the conditions established can be readily applied to PEDOT:PSS as an ECP. Research into its use as a patterned electrode in thin-film transistors has led to the development of suitable formulations for inkjet printing fine structures [192]. On a larger scale, PEDOT:PSS has been roll-to-roll coated using a slot-die coater for its use as an electrode layer in an all solution-processed organic solar cell [193]. Recently, PEDOT:PSS was used as the active material in a passive matrix display with 896 individually addressable EC pixels, as shown in Figure 5.40 [194]. It has also been used as a transparent electrode material for ECDs [171] and impregnated into spandex with the goal of developing smart textiles [195].

While PEDOT:PSS demonstrates that the use of a host polyelectrolyte is a successful strategy towards water-dispersible materials, few other noteworthy systems have employed this technique for electrochromics. Poly(thieno[3,4-b]thiophene) (PT34bT):PSS is one example that forms water-dispersible films that switch readily outside of the visible region [196]. This makes the system attractive either as an IR-switching material or as a transparent counter electrode material for an ECD. Alternatively, the polyanion species can be altered to introduce improved properties or additional functionalities. For instance, one drawback to PEDOT:PSS is related to its relatively low thermal stability above 200 °C. To improve upon this, PSS has been replaced with sulfonated polyimide (SPI), a material known for its high thermal performance. Although

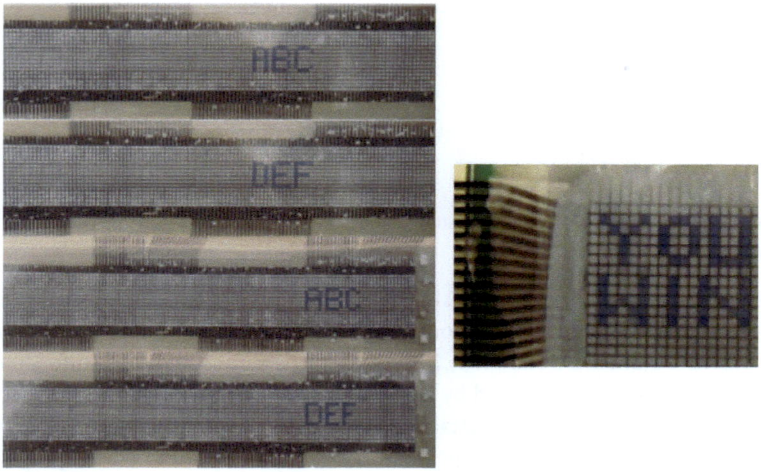

Figure 5.40 Addressable pixels (896) in a PEDOT:PSS passive display. Reproduced from Ref. [194].

electrochromic properties were not evaluated, PEDOT:SPI, PANI:SPI and PPy:SPI were all observed to have improved thermal properties compared with their PSS analogues [197]. As an example where additional functionality can be introduced through the host polyelectrolyte, DNA has been used as the polyelectrolyte in conjunction with PEDOT to introduce a non-acidic host as well as biocompatibility, while preserving the electrochromic properties of PEDOT [198]. While the polyelectrolyte in PEDOT:PSS serves to tune the processability of the polymer, in the case of PANI, polyelectrolytes that are typically also polyacids serve the additional purpose of making it possible to access the full polyelectrochromism of the polymer within stable operational voltages without resorting to low pH, as is required when small dopant acids are used. For example, when poly(2-acrylamido-2-methyl-1-propanesulfonic acid) (PAMPS) is used as the counterion, PANI switches reversibly between the fully reduced and transparent leucoemeraldine base (LB) state, the partially oxidised, green ES state and the fully oxidised, violet PB state (pernigraniline base) [199]. Subsequent solvent annealing with dichloroacetic acid further improves PANI-PAMPS electrochromic properties, for example, by enhancing its switching speeds, as well as improving the electrical conductivity by 2 orders of magnitude [199].

5.3.3.2 **Ion Functionalised Polymers**

Utilising the same strategy employed to solubilise ECPs in organic solvents, ECPs can be rendered water or alcohol soluble by appropriate functionalisation of the polymer repeat unit (pre- or post-polymerisation). Given the high dielectric constants of polar media, functionalities are often designed to have ionic pendant groups, such as $-SO_3^-$, $-CO_2^-$, $-PO_3^-$ and $-NR_3^+$ to result in water-soluble polymers such as those shown in Figure 5.41 [200–204]. The earliest examples utilising this design approach proceeded by first functionalising the monomers

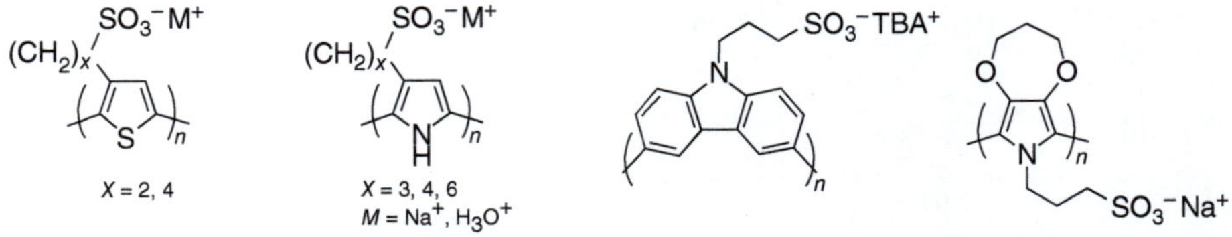

Figure 5.41 Repeat unit, for example, water-soluble ECPs achieved through electropolymerisation.

and then electrochemically polymerising them into their corresponding polymer. Water-soluble PThs were first introduced by Heeger, Wudl and co-workers, who obtained sulfonated PThs [205, 206] around the same time that sulfonated polypyrroles (PPys) were reported by Reynolds *et al.* [207, 208]. These polymers possess an ionic group covalently attached to the repeat unit, which is capable of self-doping the polymer. The self-doped nature of sulfonate functionalised poly(3-butylsulfonate pyrrole) and of sulfonated terthiophene was unequivocally proved by Havinga *et al.* [209], who electrochemically polymerised these sulfonated monomers in ACN without any additional electrolyte salt. The corresponding polymers were deposited on a Pt electrode in ACN and could then be dissolved in water or methanol, albeit to a limited extent ($<0.5\,g\,l^{-1}$). The stability of a self-doped PProDOP was evaluated under repeated potential cycling in propylene carbonate (PC), where after 10 000 cycles, the film retained over 70% of its initial charge. While electropolymerisation is an effective and understood process for forming water-soluble ECPs, limited amounts of the polymer can be generated in this fashion.

Ion functionalised polymers can also be chemically synthesised, most typically through polymerisation of the ion functionalised monomer. In the conjugated polymer literature, there are numerous examples of chemically synthesised ion functionalised polymers [201, 203]; however, there are few examples specifically designed for electrochromic applications. Some of the examples that do exist are shown in Figure 5.42 and serve to illustrate the feasibility of these materials as ECPs.

While the examples illustrated here demonstrate the ability to synthesise water-processable polymers, films of these polymers cannot be switched in water due to their solubility, which still necessitates the use of non-aqueous switching and device electrolytes. One approach towards an ionic polymer that retains the

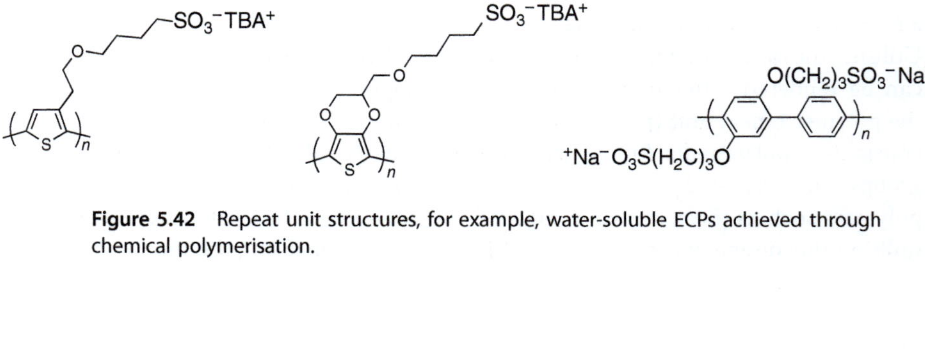

Figure 5.42 Repeat unit structures, for example, water-soluble ECPs achieved through chemical polymerisation.

benefit of water processability with the additional benefit of water switchability is to electrochemically copolymerise EDOT and EDOT-S (see middle structure Figure 5.42) which yield, a water insoluble polymer [210].

Another strategy revolves around using a layer-by-layer (LbL) technique to convert water-soluble polyelectrolytes to water-switchable polyelectrolytes. This approach can be used not only to obtain aqueous switching but also as a strategy for colour mixing. The earliest papers on for polyelectrolytes were reported by Decher *et al.* [211–213] LbL deposition proceeds by casting alternating layers of charged materials with wash steps in between. In these earliest examples, alternating layers of an anionic PSS and a cationic tetraalkylammonium polystyrene (PS) analogue were adsorbed to over 100 layers, demonstrating the use of this technology to fabricate films with a large range of thicknesses. One example of LbL deposition for electrochromics utilised alternating layers of PEDOT:PSS and linear poly(ethylene imine) (LPEI) as the cathodically colouring material. It was then used in a device setup where PANI:PAMPS was used as the anodically colouring system to afford a colour-mixed device that switched from a blue green to a yellow state [214]. In another example, PEDOT-S and polyallylamine hydrochloride (PAH) were deposited using LbL on unmodified ITO glass [215, 216]. This example illustrated the trade-off between coloration of the film and the switching times. Utilising fewer numbers of layers gave rise to faster switching times, but lower contrasts. Larger numbers of layers gave rise to improved contrast, but slower switching times. In contrast, utilising PProDOT-S and PAH, the switching times were found to be almost independent of the number of layers, giving high contrast within 1 s switching times, which the authors attributed to the improved regiosymmetry of functionalised PProDOTs compared with functionalised PEDOTs [217]. A final example utilised alternating layers of PANI:polyhedral oligomeric silsesquioxane (POSS) with PANI-S where, in contrast to the systems described so far, both layers are electrochromic. Having two electrochromic layers improved the electrochemical properties as well as coloration of the material, giving rise to high contrast devices [218].

5.3.3.3 Organic Processing to Achieve Water Solubility and Water Switchability

A final route towards water-soluble polymers is to first synthesise organic-soluble polymers bearing pendant groups that can readily be functionalised or defunctionalised to afford water-processable analogues of the polymer. This route combines the advantages of organic-soluble (ease of synthesis, purification and characterisation) and water-processable systems (environmentally benign processing conditions). While this technique has not seen widespread use in electrochromics, this strategy is commonly employed in the synthesis of conjugated polymers for sensors and other optoelectronic applications, and the syntheses developed there can be readily adopted for the design of electrochromic materials [204, 219].

Several examples include synthetic strategies to convert from organic- to water-soluble polymers such as pendant amine groups that are readily quaternised [220] post-polymerisation functionalisation via click chemistry [221] or conversion of an organic-soluble ester into an ionic carboxylate salt under basic conditions

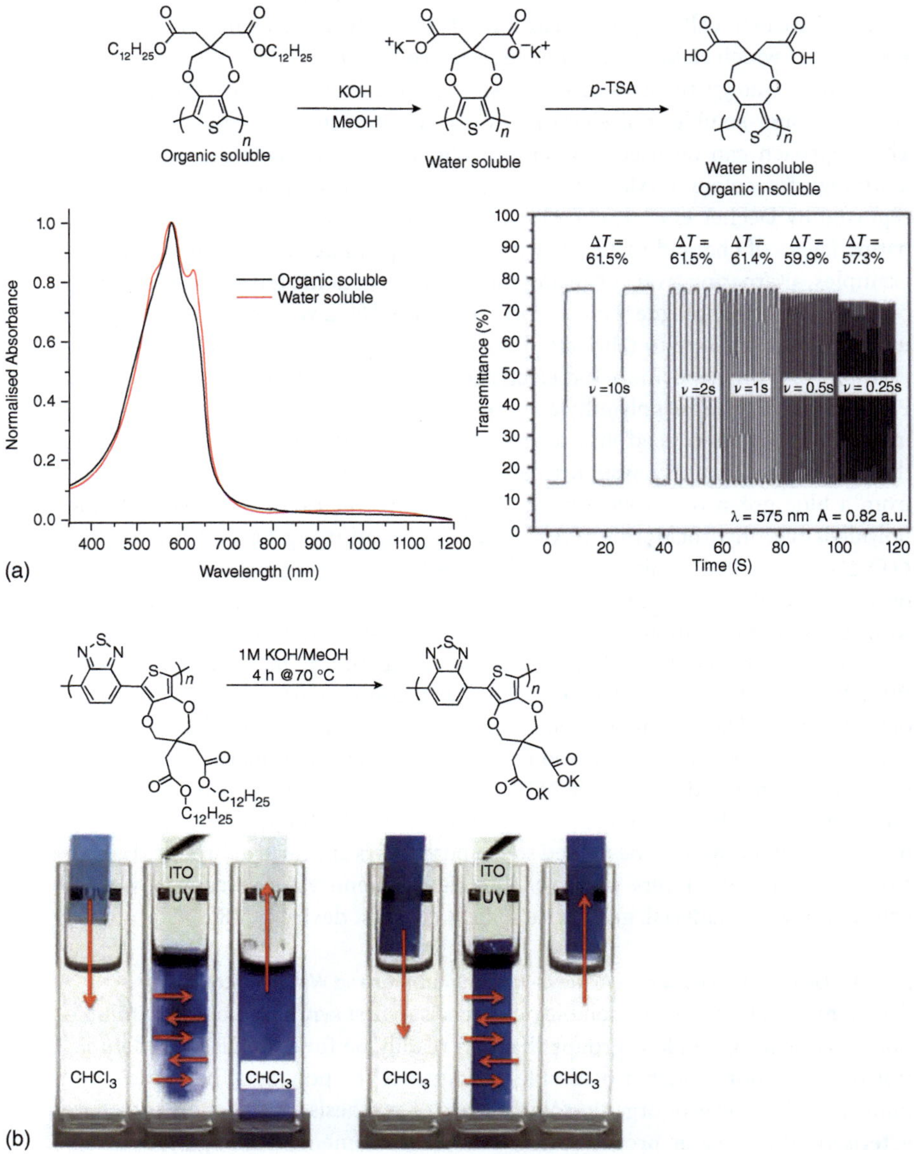

Figure 5.43 (a) Repeat unit structure and steps used to convert an organic-soluble PProDOT into a water-processable version, followed by conversion to a film insoluble in aqueous and non-aqueous solvents (top). Solution spectra of the organic-soluble and aqueous-soluble versions (left) and chronoabsorptometry experiment demonstrating the rapid switching possible for the polymer film when switched in an aqueous electrolyte. (b) Chemical defunctionalisation of an organic-soluble alkyl-ester PProDOT-BTD to an insoluble polyanion and photographs of the films on submersion in chloroform. Reproduced from Ref. [83].

[83, 200, 222]. In this latter example, an organic-soluble/processable PProDOT-ester is treated with potassium hydroxide and the ester groups are cleaved, leaving the potassium carboxylate that is soluble in water [100]. The resulting ionic polymer can then be spray-cast from aqueous solution with little change in the polymer optical properties on conversion from the organic-soluble version as shown by the solution spectra in Figure 5.43a. Further treatment of the polymer film with an acid (*p*-toluene sulfonic acid in this case) converts the carboxylate salt into carboxylic acid groups, leaving the polymer insoluble in both organic and aqueous solvents. The advantage is that the polymer can be processed from water, then treated with acid, allowing electrochromic switching in aqueous electrolyte solutions. The aqueous electrolytes offer the ability for faster switching speeds as the ionic conductivities of these electrolytes are often higher than those of non-aqueous electrolytes. This can be seen by the chronoabsorptometry experiment in Figure 5.43a (lower right) where there is only a 4.2% drop in transmittance difference for the polymer film, switched in KNO_3 aqueous solution, when the switching time was decreased from 10 to 0.25 s (there was a 15.7% drop in transmittance difference for the film switched in non-aqueous electrolyte). However, it was found that there is a limit in water solubility and a minimum in the number of ionic groups per repeat unit is required [83]. For example, utilising the same approach, a D–A polymer of ProDOT–BTD functionalised with ester groups can be spray-processed from chloroform as shown in Figure 5.43b. However, treatment of the film with base renders the polymer film insoluble in chloroform, but there is insufficient ionic content to allow solubility in water. Additional efforts by the same group circumvented this issue by attachment of multiple reactive pendant groups onto the ProDOT unit [200].

5.3.4 Methods for Patterning

The processing of electrochromic polymers differs from the processing methods of other electrochromic materials, such as their inorganic counterparts. Due to the ease with which polymers can be made to dissolve in a wide range of solvents, and their ability to form films with desired mechanical properties, these materials can take full advantage of the wide range of solution processing techniques available. This in turn allows them to be cast onto substrates of all shapes, sizes and flexibility. Without the need to deposit under ultrahigh vacuum, or to use high-temperature annealing for performance enhancement, the processing of electrochromic polymers is relatively inexpensive and can be divided into two sub-categories: coating and printing. Coating techniques are typically used for full-layer depositions with minimal direct patterning possibilities, whereas printing methods allows for intricate, high-resolution patterning. Coating methods can include spin, spray, blade, slot die and rod coating, while printing methods can include inkjet, screen, flexo, gravure and offset printing. For more detailed discussion on the techniques listed, the reader is referred to comprehensive reviews written by Krebs *et al.* and Berggren *et al.* and references therein [5, 223, 224]. Coating and printing

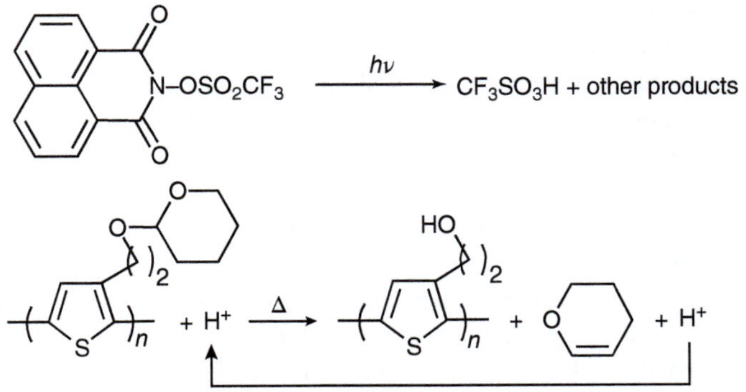

Figure 5.44 Reaction scheme for chemically amplified photolithography route to pattern polythiophene film.

methods specifically explored for preparation of electrochromic polymer films include spray coating [73, 184], slot die coating [225], inkjet printing [226, 227] and screen printing [186, 228, 229].

In many of the examples of selective patterning of ECPs, functionalisation of the polymer is required to allow for selective patterning with the ultimate goal of the polymer colour and electrochemical properties being unchanged after patterning has taken place. One of the most common routes to achieve intricate patterns is photolithography. In one example, as shown in Figure 5.44, a film of tetrahydropyranyl (THP) functionalised polythiophene (PTh) is cast in the presence of a photoacid generator [230]. When irradiated with UV–Vis light through a mask, an acid-catalysed deprotection of the functionalised polythiophene occurs, rendering the exposed areas insoluble to common organic solvents. Rinsing the film with THF removed the unradiated portions, and line widths as small as 15 μm were fabricated. This photolithographic technique takes advantage of the chemically amplified nature of the acid catalysed reaction, and low doses of 65 mJ cm^{-2} were sufficient to pattern the film with minimal photobleaching of the polymer. The aforementioned THP-protected PTh can also be deprotected at high temperatures [231]. Thermal treatment was employed onto confined regions using an NIR laser with doses of 600 mJ cm^{-2} to deprotect the underlying film and render the treated areas insoluble. After washing, the patterned lines were exposed with resolutions as fine as 10 μm. Because laser scan speeds can reach 0.6 m s^{-1}, relatively large areas and high throughput can be afforded with this method.

It has also been demonstrated that alkyl-functionalised PTh's themselves undergo cross-linking upon irradiation to alter their solubility, without the need for further additives [232]. A drawback to this method, however, is that the conductivity, as well as colour properties, of the polymer is sensitive to irradiation time. Nevertheless, irradiation times of 15 s (with a 150 W mercury lamp) were sufficient to render the film insoluble to toluene with minimal decrease in conductivity and coloration. After irradiating through a mask and washing away the unexposed areas, P3HT was still able to switch between its red neutral and blue oxidised states.

To reduce the irradiation intensity required, several examples exist whereby electrochromic polymers bearing a pendant methacrylate group are synthesised. The polymer film is typically blended with a photoinitiator and irradiated through a mask. In one example, no photoinitiator was added but either singlet oxygen or residual iron trichloride from the polymerisation synthesis acted in that role [233]. Upon irradiation, the methacrylate functionality cross-links, rendering the exposed areas insoluble. A similar strategy was employed whereby PProDOT functionalised with methacrylate groups was synthesised. An azobisisobutyronitrile (AIBN) photoinitiator was added to solutions of the polymer, which were then spray-cast to form continuous films. Irradiation through a mask was then performed at low doses, and the resulting film washed with toluene to remove un-cross-linked material as shown in Figure 5.45. The electrochromic properties were compared before and after cross-linking with no change observed as can be seen by the colorimetry provided in Figure 5.45.

The methacrylate functionality has also been used to polymerise patterned films, rather than perform a post-polymerisation cross-linking as described earlier. Methylacrylate-functionalised EDOT and ProDOT monomers have been synthesised and films prepared in the presence of iron(III) tosylate catalyst onto flexible substrates. Oxidative polymerisation was initiated thermally to afford

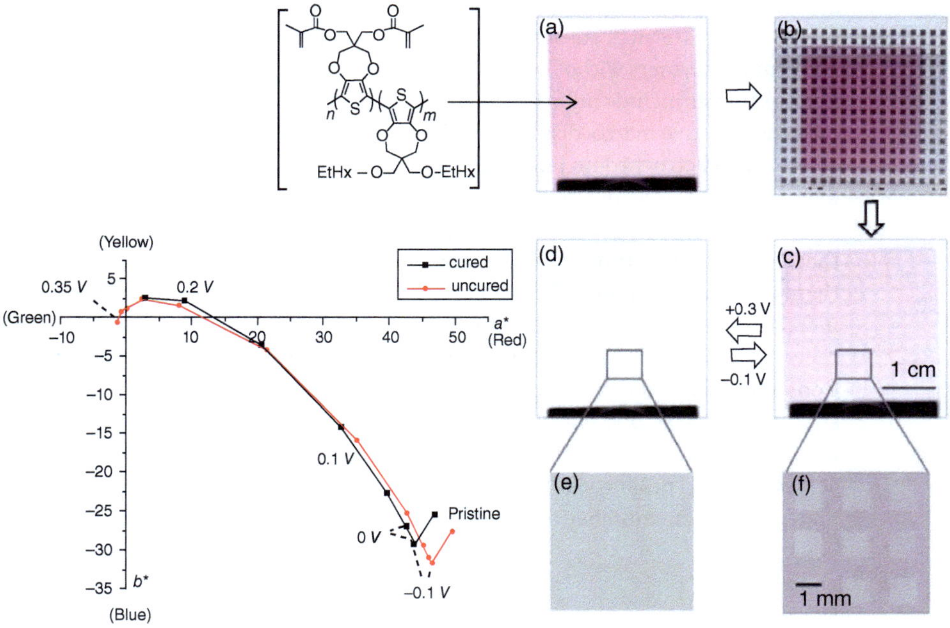

Figure 5.45 Direct photo-patterning of methacrylate-substituted ECP-Magenta. (a) Polymer-coated ITO/glass. (b) Application of photomask followed by irradiation. (c,f) Patterned film in neutral state. (d,e) Patterned film on oxidised state. Bottom left shows measured colorimetry of the polymer film before and after cross-linking. Reproduced from Ref. [234].

polymer films, which were then patterned via cross-linking under UV irradiation through a photomask [235]. The irradiation through a mask and washing afforded submicron line widths and high contrast films. As shown in Figure 5.9, photolithographic patterning via polymerisation of acrylate functionalities has also been utilised for conjugated oligomers [49]. Films of the functionalised acrylates were spray-cast in the presence of a photoinitiator and then irradiated through a photomask with UV light. The films were rinsed, leaving a patterned layer on ITO, which could be switched between various coloured states.

Fine patterning of films can also be physically induced through imprint lithography. In this method, a stamp with patterned gratings is pressed into a film, which then adopts the same patterning. This concept was demonstrated using a film of a precursor polymer containing acetate and terthiophene functionalised with norbornylene. The polymer was heated above its glass transition temperature, and a mould patterned with 20 nm wide and 80 nm deep channels was pressed into the film. After cooling, the mould was removed, and AFM images of the resulting film demonstrated lines of the precursor polymer were formed with average widths of 20 nm and heights of 80 nm. After electrochemically converting from the precursor polymer to a polythiophene, the resulting electrochromic lines switched readily between a coloured and a bleached state [236]. While the aforementioned examples utilise electrochromic polymers, or conjugated precursors, appended with a functionality to allow for printing and an initiated patterning method to create intricate active polymer layers, there of course exist examples of patterning polymers without the need for an added synthetic step or functionality. Such examples include imprint lithography of PEDOT:PSS layers [237], multilayer printing of active matrix displays using PEDOT:PSS as the transparent conductor and active electrochrome [238] and patterning PEDOT:PSS layers with selective over-oxidation to define inactive and active portions of a film [239].

One of the many advantages of conjugated electrochromic polymers is the synthetic flexibility to tune solubility and processability. As we have listed earlier, there are a large number of possibilities to deposit, print and pattern these polymers as thin films on a variety of conductors (rigid, flexible, transparent and/or reflective). While electropolymerisation is one of the earliest examples of preparation of thin films of conjugated polymers for electrochromics, the field has grown substantially with possibilities being demonstrated for films prepared via high-throughput roll-to-roll methods in addition to intricate patterning techniques to achieve desired properties. The possibilities remain endless as these are polymeric materials and any number of polymer processing methods can be adopted.

5.4
Summary and Perspective

An immense amount of organic and polymer chemistry has been employed in the development of ECPs as these materials have evolved from electrochemically polymerised lab curiosities to soluble and processable materials on the verge of commercialisation. The ability to control the colour, the so-called Completion of the Colour Palette, in a full family of polymers switching from vibrant hues to highly transmissive forms paves the way for applications in both absorptive/transmissive (window-type) and reflective (mirror and display) devices. The extensive fundamental synthetic, electrochemical, spectroscopic and colorimetric studies reported throughout the literature allow the electrochromic polymer scientist to create a system of essentially any colour. Structural modifications through copolymerisation, side-chain incorporation, pendant ion effects, cross-linking and composite formation (to name just a few of the available structural variables that can be adjusted) lead to a broad class of materials that can be processed by a variety of technologically useful methods. Considering these materials in the greater context of electrochromism in general, the colour versatility and ambient solution processability, combined with electrochromic memory and thermo- and photo-oxidative stability when correctly handled, now allow useful implementation for commercial applications.

Acknowledgements

We thank BASF and the Air Force Office of Scientific Research for their support of the electrochromic polymers and device programs at the University of Florida and Georgia Institute of Technology. JRR thanks the many students, postdocs, visiting scientists and external collaborators who have contributed to our electrochromics efforts over the years.

References

1. Amb, C.M., Dyer, A.L., and Reynolds, J.R. (2011) Navigating the color palette of solution-processable electrochromic polymers. *Chem. Mater.*, **23** (3), 397–415.
2. Beaujuge, P.M. and Reynolds, J.R. (2010) Color control in π-conjugated organic polymers for use in electrochromic devices. *Chem. Rev.*, **110** (1), 268–320.
3. Mortimer, R.J., Dyer, A.L., and Reynolds, J.R. (2006) Electrochromic organic and polymeric materials for display applications. *Displays*, **27** (1), 2–18.
4. Mortimer, R.J. (2013) Switching colors with electricity. *Ame. Sci.*, **101** (1), 38–45.
5. Dyer, A.L. and Reynolds, J.R. (2007) Electrochromism in conjugated conducting polymers, in *Handbook of Conducting Polymers*, 3rd edn (eds T.A. Skotheim and J.R. Reynolds), CRC Press, Boca Raton, FL and London.
6. Gunbas, G. and Toppare, L. (2012) Electrochromic conjugated polyheterocycles and derivatives-highlights from the last decade towards realization of long lived aspirations. *Chem. Commun. (Camb)*, **48** (8), 1083–1101.
7. Dyer, A.L., Craig, M.R., Babiarz, J.E., Kiyak, K., and Reynolds, J.R. (2010) Orange and red to transmissive electrochromic polymers based on electron-rich dioxythiophenes. *Macromolecules*, **43** (10), 4460–4467.
8. Monk, P.M.S., Mortimer, R.J., and Rosseinsky, D.R. (2007) in *Electrochromism and Electrochromic*

Devices (eds P.M.S. Monk, R.J. Mortimer, and D.R. Rosseinsky), Cambridge University Press, Cambridge, UK, pp. 312–335.

9. Ritter, S.K., Noftle, R.E., and Ward, A.E. (1993) Synthesis, characterization, and oxidative polymerization of 3-(fluoromethyl)thiophenes. *Chem. Mater.*, **5** (6), 752–754.
10. Kumar, A., Welsh, D.M., Morvant, M.C., Piroux, F., Abboud, K.A., and Reynolds, J.R. (1998) Conducting poly(3,4-alkylenedioxythiophene) derivatives as fast electrochromics with high-contrast ratios. *Chem. Mater.*, **10** (3), 896–902.
11. Walczak, R.M. and Reynolds, J.R. (2006) Poly (3,4-alkylenedioxypyrroles): the PXDOPS as versatile yet underutilized electroactive and conducting polymers. *Adv. Mater.*, **18** (9), 1121–1131.
12. Gaupp, C.L., Zong, K.W., Schottland, P., Thompson, B.C., Thomas, C.A., and Reynolds, J.R. (2000) Poly(3,4-ethylenedioxypyrrole): organic electrochemistry of a highly stable electrochromic polymer. *Macromolecules*, **33** (4), 1132–1133.
13. Schottland, P., Zong, K., Gaupp, C.L., Thompson, B.C., Thomas, C.A., Giurgiu, I., Hickman, R., Abboud, K.A., and Reynolds, J.R. (2000) Poly(3,4-alkylenedioxypyrrole)s: highly stable electronically conducting and electrochromic polymers. *Macromolecules*, **33** (19), 7051–7061.
14. Thomas, C.A., Zong, K., Schottland, P., and Reynolds, J.R. (2000) Poly(3,4-alkylenedioxypyrrole)s as highly stable aqueous-compatible conducting polymers, with biomedical implications. *Adv. Mater.*, **12** (3), 222–225.
15. Roncali, J. (1997) Synthetic principles for bandgap control in linear π-conjugated systems. *Chem. Rev.*, **97** (1), 173–205.
16. deLeeuw, D.M., Simenon, M.M.J., Brown, A.R., and Einerhand, R.E.F. (1997) Stability of n-type doped conducting polymers and consequences for polymeric microelectronic devices. *Synth. Met.*, **87** (1), 53–59.
17. Yasuda, T., Sakai, Y., Aramaki, S., and Yamamoto, T. (2005) New coplanar (ABA)(n)-type donor-acceptor π-conjugated copolymers constituted of alkylthiophene (unit A) and pyridazine (unit B): synthesis using hexamethylditin, self-organized solid structure, and optical and electrochemical properties of the copolymers. *Chem. Mater.*, **17** (24), 6060–6068.
18. Audebert, P., Sadki, S., Miomandre, F., and Clavier, G. (2004) First example of an electroactive polymer issued from an oligothiophene substituted tetrazine. *Electrochem. Commun.*, **6** (2), 144–147.
19. Salzner, U. (2002) Does the donor-acceptor concept work for designing synthetic metals? 1. Theoretical investigation of poly(3-cyano-3'-hydroxybithiophene). *J. Phys. Chem. B*, **106** (36), 9214–9220.
20. Salzner, U. and Kose, M.E. (2002) Does the donor-acceptor concept work for designing synthetic metals? 2. Theoretical investigation of copolymers of 4-(dicyanomethylene)-4H-cyclopenta[2,1-b: 3,4-b ']dithiophene and 3,4-(ethylenedioxy)thiophene. *J. Phys. Chem. B*, **106** (36), 9221–9226.
21. Ellinger, S., Graham, K.R., Shi, P.J., Farley, R.T., Steckler, T.T., Brookins, R.N., Taranekar, P., Mei, J.G., Padilha, L.A., Ensley, T.R., Hu, H.H., Webster, S., Hagan, D.J., Van Stryland, E.W., Schanze, K.S., and Reynolds, J.R. (2011) Donor-acceptor-donor-based π-conjugated oligomers for nonlinear optics and near-IR emission. *Chem. Mater.*, **23** (17), 3805–3817.
22. Arbizzani, C., Bongini, A., Mastragostino, M., Zanelli, A., Barbarella, G., and Zambianchi, M. (1995) Polyalkylthiophenes as electrochromic materials – a comparative-study of poly(3-Methylthiophenes) and poly(3-Hexylthiophenes). *Adv. Mater.*, **7** (6), 571–574.
23. Collombdunandsauthier, M.N., Langlois, S., and Genies, E. (1994) Spectroelectrochemical behavior of

poly(3-Octylthiophene) – application to electrochromic windows with polyaniline and iridium oxide. *J. Appl. Electrochem.*, **24** (1), 72–77.
24. Osaka, I. and McCullough, R.D. (2008) Advances in molecular design and synthesis of regioregular polythiophenes. *Acc. Chem. Res.*, **41** (9), 1202–1214.
25. Mastragostino, M., Arbizzani, C., Bongini, A., Barbarella, G., and Zambianchi, M. (1993) Polymer-based electrochromic devices. 1. Poly(3-methylthiophenes). *Electrochim. Acta*, **38** (1), 135–140.
26. Blohm, M.L., Pickett, J.E., and Vandort, P.C. (1993) Synthesis, characterization, and stability of poly(3,4-dibutoxythiophenevinylene) copolymers. *Macromolecules*, **26** (11), 2704–2710.
27. Havinga, E.E., Mutsaers, C.M.J., and Jenneskens, L.W. (1996) Absorption properties of alkoxy-substituted thienylene-vinylene oligomers as a function of the doping level. *Chem. Mater.*, **8** (3), 769–776.
28. Jen, K.Y., Eckhardt, H., Jow, T.R., Shacklette, L.W., and Elsenbaumer, R.L. (1988) Optical, electrochemical, and conductive properties of poly(3-Alkoxy-2,5-Thienylene Vinylenes). *J. Chem. Soc., Chem. Commun.*, (3), 215–217.
29. Martinez, M., Reynolds, J.R., Basak, S., Black, D.A., Marynick, D.S., and Pomerantz, M. (1988) Electrochemical synthesis and optical analysis of poly[(2,2'-Dithienyl)-5,5'-Diylvinylene]. *J. Polym. Sci., Part B: Polym. Phys.*, **26** (4), 911–920.
30. Sotzing, G.A., Reynolds, J.R., and Steel, P.J. (1996) Electrochromic conducting polymers via electrochemical polymerization of bis(2-(3,4-ethylenedioxy)thienyl) monomers. *Chem. Mater.*, **8** (4), 882–889.
31. Fu, Y.P. and Elsenbaumer, R.L. (1995) Structure-property relationships in conjugated polymers – oxidation potentials, reduction potentials, and band-gaps for a series of regiospecifically methoxy substituted thiophene vinylenes and poly(bithiophene vinylenes). Abstracts of Papers of the American Chemical Society, 209, 183-PMSE.
32. Bredas, J.L., Heeger, A.J., and Wudl, F. (1986) Towards organic polymers with very small intrinsic band-gaps.1. Electronic-structure of polyisothianaphthene and derivatives. *J. Chem. Phys.*, **85** (8), 4673–4678.
33. Wudl, F., Kobayashi, M., and Heeger, A.J. (1984) Poly(Isothianaphthene). *J. Org. Chem.*, **49** (18), 3382–3384.
34. Yashima, H., Kobayashi, M., Lee, K.B., Chung, D., Heeger, A.J., and Wudl, F. (1987) Electrochromic switching of the optical-properties of polyisothianaphthene. *J. Electrochem. Soc.*, **134** (1), 46–52.
35. Kobayashi, M., Colaneri, N., Boysel, M., Wudl, F., and Heeger, A.J. (1985) The electronic and electrochemical properties of poly(isothianaphthene). *J. Chem. Phys.*, **82** (12), 5717–5723.
36. Lee, K. and Sotzing, G.A. (2001) Poly(thieno[3,4-b]thiophene). A new stable low band gap conducting polymer. *Macromolecules*, **34** (17), 5746–5747.
37. Sotzing, G.A. and Lee, K.H. (2002) Poly(thieno[3,4-b]thiophene): a p- and n-dopable polythiophene exhibiting high optical transparency in the semiconducting state. *Macromolecules*, **35** (19), 7281–7286.
38. Kumar, A., Buyukmumcu, Z., and Sotzing, G.A. (2006) Poly(thieno[3,4-b]furan). A new low band gap conjugated polymer. *Macromolecules*, **39** (8), 2723–2725.
39. Kumar, A., Bokria, J.G., Buyukmumcu, Z., Dey, T., and Sotzing, G.A. (2008) Poly(thieno[3,4-b]furan), a new low band gap polymer: experiment and theory. *Macromolecules*, **41** (19), 7098–7108.
40. Ferraris, J.P. and Lambert, T.L. (1991) Narrow bandgap polymers – poly-4-dicyanomethylene-4h-cyclopenta[2,1-B-3,4-B']dithiophene (Pcdm). *J. Chem. Soc., Chem. Commun.*, (18), 1268–1270.
41. Ferraris, J.P., Henderson, C., Torres, D., and Meeker, D. (1995) Synthesis, spectroelectrochemistry and application in electrochromic devices of an N-dopable

and P-dopable conducting polymer. *Synth. Met.*, **72** (2), 147–152.
42. Ogawa, K. and Rasmussen, S.C. (2006) N-functionalized poly(dithieno[3,2-b: 2',3'-d]pyrrole)s: highly fluorescent materials with reduced band gaps. *Macromolecules*, **39** (5), 1771–1778.
43. Mert, O., Sahin, E., Ertas, E., Ozturk, T., Aydin, E.A., and Toppare, L. (2006) Electrochromic properties of poly(diphenyldithieno[3,2-b; 2',3'-d]thiophene). *J. Electroanal. Chem.*, **591** (1), 53–58.
44. Dyer, A.L., Thompson, E.J., and Reynolds, J.R. (2011) Completing the color palette with spray-processable polymer electrochromics. *ACS Appl. Mater. Interfaces*, **3** (6), 1787–1795.
45. Xu, C.X., Zhao, J.S., Yu, J.S., and Cui, C.S. (2013) Ethylenedioxythiophene derivatized polynapthalenes as active materials for electrochromic devices. *Electrochim. Acta*, **96**, 82–89.
46. Oguzhan, E., Bilgili, H., Koyuncu, F.B., Ozdemir, E., and Koyuncu, S. (2013) A new processable donor-acceptor polymer displaying neutral state yellow electrochromism. *Polymer*, **54** (23), 6283–6292.
47. Icli-Ozkut, M., Oztas, Z., Algi, F., and Cihaner, A. (2011) A neutral state yellow to navy polymer electrochrome with pyrene scaffold. *Org. Electron.*, **12** (9), 1505–1511.
48. Dey, T., Invernale, M.A., Ding, Y., Buyukmumcu, Z., and Sotzing, G.A. (2011) Poly(3,4-propylenedioxythiophene)s as a single platform for full color realization. *Macromolecules*, **44** (8), 2415–2417.
49. Nielsen, C.B., Angerhofer, A., Abboud, K.A., and Reynolds, J.R. (2008) Discrete photopatternable π-conjugated oligomers for electrochromic devices. *J. Am. Chem. Soc.*, **130** (30), 9734–9746.
50. Amb, C.M., Kerszulis, J.A., Thompson, E.J., Dyer, A.L., and Reynolds, J.R. (2011) Propylenedioxythiophene (ProDOT)-phenylene copolymers allow a yellow-to-transmissive electrochrome. *Polym. Chem.*, **2** (4), 812–814.
51. Kerszulis, J.A., Amb, C., Dyer, A., and Reynolds, J. (2014) Follow the yellow brick road—structural optimization of vibrant yellow-to-transmissive electrochromic conjugated polymers. *Macromolecules*, **47** (16), 5462–5469, Manuscript ID: ma-2014-01080u.R01081.
52. Thompson, B.C., Schottland, P., Zong, K.W., and Reynolds, J.R. (2000) In situ colorimetric analysis of electrochromic polymers and devices. *Chem. Mater.*, **12** (6), 1563–1571.
53. Zong, K. and Reynolds, J.R. (2001) 3,4-alkylenedioxypyrroles: functionalized derivatives as monomers for new electron-rich conducting and electroactive polymers. *J. Org. Chem.*, **66** (21), 6873–6882.
54. Merz, A., Schropp, R., and Dotterl, E. (1995) 3,4-Dialkoxypyrroles and 2,3,7,8,12,13,17,18-Octaalkoxyporphyrins. *Synthesis (Stuttgart)*, (7), 795–800.
55. Zotti, G., Zecchin, S., Schiavon, G., and Groenendaal, L. (2000) Conductive and magnetic properties of 3,4-dimethoxy- and 3,4-ethylenedioxy-capped polypyrrole and polythiophene. *Chem. Mater.*, **12** (10), 2996–3005.
56. Jonas, F.D., Heywang, G.D., Schmidtberg, W., Heinze, J. P. D. and Dietrich, M. (1989) Polythiophenes, process for their preparation and their use. EP Patent 0339340 (A2), filed Apr. 8, 1989 and issued Nov. 2, 1989.
57. Heywang, G. and Jonas, F. (1992) Poly(Alkylenedioxythiophene)S – new, very stable conducting polymers. *Adv. Mater.*, **4** (2), 116–118.
58. Jonas, F. and Schrader, L. (1991) Conductive modifications of polymers with polypyrroles and polythiophenes. *Synth. Met.*, **41** (3), 831–836.
59. Elschner, A., Kirchmeyer, S., Lovenich, W., Merker, U., and Reuter, K. (2010) *PEDOT: Principles and Applications of an Intrinsically Conductive Polymer*, Taylor & Francis.
60. Groenendaal, L., Zotti, G., Aubert, P.H., Waybright, S.M., and Reynolds, J.R. (2003) Electrochemistry of

poly(3,4-alkylenedioxythiophene) derivatives. *Adv. Mater.*, **15** (11), 855–879.

61. Heuer, H.W., Wehrmann, R., and Kirchmeyer, S. (2002) Electrochromic window based on conducting poly (3,4-ethylenedioxythiophene)poly(styrene sulfonate). *Adv. Funct. Mater.*, **12** (2), 89–94.
62. Sonmez, G. *et al* (2004) Red, green, and blue colors in polymeric electrochromics. *Adv. Mater.*, **16**, 1905.
63. Kumar, A. and Reynolds, J.R. (1996) Soluble alkyl-substituted poly(ethylenedioxythiophenes) as electrochromic materials. *Macromolecules*, **29** (23), 7629–7630.
64. Sankaran, B. and Reynolds, J.R. (1997) High-contrast electrochromic polymers from alkyl-derivatized poly(3,4-ethylenedioxythiophenes). *Macromolecules*, **30** (9), 2582–2588.
65. Sapp, S.A., Sotzing, G.A., and Reynolds, J.R. (1998) High contrast ratio and fast-switching dual polymer electrochromic devices. *Chem. Mater.*, **10** (8), 2101–2108.
66. Welsh, D.M., Kumar, A., Morvant, M.C., and Reynolds, J.R. (1999) Fast electrochromic polymers based on new poly(3,4-alkylenedioxythiophene) derivatives. *Synth. Met.*, **102** (1-3), 967–968.
67. Dietrich, M., Heinze, J., Heywang, G., and Jonas, F. (1994) Electrochemical and spectroscopic characterization of polyalkylenedioxythiophenes. *J. Electroanal. Chem.*, **369** (1-2), 87–92.
68. Krishnamoorthy, K., Ambade, A.V., Kanungo, M., Contractor, A.Q., and Kumar, A. (2001) Rational design of an electrochromic polymer with high contrast in the visible region: dibenzyl substituted poly(3,4-propylenedioxythiophene). *J. Mater. Chem.*, **11** (12), 2909–2911.
69. Welsh, D.M., Kumar, A., Meijer, E.W., and Reynolds, J.R. (1999) Enhanced contrast ratios and rapid switching in electrochromics based on poly(3,4-propylenedioxythiophene) derivatives. *Adv. Mater.*, **11** (16), 1379–1382.
70. Gaupp, C.L., Welsh, D.M., and Reynolds, J.R. (2002) Poly(ProDOT-Et-2): a high-contrast, high-coloration efficiency electrochromic polymer. *Macromol. Rapid Commun.*, **23** (15), 885–889.
71. Welsh, D.M., Kloeppner, L.J., Madrigal, L., Pinto, M.R., Thompson, B.C., Schanze, K.S., Abboud, K.A., Powell, D., and Reynolds, J.R. (2002) Regiosymmetric dibutyl-substituted poly(3,4-propylenedioxythiophene)s as highly electron-rich electroactive and luminescent polymers. *Macromolecules*, **35** (17), 6517–6525.
72. Cirpan, A., Argun, A.A., Grenier, C.R.G., Reeves, B.D., and Reynolds, J.R. (2003) Electrochromic devices based on soluble and processable dioxythiophene polymers. *J. Mater. Chem.*, **13** (10), 2422–2428.
73. Reeves, B.D., Grenier, C.R.G., Argun, A.A., Cirpan, A., McCarley, T.D., and Reynolds, J.R. (2004) Spray coatable electrochromic dioxythiophene polymers with high coloration efficiencies. *Macromolecules*, **37** (20), 7559–7569.
74. Aqad, E., Lakshmikantham, M.V., and Cava, M.P. (2001) Synthesis of 3,4-ethylenedioxyselenophene (EDOS): a novel building block for electron-rich π-conjugated polymers. *Org. Lett.*, **3** (26), 4283–4285.
75. Patra, A., Wijsboom, Y.H., Zade, S.S., Li, M., Sheynin, Y., Leitus, G., and Bendikov, M. (2008) Poly(3,4-ethylenedioxyselenophene). *J. Am. Chem. Soc.*, **130** (21), 6734–6736.
76. Li, M., Sheynin, Y., Patra, A., and Bendikov, M. (2009) Tuning the electrochromic properties of poly(alkyl-3,4-ethylenedioxyselenophenes) having high contrast ratio and coloration efficiency. *Chem. Mater.*, **21** (12), 2482–2488.
77. Li, M., Patra, A., Sheynin, Y., and Bendikov, M. (2009) Hexyl-derivatized poly(3,4-ethylenedioxyselenophene): novel highly stable organic electrochromic material with high contrast

ratio, high coloration efficiency, and low-switching voltage. *Adv. Mater.*, **21** (17), 1707–1711.

78. Ozkut, M.I., Atak, S., Onal, A.M., and Cihaner, A. (2011) A blue to highly transmissive soluble electrochromic polymer based on poly(3,4-propylenedioxyselenophene) with a high stability and coloration efficiency. *J. Mater. Chem.*, **21** (14), 5268–5272.
79. Sotzing, G.A., Thomas, C.A., Reynolds, J.R., and Steel, P.J. (1998) Low band gap cyanovinylene polymers based on ethylenedioxythiophene. *Macromolecules*, **31** (11), 3750–3752.
80. Thomas, C.A., Zong, K.W., Abboud, K.A., Steel, P.J., and Reynolds, J.R. (2004) Donor-mediated band gap reduction in a homologous series of conjugated polymers. *J. Am. Chem. Soc.*, **126** (50), 16440–16450.
81. Thompson, B.C., Kim, Y.G., McCarley, T.D., and Reynolds, J.R. (2006) Soluble narrow band gap and blue propylenedioxythiophene-cyanovinylene polymers as multifunctional materials for photovoltaic and electrochromic applications. *J. Am. Chem. Soc.*, **128** (39), 12714–12725.
82. Galand, E.M., Kim, Y.G., Mwaura, J.K., Jones, A.G., McCarley, T.D., Shrotriya, V., Yang, Y., and Reynolds, J.R. (2006) Optimization of narrow band-gap propylenedioxythiophene: cyanovinylene copolymers for optoelectronic applications. *Macromolecules*, **39** (26), 9132–9142.
83. Amb, C.M., Beaujuge, P.M., and Reynolds, J.R. (2010) Spray-processable blue-to-highly transmissive switching polymer electrochromes via the donor-acceptor approach. *Adv. Mater.*, **22** (6), 724–728.
84. Icli, M., Pamuk, M., Algi, F., Onal, A.M., and Cihaner, A. (2010) Donor-acceptor polymer electrochromes with tunable colors and performance. *Chem. Mater.*, **22** (13), 4034–4044.
85. Neo, W.T., Loo, L.M., Song, J., Wang, X.B., Cho, C.M., Chan, H.S.O., Zong, Y., and Xu, J.W. (2013) Solution-processable blue-to-transmissive electrochromic benzotriazole-containing conjugated polymers. *Polym. Chem.*, **4** (17), 4663–4675.
86. Hellstrom, S., Henriksson, P., Kroon, R., Wang, E.G., and Andersson, M.R. (2011) Blue-to-transmissive electrochromic switching of solution processable donor-acceptor polymers. *Org. Electron.*, **12** (8), 1406–1413.
87. Akpinar, H., Balan, A., Baran, D., Unver, E.K., and Toppare, L. (2010) Donor-acceptor-donor type conjugated polymers for electrochromic applications: benzimidazole as the acceptor unit. *Polymer*, **51** (26), 6123–6131.
88. Wu, C.G., Lu, M.I., Chang, S.J., and Wei, C.S. (2007) A solution-processable high-coloration-efficiency low-switching-voltage electrochromic polymer based on polycyclopentadithiophene. *Adv. Funct. Mater.*, **17** (7), 1063–1070.
89. Chen, H.Y., Hou, J.H., Hayden, A.E., Yang, H., Houk, K.N., and Yang, Y. (2010) Silicon atom substitution enhances interchain packing in a thiophene-based polymer system. *Adv. Mater.*, **22** (3), 371–375.
90. Lee, B., Seshadri, V., and Sotzing, G.A. (2005) Water dispersible low band gap conductive polymer based on thieno[3,4-b]thiophene. *Synth. Met.*, **152** (1-3), 177–180.
91. Turbiez, M., Frere, P., Leriche, P., Mercier, N., and Roncali, J. (2005) Poly(3,6-dimethoxy-thieno[3,2-b]thiophene): a possible alternative to poly(3,4-ethylenedioxythiophene) (PEDOT). *Chem. Commun. (Camb)*, (9), 1161–1163.
92. Sonmez, G., Sonmez, H.B., Shen, C.K.F., Jost, R.W., Rubin, Y., and Wudl, F. (2005) A processable green polymeric electrochromic. *Macromolecules*, **38** (3), 669–675.
93. Dubois, C.J. and Reynolds, J.R. (2002) 3,4-ethylenedioxythiophene-pyridine-based polymers: redox or n-type electronic conductivity? *Adv. Mater.*, **14** (24), 1844–1846.

94. DuBois, C.J., Abboud, K.A., and Reynolds, J.R. (2004) Electrolyte-controlled redox conductivity and n-type doping in poly(bis-EDOT-pyridine)s. *J. Phys. Chem. B*, **108** (25), 8550–8557.
95. Raimundo, J.M., Blanchard, P., Brisset, H., Akoudad, S., and Roncali, J. (2000) Proquinoid acceptors as building blocks for the design of efficient π-conjugated fluorophores with high electron affinity. *Chem. Commun. (Camb)*, (11), 939–940.
96. Blanchard, P., Raimundo, J.M., and Roncali, J. (2001) New synthetic strategies towards conjugated NLO-phores and fluorophores. *Synth. Met.*, **119** (1-3), 527–528.
97. Aldakov, D., Palacios, M.A., and Anzenbacher, P. (2005) Benzothiadiazoles and dipyrrolyl quinoxalines with extended conjugated chromophores-fluorophores and anion sensors. *Chem. Mater.*, **17** (21), 5238–5241.
98. Durmus, A., Gunbas, G.E., Camurlu, P., and Toppare, L. (2007) A neutral state green polymer with a superior transmissive light blue oxidized state. *Chem. Commun. (Camb)*, (31), 3246–3248.
99. Durmus, A., Gunbas, G.E., and Toppare, L. (2007) New, highly stable electrochromic polymers from 3,4-ethylenedioxythiophene-bis-substituted quinoxalines toward green polymeric materials. *Chem. Mater.*, **19** (25), 6247–6251.
100. Beaujuge, P.M., Ellinger, S., and Reynolds, J.R. (2008) Spray processable green to highly transmissive electrochromics via chemically polymerizable donor-acceptor heterocyclic pentamers. *Adv. Mater.*, **20** (14), 2772–2776.
101. Beaujuge, P.M., Ellinger, S., and Reynolds, J.R. (2008) The donor-acceptor approach allows a black-to-transmissive switching polymeric electrochrome. *Nat. Mater.*, **7** (10), 795–799.
102. Gunbas, G.E., Durmus, A., and Toppare, L. (2008) A unique processable green polymer with a transmissive oxidized state for realization of potential RGB-based electrochromic device applications. *Adv. Funct. Mater.*, **18** (14), 2026–2030.
103. Beaujuge, P.M., Vasilyeva, S.V., Ellinger, S., McCarley, T.D., and Reynolds, J.R. (2009) Unsaturated linkages in dioxythiophene-benzothiadiazole donor-acceptor electrochromic polymers: the key role of conformational freedom. *Macromolecules*, **42** (11), 3694–3706.
104. Cihaner, A. and Algi, F. (2008) A novel neutral state green polymeric electrochromic with superior n- and p-doping processes: Closer to Red-Blue-Green (RGB) display realization. *Adv. Funct. Mater.*, **18** (22), 3583–3589.
105. Gunbas, G.E., Durmus, A., and Toppare, L. (2008) Could green be greener? Novel donor-acceptor-type electrochromic polymers: towards excellent neutral green materials with exceptional transmissive oxidized states for completion of RGB color space. *Adv. Mater.*, **20** (4), 691–695.
106. Hellstrom, S., Cai, T.Q., Inganas, O., and Andersson, M.R. (2011) Influence of side chains on electrochromic properties of green donor-acceptor-donor polymers. *Electrochim. Acta*, **56** (10), 3454–3459.
107. Ozdemir, S., Balan, A., Baran, D., Dogan, O., and Toppare, L. (2011) Green to highly transmissive switching multicolored electrochromes: Ferrocene pendant group effect on electrochromic properties. *React. Funct. Polym.*, **71** (2), 168–174.
108. Celebi, S., Balan, A., Epik, B., Baran, D., and Toppare, L. (2009) Donor acceptor type neutral state green polymer bearing pyrrole as the donor unit. *Org. Electron.*, **10** (4), 631–636.
109. Unur, E., Beaujuge, P.M., Ellinger, S., Jung, J.H., and Reynolds, J.R. (2009) Black to transmissive switching in a pseudo three-electrode electrochromic device. *Chem. Mater.*, **21** (21), 5145–5153.
110. Vasilyeva, S.V., Beaujuge, P.M., Wang, S.J., Babiarz, J.E., Ballarotto, V.W., and

Reynolds, J.R. (2011) Material strategies for black-to-transmissive window-type polymer electrochromic devices. *ACS Appl. Mater. Interfaces*, **3** (4), 1022–1032.

111. Shin, H., Kim, Y., Bhuvana, T., Lee, J., Yang, X., Park, C., and Kim, E. (2012) Color combination of conductive polymers for black electrochromism. *ACS Appl. Mater. Interfaces*, **4** (1), 185–191.

112. Shi, P.J., Amb, C.M., Knott, E.P., Thompson, E.J., Liu, D.Y., Mei, J.G., Dyer, A.L., and Reynolds, J.R. (2010) Broadly absorbing black to transmissive switching electrochromic polymers. *Adv. Mater.*, **22** (44), 4949–4953.

113. Sassi, M., Salamone, M.M., Ruffo, R., Mari, C.M., Pagani, G.A., and Beverina, L. (2012) Gray to colorless switching, crosslinked electrochromic polymers with outstanding stability and transmissivity from naphthalenediimmide-functionalized EDOT. *Adv. Mater.*, **24** (15), 2004–2008.

114. Oktem, G., Balan, A., Baran, D., and Toppare, L. (2011) Donor-acceptor type random copolymers for full visible light absorption. *Chem. Commun. (Camb)*, **47** (13), 3933–3935.

115. Icli, M., Pamuk, M., Algi, F., Onal, A.M., and Cihaner, A. (2010) A new soluble neutral state black electrochromic copolymer via a donor-acceptor approach. *Org. Electron.*, **11** (7), 1255–1260.

116. Ding, G.Q., Cho, C.M., Chen, C.X., Zhou, D., Wang, X.B., Tan, A.Y.X., Xu, J.W., and Lu, X.H. (2013) Black-to-transmissive electrochromism of azulene-based donor-acceptor copolymers complemented by poly(4-styrene sulfonic acid)-doped poly(3,4-ethylenedioxythiophene). *Org. Electron.*, **14** (11), 2748–2755.

117. Lee, K.R. and Sotzing, G.A. (2013) Color tuning of black for electrochromic polymers using precursor blends. *Chem. Commun. (Camb)*, **49** (45), 5192–5194.

118. Siove, A., Ades, D., Ngbilo, E., and Chevrot, C. (1990) Chain-length effect on the electroactivity of poly(N-Alkyl-3,6-Carbazolediyl) thin-films. *Synth. Met.*, **38** (3), 331–340.

119. Chevrot, C., Ngbilo, E., Kham, K., and Sadki, S. (1996) Optical and electronic properties of undoped and doped poly(N-alkylcarbazole) thin layers. *Synth. Met.*, **81** (2-3), 201–204.

120. Tran-Van, F., Henri, T., and Chevrot, C. (2002) Synthesis and electrochemical properties of mixed ionic and electronic modified polycarbazole. *Electrochim. Acta*, **47** (18), 2927–2936.

121. Qiu, Y.J. and Reynolds, J.R. (1990) Poly[3,6-(Carbaz-9-Yl)propanesulfonate] – a self-doped polymer with both cation and anion-exchange properties. *J. Electrochem. Soc.*, **137** (3), 900–904.

122. Verghese, M.M., Ram, M.K., Vardhan, H., Malhotra, B.D., and Ashraf, S.M. (1997) Electrochromic properties of polycarbazole films. *Polymer*, **38** (7), 1625–1629.

123. Scherf, U. and Neher, D. (2008) *Polyfluorenes*, Springer.

124. Bezgin, B., Cihaner, A., and Onal, A.M. (2008) Electrochemical polymerization of 9-fluorenecarboxylic acid and its electrochromic device application. *Thin Solid Films*, **516** (21), 7329–7334.

125. Poriel, C., Liang, J.J., Rault-Berthelot, J., Barriere, F., Cocherel, N., Slawin, A.M.Z., Horhant, D., Virboul, M., Alcaraz, G., Audebrand, N., Vignau, L., Huby, N., Wantz, G., and Hirsch, L. (2007) Dispirofluorene-indenofluorene derivatives as new building blocks for blue organic electroluminescent devices and electroactive polymers. *Chem. Eur. J.*, **13** (36), 10055–10069.

126. Yen, H.J. and Liou, G.S. (2012) Solution-processable triarylamine-based electroactive high performance polymers for anodically electrochromic applications. *Polym. Chem.*, **3** (2), 255–264.

127. Yen, H.J. and Liou, G.S. (2009) Solution-processable novel near-infrared electrochromic aromatic polyamides based on electroactive

tetraphenyl-p-phenylenediamine moieties. *Chem. Mater.*, **21** (17), 4062–4070.

128. Chang, C.W., Liou, G.S., and Hsiao, S.H. (2007) Highly stable anodic green electrochromic aromatic polyamides: synthesis and electrochromic properties. *J. Mater. Chem.*, **17** (10), 1007–1015.

129. Natera, J., Otero, L., Sereno, L., Fungo, F., Wang, N.S., Tsai, Y.M., Hwu, T.Y., and Wong, K.T. (2007) A novel electrochromic polymer synthesized through electropolymerization of a new donor-acceptor bipolar system. *Macromolecules*, **40** (13), 4456–4463.

130. Sonmez, G., Schwendeman, I., Schottland, P., Zong, K.W., and Reynolds, J.R. (2003) N-substituted poly(3,4-propylenedioxypyrrole)s: high gap and low redox potential switching electroactive and electrochromic polymers. *Macromolecules*, **36** (3), 639–647.

131. Walcak, R.M., Jung, J.H., Cowart, J.S., and Reynolds, J.R. (2007) 3,4-alkylenedioxypyrrole-based conjugated polymers with finely tuned electronic and optical properties via a flexible and efficient N-functionalization method. *Macromolecules*, **40** (22), 7777–7785.

132. Walczak, R.M., Leonard, J.K., and Reynolds, J.R. (2008) Processable, electroactive, and aqueous compatible poly(3,4-alkylenedioxypyrrole)s through a functionally tolerant delodination condensation polymerization. *Macromolecules*, **41** (3), 691–700.

133. Schwendeman, I., Hickman, R., Sonmez, G., Schottland, P., Zong, K., Welsh, D.M., and Reynolds, J.R. (2002) Enhanced contrast dual polymer electrochromic devices. *Chem. Mater.*, **14** (7), 3118–3122.

134. Knott, E.P., Craig, M.R., Liu, D.Y., Babiarz, J.E., Dyer, A.L., and Reynolds, J.R. (2012) A minimally coloured dioxypyrrole polymer as a counter electrode material in polymeric electrochromic window devices. *J. Mater. Chem.*, **22** (11), 4953–4962.

135. Arroyave, F.A. and Reynolds, J.R. (2012) Dioxypyrrole-based polymers via dehalogenation polycondensation using various electrophilic halogen sources. *Macromolecules*, **45** (15), 5842–5849.

136. Vasilyeva, S.V., Unur, E., Walczak, R.M., Donoghue, E.P., Rinzler, A.G., and Reynolds, J.R. (2009) Color purity in polymer electrochromic window devices on indium-tin oxide and single-walled carbon nanotube electrodes. *ACS Appl. Mater. Interfaces*, **1** (10), 2288–2297.

137. Bulloch, R.H., Kerszulis, J.A., Dyer, A.L., and Reynolds, J.R. (2014) Mapping the broad CMY subtractive primary color gamut using a dual-active electrochromic device. *ACS Appl. Mater. Interfaces*, **6** (9), 6623–6630.

138. Druy, M.A. and Seymour, R.J. (1983) Poly (2,2' – Bithiophene): an electrochromic conducting polymer. *J. Phys. Colloqyes*, **44** (C3), C3-595–C3-598.

139. Du Bois, C.J., Larmat, F., Irvin, D.J., and Reynolds, J.R. (2001) Multi-colored electrochromic polymers based on BEDOT-pyridines. *Synth. Met.*, **119** (1-3), 321–322.

140. Ouyang, M., Wang, G., Zhang, Y., Hua, C., and Zhang, C. (2011) Multicolored electrochromic copolymer based on 1,4-di(thiophen-3-yl)benzene and 3,4-ethylenedioxythiophene. *J. Electroanal. Chem.*, **653** (1-2), 21–26.

141. Hyodo, K. and Omae, M. (1990) Colorimetric study of the electrochromic properties of a conducting polymer. *J. Electroanal. Chem.*, **292** (1-2), 93–102.

142. Macdiarmid, A.G. and Epstein, A.J. (1989) Polyanilines – a novel class of conducting polymers. *Faraday Discuss.*, **88**, 317–332.

143. Diaz, A.F. and Logan, J.A. (1980) Electroactive polyaniline films. *J. Electroanal. Chem.*, **111** (1), 111–114.

144. Macdiarmid, A.G., Yang, L.S., Huang, W.S., and Humphrey, B.D. (1987) Polyaniline – electrochemistry and application to rechargeable batteries. *Synth. Met.*, **18** (1-3), 393–398.

145. Kobayashi, T., Yoneyama, H., and Tamura, H. (1984) Electrochemical reactions concerned with electrochromism of polyaniline film-coated electrodes. *J. Electroanal. Chem.*, **177** (1-2), 281–291.
146. Kobayashi, T., Yoneyama, H., and Tamura, H. (1984) Polyaniline film-coated electrodes as electrochromic display devices. *J. Electroanal. Chem.*, **161** (2), 419–423.
147. Leclerc, M., Guay, J., and Dao, L.H. (1989) Synthesis and characterization of poly(alkylanilines). *Macromolecules*, **22** (2), 649–653.
148. Chevalier, J.W., Bergeron, J.Y., and Dao, L.H. (1992) Synthesis, characterization, and properties of poly(N-alkylanilines). *Macromolecules*, **25** (13), 3325–3331.
149. Lindfors, T. and Ivaska, A. (2002) pH sensitivity of polyaniline and its substituted derivatives. *J. Electroanal. Chem.*, **531** (1), 43–52.
150. Daprano, G., Leclerc, M., Zotti, G., and Schiavon, G. (1995) Synthesis and characterization of polyaniline derivatives – poly(2-alkoxyanilines) and poly(2,5-dialkoxyanilines). *Chem. Mater.*, **7** (1), 33–42.
151. Jiang, Y. and Epstein, A.J. (1990) Synthesis of self-doped conducting polyaniline. *J. Am. Chem. Soc.*, **112** (7), 2800–2801.
152. Tarver, J., Yoo, J.E., and Loo, Y.L. (2010) Polyaniline exhibiting stable and reversible switching in the visible extending into the near-IR in aqueous media. *Chem. Mater.*, **22** (7), 2333–2340.
153. Gaupp, C.L. and Reynolds, J.R. (2003) Multichromic copolymers based on 3,6-bis(2-(3,4-ethylenedioxythiophene))-N-alkylcarbazole derivatives. *Macromolecules*, **36** (17), 6305–6315.
154. Yiğit, D., Udum, Y.A., Güllü, M., and Toppare, L. (2014) Electrochemical and optical properties of novel terthienyl based azobenzene, coumarine and fluorescein containing polymers: multicolored electrochromic polymers. *J. Electroanal. Chem.*, **712**, 215–222.
155. Sicard, L., Navarathne, D., Skalski, T., and Skene, W.G. (2013) On-substrate preparation of an electroactive conjugated polyazomethine from solution-processable monomers and its application in electrochromic devices. *Adv. Funct. Mater.*, **23** (28), 3549–3559.
156. Irvin, D.J., DuBois, C.J., and Reynolds, J.R. (1999) Dual p- and n-type doping in an acid sensitive alternating bi(ethylenedioxythiophene) and pyridine polymer. *Chem. Commun. (Camb)*, (20), 2121–2122.
157. Camurlu, P. and Karagoren, N. (2013) Both p and n-dopable, multichromic, napthalineimide clicked poly(2,5-dithienylpyrrole) derivatives. *J. Electrochem. Soc.*, **160** (9), H560–H567.
158. Ferraris, J.P. and Skiles, G.D. (1987) Substitutional alloys of organic polymeric conductors. *Polymer*, **28** (2), 179–182.
159. Ferraris, J.P., Andrus, R.G., and Hrncir, D.C. (1989) Steric effects on the optical and electrochemical properties of n-substituted pyrrole thiophene monomers and polymers. *J. Chem. Soc., Chem. Commun.*, (18), 1318–1320.
160. Varis, S., Ak, M., Tanyeli, C., Akhmedov, I.M., and Toppare, L. (2006) A soluble and multichromic conducting polythiophene derivative. *Eur. Polym. J.*, **42** (10), 2352–2360.
161. Yildiz, E., Camurlu, P., Tanyeli, C., Akhmedov, I., and Toppare, L. (2008) A soluble conducting polymer of 4-(2,5-di(thiophen-2-yl)-1H-pyrrol-1-yl)benzenamine and its multichromic copolymer with EDOT. *J. Electroanal. Chem.*, **612** (2), 247–256.
162. Song, H.K., Lee, E.J., and Oh, S.M. (2005) Electrochromism of 2,2'-azinobis(3-ethylbenzothiazoline-6-sulfonate) incorporated into conducting polymer as a dopant. *Chem. Mater.*, **17** (9), 2232–2233.
163. Fei, J.F., Lim, K.G., and Palmore, G.T.R. (2008) Polymer composite with three electrochromic states. *Chem. Mater.*, **20** (12), 3832–3839.
164. Heinze, J., Frontana-Uribe, B.A., and Ludwigs, S. (2010) Electrochemistry of conducting polymers-persistent models

and new concepts. *Chem. Rev.*, **110** (8), 4724–4771.

165. Sadki, S., Schottland, P., Brodie, N., and Sabouraud, G. (2000) The mechanisms of pyrrole electropolymerization. *Chem. Soc. Rev.*, **29** (5), 283–293.

166. Poverenov, E., Li, M., Bitler, A., and Bendikov, M. (2010) Major effect of electropolymerization solvent on morphology and electrochromic properties of PEDOT films. *Chem. Mater.*, **22** (13), 4019–4025.

167. Sonmez, G., Meng, H., and Wudl, F. (2004) Organic polymeric electrochromic devices: polychromism with very high coloration efficiency. *Chem. Mater.*, **16** (4), 574–580.

168. Gaupp, C.L., Welsh, D.M., Rauh, R.D., and Reynolds, J.R. (2002) Composite coloration efficiency measurements of electrochromic polymers based on 3,4-alkylenedioxythiophenes. *Chem. Mater.*, **14** (9), 3964–3970.

169. Shen, D.E., Estrada, L.A., Österholm, A.M., Salazar, D.H., Dyer, A.L., and Reynolds, J.R. (2014) Understanding the effects of electrochemical parameters on the areal capacitance of electroactive polymers. *J. Mater. Chem. A*, **2** (20), 7509.

170. Zhu, Y., Otley, M.T., Alamer, F.A., Kumar, A., Zhang, X., Mamangun, D.M.D., Li, M., Arden, B.G., and Sotzing, G.A. (2014) Electrochromic properties as a function of electrolyte on the performance of electrochromic devices consisting of a single-layer polymer. *Org. Electron.*, **15** (7), 1378–1386.

171. Argun, A.A., Cirpan, A., and Reynolds, J.R. (2003) The first truly all-polymer electrochromic devices. *Adv. Mater.*, **15** (16), 1338–1341.

172. Seshadri, V., Padilla, J., Bircan, H., Radmard, B., Draper, R., Wood, M., Otero, T.F., and Sotzing, G.A. (2007) Optimization, preparation, and electrical short evaluation for 30cm^2 active area dual conjugated polymer electrochromic windows. *Org. Electron.*, **8** (4), 367–381.

173. Argun, A.A., Aubert, P.H., Thompson, B.C., Schwendeman, I., Gaupp, C.L., Hwang, J., Pinto, N.J., Tanner, D.B., MacDiarmid, A.G., and Reynolds, J.R. (2004) Multicolored electrochromism polymers: structures and devices. *Chem. Mater.*, **16** (23), 4401–4412.

174. Inganas, O., Carlberg, C., and Yohannes, T. (1998) Polymer electrolytes in optical devices. *Electrochim. Acta*, **43** (10-11), 1615–1621.

175. Ding, Y., Invernale, M.A., Mamangun, D.M.D., Kumar, A., and Sotzing, G.A. (2011) A simple, low waste and versatile procedure to make polymer electrochromic devices. *J. Mater. Chem.*, **21** (32), 11873.

176. Invernale, M.A., Ding, Y., Mamangun, D.M., Yavuz, M.S., and Sotzing, G.A. (2010) Preparation of conjugated polymers inside assembled solid-state devices. *Adv. Mater.*, **22** (12), 1379–1382.

177. Alamer, F.A., Otley, M.T., Ding, Y., and Sotzing, G.A. (2013) Solid-state high-throughput screening for color tuning of electrochromic polymers. *Adv. Mater.*, **25** (43), 6256–6260.

178. Otley, M.T., Alamer, F.A., Zhu, Y., Singhaviranon, A., Zhang, X., Li, M., Kumar, A., and Sotzing, G.A. (2014) Acrylated poly(3,4-propylenedioxythiophene) for enhancement of lifetime and optical properties for single-layer electrochromic devices. *ACS Appl. Mater. Interfaces*, **6** (3), 1734–1739.

179. İçli, M., Pamuk, M., Algi, F., Önal, A.M., and Cihaner, A. (2010) Donor–acceptor polymer electrochromes with tunable colors and performance. *Chem. Mater.*, **22** (13), 4034–4044.

180. Elsenbaumer, R.L., Jen, K.Y., and Oboodi, R. (1986) Processible and environmentally stable conducting polymers. *Synth. Met.*, **15** (2-3), 169–174.

181. Jen, K.Y., Miller, G.G., and Elsenbaumer, R.L. (1986) Highly conducting, soluble, and environmentally-stable poly(3-alkylthiophenes). *J. Chem. Soc., Chem. Commun.*, (17), 1346–1347.

182. Hotta, S., Rughooputh, D.D.V., Heeger, A.J., and Wudl, F. (1987) Spectroscopic studies of soluble poly(3-Alkylthienylenes). *Macromolecules*, **20** (1), 212–215.

183. Groenendaal, B.L., Jonas, F., Freitag, D., Pielartzik, H., and Reynolds, J.R. (2000) Poly(3,4-ethylenedioxythiophene) and its derivatives: past, present, and future. *Adv. Mater.*, **12** (7), 481–494.
184. Mortimer, R.J., Graham, K.R., Grenier, C.R., and Reynolds, J.R. (2009) Influence of the film thickness and morphology on the colorimetric properties of spray-coated electrochromic disubstituted 3,4-propylenedioxythiophene polymers. *ACS Appl. Mater. Interfaces*, **1** (10), 2269–2276.
185. Andersson, P., Nilsson, D., Svensson, P.O., Chen, M.X., Malmstrom, A., Remonen, T., Kugler, T., and Berggren, M. (2002) Active matrix displays based on all-organic electrochemical smart pixels printed on paper. *Adv. Mater.*, **14** (20), 1460–1464.
186. Andersson, P., Forchheimer, R., Tehrani, P., and Berggren, M. (2007) Printable all-organic electrochromic active-matrix displays. *Adv. Funct. Mater.*, **17** (16), 3074–3082.
187. Berggren, M., Nilsson, D., and Robinson, N.D. (2007) Organic materials for printed electronics. *Nat. Mater.*, **6** (1), 3–5.
188. Freund, M.S., Svenda, N. and Deore, B.A. (2012) In situ polymerization of conducting poly(3,4-ethylenedioxythiophene). EP Patent 2451850 A1, filed Jul. 9, 2010 and issued May 16, 2012.
189. Jonas, F.D. and Krafft, W.D. (1996) New polythiophene dispersions, their preparation and their use. EP Patent 0440957 B1, filed Dec. 20, 1990 and issued Mar. 27, 1996.
190. Krafft, W.D., Jonas, F.D., Muys, B. and Quintens, D. D. (1993) Antistatic Plastic Parts: Google Patents. US Patent 5370981 A, filed Mar. 24, 1993 and issued Dec 6, 1994.
191. Jonas, F., Krafft, W., and Muys, B. (1995) Poly(3,4-Ethylenedioxythiophene) – conductive coatings, technical applications and properties. *Macromol. Symp.*, **100**, 169–173.
192. Lim, J.A., Cho, J.H., Park, Y.D., Kim, D.H., Hwang, M., and Cho, K. (2006) Solvent effect of inkjet printed source/drain electrodes on electrical properties of polymer thin-film transistors. *Appl. Phys. Lett.*, **88** (8), 082102.
193. Krebs, F.C. (2009) All solution roll-to-roll processed polymer solar cells free from indium-tin-oxide and vacuum coating steps. *Org. Electron.*, **10** (5), 761–768.
194. Ersman, P.A., Kawahara, J., and Berggren, M. (2013) Printed passive matrix addressed electrochromic displays. *Org. Electron.*, **14** (12), 3371–3378.
195. Ding, Y., Invernale, M.A., and Sotzing, G.A. (2010) Conductivity trends of PEDOT-PSS impregnated fabric and the effect of conductivity on electrochromic textile. *ACS Appl. Mater. Interfaces*, **2** (6), 1588–1593.
196. Lee, B., Seshadri, V., and Sotzing, G.A. (2005) Poly(thieno[3,4-b]thiophene)-poly(styrene sulfonate): a low band gap, water dispersible conjugated polymer. *Langmuir*, **21** (23), 10797–10802.
197. Somboonsub, B., Thongyai, S., Scola, D.A., Sotzing, G.A., and Praserthdam, P. (2012) Sulfonated polyimide as a thermally stable template for water processable conductive polymers. *Synth. Met.*, **162** (11-12), 941–947.
198. Ner, Y., Invernale, M.A., Grote, J.G., Stuart, J.A., and Sotzing, G.A. (2010) Facile chemical synthesis of DNA-doped PEDOT. *Synth. Met.*, **160** (5-6), 351–353.
199. Tarver, J. and Loo, Y.-L. (2011) Mesostructures of polyaniline films affect polyelectrochromic switching. *Chem. Mater.*, **23** (19), 4402–4409.
200. Shi, P., Amb, C.M., Dyer, A.L., and Reynolds, J.R. (2012) Fast switching water processable electrochromic polymers. *ACS Appl. Mater. Interfaces*, **4** (12), 6512–6521.
201. Hoven, C.V., Garcia, A., Bazan, G.C., and Nguyen, T.-Q. (2008) Recent applications of conjugated polyelectrolytes in optoelectronic devices. *Adv. Mater.*, **20** (20), 3793–3810.
202. Huang, F., Wu, H., and Cao, Y. (2010) Water/alcohol soluble conjugated polymers as highly efficient electron

transporting/injection layer in optoelectronic devices. *Chem. Soc. Rev.*, **39** (7), 2500–2521.
203. Zhu, C., Liu, L., Yang, Q., Lv, F., and Wang, S. (2012) Water-soluble conjugated polymers for imaging, diagnosis, and therapy. *Chem. Rev.*, **112** (8), 4687–4735.
204. Jiang, H., Taranekar, P., Reynolds, J.R., and Schanze, K.S. (2009) Conjugated polyelectrolytes: synthesis, photophysics, and applications. *Angew. Chem. Int. Ed.*, **48** (24), 4300–4316.
205. Patil, A.O., Ikenoue, Y., Basescu, N., Colaneri, N., Chen, J., Wudl, F., and Heeger, A.J. (1987) Self-doped conducting polymers. *Synth. Met.*, **20** (2), 151–159.
206. Patil, A.O., Ikenoue, Y., Wudl, F., and Heeger, A.J. (1987) Water-soluble conducting polymers. *J. Am. Chem. Soc.*, **109** (6), 1858–1859.
207. Sundaresan, N.S., Basak, S., Pomerantz, M., and Reynolds, J.R. (1987) Electroactive copolymers of pyrrole containing covalently bound dopant ions – poly(Pyrrole-Co-[3-(Pyrrol-1-Yl)propanesulphonate]). *J. Chem. Soc., Chem. Commun.*, (8), 621–622.
208. Reynolds, J.R., Sundaresan, N.S., Pomerantz, M., Basak, S., and Baker, C.K. (1988) Self-doped conducting copolymers – a charge and mass-transport study of poly(pyrrole-Co[3-(pyrrol-1-Yl)propanesulfonate]). *J. Electroanal. Chem.*, **250** (2), 355–371.
209. Havinga, E.E., Vanhorssen, L.W., Tenhoeve, W., Wynberg, H., and Meijer, E.W. (1987) Self-doped water-soluble conducting polymers. *Polym. Bull.*, **18** (3), 277–281.
210. Stephan, O., Schottland, P., Le Gall, P.Y., Chevrot, C., Mariet, C., and Carrier, M. (1998) Electrochemical behaviour of 3,4-ethylenedioxythiophene functionalized by a sulphonate group. Application to the preparation of poly(3,4-ethylenedioxythiophene) having permanent cation-exchange properties. *J. Electroanal. Chem.*, **443** (2), 217–226.
211. Decher, G., Hong, J.D., and Schmitt, J. (1992) Buildup of ultrathin multilayer films by a self-assembly process. 3. Consecutively alternating adsorption of anionic and cationic polyelectrolytes on charged surfaces. *Thin Solid Films*, **210** (1-2), 831–835.
212. Decher, G. (1997) Fuzzy nanoassemblies: toward layered polymeric multicomposites. *Science*, **277** (5330), 1232–1237.
213. Decher, G., Lvov, Y., and Schmitt, J. (1994) Proof of multilayer structural organization in self-assembled polycation polyanion molecular films. *Thin Solid Films*, **244** (1-2), 772–777.
214. DeLongchamp, D. and Hammond, P.T. (2001) Layer-by-layer assembly of PEDOT/polyaniline electrochromic devices. *Adv. Mater.*, **13** (19), 1455–1459.
215. Cutler, C.A., Bouguettaya, M., and Reynolds, J.R. (2002) PEDOT polyelectrolyte based electrochromic films via electrostatic adsorption. *Adv. Mater.*, **14** (9), 684–688.
216. Cutler, C.A., Bouguettaya, M., Kang, T.S., and Reynolds, J.R. (2005) Alkoxysulfonate-functionalized PEDOT polyelectrolyte multilayer films: electrochromic and hole transport materials. *Macromolecules*, **38** (8), 3068–3074.
217. Jain, V., Sahoo, R., Mishra, S.P., Sinha, J., Montazami, R., Yochum, H.M., Heflin, J.R., and Kumar, A. (2009) Synthesis and characterization of regioregular water-soluble 3,4-propylenedioxythiophene derivative and its application in the fabrication of high-contrast solid-state electrochromic devices. *Macromolecules*, **42** (1), 135–140.
218. Jia, P., Argun, A.A., Xu, J., Xiong, S., Ma, J., Hammond, P.T., and Lu, X. (2010) High-contrast electrochromic thin films via layer-by-layer assembly of starlike and sulfonated polyaniline. *Chem. Mater.*, **22** (22), 6085–6091.
219. Duarte, A., Pu, K.-Y., Liu, B., and Bazan, G.C. (2011) Recent advances in

conjugated polyelectrolytes for emerging optoelectronic applications. *Chem. Mater.*, **23** (3), 501–515.
220. Huang, F., Wu, H.B., Wang, D., Yang, W., and Cao, Y. (2004) Novel electroluminescent conjugated polyelectrolytes based on polyfluorene. *Chem. Mater.*, **16** (4), 708–716.
221. Sinha, J., Sahoo, R., and Kumar, A. (2009) Processable, regioregular, and "Click"able monomer and polymers based on 3,4-propylenedioxythiophene with tunable solubility. *Macromolecules*, **42** (6), 2015–2022.
222. Beaujuge, P.M., Amb, C.M., and Reynolds, J.R. (2010) A side-chain defunctionalization approach yields a polymer electrochrome spray-processable from water. *Adv. Mater.*, **22** (47), 5383–5387.
223. Søndergaard, R.R., Hösel, M., and Krebs, F.C. (2013) Roll-to-Roll fabrication of large area functional organic materials. *J. Polym. Sci., Part B: Polym. Phys.*, **51** (1), 16–34.
224. Krebs, F.C. (2009) Fabrication and processing of polymer solar cells: a review of printing and coating techniques. *Sol. Energy Mater. Sol. Cells*, **93** (4), 394–412.
225. Jensen, J., Dam, H.F., Reynolds, J.R., Dyer, A.L., and Krebs, F.C. (2012) Manufacture and demonstration of organic photovoltaic-powered electrochromic displays using roll coating methods and printable electrolytes. *J. Polym. Sci., Part B: Polym. Phys.*, **50** (8), 536–545.
226. Small, W.R., Masdarolomoor, F., Wallace, G.G., and in het Panhuis, M. (2007) Inkjet deposition and characterization of transparent conducting electroactive polyaniline composite films with a high carbon nanotube loading fraction. *J. Mater. Chem.*, **17** (41), 4359.
227. Shim, G.H., Han, M.G., Sharp-Norton, J.C., Creager, S.E., and Foulger, S.H. (2008) Inkjet-printed electrochromic devices utilizing polyaniline–silica and poly(3,4-ethylenedioxythiophene)–silica colloidal composite particles. *J. Mater. Chem.*, **18** (5), 594.
228. Brotherston, I.D., Mudigonda, D.S.K., Osborn, J.M., Belk, J., Chen, J., Loveday, D.C., Boehme, J.L., Ferraris, J.P., and Meeker, D.L. (1999) Tailoring the electrochromic properties of devices via polymer blends, copolymers, laminates and patterns. *Electrochim. Acta*, **44** (18), 2993–3004.
229. Tehrani, P., Isaksson, J., Mammo, W., Andersson, M.R., Robinson, N.D., and Berggren, M. (2006) Evaluation of active materials designed for use in printable electrochromic polymer displays. *Thin Solid Films*, **515** (4), 2485–2492.
230. Yu, J.F., Abley, M., Yang, C., and Holdcroft, S. (1998) Chemically amplified photolithography of a conjugated polymer. *Chem. Commun. (Camb)*, (15), 1503–1504.
231. Gordon, T.J., Yu, J.F., Yang, C., and Holdcroft, S. (2007) Direct thermal patterning of a π-conjugated polymer. *Chem. Mater.*, **19** (9), 2155–2161.
232. Abdou, M.S.A., Diazguijada, G.A., Arroyo, M.I., and Holdcroft, S. (1991) Photoimaging of electronically conducting polymeric networks. *Chem. Mater.*, **3** (6), 1003–1006.
233. Lowe, J. and Holdcroft, S. (1995) Synthesis and photolithography of polymers and copolymers based on poly(3-(2-(methacryloyloxy)ethyl)thiophene). *Macromolecules*, **28** (13), 4608–4616.
234. Jensen, J. *et al.* (2013) Standard direct photopatterning of electrochromic polymers. *Adv. Funct. Mater.*, **23**, 3728.
235. Kim, J., You, J., and Kim, E. (2010) Flexible conductive polymer patterns from vapor polymerizable and photo-cross-linkable EDOT. *Macromolecules*, **43** (5), 2322–2327.
236. Choi, J., Kumar, A., and Sotzing, G.A. (2007) Nanopatterned electrochromic conjugated poly(terthiophene)s via thermal nanoimprint lithography of precursor polymer. *J. Macromol. Sci., Part A*, **44** (12), 1305–1309.
237. Admassie, S. and Inganäs, O. (2004) Electrochromism in diffractive

conducting polymer gratings. *J. Electrochem. Soc.*, **151** (6), H153.

238. Takamatsu, S., Takahata, T., Matsumoto, K., and Shimoyama, I. (2011) Micro-patterning of a conductive polymer and an insulation polymer using the Parylene lift-off method for electrochromic displays. *J. Micromech. Microeng.*, **21** (7), 075021.

239. Tehrani, P., Hennerdal, L.-O., Dyer, A.L., Reynolds, J.R., and Berggren, M. (2009) Improving the contrast of all-printed electrochromic polymer on paper displays. *J. Mater. Chem.*, **19** (13), 1799.

6
Electrochromism within Transition-Metal Coordination Complexes and Polymers

Yu-Wu Zhong

6.1
Electronic Transitions and Redox Properties of Transition-Metal Complexes

Transition-metal complexes are made of metal ions and ligands. They often possess very rich electronic transitions, including the metal-localised transitions, the ligand-localised transitions and those resulting from the mutual interaction of metal and ligands. Figure 6.1a shows a schematic molecular orbital diagram for an octahedral monometallic complex. Such a picture is very useful for the understanding of the various natures of the electronic transitions of transition-metal complexes [1]. The σ^*_{M} (M stands for metal) molecular orbital is mainly from the p and s atomic orbitals of the metal ion. The $\sigma^*_{\mathrm{M}}(e_{\mathrm{g}})$ and $\pi_{\mathrm{M}}(t_{2\mathrm{g}})$ orbitals are from the d orbital splitting of the metal ion with an octahedral configuration. The π^*_{L} (L stands for ligand) and π_{L} molecular orbitals have major contributions from the ligand π orbital, and the σ_{L} molecular orbital mainly originates from the ligand σ orbital. In the ground state, the π_{L} and σ_{L} orbitals are usually filled, and the π^*_{L}, $\sigma^*_{\mathrm{M}}(e_{\mathrm{g}})$ and σ^*_{M} orbitals are usually empty. The $\pi_{\mathrm{M}}(t_{2\mathrm{g}})$ orbital is partially or fully filled. Three kinds of electronic transitions can be classified from these orbitals, including the metal-centred (MC) transitions (also called d-d transitions), the ligand-centred (LC) transitions (also called intraligand, IL, transitions) and the charge-transfer transitions (CT) between metal and ligand. The latter transition can be further distinguished into ligand-to-metal charge-transfer (LMCT) and metal-to-ligand charge-transfer (MLCT) transitions. In addition to these transitions, ligand-to-ligand charge-transfer (LLCT) between two different ligands is possible. Transition-metal complexes-based electrochromism usually involves the MLCT absorption spectral changes in the visible region. However, it should be kept in mind that the d orbital splitting and energy orderings of different complexes could differ significantly and these complexes should be treated on a case-by-case basis.

In addition to monometallic complexes, dimetallic mixed-valence complexes [2, 3] with moderate-to-strong metal–metal electronic coupling have been used as the active components for electrochromism. This kind of complex can be represented by a general formula $[\mathrm{M}_a^n - \mathrm{L} - \mathrm{M}_b^{n+1}]$, where M_a and M_b stand for two

Electrochromic Materials and Devices, First Edition.
Edited by Roger J. Mortimer, David R. Rosseinsky and Paul M. S. Monk.

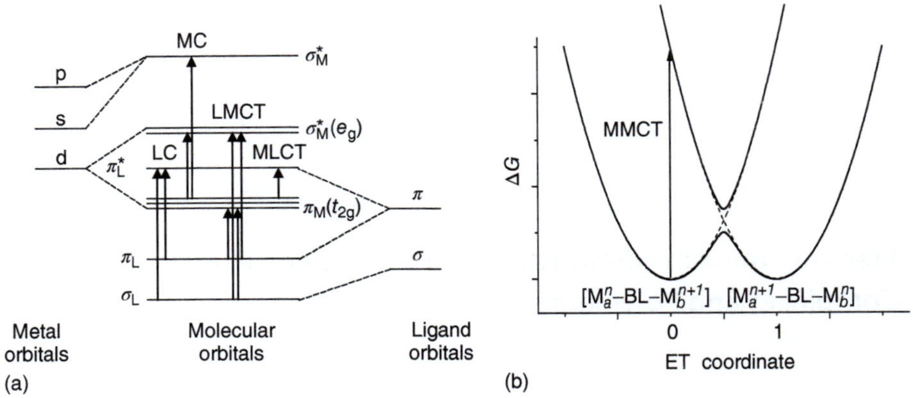

Figure 6.1 (a) Schematic molecular orbital diagram and electronic transitions for an octahedral monometallic complex. (b) Potential energy surfaces of mixed-valence compounds with MMCT transitions. Dashed curves are diabatic states; solid curves are adiabatic states.

metal centres and L for the bridging ligand. Because of the presence of strong coupling between metal centres, they display two well-separated redox waves. In the mixed-valence state, a strong metal-to-metal charge-transfer (MMCT) transition (or intervalence charge-transfer, IVCT) is often observed in the near-infrared (NIR) region. However, the MMCT band is not present in the low or high homovalent states, $[M_a^n - L - M_b^n]$ and $[M_a^{n+1} - L - M_b^{n+1}]$, which forms the basis for the uses of these complexes in electrochromism [4]. Figure 6.1b shows a schematic diagram of the adiabatic free energy surfaces (solid lines) of a mixed-valence complex, according to the classical two-state Marcus–Hush theory [5, 6]. The interaction of two localised diabatic states (dashed curves), $[M_a^n - L - M_b^{n+1}]$ and $[M_a^{n+1} - L - M_b^n]$, gives rise to two new adiabatic potential energy surfaces (solid curves) along the electron transfer (ET) coordinate. The optically induced vertical transition from the energy minimum of the adiabatic ground state surface to the excited-state potential surface is the MMCT transition.

Transition-metal complexes often possess multiple well-defined redox processes and the electronic absorptions are largely dependent on the redox states. This is one of the main reasons why transition-metal complexes are appealing for electrochromism. The redox processes of coordination complexes can localise largely on either the metal centre or the surrounding ligands. For instance, the prototype polypyridyl complex, $[Ru(tpy)_2]^{2+}$ (tpy = 2,2′:6′,2″-terpyridine), exhibits a metal-localised redox couple ($Ru^{III/II}$) at +1.32 V and two ligand-localised redox couple ($tpy^{0/-}$) at −1.22 and −1.46 V versus Ag/AgCl [7]. However, in some special complexes, the oxidation state is unclear and the redox reactions are either metal- or ligand-localised, or both metal and ligand are responsible for a single redox process. These ligands are thus redox non-innocent, which is largely caused by the strong orbital overlap between metal and ligand. Two types of non-innocent ligands are well known. One class is 1,2-dioxolene, 1,2-dithiolene, *o*-phenylenediamine and other electronically similar ligands [8]. The other is the anionic carbon ligand, such as phenylacetylene, phenylvinylene or phenyl group, when connected to a metal centre with M–C bond [9]. These complexes are very useful in electrochromism, owing to the low redox potentials and strong absorptions in the visible and NIR regions. This will be discussed later in this chapter.

In order to develop solid-state devices for practical use, electrochromic materials must be deposited onto electrodes as thin films. The following sections are thus arranged according to the difference in film-forming methods. The electrochromism of complexes as small molecules dissolved in solution will be briefly mentioned, if necessary. Electropolymerisation is a very useful means for preparing thin films, which will be the topic of Sections 6.2 and 6.3. The difference between them is the mechanism of the electropolymerisation: reductively versus oxidatively initiated. Multilayer films simply prepared by self-assembly on electrode surfaces or adsorbing on semiconductor surfaces are discussed in Section 6.4. Section 6.5 is devoted to the electrochromism in thin films prepared by spin coating or drop casting. Finally, a comparison of these film-forming methods for electrochromism will be discussed.

6.2
Electrochromism in Reductively Electropolymerised Films of Polypyridyl Complexes

The reductive electropolymerisation of vinyl-substituted polypyridyl complexes was pioneered by Murray, Meyer and Abruña in the 1980s [10, 11]. To date, a large number of transition-metal complexes with vinyl-substituted ligands have been synthesised and used for the preparations of electropolymerised films. These films are useful in many applications such as electrode coatings, electrocatalysis, electrogenerated chemiluminescence and electrochromism [12]. The mechanism of the reductive polymerisation of vinyl-substituted polypyridyl complexes is believed to be anionically initiated, followed by radical–radical chain propagation [13]. The metal coordination sphere remains essentially intact upon polymerisation. In general, at least two vinyl groups are needed to effect highly efficient polymerisation, reflecting the importance of cross-linking afforded by multiple vinyl groups.

Figure 6.2 shows some vinyl-substituted polypyridyl ligands, which have been frequently used in electropolymerisation, including 4-vinylpyridine (vpy) [10], 4-methyl-4′-vinyl-2,2′-bipyridine (vbpy) [11], 4′-vinyl-2,2′:6′,2″-terpyridine (vtpy) [14], 5,5′-divinyl-2,2′-bipyridine (dvbpy) [15] and 5,5″-divinyl-4′-tolyl-2,2′:6′,2″-terpyridine (dvtpy) [16]. The Ligand vpy is commercially available at a reasonable price. However, other ligands need to be synthesised at individual laboratories. A common synthesis of these vinyl-substituted compounds is via a Wittig reaction from corresponding aldehydes [12]. A more convenient method is via the Suzuki coupling between potassium vinyltrifluoroborate and corresponding bromo-substituted polypyridines [17]. In addition to the vinyl-substituted polypyridines, ligands with internal C–C double bonds, such as 4,4′-distyryl-2,2′-bipyridine (DSB) [18] and 4,4′-di(3,4-dimethoxystyryl)-2,2′-bipyridine

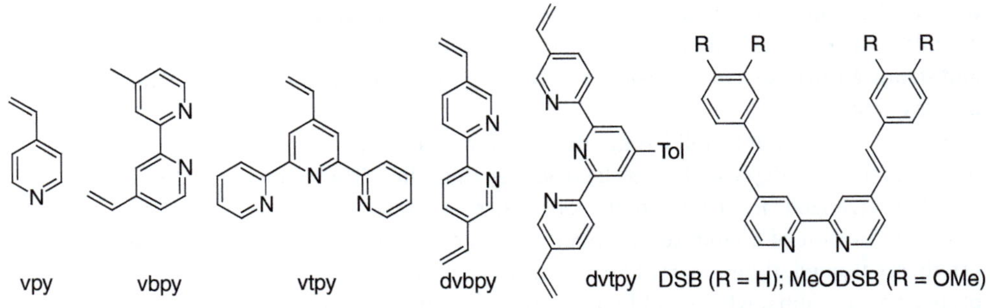

Figure 6.2 Polypyridyl ligands for reductive electropolymerisation. Tol is the *p*-tolyl group.

(MeODSB) [19, 20], have been used for the reductive electropolymerisation. The polymeric films of the ruthenium complexes of these ligands have been studied for electrochromism, where a switch between the colour and bleached states can be realised by manipulating the MLCT absorptions [16, 19, 21].

Depending on the nature of the MLCT absorptions and redox properties, the operating wavelengths, potentials and colours of the films will differ significantly. For instance, poly-$[(dvtpy)Ru(dpb)]^+$ (dpb, that is, 1,3-di(pyrid-2-yl)benzene) displays broad MLCT absorptions around 460–660 nm (Figure 6.3). This absorption decreased upon varying the potential from +0.1 to +0.8 V versus Ag/AgCl. Accordingly, the purple colour of the ITO film was bleached. In contrast, the MLCT transitions of poly-$[(dvtpy)Ru(Mebip)]^{2+}$ (Mebip is *bis(N*-methylbenzimidazolyl)pyridine) film are sharper and blue-shifted with respect to poly-$[(dvtpy)Ru(dpb)]^+$. The ruthenium ion of poly-$[(dvtpy)Ru(Mebip)]^{2+}$ was oxidised when the applied potential was increased from +0.7 to +1.2 V, and the orange colour of the film was bleached. The shallow absorptions between 680 and 900 nm are LMCT transitions [16]. The difference between these two polymers is caused by the presence of a Ru–C bond in $[(dvtpy)Ru(dpb)]^+$ (the so-called cyclometalated ruthenium complex), which significantly decreases the Ru(III/II) potential and the energy of the MLCT transition.

Spatial electrochromism in metallopolymeric films has been reported by Meyer and co-workers [22]. Thin films of poly-$[Ru(vbpy)_2(py)_2]^{2+}$ (py = pyridine) were prepared on Pt disc electrodes by reductive electropolymerisation. Approximately half of the electrode surface was covered with an optical mask. Photolysis of the films in the presence of chloride ion led to photochemical substitution of the pyridine ligands by Cl atoms in the region without optical mask. The resulting film thus consists of spatially segregated poly-$[Ru(vbpy)_2(py)_2]^{2+}$ and poly-$[Ru(vbpy)_2(py)Cl]^+$. These two components display significantly different absorption spectra and redox potentials (Figure 6.4), which provides a basis for the fabrication of bicomponent spatial electrochromism.

The reductive electropolymerisation of tridentate bisruthenium complexes $[Ru_2(vtpy)_2(tpb)]^{2+}$ (tpb = 1,3,4,6-tetra(pyrid-2-yl)benzene) has recently been reported by Zhong and co-workers [23, 24] (Figure 6.5). The resulting metallopolymeric films exhibit three-step electrochromism in the NIR region. These

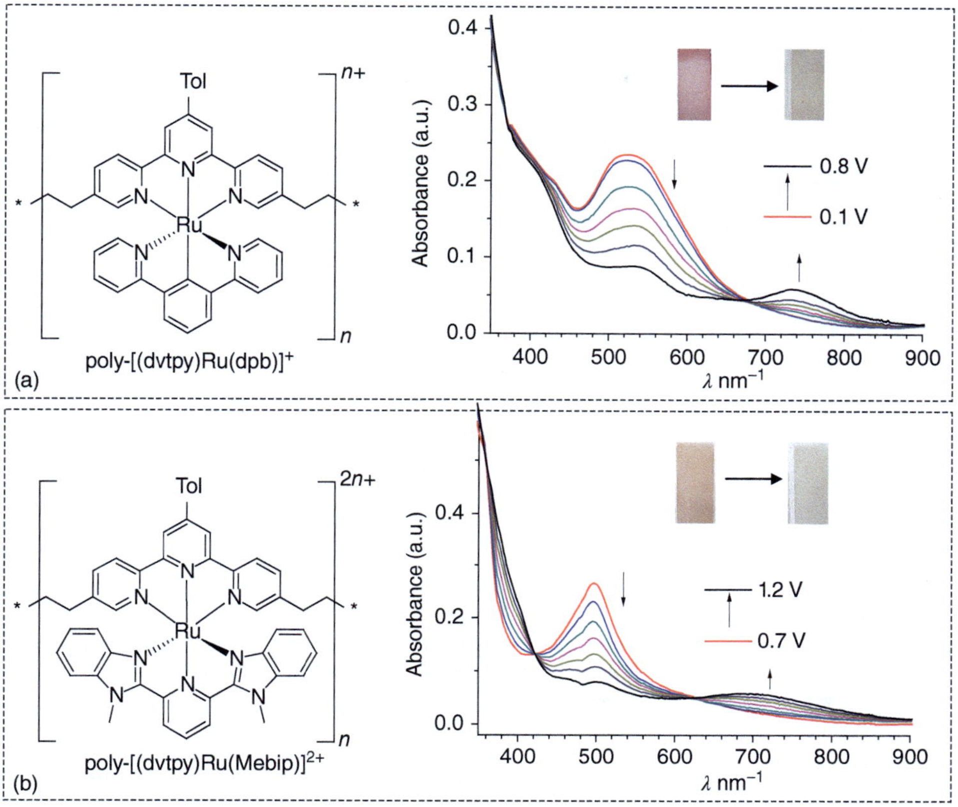

Figure 6.3 Electrochromism of (a) poly-[(dvtpy)Ru(dpb)]$^{2+}$ and (b) poly-[(dvtpy)Ru(Mebip)]$^{2+}$.

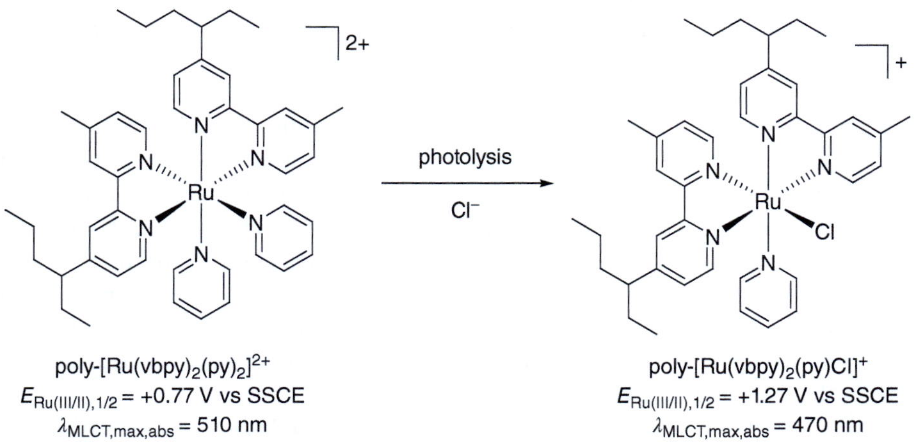

Figure 6.4 Polymers for spatial electrochromism.

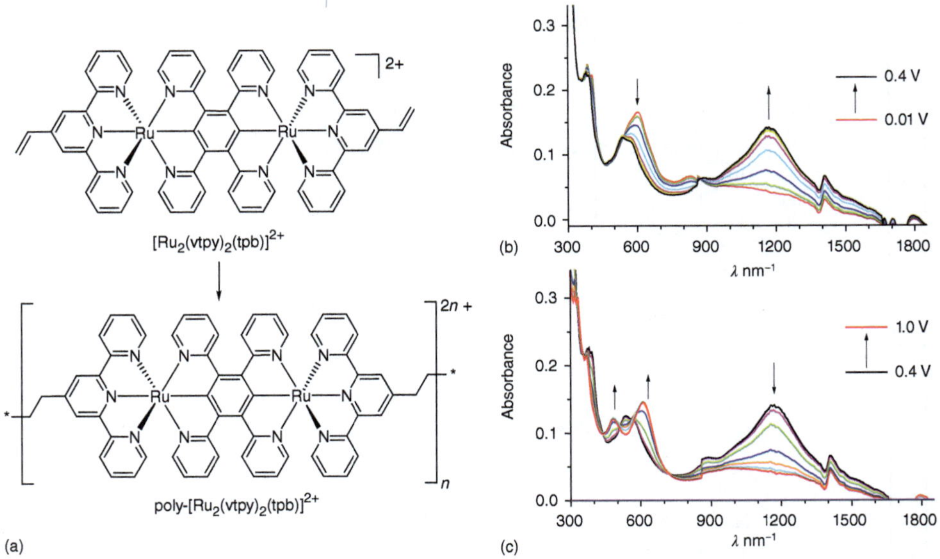

Figure 6.5 (a) Polymerisation of $[Ru_2(vtpy)_2(tpb)]^{2+}$. (b,c) Absorption spectral changes of the film upon stepwise applying potential from (b) +0.01 to +0.4 V and (c) +0.4 to +1.0 V versus Ag/AgCl, respectively.

polymers have two consecutive Ru(III/II) redox couples at +0.15 and +0.56 V versus Ag/AgCl. When the applied potential was gradually increased to +0.4 V to transform the uniform-valence dimetallic unit into the mixed-valence state, the MMCT transitions in the NIR region (centred at 1165 nm) increased and the deep blue colour of the original film changed to pink. Upon further increasing the potential to +1.0 V, the MMCT band at 1165 nm decreased. As a result, the film changed from pink to green. This tricolour change (blue → pink → green) is fully reversible when the applied potential was reversed. In the single-oxidation step, a contrast ratio ($\Delta T\%$) of 40% and coloration efficiency of 250 $cm^2\ C^{-1}$ were achieved at 1165 nm. The response time for the contrast ratio to reach over 90% of its maximum is around 5 s. These films exhibit good electrochemical stability. Well-defined redox couples retain most of their redox ability when the potential is scanned for over 1000 cycles, the peak current dropping to 76% of its original state. Notably, these films exhibit promising optical memory effect (the ability to retain the transmittance after the applied potential is turned off). The pink colour of the film after single oxidation lasts at least for several days, which means that these materials are virtually bistable.

When the ligand was changed from tpb to 1,3,6,8-tetra(2-pyridyl)pyrene (tppyr) with a pyrene backbone, the resulting poly-$[Ru_2(vtpy)_2(tppyr)]^{2+}$ films show similar three-step electrochromic behaviour [25, 26]. However, the $Ru^{II/III}$

redox potentials and the absorption maximum of the MMCT band are different. A typical poly-$[Ru_2(vtpy)_2(tppyr)]^{2+}$ film on ITO/glass displays two Ru(III/II) waves at +0.47 and +0.67 V versus Ag/AgCl. The MMCT band in the mixed-valence state shows an absorption maximum at 1900 nm. The appearance and disappearance of the MMCT band can be observed by applying potentials to stepwise oxidise two ruthenium sites. The colour of the film changed from blue to brown and then to orange during this two-step process.

NIR electrochromic materials are useful in optical telecommunications as variable optical attenuators, since the glass optical fibre has the lowest energy loss in the NIR window [27, 28]. The use of NIR electrochromic materials in other fields is rare. Zhong and co-workers have recently reported on the use of the NIR electrochromism of poly-$[(dvtpy)Ru(dpba)]^+$ (dpba = 3,5-di(pydi-2-yl)-di-*p*-anisylaminobenzene, Figure 6.6) for optical information storage [29]. The metallopolymer film undergoes two well-defined redox processes at +0.32 and +0.68 V versus Ag/AgCl, leading to three well-separated redox states. The singly and doubly oxidised states of the film possess intense absorption at 1070 and 700 nm, respectively. These polymeric films exhibit promising three-stage NIR electrochromism and the best contrast ratios ($\Delta T\%$) of 52% at 1070 nm and 76% at 700 nm have been achieved. The colours of the film at three redox stages are purple, brown and sky blue, respectively. The electrochromic states have long retention times at all three stages (infinity at −0.20 V, 4 h at +0.55 V and 30 min at +1.05 V, respectively). The singly and doubly oxidised states of a film about 10 nm

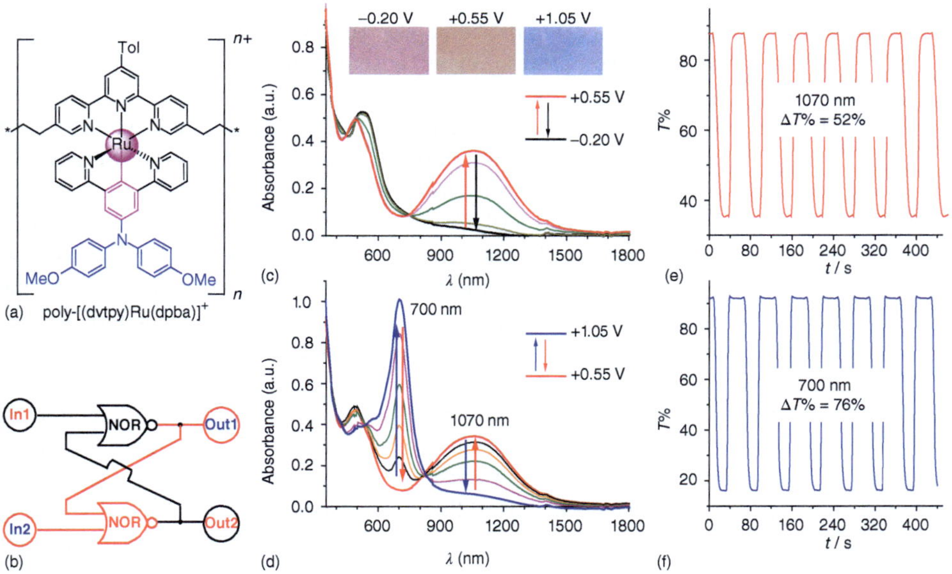

Figure 6.6 (a) Poly-$[(dvtpy)Ru(dpba)]^+$. (b) Logic circuit for a Set/Reset flip-flop. In1 = +0.55 V; In2 = +1.05 V; Out1 = 700 nm; Out2 = 1070 nm. (c–f) Electrochromism at two wavelengths.

thick have been used to build a surface-confined Set/Reset flip-flop memory with two electrochemical inputs and two NIR optical outputs. By using another intermediate state between the singly and doubly oxidised states, a multi-valued logic system has been demonstrated.

In addition to vinyl- or olefin-substituted polypyridine complexes, complexes with chlorophenanthroline or related halogen-substituted polypyridines have been reported to be reductively polymerised [30, 31]. This chemistry relies upon the electrochemically initiated carbon–halide bond cleavage and radical generation, and the resulting films show electrochromism based on the MLCT transitions.

6.3 Electrochromism in Oxidatively Electropolymerised Films of Transition-Metal Complexes

Compared to reductive electropolymerisation, preparation of thin films by oxidative electropolymerisation has received much more attention. Metal complexes substituted with aniline [32], phenol [33], pyrrole [34], carbazole [35], triphenylamine [36] and thiophene derivatives [37] have long been known to undergo oxidative electropolymerisation with retention of the coordination sphere, as has been discussed in a recent review article [38]. Some representative examples with various polymerisable monomers and recent progresses focusing on the electrochromism of the resulting metallopolymeric films are presented in this section.

Aniline has been used for oxidative electropolymerisation in many organic materials and metal complexes. For instance, polypyridyl complexes $[M(NH_2\text{-}ttpy)_2]^{2+}$ (NH_2-ttpy, that is, 4′-*m*-anilinyl-2,2′:6′,2″-terpyridine, M = Fe, Ru, Figure 6.7) have been polymerised, and the resulting filmed show MLCT-based electrochromism [33, 39]. In addition, the polymeric film of $[Fe(\textit{bis}\text{-anilbpy})_3]^{2+}$ (*bis*-anilbpy, that is, 4,4′-*bis*(pheynylaminomethylene)-2,2′-bipyridine) exhibited multi-stage electrochromism [40]. Oxidation of the film by a series of potential steps to +0.85 V versus SCE resulted in generation of an absorption band at 720 nm, which was ascribed to the radical cation of a *para*-diaminobenzene structural motif formed during electropolymerisation. Application of further positive potentials where the metal centre was oxidised resulted in the decrease of the MLCT band.

Compared to aniline, triphenylamine has received much less attention in electropolymerisation. Triphenylamine usually undergoes dimerisation upon oxidation, but not polymerisation. After dimeric coupling, a tetraphenylbenzidine unit is generated, which is very difficult for further electrochemical coupling because of charge delocalisation. However, if a molecule contains two triphenylamines on distal positions and there is little electronic coupling between them, the formation of polymeric films through the head-to-tail electropolymerisation is possible. For instance, Leung and co-workers have demonstrated that the complex $[Ru(\textit{bis}\text{-}NPh_3phen)(phen)_2]^{2+}$ (phen = 1,10-phenanthroline, Figure 6.8) can be

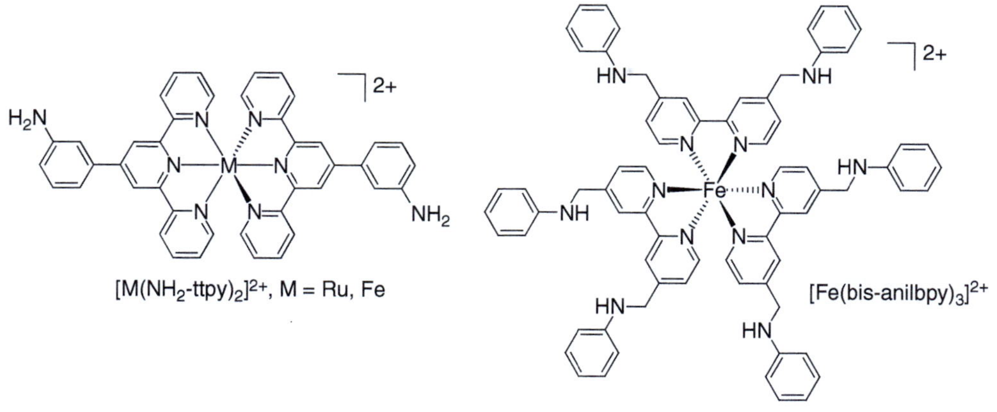

Figure 6.7 $[M(NH_2\text{-ttpy})_2]^{2+}$ and $[Fe(\textit{bis}\text{-anilbpy})_3]^{2+}$ for oxidative electropolymerisation.

smoothly polymerised by oxidation [36]. Figure 6.8b shows the cyclic voltammogram (CV) recorded during electropolymerisation. The appearance of two new redox waves between 100 and 700 mV versus $Fc^{+/0}$ is from the tetraphenylbenzidine unit generated during electropolymerisation. Electrochromism of the resulting film on ITO/glass has been examined. In the neutral state, the orange polymer shows an intense absorption at 452 nm. On oxidation, the film gives a new absorption signal in the NIR region, ranging from 1000 to 2000 nm. Further oxidation of the film leads to absorption at around 800 nm, and the film turns a deep blue–green colour. In addition to this progress, bistridentate complex $[Ru(NPh_3tpy)_2]^{2+}$ (NPh_3tpy is 4′-triphenylamino-2,2′:6′,2″-terpyridine) has been reported to undergo oxidative electropolymerisation [41]. However, no electrochromism has been examined.

By using a similar strategy, Zhong and co-workers have recently performed the oxidative electropolymerisation of the cyclometalated bisruthenium complex $[(Nptpy)_2Ru_2(tppyr)]^{2+}$ (Figure 6.9), where Nptpy is 4′-(*p*-*N*,*N*-diphenylamino)phenyl-2,2′:6′,2″-terpyridine [42]. The resulting polymer is composed of two types of alternating constituent units: tppyr-bridged cyclometalated bisruthenium units and biphenyl-bridged *bis*-triarylamine segments. These films exhibited four well-defined anodic redox couples as a result of the stepwise oxidations of these two units. By manipulating the IVCT transitions of mixed-valent bisruthenium and *bis*-triarylamine units, five-stage NIR electrochromism with stepwise colour changes accompanied by good contrast ratio and coloration efficiency has been realised in these films. Similarly, the complex $[(Nptpy)_2Ru_2(tpb)]^{2+}$ has been polymerised to yield five-stage NIR electrochromism.

The carbazole group has been used for oxidative electropolymerisation of a number of organic materials, as pioneered by Ma and co-workers [43]. However, little work has been devoted to electropolymerisation of metal complexes with carbazole [35]. Shi and co-workers have recently reported that the

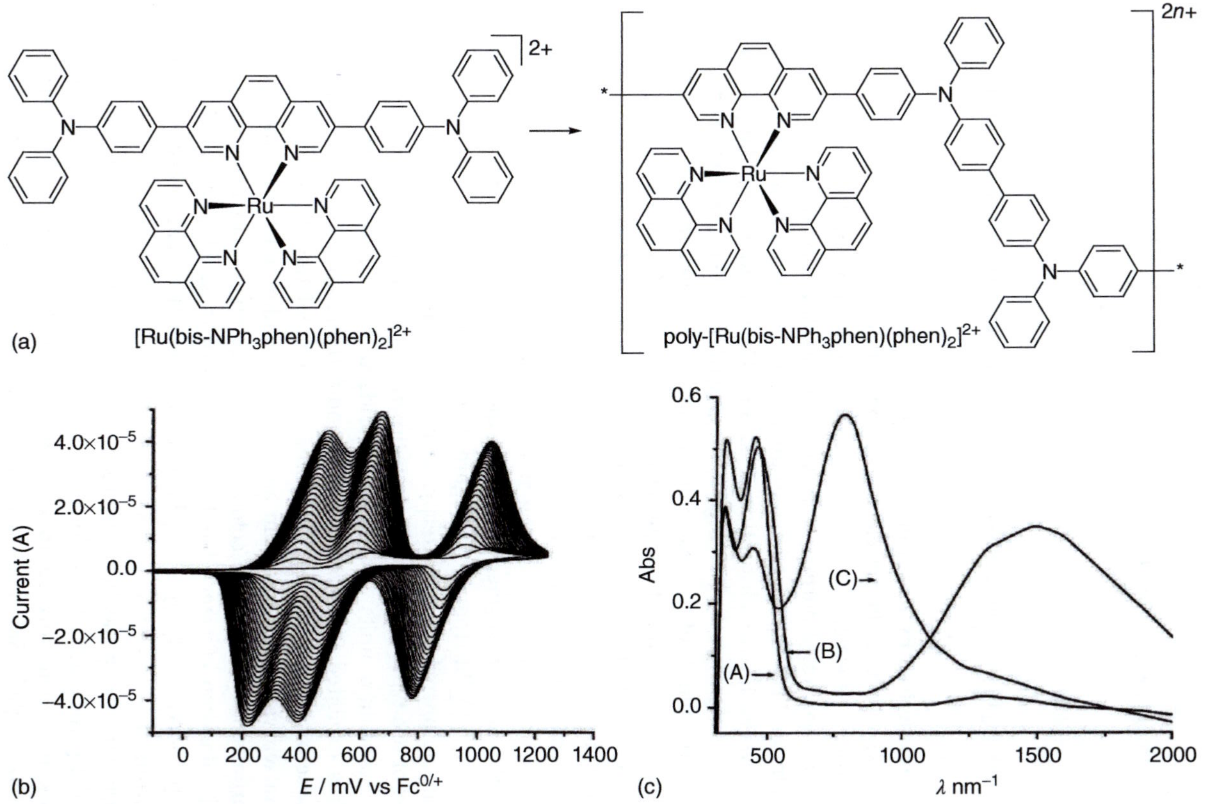

Figure 6.8 (a) Electropolymerisation of $[Ru(bis\text{-}NPh_3\text{-}phen)(Phen)_2]^{2+}$. (b) CV recorded during electropolymerisation. (c) Absorption spectra of the film at (A) 0, (B) +0.9 and (C) +1.3 V versus Ag/AgCl.

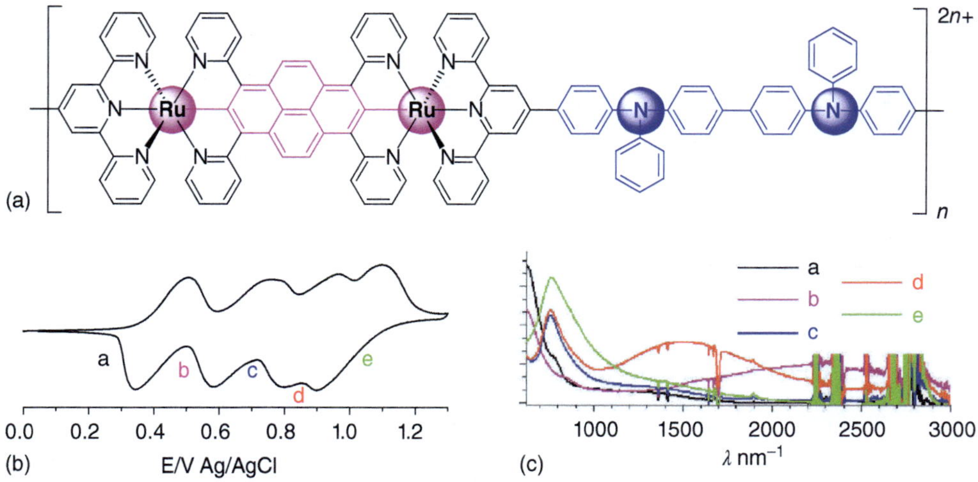

Figure 6.9 (a) Poly-[(Nptpy)$_2$Ru$_2$(tppyr)]$^{2+}$, (b) CV of poly-[(Nptpy)$_2$Ru$_2$(tppyr)]$^{2+}$ and (c) absorption spectra of the film at five different redox stages as shown in (b).

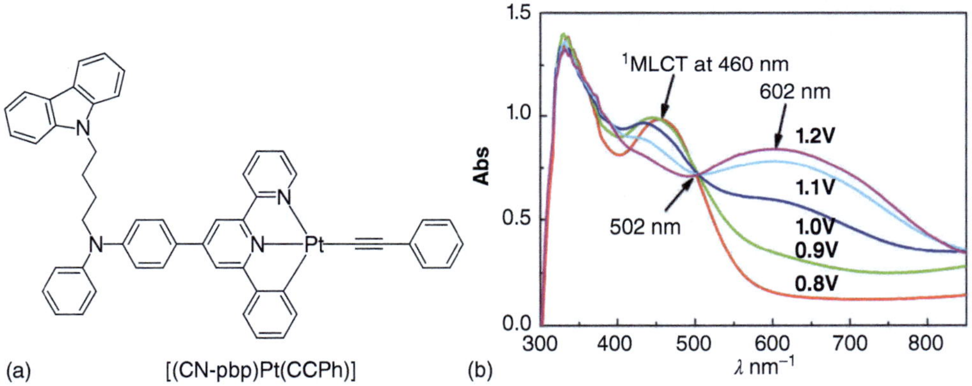

Figure 6.10 (a) [(CN-pbp)Pt(CCPh)] and (b) Absorption spectral changes of poly-[(CN-pbp)Pt(CCPh)] film upon stepwise oxidation versus SCE.

cyclometalated platinum complex [(CN-pbp)Pt(CCPh)] could be polymerised by oxidation (Figure 6.10), where CN-pbp was the NNC-type ligand 4-{p-[N-(4-(9-carbazole)butyl)-N-phenyl]aniline}-6-phenyl-2,2′-bipyridine [44]. The resulting film underwent a colour change from orange to bluish-black after oxidation. A contrast ratio of 67% around 600 nm has been achieved. The response time is 3.1 s, which is shorter relative to the poly-CN-pbp film without platinum. This is ascribed to the enhanced conductance by the presence of the metal site and the planar configuration of the [Pt-NNC] moieties in the metallopolymers.

Another useful polymerisable functional group is thiophene and its derivatives, and they have been used for the oxidative electropolymerisation of many organic materials and metal complexes [45, 46]. In particular, ethylenedioxythiophene

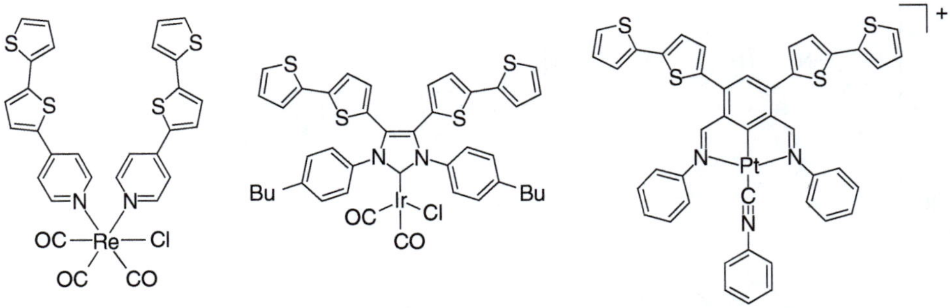

$[Re(btpy)_2(CO)_3Cl]$ $[Ir(bt\text{-}NHC)(CO)_2Cl]$ $[Pt(bt\text{-}dpib)(PhNC)]^+$

Figure 6.11 Complexes with bithiophene groups for oxidative electropolymerisation.

(EDOT) and bithiophene-based monomers are often employed because they can be polymerised at relative low potentials. Figure 6.11 shows some representative monomers for oxidative electropolymerisation. Farrell and co-workers have prepared the rhenium complex $[Re(btpy)_2(CO)_3Cl]$, where btpy is 4-bithiophenylpyridine [47]. This complex could be polymerised and the resulting polymeric film showed interesting multi-stage electrochromism. In the neutral state, the film is orange. It turns green upon reduction at −2.0 V versus $Fc^{+/0}$ (Fc = ferrocene). When held at +1.0 V, the film is purple. This polyelectrochromism is likely to represent cooperative behaviour between the organic conjugated system and the rhenium metal centres in the metallopolymers.

Where bt-NHC is a bithiophene-substituted *N*-heterocyclic carbene ligand, as shown in the centre of Figure 6.11, Cowley and co-workers have found that polymeric films of $[Ir(bt\text{-}NHC)(CO)_2Cl]$ show significant absorption at 692 and 1092 nm upon oxidation [48]. These two bands are ascribed to the formation of polarons and bipolarons along the quaterthiophene chains of the polymers, respectively. These films have a high degree of electrochromic reversibility, and they are promising for NIR electrochromism. However, the bipolaron absorption at 1092 nm has not been observed for films of gold and silver complexes with the same NHC ligand. In a similar way, the polymeric films of $[Pt(bt\text{-}dpib)(PhNC)]^+$ exhibit electrochromism of the polaron- and bipolaron-based absorptions, where bt-dpib is 4,6-*bis*(2,2′-bithiophen-5-yl)-1,3-di((*N*-phenyl)imino)benzene [49]. Owing to the synthetic feasibility and well-established electrochemistry, the oxidative electropolymerisation of monomers with thiophene derivatives is expected to produce more metallopolymeric films for electrochromism [50, 51].

6.4 Electrochromism in Self-Assembled or Self-Adsorbed Multilayer Films of Transition-Metal Complexes

Self-assembled monolayers (SAMs) of transition-metal complexes are well known. However, the change of absorbance and colour is quite small to see by the naked eye. Two methods are often used to enhance the absorptivity. One is by forming self-assembled multilayer films. The other one is by self-adsorbing the molecules onto films of nanocrystalline metal oxides which are pre-coated on electrode surfaces. Both methods generally need molecules with functional anchoring groups. Films prepared by these two methods have received much attention in electrochromism.

In 2003, Haga *et al.* reported on the SAMs or multilayers of diruthenium complexes with phosphonate groups immobilised on ITO/glass electrodes [52]. The electrochromic response at 550 nm of these films has been examined. Two years later, Kurth and co-workers prepared electrochromic films by sequential electrostatic deposition of polystyrenesulfonate and a pre-formed Co(II)-metalloviologen-based coordination polyelectrolyte [53]. The films showed electrochromic response at 643 nm during double-potential steps between 0 and −1.1 V versus Ag/AgCl.

Tieke and co-workers have reported on multilayer films of coordination polymer network structures which are built by using alternate layer-by-layer assembly of polyiminoarylenes P1–P3 with terpyridine substituent groups (acting as polytopic ligands (containing at least two coordination sites)) and various metal ions (Figure 6.12) [54]. The stepwise oxidations of nitrogen atoms in the polymeric backbone lead to the absorption spectral and colour changes of the films (Figure 6.12c,d). However, the colours at different redox states are strongly dependent on the polytopic ligand (P1–P3) and metal ions [55]. A contrast ratio of 18% at 800 nm and switching time of 450 ms have been achieved for the film with P1 and Zn ions [54]. The same authors have recently reported that, when electrochromic anions, such as poly((4-sulfonatophenyl)iminofluorene), are used, the films exhibit faster response times and higher contrast [56].

A good film-formation method is very important for electrochromism. In 2009, van der Boom and co-workers reported multilayer molecular networks based on polypyridyl osmium complexes $[Os(pybpy)_3]^{2+}$ (pybpy = 4-((pyrid-4-yl)vinyl)-4′-methyl- 2,2′-bipyridine) cross-linked via $PdCl_2$ (Figure 6.13) [57]. Formation of this porous film involves inward and outward diffusion of $PdCl_2$ and the films grows in an exponential manner. The low oxidation potential (+0.89 V vs Ag/AgCl) of the $Os^{III/II}$ process and the strong MLCT absorption of this film are advantageous for electrochromic studies. A contrast ratio of 35% at 510 nm has been achieved for these films. Only 5% signal loss was observed in the contrast ratio after 1000 redox cycles.

In recent years, metal-organic framework (MOF) materials have received tremendous interest in various fields. In general, MOF are obtained as bulk materials. The preparations of MOF films on electrode surfaces have been reported very recently. For instance, Dinca and co-workers described the solvothermal deposition of porous, micron-thick Zn-pyrazolate MOF films containing redox-active naphthalene diimide (NDI) linkers, H_2NDI-X (X = H, SEt, NHEt, Figure 6.13) [58]. These films exhibit fast and reversible colour switching based on the redox chemistry of the NDI component and the switching wavelength can be tuned by the X substituents. In addition, Hupp and co-workers prepared uniform and crack-free MOF thin film of pyrenetetrabenzoic acid (Figure 6.13) and Zn

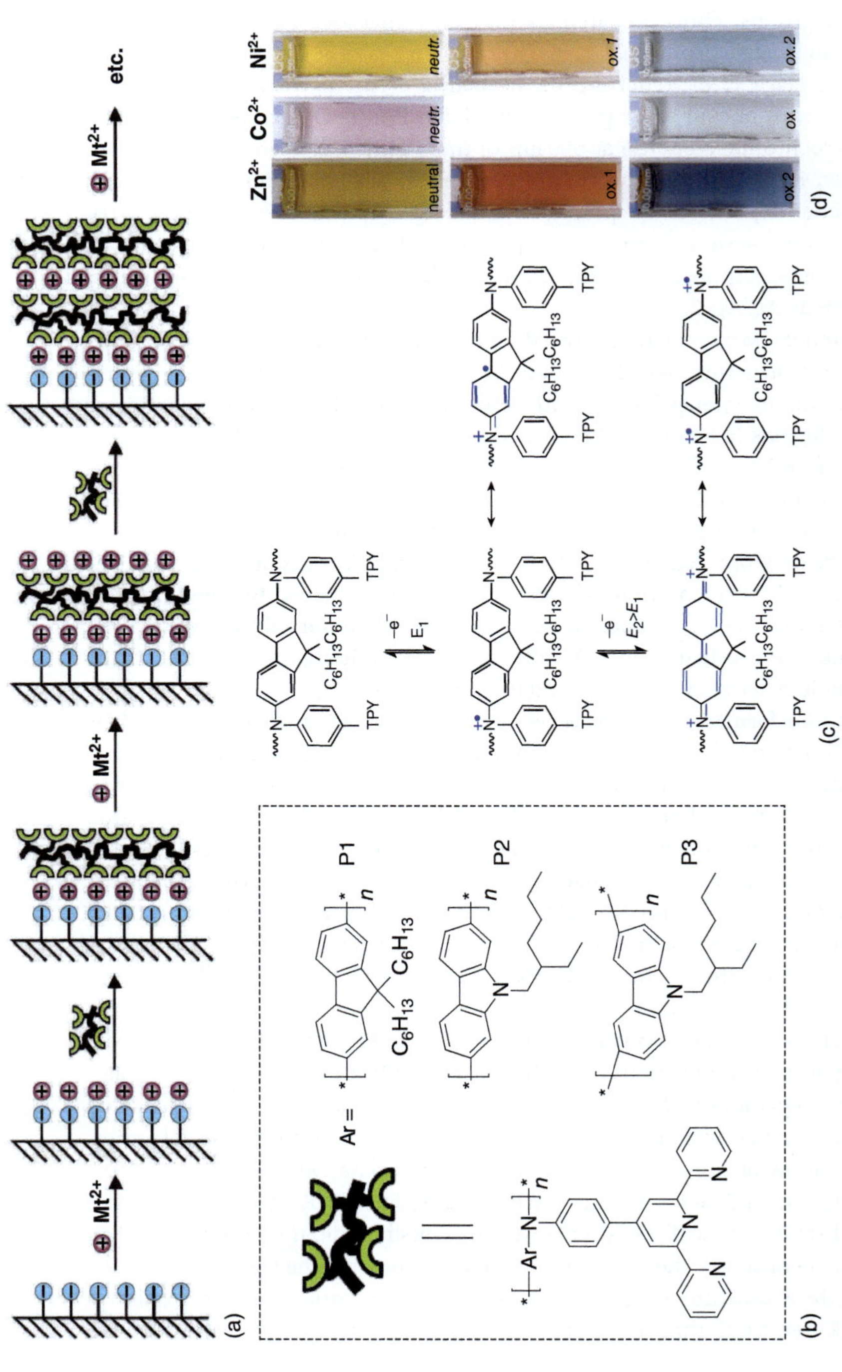

Figure 6.12 (a) Schematic diagram of layer-by-layer assembly. (b) Molecular structures of polyiminoarylenes with terpyridine substituent groups. (c,d) Different electrochemical oxidation states and colours of complex films with P1 and different metal ions.

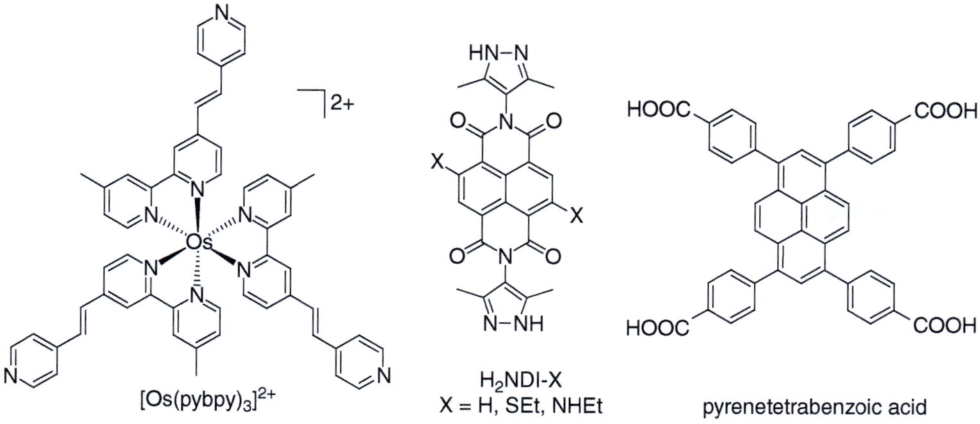

Figure 6.13 $[Os(pybpy)_3]^{2+}$ and H_2NDI-X for multilayer assembly.

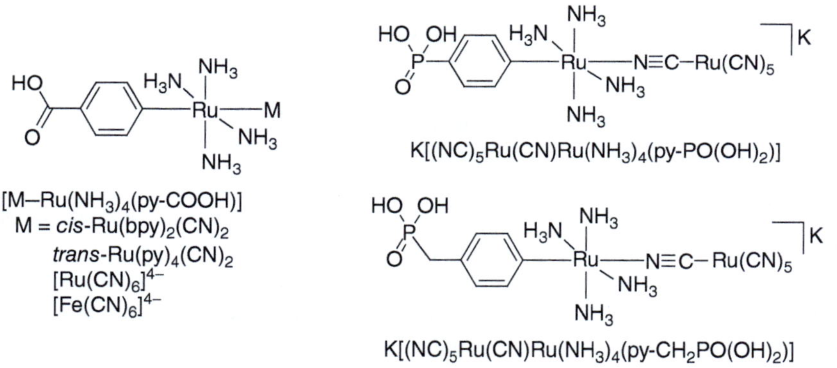

Figure 6.14 Binuclear complexes with carboxylic or phosphonic acid groups.

ions [59]. The film exhibits electrochromic switching between yellow and deep blue by means of a one-electron redox reaction at the pyrene-based linkers. The rigid MOF stabilises the radical cations of the pyrene linkers at positive applied potential, resulting in reversible colour change of the MOF film. The use of MOF materials for electrochromic purposes has remained relatively underdeveloped. Research in this area is expected to grow in the near future, as long as good film formation method has been established for MOF.

In addition to self-assembled multilayers, electrochromic materials can be adsorbed on films of nanocrystalline metal oxides such as TiO_2 or Sn-doped TiO_2 in order to gain absorptivity [60]. A similar method has been successfully used in dye-sensitized solar cells to increase the dye loadings. In 2003, Bignozzi reported a series of cyano-bridged binuclear mixed-valence complexes of the general formula $[M–Ru(NH_3)_4(py\text{-}COOH)]$ (py-COOH is isonicotinic acid; M stands for various metal components as shown in Figure 6.14) [61]. These complexes exhibit reversible electrochromic behaviour in the range of applied potential from −0.5 to

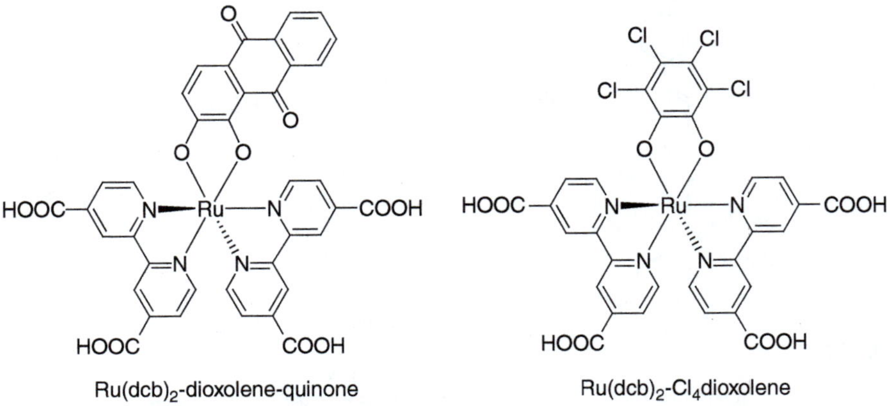

Figure 6.15 Ruthenium dioxolene with carboxylic acid groups.

+0.5 V versus SCE. The MLCT and MMCT transitions can be tuned through the changes of the metal moiety M. Analogous spectral changes were observed when these complexes were anchored through the carboxylic function to nanocrystalline TiO_2 or SnO_2 electrodes. When phosphonic acid groups were used as the anchor groups, for example, in complexes $K[(NC)_5Ru(CN)Ru(NH_3)_4(py\text{-}PO(OH)_2)]$ and $K[(NC)_5Ru(CN)Ru(NH_3)_4(py\text{-}CH_2PO(OH)_2)]$ [62], the stability could be further enhanced. In addition, electrochromic devices based on semiconductors functionalised with mononuclear charge transfer compounds or triruthenium clusters have been reported [63, 64].

The redox processes in the aforementioned materials are associated with the metal ions. Redox non-innocent complexes have been adsorbed onto semiconductor films for electrochromic studies. A series of $Ru(dcb)_2$-dioxolene (dcb, that is, 2,2′-bipyridine-4,4′-dicarboxylic acid) complexes with carboxylic acid anchor groups have been prepared by Schwab *et al.* [65]. These complexes show reversible electrochemical behaviour accompanied by a strong NIR absorption in their semiquinone forms due to a MLCT band. The NIR absorption is completely absent upon oxidation to the quinone form and upon reduction to the catechol form. The colour of the systems can be tuned by using a wide range of ligands. Complex $Ru(dcb)_2$-dioxolene-quinone (Figure 6.15) has been adsorbed on SbO_2 films and it exhibits reversibly switchable electrochromic behaviour in the NIR region. Similarly, Peter, Ward and co-workers have reported the NIR electrochromism based on complex $Ru(dcb)_2Cl_4$ dioxolene [66].

6.5
Electrochromism in Spin-Coated or Drop-Cast Thin Films of Transition-Metal Complexes

The simplest method to obtain a thin film is by spin coating or drop casting of a solution of the materials under study and allowing the solvent to evaporate. However, since most electrochromic studies are carried out by immersing the film into an electrolyte solution, the film must be insoluble in the solvent tested. Two methods have been used to deal with the solubility issue. One is to form films by using a good solvent and perform the measurement in a film-preserving solution. Another is first to synthesise materials with polymerisable groups and form films by spin coating or drop casting, which are then subjected to chemical polymerisation by a physical method (e.g. by heating) to form a cross-linked network. The resulting film has thus limited solubility in most solvents.

Metal dioxolene complexes are well-known redox non-innocent electrochromic materials [8]. They display significantly different visible and NIR absorptions in different oxidation states. The electrochromism of ruthenium dioxolene complexes in solution has been discussed in a review article by Ward [67]. In the previous section, the adsorption of ruthenium dioxolene complexes onto semiconductor films for electrochromic studies was discussed. Alternatively, the preparation of the electrochromic films could be simply achieved by drop casting of a solution of the dinuclear or trinuclear ruthenium dioxolene complexes, $[Ru_2(bpy)_4($*bis*-dioxlene$)]^+$ (bpy = 2,2′-bipyridine), $[Ru_2(dtbpy)_4($*bis*-dioxlene$)]^+$, $[Ru_3(bpy)_6($*tris*-dioxlene$)]^{3+}$ and $[Ru_3(dtbpy)_6($*tris*-dioxlene$)]^{3+}$ (dtbpy, that is, 4,4′-di-*tert*-butyl-2,2′-bipyridine, Figure 6.16), in CH_3CN onto ITO/glass electrodes and letting the solution evaporate evenly [68, 69]. The NIR electrochromic studies were then performed by immersing the film into aqueous KPF_6 solution. These films are reasonably stable in aqueous solutions, especially for complexes with peripheral hydrophobic *tert*-butyl groups.

The aforementioned films are composed of small-molecule materials. Spin-coated films of coordination polymers have been used for electrochromic

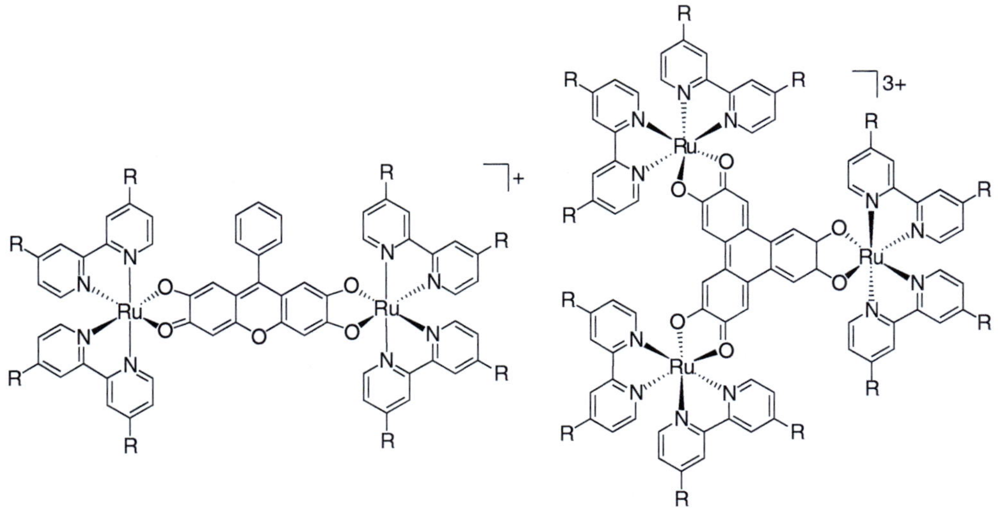

R = H: $[Ru_2(bpy)_4($*bis*-dioxolene$)]^+$
R = *t*-Bu: $[Ru_2(dtbpy)_4($*bis*-dioxolene$)]^+$

R = H: $[Ru_3(bpy)_6($*tris*-dioxolene$)]^{3+}$
R = *t*-Bu: $[Ru_3(dtbpy)_6($*tris*-dioxolene$)]^{3+}$

Figure 6.16 Multinuclear ruthenium dioxolene complexes.

studies [70, 71]. In particular, Higuchi, Kurth and co-workers have carried out systematic studies on metallo-supramolecular polymers based on functionalised ditopic ligands (ligands capable of coordination at two separate sites) as electrochromic materials [72, 73]. Coordination polymers have been prepared from the *bis*-terpyridine ligands tpy-linker-tpy (Figure 6.17) with $Fe(OAc)_2$, $Co(OAc)_2$ or $RuCl_2(DMSO)_4$. The resulting polymers exhibit moderate-to-good solubility in H_2O, MeOH, EtOH and HOAc. Thin films of these polymers were prepared by spin coating from a methanol solution. The resulting films were typically measured in an acetonitrile solution of n-Bu_4ClO_4. These films have good stability owing to the limited solubility of these polymers in acetonitrile. Reversible electrochromism of the MLCT absorptions has been achieved for all films. Polymeric films assembled from the OMe-modified ligands exhibit higher switching reversibility and stability and show a lower switching potential with respect to the unsubstituted and Br-substituted analogues. Optical memory of 15 min has been achieved with a Fe polymer. By using polymer gel electrolyte, electrochromic solid-state devices with these coordination polymers have been reported [74]. Very recently, Higuchi prepared a green coordination polymer from the bisphenanthroline ligand phen-linker-phen (Figure 6.17) and $Cu(ClO_4)_2$ [75]. A spin-coated film of this polymer showed electrochromic switching between the green and bleached state in n-Bu_4ClO_4/CH_3CN.

The success of the aforementioned film formation and electrochromic studies relies heavily on the solubility of the materials. For instance, the coordination polymers with tpy-linker-tpy are soluble in MeOH but not in CH_3CN. Thus, MeOH is used for the film formation step and CH_3CN is used for the electrochromic study step. However, when the counter anions OAc^- or Cl^- are replaced by PF_6^-, these

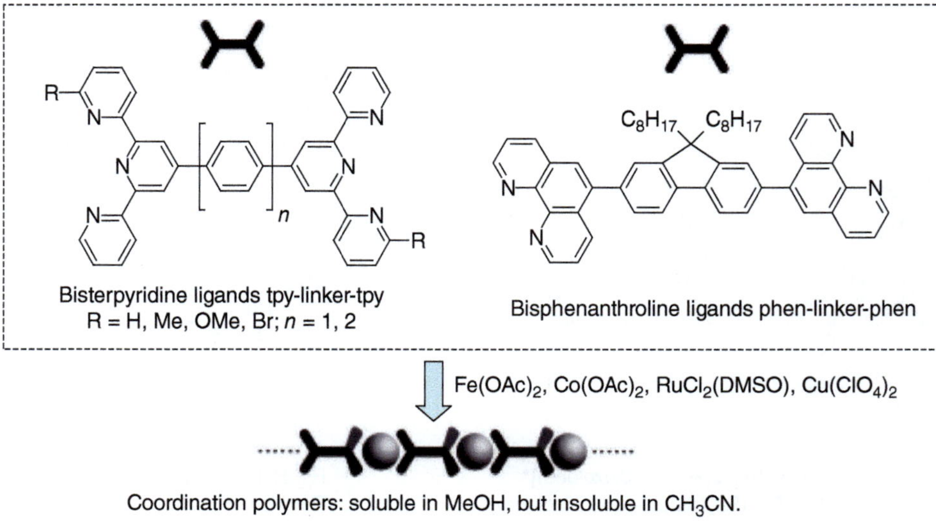

Figure 6.17 Ditopic ligands for preparations of coordination polymers.

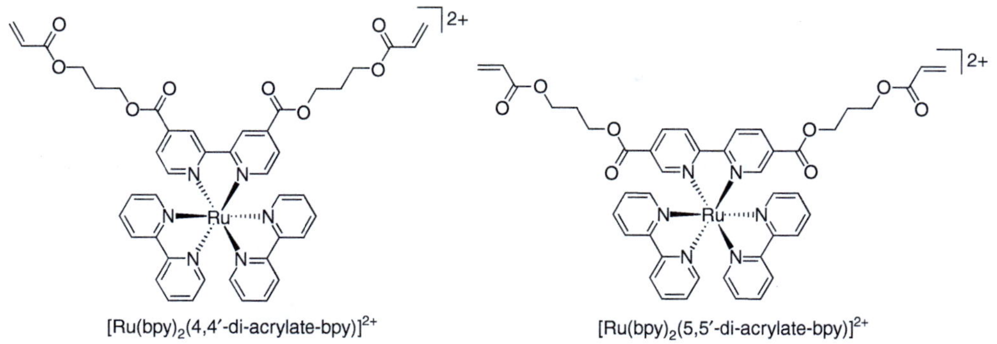

Figure 6.18 Acrylate-containing $[Ru(bpy)_3]$-type complexes.

polymers are basically insoluble in most organic solvents and thus not possible for forming a solution for spin coating [73].

Another strategy to deal with the solubility issue is first to make films of polymerisable monomers. For instance, Elliott and co-workers prepared a $[Ru(bpy)_3]$-type complex with vinyl substituents on the bpy ligand [76]. A spin-coated film of this complex was thermally polymerised to form an electroactive film which was stable under a variety of conditions. Using the same strategy, Reynolds and co-workers recently reported a family of acrylate-containing $[Ru(bpy)_3]$-type complexes [77] (Figure 6.18). The thermally cross-linked films of these complexes exhibit dual electrogenerated chemiluminescence and multicolour electrochromism. Films of $[Ru(bpy)_2(4,4'\text{-di-acrylate-bpy})]^{2+}$ and $[Ru(bpy)_2(5,5'\text{-di-acrylate-bpy})]^{2+}$, having different substitution positions of the acrylate groups, show similar MLCT bleaching upon oxidation. However, upon reduction of the 2+ state, spectral changes for each of the complexes became significantly different. Poly-$[Ru(bpy)_2(4,4'\text{-di-acrylate-bpy})]^{2+}$ demonstrated only a small increase in absorption intensity as well as a 10 nm red shift for λ_{max} at the 1+ oxidation state, resulting in a dark red-orange colour. Poly-$[Ru(bpy)_2(5,5'\text{-di-acrylate-bpy})]^{2+}$ exhibited an additional absorption band with the maximum at $\lambda_{abs} = 675$ nm in addition to substantially pronounced absorption increase at each reduced state, resulting in corresponding dark blue-green and subsequent dark orange colours for the 1+ and 0 oxidation states, respectively. A blend of two complexes has also been examined for multi-stage electrochromism.

One of the classical examples of cross-linked electrochromic films prepared by spin coating followed by chemical polymerisation is shown in Figure 6.19. The early work by Kaim and co-workers has shown that ruthenium-dicarbonyl-hydrazido complexes are electrochromic in the NIR region [78] and these complexes received much interest in electrochromism [79, 80]. Wang and co-workers prepared a ruthenium dendrimer containing three diruthenium-dicarbonyl-hydrazido units and three phenol groups (Figure 6.19) [27, 28]. A spin-coated film of the ruthenium dendrimer and a tri-isocyanate compound was heated to form a tightly bound cross-linked polymer network. This film is

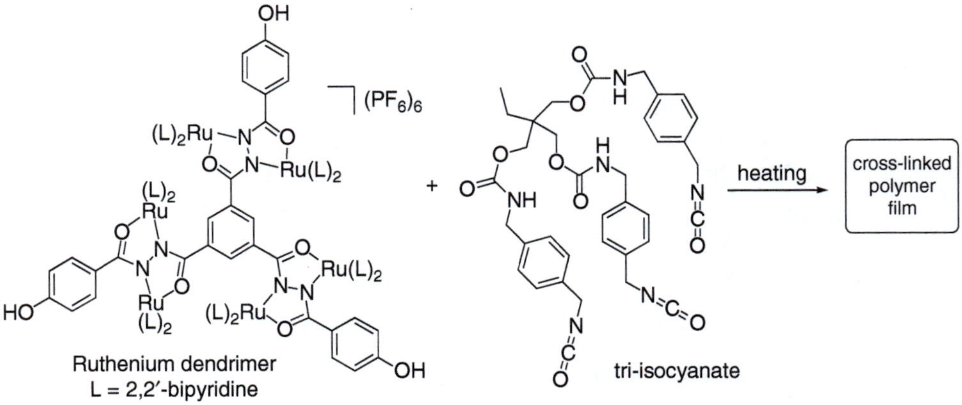

Figure 6.19 Cross-linked polymer film formation.

insoluble in most organic solvents, and it shows promising electrochromism at 1550 nm. An optical attenuation of 5.4 dB $\mu^{-1}m^{-1}$ with a switching time of 2 s has been achieved [81].

6.6 Conclusion and Outlook

Transition-metal complexes possess well-defined redox processes and rich absorption spectra spanning from UV to NIR region. They are excellent electrochromic materials. In conjugation with good film-formation methods, promising electrochromic devices can be built. These devices are useful in electronic display, smart window, optical telecommunication and information storage, among other uses. The electrochromic behaviour of these films in many cases arises from the redox state changes of metal ions, for example, in polypyridyl complexes. However, it is also common for the organic ligands to be mainly associated with the redox reactions involved, for example, in redox non-innocent complexes. Another example is the metallopolymeric films prepared by the oxidative electropolymerisation of complexes containing bithiophene units. The NIR bands arise from the polaron and bipolaron absorptions in the tetrathiophene segments of the polymers [48, 49]. The function of the metal ions is to regulate (either enhance or attenuate) these absorptions, in addition to behaving as linkers of the polymer backbones.

Some main film-formation methods for electrochromic studies have been introduced in the previous sections. These methods all have pros and cons. Electropolymerisation (either reductive or oxidative) is a well-established method, with the *in situ* polymer formation and film deposition. The thickness can be well controlled. However, the morphology of the film is difficult to control and the composition of the film is not fixed as a result of the ion exchange with electrolyte solution. Another disadvantage is that it requires the material molecules to be functionalised with appropriate polymerisable substituents. This point also applies to the films prepared by chemical polymerisation. Self-assembled multilayers possess better morphology. However, preparation is a rather time-consuming procedure, since it needs repetitive adsorption in different solutions of individual composition. Self-adsorbed films on semiconductor surfaces are well established for solar cell studies. Some common features are shared by solar cell and electrochromism studies, including the strong adhesion of the dye and the control of dye loadings. The scattering of light at longer wavelengths by the metal oxide particles is disadvantageous, which reduces the transparency of the window. Spin coating is a very simple method for making thin films, if the solubility is not a problem for the materials under study. However, the reproducibility of the electrochromic behaviour needs to be tested.

In addition to the aforementioned film formation methods, vacuum-sublimed thin films [82, 83] and Langmuir–Blodgett films [84, 85] of phthalocyanine compounds have been reported for electrochromism studies. This topic was discussed previously [86]. In addition, some excellent studies on electrochromism in solution have been reported, including square-planar *bis*-dithiolene complexes [87, 88] and Ward's oxomolybdenum complexes [89, 90]. Incorporation of these complexes into films or conducting solids has yet to be exploited.

Acknowledgements

Financial support from the National Natural Science Foundation of China (grants 91227104, 21271176, 21221002, and 21472196), the National Basic Research 973 program of China (grant 2011CB932301) and Institute of Chemistry, Chinese Academy of Sciences ('100 Talent' Program and grant XDB 12010400) are acknowledged.

References

1. Balzani, V., Juris, A., Venturi, M., Campagna, S., and Serroni, S. (1996) Luminescent and redox-active polynuclear transition-metal complexes. *Chem. Rev.*, **96**, 759–833.
2. Aguirre-Etcheverry, P. and O'Hare, D. (2010) Electronic communication through unsaturated hydrocarbon bridges in homobimetallic organometallic complexes. *Chem. Rev.*, **110**, 4839–4864.
3. Kaim, W. and Lahiri, G.K. (2007) Unconventional mixed-valent complexes of ruthenium and osmium. *Angew. Chem. Int. Ed.*, **46**, 1778–1796.
4. Kaim, W. (2011) Concepts for metal complex chromophores absorbing in the near infrared. *Coord. Chem. Rev.*, **255**, 2503–2513.
5. Hush, N.S. (1985) Distance dependence of electron transfer rates. *Coord. Chem. Rev.*, **64**, 135–157.
6. D'Alessandro, D.M. and Keene, F.R. (2006) Current trends and future challenges in the experimental, theoretical and computational analysis of intervalence charge transfer (IVCT) transitions. *Chem. Soc. Rev.*, **35**, 424–440.
7. Yao, C.-J., Yao, J., and Zhong, Y.-W. (2011) Electronic communication between two amine redox centers bridged by a bis(terpyridine)ruthenium(II) complex. *Inorg. Chem.*, **50**, 6847–6849.

8. Kaim, W. (2011) Manifestations of non-innocent ligand behavior. *Inorg. Chem.*, **50**, 9752–9765.
9. Costuas, K. and Rigaut, S. (2011) Polynuclear carbon-rich organometallic complexes: clarification of the role of the bridging ligand in the redox properties. *Dalton Trans.*, **40**, 5643–5658.
10. Abruña, H.D., Denisevich, P., Umaña, M., Meyer, T.J., and Murray, R.W. (1981) Rectifying interfaces using two-layer films of electrochemically polymerized vinylpyridine and vinylbipyridine complexes of ruthenium and iron on electrodes. *J. Am. Chem. Soc.*, **103**, 1–5.
11. Denisevich, P., Abruña, H.D., Leidner, C.R., Meyer, T.J., and Murray, R.W. (1982) Electropolymerization of vinylpyridine and vinylbipyridine complexes of iron and ruthenium: homopolymers, copolymers, reactive polymers. *Inorg. Chem.*, **21**, 2153–2161.
12. Zhong, Y.-W., Yao, C.-J., and Nie, H.-J. (2013) Electropolymerized films of vinyl-substituted polypyridine complexes: synthesis, characterization, and applications. *Coord. Chem. Rev.*, **257**, 1357–1372.
13. Calvert, J.M., Schmehl, R.H., Sullivan, B.P., Facci, J.S., Meyer, T.J., and Murray, R.W. (1983) Synthetic and mechanistic investigations of the reductive electrochemical polymerization of vinyl-containing complexes of iron(II), ruthenium(II), and osmium(II). *Inorg. Chem.*, **22**, 2151–2162.
14. Potts, K.T., Usifer, D.A., Guadalupe, A.R., and Abruña, H.D. (1987) 4-vinyl-, 6-vinyl-, and 4′-vinyl-2,2′:6′,2″-terpyridinyl ligands: their synthesis and the electrochemistry of their transition-metal coordination complexes. *J. Am. Chem. Soc.*, **109**, 3961–3967.
15. Nie, H.-J., Shao, J.-Y., Wu, J., Yao, J., and Zhong, Y.-W. (2012) Synthesis and reductive electropolymerization of metal complexes with 5,5′-divinyl-2,2′-bipyridine. *Organometallics*, **31**, 6952–6959.
16. Cui, B.-B., Nie, H.-J., Yao, C.-J., Shao, J.-Y., Wu, S.-H., and Zhong, Y.-W. (2013) Reductive electropolymerization of bis-tridentate ruthenium complexes with 5,5″-divinyl-4′-tolyl-2,2′:6′,2″-terpyridine. *Dalton Trans.*, **42**, 14125–14133.
17. Nie, H.-J., Yao, J., and Zhong, Y.-W. (2011) Synthesis of vinyl-substituted polypyridinyl ligands through Suzuki-Miyauru cross-coupling of potassium vinyltrifluoroborate with bromopolypyridines. *J. Org. Chem.*, **76**, 4771–4775.
18. Leidner, C.R., Sulivan, B.P., Reed, R.A., White, B.A., Crimmins, M.T., Murray, R.W., and Meyer, T.J. (1987) Synthesis and electropolymerization of distyrylbipyridine and methyldistyrylbipyridine complexes of iron, ruthenium, osmium, rhenium, and cobalt. *Inorg. Chem.*, **26**, 882–891.
19. Beer, P.D., Kocian, O., Mortimer, R.J., and Ridgway, C. (1993) Cyclic voltammetry of benzo-15-crown-5 ether-vinyl-bipyridyl ligands, their ruthenium(II) complexes and bismethoxyphenyl-vinyl-bipyridyl ruthenium(II) complexes. *J. Chem. Soc., Faraday Trans.*, **89**, 333–338.
20. Aranyos, V., Hjelm, J., Hagfeldt, A., and Grennberg, H. (2001) Electropolymerisable bipyridine ruthenium(II) complexes. Synthesis and electrochemical characterisation of 4-(3-methoxystyryl)- and 4,4′-di(3-methoxystyryl)-2,2′-bipyridine ruthenium complexes. *J. Chem. Soc. Dalton Trans.*, 1319–1325.
21. Beer, P.D., Kocian, O., Mortimer, R.J., Ridgway, C., and Stradiotto, N.R. (1996) Electrochemical polymerization studies of aza-15-crown-5 vinyl-2,2′-bipyridine ruthenium(II) complexes. *J. Electroanal. Chem.*, **408**, 61–66.
22. Leasure, R.M., Ou, W., Moss, J.A., Linton, R.W., and Meyer, T.J. (1996) Spatial electrochromism in metallopolymeric films of ruthenium polypyridyl complexes. *Chem. Mater.*, **8**, 264–273.
23. Yao, C.-J., Zhong, Y.-W., and Yao, J. (2011) Charge delocalization in a cyclometalated bisruthenium complex bridged by a noninnocent 1,2,4,5-tetra(2-pyridyl)benzene ligand. *J. Am. Chem. Soc.*, **133**, 15697–15706.
24. Yao, C.-J., Zhong, Y.-W., Nie, H.-J., Abruna, H.D., and Yao, J. (2011) Near-IR electrochromism in electropolymerized

films of a biscyclometalated ruthenium complex bridged by 1,2,4,5-tetra(2-pyridyl) benzene. *J. Am. Chem. Soc.*, **133**, 20720–20723.
25. Yao, C.-J., Sui, L.-Z., Xie, H.-Y., Xiao, W.-J., Zhong, Y.-W., and Yao, J. (2010) Electronic coupling between two cyclometalated ruthenium centers bridged by 1,3,6,8-tetra(2-pyridyl)-pyrene (tppyr). *Inorg. Chem.*, **49**, 8347–8350.
26. Yao, C.-J., Yao, J., and Zhong, Y.-W. (2012) Metallopolymeric films based on a biscyclometalated ruthenium complex bridged by 1,3,6,8-tetra (2-pyridyl) pyrene: applications in near-infrared electrochromic windows. *Inorg. Chem.*, **51**, 6259–6263.
27. Qi, Y.-H., Desjardins, P., Meng, X.-S., and Wang, Z.-Y. (2002) Electrochromic ruthenium complex materials for optical attenuation. *Opt. Mater.*, **21**, 255–263.
28. Qi, Y., Desjardins, P., and Wang, Z.Y. (2002) Novel near-infrared active dinuclear ruthenium complex materials: effects of substituents on optical attenuation. *J. Opt. A: Pure Appl. Opt.*, **4**, S273–S277.
29. Cui, B.-B., Yao, C.-J., Yao, J., and Zhong, Y.-W. (2014) Electropolymerized films as a molecular platform for volatile memory devices with two near-infrared outputs and long retention time. *Chem. Sci.*, **5**, 932–941.
30. Fussa-Rydel, O., Zhang, H.-T., Hupp, J.T., and Leidner, C.R. (1989) Electrochemical assembly of metallopolymeric films via reduction of coordinated 5-chlorophenanthroline. *Inorg. Chem.*, **28**, 1533–1537.
31. Zhang, H.-T., Subramanian, P., Fussa-Rydel, O., Bebel, J.C., and Hupp, J.T. (1992) Electrochromic devices based on thin metallopolymeric films. *Sol. Energy Mater. Sol. Cells*, **25**, 315–325.
32. Ellis, C.D., Margerum, L.D., Murray, R.W., and Meyer, T.J. (1983) Oxidative electropolymerization of polypyridyl complexes of ruthenium. *Inorg. Chem.*, **22**, 1283–1291.
33. Hanabusa, K., Nakamura, A., Koyama, T., and Shirai, H. (1994) Electropolymerization and characterization of terpyridinyl iron(II) and ruthenium(II) complexes. *Polym. Int.*, **35**, 231–238.
34. Deronzier, A. and Moutet, J.-C. (1996) Polypyrrole films containing metal complexes: syntheses and applications. *Coord. Chem. Rev.*, **147**, 339–371.
35. Zhu, Y., Gu, C., Tang, S., Fei, T., Gu, X., Wang, H., Wang, Z., Wang, F., Lu, D., and Ma, Y. (2009) A new kind of peripheral carbazole substituted ruthenium(II) complexes for electrochemical deposition organic light-emitting diodes. *J. Mater. Chem.*, **19**, 3941–3949.
36. Leung, M.-k., Chou, M.-Y., Su, Y.O., Chiang, C.L., Chen, H.-L., Yang, C.F., Yang, C.-C., Lin, C.-C., and Chen, H.T. (2003) Diphenylamino group as an effective handle to conjugated donor-acceptor polymers through electropolymerization. *Org. Lett.*, **5**, 839–842.
37. Wolf, M.D. (2001) Transition-metal-polythiophene hybrid materials. *Adv. Mater.*, **13**, 545–553.
38. Friebe, C., Hager, M.D., Winter, A., and Schubert, U.S. (2012) Metal-containing polymers via electropolymerization. *Adv. Mater.*, **24**, 332–345.
39. Junior, G.C. and Silva, A.P.S. (2008) Spectroelectrochemistry and investigation of charge transport mechanisms of iron poly(pyridyl) redox polymers. *Polyhedron*, **27**, 1860–1866.
40. Horwitz, C.P. and Zuo, Q. (1992) Oxidative electropolymerization of iron and ruthenium complexes containing aniline-substituted 2,2′-bipyridine ligands. *Inorg. Chem.*, **31**, 1607–1613.
41. Qiu, D., Zhao, Q., Bao, X., Liu, K., Wang, H., Guo, Y., Zhang, L., Zeng, J., and Wang, H. (2011) Electropolymerization and characterization of an alternatively conjugated donor-acceptor metallopolymer: poly-[Ru(4′-(4-(diphenylamino)phenyl)-2,2′:6′,2″-terpyridine)$_2$]$^{2+}$. *Inorg. Chem. Commun.*, **14**, 296–299.
42. Yao, C.-J., Zhong, Y.-W., and Yao, J. (2013) Five-stage near-infrared electrochromism in electropolymerized films composed of alternating cyclometalated bisruthenium and bis-triarylamine segments. *Inorg. Chem.*, **52**, 10000–10008.
43. Gu, C., Zhang, Z., Sun, S., Pan, Y., Zhong, C., Lv, Y., Li, M., Ariga, K., Huang, F., and Ma, Y. (2012) In situ electrochemical deposition and doping

of C_{60} films applied to high-performance inverted organic photovoltaics. *Adv. Mater.*, **24**, 5727–5731, and references therein.

44. Qiu, D., Bao, X., Feng, Y., Liu, K., Wang, H., Shi, H., Guo, Y., Huang, Q., Zeng, J., Zhou, J., and Xing, Z. (2012) Electrochromic films with high optical contrast prepared by oxidative electropolymerization of a novel multi-functionalized cyclometalating ligand and its neutral-charged Pt(II) complexes. *Electrochim. Acta*, **60**, 339–346.
45. Hjelm, J., Handel, R.W., Hagfeldt, A., Constable, E.C., Housecroft, C.E., and Forster, R.J. (2005) Conducting polymers containing in-chain metal complexes: electropolymerization of oligothienyl-substituted $\{M(tpy)_2\}$ complexes and in situ conductivity studies, M = Os(II), Ru(II). *Inorg. Chem.*, **44**, 1073–1081.
46. Heinze, J., Frontana-Uribe, B.A., and Ludwigs, S. (2010) Electrochemistry of conducting polymers-persistent models and new concepts. *Chem. Rev.*, **110**, 4724–4771.
47. Farrell, J.R., Lavoie, D.P., Pennell, R.T., Cetin, A., Shaw, J.L., and Ziegler, C.J. (2007) Electrochromic polymer films containing Re(I) and Pt(II) metal centers. *Inorg. Chem.*, **46**, 6840–6842.
48. Powell, A.P., Bielawski, C.W., and Cowley, A.H. (2010) Design, synthesis, and study of main chain poly(N-heterocyclic carbene) complexes: applications in electrochromic devices. *J. Am. Chem. Soc.*, **132**, 10184–10195.
49. Milum, K.M., Kim, Y.N., and Holliday, B.J. (2010) Pt-[NCN] pincer conducting metallopolymers that display redox-attenuated metal-ligand interactions. *Chem. Mater.*, **22**, 2414–2416.
50. Liu, W., Huang, W., Chen, C.-H., Pink, M., and Lee, D. (2012) Charge injection and transport in metal-containing conducting polymers: spectroelectrochemical mapping of redox activities. *Chem. Mater.*, **24**, 3650–3658.
51. Matsuse, R., Abe, M., Tomiyasu, Y., Inatomi, A., Yonemura, H., Yamada, S., and Hisaeda, Y. (2013) Metallopolymer films exhibiting three-color electrochromism in the UV/Vis and near-IR region: remarkable utility of trimetallic clusters bearing thienyl pendants and their mixed-valent charge transfer transitions. *J. Inorg. Organomet. Polym.*, **23**, 136–146.
52. Haga, M.-a., Takasugi, T., Tomie, A., Ishizuya, M., Yamada, T., Hossain, M.D., and Inoue, M. (2003) Molecular design of a proton-induced molecular switch based on rod-shaped Ru dinuclear complexes with bis-tridentate 2,6-bis(benzimidazol-2-yl)pyridine derivatives. *Dalton Trans.*, 2069–2079.
53. Kurth, D.G., Lopez, J.P., and Dong, W.-F. (2005) A new Co(II)-metalloviologen-based electrochromic material integrated in thin multilayer films. *Chem. Commun.*, 2119–2121.
54. Maier, A., Rabindranath, A.R., and Tieke, B. (2009) Fast-switching electrochromic films of zinc polyiminofluorene-terpyridine prepared upon coordinative supramolecular assembly. *Adv. Mater.*, **21**, 959–963.
55. Maier, A., Rabindranath, A.R., and Tieke, B. (2009) Coordinative supramolecular assembly of electrochromic films based on metal ion complexes of polyiminofluorene with terpyridine substituent groups. *Chem. Mater.*, **21**, 3668–3676.
56. Maier, A., Cheng, K., Savych, J., and Tieke, B. (2011) Double-electrochromic coordination polymer network films. *ACS Appl. Mater. Interfaces*, **21**, 3668–3676.
57. Motiei, L., Lahav, M., Freeman, D., and van der Boom, M.E. (2009) Electrochromic behavior of a self-propagating molecular-based assembly. *J. Am. Chem. Soc.*, **131**, 3468–3469.
58. Wade, C.R., Li, M., and Dinca, M. (2013) Facile deposition of multicolored electrochromic metal-organic framework thin films. *Angew. Chem. Int. Ed.*, **52**, 13377–13381.
59. Kung, C.-W., Wang, T.C., Mondloch, J.E., Fairen-Jimenez, D., Gardner, D.M., Bury, W., Klingsporn, J.M., Barnes, J.C., Duyne, R.V., Stoddart, J.F., Wasielewski, M.R., Farha, O.K., and Hupp, J.T. (2013) Metal-organic framework thin films composed of free-standing acicular nanorods exhibiting reversible

electrochromism. *Chem. Mater.*, **25**, 5012–5017.

60. Argazzi, R., Iha, N.Y.M., Zabri, H., Odobel, F., and Bignozzi, C.A. (2004) Design of molecular dyes for application in photoelectrochemical and electrochromic devices based on nanocrystalline metal oxide semiconductors. *Coord. Chem. Rev.*, **248**, 1299–1316.
61. Biancardo, M., Schwab, P.F.H., Argazzi, R., and Bignozzi, C.A. (2003) Electrochromic devices based on binuclear mixed valence compounds adsorbed on nanocrystalline semiconductors. *Inorg. Chem.*, **42**, 3966–3968.
62. Ferri, C., Costa, E., Biancardo, M., Argazzi, R., and Bignozzi, C.A. (2007) Electrochromic properties of mixed valence binuclear ruthenium complexes adsorbed on nanocrystalline SnO_2 films. *Inorg. Chim. Acta*, **360**, 1131–1137.
63. Biancardo, M., Argazzi, R., and Bignozzi, C.A. (2006) Electrochromic devices based on wide band-gap nanocrystalline semiconductors functionalized with mononuclear charge transfer compounds. *Displays*, **27**, 19–23.
64. Toma, S.H. and Toma, H.E. (2006) Versatile electrochromic displays based on TiO_2 nanoporous films modified with triruthenium clusters. *Electrochem. Commun.*, **8**, 1628–1632.
65. Schwab, P.F.H., Diegoli, S., Biancardo, M., and Bignozzi, C.A. (2003) Novel Ru-dioxolene complexes as potential electrochromic materials and NIR dyes. *Inorg. Chem.*, **42**, 6613–6615.
66. Garcia-Canadas, J., Meacham, A.P., Peter, L.M., and Ward, M.D. (2003) A near-infrared electrochromic window based on an Sb-doped SnO_2 electrode modified with a Ru-dioxolene complex. *Angew. Chem. Int. Ed.*, **42**, 3011–3014.
67. Ward, M.D. (2005) Near-infrared electrochromic materials for optical attenuation based on transition-metal coordination complexes. *J. Solid State Electrochem.*, **9**, 778–787.
68. Vickers, S.J. and Ward, M.D. (2005) Facile preparation of a visible- and near-infrared-active electrochromic film by direct deposition of a ruthenium dioxolene complex on an ITO/glass surface. *Electrochem. Commun.*, **7**, 389–393.
69. Grange, C.S., Meijer, A.J.H., and Ward, M.D. (2010) Trinuclear ruthenium dioxolene complexes based on the bridging ligand hexahydroxytriphenylene: electrochemistry, spectroscopy, and near-infrared electrochromic behavior associated with a reversible seven-membered redox chain. *Dalton Trans.*, **39**, 200–211.
70. Bernhard, S., Goldsmith, J.I., Takada, K., and Abruna, H.D. (2003) Iron(II) and copper(I) coordination polymers: electrochromic materials with and without chiroptical properties. *Inorg. Chem.*, **42**, 4389–4393.
71. Zeng, Q., McNally, A., Keyes, T.E., and Forster, R.J. (2008) Three colour electrochromic metallopolymer based on a ruthenium phenolate complex bound to poly(4-vinyl)pyridine. *Electrochem. Commun.*, **10**, 466–470.
72. Han, F.S., Higuchi, M., and Kurth, D.G. (2007) Metallo-supramolecular polymers based on functionalized bis-terpyridines as novel electrochromic materials. *Adv. Mater.*, **19**, 3928–3931.
73. Han, F.S., Higuchi, M., and Kurth, D.G. (2008) Metallosupramolecular polyelectrolytes self-assembled from various pyridine ring-substituted bis-terpyridines and metal ions: photophysical, electrochemical, and electrochromic properties. *J. Am. Chem. Soc.*, **130**, 2073–2081.
74. Higuchi, M., Akasaka, Y., Ikeda, T., Hayashi, A., and Kurth, D.G. (2009) Electrochromic solid-state devices using organic-metallic hybrid polymers. *J. Inorg. Organomet. Polym.*, **19**, 74–78.
75. Hossain, M.D., Sato, T., and Higuchim, M. (2013) A green copper-based metallo-supramolecular polymer: synthesis, structure, and electrochromic properties. *Chem. Asian J.*, **8**, 76–79.
76. Elliott, C.M. and Redepenning, J.G. (1986) Stability and response studies of multicolor electrochromic polymer modified electrodes prepared from tris(5,5′-dicarboxyester- 2,2′-bipyridine)ruthenium(II). *J. Electroanal. Chem.*, **197**, 219–232.
77. Puodziukynaite, E., Oberst, J.L., Dyer, A.L., and Reynolds, J.R. (2012) Establishing dual electrogenerated chemiluminescence and multicolor electrochromism

in functional ionic transition-metal complexes. *J. Am. Chem. Soc.*, **134**, 968–978.

78. Kasack, V., Kaim, W., Binder, H., Jordanov, J., and Roth, E. (1995) When is an odd-electron dinuclear complex a mixed-valent species? Tuning of ligand-to-metal spin shifts in diruthenium(III,II) complexes of noninnocent bridging ligands OC(R)NNC(R)O. *Inorg. Chem.*, **34**, 1924–1933.
79. Wang, S., Li, X., Xun, S., Wan, X., and Wang, Z.Y. (2006) Near-infrared electrochromic and electroluminescent polymers containing pendant ruthenium complex groups. *Macromolecules*, **39**, 7502–7507.
80. Zhang, J., Wu, X., Yu, H., Yan, D., and Wang, Z. (2005) Near infrared electrochromic variable optical attenuator based on ruthenium complex and polycrystalline tungsten oxide. *Chin. Sci. Bull.*, **50**, 2688–2690.
81. Qi, Y. and Wang, Z.Y. (2003) Dendritic mixed-valence dinuclear ruthenium complexes for optical attenuation at telecommunication wavelengths. *Macromolecules*, **36**, 3146–3151.
82. Collins, G.C.S. and Schiffrin, D.J. (1982) The electrochromic properties of lutetium and other phthalocyanines. *J. Electroanal. Chem.*, **139**, 335–369.
83. Nicholson, M.M. (1982) Lanthanide diphthalocyanines – electrochemistry and display applications. *Ind. Eng. Chem. Prod. Res. Dev.*, **21**, 261–266.
84. Granito, C., Goldenberg, L.M., Bryce, M.R., Monkman, A.P., Troisi, L., Pasimeni, L., and Petty, M.C. (1996) Optical and electrochemical properties of metallophthalocyanine derivative Langmuir–Blodgett films. *Langmuir*, **12**, 472–476.
85. Rodriguez-Mendez, M.L., Souto, J., de Saja, J.A., and Aroca, R. (1995) Electrochromic display based on Langmuir Blodgett films of praseodymium bisphthalocyanine. *J. Mater. Chem.*, **5**, 639–642.
86. Monk, P.M.S., Mortimer, R.J., and Rosseinsky, D.R. (2007) *Electrochromism and Electrochromic Devices*, Cambridge University Press, Cambridge, pp. 258–265.
87. McCleverty, J.A. (1968) Metal 1,2-dithiolene and related complexes. *Prog. Inorg. Chem.*, **10**, 49–221.
88. Bonneval, B.G., Ching, K.I.M.-C., Alary, F., Bui, T.-T., and Valade, L. (2010) Neutral d^8 metal bis-dithiolene complexes: synthesis, electronic properties and applications. *Coord. Chem. Rev.*, **254**, 1457–1467.
89. Lee, S.-M., Marcaccio, M., McCleverty, J.A., and Ward, M.D. (1998) Dinuclear complexes containing ferrocenyl and oxomolybdenum(V) groups linked by conjugated bridges: a new class of electrochromic near-infrared dye. *Chem. Mater.*, **10**, 3272–3274.
90. McDonagh, A.M., Bayly, S.R., Riley, D.J., Ward, M.D., McCleverty, J.A., Cowin, M.A., Morgan, C.N., Verrazza, R., Penty, R.V., and White, I.H. (2000) A variable optical attenuator operating in the near-infrared region based on an electrochromic molybdenum complex. *Chem. Mater.*, **12**, 2523–2524.

7
Organic Near-Infrared Electrochromic Materials

Bin Yao, Jie Zhang, and Xinhua Wan

7.1
Introduction

Electrochromic materials exhibit reversible optical changes in absorption or reflection upon electrochemical oxidation or reduction. Since Deb studied the electrochromism of amorphous WO_3 film [1], numerous efforts have been paid to develop various electrochromic materials. Traditionally electrochromic materials are mainly focused on colour changes in the visible region (e.g. 400–780 nm), which find wide potential and practical applications such as e-papers, optical switching devices, smart windows and camouflage materials [2–4]. Over the last two decades, with the comprehensive development of photonics, increasing interest has been shifting to the near-infrared region (NIR, e.g. 780–2500 nm), which could be exploitable for optical communication, data storage and thermal control in buildings and spacecraft [5–7]. Especially for the communication wavelengths (e.g. 1310 and 1550 nm), NIR electrochromic materials emerge as the core components in variable optical attenuators (VOAs) owing to the easily modulated optical signals by externally applied small voltages [8, 9].

Plenty of materials including both inorganic and organic compounds exhibit NIR electrochromic behaviours. Inorganic NIR electrochromic materials mainly include transition metal oxides (e.g. WO_3, NiO, MoO_3) [10–13] and transition metal complexes (e.g. Ni complexes, Mo complexes, Ru complexes) [14, 15]. They usually exhibit excellent electrochromic stability and reversibility. Organic NIR electrochromic materials outdo inorganic ones in terms of low cost, large-area flexibility, good solution processability and high colour contrast [7]. Many kinds of organic compounds are known to exhibit electrochromism in the NIR region, and they can be divided into cathodically and anodically colouring electrochromic materials in terms of addition or subtraction of electrons in operation. Aromatic quinones and aromatic imides are the most well-known cathodically colouring electrochromic materials [16, 17]. To improve processability and stability, they have also been incorporated into polymer backbones or attached to polymer backbones as pendant groups. Anodically colouring electrochromic materials primarily include poly(triarylamine)s [7] and conjugated polymers [18]. Polymeric

Electrochromic Materials and Devices, First Edition.
Edited by Roger J. Mortimer, David R. Rosseinsky and Paul M. S. Monk.

NIR electrochromic materials possess advantages of good film-forming capability, long-term stability, high colour contrast and polyelectrochromism.

Several mechanisms have been proposed to achieve NIR electrochromism in organic materials, including both intramolecular and intermolecular interactions. The $\pi^*-\pi^*$ transition [19] is one of the most common intramolecular transitions, and the absorption position and intensity are often independent of solvent, concentration and temperature. Inter-valence charge transfer (IVCT) [20] is another mode of intramolecular interactions. Compared with $\pi^*-\pi^*$ transitions, IVCT involves a localised electronic structure, which always needs a proper connection pattern and distance between different redox centres. As for some conducting materials like thiophene-based conjugated polymers, when electrons are injected or extracted, the resultant polymers display intense NIR absorption bands because of the formation of charge carriers such as polarons and bipolarons [21]. The donor–acceptor (D-A) approach is considered the most successful method to modulate the electrochromic properties of conjugated polymers. With a delicate macromolecular design, D-A type conjugated polymers could behave with either p- or n-dopable characteristics, and as high as 99% colour contrast ratio has been reported in the NIR region [22]. For some particular compounds, although no NIR electrochromism is observed in the molecularly dispersed state, 'optical conduction bands' covering the NIR region can be induced by the ordered stacking of molecules [23]. Such molecules usually possess extensive π structures like naphthalene diimides (NDIs) [24]. It is worth noting that these interactions are not independent of each other, and different interactions could exist in the same molecule [25].

This chapter describes several kinds of electrochromic materials according to their chemical structures, mainly including aromatic quinones, aromatic imides, poly(triarylamine)s, conjugated polymers and some other electroactive small organic molecules.

7.2 Aromatic Quinones

Aromatic quinones have been used as dyes and electron-deficient materials for a long time. Investigations of NIR electrochromism in them began in the 1990s by Miller and co-workers, who synthesised a few types of aromatic quinones and systematically investigated their structure–property relationships. The effect of quinone moiety number on electrochromism was first investigated (Chart 7.1) [26].

When reduced to radical anionic states, the compounds **5–7**, which contain only a single quinone moiety, just display absorptions in the visible region (Table 7.1), whereas the others with more than one quinone moiety show NIR absorptions. Especially for the radical anion 1^-, intense NIR absorption bands centred at 1730 nm appear independent of solvent and concentration due to $\pi^*-\pi^*$ transitions (Figure 7.1).

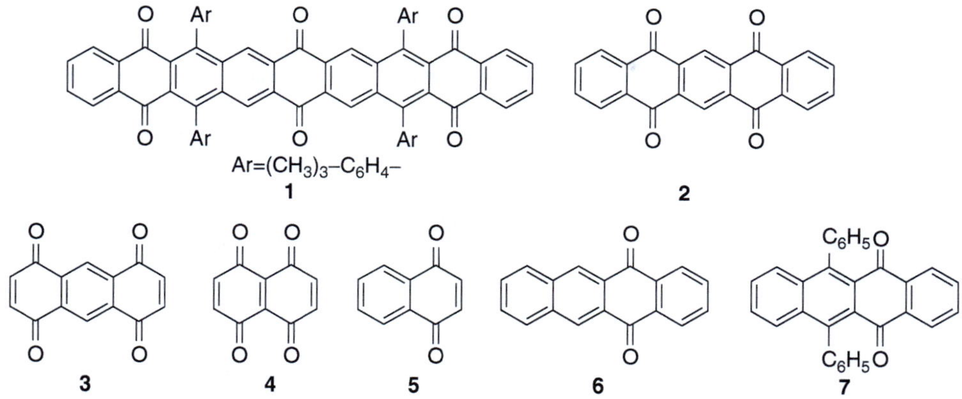

Chart 7.1 Chemical structures of quinone derivatives **1–7**.

Table 7.1 Physical properties of semiquinones **1–7**.

Compounds	$-E^{\circ}$ (mV, SCE)	λ_{max} (nm) (log ε)
1^-	665, 981, 1265	1730 (4.43)
2^-	420, 920	1335 (4.50)
3^-	256, 731	1560 (4.21)
4^-	−142, 586	900 (4.66)
5^-	608, 1412	480 (3.89)
6^-	960, 1565	595 (4.03)
7^-	1048, 1655	650 (3.95)

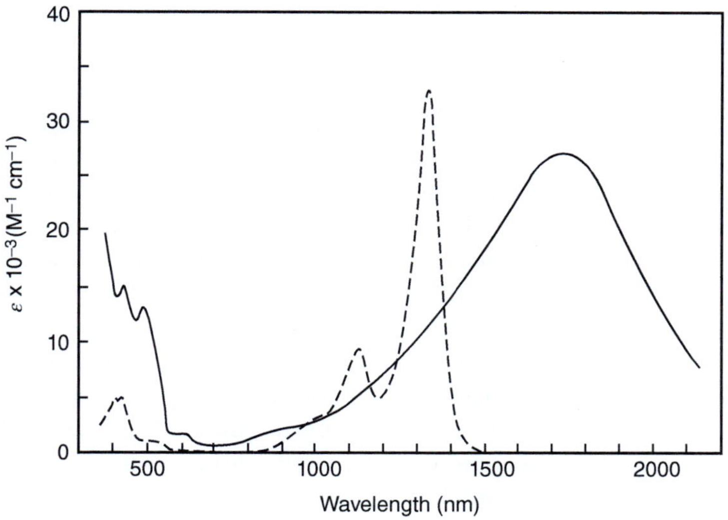

Figure 7.1 Absorption spectra for **1**$^-$ (–) and **2**$^-$ (--) in *N*-dimethylformamide (DMF).

Table 7.2 Spectroscopic and electrochemical data of diquinone derivatives.

Compounds	Structure	λ_{max} a) (nm)	$E°$ X b) (mV)	Compounds	Structure	λ_{max} (nm)	$E°$ (mV)
2	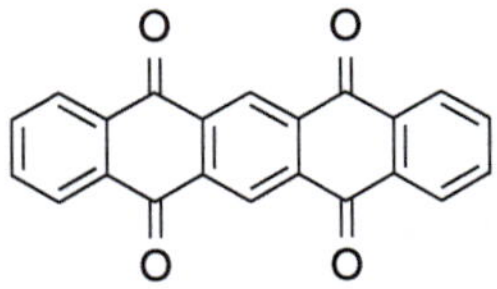	1335	−0.42	11	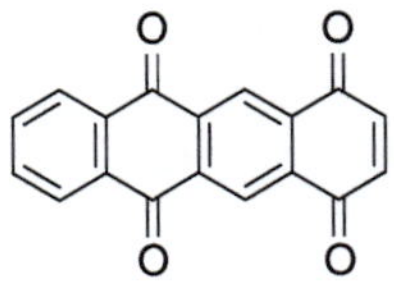	1200–1650	−0.30
8	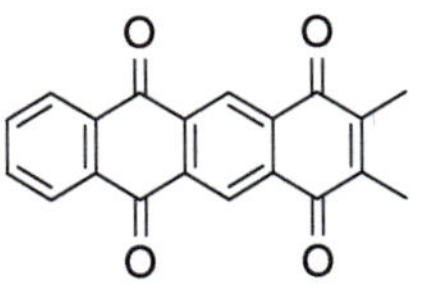	1400	−0.41	3	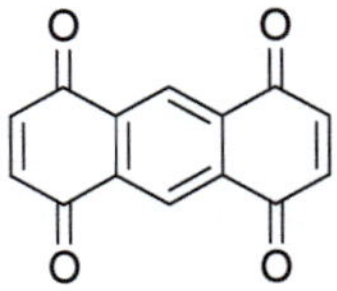	1560	−0.25
9	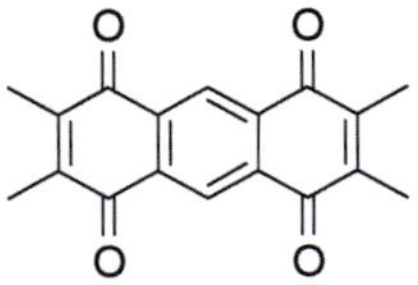	1480	−0.41	12	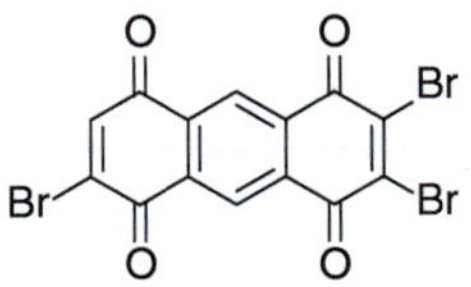	1615	−0.05
10	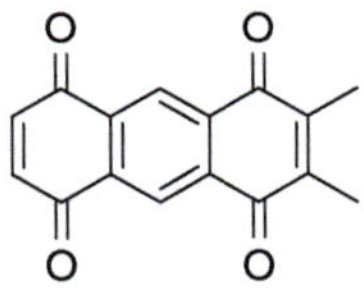	1500	−0.31	13	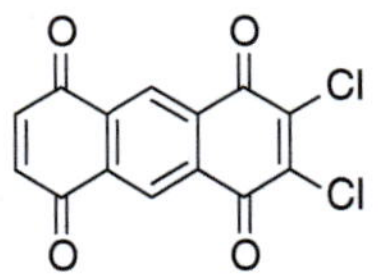	1200–1700	−0.08

a) Only the wavelengths of maximum absorption of radical anions are listed.
b) Only the first reduction potentials are listed versus saturated calomel electrode (SCE).

Subsequently, the substitution effect was investigated based on the diquinone structure of compound **3** [27]. It is evident from Table 7.2 that a methyl substituent and fused benzene at the periphery mainly serving as electron-donating groups make the reduction of molecules more difficult and cause the blue-shift of the maximum absorption wavelengths of radical anions. On the contrary, the electron-withdrawing substituents like bromine and chlorine make the reduction easier. As a result, the wavelengths of maximum absorption are red-shifted.

Both **3** and **4** display intense NIR electrochromism arising from $\pi^*-\pi^*$ transitions. As two quinone moieties are separated by more fused benzenes (**15** and **16**), the NIR electrochromic properties of the resultant molecules are quite distinct [28]. As displayed in Figure 7.2, when reduced to radical anionic states, both **15** and **16** show broad and featureless optical absorption bands covering the entire visible and NIR regions, which are characteristics of mixed-valence compounds. In addition to the long NIR absorptions, the separations (ΔE) between the first and second reduction peaks of compounds **15** and **16** are close to model compound

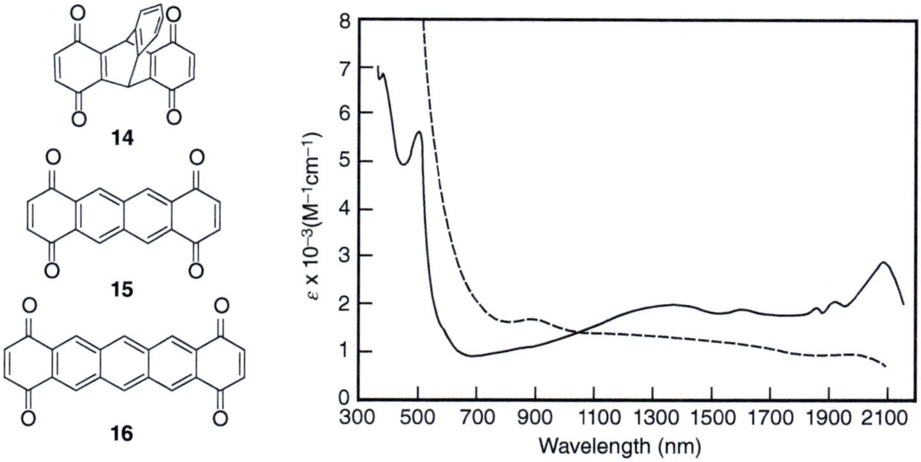

Figure 7.2 Chemical structures of **14–16** and vis-NIR spectra of **15**$^-$ (–) and **16**$^-$ (--) in DMF.

14 and much smaller than those of compounds **3** and **4**. It is therefore considered that, owing to the extraordinary polarisability of these radical anionic species, a little perturbation from solvent or ion-pairing effect renders the electron to be localised at one side of **15**$^-$ and **16**$^-$; thus, the broad NIR absorption bands come from IVCT. Electron spin resonance (ESR) investigation [29] and theoretical computation [30] further confirmed the localised character of these compounds, and the hopping rate from one quinone to the other was also measured.

For low-molecular-mass NIR electrochromic materials, device fabrication requires deposition on the electrode surface under high vacuum or blending with polymers. The crystallisation and phase separation of small molecules due to their incompatibility with polymer in the active layer would influence the property and stability of devices. To solve this problem, pentacenediquinone was introduced into the polymer main chain by nucleophilic substitution (Scheme 7.1) [31]. All

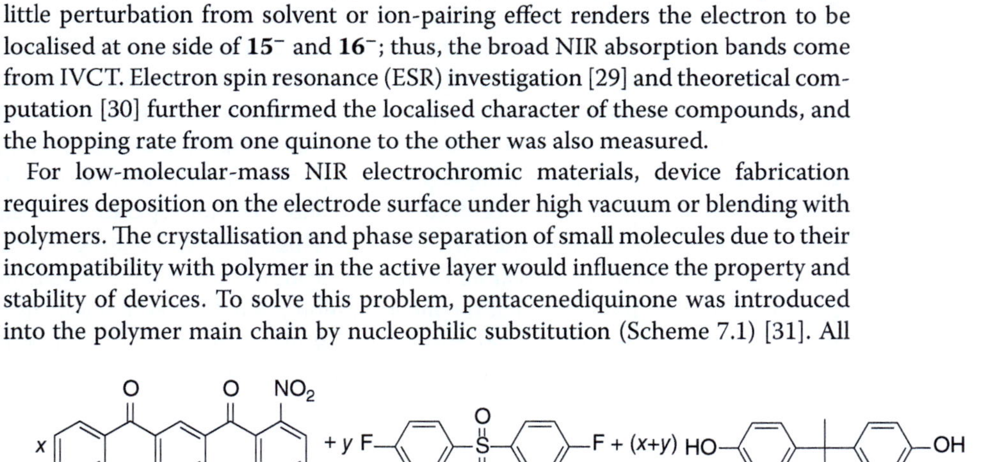

Scheme 7.1 Corresponding polymers **17a–d** based on pentacenediquinone.

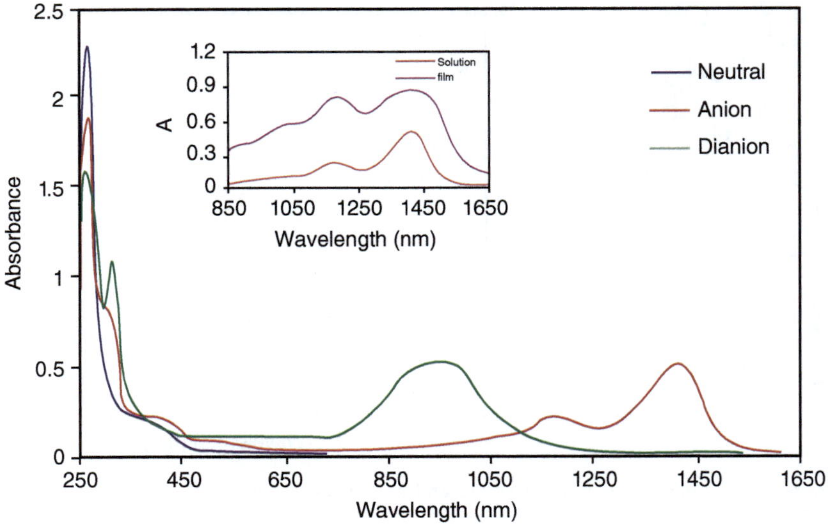

Figure 7.3 UV/vis/NIR spectra of polymer **17c** in DMF containing tetrabutylammonium perchlorate (TBAP) in its neutral, anionic and dianionic states. The inset showed the NIR absorptions in solution and in the solid state.

the polymers **17a–d** have excellent thermal stability and good solubility in various solvents, such as tetrahydrofuran (THF) and $CHCl_3$. A cyclic voltammetric investigation indicates that all polymers show two reversible redox peaks in −0.40 and −0.85 V, which correspond to the formation of radical anion and dianion, respectively. Taking **17c** as an example (Figure 7.3), in its radical anionic state, **17c** exhibits NIR absorption bands centred at 1170 and 1396 nm separately. When further reduced to dianion, the NIR absorption bands corresponding to the radical anion gradually disappear and a new absorption band peaking at 880 nm appears. These results reveal that the incorporation of pentacenediquinone into the polymer main chain does not affect the NIR electrochromic property too much, while the solution-processable performance is effectively improved.

7.3 Aromatic Imides

Aromatic imides are common optoelectronic building blocks that are widely used in solar cells and n-channel thin-film transistors. In addition, they are also potential materials that exhibit NIR electrochromic activities when reduced to different states. Aromatic imides showing NIR electroactivity chiefly include NDIs, pyrene diimides (PDIs) and anthraquinone imides (AQIs) (Chart 7.2).

The maximum absorption wavelengths of PDI radical anions have been measured in the NIR region in both solutions and solid states [32, 33]. NDIs are known to undergo electrical colour change in the visible region between three redox states

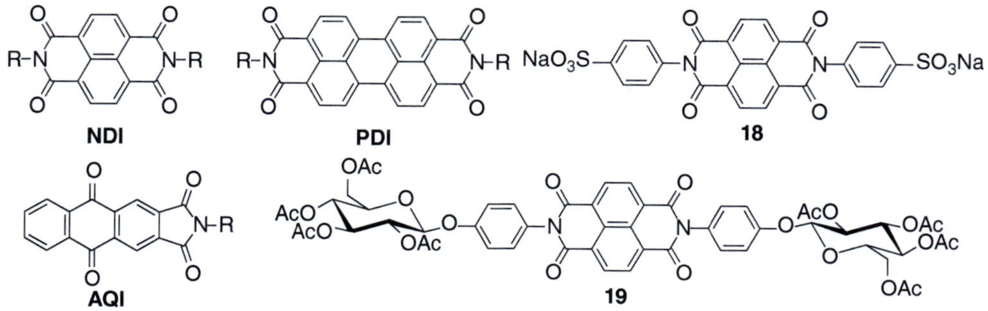

Chart 7.2 Chemical structures of NDIs, PDIs, AQIs, **18** and **19**.

(neutral, radical anion and dianion) in dilute solution. Recently, it has been proved that the stacked NDIs, upon reduction, display an 'optical conduction band' in the NIR region arising from excitation of π electrons along the stacks [23, 34]. Based on this interesting finding, layer-by-layer (LbL) assembly of Naph-SO_3Na (**18**) and poly(diallyldimethylammonium) chloride was employed to fabricate an NIR electrochromic VOA [35]. Upon one-electron reduction, the multi-layer film exhibits a broad absorption band from 1200 to 1900 nm. Owing to the improved ionic conductivity, the optical attenuation at 1550 nm of the film shows a rapid response time that reaches 1.3 dB μm^{-1} within 5 s. By fine design, an LbL film assembly of Naph-SO_3Na and its complementary NIR material probably expands the attenuation range further.

Chirality is an important component of many fields such as material science and life science, and modulation of chiral signal in the NIR region plays an important role in the communication field. By having NDI molecules stacked asymmetrically in a chiral aggregate, the induced circular dichroism (CD) and other chiroptical properties could be expected in both the visible and NIR regions and be electrochemically modulated between different redox states. NDI bearing two chiral acetylated glucoses at amino terminals (**19**) forms a chiral low-molecular-weight organogel in n-butanol [24]. Under different external voltages, the NDI core undergoes redox reactions without destroying the asymmetric structure of gel. As a result, reversible and drastic changes of CD signals in both the visible and NIR regions between three redox states are observed (Figure 7.4). A large CD signal at 1310–1550 nm of the communication wavelengths is induced when the stacked NDI chromophores in the chiral gel are reduced to radical anionic states.

Two apparent drawbacks of low-molecular-weight organogel are low mechanical strength and instability over a long time. For improving mechanical strength and long-term stability of electrochromic devices, NDI and PDI structures have been attempted to be incorporated into polymers (**20–22** in Chart 7.3) [36–38]. However, relatively low colour contrast in the NIR region hinders their further practical applications as NIR electrochromic materials.

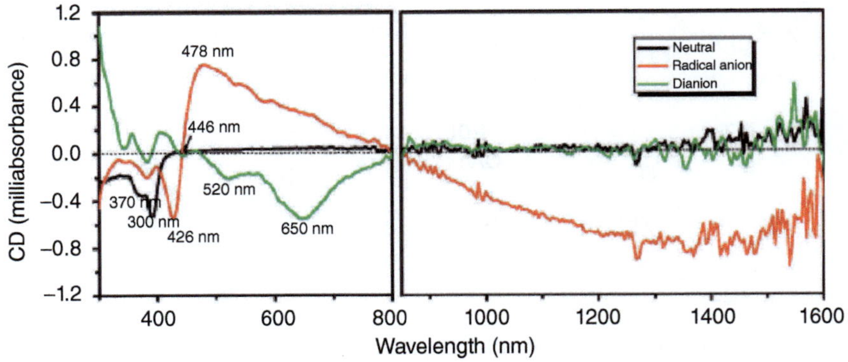

Figure 7.4 CD spectra of films of gels of **19** in its neutral, radical anionic and dianionic states.

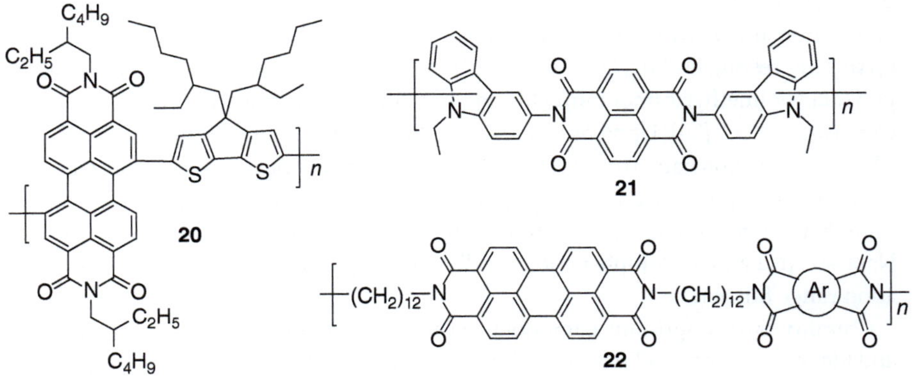

Chart 7.3 NIR electrochromic polymers **20–22** based on NDI and PDI.

7.4 Anthraquinone Imides

AQIs are a unique type of cathodically colouring electrochromic material with both quinone and imide moieties. By introducing an imide group, AQIs not only maintain the NIR electrochromism of quinones but also show good solubility and processability. The radical anions of unsubstituted AQIs absorb strongly in the range 700–1200 nm, which is far shorter than communication wavelength. Wan *et al.* reported a facile route to synthesise AQIs substituted at the six-position (Scheme 7.2) [39]. In this way, functional groups are readily introduced and efficiently tune the energy gap and facilitate further derivation [40, 41].

The structure–property relationships between different AQI structures and their NIR electrochromism have been systematically investigated. When forming radical anions, both **23b** and **23c** exhibit strong absorption bands in the NIR region. Interestingly, the NIR absorption bands vary a lot between different substituents. Polymer **23c**$^-$ shows NIR absorptions at 860 and 1260 nm, while

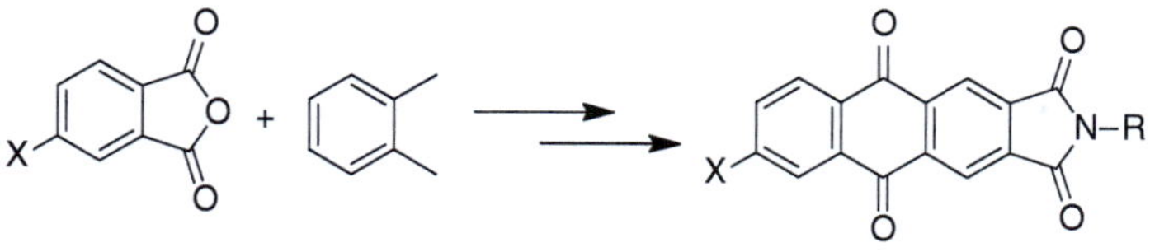

Scheme 7.2 A novel synthetic route of substituted AQIs.

23b$^-$ shows NIR absorptions peaked at 840 nm with a shoulder peak at 1000 nm. It is considered that the stronger electron-withdrawing substituent makes the absorption wavelength of the radical anion red-shift. By means of a Gaussian calculation to guide molecular design, the nitro moiety is further replaced by a dicyanovinyl group (**24** in Figure 7.5), and electrochromic absorption of the resultant molecule further red-shifts to 1400 nm (Figure 7.5) [42]. This is an interesting example that combines theoretical calculation and experiment to design a molecule authoritatively (Figure 7.6). It should be noted that the first reduction centre of AQIs often corresponds to the quinone moiety [16], which is different from NDIs and PDIs where the first reduction is attributed to aromatic imide.

For the purpose of modulating the NIR chiral signal, an attempt has been carried out to introduce AQI into a helical polymer. By the modification of the imide portion, the AQI moiety was successfully introduced into the side chain of poly(*N*-propargylamides)s (**25** in Figure 7.7) [43]. Due to the chiral centre of the amino acid and the strong interactions including H-bonding as well as $\pi-\pi$ stacking, **25** takes a helical conformation of backbone and skewed packing of pendant chromophores. Upon application of different voltages, intense CD signals centred at 780 and 530 nm appear and are assigned to the radical anionic and dianionic intermediates, respectively (Figure 7.7). A further voltage-controlled reversible Cotton effect demonstrates that the polymers are potential electrochemically driven chiroptical-switching materials operating in the NIR region. However, the reversibility should be improved and the electrochromic wavelength should also be extended to the communication wavelengths.

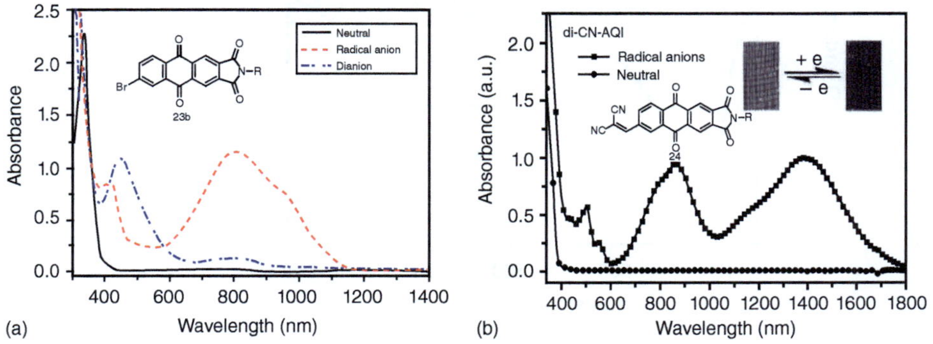

Figure 7.5 UV/vis/NIR spectra of **23b** (a) and **24** (b) in different states in CH_2Cl_2 solution.

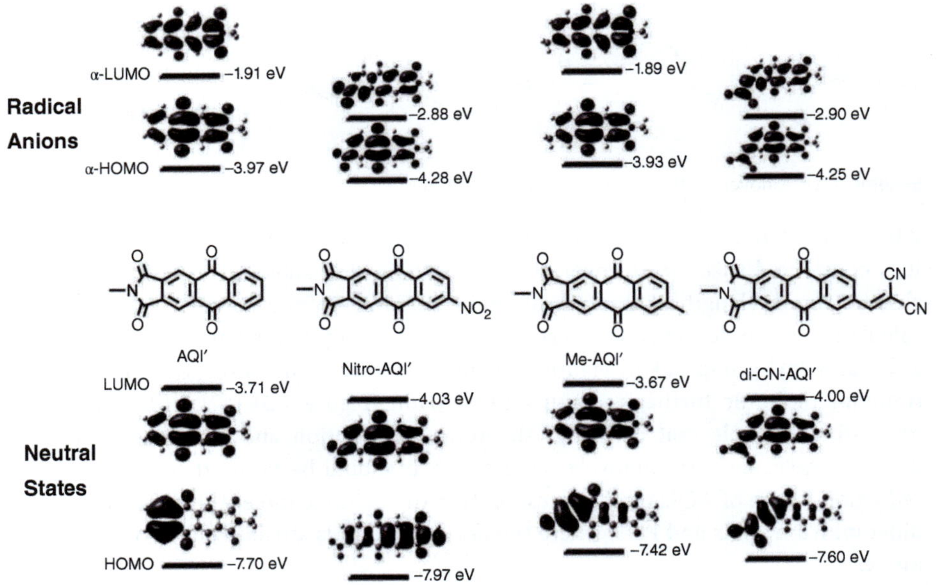

Figure 7.6 Selected frontier molecular orbitals of unsubstituted and six-substituted AQIs in their neutral and radical anionic states obtained at the UB3LYP/6-31G+(d, p) level.

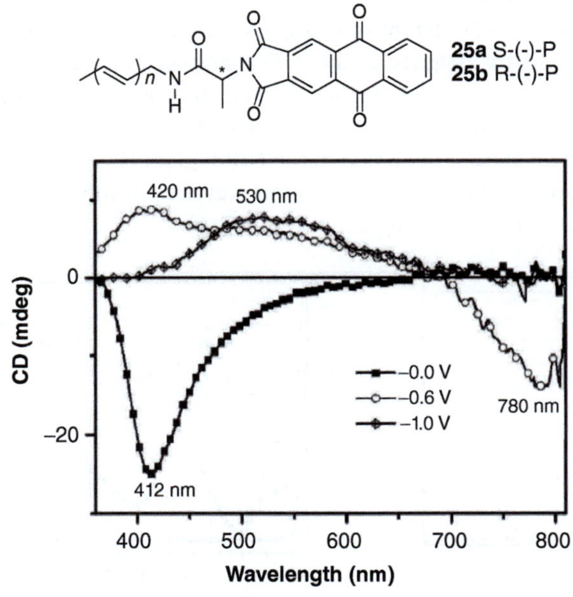

Figure 7.7 CD spectra of **25a** film at 0.0, −0.6 and −1.0 V with stepping potentials between −2 and 0 V.

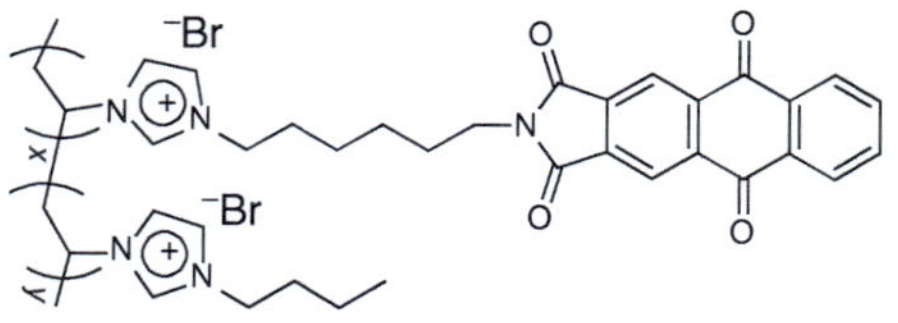

	x	y
26a:	1	0
26b:	0.14	0.86
26c:	0.11	0.89

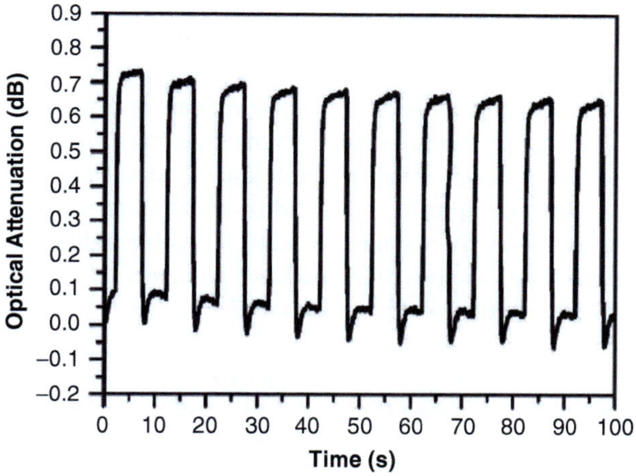

Figure 7.8 Optical attenuation of **26a** film at 810 nm.

Common NIR electrochromic devices are multi-layer configurations which always contain an electrochromic layer, electrolyte, transparent electrode and background sheet. However, such multi-layer configurations are not convenient for large-area applications. Besides, the interface defects between different layers also remarkably affect the cell lifetime. To simplify the device structure, an AQI derivative is copolymerised with ionic liquid to construct a single all-solid-state device (**26** in Figure 7.8) [44]. This electrochromic device is assembled in the air with a simple configuration ITO|**26**|ITO. The device requires only 1 s at −2 V for switching transmittance at 810 nm and 1 s for bleaching (Figure 7.8). Such a simple device demonstrates a rapid response in the NIR region to an externally applied voltage. However, the optical contrast of the device is low compared with those observed in thin films dipped into electrolyte solutions; perhaps more design factors should be taken into account.

7.5 Poly(triarylamine)s

The aforementioned materials are mainly low-molecular-mass cathodically colouring electrochromic materials with poor processability. Even though they are successfully incorporated into polymers, the electrochromic devices always exhibit low colour contrast and poor reversibility. In short, cathodically colouring

NIR electrochromism of both organic small molecules and polymeric materials are scarce and difficult to obtain, in part due to the relatively poor stability of their charged states to oxygen and water [44].

In contrast with cathodically colouring electrochromic materials, anodically colouring electrochromic materials are not very sensitive to oxygen and water and display very good electrochromic stability. Besides, anodically colouring electrochromic materials are often easy to be derivatised to form high-molecular-weight polymers. Compared with small molecules, polymeric electrochromic materials possess several advantages, such as low driving voltage, rapid response time, high colour contrast, long cycle life, good environmental stability, easy solution processing and polyelectrochromism with the same material.

Polyaniline (PANI, **27a** in Figure 7.9a) is a well-known anodically colouring electrochromic material operating in both the visible and NIR regions [45, 46]. PANI shows electrochromism in three states, including leucoemeraldine base (LB, fully reduced), pernigraniline base (PB, fully oxidised) and emeraldine base (EB, protonation of PANI's intermediate oxidation state) (Figure 7.9a) [47]. However, switching across the full range of PANI is not easy to achieve in aqueous media due to hydrolytic degradation and loss of electroactivity of the PB state in the presence of water. A recent study shows that this can be avoided by template synthesised with poly(2-acrylamido-2-methyl-1-propanesulfonic acid) (PAAMPSA) [48]. The resultant PANI-PAMS particles show enhanced stability at elevated pH, and no hydrolytic degradation is found at oxidation potentials. The structural rearrangement caused by the further treatment with dichloroacetic acid (DCA) allows quick and complete access to the fully oxidised and reduced forms of PANI [47]. The corresponding device switches across the full range of PANI among different potentials quickly and reversibly (Figure 7.9b), and the colour contrast at 1063 nm reaches 50% between LB and PB states.

Unlike PANI, which needs doping with small-molecule acids, such as HCl or H_2SO_4, to act as an electrochromic material, electron-rich triphenylamines

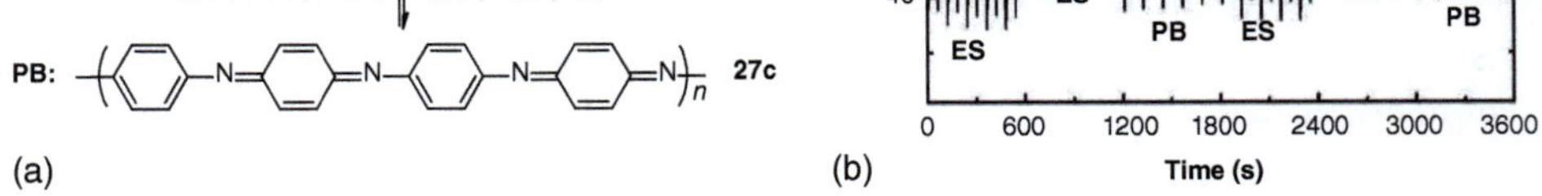

Figure 7.9 (a) Reaction sequence shows redox transitions between LB, emeraldine salt (ES) and PB (pernigraniline base) states, where A^- represents the counter ion. (b) DCA-treated PANI-PAAMPSA films on bisphosphonic-acid-modified indium tin oxide (ITO) to series of switching potentials.

(TPAs) can be easily oxidised to form stable radical cations, and meanwhile, the oxidation processes are always associated with significant absorption changes in both the visible and NIR regions. Besides, TPAs can be derivatised easily and yield different topologies, providing researchers with great convenience of good solution processability and flexible structure regulation [7].

Electron-rich TPAs can be easily polymerised by electrochemical oxidation. However, the molecular weight of the polymer product is not easy to control. Most of the time, polymers after electrochemical polymerisations are insoluble in common solvents, which hampers their characterisation and application for large-area devices [49]. Various coupling reactions are applied to acquire solution-processable polymers based on TPAs, such as Suzuki coupling, Yamamoto coupling and oxidative coupling (Scheme 7.3) [50–52]. For example, polymer **28** from Suzuki coupling shows good solubility in many common organic solvents, such as *N*-methyl-2-pyrrolidone (NMP), THF, chloroform and toluene at room temperature [50]. Polymer **28** thin film shows rapid and reversible UV-vis-NIR absorption changes at different potentials, and it exhibits excellent stability of electrochromic characteristics after continuous cyclic scans over 900 s.

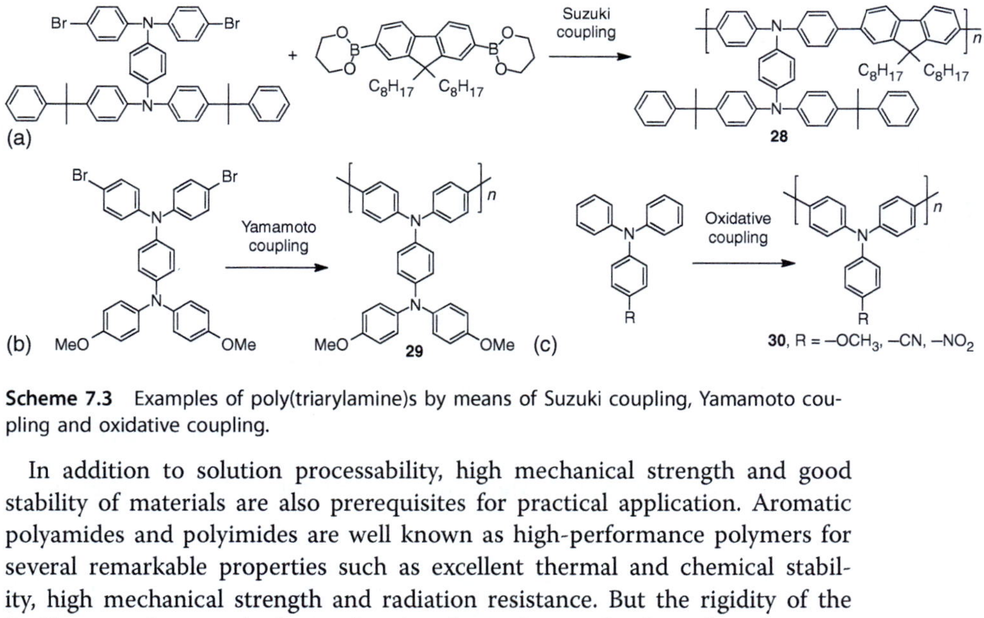

Scheme 7.3 Examples of poly(triarylamine)s by means of Suzuki coupling, Yamamoto coupling and oxidative coupling.

In addition to solution processability, high mechanical strength and good stability of materials are also prerequisites for practical application. Aromatic polyamides and polyimides are well known as high-performance polymers for several remarkable properties such as excellent thermal and chemical stability, high mechanical strength and radiation resistance. But the rigidity of the backbone and strong hydrogen bonding limit their technological applications because of high T_g and poor solubility in most organic solvents. Combining TPAs with polyamides or polyimides will be beneficial. On the one hand, with propeller-shaped TPAs as packing-disruptive units introduced into the polymer backbones, polyamides and polyimides show good solubility and good film-forming capability without sacrificing high stability too much. On the other hand, the obtained high-performance polyamides and polyimides exhibit excellent NIR electrochromism by virtue of the TPA units. Liou, Hsiao and co-workers have done a lot of work on the NIR electrochromism of aromatic polyamides and polyimides based on TPAs [7].

In 2005, Liou *et al.* prepared electrochromic aromatic polyimides **31** (Chart 7.4) with a pendant TPA derivative [53]. All the polyimides **31** are amorphous showing high solubility in many organic solvents and can be solution-cast into transparent, tough and flexible films with good mechanical properties. The CVs of the polymer films indicate that all the polymers display two reversible oxidation redox couples at 0.78 and 1.14 V corresponding to the cationic radical and dication, respectively. All the films of **31** show good NIR electrochromism. Taking **31a** as an example, when the applied potential is increased positively from 0.65 to 0.98 V, the characteristic NIR absorption band peaking at 879 nm increases gradually and is assigned as that of the cationic radical (Figure 7.10a). When the visible and NIR colour switchings between the neutral and the cationic radical state are conducted, **31a** exhibits no deviation in electrochromic properties in both regions during five continuous cyclic scans (Figure 7.10b).

Figure 7.10b shows that polymer **31a** thin film presents reversible colour switching between different states during five continuous cyclic scans. But with a further

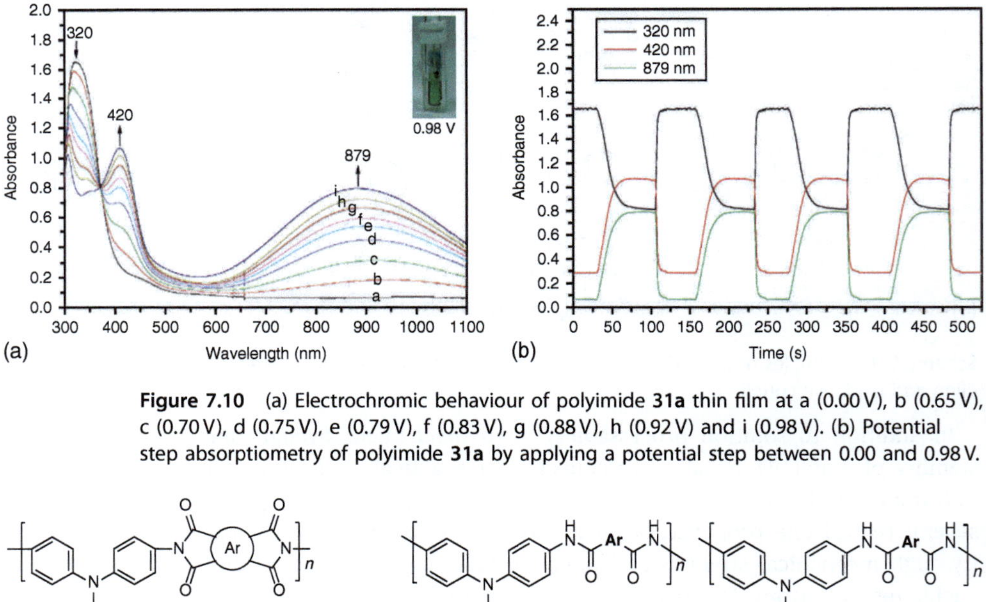

Figure 7.10 (a) Electrochromic behaviour of polyimide **31a** thin film at a (0.00 V), b (0.65 V), c (0.70 V), d (0.75 V), e (0.79 V), f (0.83 V), g (0.88 V), h (0.92 V) and i (0.98 V). (b) Potential step absorptiometry of polyimide **31a** by applying a potential step between 0.00 and 0.98 V.

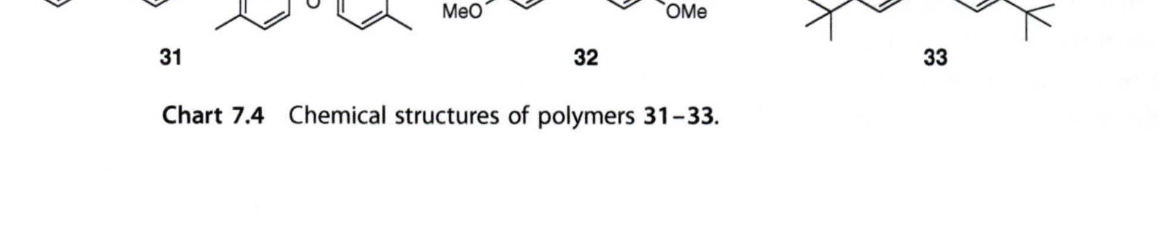

31 **32** **33**

Chart 7.4 Chemical structures of polymers **31–33**.

increase in numbers of switching scans, both the corresponding redox current in the CV and electrochromic contrast in spectroelectrochemistry decrease gradually. Long-term stability is a crucial factor for electrochromic materials. In this case, the decreased stability may be due to the instability of the electrogenerated cationic radical of TPA and the formation of tetraphenylbenzidine by tail-to-tail coupling reaction (Scheme 7.4) [54]. Incorporation of electron-donating substituents (e.g. **32** and **33** with methoxy and tert-butyl groups, Chart 7.4) into TPAs at the para-position of benzene not only efficiently blocks the reaction sites but also lowers the oxidation potential to improve stability [55, 56]. As a result, polymers **32** and **33** exhibit excellent stability and reversibility of NIR electrochromic properties after several hundreds of switching times.

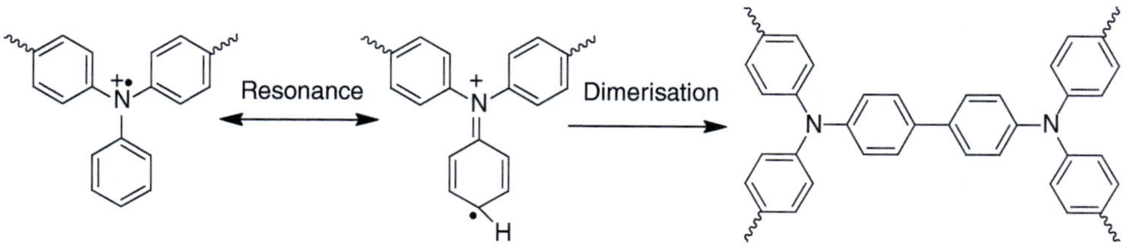

Scheme 7.4 Schematic diagram of the dimerisation of radical cations of TPAs.

In the following years, more polyimides, polyamides and poly(amide-imide)s based on TPA derivatives have been designed and synthesised to modulate the electrochromism of poly(triarylamine)s with low driving voltage, rapid switching time, high colour contrast, high coloration efficiency and long-term stability [7]. Overall, almost all the polymers are soluble in polar organic solvents and show good film-forming capability with high chemical and thermal stability. Generally, the polymers could be divided into several classes according to the method applied.

The first class involves modification of diamines with different functional groups while the diacids are not electroactive. Various functional groups are introduced into diamine units of poly(triarylamine)s through rational design, such as anthraquinone (**34**), pyrene (**35**), carbazole (**36**), phenothiazine (**37**) and so on (Chart 7.5) [57–60]. It is worth noting that poly(triarylamine)s

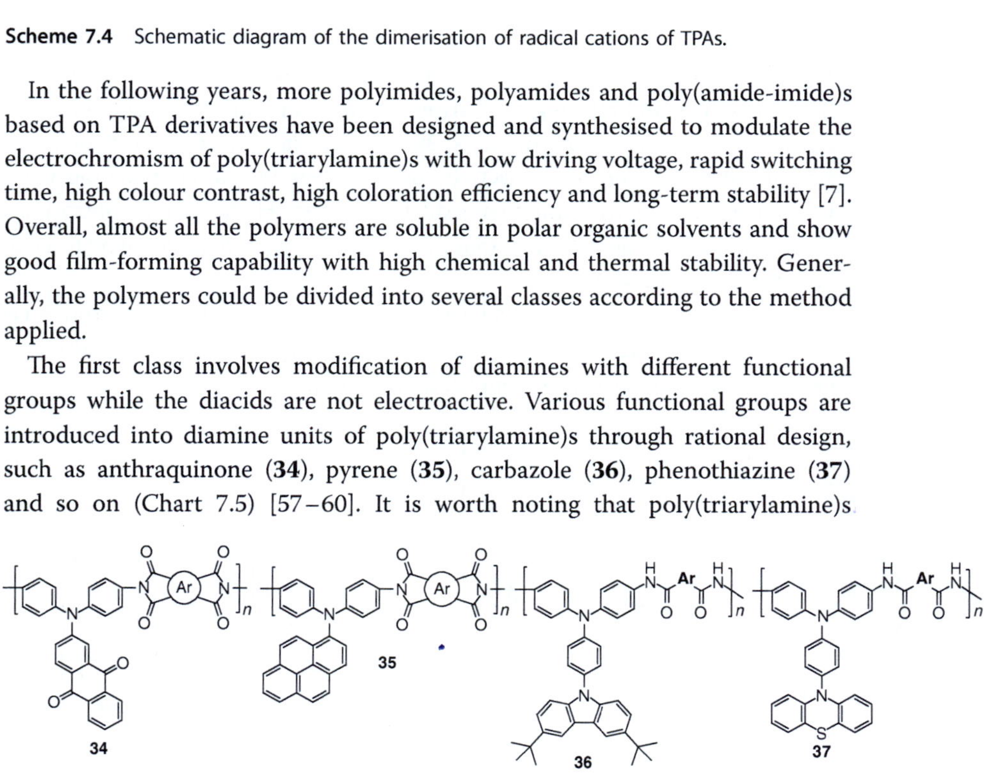

Chart 7.5 Examples of poly(triarylamine)s with different functional groups.

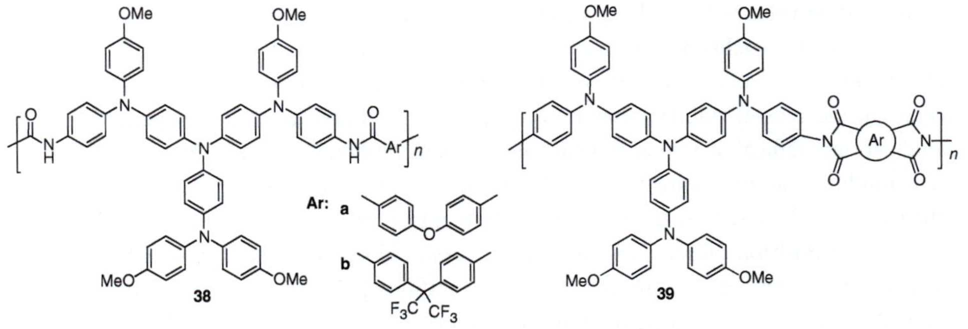

Chart 7.6 Examples of poly(triarylamine)s with starburst triarylamine.

containing starburst diamine display excellent electrochromism owing to its special structural and electronic character (Chart 7.6) [61–63].

Taking **38** as a good example [63], four redox centres are incorporated into a starburst diamine. As shown in Figure 7.11, the starburst structure provides more electroactive sites and higher solution processability. The incorporation of electron-donating methoxy moieties to the para-position of benzene lowers the oxidation potential and enhances electrochromic stability. The para-phenylenediamine-based core leads to IVCT absorption bands in the NIR region. With these careful molecular designs, **38a** exhibits high colour contrasts (85%) in both the visible and NIR regions between neutral and different oxidation states. In addition to high colour contrasts, **38a** also displays excellent electrochromic stability that maintains electroactivity of more than 99% after switching 10 000 times between neutral and first oxidation states. High colour contrast and long-term stability demonstrate that this polymer is a good potential material for commercial application.

Interestingly, for polyimides **39** bearing starburst triarylamine, apart from excellent NIR electrochromic characteristics, the D-A type polyimides are also good candidates for memory device application [61, 62].

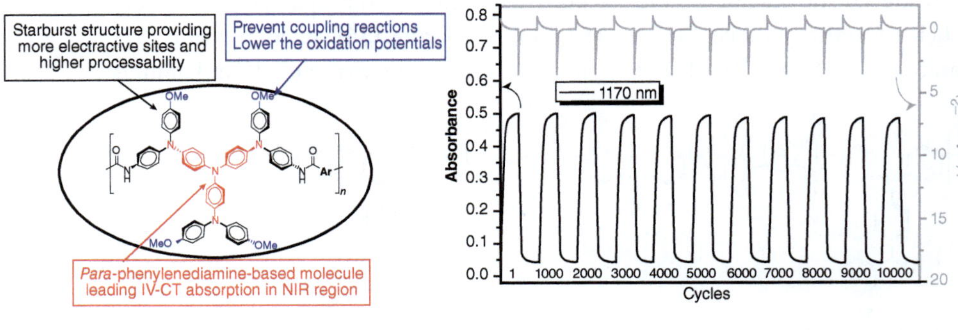

Figure 7.11 Molecular design and electrochromic switching between 0.00 and 0.55 V of **38a** thin film on ITO-coated glass substrate.

The second class involves polymerisation of TPA-based diacids and TPA-based diamines (Scheme 7.5) [64, 65]. Polymerisation of TPA-based diacids and normal diamines leads to results similar to those of the first class, which possesses only one monomer to be tuned. Besides, TPA-based diacids are often more difficult to be synthesised and purified than TPA-based diamines. Polymerisation of TPA-based diacids and TPA-based diamines would make a big difference. As two polymerised monomers are both electroactive, the resultant polymers may involve diverse redox stages to modulate. For example, polymer **40d** is composed of two similar monomers and CV investigation shows that **40d** reveals four reversible oxidation redox couples at 0.35, 0.64, 0.84 and 0.99 V corresponding to successive one-electron removal from four nitrogen atoms. As expect, four new absorption bands at 1030, 999, 831 and 800 nm correspond to these four oxidation states. With methoxy protection group, polymer **40d** film shows excellent electrochromic stability after 3000 cyclic switches.

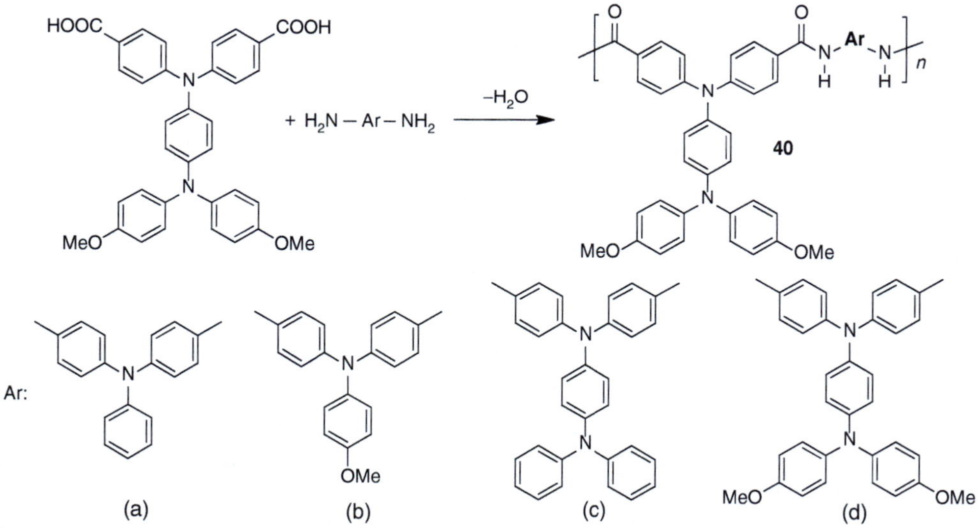

Scheme 7.5 Examples of polymers from TPA-based diacids and TPA-based diamines.

Finally, random copolymerisation of different diamine monomers is also an alternative method, such as *N,N,N′,N′*-tetraphenylbenzidine (TPB) and *N,N,N′,N′*-tetraphenyl-p-phenylenediamine (TPPA) (Chart 7.7) [66]. This method is very effective in controlling the absorption bands of polymers in both neutral and charged states. Nevertheless, the polymerisation activity of different diamines should not vary too much.

In recent years, in contrast with adopting sophisticated routes to synthesise different monomers, more attempts are made to construct electrochromic materials by means of economic and effective methods. Epoxy resins are commercially available materials showing high performance like high toughness, low shrinkage, excellent adhesion and good corrosion resistance. Amines are common

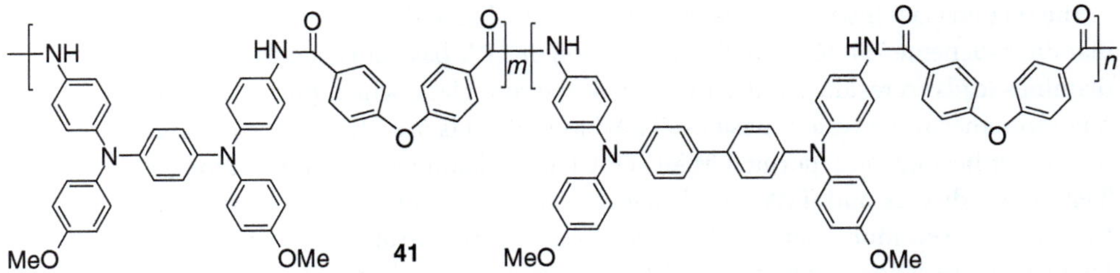

41

Chart 7.7 Copolymer of poly(triarylamine)s from TPB and TPPA.

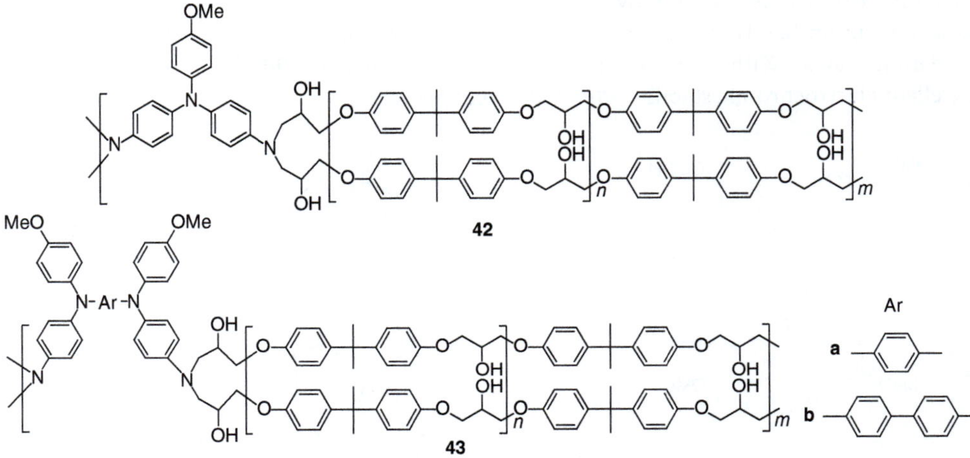

42

43

Chart 7.8 Chemical structures of the electrochromic thermosetting epoxies.

hardeners used in curing epoxy resins. Therefore, novel multi-electrochromic TPA-containing thermosetting epoxies have been readily prepared by using triarylamine-containing diamines as curing agents (Chart 7.8) [67]. These TPA-based epoxy materials show polyelectrochromism in both the visible and NIR regions with good stability and high optical contrast. As different types of NIR absorption bands could be modulated by adjusting the connection modes and N–N distances in bridge TPA systems [68], more practical TPA-based electrochromic thermoset epoxy materials could appear in the future.

7.6 Conjugated Polymers

Conjugated polymers have attracted a great amount of attention since their discovery owing to their applications in various fields, such as solar cells, light-emitting diodes, field effect transistors and sensors [69, 70]. In recent years, with the development of display technology and communication technology, the application of conjugated polymers in electrochromism has also made considerable progress [18, 71, 72]. The colour control of electrochromic conjugated conducting polymers in the visible region has been well expounded in Chapter 5, and this chapter pays more attention to the NIR electrochromic characteristics of conjugated polymers.

Thiophene- and pyrrole-based conjugated polymers are good candidates for electrochromic applications because of their good electrochromic stability, outstanding coloration efficiency, high colour contrast, fast switching time and tunable chemical structure [71]. Poly(3,4-ethylenedioxythiophene) (PEDOT, **44**), one of the most important advances in electrochromic conjugated polymers, has been known as the blue component for realisation of (red, green and blue) RGB-based display devices [73, 74]. Apart from considerable visible electrochromism, PEDOT also reveals intense electrochromism in the NIR region. As shown in Figure 7.12 [74], with an increase of external voltage, the absorption band in the visible region gradually decreases, and a new absorption band peaking at 900 nm gradually increases, which is ascribed to the optical band of a polaron charge carrier. The NIR absorption band at longer wavelength also begins to increase, which is attributed to a bipolaron charge carrier.

Pozo-Gonzalo *et al.* reported a new all-plastic electrochromic device with an optical contrast of 44% at 1971 nm based on PEDOT formulations and ionic liquid blends as electrolytes [75, 76]. This device could be useful for NIR applications

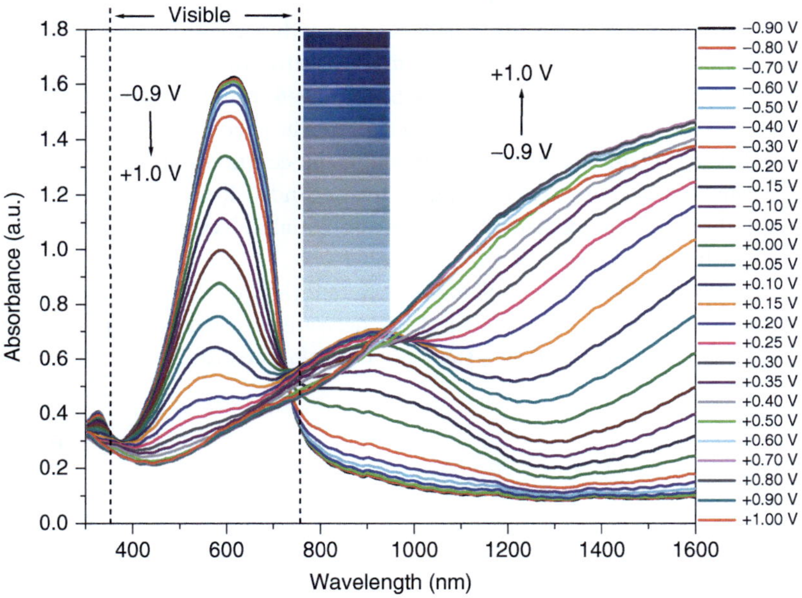

Figure 7.12 Spectroelectrochemistry of a PEDOT thin film in different oxidation states with inset pictures of variation in the blue colour at different oxidation levels.

such as spacecraft for thermal control, replacing the existing systems, adding lightness and flexibility to the final device.

As for the TPA-based polymers research, in order to achieve the purpose of practical application of conjugated polymers, several factors are crucial, including long-term stability, low redox potential, high colour contrast, high coloration efficiency, fast response time and polyelectrochromism. There are several strategies that are employed to modulate the performance of the electrochromic stages of conjugated polymers.

The first one is to modify the side chain of polythiophene and polypyrrole while keeping the main chain unchanged (Chart 7.9). Several examples are listed, such as the change of ring sizes in poly(3,4-alkylenedioxypyrrole)s (PXDOPs, **45**) and poly(3,4-alkylenedioxythiophene)s, (PXDOTs, **46**), replacement of the oxygen atom with a sulfur atom in PEDOT (**47**), introduction of side chains into polythiophene (**48**) and polypyrrole (**49**), and fusion of another thiophene (**50**) or furan ring (**51**) into polythiophene [18, 71, 77, 78]. It is worth noting that poly(3,4-propylenedioxythiophene)s (PProDOTs, **52**) have attracted much attention because of their excellent optoelectronic properties. Reynolds *et al.* have done a lot of interesting and meaningful work based on this structure [71, 79–81].

When incorporated into a reflective device platform, disubstituted PProDOTs (PProDOT-$(CH_2OEtHx)_2$, i.e. R = ethylhexyloxyl) [80] reveal relatively no colour change in the visible region but a large change (70–90%) in the NIR region between −0.8 and 0.4 V. Interestingly, this unsymmetrical switching is thickness-dependent and does not exist in transmission mode (Figure 7.13). This phenomenon is attributed to the penetration depth reached by the NIR light, which allows light from that region to be more sensitive to optical changes occurring in the polymer film as a conductive front moved from the electrode surface to the top of the film. A VOA device of the material has been constructed, and the device displays 12 dB of optical attenuation at the communication wavelengths with minimal optical loss (<0.2 dB) and a sub-second switching time. The result indicates that these conjugated polymers are potentially electrochromic materials in VOA with further optimisation.

The second approach is to copolymerise different electron-donating monomers. In addition to thiophene, pyrrole and their derivatives, other electron-donating building blocks are also involved, such as carbazole, substituted benzene and so on (Chart 7.10) [82–86].

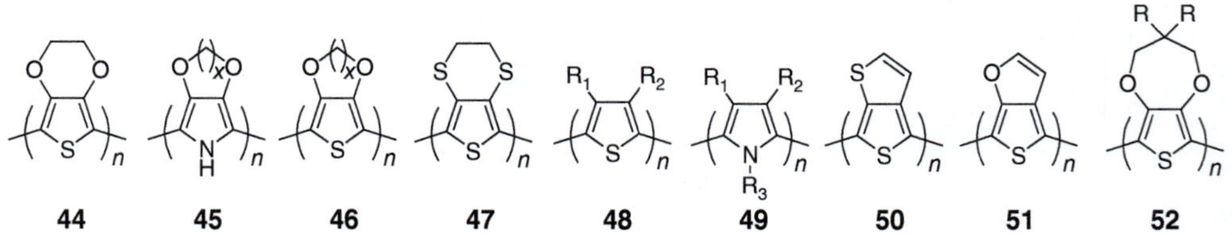

44 45 46 47 48 49 50 51 52

Chart 7.9 Modifications of the side chains of polythiophene and polypyrrole.

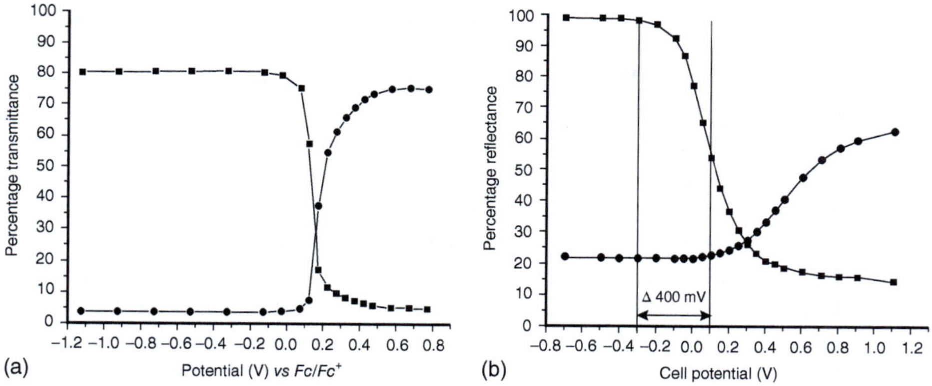

Figure 7.13 (a) Percentage transmittance for a 500 nm thick film of PProDOT-$(CH_2OEtHx)_2$ on ITO|glass as a function of applied potential at 1.55 μm (■) and 550 nm (•). (b) Percentage total reflectance for a 500-nm-thick PProDOT-$(CH_2OEtHx)_2$ reflective electrochromic device (ECD) as a function of applied potential at 1.55 μm (■) and 550 nm (•).

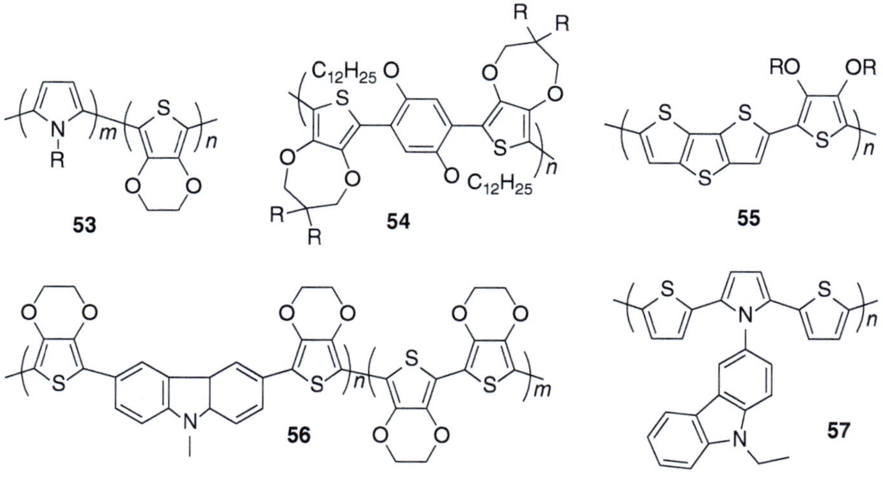

53 54 55 56 57

Chart 7.10 Examples of copolymers of different electron-donating monomers.

Taking polymer **57** as an example, the optical contrast is found to be 50% at 1000 nm [84]. A thin film of **57** shows excellent electrochromic stability with maintaining 97.6% optical activity after 5000 switching cycles. Besides, the colouring and bleaching times are only 1.4 and 1.2 s, respectively. High contrast ratio in the NIR region and rapid switching time are advantages of polymer **57** for NIR electrochromic application. An electrochromic device with a sandwich configuration, ITO|**57**|gel electrolyte|PEDOT|ITO, was assembled from **57**. The device exhibits reversible electrochemical oxidation, low driving voltage, high coloration efficiency and high redox stability. However, the optical contrast of the **57**|PEDOT device at 1000 nm is only 16%, which is much smaller than that on an ITO|glass surface. Maybe more design factors of the device structure should be considered.

Low-band-gap polymers are greatly desired for electrochromic applications since they could be utilised as cathodically colouring polymers with highly transparent oxidised states. Besides, low-band-gap polymers often bring about low oxidation or reduction potentials; thus, polymer degradation can be greatly avoided at higher numbers of switching cycles [87]. Modification of the side chains of polythiophene and polypyrrole and copolymerisation of different donating monomers can only modulate the energy gap of polymers to a limited degree. Reduction of bond-length alternation and syntheses of highly planar systems make a certain contribution, but numerous experiments have confirmed that the D-A approach is the most effective method applied to obtain low-band-gap polymers [18, 71].

Thus, the most successful class is the application of an alternating sequence of donor (electron-rich) and acceptor (electron-deficient) units in the polymer chain. Common donor units are based on thiophene, pyrrole and their derivatives. The challenge of the D-A approach lies in choosing appropriate acceptors as well as their combinations with different donors. Toppare *et al.* have done a lot of important work on NIR electrochromism based on D-A type conjugated polymers [18].

Various electron-deficient acceptors have been employed to build D-A conjugated polymers for electrochromic researches, such as benzothiadiazole, benzoquinoxaline, thienopyrazine, benzotriazole, benzimidazole and so on (Chart 7.11) [86, 88–91]. In addition to making a big contribution to tune the colour change in the visible region, great achievements have also been realised in the NIR region based on these D-A conjugated polymers.

To date, the majority of the electroactive organic materials that have been investigated are p-dopable materials. However, n-dopable polymers are only limited to a small number of polymers. Polymers containing stable n-dopable states are desirable for their potential applications in many complex device designs such as smart windows and mirrors [92]. The D-A approach provides an effective way to obtain polymers possessing p- and n-dopable states at the same time.

A variety of low-band-gap, p- and n-dopable polymers have been synthesised by the D-A approach, such as polymers **58–61** (Chart 7.12) [87, 92–95]. It should be noted that n-doping of a conjugated polymer system is not just an electrochemical formation of a reduced state. As shown in Figure 7.14, except for a reversible redox

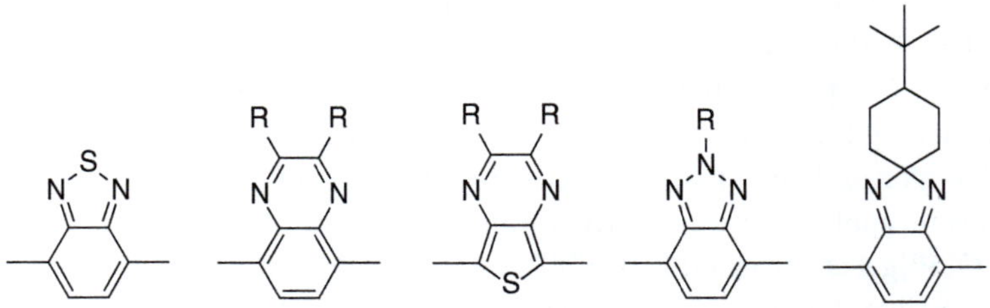

Chart 7.11 Common acceptor units in D-A type electrochromic conjugated polymers.

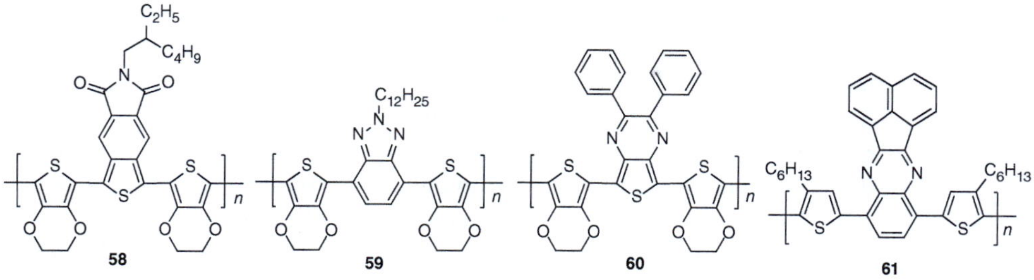

Chart 7.12 Structures of both p- and n-dopable D-A polymers **58–61**.

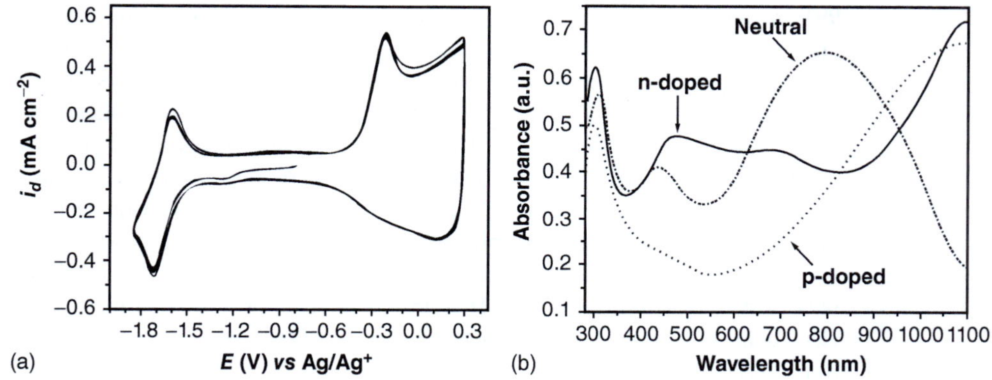

Figure 7.14 (a) p- and n-dopable **58** film in monomer-free electrolyte solution. (b) Spectroelectrochemistry of **58** film in three different states: neutral, n-dopable and p-dopable.

couple at negative potential, a drastic spectral change in NIR region is also needed to prove the true n-type doping process. Interestingly, in addition to displaying p- and n-dopable properties simultaneously, **58** film also shows unusual spectroelectrochemistry with small changes in the transparency and colour of the polymer in the visible region but substantial changes in the NIR region in transmission mode. Even though the reason is unknown, it could be very useful for NIR devices [94].

As mentioned earlier, materials exhibiting electrochromism with high colour contrast at 1310 and 1550 nm are of great possible applications in VOAs in fibre-optical communications. Regulation of the colour contrast, response time and long-term stability at these wavelengths are important issues in NIR

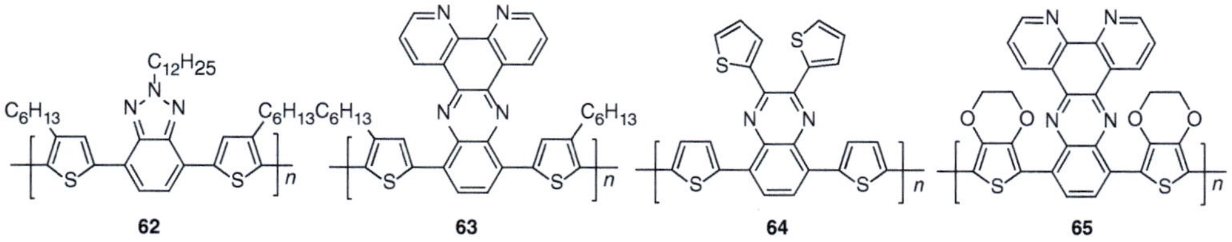

Chart 7.13 Chemical structures of D-A conjugated polymers **62–65**.

electrochromic researches. Polymers **62** and **63** (Chart 7.13) are two D-A type conjugated polymers for this particular application [96, 97]. The colour contrasts of **62** and **63** between different states are 56% at 1310 nm and 65% at 1550 nm (Figure 7.15), and the response times upon switching of **63** thin film between neutral and oxidised states are only 1.5 and 0.85 s, respectively. High colour contrast and rapid response time at communication wavelengths provide favourable conditions for their further telecommunication application.

The D-A type conjugated polymers also provide the highest optical contrasts in the NIR region reported to date. Polymer **64** shows an optical contrast of 98% at 1800 nm with a fast switching time of 0.6 s and no significant charge loss is observed after 1000 cycles [98]. Polymer **65** has an excellent optical contrast of 99% in the NIR region (1460 nm) between the neutral and oxidised states, and only 1 s is needed to achieve the optical contrast [22]. This is the highest optical contrast reported for a conjugated polymer system. Such large contrasts allow the polymers to be utilised as active electrochromic materials for NIR application.

Thiophene, pyrrole and their derivatives are widely investigated as building blocks for electrochromic conjugated polymers, yet selenophene and its derivatives are paid scarce attention. The higher electron-donating character of selenium atom leads selenophene to have a lower oxidation potential in comparison with thiophene and pyrrole; thus, degradation of polymers through over oxidation is minimised [99]. Therefore, another alternative method to adjust electrochromic properties of conjugated polymers is to incorporate selenophene or its derivatives into conjugated polymers.

Chart 7.14 displays several examples of conjugated polymers based on selenophene derivatives. 3,4-(Ethylenedioxy)selenophene (EDOS) was first synthesised and electrochemically and chemically polymerised by Cava (**66**) [100]. Polymer **67** shows not only high contrast visible in NIR electrochromism but also has a useful photothermal effect [101]. Based on this unusual photothermal effect, a transparent photo-thermo-electric converter has been realised [102]. Polymer **68** shows a very subtle contrast change (0.3%) in the visible region

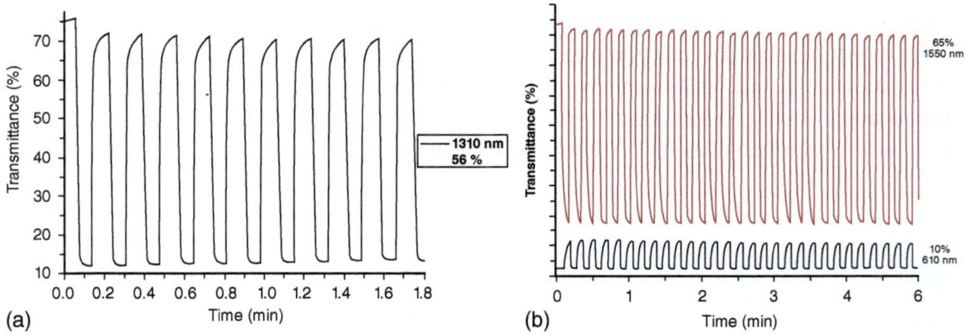

Figure 7.15 (a) Optical transmittance changes of **62** monitored at 1310 nm while switching the potentials between its oxidised and reduced states. (b) Electrochromic switching, optical absorbance change monitored of at 1550 nm for **63** between oxidised and neutral states.

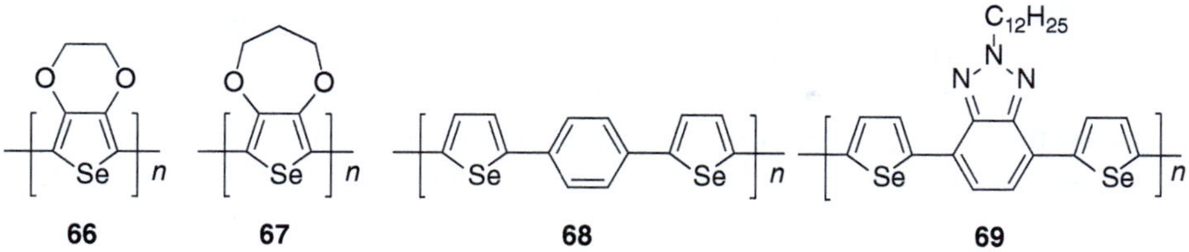

Chart 7.14 Examples of polymers based on selenophene derivatives.

while a 25% optical contrast in the NIR region (1240 nm) with a short switching time of 0.6 s [99]. Polymer **69** displays both p- and n-dopable properties, and a 56% optical contrast is achieved in the NIR region with a 0.4 s switching time between neutral and oxidised states [103]. Besides, this is the first example of an n-dopable selenophene-containing polymer. These results indicate that selenophene-based conjugated polymers could also be promising materials as NIR active electrochromic materials with further optimisation.

7.7 Other NIR Electrochromic Materials

Except for the aforementioned electrochromic materials, there are also other kinds of materials that exhibit NIR electroactivities (Chart 7.15). They are important in opening researchers' minds to look for new materials and find applications of NIR electrochromic materials.

Lutetium phthalocyanine (**70**, $LuPc_2$) has seven redox states containing both oxidations and reductions and is therefore a potential polyelectrochromic material [104, 105]. Apart from colour change in the visible region, it also exhibits NIR electrochromism. The response time of a device based on **70** thin film has reached about only 1 s, which is very fast among the known NIR electrochromic devices

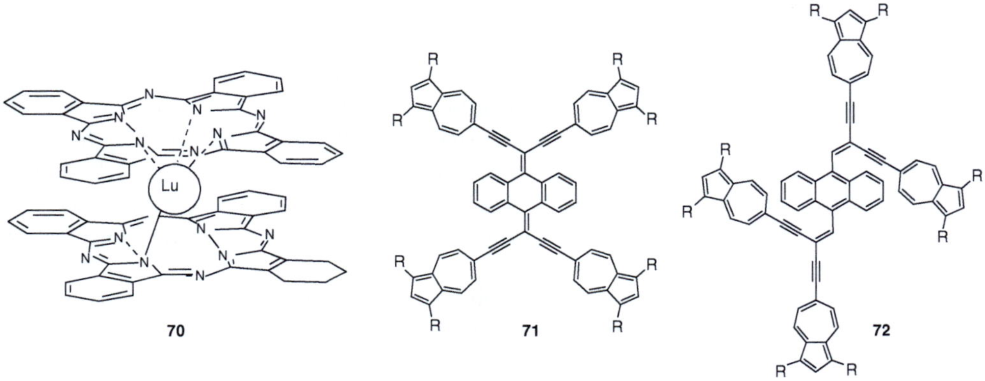

Chart 7.15 Structures of $LuPc_2$(**70**) and azulene-based compounds **71** and **72**.

[105]. However, both NIR absorption and attenuation are relatively low and are needed to be further improved.

Enediyne structures are common molecular scaffolds that reveal good electron-withdrawing properties, and the azulenyl group has a strong ability to stabilise anionic species owing to the remarkable polarisability of the azulene ring. Combining the enediyne unit with the azulenyl group provides redox systems that exhibit stabilised NIR electrochromic materials (e.g. **71** and **72**) [106–109]. Compound **71** exhibits two one-step, two-electron reductions (Scheme 7.6) and affords strong absorption bands in the NIR region in its two-electron-reduced state [107]. Azulene-containing conjugated polymers have also been synthesised, but the results are not satisfactory since the colouring and bleaching times are both much slower than those of common conjugated polymers [110].

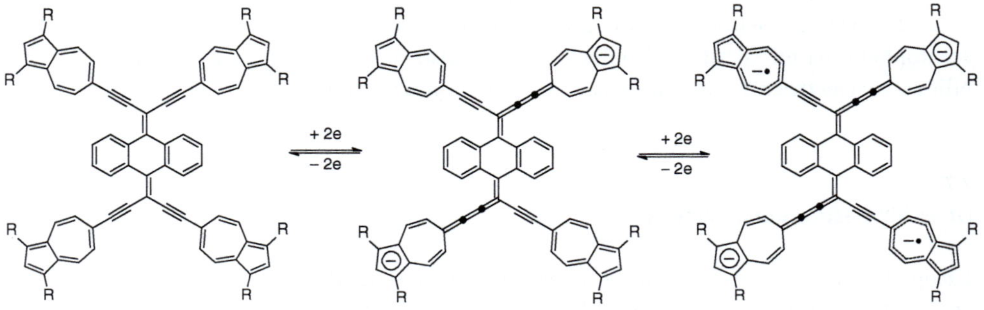

Scheme 7.6 Redox-system of azulene-substituted enediynes **71**.

7.8 Conclusion

This chapter dealt with NIR electrochromism of several kinds of organic redox systems with both low and high molecular mass.

Aromatic quinones, aromatic imides and AQIs are cathodically active and exhibit NIR absorption bands upon reduction to radical anionic states. The relationship between molecular structure and NIR electrochromic property has been well investigated. Their incorporation into polymer structures remarkably improves their processability and stability. NIR chiroptical switching has been realised by placing chromophores in chiral environments, such as helical polymers and chiral low-molecular-mass organogels. Long-term stability, high colour contrast and rapid switching time are crucial issues to be resolved for practical applications.

Poly(triarylamine)s are anodically colouring electrochromic materials and display polyelectrochromism in both the visible and NIR regions. Delicately designed TPA-based polyimides and polyamides based on TPAs show good solubility, high thermal stability as well as good film-forming capability. Derivation of TPA with various functional groups is able to tune the electrochromic properties to a large extent. The optimisation of device configuration may help them to better meet the various application requirements.

Conjugated polymers based on thiophene and pyrrole derivatives could display both cathodic and anodic electrochromism in the NIR region depending on their structures. The D-A approach is considered as the most successful method to adjust the band gap of conjugated polymers. A 99% colour contrast ratio in the NIR region has been achieved by a D-A type polymer. Small changes in the visible region and drastic changes in the NIR region have also been realised. Conjugated polymers are the most potential NIR electrochromic materials for practical applications owing to their several excellent performances, including tunable chemical structure, high conductivity, rapid switching time, high colour contrast, good film-forming capability and low cost.

References

1. Deb, S.K. (1969) *Appl. Opt.*, **8**, 192–195.
2. Monk, P.M.S., Mortimer, R.J., and Rosseinsky, D.R. (2007) *Electrochromism and Electrochromic Devices*, Cambridge University Press, Cambridge.
3. Monk, P.M.S., Mortimer, R.J., and Rosseinsky, D.R. (1995) *Electrochromism: Fundamentals and Applications*, Wiley-VCH Verlag GmbH, Weinheim.
4. Mortimer, R.J. (2011) *Annu. Rev. Mater. Res.*, **41**, 147–156.
5. Fabian, J., Nakazumi, H., and Matsuoka, M. (1992) *Chem. Rev.*, **92**, 1197–1226.
6. Qian, G. and Wang, Z.Y. (2010) *Chem. Asian J.*, **5**, 1006–1029.
7. Yen, H.J. and Liou, G.S. (2012) *Polym. Chem.*, **3**, 255–264.
8. McDonagh, A.M., Bayly, S.R., Riley, D.J., Ward, M.D., McCleverty, J.A., Cowin, M.A., Morgan, C.N., Varrazza, A.R., Penty, R.V., and White, I.H. (2000) *Chem. Mater.*, **12**, 2523–2524.
9. Zhang, J.D., Yu, H.A., Wu, X.G., and Wang, Z.Y. (2004) *Opt. Mater.*, **27**, 265–268.
10. Granqvist, C.G. (2012) *Sol. Energy Mater. Sol. Cells*, **99**, 1–13.
11. Franke, E.B., Trimble, C.L., Hale, J.S., Schubert, M., and Woollam, J.A. (2000) *J. Appl. Phys.*, **88**, 5777–5784.
12. Beydaghyan, G., Boudreau, M., and Ashrit, P.V. (2011) *J. Mater. Res.*, **26**, 55–61.
13. Hale, J.S. and Woollam, J.A. (1999) *Thin Solid Films*, **339**, 174–180.
14. Ward, M.D. (2005) *J. Solid State Electrochem.*, **9**, 778–787.
15. Zhong, Y.W., Yao, C.J., and Nie, H.J. (2013) *Coord. Chem. Rev.*, **257**, 1357–1372.
16. Rak, S.F., Jozefiak, T.H., and Miller, L.L. (1990) *J. Org. Chem.*, **55**, 4794–4801.
17. Mackinnon, S.M. and Wang, Z.Y. (2000) *J. Polym. Sci., Part A: Polym. Chem.*, **38**, 3467–3475.
18. Gunbas, G. and Toppare, L. (2012) *Chem. Commun.*, **48**, 1083–1101.
19. Todd, E.K., Wang, S., Wan, X.H., and Wang, Z.Y. (2005) *Tetrahedron Lett.*, **46**, 587–590.
20. Heckmann, A. and Lambert, C. (2012) *Angew. Chem. Int. Ed.*, **51**, 326–392.
21. Neo, W.T., Loo, L.M., Song, J., Wang, X.B., Cho, C.M., Chan, H.S.O., Zong, Y., and Xu, J.W. (2013) *Polym. Chem.*, **4**, 4663–4675.
22. Tarkuc, S., Unver, E.K., Udum, Y.A., and Toppare, L. (2010) *Eur. Polym. J.*, **46**, 2199–2205.
23. Miller, L.L. and Mann, K.R. (1996) *Acc. Chem. Res.*, **29**, 417–423.
24. Zheng, J., Qiao, W.Q., Wan, X.H., Gao, J.P., and Wang, Z.Y. (2008) *Chem. Mater.*, **20**, 6163–6168.

25. Lambert, C., Nöll, G., Zabel, M., Hampel, F., Schmälzlin, E., Bräuchle, C., and Meerholz, K. (2003) *Chem. Asian J.*, **9**, 4232–4239.
26. Jozefiak, T.H. and Miller, L.L. (1987) *J. Am. Chem. Soc.*, **109**, 6560–6561.
27. Almlöf, J.E., Feyereisen, M.W., Jozefiak, T.H., and Miller, L.L. (1990) *J. Am. Chem. Soc.*, **112**, 1206–1214.
28. Jozefiak, T.H., Almlöf, J.E., Feyereisen, M.W., and Miller, L.L. (1989) *J. Am. Chem. Soc.*, **111**, 4105–4106.
29. Rak, S.F. and Miller, L.L. (1992) *J. Am. Chem. Soc.*, **114**, 1388–1394.
30. Renz, M. and Kaupp, M. (2012) *J. Phys. Chem. A*, **116**, 10629–10637.
31. Wang, S., Todd, E.K., Birau, M., Zhang, J.D., Wan, X.H., and Wang, Z.Y. (2005) *Chem. Mater.*, **17**, 6388–6394.
32. Kim, S.H., Ko, H.C., Moon, B.J., and Lee, H.S. (2006) *Langmuir*, **22**, 9431–9435.
33. Gawrys, P., Louarn, G., Zagorska, M., and Pron, A. (2011) *Electrochim. Acta*, **56**, 3429–3435.
34. Penneau, J.F., Stallman, B.J., Kasai, P.H., and Miller, L.L. (1991) *Chem. Mater.*, **3**, 791–796.
35. Zheng, J., Zheng, Y.J., and Wan, X.H. (2011) *Chin. J. Polym. Sci.*, **29**, 117–123.
36. Koyuncu, F.B. (2012) *Electrochim. Acta*, **68**, 184–191.
37. Hou, J.H., Zhang, S.Q., Chen, T.L., and Yang, Y. (2008) *Chem. Commun.*, **44**, 6034–6036.
38. Lu, W., Gao, J.P., Wang, Z.Y., Qi, Y., Sacripante, G.G., Duff, J.D., and Sundararajan, P.R. (1999) *Macromolecules*, **32**, 8880–8885.
39. Qiao, W.Q., Zheng, J., Wang, Y.F., Zheng, Y.J., Song, N.H., Wan, X.H., and Wang, Z.Y. (2008) *Org. Lett.*, **10**, 641–644.
40. Qiao, W.Q. (2012) *Chem. Res. Chin. Univ.*, **28**, 238–241.
41. Chen, F.K., Fu, X.Y., Zhang, J., and Wan, X.H. (2013) *Electrochim. Acta*, **99**, 211–218.
42. Chen, F.K., Zhang, J., Jiang, H., and Wan, X.H. (2013) *Chem. Asian J.*, **8**, 1497–1503.
43. Zheng, Y.J., Cui, J.X., Zheng, J., and Wan, X.H. (2010) *J. Mater. Chem.*, **20**, 5915–5922.
44. Zheng, Y.J., Zheng, J., Dou, L.T., Qiao, W.Q., and Wan, X.H. (2009) *J. Mater. Chem.*, **19**, 8470–8477.
45. Gervais, F., Petit, N., Popon, C., and Buvat, P. (2003) *Eur. Phys. J. B*, **31**, 47–52.
46. Topart, P. and Hourquebie, P. (1999) *Thin Solid Films*, **352**, 243–248.
47. Tarver, J., Yoo, J.E., and Loo, Y.L. (2010) *Chem. Mater.*, **22**, 2333–2340.
48. Tarver, J., Yoo, J.E., Dennes, T.J., Schwartz, J., and Loo, Y.L. (2009) *Chem. Mater.*, **21**, 280–286.
49. Otero, L., Sereno, L., Fungo, F., Liao, Y.L., Lin, C.Y., and Wong, K.T. (2006) *Chem. Mater.*, **18**, 3495–3502.
50. Wu, H.Y., Wang, K.L., Jiang, J.C., Liaw, D.J., Lee, K.R., Lai, J.Y., and Chen, C.L. (2010) *J. Polym. Sci., Part A: Polym. Chem.*, **48**, 3913–3923.
51. Chang, H.W., Lin, K.H., Chueh, C.C., Liou, G.S., and Chen, W.C. (2009) *J. Polym. Sci., Part A: Polym. Chem.*, **47**, 4037–4050.
52. Lin, H.Y. and Liou, G.S. (2009) *J. Polym. Sci., Part A: Polym. Chem.*, **47**, 285–294.
53. Cheng, S.H., Hsiao, S.H., Su, T.H., and Liou, G.S. (2005) *Macromolecules*, **38**, 307–316.
54. Yen, H.J. and Liou, G.S. (2010) *Org. Electron.*, **11**, 299–310.
55. Liou, G.S. and Chang, C.W. (2008) *Macromolecules*, **41**, 1667–1674.
56. Hsiao, S.H., Liou, G.S., and Wang, H.M. (2009) *J. Polym. Sci., Part A: Polym. Chem.*, **47**, 2330–2343.
57. Wang, H.M., Hsiao, S.H., Liou, G.S., and Sun, C.H. (2010) *J. Polym. Sci., Part A: Polym. Chem.*, **48**, 4775–4789.
58. Yen, H.J. and Liou, G.S. (2010) *J. Mater. Chem.*, **20**, 9886–9894.
59. Hu, Y.C., Chen, C.J., Yen, H.J., Lin, K.Y., Yeh, J.M., Chen, W.C., and Liou, G.S. (2012) *J. Mater. Chem.*, **22**, 20394–20402.
60. Kung, Y.C. and Hsiao, S.H. (2011) *J. Mater. Chem.*, **21**, 1746–1754.
61. Chen, C.J., Yen, H.J., Hu, Y.C., and Liou, G.S. (2013) *J. Mater. Chem. C*, **1**, 7623–7634.

62. Yen, H.J., Chen, C.J., and Liou, G.S. (2013) *Adv. Funct. Mater.*, **23**, 5307–5316.
63. Yen, H.J., Lin, H.Y., and Liou, G.S. (2011) *Chem. Mater.*, **23**, 1874–1882.
64. Chang, C.W. and Liou, G.S. (2008) *J. Mater. Chem.*, **18**, 5638–5646.
65. Wang, H.M. and Hsiao, S.H. (2010) *Polym. Chem.*, **1**, 1013–1023.
66. Yen, H.J., Lin, K.Y., and Liou, G.S. (2011) *J. Mater. Chem.*, **21**, 6230–6237.
67. Chuang, Y.W., Yen, H.J., and Liou, G.S. (2013) *Chem. Commun.*, **49**, 9812–9814.
68. Yen, H.J., Guo, S.M., Liou, G.S., Chung, J.C., Liu, Y.C., Lu, Y.F., and Zeng, Y.Z. (2011) *J. Polym. Sci., Part A: Polym. Chem.*, **49**, 3805–3816.
69. Cheng, Y.J., Yang, S.H., and Hsu, C.S. (2009) *Chem. Rev.*, **109**, 5868–5923.
70. Zhao, Y., Guo, Y.L., and Liu, Y.Q. (2013) *Adv. Mater.*, **25**, 5372–5391.
71. Beaujuge, P.M. and Reynolds, J.R. (2010) *Chem. Rev.*, **110**, 268–320.
72. Amb, C.M., Dyer, A.L., and Reynolds, J.R. (2011) *Chem. Mater.*, **23**, 397–415.
73. Sonmez, G., Shen, C.K.F., Rubin, Y., and Wudl, F. (2004) *Angew. Chem.*, **116**, 1524–1528.
74. Sonmez, G., Sonmez, H.B., Shen, C.K.F., and Wudl, F. (2004) *Adv. Mater.*, **16**, 1905–1908.
75. Pozo-Gonzalo, C., Mecerreyes, D., Pomposo, J.A., Salsamendi, M., Marcilla, R., Grande, H., Vergaz, R., Barrios, D., and Sánchez-Pena, J.M. (2008) *Sol. Energy Mater. Sol. Cells*, **92**, 101–106.
76. Vergaz, R., Barrios, D., Pena, J.M.S., Marcos, C., Pozo, C., and Pomposo, J.A. (2008) *Sol. Energy Mater. Sol. Cells*, **92**, 107–111.
77. Schottland, P., Zong, K., Gaupp, C.L., Thompson, B.C., Thomas, C.A., Giurgiu, I., Hickman, R., Abboud, K.A., and Reynolds, J.R. (2000) *Macromolecules*, **33**, 7051–7061.
78. Sotzing, G.A. and Lee, K. (2002) *Macromolecules*, **35**, 7281–7286.
79. Cirpan, A., Argun, A.A., Grenier, C.R.G., Reeves, B.D., and Reynolds, J.R. (2003) *J. Mater. Chem.*, **13**, 2422–2428.
80. Dyer, A.L., Grenier, C.R.G., and Reynolds, J.R. (2007) *Adv. Funct. Mater.*, **17**, 1480–1486.
81. Schwendeman, I., Hwang, J., Welsh, D.M., Tanner, D.B., and Reynolds, J.R. (2001) *Adv. Mater.*, **13**, 634–637.
82. Galand, E.M., Mwaura, J.K., Argun, A.A., Abboud, K.A., McCarley, T.D., and Reynolds, J.R. (2006) *Macromolecules*, **39**, 7286–7294.
83. Gaupp, C.L. and Reynolds, J.R. (2003) *Macromolecules*, **36**, 6305–6315.
84. Sefer, E., Koyuncu, F.B., Oguzhan, E., and Koyuncu, S. (2010) *J. Polym. Sci., Part A: Polym. Chem.*, **48**, 4419–4427.
85. Silva, A.J.C., Ferreira, S.M.F., Santos, D.d.P., Navarro, M., Tonholo, J., and Ribeiro, A.S. (2012) *Sol. Energy Mater. Sol. Cells*, **103**, 108–113.
86. Neo, W.T., Cho, C.M., Song, J., Chin, J.M., Wang, X.B., He, C.B., Chan, H.S.O., and Xu, J.W. (2013) *Eur. Polym. J.*, **49**, 2446–2456.
87. Balan, A., Gunbas, G., Durmus, A., and Toppare, L. (2008) *Chem. Mater.*, **20**, 7510–7513.
88. Durmus, A., Gunbas, G.E., Camurlu, P., and Toppare, L. (2007) *Chem. Commun.*, **43**, 3246–3248.
89. Namal, I., Ozelcaglayan, A.C., Udum, Y.A., and Toppare, L. (2013) *Eur. Polym. J.*, **49**, 3181–3187.
90. Tarkuc, S., Udum, Y.A., and Toppare, L. (2012) *Thin Solid Films*, **520**, 2960–2965.
91. Beaujuge, P.M., Vasilyeva, S.V., Ellinger, S., McCarley, T.D., and Reynolds, J.R. (2009) *Macromolecules*, **42**, 3694–3706.
92. Udum, Y.A., Durmus, A., Gunbas, G.E., and Toppare, L. (2008) *Org. Electron.*, **9**, 501–506.
93. Gunbas, G.E., Camurlu, P., Akhmedov, İ.M., Tanyeli, C., Önal, A.M., and Toppare, L. (2008) *J. Electroanal. Chem.*, **615**, 75–83.
94. Sonmez, G., Meng, H., and Wudl, F. (2003) *Chem. Mater.*, **15**, 4923–4929.
95. Meng, H., Tucker, D., Chaffins, S., Chen, Y.S., Helgeson, R., Dunn, B., and Wudl, F. (2003) *Adv. Mater.*, **15**, 146–149.

96. Baran, D., Balan, A., Celebi, S., Esteban, B.M., Neugebauer, H., Sariciftci, N.S., and Toppare, L. (2010) *Chem. Mater.*, **22**, 2978–2987.

97. Esmer, E.N., Tarkuc, S., Udum, Y.A., and Toppare, L. (2011) *Mater. Chem. Phys.*, **131**, 519–524.

98. Udum, Y.A., Yildiz, E., Gunbas, G., and Toppare, L. (2008) *J. Polym. Sci., Part A: Polym. Chem.*, **46**, 3723–3731.

99. Aydemir, K., Tarkuc, S., Durmus, A., Gunbas, G.E., and Toppare, L. (2008) *Polymer*, **49**, 2029–2032.

100. Aqad, E., Lakshmikantham, M.V., and Cava, M.P. (2001) *Org. Lett.*, **3**, 4283–4285.

101. Kim, B., Kim, J., and Kim, E. (2011) *Macromolecules*, **44**, 8791–8797.

102. Kim, B., Shin, H., Park, T., Lim, H., and Kim, E. (2013) *Adv. Mater.*, **25**, 5483–5489.

103. Çetin, G.A., Balan, A., Durmuş, A., Gunbaş, G., and Toppare, L. (2009) *Org. Electron.*, **10**, 34–41.

104. Arící, M., Bozoğlu, C., Erdoğmuş, A., Uğur, A.L., and Koca, A. (2013) *Electrochim. Acta*, **113**, 668–678.

105. Zhang, J.D., Lu, F.L., Huang, H.C., Wang, J., Yu, H.A., Jiang, J.Z., Yan, D.H., and Wang, Z.Y. (2005) *Synth. Met.*, **148**, 123–126.

106. Ito, S., Iida, T., Kawakami, J., Okujima, T., and Morita, N. (2009) *Eur. J. Org. Chem.*, **2009**, 5355–5364.

107. Ito, S., Inabe, H., Morita, N., and Tajiri, A. (2004) *Eur. J. Org. Chem.*, **2004**, 1774–1780.

108. Ito, S. and Morita, N. (2009) *Eur. J. Org. Chem.*, **2009**, 4567–4579.

109. Shoji, T., Shimomura, E., Maruyama, M., Ito, S., Okujima, T., and Morita, N. (2013) *Eur. J. Org. Chem.*, **2013**, 957–964.

110. Tang, T., Ding, G.Q., Lin, T.T., Chi, H., Liu, C., Lu, X.H., Wang, F.K., and He, C.B. (2013) *Macromol. Rapid Commun.*, **34**, 431–436.

8
Metal Hydrides for Smart-Window Applications

Kazuki Yoshimura

8.1
Switchable-Mirror Thin Films

A new chromogenic material which can be switched between transparent and mirror states was discovered in 1996 and its application named 'switchable mirror' [1]. A certain metal thin film capped by Pd is changed into a transparent metal hydride state by hydrogenation.

This phenomenon was first found in the thin films of rare metals such as yttrium and lanthanum, several new materials being found later (Table 8.1). Rare earth and transition metal alloys have better colour neutrality in the transparent state [2–4]. The alloy of magnesium and a transition metal, found in 2001 [5–7], was considered to be a better material for window applications, because it was less expensive and was thought to have higher durability than the rare earth metals. Most work on a switchable mirror for practical applications has been focused on the Mg–Ni alloy thin film [8]. Recently, the Mg–Y alloy thin film has attracted much attention regarding durability and optical properties [9, 10].

Figure 8.1a shows the basic structure of the switchable mirror composed of the magnesium-alloy thin film as a switching layer, about 40 nm thick, which is coated with a thin catalyst layer of Pd about 4 nm thick. In the state as deposited, the magnesium alloy layer is a mirror since it is metallic, but when it is exposed to a diluted hydrogen atmosphere, the switching layer becomes hydrogenated by palladium catalysis, turning it into an insulator that is transparent. When this is next exposed to an atmosphere containing oxygen, the H in the hydride reacts with oxygen by the palladium catalysis and is pulled out by forming H_2O, the film thus returning to the metallic mirror state.

The Pd layer plays two roles, in catalysis and in protection of the Mg alloy layer from oxidation. The TEM image of the cross section of the prepared switchable-mirror thin film shows that although it is very thin (Figure 8.1b), the Pd layer is flat and complete, which is appropriate for protection.

Electrochromic Materials and Devices, First Edition.
Edited by Roger J. Mortimer, David R. Rosseinsky and Paul M. S. Monk.

Table 8.1 Variety of switchable-mirror materials.

First generation	Rare earth metal: Y [1], La [1] and so on, Amsterdam Free University (1996)
Second generation	Rare earth metal and transition metal: Gd–Mg [2], Sm–Mg [3], Y–Mg [4] and so on, Philips UK Ltd (1997)
Third generation	Magnesium and transition metal: Mg–Ni [5], Mg–Ti [6], Mg–Co [7], Mg–Mn [7] and so on, Lawrence Berkeley National Laboratory (2001)

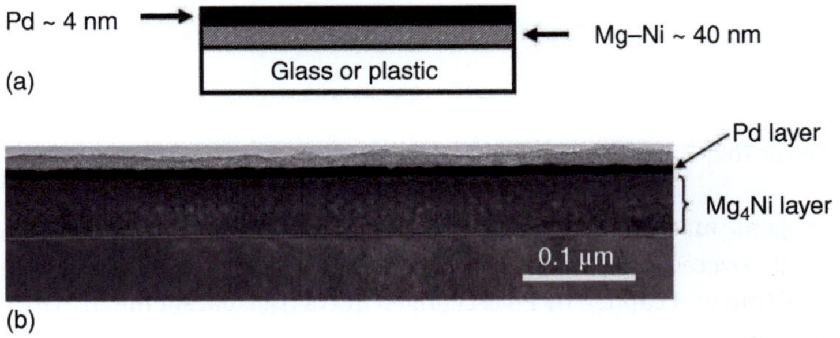

Figure 8.1 Basic structure (a) and cross-sectional TEM image of switchable-mirror thin film (b).

8.2 Optical Switching Property

Thin-film switchable mirrors are commonly prepared by using sputtering or e-beam deposition, and the magnetron sputtering method is especially suitable because the large gettering effect of Mg maintains a good vacuum. There are two methods to switch such mirror materials. One is electrochemical, using liquid or solid electrolyte with electrochromic materials. The other one is to use hydrogen- and oxygen-containing gas, in the gasochromic method, as follows.

Figure 8.2 shows an example of gasochromic switching of Pd/Mg–Ni thin film [11]. The as-deposited sample has a shiny metallic surface. Passage of diluted H_2 gas causes a drastic change from the metallic to the transparent state by hydridation of the Mg–Ni alloy, and its transmittance increases rapidly. Then, dry air introduction causes a slow change from the transparent to the metallic state by dehydridation, and the transmittance decreases gradually. The spectral changes of a Pd/Mg_6Ni thin film by gasochromic switching is shown in Figure 8.3 [8]. The transmittance and reflectance changes show that a switchable mirror provides efficient reflective control.

Such a gasochromic switching response has a strong dependence on the composition of the film. As shown in Figure 8.2, hydridation speed from the mirror state to the transparent state is almost the same for Mg_2Ni, Mg_4Ni and Mg_6Ni, taking several seconds for the change. By contrast, the dehydridation time gets longer with the increase of Mg content in the films.

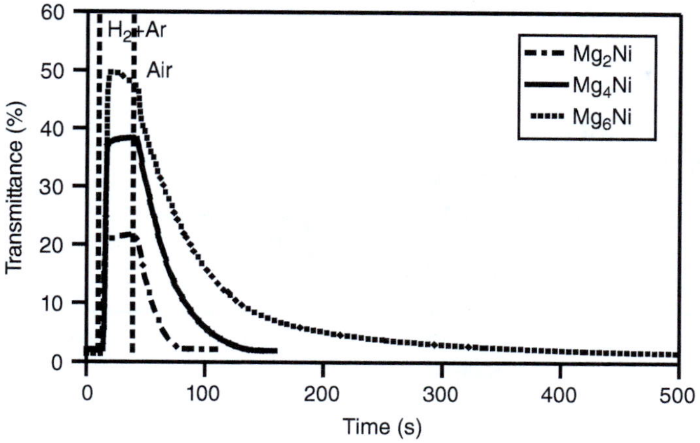

Figure 8.2 Transmittance change of Mg–Ni alloy switchable mirrors with different compositions following gasochromic switching.

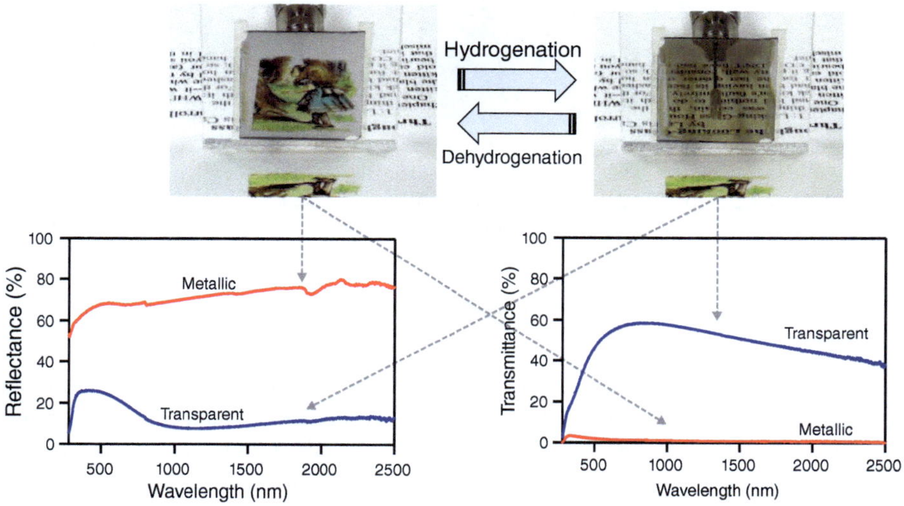

Figure 8.3 Mirror and transparent states of switchable mirror and its transmittance and reflection spectra.

8.3 Switching Durability

Switchable mirrors based on Mg–Ni initially show excellent optical switching. However, significant degradation occurs on repeated gasochromic cycling (alternate exposure to 4% H_2 in Ar and dry air of 0.1 MPa) [12]. The optical modulation between the metallic and transparent states decreases to 10% after 170 cycles (Figure 8.4a). One cause of degradation appears to be the migration of Mg to the outermost surface on repeated switching. This can be prevented by insertion of

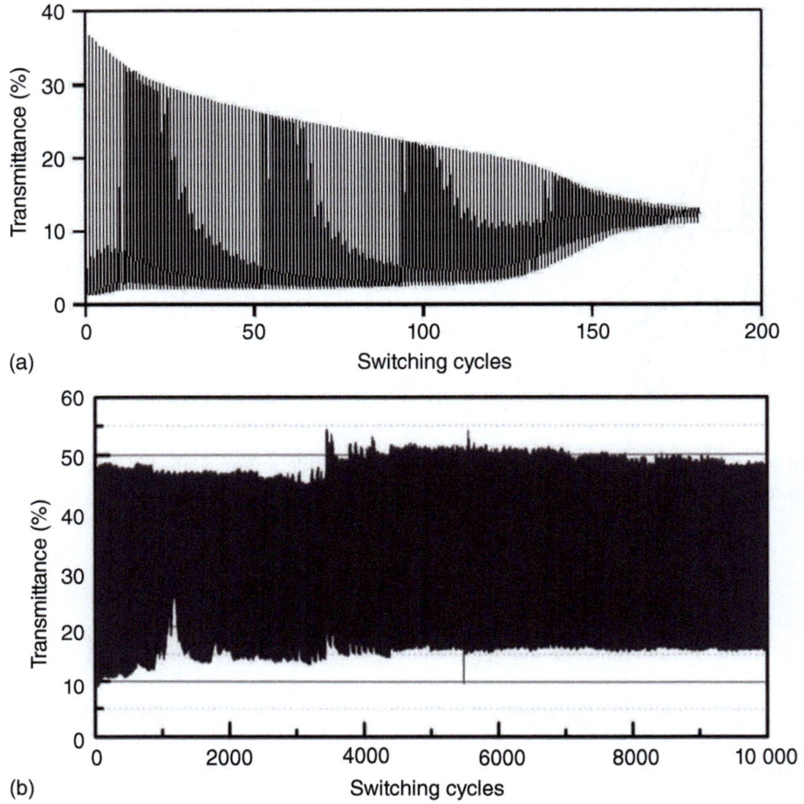

Figure 8.4 Switching durability of Mg–Ni switchable mirror (a) and Mg–Y switchable mirror (b).

a metal buffer layer, such as Ti, between the Pd and Mg–Ni layers [13]. Another cause is the deformation of the Pd layer due to volumetric changes that occur during hydrogenation and dehydrogenation. Apparently, a protective coating such as PTFE lessens the degradation, thereby improving the switching durability [14].

Recently, plastic distortions were shown to play an important role in the deformation process in experiments using an Mg–Y alloy thin film, in which 10 000 switching cycles were accomplished (Figure 8.4b) [9]. Due to the YH_2-to-YH_3 transition, the expansion of the lattice is limited to ~9% (for details, see the citation [9]).

8.4 Colour in the Transparent State

The colour in the transparent sate is very important for window applications. Although Mg–Ni thin films show good optical switching, this film has a yellowish colour in the transparent state, a colour not favoured for architectural

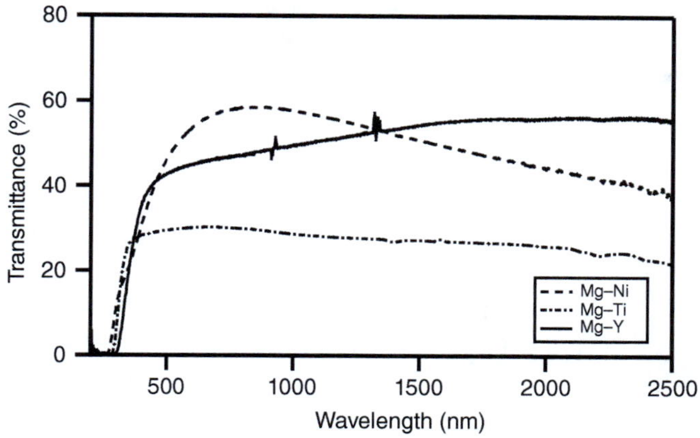

Figure 8.5 Transmittance spectra of Pd/Mg–Ni, Pd/Mg–Ti and Pd/Ta/Mg–Y thin films.

applications. Using an Mg–Ti alloy thin film, a colour-neutral switchable mirror can be realised [15]. However, the transmittance level in the transparent state is low compared to that of Mg–Ni alloy (Figure 8.5). Recent investigations indicate that the Mg–Y alloy switchable mirror might show high transmittance and colour neutrality by adjustment of the thickness of the Pd layer, together with insertion of a Ta buffer layer between Pd and Mg–Y layers, as shown in Figure 8.5 [10].

8.5 Electrochromic Switchable Mirror

Although the gasochromic devices have the advantage of a simple layer structure, there is a strong restriction: they can only be used in a double glass pane such that the hydrogen is retained. Hence, electrochromic switching is desired for some applications. In this case, the optical properties are switched by applying a voltage to the multi-layer structured switching layer.

After the first switchable-mirror experiments by gasochromic switching were reported, several papers appeared showing the possibility of electrochromic loading [16–18], although the switching durability of these devices is poor.

An all-solid-state switchable-mirror device was developed that shows promising cycling properties by using an Mg–Ni alloy thin film as a switching layer [19]. It consists of five layers, ITO (tin-doped indium oxide), H_xWO_3, Ta_2O_5, Pd and Mg–Ni, sequentially deposited on a glass substrate (Figure 8.6). By applying −5 V to the Mg–Ni layer with respect to the ITO, protons move from the H_xWO_3 to the Mg–Ni layer, which then becomes transparent. On reversing the polarity, protons return to the H_xWO_3, and the device reverts to the mirror state. The Ta_2O_5 layer acts as a solid-state electrolyte, and the Pd layer stimulates proton insertion and extraction.

The quality of the electrolyte is essential for device functionality. The Ta_2O_5 layer is prepared by reactive sputtering of a metal Ta target, resulting in the required

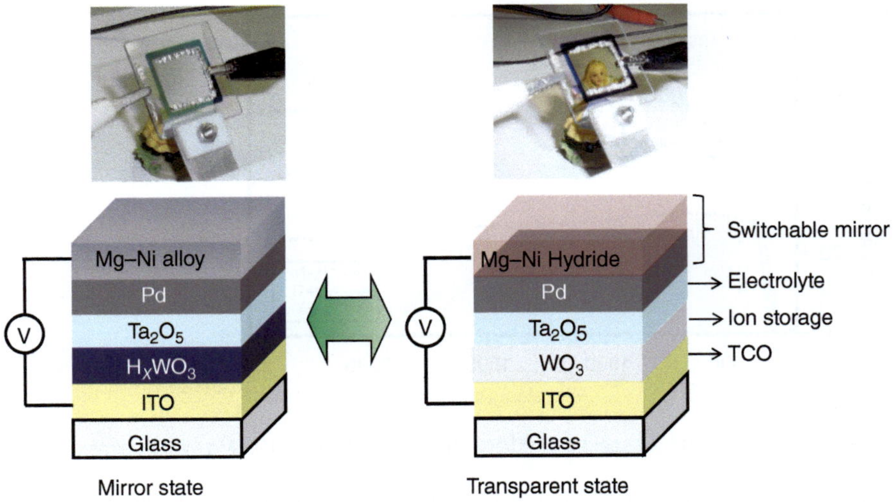

Figure 8.6 Structure of all-solid-state switchable-mirror device and its mirror and transparent states.

high electronic resistivity but high ionic conductivity. The density of the Ta_2O_5 layer determines, to a large extent, its switching performance [20].

It is very difficult for an electrochromic switchable mirror to allow stable switching of a large area. Further improvement is necessary to realise a practical electrochromic switchable-mirror device.

8.6 Smart-Window Application

One of the most promising applications of switchable-mirror thin films is the 'smart window'. Smart windows promise an important technology to decrease energy consumption in buildings and vehicles. One current state-of-the-art smart window is the purely electrochromic window using WO_3 as the switching layer, with its transparent and dark blue states, depending on the electron content, but appreciable energy absorbed when in the blue state is re-emitted as heat. Hence, a device that uses a change in reflective properties is desirable, which is an important advantage of metal hydrides, making them highly attractive for smart-window applications.

The energy performance of the switchable-mirror window is important in its application. To evaluate its energy performance, a prototype gasochromic switchable-mirror window of a size of $0.8\,m \times 1.2\,m$ was fabricated [21] (Figure 8.7). Two transparent glass panes with a thickness of 5 mm make up the double glazing. Mg_4Ni alloy and Pd thin films are deposited to form the switching layer on the inner side of the outer glass. Inlet and outlet pipes are connected to an inner glass pane for gas introduction and extraction. Thin-film deposition was

Figure 8.7 Structure of switchable-mirror window of practical size and its mirror and transparent states.

effected using a large-scale sputtering machine with an Mg target onto which sufficient Ni pieces were added to reproduce the desired composition.

The cooling required to keep a room at constant temperature when equipped with a switchable-mirror window has been compared to that of a room equipped with a normal transparent double window (Figure 8.7). It was found that the smart window reduces the required cooling power by more than 30% on a hot summer day [21].

References

1. Huiberts, J.N., Griessen, R., Rector, J.H., Wijngaarden, R.J., Dekker, J.P., de Groot, D.G., and Koeman, N.J. (1996) *Nature*, **380**, 231.
2. van der Sluis, P., Ouwerkerk, M., and Duine, P.A. (1997) *Appl. Phys. Lett.*, **70**, 3356.
3. Ouwerkerk, M. (1998) *Solid State Ionics*, **113–115**, 431.
4. Nagengast, D.G., van Gogh, A.T.M., Kooij, E.S., Dam, B., and Griessen, R. (1999) *Appl. Phys. Lett.*, **75**, 2050.
5. Richardson, T.J., Slack, J.L., Armitage, R.D., Kostecki, R., Farangis, B., and Rubin, M.D. (2001) *Appl. Phys. Lett.*, **78**, 3047.
6. Farangis, B., Nachimuthu, P., Richardson, T.J., Slack, J.L., Meyer, B.K., Perera, R.C.C., and Rubin, M. (2003) *Solid State Ionics*, **165**, 309.
7. Richardson, T.J., Slack, J.L., Farangis, B., and Rubin, M.D. (2002) *Appl. Phys. Lett.*, **80**, 1349.
8. Yoshimura, K., Yamada, Y., and Okada, M. (2002) *Appl. Phys. Lett.*, **81**, 4709.
9. Yamada, Y., Miura, M., Tajima, K., Okada, M., and Yoshimura, K. (2013) *Sol. Energy Mater. Sol. Cells*, **117**, 396.
10. Yamada, Y., Miura, M., Tajima, K., Okada, M., and Yoshimura, K. (2013) *Sol. Energy Mater. Sol. Cells*, **117**, 396.
11. Yoshimura, K., Bao, S., Yamada, Y., and Okada, M. (2006) *Vacuum*, **80**, 684.
12. Yoshimura, K., Yamada, Y., Bao, S., and Okada, M. (2007) *Jpn. J. Appl. Phys.*, **46**, 4260.
13. Bao, S., Yamada, Y., Okada, M., and Yoshimura, K. (2006) *Jpn. J. Appl. Phys.*, **45**, L588.
14. Bao, S., Yamada, Y., Tajima, K., Okada, M., and Yoshimura, K. (2009) *Sol. Energy Mater. Sol. Cells*, **93**, 1642.
15. Bao, S., Tajima, K., Yamada, Y., Okada, M., and Yoshimura, K. (2007) *Appl. Phys. A-Mater. Sci. Process.*, **87**, 621.

16. Armitage, R., Rubin, M., Richardson, T., O'Brien, N., and Chen, Y. (1999) *Appl. Phys. Lett.*, **75**, 1863.
17. Mercier, V.M.M. and van der Sluis, P. (2001) *Solid State Ionics*, **145**, 17.
18. Janner, A.M., van der Sluis, P., and Mercier, V. (2001) *Electrochim. Acta*, **46**, 2173.
19. Tajima, K., Yamada, Y., Bao, S., Okada, M., and Yoshimura, K. (2007) *Electrochem. Solid-State Lett.*, **10**, J52.
20. Tajima, K., Yamada, Y., Bao, S., Okada, M., and Yoshimura, K. (2008) *Appl. Phys. Lett.*, **92**, 41912.
21. Yoshimura, K., Yamada, Y., Bao, S., Tajima, K., and Okada, M. (2009) *Sol. Energy Mater. Sol. Cells*, **93**, 2138.

Part II
Nanostructured Electrochromic Materials and Device Fabrication

Electrochromic Materials and Devices, First Edition.
Edited by Roger J. Mortimer, David R. Rosseinsky and Paul M. S. Monk.

9
Nanostructures in Electrochromic Materials

Shanxin Xiong, Pooi See Lee, and Xuehong Lu

9.1
Introduction

9.1.1
Why Nanostructures?

As introduced in the previous chapters, electrochromism is a phenomenon related to the reversible changes in optical absorbance/transmittance/reflectance of a material induced by externally applied potentials. A typical complementary electrochromic device consists of an anodically colouring electrochromic layer and a cathodically colouring one separated by a transparent electrolyte, with the three layers being sandwiched between two transparent electrodes (Figure 9.1). When suitable alternating potentials are applied to the two electrochromic layers through the transparent electrodes, the electrochromic layers can undergo redox reactions, exhibiting reversible optical changes. The electrochromic process is induced by oxidation-state changes of organic molecules, conjugated polymers or transition metal compounds, the redox reactions involving gain or loss of electrons and simultaneous insertion/extraction of ions with opposite charges to balance the internally created electric fields. Thus, the performance of an electrochromic device is frequently limited by kinetics of charge transport, that is, the rates of electron conduction in the electrochromic materials as well as ion transport in the electrolyte, electrochromic materials and at their interfaces. From the electrochromic device operation principle discussed earlier, we can see that besides electrochromic characteristics of the active materials, electron and ion transport capabilities of the materials in the device also determine how fast the electrochromic process can proceed and how many active units can be effectively switched to desired redox states. Thus, the charge transport in an electrochromic device significantly influences key electrochromic properties of the device, such as optical contrast and response speed. Nanostructured electrochromic materials have an extremely short charge transport distance in

Electrochromic Materials and Devices, First Edition.
Edited by Roger J. Mortimer, David R. Rosseinsky and Paul M. S. Monk.

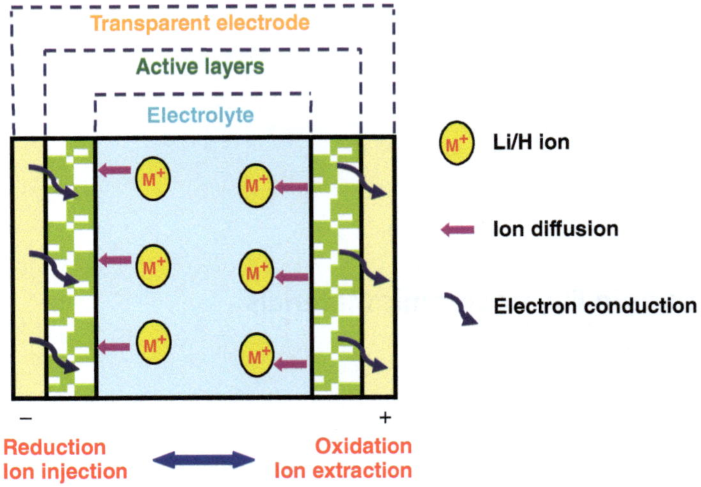

Figure 9.1 Operation principle of a typical complementary electrochromic device.

at least one dimension and an enormously large specific surface area for interaction with electrolytes and, hence, may provide fast charge transport and high electrochemical activities promising remarkable enhancement of electrochromic properties.

Nanostructured electrochromic materials can facilitate charge transport in electrochromic devices in several ways, such as promoting electron conduction in electrochromic materials through incorporation of conductive nanofillers, providing loose packing structures for facile ion motion and creating large interfacial areas between the electrochromic material and electrolyte to increase the number of accessible reaction sites. In Sections 9.2–9.5, these approaches are discussed in detail with examples.

9.1.2
Classification of Nanostructural Electrochromic Materials

Nanostructural electrochromic materials can be classified into two large categories: single-phase nanostructures and multi-phase nanostructural materials that include nanocomposites and hybrid materials with interfacial chemical bonds.

Single-phase nanostructures include zero-dimensional (0D) nanostructures, mainly nanoparticles; one-dimensional (1D) nanostructures such as nanowires, nanofibres and nanotubes; and two-dimensional (2D) nanostructures such as electrochromic thin films prepared by layer-by-layer (LbL) deposition and

Langmuir–Blodgett (LB) processes. Multi-phase electrochromic materials are composed of nanostructured materials with or without electrochromic activity dispersed in electrochromic matrices or electrochromic nanostructures dispersed in non-electrochromic matrices. In some cases, interfacial chemical bonds are also introduced into the multi-phase systems to facilitate interfacial charge transport.

9.1.3 Preparation Method

Typical nanostructural electrochromic materials and their preparation methods are summarised in Table 9.1. For single-phase electrochromic nanostructures, the major preparation routes include template-assisted approaches, self-assembly, sol–gel processes, electrospinning, LB film deposition, sputtering and so on, while for multi-phase electrochromic materials, *in situ* polymerisation, sol–gel and LbL deposition are frequently used methods.

In the following sections, the preparation methods, structures and morphologies, as well as electrochemical and electrochromic properties of different types of nanostructural electrochromic materials, are reviewed. The underlying mechanisms for their enhanced electrochromic performance are also discussed.

9.2 Nanostructures of Transition Metal Oxides (TMOs)

9.2.1 Introduction

In transition metal oxides (TMOs), there are mainly three types of electrochemical reactions that take place at the surface or interior of the active materials upon voltage biasing. The first electrochemical reaction taking place at the interface between the active material and electrolyte is described as electrical double layer (EDL) formation as shown in Figure 9.2.

There is a thin layer of specifically adsorbed ions on the surface of the TMO. This thin layer is called Helmholtz or Stern layer. It consists of the inner layer of the EDL and the locus of the inner layer which is at x_1 is called the inner Helmholtz plane (IHP). When the voltage is applied, solvated ions bearing charges opposite to that at the TMO surface would be attracted to a distance $x_{2.}$ The species of charge opposite to the IHP ions' form the outer Helmholtz plane (OHP), which carries a charge density greater than that of IHP. These two layers form the EDL via fast electrochemical kinetic behaviour involving only solution-phase charge

Table 9.1 Preparation routes of nanostructural electrochromic materials.

Nanostructures		Electrochromic materials	Preparation method	References
Single phase	Zero-dimensional nanostructures	Polyaniline (PANI)/poly(styrene sulfonate) hollow nanoparticles	Star-like PSS as template	[1]
		Prussian blue nanoparticles	Surface modification	[2]
		WO_3 nanoparticles	Ultrasonic spray deposition	[3]
	One-dimensional nanostructures	PPy nanotubes	Electrochemical synthesis	[4]
		PMProDot nanotubes	Template synthesis	[5]
		Poly(3-hexylthiophene) nanotubes	Solution wetting	[6]
		PEDOT nanotubes	Electrochemical deposition	[7]
	Two-dimensional nanostructures	3D-ordered nanostructured PANI film	Template synthesis	[8]
		Mesostructured PANI film	Solvent-induced structural relaxation	[9]
		Ultra-thin PANI film	Molecular assembly	[10]
		PEDOT nanofilm	Electrochemical synthesis	[11]
		Nanopatterned poly(3-hexylthiophene)	Soft lithography technique	[12]
		Phthalocyanine film	Langmuir–Blodgett method	[13]
Multi-phase	Electrochromic nanostructures	Polyoxometalate/PEDOT multi-layer film	LbL	[14]
		Star-like PANI/self-doped PANI	LbL	[15]
		PANI/Prussian blue	LbL	[16]
		WO_3/polyvinylimidazole	Sol–gel plus *in situ* polymerisation	[17]
		Vanadium oxide/carboxybenzylviologen	Sol–gel	[18]

	Nickel/WO_3	Reactive DC magnetron co-sputtering	[19]
	Nickel oxide/manganese oxide films	Electrochemical synthesis plus template agent	[20]
	NiO/PPy thin films	Electrochemical deposition	[21]
	PEDOT nanotubes	Electropolymerisation	[22]
	PANI-WO_3 hybrid films	Molecular assembly	[23]
	WO_3 nanorod/PANI	Surface-initiated polymerisation	[24]
Non-electrochromic nanostructures in electrochromic matrices	CNTs/Prussian blue	Electrochemical deposition	[25]
	TiO_2 nanotube|WO_3 nanocrystal	Sol–gel	[26]
	TiO_2-WO_3 core|shell nanorod	Hydrothermal and electrodeposition method	[27]
	MWCNTs|WO_3	Sol–gel	[28]
	Pt/WO_3, Au|WO_3	Co-sputtering deposition	[29]
	PEDOT|TiO_2	Electrochemical deposition	[30]
	PEDOT|graphene	Electrochemical deposition	[31]
	PANI|graphene	LbL	[32]
	PEDOT|Au|CdSe	Electrochemical polymerisation	[33]
	PANI-TiO_2	Sol–gel plus *in situ* polymerisation	[34]
	PANI-POSS	LbL	[35]

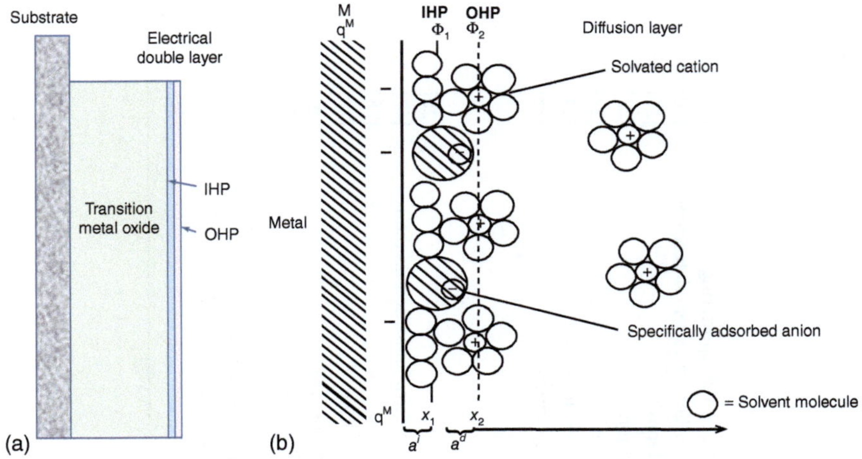

Figure 9.2 Schematic illustration of (a) EDL at the interface between TMO and electrolyte [36]. IHP and OHP are Helmholtz planes (see Section 9.2). (b) The proposed model of EDL [37]. Adopted from Refs [36, 37].

separation, like an electrostatic capacitor [38] free of ion intercalation and de-intercalation through solid, such processes disfavouring cycling stability of the electrode.

It may be useful to examine the atom-scale steps of the electrochromic reaction of oxides simply described as:

$$MeO_n + xe + xM^+ \rightleftharpoons M_xMeO_n$$

where M is the monovalence cation intercalated into the oxide and x is the extent of the electron insertion and cation intercalation.

The first redox reaction occurs at the interface between TMO and electrolyte [39] as follows:

$$MeO_n + M^+ + e \rightleftharpoons M.MeO_n$$

The electron injects into the oxide at the interface of the transition metal, complemented by the solvated-cation intercalation rendering the interface neutral.

The second redox reaction occurs within the active material. Similar to the redox reaction at the interface, electrons injected into the TMO make the negatively charged oxide attract cations in the electrolyte into the bulk surface, thus effecting neutrality in the TMO. In the reverse process, when a positive potential is applied, electrons are removed from the oxide and cations ejected from the TMO into the electrolyte. This redox reaction is a diffusion controlled reaction which in our view takes more time compared with the redox reaction at the interface. The diffusion time can be decreased by designing thin nanostructures to achieve fast kinetics, via nanoscale features of nanostructured metal oxides. Thus, the wall thickness of the mesoporous oxides is an important feature controlling the ion diffusion distance in such films [40].

Regarding the rates of different reactions, the surface charge transfer reactions in the EDL and surface redox reaction are inversely proportional to time, while the diffusion controlled reaction is inversely proportional to the square root of time [41]. Generally, the diffusion controlled redox reaction is slower than the surface dependent reactions, creating a limiting step in the kinetics.

9.2.2
Single TMO Systems

There are various components in a typical electrochromic device as shown in Figure 9.1. Tin-doped indium oxide (indium tin oxide (ITO)) or fluorinated tin oxide (FTO) is usually used as the conductive electrode because of their good conductivity and transparency in the visible range. Electrochromic materials can be categorised into anodic materials and cathodic materials, depending on the optical change with voltage application. Cathodic electrochromic TMO materials, such as WO_3, Ta_2O_5, Nb_2O_5 and TiO_2, will be coloured upon electron and ion co-injection. Correspondingly, the anodic electrochromic TMO materials, such as Cr_2O_3 MnO_2, FeO, CoO_2, NiO and RhO_2, will be coloured upon electron and ion co-extraction. These two categories of electrochromic material are hence complementary to each other and can be applied on working and counter electrode in an electrochromic device to enhance the contrast or produce different colorations.

The optical modulation of many electrochromic materials is dependent on the availability of reduced or oxidised transition metal ions and requires surface-dependent redox reaction leading to fast kinetics in electrochromic devices; therefore, it is important to prepare suitable nanostructured metal oxides with large surface area. Various structures have been demonstrated to increase the oxide–electrolyte interface and enhance the redox reaction such as three-dimensional (3D) interconnected networks of the TMO, including mesoporous [42–46] and macroporous [47, 48] oxide layers. Template-assisted synthesis has been found to be an effective method in producing these porous oxide layers. The employed template can be a soft template such as surfactant sodium dodecyl sulfate (SDS) [49] and a block co-polymer or a hard template such as polystyrene (PS) spheres. After removing the template through chemical dissolution or heating, the initial cavity occupied by the template becomes interconnected pores.

In recent literature, inorganic electrochromic materials have been tailored into nanostructures such as nanoparticles [50–52], nanorods [53–55] and nanowires [56–59] Apart from homogeneous nanostructures, there are hetero-nanostructures formed by coating electrochromic materials onto conductive nanostructures, such as ITO nanoparticles. The latter is an analogue to a core–shell structure with a conductive material as the core and a thin layer of electrochromic material as the shell [60, 61]. All these nanostructures possess two common features: large surface-to-volume ratio and at least one dimension is constrained to the nanometre scale. The advantages of employing electrochromic nanostructures as the working electrode have been demonstrated by comparison to the electrochromic performances of thin films [61, 62]. The electrochromic performances such as bleaching time, colouring time and coloration efficiency (η) have been shown to improve in electrochromic nanostructures. In general, the reported nanostructures exhibit promising bleaching and colouring time below 5 s [52, 57, 59, 61] and moderate η around 35–65 $cm^2 C^{-1}$ [50, 52, 53, 57, 58, 62, 63].

A key challenge for the effective usage of the TMO nanostructures in electrochromic devices is the large-area assembly of such nanostructures to form high-quality electrochromic thin films. Recently, to address this issue, the Lee and Ma group synthesised and employed a scalable assembly of WO_3 nanorods for electrochromic applications [64]. These uniform crystalline nanorods were synthesised without any surfactants using a hydrothermal process. The as-synthesised product possesses preferential growth direction and is almost completely horizontally aligned on the substrate. The uniform nanorods have diameters of ~100 nm and lengths of ~2 μm as shown in the Transmission electron microscope (TEM) images of Figure 9.3. The surfaces of the nanorods were very coarse, which increases their surface areas. The selected area electron diffraction pattern (inset in Figure 9.3a) was taken from single nanorods with regular diffraction spots implying that the nanorods were single crystals. For a large-area device, the WO_3 nanorods were assembled systematically into LbL superstructures as shown in Figure 9.4. An ordered anisotropic orientation resembling a 3D-nematic configuration with side-by-side alignment of

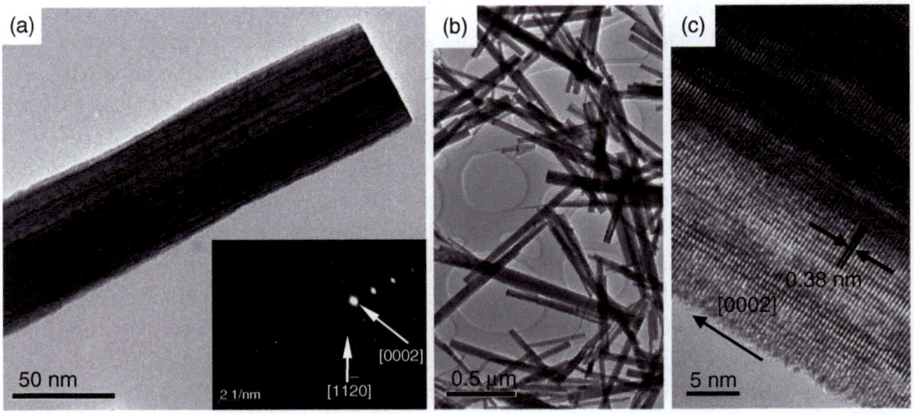

Figure 9.3 (a) Low-magnification TEM image, (b) higher-magnification TEM image showing the coarse surface of WO_3 nanorods and (c) High resolution transmission electron microscopy (HRTEM) images of the as-synthesised WO_3 nanorods. The inset shows the selected area electron diffraction pattern (SAED) takes from the single nanorods. Reproduced from Ref. [30] by Ma, J. with permission from American Chemical Society.

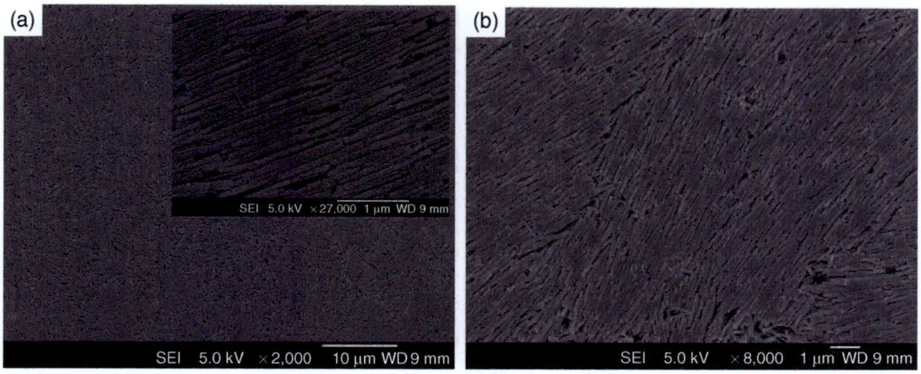

Figure 9.4 Field emission scanning electron microscope (FESEM) images of the WO_3 nanorod film at (a) low magnification and (b) higher magnification. The inset shows a high-magnification FESEM image of the highly oriented compact assembly of WO_3 nanorods. Reproduced from Ref. [30] by Ma, J. with permission from American Chemical Society.

nanorods can be found. The assembly is homogeneous and uniformly spreads over the substrates, which indicates that a centimetre-scale assembly is attainable (Figure 9.4). We believe that this approach can be extended to other nanorods with large active specific surface area, albeit optimisation of the assembly is required. The as-prepared electrochromic device exhibits satisfactory integrated electrochromic properties: high electrochromic stability (>3000 CV cycles), high contrast (~66%), compatible colour display (from colourless to green, blue and deep blue) (Figure 9.5). Furthermore, the average intercalated charge during coloration was ~133 mC cm^{-2} mg^{-1}, which is higher than other reported data. This result demonstrates the successful practical applications of nanorod films in scalable electrochromics.

Electrophoretic deposition (EPD) is another solution deposition method to coat nanostructured redox materials onto conducting substrates. Constant-voltage and constant-current EPD has been performed on a transparent conductive substrate (ITO/glass) to deposit WO_3 1D nanorods. Optimum deposition conditions were identified in correlation to its deposition mechanism. The resultant electrochromic performances such as optical modulation and switching time of the WO_3 nanorods coated substrates were also studied. Coloration time $t_c^{70\%}$ is found to be 28.8 s, which is comparable to the reported values ($t_b \sim 25$ s) [55]. However, the bleaching time $t_b^{70\%}$ of 4.5 s is found to be much superior to those of the crystalline [26] ($t_b \sim 22$ s) and amorphous WO_3 ($t_b \sim 18-36$ s) [55, 56]. This improvement is attributed to the porous nature of the oxide layer which consists of nanorods. The porous oxide layer allows electrolyte penetration and hence shortens the ionic diffusion length from the electrolyte to the centre of the oxide. On the other hand, the high surface-to-volume ratio of the nanorods facilitates the Li^+ ion movement across the oxide–electrolyte interface by providing a large amount of

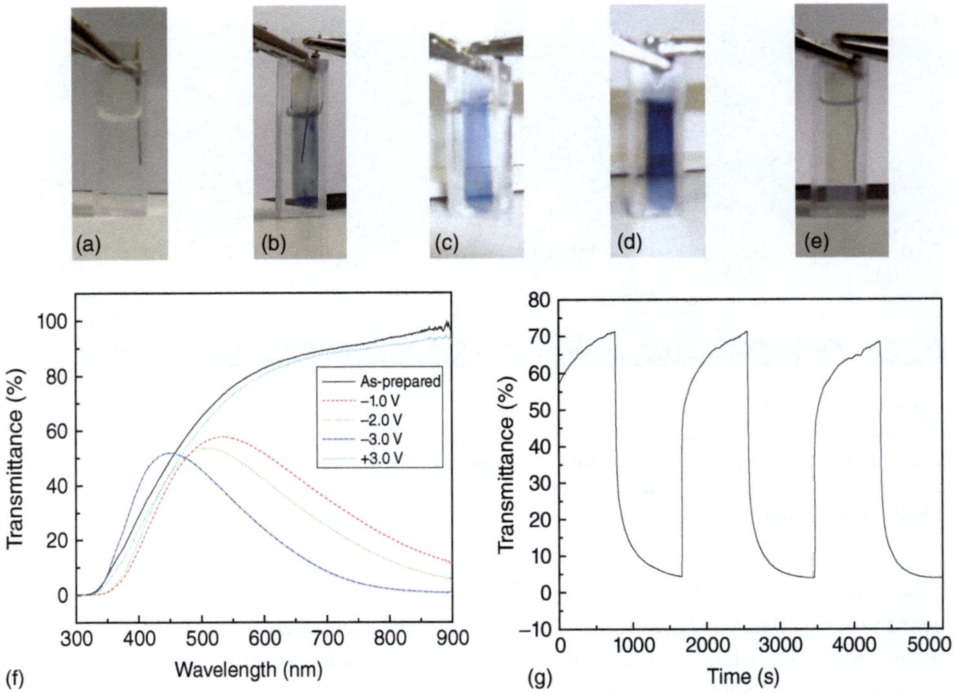

Figure 9.5 Colour changes of the WO_3 nanorod film at different voltages with 1.0 M lithium perchlorate ($LiClO_4$) in propylene carbonate (PC) as the electrolyte: (a) the as-prepared film, (b) coloured at −1.0 V, (c) coloured at −2.0 V, (d) coloured at −3.0 V, (e) bleached at 3.0 V, (f) corresponding UV-vis spectra and (g) switching time characteristics between the coloured and bleached states for the WO_3 nanorod film measured at ±3.0 V with an absorbance wavelength of 632.8 nm. Reproduced from Ref. [30] by Ma, J. with permission from American Chemical Society.

reaction sites. Hence, the shorter switching time can be a result of these two factors.

Non-porous and macroporous WO_3 films can be synthesised through electrodeposition on bare substrate and template-coated substrate, respectively. PS microspheres were coated onto the substrate as the template to form macroporous WO_3 film. The electrodeposition was carried out at −0.45 V (vs Ag/AgCl) at room temperature with same deposition duration of 200 s for both non-porous and macroporous WO_3 films to achieve the same film thickness. The formation of WO_3 can be expressed as the reduction of peroxy-tungstate complex [65]:

$$W_2O_{11}^{2-} + 2H^+ \rightarrow 2WO_3 + H_2O + 2O_2$$

The porosity of macroporous WO_3 film was determined by comparing the densities of the non-porous and macroporous WO_3 films. The porosity of macroporous WO_3 film was thereby found to be around 40%.

As shown in Figure 9.6, the current in the CV curves has been normalised to the mass of the WO_3 to study the current densities in macroporous and non-porous

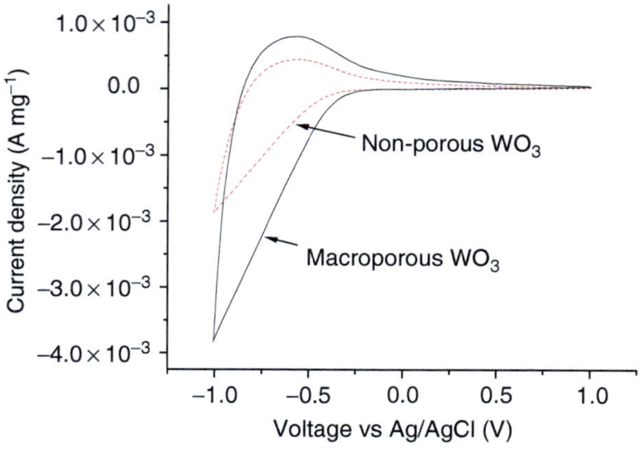

Figure 9.6 Cyclic voltammograms of the macroporous and non-porous WO_3 film at 100 mV s^{-1} scan rate. (Electrolyte: 1 M $LiClO_4$/PC)

WO_3 films. The macroporous film exhibits larger current density compared to the non-porous film. This suggests that within unit time (or under same scan rate), the macroporous WO_3 film has a higher charge transfer rate and more redox reactions occur as compared to non-porous film. Similar observations have been reported for porous WO_3 film [66, 67]. The area under the CV curve represents the total charge stored in the WO_3. The non-porous WO_3 film possesses total stored charge of 3.5 C g^{-1}, which corresponds to 0.008 Li^+ per WO_3. In contrast, the macroporous WO_3 film shows a larger area under CV curve and can accommodate up to 5.6 C g^{-1}, which corresponds to 0.013 Li^+ per WO_3, that is, 1.6 times higher than that of the non-porous film.

The enhancements should be attributed to the advantages associated with the macroporous structure, where the redox reaction has been facilitated by having more oxide–electrolyte interface and short ionic diffusion distance in the macroporous film. Both advantages can enhance the charge transfer rate by creating more reaction sites and shortening the diffusion time of Li^+. The η in the macroporous film in this work is significantly increased up to 90 $cm^2\,C^{-1}$, which is almost double that of the non-porous film η at 55 $cm^2\,C^{-1}$. The reaction kinetic has been greatly improved in the macroporous structure and leads to higher efficiency in coloration.

9.2.3 Binary TMO Systems

Binary TMOs can be readily prepared by sol–gel and sputtering methods. Usually, the component with low feed content is called dopant. Through combining various metal oxides, the optical absorption bands of the doped TMOs can be adjusted systematically by varying the ratio of the constituent oxides. Doping by a specific oxide may alter the electronic structures of the system and thereby shift the

Table 9.2 Values of the partial-molar shift parameter σ_i (shifting constant) for a variety of electrochromic metal oxide dopants in electrochromic hosts.

Dopant	Oxide host	σ_i
Binary mixtures: transitions on single ions		
CoO	NiO	0.045
CoO	WO_3	0.174
FeO	NiO	0.250
NiO	CoO	0.151
NiO	FeO	0.230
NiO	WO_3	0.211
WO_3	CoO	0.025
Intervalence transitions		
MoO_3	WO_3	0.032
WO_3	MoO_3	0.011

Reproduced from Ref. [68] by Monk, P. M. S. with permission from Elsevier B.V.

maximum optical absorption wavelength, as well as modulate the electrochromic performance. To quantify the shift, a shifting constant (σ_i) has been created to represent the ability of a metal oxide dopant i to alter the electronic environment within a solid-state oxide mixture and thereby shift the frequency maximum of the metal oxide mixture from that of the matrix. As listed in Table 9.2, cobalt oxide as a dopant is effective in shifting the W^V band in WO_3. It can also be seen that the shifting constants of FeO and NiO are significantly higher [68].

Pawlicka [69] investigated the kinetic behaviour of WO_3-doped Nb_2O_5 thin films for further understanding of the mechanisms of injection and extraction of small cations during the device operation. The switching kinetics of metal oxides usually depends on ion diffusion rate. With WO_3 incorporated, the binary system gives a higher diffusion coefficient of $1.6 \times 10^{-10}\,cm^2\,s^{-1}$ in comparison with $2.5 \times 10^{-11}\,cm^2\,s^{-1}$ for undoped Nb_2O_5.

9.3 Nanostructures of Conjugated Polymers

9.3.1 Introduction

Conjugated polymers, such as polythiophene, polyaniline (PANI), polypyrrole (PPy) and their derivatives, possess long-chain molecular structure with alternating arrangement of σ bond and π bond, which leads to π-electron delocalisation along the molecular chains and hence unique optoelectronic properties of the polymers. The ordered arrangements of conjugated polymers in nanostructures, for example, in nanotubes, can enhance the π-electron delocalisation and improve the charge transport and consequently boost the electrochromic performance of the materials. In addition to the nanostructures formed by ordered/disordered aggregation of conjugated polymer chains, namely, crystalline/semicrystalline/amorphous nanoparticles, nanofibres, nanowires, nanotubes, nanosheets and so on, large conjugated polymer molecules with non-linear molecular architectures, such as hyperbranched or star-like molecules, can also be used to improve the electron and ion transport in an electrochromic device during its operation, thereby enhancing the performance of the device.

9.3.2 Polythiophene and Its Derivatives

In the past three decades, with the development of new polythiophene derivatives, such as poly(3,4-ethylenedioxy thiophene) (PEDOT) and poly(3,4-(2-methylene)propylenedioxythiophene) (PMProDOT), where the thiophene ring is tethered with large substituent groups, the electrochromic properties of polythiophenes have been improved remarkably and they have become one of the most popular types of electrochromic materials. Much work has also been done to achieve better electrochromic properties of polythiophene derivatives through the preparation of nanostructures. Kim *et al.* [12] prepared poly(3-hexylthiophene) nanostructures by nanopatterning poly(3-hexylthiophene) via a soft lithography technique. A nanopatterned polydimethylsiloxane (PDMS) mould was used to generate 1D grating and 2D crossed line pillar patterns. The dimensions of the grating and pillars are only about 110 and 380 nm, respectively. Although the molecular structure and electronic characteristics of the nanopatterned poly(3-hexylthiophene) films are unchanged in comparison with that of the pristine poly(3-hexylthiophene) film (without patterning), their redox current and ion diffusion coefficient are increased significantly owing to the nanostructures. As a result of such enhanced charge transport, the electrochromic properties, including response speed, coloration and bleaching efficiencies, of the nanopatterned poly(3-hexylthiophene) are drastically improved. The performance enhancement can be attributed to the facile ion diffusion into the nanopatterned electrochromic films during the electrochromic process.

A template route is widely used to prepare 1D conjugated polymer nanostructures. Lee and Cho [70] prepared PEDOT nanotube arrays by electrochemical polymerisation of 3,4-ethylenedioxy thiophene (EDOT) using anodic aluminium oxide (AAO) as the template. Porous AAO film was obtained through anodisation of aluminium film on an ITO/glass substrate. The electrochemical polymerisation of EDOT was performed in the pores of the template to produce PEDOT nanotube arrays. Well-defined PEDOT nanotubes were formed, as shown in Figure 9.7. Owing to the porous nature of the PEDOT nanostructures, the ion diffusion rate is very high, resulting in very short response times, that is, 70 ms for colouring and 50 ms for bleaching.

In addition to PEDOT nanostructures, the nanostructures of another important polythiophene derivative, PMProDOT, can also be synthesised using

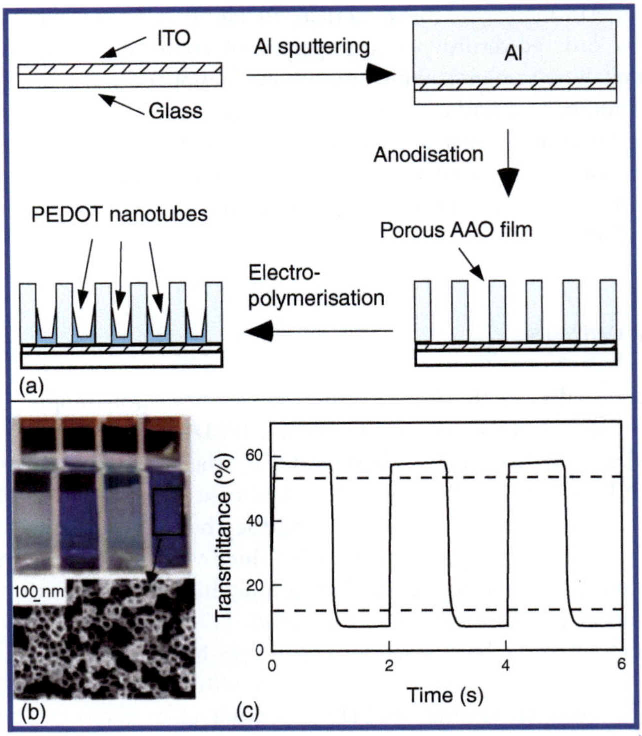

Figure 9.7 (a) Schematic diagram of the synthesis of PEDOT nanotubes in porous alumina structures on an ITO/glass substrate. (b) Electrochromic device with four windows (the first and third windows are oxidised (pale blue) while the second and fourth ones are reduced (dark blue)) and a Scanning electron microscope (SEM) image at a certain area of one window, showing the PEDOT nanotubular array structure after the removal of the template. (c) Plot of the transmittance change at 600 nm as a function of time. Reproduced from Ref. [70] by Lee, S. B. with permission from American Chemical Society.

the template approach [5]. The higher surface-area-to-volume ratio of the PMProDOT nanotubes compared to the corresponding solid thin films significantly decreases the diffusion time of the counter ions. The response time of the film made of nanotubes is less than half that of a 200 mC PMProDot film, showing that this is an effective approach to shorten the response time.

9.3.3 Polyaniline

PANI is also a widely studied conjugated polymer for electrochromic applications owing to its ease in preparation and low cost. As a popular anodically colouring electrochromic material, it is yellow to green in reduced and neutral states and turns to sky blue in oxidised state when the applied potential changes from negative to positive. Various 0D, 1D, 2D and 3D nanostructures of PANI have been prepared, giving enhanced electrochromic properties. An example for 0D nanostructures is PANI nanoparticles prepared by Lu's group [1] using a star-like sulfonated PS with polyhedral oligomeric silsesquioxane core-poly(4-styrenesulfonic acid) (PSSA) as the doping agent and template. Hollow PANI nanoparticles were formed by *in situ* polymerisation of aniline in the presence of PSSA. The hollow nanoparticle structure allows easier access of ions to PANI under external potentials. In comparison with conventional PANI with PSS as the doping agent, a much higher η, faster response speed and higher contrast are achieved with the film made of PANI hollow nanoparticles owing to its high porosity. Besides its porous morphology, the molecular chain packing also becomes more disordered than that in the common PANI thin film as the star-like geometry of the macromolecular dopant acid induces non-parallel packing of PANI chains, which also facilitate ion motion. The low-dimensional nanostructures benefit not only the electrochromic performance but also the processability of the conjugated polymers. Killard and co-authors [71] fabricated PANI-based electrochromic devices using inkjet-printed PANI nanoparticles. This approach can potentially be used for fabrication of large-area electrochromic devices with high production rate.

1D and 2D PANI nanostructures have also been prepared for electrochromic applications. Wei and co-authors [72] prepared PANI nanowires on PSS-doped PEDOT (PEDOT:PSS)/PET (polyethylene terephthalate) substrates using a dilute polymerisation process. Electrochemical impedance spectrum (EIS) measurement results show that the 45° portion in the Nyquist plots is very short, indicating that the PANI nanowire arrays can facilitate ion transport to the inner transparent electrode. 2D nanofilms can be obtained by self-assembly methods. Yang *et al.* [10] prepared self-doped PANI ultra-thin films by a molecular assembly approach. With the step potentials applied, the absorption of the film responds very fast to potentials. The coloration time and bleaching time are at the sub-second level, indicating the ultrafast switching speed of the ultra-thin films.

Li and co-authors [8] created a 3D-ordered macroporous (3DOM) PANI film using a template-assisted electrodeposition method and used the porous film as active layer in electrochromic devices. The ordered PS colloidal crystal was chosen as the template. With electropolymerisation of aniline in the gaps between PS spheres, good replication of the template was achieved. After the PS was removed by tetrahydrofuran (THF), the 3DOM PANI film was obtained. Figure 9.8 shows the morphology of the template, the macroporous PANI film and their electrochromic behaviours. In such an ordered porous film, the enhancement of ion mobility is expected. In comparison with the pristine PANI film without pores, the ion diffusion coefficient of the macroporous PANI film is increased more than three times in both the reduced and oxidised states owing to the short diffusion distance of the ions, resulting in much shorter response time. In addition, the optical contrast is also improved, as shown in Figure 9.8.

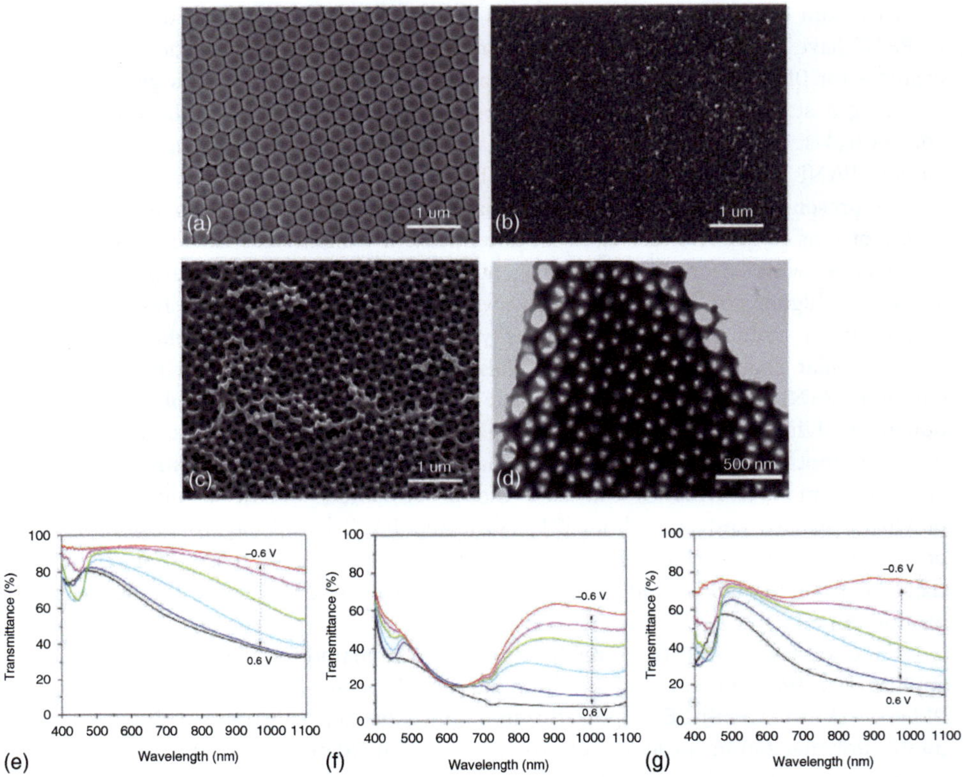

Figure 9.8 Morphologies and electrochromic properties of samples. SEM images of PS templates (a), pristine PANI films (b), 3DOM PANI films (c) and TEM photo of 3DOM PANI films (d). The diameter of PS spheres for templates was 300 nm and the electrodeposition was carried out at 0.85 V versus Ag/Ag^+ for 300 s under chronoamperometry. Transmittance switching under different potentials of (e) pristine PANI, (f) PANI/PS composites and (g) 3DOM films at the potentials of −0.6, −0.4, −0.2, 0.2, 0.4 and 0.6 V versus Ag/Ag^+. Reproduced from Ref. [8] by Li, Y. with permission from Elsevier B.V.

9.3.4 Polypyrrole

In comparison with polythiophene and PANI systems, electrochromic properties of nanostructural PPy are sparsely studied. Yamada *et al.* [4] prepared PPy nanotubes on an Au nanobrush electrode using a modified template method and the nanotubes were used as the working electrode in a reflective electrochromic device. Firstly, Au nanobrushes were plated electrically in a porous track-etched polycarbonate filter membrane. After the membrane was removed by chloroform, the PPy nanotubes were grown on the Au nanobrush electrode by electrochemical polymerisation. Owing to the special configuration of the working electrode, the electron transport and ion diffusion in the PPy nanotubes-on-Au nanobrush are much easier than that with the solid film electrode, leading to increased switching speed and enhanced stability of the device.

9.4
Nanostructures of Organic-Metal Complexes and Viologen

9.4.1
Introduction

Both organic-metal complexes and viologen are small-molecule electrochromic materials. Viologens are a class of bipyridinium compounds derived from 4,4′-bipyidine. They often work in a solution containing various anions to form self-erasing electrochromic devices, which show colour changes when a potential is applied, and bleaching occurs after the potential is removed. Organic-metal complexes (coordination compounds) consist of central cations (metal ions in most cases) and ligands (usually organic molecules/ions). In some coordination compounds, the valence states of the central cations and electronic structures of the ligands can change reversibly with external potentials, resulting in electrochromic activity. The major organic-metal complexes with electrochromic properties include Prussian blue (PB) and phthalocyanines. Some organic-metal complexes can form self-assembled structures owing to their unique molecular structures and multi-position coordination abilities.

9.4.2
Organic-Metal Complexes

PB, iron(III) hexacyanoferrate(II), is one of polynuclear transition metal hexacynometallates. In 1978, Neff [73] first prepared thin films of PB on platinum and gold electrodes and demonstrated its redox reactions accompanied by electrochromic behaviours. Typically, PB is one type of insoluble mixed-valence compounds. Through surface modification by $[Fe(CN)_6]^{4-}$ ion and alkylamine, inks composed of water-soluble and organic-solvent-soluble PB nanoparticles have been prepared [74, 75]. The inks made of nickel hexacyanoferrate and cobalt hexacyanoferrate (CoHCF) nanoparticles can be treated via similar process for assembly of a full-colour device [76].

As another type of organic-metal complexes, metallophthalocyanines exhibit unique electronic and optical properties, owing to their strong $\pi-\pi$ interactions. Rodriguez-Mendez and Antonio de Saja [13] fabricated an electrochromic thin film consisting of 20 monolayers of praseodymium bisphthalocyanine using LB technique. The *in situ* UV-vis spectra of the electrochromic device show colour changes from blue-green to red with potentials changing from −2 to +2 V. The device based on the LB film is extremely stable.

9.4.3 Viologens

Unlike most of the other electrochromic materials, the electrochromic process of viologen is realised by redox reactions of viologen and its complementary ion in a solution. Without potential applied, the device will bleach rapidly. To date, the studies on electrochromic behaviours of viologens are focused on how to confine the viologens on the electrode to improve its slow response speed caused by the situation that most of the chromophores are dissolved in liquid electrolytes and hence not in contact with electrodes. Grafting viologen groups onto macromolecules is one of the efficient ways. Suh and co-authors [77] grafted the viologen moiety onto a monodispersed PS microsphere for assembly of a reflective electrochromic device. With the reduced mobility of the chromophores, the coloration time and bleaching time are decreased to some extent.

Wang and co-authors [78] tethered viologen groups at the chain-ends of a hyperbranched polyglycidol. Thanks to the amorphous, soft and hyperbranched structure of the polyether, aggregation and crystallisation of highly concentrated and ionic viologen chromophores are inhibited. The viologens grafted at the chain ends of the hyperbranched polymer form a core–shell structure. The kinetic switching tests show that the coloration time at 2.5 V is 0.52 s and the bleaching time (switch off) is ~2 s, which are less than the tens of seconds level of other polymer systems with viologen moieties. In addition, the gradual decrease in optical contrast with the increase in cycling time is not observed in this system. It is attributed to the globular conformation of the macromolecule, which suppresses the aggregation and hinders the crystallisation of the viologen.

9.5 Electrochromic Nanocomposites and Nanohybrids

9.5.1 Introduction

Although various nanostructures of electrochromic polymers, small molecules and TMOs have been prepared and show superior electrochromic properties in some aspects, the single-phase electrochromic materials do have their limitations. For example, although promising electrochromic performances such as relatively fast switching speeds have been demonstrated with some TMO nanostructures, as demonstrated in Section 9.2, relatively high contact resistance between these rigid nanostructures still poses enormous challenge to further improving their electrochromic properties, and the coloration efficiencies of these materials still fall far behind that of conjugated polymers. On the other hand, for most electrochromic conjugated polymers, their relatively poor electrochemical and environmental stabilities still remain as major obstacles for their practical applications. To meet the requirements for real-world applications, preparation of electrochromic materials with two or more phases is an efficient approach as the synergistic combination of the merits of different types of materials may provide an opportunity to deploy a novel multi-phase electrochromic material with higher η, shorter response time and outstanding stabilities.

Multi-phase electrochromic materials in general can be divided into two classes: nanocomposites with only physical interactions between their nanometre-scale phases, which are introduced in the following two subsections, and nanohybrids with interfacial chemical bonds, which are discussed in Section 9.5.4. Furthermore, according to the electrochromic activity of each individual phase, the electrochromic nanocomposites can be further divided into two categories: electrochromic/electrochromic and electrochromic/non-electrochromic nanocomposites, which are the focuses of Sections 9.5.2 and 9.5.3, respectively. The major preparation routes of the nanocomposites include *in situ* polymerisation, sol–gel, LbL and physical blending.

9.5.2 Nanocomposites of Electrochromic Materials

To design a nanocomposite composed of two kinds of immiscible electrochromic materials nanocomposite, firstly, it should be determined if the working potentials and electrochromic features of the two phases match well. Generally, the two phases should exhibit reversible optical changes in the same potential window. Otherwise, either one phase cannot be fully switched to the desired colour state or the other phase is overloaded. Also, in most cases, anodically and cathodically colouring materials should be paired separately so that when a potential is applied, the two phases will be simultaneously switched to their coloured states or bleached states. Sometimes, an anodically colouring material may also be paired with a cathodically colouring one when the absorption bands of the two electrochromic materials in their coloured states are not in the same region; such a nanocomposite thin film may be switched from one coloured state to another coloured state, instead of from a coloured state to bleached state.

9.5.2.1 Conjugated Polymer/TMO and TMO/TMO Nanocomposites

Based on the pairing principle stated in the previous section, cathodically colouring PEDOT and its derivatives can be combined with cathodically colouring TMOs such as WO_3 and V_2O_5, while anodically colouring PANI can be combined with anodically colouring IrO and NiO, to form conjugated polymer/TMO nanocomposites. In such nanocomposites, usually TMO nanostructures are dispersed in conjugated polymer matrices or in direct contact with a transparent electrode to act as a supplementary electrochromic material or improve the charge transport in the devices. Birss and co-authors [79] prepared Ir oxide/PANI composite films by an electrochemical method. PANI was deposited on a pre-formed porous IrO_x film. The resultant nanocomposites show a colour contributed by both the PANI and IrO_x phases. The colour changes from light yellow, typical of PANI in the fully reduced state, through dark green, to black for the fully oxidised IrO_x. In comparison with the composite consisting of PANI on a sputtered solid IrO_x film, the anodic peak current of PANI on the porous IrO_x film is approximately two times of those of the main IrO_x peaks, indicating much improved electrochemical activity.

As a typical cathodically colouring conjugated polymer, PEDOT is frequently paired with cathodically colouring WO_3. Furthermore, PEDOT and WO_3 show almost same colour change from colourless to blue when the external potential changes from positive to negative. Deepa *et al.* [80] prepared PEDOT/WO_3 nanocomposites via a two-step electrochemical deposition approach. Attributing to the nanostructures of the two phases and their Coulombic interactions, the composite films exhibit much higher colouring efficiency (227 $cm^2\,C^{-1}$ at 600 nm), higher redox activity and cycling stability than the corresponding neat PEDOT film. Recently, hybrid thin films composed of multi-layer PEDOT:PSS and WO_3 were electrochemically deposited on ITO glass from a one-pot solution using square-wave galvanostatic method [81], as shown in Figure 9.9. The

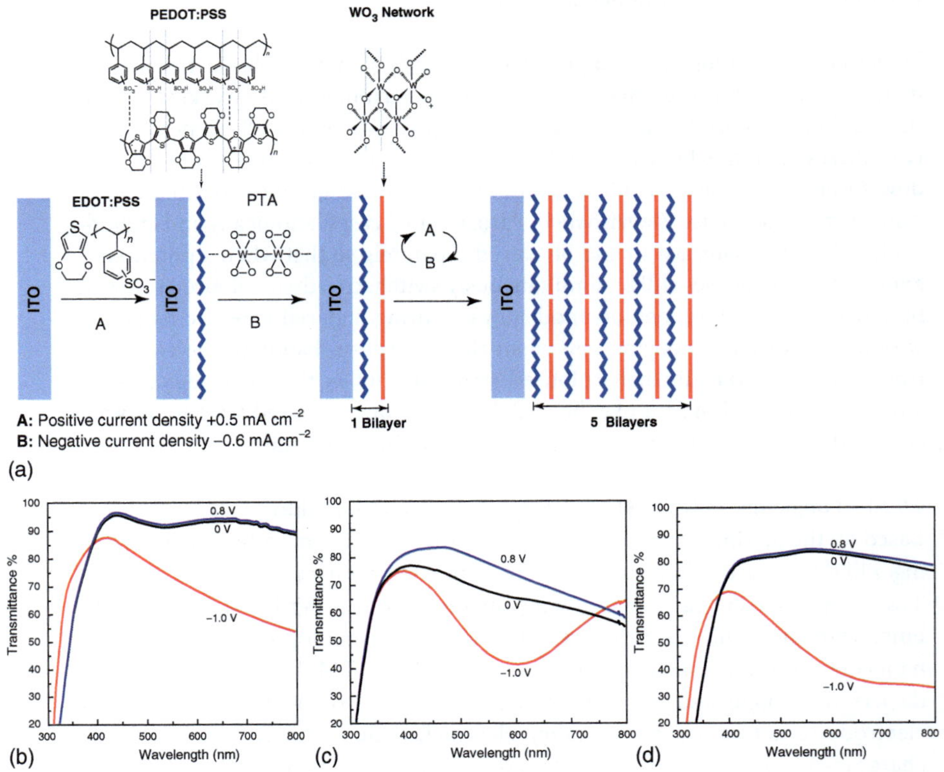

Figure 9.9 Synthesis process for the multi-layer PEDOT:PSS/WO_3 hybrid thin films (a), and UV-vis optical transmittance spectra in visible range of (b) WO_3, (c) PEDOT:PSS and (d) five-bilayer hybrid thin films under different potentials of 0.8, 0 and −1.0 V. Reproduced from Ref. [81] by Lu, X. with permission from Royal Society of Chemistry.

enhanced electrochromic properties, including the optical contrast and stability can be obtained with very short deposition times of PEDOT:PSS and WO_3 in each cycle. The improved performance is due to the favourable interactions between the two components. The surface-defect-induced anodic dissolution of WO_3 during the cycling can be prevented when the PEDOT:PSS enters the defect sites in the electrodeposited WO_3, while the surface functional groups of WO_3 may act as dopants to inhibit over-oxidation of PEDOT.

When the colour of one phase in oxidised/reduced state is not identical with the colour of the other phase under same potentials, an anodically colouring electrochromic material may still be combined with a cathodically colouring one. PANI/WO_3 is one of such systems, in which PANI shows yellow to green, sky blue and purple, while WO_3 shows blue and colourless with applied potentials from negative to positive. Zhang *et al.* [24] prepared the ultra-thin film of WO_3 nanorod-embedded PANI via surface-initiated polymerisation. The morphologies of the WO_3 nanorods obtained by solvothermal treatment, PANI and PANI/WO_3 nanocomposite are shown in Figure 9.10. Typically, the WO_3 nanorods exhibit low contrast and slow switching speed, and PANI exhibits moderate contrast and fast switching speed. By contrast, the PANI/WO_3 nanocomposite shows highest contrast and improved switching speed in comparison with that of WO_3. The fast switching speed can be attributed to the porous morphology of the PANI/WO_3 nanocomposite and enhanced conductivity in both oxidised state and reduced

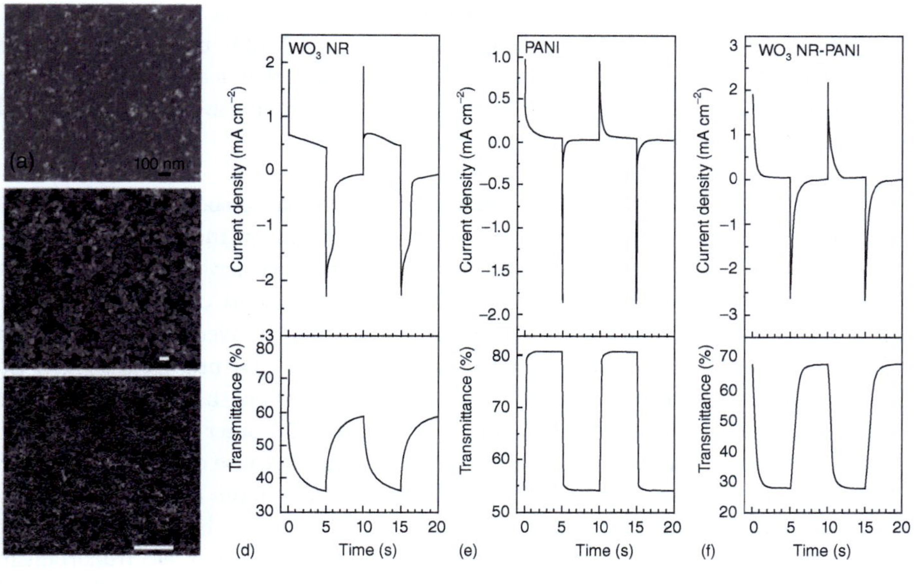

Figure 9.10 Morphologies and electrochemical behaviours of WO_3/PANI composites. SEM images of (a) WO_3 nanorods, (b) PANI and (c) WO_3 nanorod-PANI composite. Current transient response and corresponding optical switching at 550 nm for (d) WO_3 nanorods, (e) PANI and (f) WO_3 nanorod-PANI composite films in 0.5 M H_2SO_4. Reproduced from Ref. [24] by Zhang, J. with permission from Elsevier B.V.

state, owing to the high conductivity of PANI at positive potential and WO_3 at negative potential. PEDOT/NiO is another nanocomposite system consisting of both anodically and cathodically colouring materials, in which PEDOT was deposited electrochemically on the highly porous NiO film. The fast reaction kinetics with short response time of 500–600 ms was obtained owing to the fast ion diffusion in the porous NiO film [82].

Sometimes, the TMOs with weak electrochromic properties, such as ZnO and TiO_2, can also be integrated with conjugated polymers or used as ion storage layers in the electrochromic devices. Ahmadi and co-authors [22] used ZnO nanowires as template to prepare two types of PEDOT/ZnO composites using cyclic voltammetry CV and constant potential (CP) techniques. The two preparation methods, that is, PEDOT nanotube coating on ZnO nanowires via the CV route or PEDOT nanoparticles attached on the surface of ZnO nanowires via the CP route, give rise to different morphologies, as well as different electrochromic performance. The CV sample exhibits a higher contrast (45.05% vs 37.7% of the CP sample) and shorter response time (66 ms vs >300 ms of the CP sample), which can be attributed to the enhanced charge transport of the CV sample.

Different TMO nanostructures could also be integrated together to form electrochromic TMO/TMO nanocomposites. For example, the composites of mesoporous NiO and manganese oxide (MnO_2) nanotubes exhibit improved electrochromic performance as the composite film has both a high value of transmission that is the characteristic of lamellar structure of NiO and a wide potential window that is the characteristic of the MnO_2 nanotubes. It is the synergistic effect between the two characteristics that leads to quick redox reaction and optical response of the mesoporous composite of lamellar nickel oxide and MnO_2 nanotubes [20].

9.5.2.2 Conjugated Polymer/Organic Small-Molecule Nanocomposites

Electrochromic organic small molecules usually exist in the form of nanoparticles. In an electrochromic layer, the charge transport may be affected by high contact resistance between the particles. The continuous conducting polymer matrices are a promising candidate for improving the electric conduction of the organic small molecules. Chen and co-authors [83] prepared a PANI/PB bilayer film to demonstrate the multi-electrochromism of the film. Two kinds of bilayer films with inner PANI-outer PB and inner PB-outer PANI, respectively, exhibit totally different electrochromic behaviours. Because of the good electron transport property of the inner conducting PANI layer and numerous ion pathways offered by the submicron-crack open structure of the PB layer, the inner PANI-outer PB bilayer film shows intense and multicolour electrochromism (from colourless, light green, green/blue to deep blue). CoHCF was also paired with PANI nanofibres to improve electrochromic performance of the materials [84]. A CoHCF-PANI nanocomposite was prepared via a coordination reaction (for the formation of CoHCF) followed by electropolymerisation of aniline. The charge transfer, capacity and ion diffusion of the nanocomposite were found

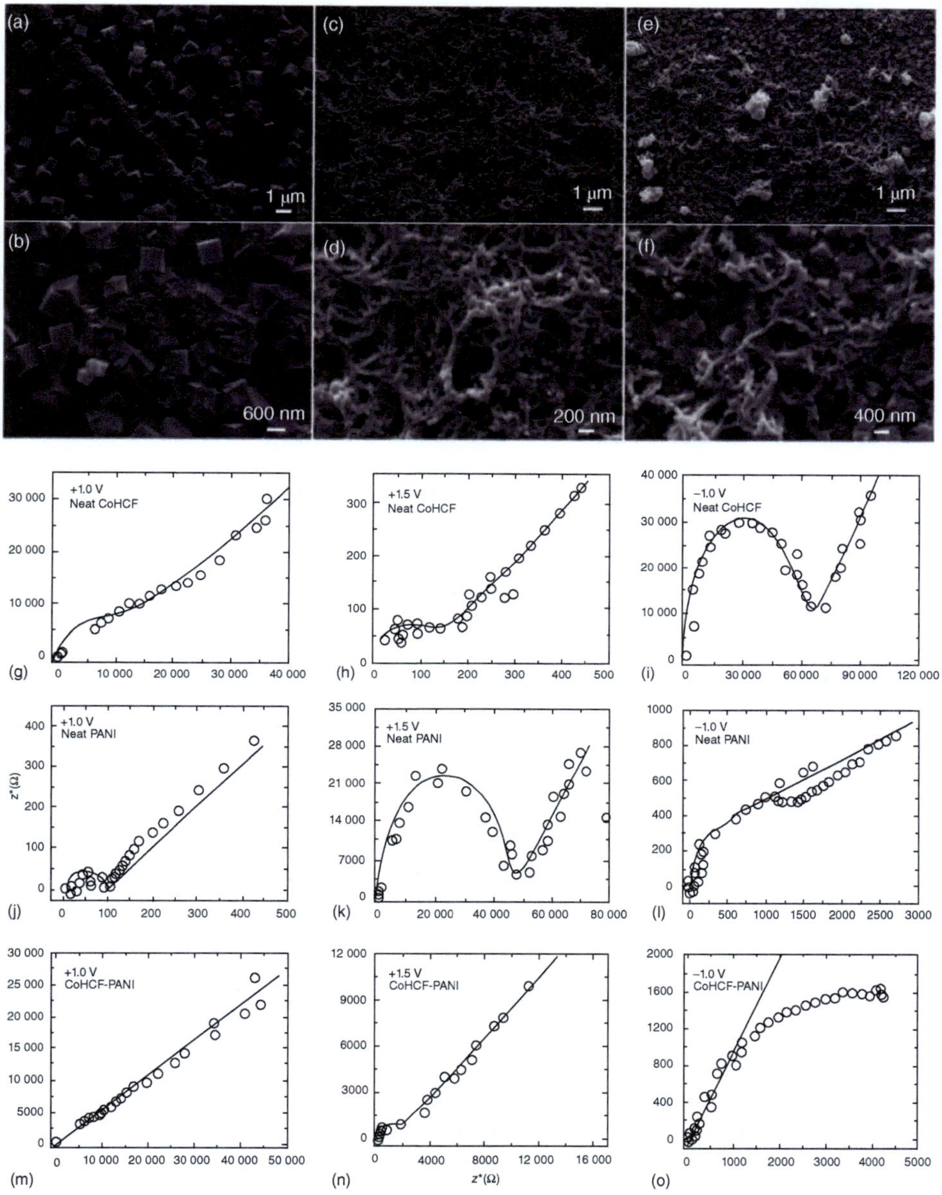

Figure 9.11 The morphologies and EIS analysis. SEM images of (a,b) neat CoHCF, (c,d) neat PANI and (e,f) hybrid CoHCF-PANI films recorded at low and high magnifications. Nyquist plots of (g–i) neat CoHCF, (j–l) neat PANI and (m–o) hybrid CoHCF-PANI films recorded at different DC potentials in the ionic liquid 1-butyl-1-methylpyrrolidinium bis(trifluoromethylsulfonyl)imide: an AC amplitude of 5 mV was superimposed over the DC potentials. The solid lines are obtained by fitting the experimental data using a Randles equivalent circuit of $R_1(C_1(R_2Z_W))$. Reproduced from Ref. [84] by Deepa, M. with permission from Elsevier B.V.

to be controlled by the film microstructures. The morphologies and EIS chart of PANI, CoHCF and the nanocomposite under different potentials are shown in Figure 9.11. It is obvious that under +1 and +1.5 V, the nanocomposite has intermediate charge transfer resistance, whereas under −1 V, the nanocomposite shows the lowest charge transfer resistance in comparison with that of neat PANI or CoHCF. The electrochromic performances are improved with a high η of 262 $cm^2 C^{-1}$ and good stability.

Another PB analogue, zinc hexacyanoferrate (ZnHCF), has the same drawback as PB in terms of high contact resistance between the particles. The Conducting polymer, PEDOT:PSS, has been used to improve the processability of ZnHCF [85]. Enhanced charge capacity, cycling life and diffusion characteristics were achieved for nano-ZnHCF through networking with PEDOT:PSS. EIS studies prove that the PEDOT:PSS not only serves as a transparent conductive binder but also acts as a pseudocapacitor that holds electric charge for facile electron transfer between the nanoparticles.

An indigo carmine (IC) dye has also been used to modulate the colour and enhance the optical contrast of PPy [86]. The enhancement in contrast, lifetime and response time can be ascribed to the better charge retention instigated by the IC molecules or structural arrangements in the PPy/IC composites.

9.5.3
Nanocomposites of Electrochromic/Non-Electrochromic Active Materials

Unlike the nanocomposites composed of two electrochromic materials, which require matching of the potential windows and colours of the two materials, the design of a nanocomposite with an electrochromic material and a non-electrochromic active material is more flexible. In such electrochromic/non-electrochromic nanocomposites, the non-electrochromic materials may consist of conducting materials, such as carbon nanotubes (CNTs), graphene, silver and gold nanoparticles or insulating/semiconductor materials, such as SiO_2, TiO_2 and ZnO. In general, the function of conducting additives is to promote the electron and ion transport, while the functions of insulating materials and semiconductors are to improve the ion transport.

9.5.3.1 Conjugated Polymers as Electrochromic Materials

Incorporation of conductive nanofillers into conjugated polymers is a common strategy to enhance electrochromic performances of the conjugated polymers. Various carbon nanostructures and metal nanoparticles have been attempted as the conductive additives in conjugated polymers. The conductive carbon nanostructures, including CNTs and graphene, can enhance the electrical conductivity of conjugated polymers at very low feed contents, which are the effective way to enhance the charge transfer in the conjugated polymers, especially when the conjugated polymers are switched to their low-conductivity states. Deepa and co-authors [87] employed multi-walled carbon nanotubes (MWCNTs) to amplify the electrochromic response and improve the redox activity of PEDOT. In addition to utilisation of the high conductivity of the MWCNTs to facilitate the charge transport in PEDOT, the interlinked bundles of nanotubular structures and porous conducting CNT network created by the incorporation of the MWCNTs can also promote the ion diffusion during the switch of the nanocomposites. Reduced graphene oxide (rGO) nanosheets have also been introduced into poly(3,4-ethylenedioxypyrrole) (PEDOP) and PEDOT to improve their electrochromic performance [31, 88]. Less-hindered electron transport and more uniform current pathways were achieved with the incorporation of rGO nanosheets. Their benefits to the electrochromic performance of the PEDOP-rGO nanocomposites are clearly reflected in their improved optical contrast, as the ease of electron ingress and egress during redox of the nanocomposites allows higher ion uptake, thus imparting higher η and faster bleaching rate.

Noble metals, such as Au and Ag, possess high electrical conductivities, as well as strong absorption enhancement effect, which are also ideal additives for conjugated polymers. Au and Ag nanoparticles have been dispersed into PEDOP, PEDOT and PANI to adjust their electrochromic behaviours [89]. The major electrochromic parameters including the optical contrast, switching speed, η and lifetime are all increased with the incorporation of Au or Ag. The enhanced electrochromic performance may be related to the decrease of charge transfer resistance. Also, the band gap of the conjugated polymers can be narrowed by noble metals from 1.83 eV for PEDOP to 1.66 eV for PEDOP-Ag and 1.6 eV for PEDOP-Au, as shown in Figure 9.12. The lower band gap would give higher absorption coefficient, reducing the potentials required to achieve certain colour states [89].

Other non-electrochromic active nanomaterials, for example, SiO_2 [90, 91], TiO_2 [30, 34, 92], ZnS and CdSe [93] nanoparticles/quantum dots, can also be utilised to modulate electrochromic properties of conjugated polymers as well as to improve their electrochemical stability. For example, the switching speed of PEDOT can be increased by SiO_2 as the packing density of PEDOT chains in the nanocomposites is decreased and the specific surface area is increased [91]. Also, owing to the suppression of formation of highly oxidised form of PANI/SiO_2, core–shell nanoparticles exhibit enhanced stability [90].

Besides the positive influences of the additives, negative influence on the electrochromic properties of a conjugated polymer was found when CdSe was used as an additive. The composite films of poly(*N*,*N*′-di[p-phenylamino(phenyl)]-1,4,5,8-naphthalene tetracarboxylic diimide)/CdSe showed bipolar properties and became more difficult to be oxidised with a widened band gap as a result of reduced conjugated length caused by the attachment of functionalised CdSe quantum dots on the polymer chains [93].

9.5.3.2 TMOs as Electrochromic Materials

Almost all additives used in conjugated polymer systems can be employed for TMOs. For example, the mechanisms for the enhancements of electrochromic

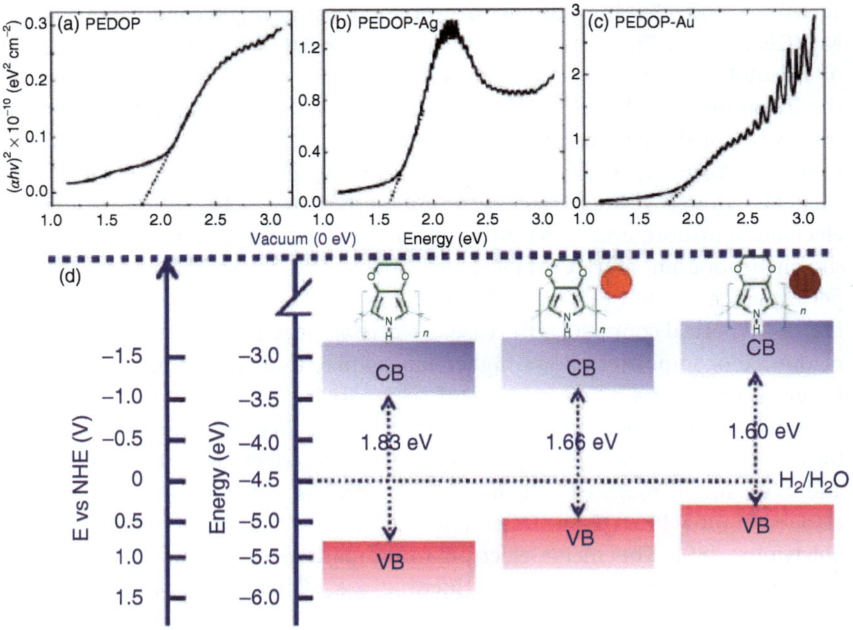

Figure 9.12 Main edge absorption spectra plotted as $(\alpha h\nu)^2$ versus photon energy ($h\nu$) for the (a) neat PEDOP, (b) PEDOT-Ag and (c) PEDOP-Au films in their neutral states. (d) Approximate levels of the highest occupied molecular orbital (HOMO)/Valence band (VB) and lowest unoccupied molecular orbital (LUMO)/Conduction band (CB), showing the narrowing of the direct band gap (E_g) in the composites. Reproduced from Ref. [89] by Deepa, M. with permission from American Chemical Society.

properties of conjugated polymers by CNTs also work for WO_3/CNT systems [28]. In a suitable range of CNT feed contents, the electrochromic performances, including contrast, reversibility, η and switching speed, of WO_3 can be enhanced with increasing CNT content. However, the overfeeding of CNTs will degrade the electrochromic performances due to the strong optical absorption of the CNTs. Noble metal nanoparticles, including that of Pt, Au and Ag, have also been incorporated into TMO electrochromic materials. The major role played by the noble metal nanoparticles is still increasing the electrical conductivity of the metal oxide films, improving the charge transport and hence electrochromic performance of the TMOs. Unlike the CNTs additives, the protons (H^+) in the Au–WO_3 and Pt–WO_3 electrodes could be modified and electrochemically affected by the Pt and Au nanophases. In other words, conducting nanophases in the electrode could affect or modify both electrochromic and electrochemical properties of WO_3 [29].

TiO_2 is a type of cathodically colouring material with weak electrochromic activity. It can be viewed as a non-electrochromic material in some cases,

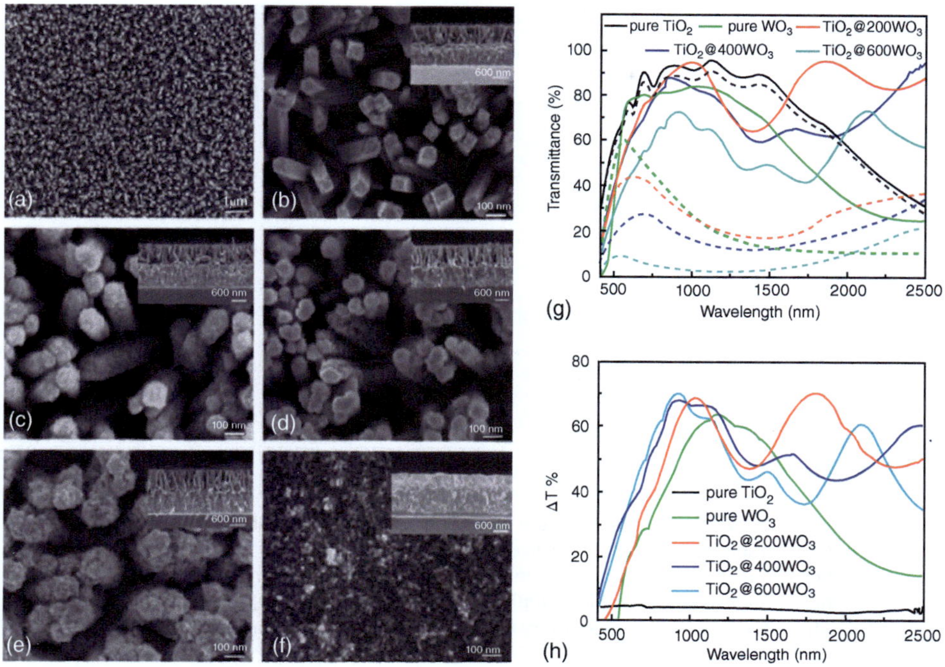

Figure 9.13 Morphologies and electrochromic properties of samples. SEM images of (a,b) TiO_2 nanorod array, (c) TiO_2-200WO_3, (d) TiO_2-400WO_3, (e) TiO_2-600WO_3 core/shell nanorod array and (f) WO_3 film (cross-sectional view presented in the inset). (g) Visible and near-infrared transmittance spectra of TiO_2, WO_3 and TiO_2-WO_3 core/shell nanorod array films (solid curves, bleached state; dashed curves, coloured state) and (h) corresponding transmittance modulations. Reproduced from Ref. [27] by Tu, J. with permission from Elsevier B.V.

depending on its preparation method. Tu and co-authors [27] prepared TiO_2-WO_3 core/shell nanorod arrays using the combination of hydrothermal and electrodeposition methods, in which the TiO_2 nanorods show almost negligible colour change. As shown in Figure 9.13, though the TiO_2 itself does not contribute much to the optical contrast, the TiO_2-WO_3 nanocomposite exhibits significantly different optical absorption properties and enhanced contrast, especially in near-infrared (NIR) region. Also, the core/shell approach shortens the response time of WO_3, which is due to the lower charge transfer resistance, as compared with that of WO_3 film, brought by the large surface area of the nanowire arrays and the space between the nanowires.

9.5.3.3 Organic Small Molecules as Electrochromic Materials

The carbon nanostructures including CNTs and graphene can also be introduced into organic small molecules to improve their electrochromic properties. With the incorporation of CNTs, the electrochromic parameters, including responding speed and stability, of PB can be improved [25]. Another type of

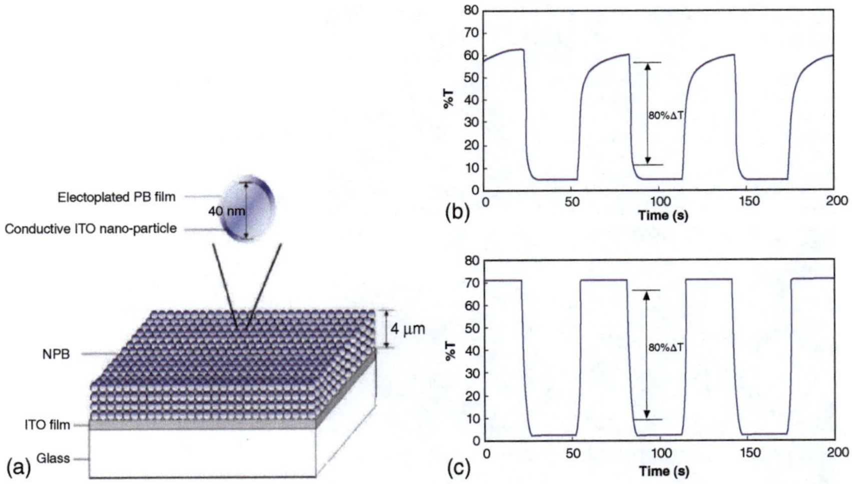

Figure 9.14 Conceptual structure of PB nanocomposite film and its electrochromic behaviour. (a) Conceptual structure of nanocomposite Prussian blue (NPB) film. After PB is electrodeposited, every ITO nanoparticle can be well covered by PB and by an NPB structure. (b,c) Transmittance spectra of coloured and bleached states of PB at 750 nm with switching potential between +0.5 and −0.5 V of normal PB film and PB/ITO composite film, respectively. Reproduced from Ref. [61] by Cheng, K.C. with permission from Elsevier B.V.

conductive nanofillers, ITO nanoparticles, have also been combined with PB to form nanocomposites. Cheng *et al.* [61] deposited a PB coating onto the film formed by ITO nanoparticles to form an ITO/PB nanocomposite with core–shell nanoparticles, as shown in Figure 9.14. The ITO nanoparticles can act as a porous conductive substrate for PB, which efficiently reduce the impedance of ITO/PB nanocomposite and the PB shell offers much larger surface area for facile redox reactions. As a result, the ITO/PB nanocomposite gives a contrast as high as 78% at 750 nm and much faster switching speed than PB film, as shown in Figure 9.14.

9.5.3.4 Electrochromic Nanohybrids with Covalent Bonds

It is well known that in structural nanocomposites, the mechanical properties of the materials are determined by the dispersion states of the nanophases in the matrices and their interfacial interactions. Similarly, for electrochromic multiphase materials, better dispersion of the nanophase in the electrochromic matrix and stronger interactions between the two phases may also lead to enhanced electrochromic performances. Creating covalent bonds at the interface is one of the efficient routes to achieve nanometre-scale, uniform dispersion of one phase into the other. Moreover, the interfacial covalent bonds may enhance the interactions between the matrix and the fillers, promoting interfacial electron conduction. By covalently attaching rigid molecular chains onto nanofillers, the packing structures of the conjugated polymers may be modified, facilitating the ion motion in the matrix. Thus, the basic physical processes involved in the operation of an electrochromic device, electron conduction and ion transport, can all be enhanced by the covalent bonding approach. Nanostructures with/without electrochromic activity including carbon nanostructures, functional POSS, silica, TiO_2, WO_3 and so on can be tethered with conjugated polymers to boost the electrochromic performances of the polymers. The major preparation methods include surface modification, *in situ* polymerisation and sol–gel process.

The function of non-active nanostructures is mainly to promote facile ion motion in the system through creating a loose arrangement of molecular chains. Xiong *et al.* [94] prepared POSS–PANI nanohybrids by copolymerisation of aniline with aniline functionalised POSS. With the incorporation of only 0.5 mol% octa(aminophenyl)silsesquioxane (OAPS), the optical contrast of POSS–PANI exhibits a 40% enhancement in comparison with that of neat PANI, as the loose packing of PANI chains caused by the star-like geometry of POSS–PANI leads to more accessible active units. In addition, the switching speed is also benefited from its loose packing structure, which is evidenced by the increase of ion conductivity by 1 order of magnitude, though the electrical conductivity is decreased simultaneously.

Unlike the highly insulating POSS cage that separates the PANI arms attached onto the same cage from one another, TiO_2 nanoparticles can interact with conjugated polymers due to their semiconductor nature. Lu's group [34] synthesised PANI–TiO_2 nanohybrids using a bridging component to tether PANI onto TiO_2. With the help of interfacial covalent bonds, the *in situ* formed TiO_2 nanoparticles are well dispersed in the hybrids. Owing to the strong interactions between TiO_2 and PANI, the operation potentials of the device could be decreased to ± 1.2 V. The low working voltages render good stability to the device during long-term cycling, while lifetime is usually the bottleneck for practical applications of conjugated polymer-based electrochromic devices. As shown in Figure 9.15, the device with the PANI-TiO_2 hybrid as the active material exhibits higher contrast (lower absorption in bleached state and higher absorption in coloration state) and excellent stability. The neat PANI-based device shows an obvious decline in contrast at about 18 000 cycles, and the contrast drops fast to less 10% at 30 000 cycles. By contrast, the PANI–TiO_2 nanohybrid-based device does not show obvious contrast decay in 50 000 cycles.

As mentioned in the previous sections, carbon nanostructures may have high electrical conductivity and strong interactions with conjugated polymers. They are promising candidates for promoting the electrochromic performances of conjugated polymers, except that the feed contents of the carbon nanostructures must be strictly controlled due to their strong optical absorption. Xiong *et al.* tethered PANI molecular chains onto fullerene balls (C_{60}) [95] and single-walled carbon nanotubes (SWCNTs) [96] by copolymerisation of aniline with *p*-phenylenediamine (PPD)-functionalised carbon nanostructures. For C_{60}–PANI nanohybrids, owing to the capability of fullerene to accept six electrons, a star-like geometry with six PANI arms can be obtained, as shown in Figure 9.16. The strong interaction between fullerene and PANI is verified by the

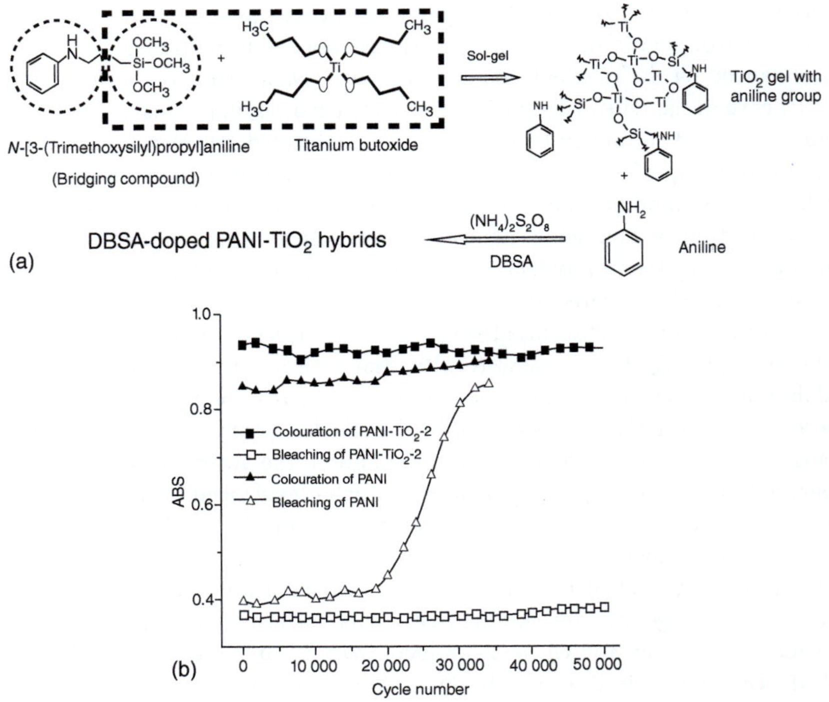

Figure 9.15 Synthesis route and performance of PANI-TIO$_2$ hybrids. (a) Synthesis route of PANI-TiO$_2$ hybrid. (b) Absorbance of the PANI/PEDOT and PANI-TiO$_2$-2/PEDOT complementary devices in their coloured and bleached states, as a function of the number of switching cycles. Absorption was measured at 680 nm for the PANI/PEDOT devices. The corresponding wavelength for the PANI-TiO$_2$-2/PEDOT device was 650 nm. Reproduced from Ref. [34] by Lu, X. with permission from American Chemical Society.

red-shift of the π-polaron transition band of the C_{60}−PANI solution, as well as the photoluminescence (PL) quench effect induced by grafting PANI onto C_{60}. With introduction of fullerene into the PANI matrix, the ionic conductivity is increased without sacrifice of electric conductivity, which can be attributed to the loose star-like molecular structure of C_{60}−PANI and the strong interaction between fullerene and PANI. The electrochromic performances of the C_{60}−PANI nanohybrids are benefited from the enhanced charge transfer, as shown in Figure 9.16.

Furthermore, a nanomaterial with electrochromic activity can also be covalently incorporated into electrochromic matrices for further enhancement of electrochromic performance. Yang *et al.* [17] prepared covalently bonded poly(vinylimidazole) (PVI)/WO_3 hybrid with core–shell structure using the chemically modified WO_3 as core. In comparison with neat WO_3 and the PVI/WO_3 nanocomposites without interfacial covalent bonds, the covalently bonded PVI/WO_3 hybrid exhibits higher contrast and faster switching speed.

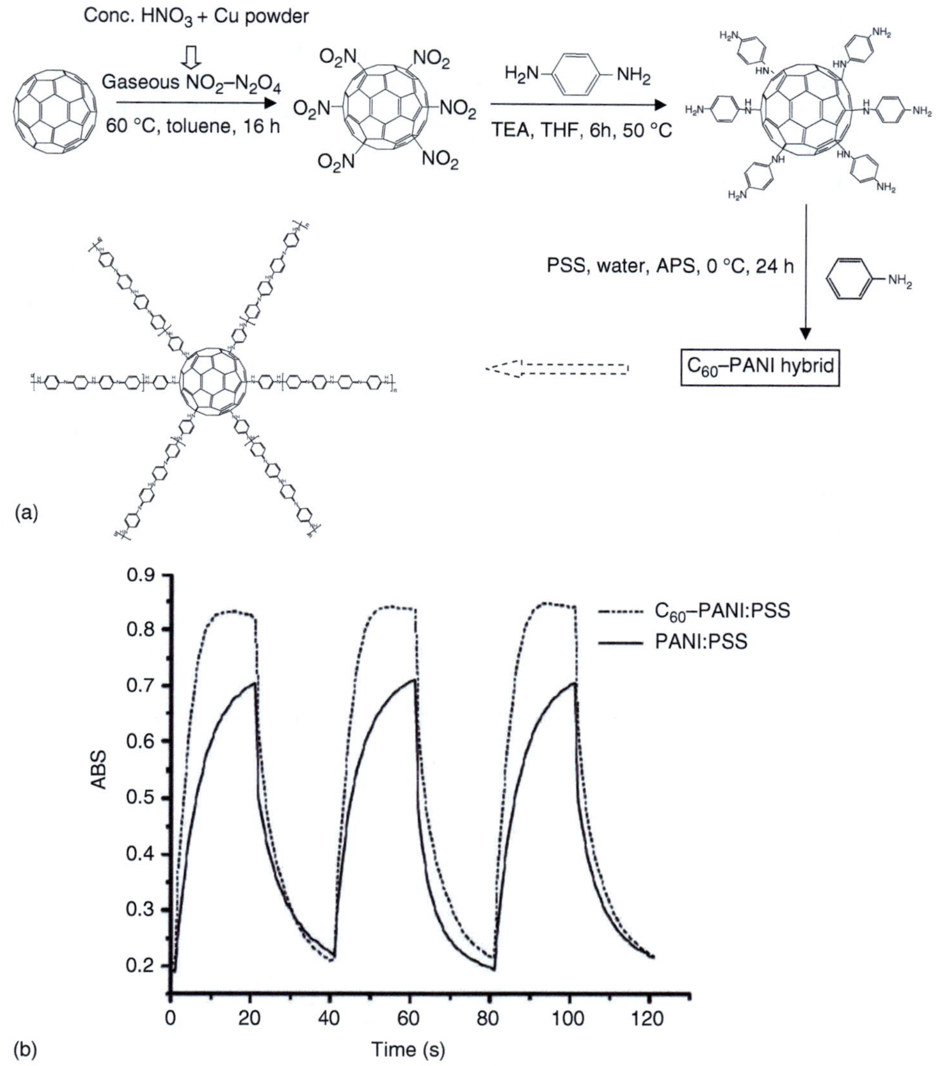

Figure 9.16 Synthesis route and performance of C_{60}–PANI. (a) Synthesis route of C_{60}–PANI hybrid. (b) Optical absorbance at λ_{max} for the devices with PANI:PSS and C_{60}–PANI:PSS as the electrochromic layers, respectively, under the step potential oscillating between +2.0 and −2.4 V with 40 s interval. Reproduced from Ref. [95] by Lu, X. with permission from Elsevier B.V.

9.6 Conclusions and Perspective

With the preparation of electrochromic nanomaterials or the addition of nanomaterials into electrochromic materials, electrochromic performances of the resultant nanostructured electrochromic materials can be significantly enhanced. The principal enhancement mechanisms the introduction of nanomaterials into electrochromic materials are that the nanostructures can improve electron conduction and ion transport in the systems, owing to either the high electrical conductivity of the incorporated nanostructures and strong interactions between the dispersed phase and matrix or the induced porous morphology/loose molecular packing structure. As individual electrochromic material usually has some shortcomings in one or more aspects, and a new class of electrochromic materials needs enormous efforts and time to be developed, multi-phase electrochromic materials are viable choices for achieving high-performance electrochromic materials. Among numerous multi-phase electrochromic systems discussed in this chapter, the hybrids with interfacial covalent bonds have unique advantages, that is, they may simultaneously exhibit enhanced optical contrast, switching speed and long cycle life.

Although nanostructured electrochromic materials were intensively studied in the past decade, some challenges still remain. Firstly, in order to further enhance the electron conduction and ion transport in multi-phase electrochromic materials, the morphology of the materials, especially the ones with interfacial covalent bonds, needs to be controlled more precisely to further realise the potential of the covalent approach. Secondly, the mechanisms for the enhanced electrochromic properties of the nanostructured electrochromic materials, in particular the nature of the interactions at the interfaces of the multi-phase materials and the roles played by the interfacial interactions in determining the mass transfer and electron conduction across the interfaces, need to be further clarified, thereby improving our understanding of electrochromism of the multi-phase systems.

References

1. Wei, J., Xiong, S., Bai, Y., Jia, P., Ma, J., and Lu, X. (2012) Polyaniline nanoparticles doped with star-like poly(styrene sulfonate): synthesis and electrochromic properties. *Sol. Energy Mater. Sol. Cells*, **99**, 141–147.
2. Gotoh, A., Uchida, H., Ishizaki, M., Satoh, T., Kaga, S., Okamoto, S., Ohta, M., Sakamoto, M., Kawamoto, T., Tanaka, H., Tokumoto, M., Hara, S., Shiozaki, H., Yamada, M., Miyake, M., and Kurihara, M. (2007) Simple synthesis of three primary colour nanoparticle inks of Prussian blue and its analogues. *Nanotechnology*, **18**, 345609.
3. Li, C.-P., Wolden, C.A., Dillon, A.C., and Tenent, R.C. (2012) Electrochromic films produced by ultrasonic spray deposition of tungsten oxide nanoparticles. *Sol. Energy Mater. Sol. Cells*, **99**, 50–55.
4. Yamada, K., Seya, K., and Kimura, G. (2009) Electrochromism of poly(pyrrole) film on Au nano-brush electrode. *Synth. Met.*, **159**, 188–193.
5. Nguyen, T.M., Cho, S., Varongchayakul, N., Yoon, D., Seog, J., Zong, K., and Lee, S.B. (2012) Electrochemical synthesis and one step modification of PMProDot nanotubes and their enhanced electrochemical properties. *Chem. Commun.*, **48**, 2725–2727.
6. Byun, J., Kim, Y., Jeon, G., and Kim, J.K. (2011) Ultrahigh density array of free-standing poly(3-hexylthiophene) nanotubes on conducting substrates via solution wetting. *Macromolecules*, **44**, 8558–8562.
7. Kumar, A., Welsh, D.M., Morvant, M.C., Abboud, K., and Reynolds,

J.R. (1998) Conducting poly(3,4-alkylenedioxythiophene) derivatives as fast electrochromics with high contrast ratios. *Chem. Mater.*, **10**, 896–902.

8. Ge, D., Yang, L., Tong, Z., Ding, Y., Xin, W., Zhao, J., and Li, Y. (2013) Ion diffusion and optical switching performance of 3D ordered nanostructured polyaniline films for advanced electrochemical/electrochromic devices. *Electrochim. Acta*, **104**, 191–197.
9. Tarver, J. and Loo, Y.-L. (2011) Mesostructures of polyaniline films affect polyelectrochromic switching. *Chem. Mater.*, **23**, 4402–4409.
10. Yang, C.-H., Huang, L.-R., Chih, Y.-K., Lin, W.-C., Liu, F.-J., and Wang, T.-L. (2007) Molecular assembled self-doped polyaniline copolymer ultra-thin films. *Polymer*, **48**, 3237–3247.
11. Lakshmi, A., Anandha Raj, J., Gopu, G., Arumugam, P., and Vedhi, C. (2013) Electrochemical, electrochromic behaviour and effects of supporting electrolyte on nano-thin film of poly (3,4-ethylenedioxy thiophene). *Electrochim. Acta*, **92**, 452–459.
12. Kim, Y., Kim, Y., Kim, S., and Kim, E. (2010) Electrochromic diffraction from nanopatterned poly(3-hexylthiophene). *ACS Nano*, **4**, 5277–5284.
13. Rodriguez-Mendez, M.L. and Antonio de Saja, J. (2009) Nanostructured thin films based on phthalocyanines: electrochromic displays and sensors. *J. Porphyrins Phthalocyanines*, **13**, 606–615.
14. Liu, S., Xu, L., Li, F., Xu, B., and Sun, Z. (2011) Enhanced electrochromic performance of composite films by combination of polyoxometalate with poly(3,4-ethylenedioxythiophene). *J. Mater. Chem.*, **21**, 1946–1952.
15. Jia, P., Argun, A.A., Xu, J., Xiong, S., Ma, J., Hammond, P.T., and Lu, X. (2010) High-contrast electrochromic thin films via layer-by-layer assembly of star-like and sulfonated polyaniline. *Chem. Mater.*, **22**, 6085–6091.
16. DeLongchamp, D.M. and Hammond, P.T. (2004) Multiple-color electrochromism from layer-by-layer-assembled polyaniline/Prussian blue nanocomposite thin films. *Chem. Mater.*, **16**, 4799–4805.
17. Yang, Z., Pu, H., and Yin, J. (2005) Preparation and electrochromic property of covalently bonded WO_3/polyvinylimidazole core-shell microspheres. *J. Colloid Interface Sci.*, **292**, 108–112.
18. Oliveira, R.S.D., Alves, W.A., and Ponzio, E.A. (2012) Spectroelectrochemical study of the hybrid between vanadium oxide and carboxybenzylviologen for application in electrochromic electrodes. *ECS Trans.*, **43** (1), 363–369.
19. Green, S.V., Pehlivan, E., Granqvist, C.G., and Niklasson, G.A. (2012) Electrochromism in sputter deposited nickel-containing tungsten oxide films. *Sol. Energy Mater. Sol. Cells*, **99**, 339–344.
20. Yoshino, T., Kobayashi, K., Araki, S., Nakamura, K., and Kobayashi, N. (2012) Electrochromic properties of electrochemically fabricated nanostructure nickel oxide and manganese oxide films. *Sol. Energy Mater. Sol. Cells*, **99**, 43–49.
21. Sonavane, A.C., Inamdara, A.I., Dalavi, D.S., Deshmukh, H.P., and Patil, P.S. (2010) Simple and rapid synthesis of NiO/PPy thin films with improved electrochromic performance. *Electrochim. Acta*, **55**, 2344–2351.
22. Kateb, M., Ahmadi, V., and Mohseni, M. (2013) Fast switching and high contrast electrochromic device based on PEDOT nanotube grown on ZnO nanowires. *Sol. Energy Mater. Sol. Cells*, **112**, 57–64.
23. Zhang, J., Tu, J., Zhang, D., Qiao, Y., Xia, X., Wang, X., and Gu, C. (2011) Multicolor electrochromic polyaniline-WO_3 hybrid thin films: one-pot molecular assembling synthesis. *J. Mater. Chem.*, **21**, 17316.
24. Zhang, J., Tu, J., Dua, G., Dong, Z., Wu, Y., Chang, L., Xie, D., Cai, G., and Wang, X. (2013) Ultra-thin WO_3 nanorod embedded polyaniline composite thin film: synthesis and electrochromic characteristics. *Sol. Energy Mater. Sol. Cells*, **114**, 31–37.
25. Nossol, E. and Zarbin, A.J.G. (2013) Electrochromic properties of carbon nanotubes/Prussian blue nanocomposite films. *Sol. Energy Mater. Sol. Cells*, **109**, 40–46.

26. Benoit, A., Paramasivam, I., Nah, Y.-C., Roy, P., and Schmuki, P. (2009) Decoration of TiO_2 nanotube layers with WO_3 nanocrystals for high-electrochromic activity. *Electrochem. Commun.*, **11**, 728–732.
27. Cai, G.F., Zhou, D., Xiong, Q.Q., Zhang, J.H., Wang, X.L., Gu, C.D., and Tu, J.P. (2013) Efficient electrochromic materials based on TiO_2@WO_3 core/shell nanorod arrays. *Sol. Energy Mater. Sol. Cells*, **117**, 231–238.
28. Lin, C.-K., Tseng, S.-C., Cheng, C.-H., Chen, C.-Y., and Chen, C.-C. (2011) Electrochromic performance of hybrid tungsten oxide films with multiwalled-CNT additions. *Thin Solid Films*, **520**, 1375–1378.
29. Park, K.-W., Song, Y.-J., Lee, J.-M., and Han, S.-B. (2007) Influence of Pt and Au nanophases on electrochromism of WO_3 in nanostructure thin-film electrodes. *Electrochem. Commun.*, **9**, 2111–2115.
30. Ma, L., Li, Y., Yu, X., Yang, Q., and Noh, C.-H. (2008) Using room temperature ionic liquid to fabricate PEDOT/TiO_2 nanocomposite electrode-based electrochromic devices with enhanced long-term stability. *Sol. Energy Mater. Sol. Cells*, **92**, 1253–1259.
31. Saxena, A.P., Deepa, M., Joshi, A.G., Bhandari, S., and Srivastava, A.K. (2011) Poly(3,4-ethylenedioxythiophene)-ionic liquid functionalized graphene/reduced graphene oxide nanostructures: improved conduction and electrochromism. *ACS Appl. Mater. Interfaces*, **3**, 1115–1126.
32. Sheng, K., Bai, H., Sun, Y., Li, C., and Shi, G. (2011) Layer-by-layer assembly of graphene/polyaniline multilayer films and their application for electrochromic devices. *Polymer*, **52**, 5567–5572.
33. Bhandari, S., Deepa, M., Sharma, S.N., Joshi, A.G., Srivastava, A.K., and Kant, R. (2010) Charge transport and electrochromism in novel nanocomposite films of poly(3,4-ethylenedioxythiophene)-Au nanoparticles-CdSe quantum dots. *J. Phys. Chem. C*, **114**, 14606–14613.
34. Xiong, S., Phua, S.L., Dunn, B.S., Ma, J., and Lu, X. (2010) Covalently bonded polyaniline-TiO2 hybrids: a facile approach to highly stable anodic electrochromic materials with low oxidation potentials. *Chem. Mater.*, **22**, 255–260.
35. Jia, P., Argun, A.A., Xu, J., Xiong, S., Ma, J., Hammond, P.T., and Lu, X. (2009) Enhanced electrochromic switching in multilayer thin films of polyaniline-tethered silsesquioxane nanocage. *Chem. Mater.*, **21**, 4434–4441.
36. Frackowiak, E. and Beguin, F. (2001) Carbon materials for the electrochemical storage of energy in capacitors. *Carbon*, **39**, 937–950.
37. Conway, B.E. (1999) *Electrochemical Supercapacitors: Scientific Fundamentals and Technological Applications*, Springer.
38. Pandolfo, A.G. and Hollenkamp, A.F. (2006) Carbon properties and their role in supercapacitors. *J. Power Sources*, **157**, 11–27.
39. Conway, B.E., Birss, V., and Wojtowicz, J. (1997) The role and utilization of pseudocapacitance for energy storage by supercapacitors. *J. Power Sources*, **66**, 1–14.
40. Sorensen, E.M., Barry, S.J., Jung, H.K., Rondinelli, J.R., Vaughey, J.T., and Poeppelmeier, K.R. (2006) Three-dimensionally ordered macroporous $Li_4Ti_5O_{12}$: effect of wall structure on electrochemical properties. *Chem. Mater.*, **18**, 482–489.
41. Lindstrom, H., Sodergren, S., Solbrand, A., Rensmo, H., Hjelm, J., Hagfeldt, A., and Lindquist, S.-E. (1997) Li^+ ion insertion in TiO_2 (anatase). 2. Voltammetry on nanoporous films. *J. Phys. Chem. B*, **101**, 7717–7722.
42. Baeck, S.H., Jaramillo, T.F., Stucky, G.D., and McFarland, E.W. (2003) Synthesis of tungsten oxide on copper surfaces by electroless deposition. *Chem. Mater.*, **15**, 3411–3413.
43. Sasi, B. and Gopchandran, K.G. (2007) Nanostructured mesoporous nickel oxide thin films. *Nanotechnology*, **18**, 115613.
44. Cheng, W., Baudrin, E., Dunn, B., and Zink, J.I. (2001) Synthesis and electrochromic properties of mesoporous tungsten oxide. *J. Mater. Chem.*, **11**, 92–97.
45. Khoo, E., Wang, J.M., Ma, J., and Lee, P.S. (2010) Electrochemical energy storage in a beta-$Na_{0.33}V_2O_5$

nanobelt network and its application for supercapacitors. *J. Mater. Chem.*, **20**, 8368–8374.

46. Baeck, S.H., Choi, K.S., Jaramillo, T.F., Stucky, G.D., and McFarland, E.W. (2003) Enhancement of photocatalytic and electrochromic properties of electrochemically fabricated mesoporous WO_3 thin films. *Adv. Mater.*, **15**, 1269–1273.
47. Hu, J., Abdelsalam, M., Bartlett, P., Cole, R., Sugawara, Y., Baumberg, J., Mahajan, S., and Denuault, G. (2009) Electrodeposition of highly ordered macroporous iridium oxide through self-assembled colloidal templates. *J. Mater. Chem.*, **19**, 3855–3858.
48. Balaji, S., Djaoued, Y., Albert, A.S., Ferguson, R.Z., and Bruning, R. (2009) Hexagonal tungsten oxide based electrochromic devices: spectroscopic evidence for the Li ion occupancy of four-coordinated square windows. *Chem. Mater.*, **21**, 1381–1389.
49. Deepa, M., Srivastava, A.K., Sood, K.N., and Agnihotry, S.A. (2006) Nanostructured mesoporous tungsten oxide films with fast kinetics for electrochromic smart windows. *Nanotechnology*, **17**, 2625–2630.
50. Deshpande, R., Lee, S.H., Mahan, A.H., Parilla, P.A., Jones, K.M., Norman, A.G., To, B., Blackburn, J.L., Mitra, S., and Dillon, A.C. (2007) Optimization of crystalline tungsten oxide nanoparticles for improved electrochromic applications. *Solid State Ionics*, **178**, 895–900.
51. Song, Y.Y., Gao, Z.D., Wang, J.H., Xia, X.H., and Lynch, R. (2011) Multistage coloring electrochromic device based on TiO_2 nanotube arrays modified with WO_3 nanoparticles. *Adv. Funct. Mater.*, **21**, 1941–1946.
52. Lee, S.H., Deshpande, R., Parilla, P.A., Jones, K.M., To, B., Mahan, A.H., and Dillon, A.C. (2006) Crystalline WO_3 nanoparticles for highly improved electrochromic applications. *Adv. Mater.*, **18**, 763–766.
53. Wang, J.M., Khoo, E., Lee, P.S., and Ma, J. (2009) Controlled synthesis of WO_3 nanorods and their electrochromic properties in H_2SO_4 electrolyte. *J. Phys. Chem. C*, **113**, 9655–9658.
54. Devan, R.S., Gao, S.Y., Ho, W.D., Lin, J.H., Ma, Y.R., Patil, P.S., and Liou, Y. (2011) Electrochromic properties of large-area and high-density arrays of transparent one-dimensional beta-Ta_2O_5 nanorods on indium-tin-oxide thin-films. *Appl. Phys. Lett.*, **98**, 133117.
55. Park, S.Y., Lee, J.M., Noh, C., and Son, S.U. (2009) Colloidal approach for tungsten oxide nanorod-based electrochromic systems with highly improved response times and color efficiencies. *J. Mater. Chem.*, **19**, 7959–7964.
56. Gubbala, S., Thangala, J., and Sunkara, M.K. (2007) Nanowire-based electrochromic devices. *Sol. Energy Mater. Sol. Cells*, **91**, 813–820.
57. Liao, C.C., Chen, F.R., and Kai, J.J. (2007) Annealing effect on electrochromic properties of tungsten oxide nanowires. *Sol. Energy Mater. Sol. Cells*, **91**, 1258–1266.
58. Yoo, S.J., Lim, J.W., Sung, Y.E., Jung, Y.H., Choi, H.G., and Kim, D.K. (2007) Fast switchable electrochromic properties of tungsten oxide nanowire bundles. *Appl. Phys. Lett.*, **90**, 173126.
59. Xiong, C.R., Aliev, A.E., Gnade, B., and Balkus, K.J. (2008) Fabrication of silver vanadium oxide and V_2O_5 nanowires for electrochromics. *ACS Nano*, **2**, 293–301.
60. Lin, S.H., Chen, F.R., and Kai, J.J. (2008) Electrochromic properties of nano-structured nickel oxide thin film prepared by spray pyrolysis method. *Appl. Surf. Sci.*, **254**, 2017–2022.
61. Cheng, K.-C., Chen, F.-R., and Kai, J.-J. (2007) Electrochromic property of nano-composite Prussian blue based thin film. *Electrochim. Acta*, **52**, 3330–3335.
62. Shim, H.S., Kim, J.W., Sung, Y.E., and Kim, W.B. (2009) Electrochromic properties of tungsten oxide nanowires fabricated by electrospinning method. *Sol. Energy Mater. Sol. Cells*, **93**, 2062–2068.
63. Zhang, J., Wang, X.L., Xia, X.H., Gu, C.D., and Tu, J.P. (2011) Electrochromic behavior of WO_3 nanotree films prepared by hydrothermal oxidation. *Sol. Energy Mater. Sol. Cells*, **95**, 2107–2112.
64. Wang, J.M., Khoo, E., Lee, P.S., and Ma, J. (2008) Synthesis, assembly, and

electrochromic properties of uniform crystalline WO_3 nanorods. *J. Phys. Chem. C*, **112**, 14306–14312.

65. Meulenkamp, E.A. (1997) Mechanism of WO_3 electrodeposition from peroxy-tungstate solution. *J. Electrochem. Soc.*, **144**, 1664–1671.

66. Hsu, C.S., Lin, C.K., Chan, C.C., Chang, C.C., and Tsay, C.Y. (2006) Preparation and characterization of nanocrystalline porous TiO_2/WO_3 composite thin films. *Thin Solid Films*, **494**, 228–233.

67. Jiang, F.S., Zheng, T., and Yang, Y. (2008) Preparation and electrochromic properties of tungsten oxide and iridium oxide porous films. *J. Non Cryst. Solids*, **354**, 1290–1293.

68. Monk, P.M.S., Akhtar, S.P., Boutevin, J., and Duffield, J.R. (2001) Toward the tailoring of electrochromic bands of metal–oxide mixtures. *Electrochim. Acta*, **46**, 2091–2096.

69. Melo, L.O., Dragunski, D.C., Avellaneda, C.O., and Pawlicka, A. (2003) Kinetic behavior of WO_3 doped Nb_2O_5 thin films. *Proc. SPIE*, **4986**, 666–672.

70. Cho, S.I. and Lee, S.B. (2008) Fast electrochemistry of conductive polymer nanotubes: synthesis, mechanism, and application. *Acc. Chem. Res.*, **41**, 699–707.

71. Morrin, A., Ngamna, O., O'Malley, E., Kent, N., Moulton, S.E., Wallace, G.G., Smyth, M.R., and Killard, A.J. (2008) The fabrication and characterization of inkjet-printed polyaniline nanoparticle films. *Electrochim. Acta*, **53**, 5092–5099.

72. Wang, K., Wu, H., Meng, Y., Zhang, Y., and Wei, Z. (2012) Integrated energy storage and electrochromic function in one flexible device: an energy storage smart window. *Energy Environ. Sci.*, **5**, 8384–8389.

73. Neff, V.D. (1978) Electrochemical oxidation and reduction of thin-films of Prussian blue. *J. Electrochem. Soc.*, **125**, 886–887.

74. Hara, S., Shiozaki, H., Omura, A., Tanaka, H., Kawamoto, T., Tokumoto, M., Yamada, M., Gotoh, A., Kurihara, M., and Sakamoto, M. (2008) Color-switchable glass and display devices fabricated by liquid processes with electrochromic nanoparticle "ink". *Appl. Phys. Express*, **1**, 104002.

75. Shiozaki, H., Kawamoto, T., Tanaka, H., Hara, S., Tokumoto, M., Gotoh, A., Satoh, T., Ishizaki, M., Kurihara, M., and Sakamoto, M. (2008) Electrochromic thin film fabricated using a water-dispersible ink of Prussian blue nanoparticles. *Jpn. J. Appl. Phys.*, **47**, 1242–1244.

76. Wang, J., Sun, X., and Jiao, Z. (2010) Application of nanostructures in electrochromic materials and devices: recent progress. *Materials*, **3**, 5029–5053.

77. Ryu, J.-H., Shin, D.-O., and Suh, K.-D. (2005) Preparation of a reflective-type electrochromic device based on monodisperse, micrometer-size-range polymeric microspheres and viologen pendants. *J. Polym. Sci., Part A: Polym. Chem.*, **43**, 6562–6572.

78. Cao, L., Mou, M., and Wang, Y. (2009) Hyperbranched and viologen-functionalized polyglycerols: preparation, photo- and electrochromic performance. *J. Mater. Chem.*, **19**, 3412–3418.

79. Elzanowska, H., Miasek, E., and Birss, V.I. (2008) Electrochemical formation of Ir oxide/polyaniline composite films. *Electrochim. Acta*, **53**, 2706–2715.

80. Deepa, M., Srivastava, A.K., Sood, K.N., and Murugan, A.V. (2008) Nanostructured tungsten oxide-poly(3,4-ethylenedioxythiophene): poly(styrenesulfonate) hybrid films: synthesis, electrochromic response, and durability characteristics. *J. Electrochem. Soc.*, **155**, D703–D710.

81. Ling, H., Lu, J., Phua, S., Liu, H., Liu, L., Huang, Y., Mandler, D., Lee, P.S., and Lu, X. (2014) One-pot sequential electrochemical deposition of multilayer poly(3,4-ethylenedioxythiophene):poly(4-styrenesulfonic acid)/tungsten trioxide hybrid films and their enhanced electrochromic properties. *J. Mater. Chem. A.*, **2**, 2708–2717.

82. Xia, X.H., Tu, J.P., Zhang, J., Huang, X.H., Wang, X.L., Zhang, W.K., and Huang, H. (2009) Multicolor and fast electrochromism of nanoporous NiO/poly(3,4-ethylenedioxythiophene)

composite thin film. *Electrochem. Commun.*, **11**, 702–705.
83. Hong, S.-F., Hwang, S.-C., and Chen, L.-C. (2008) Deposition-order-dependent polyelectrochromic and redox behaviors of the polyaniline–Prussian blue bilayer. *Electrochim. Acta*, **53**, 6215–6227.
84. Pahal, S., Deepa, M., Bhandari, S., Sood, K.N., and Srivastava, A.K. (2010) Electrochromism and redox switching of cobalt hexacyanoferrate–polyaniline hybrid films in a hydrophobic ionic liquid. *Sol. Energy Mater. Sol. Cells*, **94**, 1064–1075.
85. Hong, S.-F. and Chen, L.-C. (2012) Nano-Prussian blue analogue/PEDOT:PSS composites for electrochromic windows. *Sol. Energy Mater. Sol. Cells*, **104**, 64–74.
86. Girotto, E.M. and Paoli, M.-A.D. (1998) Polypyrrole color modulation and electrochromic contrast enhancement by doping with a dye. *Adv. Mater.*, **10**, 790–793.
87. Bhandari, S., Deepa, M., Srivastava, A.K., Joshi, A.G., and Kant, R. (2009) Poly(3,4-ethylenedioxythiophene)-multiwalled carbon nanotube composite films: structure-directed amplified electrochromic response and improved redox activity. *J. Phys. Chem. B*, **113**, 9416–9428.
88. Reddy, B.N., Deepa, M., Joshi, A.G., and Srivastava, A.K. (2011) Poly(3,4-Ethylenedioxypyrrole) enwrapped by reduced graphene oxide: how conduction behavior at nanolevel leads to increased electrochemical activity. *J. Phys. Chem. C*, **115**, 18354–18365.
89. Deepa, M., Kharkwal, A., Joshi, A.G., and Srivastava, A.K. (2011) Charge transport and electrochemical response of poly(3,4-ethylenedioxypyrrole) films improved by noble-metal nanoparticles. *J. Phys. Chem. B*, **115**, 7321–7331.
90. Hwang, T., Lee, H., Kim, H., Kim, G., and Mun, G. (2010) Enhancement of electrochromic durability of a film made of silica-polyaniline core-shell nanoparticles. *Surf. Rev. Lett.*, **17**, 39–44.
91. Lee, J.S., Choi, Y.-J., Park, H.-H., and Pyun, J.C. (2011) Electrochromic properties of poly(3,4-ethylenedioxythiophene) nanocomposite film containing SiO_2 nanoparticles. *J. Appl. Polym. Sci.*, **122**, 3080–3085.
92. Ma, L., Li, Y., Yu, X., Zhu, N., Yang, Q., and Noh, C.-H. (2008) Electrochemical preparation of PMeT/TiO_2 nanocomposite electrochromic electrodes with enhanced long-term stability. *J. Solid State Electrochem.*, **12**, 1503–1509.
93. Wei, H., Yan, X., Li, Y., Wu, S., Wang, A., Wei, S., and Guo, Z. (2012) Hybrid electrochromic fluorescent poly(DNTD)/CdSe@ZnS composite films. *J. Phys. Chem. C*, **116**, 4500–4510.
94. Xiong, S., Xiao, Y., Ma, J., Zhang, L., and Lu, X. (2007) Enhancement of electrochromic contrast by tethering conjugated polymer chains onto polyhedral oligomeric silsesquioxane nanocages. *Macromol. Rapid Commun.*, **28**, 281–285.
95. Xiong, S., Yang, F., Ding, G., Mya, K.Y., Ma, J., and Lu, X. (2012) Covalent bonding of polyaniline on fullerene: enhanced electrical, ionic conductivities and electrochromic performances. *Electrochim. Acta*, **67**, 194–200.
96. Xiong, S., Wei, J., Jia, P., Yang, L., Ma, J., and Lu, X. (2011) Water-processable polyaniline with covalently bonded single-walled carbon nanotubes: enhanced electrochromic properties and impedance analysis. *ACS Appl. Mater. Interfaces*, **3**, 782–788.

10
Advances in Polymer Electrolytes for Electrochromic Applications

Alice Lee-Sie Eh, Xuehong Lu, and Pooi See Lee

10.1
Introduction

Electrochromic (EC) devices operate as rechargeable electrochemical cells, comprising at least two electrodes that are separated by an electrolyte. At least one of the two electrodes is coated with an EC material that is an anodically or cathodically colouring electrochrome. Colour switching takes place on charge/discharge by application of an appropriate electrical potential. EC devices are designed to operate in absorptive, transmissive or reflective modes, depending on the requirements of the EC applications. The EC materials and electrolytes are the most essential active components operating synergistically. Without a good electrolyte, even a good EC material cannot operate well. Electrolyte provides an indispensable role as the prime ionic conduction medium between the electrodes, while preventing electron conduction between the two electrodes during EC operation. The important electrolyte properties greatly affecting the electrochromic performance are the electrolyte ionic conductivity, ion dissociation, transport rate of ion through bulk and interface and thermal stability [1].

Electrolytes can be in solid, liquid or gel form. The mobile ions that can be transported through the electrolyte layer in contact with the EC film can be easily prepared by mixing salts such as lithium perchlorate ($LiClO_4$), lithium hexafluorophosphate ($LiPF_6$) and lithium trifluoromethanesulfonate ($LiCF_3SO_3$) in a polar solvent (water, propylene carbonate (PC), acetonitrile, etc.) to obtain dissociated cations and anions [1–3]. These must allow simultaneous ionic uptake at each electrode. In general, electrolytes can be classified into polymer electrolytes, liquid electrolytes, ceramic electrolytes and solid inorganic electrolytes [4–8]. A number of review articles on different types of electrolytes for electrochemical applications, especially in the field of lithium ion batteries and fuel cells, give a comprehensive overview of the detailed classification of electrolytes, their strengths/weaknesses, materials designing aspects and conventional techniques employed for material, structure, thermal and ion transport characterisation studies [9, 10]. Liquid electrolytes possessing high ionic conductivity have several inevitable drawbacks such as the possibility of electrolyte leakage, low chemical

Electrochromic Materials and Devices, First Edition.
Edited by Roger J. Mortimer, David R. Rosseinsky and Paul M. S. Monk.

stability, hydrostatic pressure considerations, difficulty in assured sealing and are unsafe for practical applications especially in scaling-up processes [11]. This chapter illustrates the most significant aspects of lithium polymer electrolytes from solid polymer electrolytes (SPEs) to gel polymer electrolytes (GPEs) that are extensively used in the EC applications.

10.2 Requirements of Polymer Electrolytes in Electrochromic Applications

Transference number: The lithium ion transference number is important in the characterisation of polymer electrolytes, but this is rarely determined in the studies and characterisation of polymer electrolytes. Being the fraction of current carried by the lithium ions, it should be as high as possible. The transference number analysis can be performed by combining the direct current (DC) polarisation method and alternating current (AC) impedance measurement for devices made of solid electrolyte films sandwiched between two Li foils [12]. Transference number analysis of Li^+ cation is used to characterise the dissolution behaviour of Li salt in all the electrolytes to be used.

High ionic conductivity: Polymer electrolytes should possess good ionic conductivity and be electronically insulating ($\sigma_e < 10^{-12}$ S cm^{-1}) in order to facilitate ion transport and to minimise self-discharge. Colour change in an EC device involves the movement of ions into and out of an EC film. Higher ionic conductivity can facilitate faster switching of electrochromic states.

Functional separator: In the absence of polymer electrolytes, a porous separator is required to prevent short circuit between the electrodes. Polymer electrolytes can act as separators in EC devices, making these devices light and compact without the need for additional separators. Binding effect is another desirable property of polymer electrolytes which can aid in facilitating good electrical contact with the electrodes/good adherence to the adjoining EC layers, leading to compact device fabrication.

Robustness (mechanical stability, thermal stability, chemical and electrochemical stability): Polymer electrolytes should be robust against electrical, mechanical and thermal abuses. Good thermal stability ensures that the inevitable heat released during the functioning of the EC device does not result in the degradation of the device itself. All polymer electrolytes must possess suitable electrochemical stability within the operational voltage range that is used for switching the EC device. Polymer electrolytes chosen must have good chemical stability so as to prevent undesired chemical reactions especially during the assembly of the EC devices.

Facile processing: Applications of polymer electrolytes in the EC devices are highly driven by their ease of processing due to their flexible film-forming ability. Polymer-electrolyte films of desired thickness can be controlled and formed into very thin films, making the EC devices very light and compact.

Wide range of operating temperatures: Polymer electrolytes exhibit the ability to operate over a wide range of temperatures compared to alternatives.

Wide potential window: To ensure the inclusive stability of the EC devices upon cycling, we should consider the range of potentials used in the electrochemical window of the electrolyte. The ideal polymer electrolyte should have a wide potential window in which cycling processes can proceed without any side reactions.

High transparency in the wavelength used: Transparent polymer electrolytes are highly desirable for EC applications. Highly transmissive polymer electrolytes maximise the transparency of an EC device in the bleached state.

Low volatility: Polymer electrolytes should not be volatile. All organic solvents used during the preparation of the polymer electrolytes need to be finally removed. This is especially critical in ensuring long-term stability of an EC device as slight seepage of the polymer electrolytes will affect electrochromism leading to degradation of the EC devices.

Cation coordination ability: Polymers with good cation coordination ability are favoured as this allows for coordination bonding interactions with the metal cations.

Bond rotation: Polymers with low bond rotation energy are favoured in allowing polymer segmental motion to occur readily as the ionic conduction mechanism of polymer electrolytes is closely associated with the local segmental motions of the polymer.

Low cost and environmentally neutrality: Inexpensive, environmental friendly and readily available polymer electrolytes are highly desirable for the successful commercialisation and implementation of EC devices.

During the synthesis of polymer electrolyte, we have to take into account the prerequisites for the intended EC applications. For instance, smart windows are judged by their transmittance modulation range in the visible and whole solar spectrum, expected lifetime without (or only minor) degradation, reasonable switching time for coloration and bleaching, need to achieve window size, total energy consumption, operating voltage and operating temperature range [13]. Based on the specific requirements of the smart windows, we can tailor an ideal polymer electrolyte to complement the EC materials used.

10.3 Types of Polymer Electrolytes

Polymer electrolytes are typically made up of a dispersion of salt in a neutral polymer matrix [4–10, 14, 15]. The properties of the doping salt are important during the formulation of polymer electrolytes. Polymer electrolytes that are used in EC applications can be classified into four groups, namely SPEs, GELs, semi-SPEs, polyelectrolytes and composite polymer electrolytes (CPEs).

10.3.1 Solid Polymer Electrolytes (SPEs)

SPEs are defined as solvent-free systems that are formed by dissolving salts in a polar polymer matrix resulting in an ionic conducting phase [5, 6]. Lithium salts that are often employed in SPEs are lithium perchlorate ($LiClO_4$), lithium tetrafluorophosphate ($LiPF_4$), lithium hexafluorophosphate ($LiPF_6$), lithium methanesulfonate ($LiCF_3SO_3$), lithium bis(trifluoromethylsulfonyl)imide ($LiN(SO_2CF_3)_2$) and lithium bis(oxalate)borate ($LiB(C_2O_4)_2$) dissolved in polymer hosts such as poly(vinylidene fluoride) (PVDF), poly(methyl methacrylate) (PMMA), polyethylene oxide (PEO), poly(acrylonitrile) PAN and so on. Electrolyte leakage, considerable weight gain from electrolyte solution and the low chemical stability of the liquid electrolyte solution are sometimes drawbacks for electrochromic applications [2]. In fact, the world's best-selling EC device, the Gentex rear-view mirror actually employed liquid electrolyte in its EC system. Nevertheless, SPEs offer chemical and electrochemical stability, flexibility, mechanical robustness and ease of scalability. As has been stated earlier, the ionic conductivity mechanism is closely associated with the local segmental motions of the polymer and takes place primarily in the amorphous phase [4–6]. The ionic conduction mechanism is based on the creation of free volume in the amorphous phase due to the polymer chain's dynamics. The population of ions and electrons in the systems must be large so as to avoid problems with migration and ohmic resistance [16]. However, the ionic conductivity of these SPEs is reported in the range of 10^{-9} to 10^{-5} S cm^{-1} at room temperature which is too low for practical EC applications [2]. Conductivity greater than 10^{-4} S cm^{-1} at room temperature is typically desirable to minimise the ohmic drop across the devices.

10.3.2 Gel Polymer Electrolytes (GPEs)

GPEs have relatively higher ambient-temperature ionic conductivity compared with SPEs. GPEs are usually synthesised by incorporating liquid plasticiser and/or solvents within the polymer matrix that are capable of forming a stable gel with the polymer host structure [8]. The GPE is a two-phase system composed of ionically conducting medium entrapped in a polymer matrix host. The saturated polymer electrolyte consists of one or more salts dissolved in one or more solvents. GPEs possess both the cohesive properties of solids and the diffusive transport properties of liquids owing to their unique hybrid network structures [5]. Soft morphology, poor mechanical property and considerable viscosity of GPEs render them unsuitable for certain manufacturing processes due to safety concerns arising from internal short circuits. Additionally, GPEs tend to degrade upon extensive heating and UV irradiation. The poor mechanical property of GPEs can be ameliorated by the addition of plasticiser. Plasticisers such as PC, electrochromic (EC), diethyl carbonate (DEC) and dimethyl carbonate (DMC) are added to adjust the physical characteristics of the overall blend. Plasticisers improve the ionic conductivity by (i) increasing the amorphous phase content, (ii) dissociating ion aggregates, (iii) increasing ionic mobility within the gel electrolytes and (iv) lowering the glass transition temperature (T_g) of the system [5, 16]. In order to achieve these improvements, the plasticisers must also exhibit a high dielectric constant and low vapour pressure. Cross-linking materials can also be added to improve the mechanical property of gel electrolytes via UV, thermal, photo or electron beam radiations. However, a compromise needs to be established between the increase in ionic conductivity and the decrease in mechanical strength while plasticisation occurs. The ionic conductivity is strongly influenced by the presence of immobilised solvent in the polymer matrix. The polymer electrolytes act as the conduction medium but at the same time, act as the stiffener for the solvent which solvates the lithium salts. There have been extensive investigations of polymer matrix materials in recent years, such as PEO, PMMA, poly(vinylidene fluoride) (PVDF), poly(vinylidene fluoride-*co*-hexafluoropropylene) (PVDF-HFP) and PAN [4–10].

10.3.3
Polyelectrolytes

Polyelectrolytes differ from the SPEs and GELs. Polyelectrolytes are polymers that bear ionised units where the charged functional groups are covalently bonded to the polymer backbone [17]. The ionic conductivity contribution comes from the self ion-generating group. In most cases, the cations are mobile whereas the anions are immobile. Perfluorosulfonic ionomers are the most widely applied ionomer electrolytes, which include Nafion and its derivatives. In fact, polyelectrolytes are omnipresent in nature, for example, polysaccharides, proteins, deoxyribonucleic acid (DNA) and ribonucleic acid (RNA). Every repeating unit bears either a positively or a negatively ionisable group. Thus, such polymers form cationic or anionic polyelectrolytes. Polyampholytes are another group of polyelectrolytes which possess both positively and negatively charged monomeric units. Polyelectrolytes are of interest for exploration of novel phenomena and new applications due to their conformation and controllable interactions via a number of parameters. Polymer electrolytes and polyelectrolytes differ in the mobility of anions. In a polymer-electrolyte matrix, both cations and anions from dissolved salts are mobile. However, the anions of polyelectrolytes are attached to the polymer matrix through covalent bonding while the cations are mobile. Therefore, the polyelectrolyte is a single ion conductor and its cation transference number is close to 1.0, a crucial requirement for specific electrochemical applications. In collaboration with the Hammond group [18], our group introduced tetralayer polyelectrolytes that combined electrostatic and hydrogen bonding via layer-by-layer (LbL) assembly in the fabrication of both transmissive and reflective mode EC devices with a modulation of 30–40% in the visible and near-infrared range. The successful fabrication of the LbL-assembled electrolytes is an added advantage to the processing of solid EC devices as it offers complete flexibility, polymeric and solid-state film.

10.3.4 Composite Polymer Electrolytes (CPEs)

CPEs are synthesised by adding inorganic material or ionic liquid to polymer gel electrolytes [7]. This addition is expected to improve electrochemical and mechanical properties of the gel electrolyte, to enhance electrolyte/electrodes compatibilities and to provide better safety measure than GPEs [3]. Addition of inorganic filler materials has been found to improve the properties of GPEs for EC applications [8]. These fillers can be in the form of ionic conductive such as zeolites or neutral such as Al_2O_3, SiO_2 and MgO. The resultant electrolyte's properties are highly dependent on the particle size used and the characteristics of the fillers used. Besides improving the ionic conductivity of the electrolytes, the interfacial stability and mechanical strength are also enhanced in the composites [17]. However, the ionic conductivity at ambient temperature still remains relatively low for practical EC applications. A comprehensive understanding about the dominant conduction mechanism has yet to emerge. A major challenge in developing new CPEs lies in the lack of structure-properties models.

10.4 Polymer Hosts of Interest in Electrochromic Devices

Polymer electrolytes serve to transport ions to and from the electrochromic thin film in EC devices. To reach an acceptable dynamics of EC device, the ionic conductivity should range between 10^{-3} and 10^{-7} S cm^{-1} depending upon the intended EC applications. However, the electrical conductivity must be low preferably below 10^{-12} S cm^{-1}. There have been numerous reports in the literature based on the studies of polymer electrolytes in electrochemical devices, particularly in the development of batteries and fuel cells. This greatly facilitates the investigation of ideal polymer electrolytes in the application of EC devices. Early works on the application of polymer electrolytes in EC devices had been pioneered by Giglia *et al.* [19] and Randin [20]. Some of the most commonly incorporated polymers in electrolytes meant for EC applications are poly(ethylene oxide)/poly(ethylene glycol) (PEG), PMMA, poly(vinylidene fluoride) (PVDF), poly(vinyl alcohol) (PVA), PAN, poly(ethyl methacrylate) (PEMA) and poly(2-ethoxyethyl methacrylate) (PEOEMA), poly(vinyl chloride) (PVC), poly(vinyl sulfones) (PVSs), boronate esters polymers and water-borne polyurethane. In this review chapter, the main interest concentrates on some plasticised systems that have received much more attention from a practical viewpoint in the EC applications. This includes PEO/PEG, PMMA, PVDF and ionic liquid–based polymer electrolytes. Additionally, "green" polymer hosts that are gaining interest in EC applications are also discussed here. In the following sections, we explain the basic properties of the polymer systems and their applications in the EC device.

10.4.1 PEO/PEG-Based Polymer Electrolytes

PEO is one of the most conventional polymers in the synthesis of polymer electrolytes. PEO refers to polymers with a molecular mass above 20 000 g mol^{-1} while PEG refers to oligomers with a molecular mass below 20 000 g mol^{-1}. Both PEO and PEG are available commercially and PEO/PEG with different molecular weights serves for different purposes in different applications. Both have similar chemical properties but differ in terms of physical properties especially viscosity due to the chain length effects. PEO-based electrolyte is the earliest and the most extensively studied host for polymer electrolytes synthesis. Armand introduced PEO-based electrolytes for lithium batteries in 1978 and since then, there has been extensive research in the field of polymer electrolytes [21]. PEO has a polar ether group and hence a significant segmental mobility in the amorphous phase. The ether oxygen has sufficient electron-donor power to form coordinate bonds with cations and to compensate for the lattice energy of the salt. In addition, PEO has a very good film-forming characteristic and can be easily moulded into various sizes and shapes, thus facilitating the facile synthesis of polymer electrolytes in the laboratory at a low cost. However, the drawbacks of PEO include low melting point, limited operating temperature range, a low hydroxide-ion transference number, high crystallinity and, thus, poor interfacial characteristics [17]. Several approaches have been used to increase the ion dissociation in the PEO-based electrolytes such as the use of salts with low lattice energy, the addition of polar plasticiser, and the addition of microscopic fillers that may inhibit the formation of polymer crystals and, at the same time, aid fast ion transport paths via interaction between the fillers and the electrolyte [22].

Desai *et al.* introduced ionic liquid plasticiser in their gel electrolytes in the fabrication of EC devices [23]. The PEO-based GPE was plasticised with ethylene carbonate/PC or *N*-butyl-3-methylpyridinium trifluoromethanesulphonylimide (P_{14}TFSI) for the EC device of polypyrrole (PPy) in the following configuration: indium-tin oxide (ITO)|PPy|GPE|PEDOT|ITO. Lithium perchlorate ($LiClO_4$) and lithium trifluoromethanesulphonylimide (LiTFSI) salts were used in the GPEs to enhance the ionic conductivity and inhibit phase separation of the PEO and plasticiser. The studies demonstrated that the ionic liquid plasticised GPEs exhibited better performance compared to pure ionic liquid–based devices and volatile organic solvent plasticisers based upon EC/PC mixtures. Our group has introduced the novel incorporation of LbL assembled SPE for EC device [18]. The SPE films were assembled via LbL method by using linear poly(ethylene imine) (LPEI), PEO and poly(acrylic acid) (PAA) in the polyaniline with dodecylbenzene sulfonic acid (PANI-DBSA) EC device. The multi-layer polymer electrolyte film comprises four interbonding layers per deposition cycle combining both electrostatic and hydrogen bonding in the same structure. The ionic conductivity obtained from SPE tetralayer films can reach up to 9.1×10^{-4} S cm^{-1}. Tetralayers with low-molecular-weight PEO (20 k) films showed slightly higher ionic conductivity values compared to high-molecular-weight PEO (300 k). Overall, similar ionic conductivities are observed for both in-plane and through-plane measurement for each sample, which indicated isotropic conducting characteristics and good intermixing of the participating polymer and hence eliminated the notion of any stratified layer formation.

10.4.2
PMMA-Based Polymer Electrolytes

PMMA is also known as poly(methyl-2-methylpropenoate) which is a synthetic amorphous polymer of methyl methacrylate. PMMA is a hard thermoplastic, which possesses high light transparency and is more impact resistant than glass. PMMA has received immense attention in EC applications due to its high ionic conductivity, high transparency, excellent compatibility with liquid electrolytes, excellent environmental stability, high solvent retention ability and good gelatinising properties [24]. Other advantages of PMMA are low cost, good outdoor weatherability, good resistance to acid, high surface resistance, compatibility with most polymers, good interfacial stability towards lithium electrodes and high ability to solvate inorganic salts to form complexation between salt and polymer due to a polar functional group in its polymer chain that has high affinity for lithium ions.

Argun *et al.* introduced the very first study on all-polymeric EC devices using poly(3,4-propylenedioxythiophene) (PProDOTMe$_2$) as the cathodically colouring polymer and poly[3,6-bis(2-(3,4-ethylenedioxy)thienyl)-*N*-methylcarbazole] (PBEDOT-*N*-MeCz) as the anodically colouring polymers with PMMA as the polymer matrix [25]. The EC device demonstrated a maximum transmittance change ($\Delta\%T$) of 51% at 540 nm and was highly stable, losing only 5% contrast after 32 000 switches.

Interestingly, Sydam *et al.* synthesised PMMA electrolytes using two different approaches and applied the electrolytes to the EC devices of WO_3 PB and poly(3,4-ethylenedioxythiophene)-Prussian blue (PEDOT-PB) [26]. In the first approach, formation of a homogeneous transparent gel was carried out by immobilisation of PMMA in an ionic liquid. The second approach relied on the *in situ* thermal polymerisation of the methylmethacrylate (MMA) in the hydrophobic ionic liquid, which yielded a solidified transparent gel. The WO_3 PB (MMA)–based device showed much higher coloration efficiency of 119 cm^2 C^{-1} ($\lambda_{max} = 550$ nm) compared to WO_3 PB (PMMA)–based device (coloration efficiency (CE) = 54 cm^2 C^{-1} at the same wavelength). The film microstructure played an important role in facilitating the ion insertion–extraction kinetics and hence the open channels provided by the four/six/three coordinated voids created by the hexagonal crystal structure of the WO_3 film impacted the bleaching kinetics favourably.

10.4.3 PVDF-Based Polymer Electrolytes

Another popular candidate for hosting polymer electrolyte in EC devices is PVDF. PVDF is highly favoured over other polymer materials due to its durability, ease of fabrication, high hydrophobicity, good thermomechanical property, fairly high permittivity, thermal and chemical stabilities and chemical resistance, rendering it suitable for processing EC devices. Due to its semicrystalline nature, PVDF is highly desirable for EC applications as the crystalline phase provides thermal stability while the amorphous phase provides flexibility. In addition, the high dielectric constant of PVDF ($\varepsilon = 8.4$) assists the greater dissociation of the lithium salts, providing higher charge carrier density and a high electrochemical stability window because of the presence of strong electron-withdrawing functional groups (–C–F) that increase the anodic stability, making it a potential candidate in EC applications [5]. Additionally, PVDF is soluble in high-boiling point solvents such as *N*-methyl-2-pyrrolidone (NMP), *N*-dimethylformamide (DMF) and dimethylacetamide (DMAc), making it easier to synthesise suitable polymer electrolyte.

Despite the excellent electrochemical properties offered by the PVDF-based gel electrolytes, they are not very stable towards lithium salts due to a fluorinated polymer host leading to poor interfacial properties between lithium and fluorine [5]. This explains why PVDF has a low donor number (DN) and is expected to have no solvating ability vs. lithium salts. Thus, PVDF cannot be efficiently used in polymer-salt complexes. The use of copolymers is expected to overcome this drawback and increase the polymer electrolytes performance since the amorphous parts would favour the entrapment of a liquid electrolyte while the crystalline parts can help to maintain the mechanical property. It has been observed that electrolyte properties improved substantially when PVDF was copolymerised with HFP in the Bellcore's plastic rechargeable Li-ion batteries applications [27]. PVDF-HFP exhibited greater solubility towards organic solvents besides having lower crystallinity with a reduced glass transition temperature than the pure PVDF polymer in the gel electrolytes. PVDF-HFP [28–30] and poly(vinylidene fluoride-*co*-trifluoroethylene) (PVDF-TrFE) [1] have also been recently identified as interesting candidates for polymer electrolytes in EC applications. Our research group studied the performance of PVDF-TrFE in the polymer electrolyte in both solid and gel forms in the EC device which utilised PANI as the EC materials [1]. The average values of the semisolid and SPEs obtained were only 6.26×10^{-6} and 3.46×10^{-7} S cm^{-1} (Figure 10.1). The reduced ionic conductivity was further explained by prominent effect of the transverse internal interfacial impedance found in the behaviour of the devices with semisolid and SPEs. The ionic conductivity of the gel state devices was dictated by the amount of dissociated ions while ionic movement in the electrolyte bulk and through the electrolyte/EC material interface dictated the amount of light modulation perceived by the semisolid EC devices.

The most striking advancements in the ionic conductivity of PVDF polymer electrolyte have been attained through the incorporation of substantial amounts

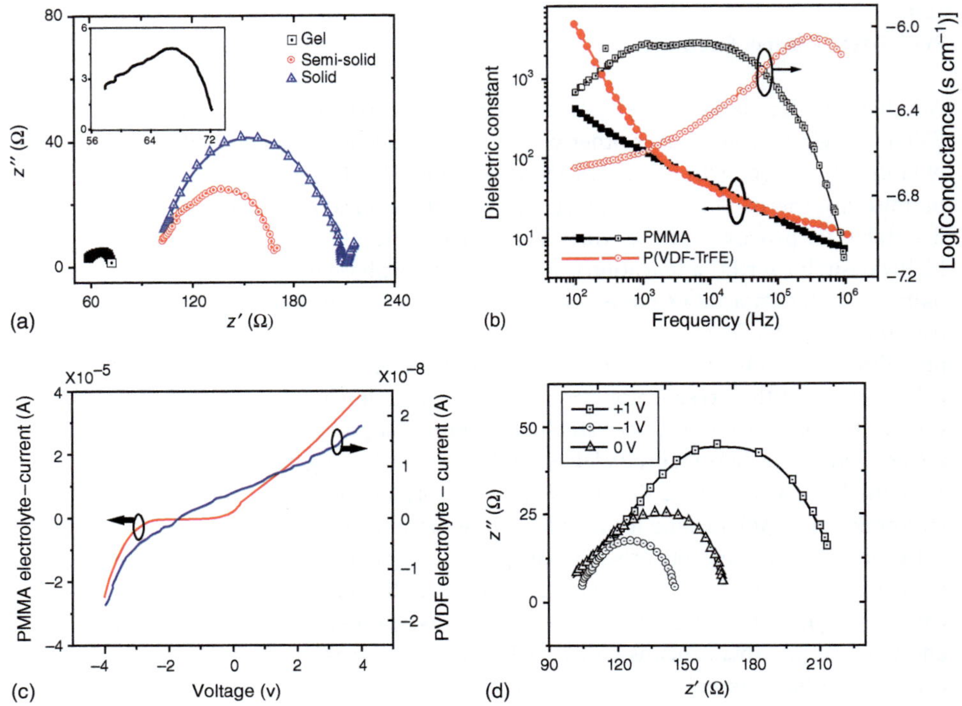

Figure 10.1 (a) Electrochemical impedance spectra of electrolyte with P(VDF-TrFE) host polymer. The inset shows fitting circuit. (b) Dielectric properties of solid PMMA and P(VDF-TrFE) electrolytes. (c) Current leakage data of solid PMMA and P(VDF-TrFE) electrolytes from same samples as (b). (d) Electrochemical impedance spectra of semisolid P(VDF-TrFE) electrolytes under various applied biases. Reprinted with permission from [1]. Copyright 2009 American Chemical Society.

of plasticisers in the formulation. Plasticisers can reduce the crystallinity of the polymer and increase the ionic conductivity of the electrolytes through segmental mobility and higher ion dissociation which allow greater numbers of charge carriers for ionic transport [17]. The most conventional plasticisers used are low molecular weight polyethers and polar organic solvents. Our research group has successfully demonstrated the high ionic conductivity of PVDF–TrFE/PEO blended polymer electrolytes in the solid EC device prepared using PANI–DBSA as anodic materials, poly(3,4-ethylene dioxythiophene)-poly(4-styrene sulfonic acid) (PEDOT:PSS) as the ion storage layer on the counter electrode [12]. We studied the effect of PEO molecular weight on the polymer electrolyte system. From the experimental result, low molecular weight of PEO caused no crystallisation. However, the crystallisation of PEO was manifested with increasing molecular weights of PEO. The use of PEO eliminated the need of incorporating liquid plasticiser while maintaining the high ionic conductivity of the polymer electrolyte (above 10^{-4} S cm^{-1}) under various environmental conditions, namely normal ambient, humid conditions and extremely hot or dry (Figure 10.2).

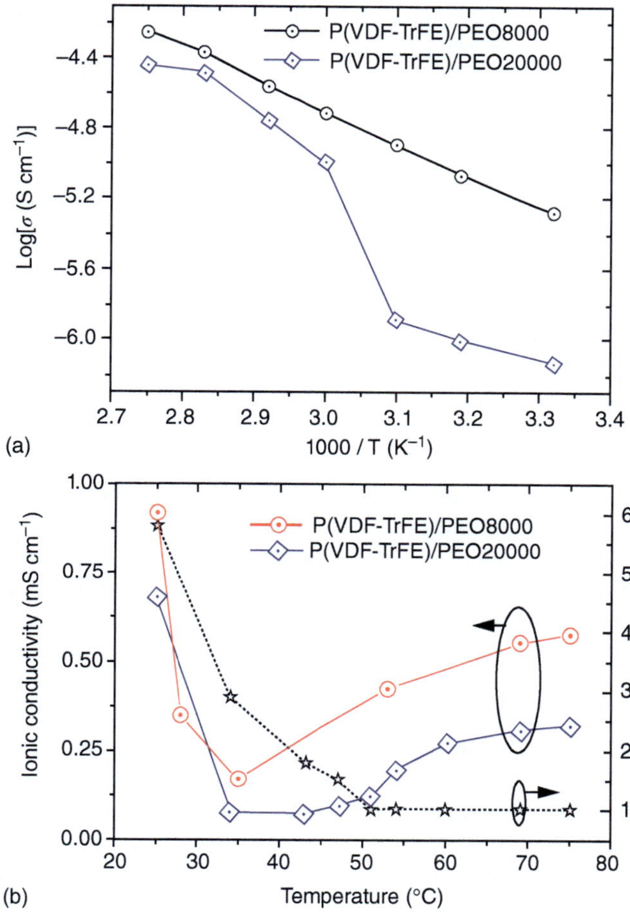

Figure 10.2 Ionic conductivity of P(VDF-TrFE)/PEO8000 and PEO20000 electrolytes measured in (a) controlled chamber with relative humidity (RH) 10% and (b) normal ambient conditions. The right axis in (b) shows the RH value recorded during each measurement [12]. Reproduced by permission of the PCCP Owner Societies.

PVDF–HFP nanofibrous membrane is another promising candidate for transparent electrolytes to be used in the transmissive EC devices. These as-spun PVDF–HFP nanofibrous membranes can be used to host ionic liquids in the electrolytes. However, it cannot retain its shape and form when swollen with a large amount of ionic liquids. By contrast, supercritical carbon dioxide ($SCCO_2$)-treated membranes showed excellent dimensional stability after swelling in ionic liquids as the treatment created interfibre junctions without destroying the nanofibrous morphology, resulting in a remarkable improvement in the mechanical properties of the membrane [29]. The good structural integrity of the ionic liquid–loaded $SCCO_2$-treated membranes allowed controlled uniformity of electrolyte thickness across a large area which is especially beneficial in the processing of a large-area EC device. With ionic liquids, the ionic conductivity of the treated membranes is very close to that of the neat ionic liquids and is nearly 1 order higher than that of the corresponding ionic liquid/PVDF gel due to the ease of ions movements in the interpenetrating pores of the membranes. Jia *et al.* studied the influences of IL-loaded $SCCO_2$-treated electrospun PVDF-HFP membrane as the polymer electrolyte in the transmissive EC device [30]. The EC device showed consistently high optical transmittance and ionic conductivity while maintaining good structural integrity.

10.4.4
Ionic Liquid-Based Polymer Electrolytes

Ionic liquids (ILs) are liquid molten salts at room temperature (<100 °C) and consist of dissociated ions with no intervening solvent [31]. In fact, IL can be used as an electrolyte without the addition of any salt since it is intrinsically ionic conductive. Nevertheless, for some practical applications where the presence of some specific ions is necessary, different kinds of salts (most frequently Li salts are preferred) can be added depending on the requirement [17]. ILs offer several advantages in terms of ionic conductivity, transference number, electrochemical window, as well as physical and chemical properties such as non-volatility, melting point, density, viscosity and observed temperature of solidification [32]. Additionally, ILs offer the possibility to recycle the solvents after usage. Room-temperature ionic liquids (RTILs) are salts of low-temperature melting points either at or below room temperature. RTILs are usually quaternary ammonium salts characterised by negligible vapour pressure, rendering them non-flammable [32]. RTILs exhibit a broad electrochemical stability window (>4 V), rendering them suitable for the application of Li-ion batteries with high-energy cathodes. These RTILs show identical physicochemical properties with high-temperature molten salts, but they are easier to handle during processing due to the liquid nature at room temperature. Consequently, the RTIL-based electrolytes could potentially exhibit higher ionic conductivity than SPEs and better long-term stability than conventional solvent-containing gel electrolytes and have no detrimental effect on the safety aspects of the electrochromic device (ECD)s.

The transparent RTILs that have been used in the electrolytes of transmissive EC devices are mostly aprotic RTILs with organic cations such as imidazolium [33–35], pyridinium [23] or pyrrolidinium [26, 36], as shown in Figure 10.3. Common anions in the RTILs include PF_6^-, BF_4^- and bis(trifluoromethanesulfonyl)imide ($[N(SO_2CF_3)_2]^-$, $TFSI^-$).

RTILs can be directly used as electrolytes for spectroelectrochemistry and switch studies of the electrochromic thin films deposited on transparent electrodes [35]. However, when the RTILs are used as liquid electrolytes in EC devices, to ensure uniform electrolyte thickness and prevent the leakage of the liquid electrolytes, it is necessary to use spacers to regulate the thickness of the electrolytes and then seal the devices tightly with a room-temperature curing or UV-hardening adhesive. Direct electrochemical preparation of the electrochromic

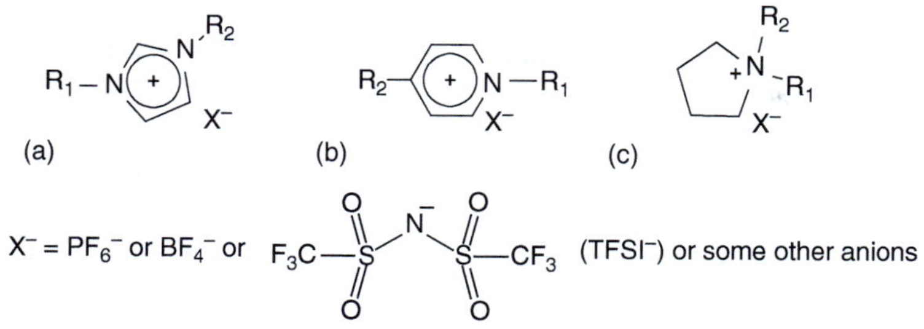

Figure 10.3 Structures of aprotic RTILs with (a) imidazolium, (b) pyridinium and (c) pyrrolidinium cations.

thin films in RTILs could reduce the moisture pick-up by the electrochromic thin films [33, 34]. For example, Ma *et al.* used 1-butyl-3-methyl-imidazolium tetrafluoroborate ([BMIM]BF_4) as both the growth medium in the electrochemical deposition of PEDOT on porous TiO_2/ITO-coated electrodes and the liquid electrolyte in the assembly of the devices [33]. The EC devices were found to exhibit good long-term stability, which could retain up to 95% of their optical response after 10 000 cycles. The RTIL-based electrolytes are more common in the form of gels, which do not need any spacer to regulate the electrolyte thickness, making the device assembly processes much simpler. The polymers that have been used to host the RTILs in ECDs include PEO [23], PVA [36], PMMA [26], PVDF-HFP [37] and poly(propylene carbonate) (PPC) [38]. The use of polymeric ionic liquids (PILs) as the polymer hosts has also been reported [39]. The main advantage of the PILs is that they have good compatibility with the RTILs of similar structures, facilitating the incorporation of the RTILs into the PIL matrices. Marcilla *et al.* mixed 1-butyl-3-methylimidazolium *bis*(trifluoromethanesulfonimide) [bmim][Tf_2N], 1-butyl-3-methylimidazolium tetra-fluoroborate [bmim][BF_4] and 1-butyl-3-methylimidazolium bromide [bmim][Br], with poly[1-vinyl-ethyl-imidazolium] bearing similar counter-anions [Tf_2N], [BF_4] and [Br]. The reported ionic conductivity varied between 10^{-2} and 10^{-5} S cm^{-1} at room temperature and showed higher values of conductivity with increasing IL content. The EC devices demonstrated enhanced cycle life up to 70 000 cycles with the tailor-made polymer electrolytes based upon ILs compared to previously used PEO-Li electrolytes.

In RTIL-based gel electrolytes, the RTILs play the roles of both conductive media and non-volatile plasticisers. Some RTIL-based electrolyte formulations may contain residue solvents as plasticisers, which improve the mobility of the ions and hence the electrochromic performances of the EC devices, whereas the performances of the EC devices may gradually deteriorate as such solvents may slowly diffuse out through polymeric sealant materials. To increase the ionic conductivities of the RTIL-based electrolytes without sacrificing their long-term stability, a possible approach is to incorporate lithium salts into the RTIL-based electrolytes as some RTILs are non-reactive towards some lithium salts [38]. The interactions between lithium ions and RTIL molecules may have positive influences on lithium ion mobility in such RTIL/lithium salt composite electrolytes [40]. More importantly, the size of lithium ions is much smaller than the large cations in RTILs, and hence the mobile lithium ions may diffuse into some electrochromic materials more easily than the large cations.

A drawback of RTIL-based electrolytes is that most RTILs are fairly expensive. Although $SCCO_2$-treated electrospun PVDF-HFP mats have been successfully demonstrated as RTIL hosts in ECDs, such mats only have fairly weak inter-fibre junctions. In order to handle such electrospun mats easily in the EC device fabrication process, the thickness has to be above 20 μm and hence the mats would absorb a considerably large amount of RTILs. To reduce the cost of RTIL-based electrolytes, recently, Lu's group reported a facile method to prepare ultra-thin free-standing electrospun PVDF-HFP mats using 1,3-diaminopropane (DAP) as the cross-linking agent [41]. When cross-linked with 1 wt% DAP, the $[BMIM]BF_4$-loaded electrospun PVDF-HFP mat can be a free-standing mat with good structural integrity at the thickness of 2 μm. It shows a high ionic conductivity of $2.6\,mS\,cm^{-1}$, and the complementary ECD using this mat as the electrolyte layer shows good performance. More importantly, the amount of the RTIL hosted in the ultra-thin free-standing mat is less than $5\,g\,m^{-2}$, which shows the great potential of this approach for practical EC device applications.

10.4.5 Poly(propylene carbonate) (PPC)-Based Polymer Electrolytes

PPC has been attracting attention due to its biodegradable property. With biodegradable polymer electrolyte, the environmental pollution resulted from battery disposal can be alleviated. PPC is an amorphous alternating copolymer, which can be obtained from the copolymerisation of CO_2 and propylene oxide. PPC possesses a structure similar to those of carbonate-based organic solvents that are conventionally used in gel electrolytes. This extends its potential to be used in gel electrolytes as it offers compatibility with lithium salts and good interfacial contact with electrodes in electrochemical devices [38]. PPC-based electrolytes have high transparency due to the amorphous nature of PPC. PPC has a T_g close to room temperature and can thus facilitate the motion of lithium ions in the polymer by incorporating a small amount of ionic liquid. Yu *et al.* first reported on the new $PEO/50\%PPC/10\%LiClO_4$ polymer electrolytes which possess ionic conductivity of $6.83\times10^{-5}\,S\,cm^{-1}$ for lithium-ion battery application at room temperature and good electrochemical stability up to 4.5 V [42]. Zhou *et al.* synthesised solvent-free PPC-based electrolytes ($PPC/LiClO_4/BMIM^+BF_4^-=1/0.2/3$) which exhibited high ionic conductivity of $1.5\,mS\,cm^{-1}$ at room temperature [38]. The PPC-based EC device exhibited good electrochromic performance with high contrast ($\Delta A = 0.74$), short coloration (3.3 s) and bleaching time (10.9 s) at low operation potentials (±2.5 V), demonstrating the potential of PPC in EC devices.

10.5 Recent Trends in Polymer Electrolytes

10.5.1 Flexible, Imprintable, Bendable and Shape-Conformable Polymer Electrolytes

One of the polymer-electrolyte emerging technologies is the capability to obtain imprintable, bendable and shape-conformable electrolyte. It is challenging to secure dimensional stability (in terms of the solid form) of polymer electrolytes during preparation of electrolyte and the cell assembly process without impairing their electrochemical performances. Kil *et al.* had successfully synthesised highly ion-conductive, bendable polymer electrolytes which can also conform to 3D micropatterned architectures of electrodes over large areas [43]. Other advantages offered by the polymer electrolytes were capability of being directly writable or printable onto complex, contoured substrates, owing to the structural uniqueness and well-tuned rheological characteristics. However, the major challenge still lies in the maintenance of good contact between the polymer electrolytes and the electrodes in the development of 3D-structured or flexible devices. For EC applications, the interfaces between electrolyte and the two electrodes in the EC device dictate the overall performance [17]. The adhesion between the electrolytes and the two electrodes must be good in order to have an excellent electrochemical stability. Our group has demonstrated a mechanically robust WO_3 EC device which can be stretched, twisted, folded and crumpled with minimal performance failure [44]. Short coloration (1 s) and bleaching (4 s) times were reported with good cyclic stability (retention of 81% after 100 cycles) at relaxed state. These WO_3 EC devices could also be implanted on the cotton textile substrates, demonstrating their potential applications in wearable electronics. To further enhance their applicability and durability in areas of flexible and wearable EC devices, mechanically robust polymer electrolytes should be employed instead of the proton-based H_2SO_4 electrolyte. Unlike liquid electrolyte, polymer electrolytes are highly advantageous in terms of enhanced protection and stability, can act as a novel separator, wider potential window, chemical and electrochemical stability and confinement to specific area.

10.5.2 Potentially 'Green' Biodegradable Polymer Electrolytes Using Naturally Available Polymer Host

From a 'green' chemistry point of view, many researchers are gaining keen interest in biodegradable polymer electrolytes that have great potential for applications in rechargeable lithium batteries, solar cells, dye-sensitised solar cells, supercapacitors and fuel cells [45]. Many other alternatives/polymer hosts that are used in lithium-ion batteries and electrochemical applications are in fact viable for applications in electrochromics on the basis that the polymer hosts chosen must satisfy/fulfil the requirements of electrochromics applications. The ideal EC devices are aimed at their lightness, durability, high contrast, high energy efficiency at a low cost while preserving the green chemistry idea. Recent attention has been shifted to the use of natural polymers (e.g. hydroxyethyl cellulose, starch, chitosan and gelatin) which offer eco-friendliness, biodegradability, low production cost and good physical and chemical properties. These natural polymers are commonly employed in the cosmetic, pharmaceutical and food industries but can be used in the synthesis of polymer electrolytes as well.

Gelatin is a polypeptide which consists mostly of proline, hydroxyproline and glycine. It can be extracted from denatured collagen and dissolves in boiling water to form a pale yellow, semi-transparent, viscous solution. Avellaneda *et al.* assembled $6 \times 8\,cm^2$ all-solid-state EC device using Nb_2O_5:Mo as the EC layer and CeO_2-TiO_2 as the ion-storage layer with the incorporation of Li-gelatin-glycerol electrolyte [46]. The EC device exhibited a long-term stability of more than 25 000 switching cycles ($\pm 2.0\,V/90\,s$) with a coloration efficiency increasing from $10\,cm^2\,C^{-1}$ (initial cycle) to $23\,cm^2\,C^{-1}$ (25 000th cycle). Agar is a heterogeneous mixture of two polysaccharides: agaropectin and agarose. Raphael *et al.* reported assembly of EC devices (WO_3|agar-$LiClO_4$|CeO_2 TiO_2) using polymer gel electrolytes based on agar and $LiClO_4$ [16]. The results revealed that the $LiClO_4$ concentration influenced the ionic conductivity of electrolytes with optimum ionic conductivity of $6.5 \times 10^{-5}\,S\,cm^{-1}$ at room temperature when employing 17 wt% of $LiClO_4$. The coloration efficiency obtained was $25\,cm^2\,C^{-1}$ during the 5000th chronoamperometric cycles. Cellulose has been used widely in the development of green materials due to its non-toxic nature, environmentally friendly nature, low cost, availability of renewable resources, biocompatibility and biodegradable characteristics. Cellulose has been used to prepare separator in lithium-ion batteries. Lalia *et al.* prepared nanocrystalline cellulose (NCC)-reinforced PVDF-HFP composite mat (via electrospinning method) that was capable of high uptake of liquid electrolyte (1 M LiTFSI + BMPyTFSI) resulting in high conductivity ($4 \times 10^{-4}\,S\,cm^{-1}$ at 30 °C) [47]. Another worth considering polymer matrix is chitosan. Chitosan consists of glucosamine repeat units and can be derived by partial or full deacetylation of chitin, which is obtained from the shells of shrimps, crabs and lobsters. The complexation between chitosan and salt occurred between the lone pair electron of oxygen or nitrogen atoms of chitosan and the conducting charge species of salt. PEO is ideal for cationic solvation due to the presence of the ether group along its backbone. The amorphous chains of chitosan can reduce the crystallinity of PEO while the flexibility of the PEO chains may compensate the rigidity of chitosan [48]. Mobarak *et al.* have studied the potential of using chitosan derivatives, carboxymethyl chitosan as a green polymer host for electrolyte synthesis [49]. The chemical modification of chitosan has the ability to enhance the amorphous properties and thus increase the ionic conductivity of carboxymethyl chitosan–based electrolyte by two-magnitude higher.

Starch has a wide range of applications in food packaging, drug delivery and medicine [45]. Some of the unique properties of starch are excellent thermal and steel adhesion properties, superior biocompatibility, good solubility, high recrystallisation stability of the amorphous phase and good capability of film formation [50]. Polymer electrolytes based on biodegradable rice starch (65 wt%) and lithium iodide (35 wt%) were prepared using a solution casting method. The highest ionic conductivity achieved was $4.68 \times 10^{-5}\ S\ cm^{-1}$ [45]. Rubber-based polymer electrolytes have been claimed to acquire several advantages such as free standing, flexibility and good elasticity [51]. Modified natural rubber is an alternative to the use of ceramic fillers or blends in the synthesis of polymer electrolytes [52]. The modification is intended to lower the glass transition temperature and provide good elasticity and adhesion. The incorporation of plasticisers in the modified natural (ENR-25/ENR-50)/PEO rubbers with EC/PC/$LiCF_3SO_3$ salt solution significantly increased the ionic conductivity to $10^{-4}\ S\ cm^{-1}$ at room temperature. Another green polymer host for polymer electrolyte is nata-de-coco. Radiman and Rifathin applied microwave-assisted phosphorylation as an effective method to prepare an electrolyte membrane from nata-de-coco, which used environmentally friendly chemicals and techniques [53]. Sudhakar *et al.* employed a novel GPE system based on biodegradable renewable resources in the fabrication of their supercapacitors [54]. The biodegradable GPE was synthesised using guar gum as the polymer matrix, $LiClO_4$ as the doping salt and glycerol as the plasticiser. Guar gum is a hydrophilic, non-ionic polysaccharide extracted from the endospermic seed of *Cyamopsis tetragonolobus* plant. All the aforementioned green polymer hosts have the potential to serve as polymer electrolytes meant for electrochromics applications. However, further studies need to be conducted in order to tune the desirable properties for electrochromics applications.

10.6 Future Outlook

10.6.1 Recent Trends in Electrochromic Devices

It is evident that EC devices hold high promise in the 'smart' glass market. According to BCC Research, the global market for smart glass-based products increased from \$883.1 million in 2009 to \$1.2 billion in 2010 and was valued nearly \$1.6 billion in 2011 [55]. The market is forecasted to grow at a compound annual growth rate (CAGR) of 21.6% from 2011 to 2016 and is predicted to reach global revenues of nearly \$4.2 billion in 2016. The transportation and aerospace sector accounted for 91.4% of the smart glass market in 2011 with revenues of \$1.4 billion. This sector is estimated to expand at a CAGR of 22.1% in the next 5 years and shall reach \$3.9 billion by 2016. The Pike Research report on global demand for smart glass stated that the smart glass output will exceed 3 million square metres annually [56]. In 2012, Pike Research estimated that the market value for architectural and transportation smart glass (excluding auto-dimming rear-view mirrors) was $84 million. It is estimated that, by 2020, the market will grow to $693 million worldwide at a CAGR of 30.1%. Smart glass can satisfy both the design aspirations for increased glazing and the push towards lower energy consumption.

EC devices can work concurrently with other electrochemical devices. For instance, when sunlight illuminates dye-sensitised solar cells (DSSCs), they will generate electrical power and turn the EC device into a darker colour state, dimming the light transmitted. The connection polarity of the DSSCs can be switched to regain the lighter colour of the EC device, thus implying the potential application of the self-powered colour-changing system for the next-generation sun glasses and smart windows [57]. Österholm *et al.* demonstrated the efficacies of using PEDOT Type I electrochemical supercapacitors (ESCs) as a power source for a high-contrast and fast-switching EC device [58]. Yang *et al.* demonstrated a self-powered EC device driven by a nanogenerator [59]. The EC material used was WO_3 nanoparticles, which were radio frequency (RF)-sputtered on

Figure 10.4 Important considerations in versatile polymer electrolyte (PE) synthesis.

the fluorinated tin oxide (FTO)/glass. The average visible optical modulation obtained was 15.3% with electrochromic response time of 10 s and coloration efficiency of 58.7 $cm^2\ C^{-1}$ using the PMMA as the polyelectrolyte.

10.6.2
Challenges in Creating Versatile Polymer Electrolytes for EC Devices

The Electrolyte is the key component in any electrochemical device because it determines the voltage and temperature range, charge/discharge rates and stability of the device [58]. In any EC device, the main concerns in versatile polymer electrolyte synthesis are maintaining high ionic conductivity, high optical transparency, high stability, wide potential window, wide range of working temperatures, act as novel separator, ease of processing, good chemical and electrochemical stabilities, good mechanical strength and robustness (Figure 10.4). All in all, these concerns are dependent on the intended application fields and purposes. Due to the recent active development of electrochemical devices, versatile polymer electrolytes should not be restricted to electrochromics applications alone. The polymer electrolytes synthesised should be versatile for various applications such as organic photovoltaic (OPV)/EC devices, supercapacitor/EC devices and nanogenerator/EC devices. With more applications in a single flexible device, more challenges need to be addressed in order to provide fully functional application.

References

1. Nguyen, C.A., Xiong, S.X., Ma, J., Lu, X.H., and Lee, P.S. (2009) Toward electrochromic device using solid electrolyte with polar polymer host. *J. Phys. Chem. B*, **113**, 8006–8010.
2. Su, L.Y., Wang, H., and Lu, Z.H. (1998) All solid-state electrochromic smart window of electrodeposited WO_3 and Prussian blue film with PVC gel electrolyte. *Supramol. Sci.*, **5**, 657–659.
3. Gong, Y.F., Fu, X.K., Zhang, S.P., and Jiang, Q.L. (2007) Preparation of a star network PEG-based gel polymer electrolyte and its application to electrochromic devices. *Chin. J. Chem.*, **25**, 1743–1747.
4. Choudhury, N.A., Sampath, S., and Shukla, A.K. (2009) Hydrogel-polymer electrolytes for electrochemical capacitors: an overview. *Energy Environ. Sci.*, **2**, 55–67.
5. Agrawal, R.C. and Pandey, G.P. (2008) Solid polymer electrolytes: materials designing and all-solid-state battery applications: an overview. *J. Phys. D: Appl. Phys.*, **41**, 223001.
6. Meyer, W.H. (1998) Polymer electrolytes for lithium-ion batteries. *Adv. Mater.*, **10**, 439–448.
7. Stephan, A.M. and Nahm, K.S. (2006) Review on composite polymer electrolytes for lithium batteries. *Polymer*, **47**, 5952–5964.
8. Gonçalves, A., Costa, C., Pereira, S., Correia, N., Silva, M.M., Barbosa, P.C., Rodrigues, L.C., Henriques, I., Martins, R., and Fortunato, E. (2012) Study of electrochromic devices with nanocomposites polymethacrylate hydroxyethylene resin based electrolyte. *Polym. Adv. Technol.*, **23**, 791–795.
9. Dias, F.B., Plomp, L., and Veldhuis, J.B.J. (2000) Trends in polymer electrolytes for secondary lithium batteries. *J. Power Sources*, **88**, 169–191.
10. Pérez, L.C., Brandão, L., Sousa, J.M., and Mendes, A. (2011) Segmented polymer electrolyte membrane fuel cells – a

review. *Renewable Sustainable Energy Rev.*, **15**, 169–185.

11. Wu, T.Y., Li, W.B., Kuo, C.W., Chou, C.F., Liao, J.W., Chen, H.R., and Tseng, C.G. (2013) Study of poly (methyl methacrylate)-based gel electrolyte for electrochromic device. *Int. J. Electrochem. Sci.*, **8**, 10720–10732.
12. Nguyen, C.A., Xiong, S.X., Ma, J., Lu, X.H., and Lee, P.S. (2011) High ionic conductivity P(VDF-TrFE)/PEO blended polymer electrolytes for solid electrochromic devices. *Phys. Chem. Chem. Phys.*, **13**, 13319–13326.
13. Baetens, R., Jelle, B.P., and Gustavsen, A. (2010) Properties, requirements and possibilities of smart windows for dynamic daylight and solar energy control in buildings: a state-of-the-art review. *Sol. Energy Mater. Sol. Cells*, **94**, 87–105.
14. Neburchilov, V., Martin, J., Wang, H., and Zhang, J. (2007) A review of polymer electrolyte membranes for direct methanol fuel cells. *J. Power Sources*, **169**, 221–238.
15. Miller, M. and Bazylak, A. (2011) A review of polymer electrolyte membrane fuel cell stack testing. *J. Power Sources*, **196**, 601–613.
16. Raphael, E., Avellaneda, C.O., Aegerter, M.A., Silva, M.M., and Pawlicka, A. (2012) Agar-based gel electrolyte for electrochromic device application. *Mol. Cryst. Liq. Cryst.*, **554**, 264–272.
17. Thakur, V.K., Ding, G.Q., Ma, J., Lee, P.S., and Lu, X.H. (2012) Hybrid materials and polymer electrolytes for electrochromic device applications. *Adv. Mater.*, **24**, 4071–4096.
18. Nguyen, C.A., Argun, A.A., Hammond, P.T., Lu, X.H., and Lee, P.S. (2011) Layer-by-layer assembled solid polymer electrolyte for electrochromic devices. *Chem. Mater.*, **23**, 2142–2149.
19. Giglia, R.D., Sedlak, J.A. and Lipp, D.W. (1979) Electrochromic devices with polymeric electrolytes. US Patent 4,174,152.
20. Randin, J.P. (1982) Ion-containing polymers as semisolid electrolytes in WO_3-based electrochromic devices. *J. Electrochem. Soc.*, **129**, 1215–1220.
21. Armand, M., Chabagno, J.M. and Duclot, M. (1978) Polymeric solid electrolytes. Second International Meeting on Solid Electrolytes Extended Abstracts, St Andrews, Scotland, September 20–22, 1978.
22. Pitawala, H.M.J.C., Dissanayake, M.A.K.L., and Seneviratne, V.A. (2007) Combined effect of Al_2O_3 nano-fillers and EC plasticizer on ionic conductivity enhancement in the solid polymer electrolyte $(PEO)_9$LiTf. *Solid State Ionics*, **178**, 885–888.
23. Desai, S., Shepherd, R.L., Innis, P.C., Murphy, P., Hall, C., Fabretto, R., and Wallace, G.G. (2011) Gel electrolytes with ionic liquid plasticiser for electrochromic devices. *Electrochim. Acta*, **56**, 4408–4413.
24. Andersson, A.M., Granqvist, C.G., and Stevens, J.R. (1989) Electrochromic Li_xWO_3/polymer laminate/$Li_yV_2O_5$ device: toward an all-solid-state smart window. *Appl. Opt.*, **28**, 3295–3302.
25. Argun, A.A., Cirpan, A., and Reynolds, J.R. (2003) The first truly all-polymer electrochromic devices. *Adv. Mater.*, **16**, 1338–1341.
26. Sydam, R., Deepa, M., and Srivastava, A.K. (2012) Electrochromic device response controlled by an *in situ* polymerized ionic liquid based gel electrolyte. *RSC Adv.*, **2**, 9011–9021.
27. Tarascon, J.M., Gozdz, A.S., Schmutz, C., Shokoohi, F., and Warren, P.C. (1996) Performance of Bellcore's plastic rechargeable Li-ion batteries. *Solid State Ionics*, **86-88**, 49–54.
28. Xiong, S.X., Wei, J., Jia, P.T., Yang, L.P., Ma, J., and Lu, X.H. (2011) Water-processable polyaniline with covalently bonded single-walled carbon nanotubes: enhanced electrochromic properties and impedance analysis. *ACS Appl. Mater. Interfaces*, **3**, 782–788.
29. Yee, W.A., Xiong, S.X., Ding, G.Q., Nguyen, C.A., Lee, P.S., Ma, J., Kotaki, M., Liu, Y., and Lu, X.H. (2010) Supercritical carbon dioxide-treated electrospun poly(vinylidene fluoride) nanofibrous membranes: morphology, structures and properties as an ionic-liquid host. *Macromol. Rapid Commun.*, **31**, 1779–1784.
30. Jia, P.J., Yee, W.A., Xu, J.W., Toh, C.L., Ma, J., and Lu, X.H. (2011) Thermal stability of ionic liquid-loaded electrospun

poly(vinylidene fluoride) membranes and its influences on performance of electrochromic devices. *J. Membr. Sci.*, **376**, 283–289.

31. Čolović, M., Jerman, I., Gaberšček, M., and Orel, B. (2011) POSS based ionic liquid as an electrolyte for hybrid electrochromic devices. *Sol. Energy Mater. Sol. Cells*, **95**, 3472–3481.
32. Lewandowski, A. and Świderska-Mocek, A. (2009) Ionic liquids as electrolytes for Li-ion batteries—an overview of electrochemical studies. *J. Power Sources*, **194**, 601–609.
33. Ma, L., Li, Y., Yu, X., Yang, Q., and Noh, C.-H. (2008) Using room temperature ionic liquid to fabricate PEDOT/TiO_2 nanocomposite electrode-based electrochromic devices with enhanced long-term stability. *Sol. Energy Mater. Sol. Cells*, **92**, 1253–1259.
34. Ma, L.J., Li, Y.X., Yu, X.F., Yang, Q.B., and Noh, C.H. (2009) Fabricating red–blue-switching dual polymer electrochromic devices using room temperature ionic liquid. *Sol. Energy Mater. Sol. Cells*, **93**, 1253–1259.
35. Pang, Y.H., Xu, H., Li, X.Y., Ding, H.L., Cheng, Y.X., Shi, G.Y., and Jin, L.T. (2006) Electrochemical synthesis, characterization, and electrochromic properties of poly(3-chlorothiophene) and its copolymer with 3-methylthiophene in a room temperature ionic liquid. *Electrochem. Commun.*, **8**, 1757–1763.
36. Deepa, M., Awadhia, A., and Bhandari, S. (2009) Electrochemistry of poly(3,4-ethylenedioxythiophene)-polyaniline/Prussian blue electrochromic devices containing an ionic liquid based gel electrolyte film. *Phys. Chem. Chem. Phys.*, **11**, 5674–5685.
37. Reddy, B.N. and Deepa, M. (2013) Electrochromic switching and nanoscale electrical properties of a poly(5-cyano indole)-poly(3,4-ethylenedioxypyrrole) device with a free standing ionic liquid electrolyte. *Polymer*, **54**, 5801–5811.
38. Zhou, D., Zhou, R., Chen, C.X., Yee, W.A., Kong, J.H., Ding, G.Q., and Lu, X.H. (2013) Non-volatile polymer electrolyte based on poly(propylene carbonate), ionic liquid, and lithium perchlorate for electrochromic devices. *J. Phys. Chem. B*, **117**, 7783–7789.
39. Marcilla, R., Alcaide, F., Sardon, H., Pomposo, J.A., Pozo-Gonzalo, C., and Mecerreyes, D. (2006) Tailor-made polymer electrolytes based upon ionic liquids and their application in all-plastic electrochromic devices. *Electrochem. Commun.*, **8**, 482–488.
40. Echelmeyer, T., Meyer, H.W., and van Wuellen, L. (2009) Novel ternary composite electrolytes: Li ion conducting ionic liquids in silica glass. *Chem. Mater.*, **21**, 2280–2285.
41. Zhou, R., Liu, W.S., Kong, J.H., Zhou, D., Ding, G.Q., Leong, Y.W., Pallathadka, P.K., and Lu, X.H. (2014) Chemically cross-linked ultrathin electrospun poly(vinylidene fluoride-co-hexafluoropropylene) nanofibrous mats as ionic liquid host in electrochromic devices. *Polymer*, **55**, 1520–1526.
42. Yu, X.Y., Xiao, M., Wang, S.J., Zhao, Q.Q., and Meng, Y.Z. (2010) Fabrication and characterization of PEO/PPC polymer electrolyte for lithium-ion battery. *J. Appl. Polym. Sci.*, **115**, 2718–2722.
43. Kil, E.H., Choi, K.H., Ha, H.J., Xu, S., Rogers, J.A., Kim, M.R., Lee, Y.G., Kim, K.M., Cho, K.Y., and Lee, S.Y. (2013) Imprintable, bendable, and shape-conformable polymer electrolytes for versatile-shaped lithium-ion batteries. *Adv. Mater.*, **25**, 1395–1400.
44. Yan, C.Y., Kang, W.B., Wang, J.X., Cui, M.Q., Wang, X., Foo, C.Y., Chee, K.J.Z., and Lee, P.S. (2013) Stretchable and wearable electrochromic devices. *ACS Nano*, **1**, 316–322.
45. Khanmirzaei, M.H. and Ramesh, S. (2013) Ionic transport and FTIR properties of LiI doped biodegradable rice starch based PE. *Int. J. Electrochem. Sci.*, **8**, 9977–9991.
46. Avellaneda, C.O., Vieira, D.F., Al-Kahlout, A., Heusing, S., Leite, E.R., Pawlicka, A., and Aegerter, M.A. (2008) All solid-state electrochromic devices with gelatin-based electrolyte. *Sol. Energy Mater. Sol. Cells*, **92**, 228–233.

47. Lalia, B.S., Samad, Y.A., and Hashaikeh, R. (2012) Nanocrystalline cellulose-reinforced composite mats for lithium-ion batteries: electrochemical and thermomechanical performance. *J. Solid State Electrochem.*, **17**, 575–581.
48. Rakkapao, N., Vao-soongnern, V., Masubuchi, Y., and Watanabe, H. (2011) Miscibility of chitosan/poly(ethylene oxide) blends and effect of doping alkali and alkali earth metal ions on chitosan/PEO interaction. *Polymer*, **52**, 2618–2627.
49. Mobarak, N.N., Ahmad, A., Abdullah, M.P., Ramli, N., and Rahman, M.Y.A. (2013) Conductivity enhancement via chemical modification of chitosan based green polymer electrolyte. *Electrochim. Acta*, **92**, 161–167.
50. Liew, C.W., Ramesh, S., Ramesh, K., and Arof, A.K. (2012) Preparation and characterization of lithium ion conducting ionic liquid-based biodegradable corn starch polymer electrolytes. *J. Solid State Electrochem.*, **16**, 1869–1875.
51. Gesheva, K.A., Ivanova, T.M., and Bodurov, G. (2012) Transition metal oxide films: technology and "smart windows" electrochromic device performance. *Prog. Org. Coat.*, **74**, 635–639.
52. Glasse, M.D., Idris, R., Latham, R.J., Linford, R.G., and Schlindwein, W.S. (2002) Polymer electrolytes based on modified natural rubber. *Solid State Ionics*, **147**, 289–294.
53. Radiman, C.L. and Rifathin, A. (2013) Preparation of phosphorylated nata-de-coco for polymer electrolyte membrane applications. *J. Appl. Polym. Sci.*, **130**, 399–405.
54. Sudhakar, Y.N., Selvakumar, M., and Bhat, D.K. (2014) Tubular array, dielectric, conductivity and electrochemical properties of biodegradable gel polymer electrolyte. *Mater. Sci. Eng., B*, **180**, 12–19.
55. Gagliardi, M. (2012) Global Markets and Technologies for Smart Glass, *BCC Research*, May 2012.
56. Pike Research, S.G.E. (2012) Suspended Particle, Thermochromic, and Liquid Crystal Glass Technologies for Architectural and Transportation Applications, Q2 2012, *http://www.pikeresearch.com/research/smart-glass* (accessed 23 February 2015).
57. Qiang, P.F., Chen, Z.W., Yang, P.H., Cai, X., Tan, S.Z., Liu, P.Y., and Mai, W.J. (2013) TiO_2 nanowires for potential facile integration of solar cells and electrochromic devices. *Nanotechnology*, **24**, 435403.
58. Österholm, A.M., Shen, D.E., Dyer, A.L., and Reynolds, J.R. (2013) Optimization of PEDOT films in ionic liquid supercapacitors: demonstration as a power source for polymer electrochromic devices. *ACS Appl. Mater. Interfaces*, **5**, 13432–13440.
59. Yang, X.H., Zhu, G., Wang, S.H., Zhang, R., Lin, L., Wu, W.Z., and Wang, Z.L. (2012) A self-powered electrochromic device driven by a nanogenerator. *Energy Environ. Sci.*, **5**, 9462–9466.

11
Gyroid-Structured Electrodes for Electrochromic and Supercapacitor Applications

Maik R.J. Scherer and Ullrich Steiner

11.1
Introduction to Nanostructured Electrochromic Electrodes

The reversible, persistent change in colour or transparency controlled by a temporarily applied electrical potential displayed by electrochromic materials is truly fascinating [1, 2]. Especially in view of global warming, the low energy consumption of electrochromic devices makes them even more attractive. While these materials have been employed in smart windows and display applications, only very few electrochromic products are commercially available [3], despite extensive research for over 40 years [4]. This is mainly due to low chromatic contrasts, long switching times, stability issues and unsatisfying coloration efficiencies [5]. The chromatic contrast of conventional unstructured, compact electrode devices is limited by finite ion penetration depths, whereas the switching time is restricted by the long diffusion distances required to achieve a satisfactory colour saturation [6]. Only with the recent advances in nanotechnology, especially in terms of structure control on the nanometre scale, progress was made concerning these shortcomings. Nanostructuring of the electrode material has proved to be a promising strategy for improving several of the electrochromic figures of merit at once. The enhanced and unique properties that are present neither in the bulk nor in structures with dimensions on larger length scales often arise from the ability to access both bulk and surface/interface properties likewise. Moreover, nanostructured electrochromic materials offer new opportunities to study the involved fundamental surface processes in a controlled manner.

Over the last two decades, several hundred studies on nanostructured electrochromic electrodes have been published; see Chapter 9. The discussion of all employed structures and material systems lies well outside the scope of this chapter. Here the focus is directed at ordered three-dimensional nanostructures with structural features ranging in size from 1 to 100 nm. An especially interesting subset of these are the mesoporous nanostructures that exhibit interconnected pores with diameters between 2 and 50 nm and therefore high surface-to-volume ratios.

Electrochromic Materials and Devices, First Edition.
Edited by Roger J. Mortimer, David R. Rosseinsky and Paul M. S. Monk.

The most important feature of such mesoporous electrodes when applied in electrochromic devices is their extremely large accessible internal surface area. The advantages and consequential performance gains arising from an increased surface for electrochromic systems where the redox active species, such as a viologen or a discrete transition metal complex, is adsorbed and anchored as monolayer to a mesoporous transparent conducting oxide are evident [7–10]. In contrast, the implications on the material properties of electrochromic transient metal oxides or conjugated polymers affected by nanostructuring are more far-reaching. In these materials, the electrochromism is inherently linked to the intercalation/deintercalation or insertion/extraction of ions that ensure the maintenance of charge neutrality during the redox reaction. Due to intrinsically low ion diffusion coefficients and relatively short timescales being of interest, just a thin layer or shell of typically a few nanometre thickness participates in the electrochemical reaction and contributes to the electrochromic effect while the majority of the bulk material remains inactive, see Figure 11.1. On nanostructuring, the inactive part may be fully transformed into accessible, active electrode material given that at least one structural feature size is sufficiently small. Another very important advantage of three-dimensionally nanostructured electrodes compared to bulk material is their ability to freely expand and shrink during the insertion and extraction of ions on electrochemical cycling. Unstructured electrodes tend to crack and therefore exhibit poor device lifetimes, whereas the nanostructured counterparts seem to cope much better with the substantial volume changes.

In summary, an effective and efficient mesoporous three-dimensional nanostructure concerning electrochromism should exhibit the following features:

1) an extremely large accessible internal surface area that makes both bulk and surface/interface properties accessible,
2) a three-dimensionally continuous structural network with a constant strut diameter that ensures a very rapid and efficient transport of charge carriers through the electrochromic material itself,

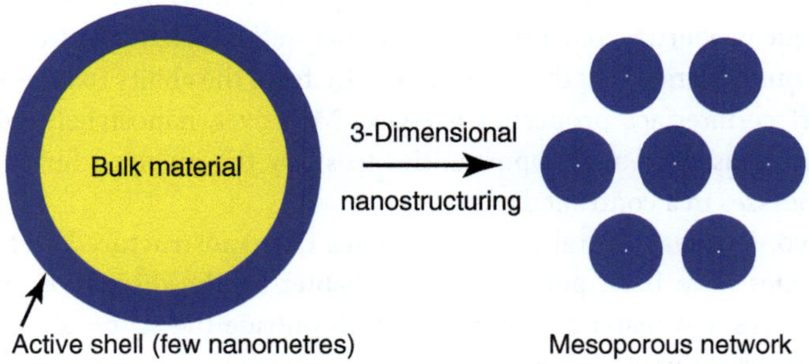

Figure 11.1 The majority of bulk material (yellow) remains inactive during electrochemical cycling and only a thin active shell (blue) that is accessible for ion insertion contributes to the electrochromic effect. On three-dimensional nanostructuring, ions can easily penetrate the entire electrode material which leads to a significant enhancement of the electrochromic properties.

3) an interconnected network of porous channels with a constant pore diameter that enables an efficient electrolyte infiltration and a very rapid transport of charge carriers through the electrolyte-filled pores,
4) a high surface-to-volume ratio and a highly efficient interfacial charge transfer between the electrochromic material and the electrolyte,
5) an interconnected and self-supporting architecture that exhibits excellent mechanical stability and robustness,
6) and last but not least, sufficient porous space that can accommodate the recurring volume expansion/contraction on ion insertion/extraction.

One such nanomorphology that meets all listed requirements is the so-called double-gyroid (DG) with $Ia\bar{3}d$ symmetry arising from block-copolymer self-assembly [11]. Before discussing the DG morphology, its synthesis and applications in electrochromic and supercapacitor devices in more detail, other techniques of modern nanotechnology employed in the three-dimensional nanopatterning of electrochromic materials are briefly introduced.

11.1.1 Three-Dimensional Nanostructuring Strategies

Currently, the fabrication of well-defined nanostructures for industrial purposes mainly relies on top-down nanolithographic processes, such as nanoimprint, X-ray, extreme ultraviolet and electron-beam lithography [12, 13]. Besides being very time-consuming and requiring expensive production facilities, the main disadvantage of these top-down techniques is their rather limited capability concerning the fabrication of complex three-dimensional structures. Recently, bottom-up processes have been developed that could become the next generation of nanofabrication, as they overcome the problems associated with nanolithographic techniques: whether the novel bottom-up approaches are going to leverage the commercialisation of electrochromic devices and their large-scale architectural applications will largely depend on their scalability, the number and complexity of the involved synthesis steps and the resultant manufacturing costs.

The bottom-up approaches employed in the three-dimensional nanopatterning of electrochromic materials and transparent conducting oxides for electrochromic devices can be broadly divided into three categories: firstly, the synthesis of nanoparticles and their aggregation into mesoporous electrodes [14, 15]; secondly, the direct formation of mesoporous electrodes during the synthesis of electrode materials [16–22]; and finally, the replication of porous templates with electrode material [23–31]. The order of the listing roughly follows the degree of complexity and effort involved in the corresponding synthesis process.

The template-assisted nanostructuring approach splits the fabrication process into two successive steps, namely the synthesis of mesoporous templates and their subsequent replication into electrode materials, as illustrated in Figure 11.2. Compared to the other approaches, this strategy has the major advantage that it allows for the independent optimisation of the structure formation, that is, the template

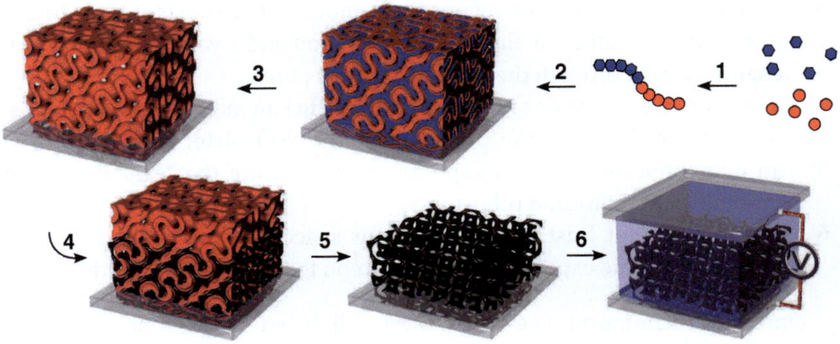

Figure 11.2 Template-assisted three-dimensional nanopatterning approach. This strategy is particularly versatile since it separates the template fabrication (top row) from the electrode material synthesis (bottom row). The scheme illustrates how block-copolymer self-assembly can be utilised for the template synthesis: **1.** Copolymer synthesis. **2.** Microphase separation into the DG morphology. **3.** Selective degradation of one block yields a mesoporous template. **4.** Template refilling with an electrochromic material. **5.** Removal of the polymeric matrix. **6.** Assembly of electrochromic devices. (Reproduced with permission from [39].)

preparation, and the actual synthesis of the electrochromic material. A variety of templates have been used to structure electrochromic electrodes, including colloidal templates [23–27], anodic aluminium membranes [28–30], and track-etched poly(carbonate) membranes [31]. Since the same template can be used to structure various electrode materials employing different replication techniques, such as sol-gel, atomic layer deposition and electrochemical deposition, it is considered as an extremely versatile approach [32–35].

An especially elegant technique utilises block-copolymer self-assembly to fabricate mesoporous templates in a potentially inexpensive and scalable manner. After adoption of an appropriate nanomorphology during microphase separation, the selective removal of one polymeric block yields the desired mesoporous template, see Figure 11.2. Crossland *et al.* demonstrated the fabrication of titanium dioxide nanowires orientated perpendicular to the substrate's surface using a cylinder-forming diblock copolymer [36]. The reported collapse of the nanowires after dissolution of the polymeric template anticipated the application in electrochromic devices. This led to the investigation of another diblock copolymer morphology, namely the DG, which is distinguished by its truly three-dimensional and self-supporting network architecture. The remaining part of this chapter explores the fascinating gyroidal intermaterial dividing surfaces formed by self-assembling systems and their application in electrochromic and supercapacitor devices [37–40].

The reader might be of the opinion that the template-assisted approach is currently too complex and costly for the industrial mass production of three-dimensionally nanostructured electrochromic electrodes. However, from a scientific point of view, these are especially interesting since they are highly versatile and can often be applied to electrochromic materials that do not show any self-organisation properties and are incompatible with self-assembly. Furthermore, the resulting well-defined and highly ordered electrode morphologies are particularly suitable for the characterisation and study of the nanostructuring effects on the fundamental electrochromic processes.

11.2 Polymer Self-Assembly and the Gyroid Nanomorphology

11.2.1 Copolymer Microphase Separation

Polymers are macromolecules that consist of repeating molecular units known as monomers covalently bonded during polymerisation. Homopolymers contain only one single monomer species, whereas block copolymers are made up of a sequence of chemically distinct homopolymers. A remarkable property of block copolymers is their ability to self-assemble, driven by thermodynamic incompatibilities of the constituent blocks that result in intramolecular repulsive forces. Being covalently bonded, the immiscible blocks are prevented from separating on a macroscopic level as the corresponding homopolymer blend would do. Even for the most simple copolymer architecture, namely diblocks, this gives rise to the formation of a variety of highly ordered morphologies with nanoscale periodicities [41].

The volume ratio of the copolymer blocks f primarily determines which structure is adopted at equilibrium, whereas its dimensions are dependent on the degree of polymerisation N, as shown in Figure 11.3. Symmetric diblock copolymers with equal block volumes arrange into a lamellar morphology with alternating layers of the constituent blocks. An increase in the red block's volume fraction leads to the formation of morphologies where the minority component (blue) is enclosed by the matrix phase (red). Between the lamellar and the hexagonally packed cylinders, the DG and *Fddd* are observed. A further increase in asymmetry leads to the formation of body-centred cubic and close-packed spheres.

Most copolymer melts tend to segregate on lowering the temperature, although mixtures of the corresponding monomers form a single homogeneous phase. The Flory–Huggins mean field theory provides a framework which explains this discrepancy [42]. It considers the two competing terms of enthalpic and entropic nature which are related to the interaction parameter χ and the degree of polymerisation N, respectively. The dimensionless χ parameter, which is usually found to vary linearly with the inverse temperature T, captures the energetic contributions arising from the monomer interactions. At sufficiently high temperatures (small χ), the entropic terms, which scale with N^{-1}, exceed the energetic interactions, and the chains are mixed homogeneously [43]. As the temperature is reduced (large χ), the repulsive enthalpic term increases and the system tends to minimise the unfavourable interface between the incompatible blocks via formation of intermaterial dividing surfaces (IMDS). Note that during

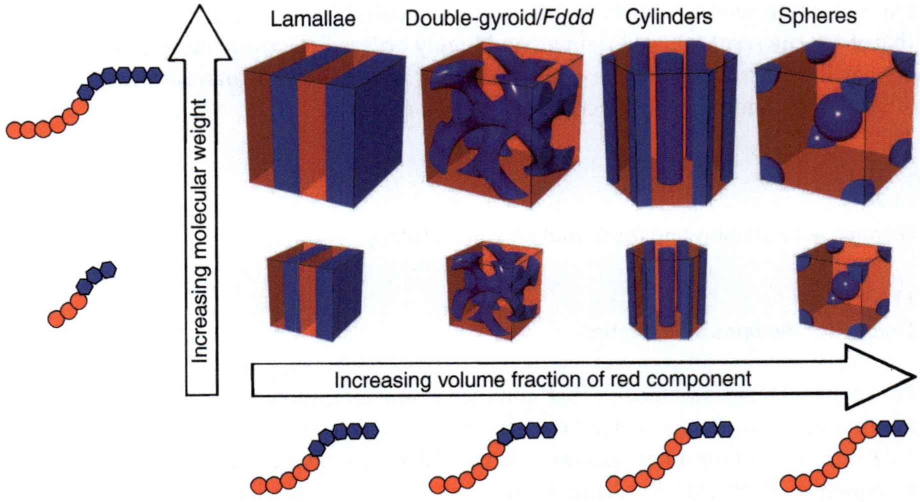

Figure 11.3 Diblock-copolymer equilibrium phase morphologies. Inverting the block length ratio yields colour-inverted structures. (Reproduced with permission from [39].)

this process, the polymer chains located in close proximity of the IMDS are significantly stretched and the resulting extended chain conformation gives rise to additional entropic forces. For any given segregation strength χN and copolymer composition f, the melt adopts the IMDS that minimises the free energy.

Due to the distinct temperature dependency of χ, it is possible to induce morphology transitions through thermal annealing, which corresponds to an almost vertical path in the phase diagram; see Figure 11.4. Given an inverse temperature dependence, an increase in temperature will result in a shift towards smaller χN values in the phase diagram.

Not all morphologies are equally well suited for the template-assisted nanostructuring process, as the work by Crossland *et al.* has demonstrated [36]. To ensure that neither the porous template nor its replica collapses upon removal of the sacrificial phase or the template, respectively, both segregated copolymer phases need to be continuous and thus independently self-supporting. A further advantage of such multicontinuous morphologies when applied in thin-film architectures is the higher likelihood to be porous at the interfaces. The only ordered multicontinuous microdomain structures adopted by diblock copolymers are the *Fddd* and the DG. The *Fddd*'s region in the phase diagram is extremely small, leaving the DG as the only experimentally practicable [45].

11.2.2
Double-Gyroid

In 1986, Thomas *et al.* were the first to discover the DG in strongly segregated multiarm star diblock copolymers of poly(isoprene) and poly(styrene) (PS),

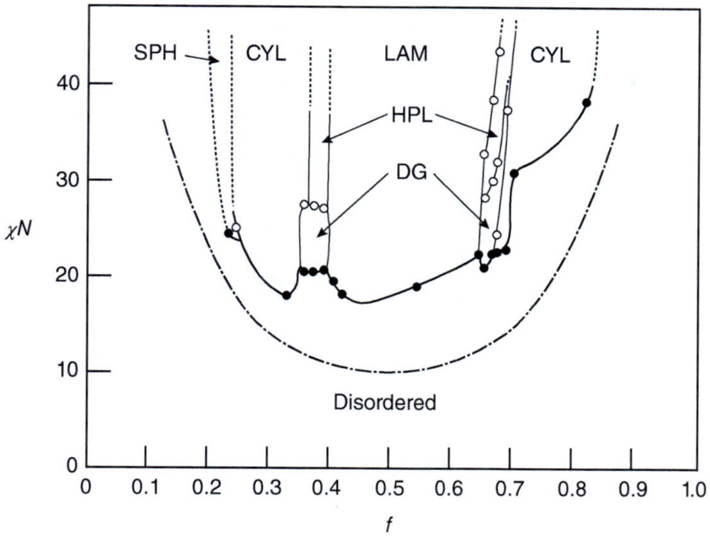

Figure 11.4 Experimentally determined phase diagram of poly(styrene-*b*-isoprene) measured by Khandpur *et al.*, with $\chi N \propto T^{-1}$ the segregation strength and f the volume ratio of the poly(isoprene) block. (Lamellar, LAM; double-gyroid, DG; cylinders, CYL; spheres, SPH; and hexagonally perforated lamellar, HPL.) (Reproduced with permission from [44].)

but incorrectly identified the new morphology, lying between the lamellar and cylindrical phase, as the double-diamond [46]. Eight years later, in 1994, two groups independently characterised the IMDS observed in a low molecular diblock copolymer melt and in a diblock copolymer blend correctly as DG with $Ia\bar{3}d$ (No. 230) symmetry [47, 48]. These results suggested a re-examination of the star diblock structure with improved X-ray measurements, which led to a re-classification as DG [49]. Thereafter, the DG has been observed in various diblock copolymers [50, 51], in nanoporous silica films [52], in ABA triblock copolymers [53, 54] and in ABC triblock copolymers where the A and C blocks have identical volume fractions [55].

The DG structure comprises two interwoven single-gyroid (SG) surfaces with $I4_132$ symmetry that are related by inversion. An approximation of the gyroidal networks is given by a level surface function,

$$f(x,y,z) = \sin\frac{2\pi}{L}x\cos\frac{2\pi}{L}y + \sin\frac{2\pi}{L}y\cos\frac{2\pi}{L}z + \sin\frac{2\pi}{L}z\cos\frac{2\pi}{L}x = \pm c \tag{11.1}$$

where L is the cubic unit cell length and c controls the volume fraction occupied by the network phase [56, 57]. In the interval $0 < c \leq 1.413$, the level surface effectively divides space into three triply periodic, embedded, continuous, self-supporting and freestanding subvolumes. Each of the two SG networks consists of tubular segments interlinked by nodes with threefold connectivity that form two

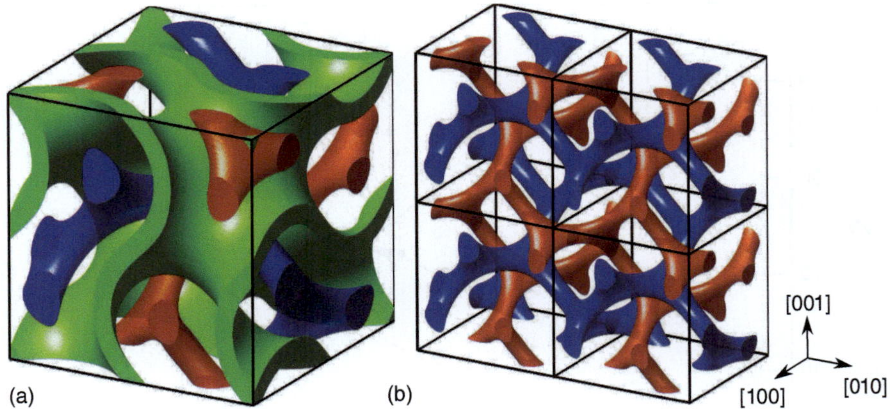

Figure 11.5 (a) The DG cubic unit cell showing the matrix phase (green) and the two networks (blue and red). (b) Four cubic unit cells of the DG network with a volume fraction of 12%($c = \pm 1.3$). (Reproduced with permission from [39])

adjoining spirals of opposite chirality along both the ⟨100⟩ and ⟨111⟩ directions. The third phase, which is located between the SG surfaces, centred on the minimal gyroid or Schoen G surface ($c = 0$), forms the so-called matrix phase through which the two networks run [58]. The matrix phase volume increases with increasing values of c, thereby gradually reducing the strut diameters of the two enclosed networks.

The matrix is also called the majority phase, because it takes up about 63–66% of the volume in DG morphologies formed by diblock polymers, and accordingly the networks together are called the minority phase. The cubic DG unit cell showing the matrix phase together with the two minority networks is presented in Figure 11.5a. To illustrate that the two distinct intertwined networks are non-intersecting, an arrangement of four unit cells with the matrix phase removed and the two networks coloured blue and red is shown in Figure 11.5b.

11.2.3 Synthesis of Mesoporous DG Templates

The first step of the DG template synthesis is the selection of a suitable diblock copolymer. For the template-forming block, PS (or partially fluorinated PS) is especially suitable, since it is cheap, cross-linkable, chemically resistant against most bases and acids and has a suitable glass transition temperature of 104°C, while being easily removable after replication by combustion or dissolution in organic solvents. As sacrificial block, the readily degradable and low-cost biopolymer poly(D,L-lactide) (PLA) is of particular interest.

Equally important is the selection of the synthesis route, and we chose the following: first, PS is synthesised by atom transfer radical polymerisation (ATRP), followed by the addition of PLA via the recently developed

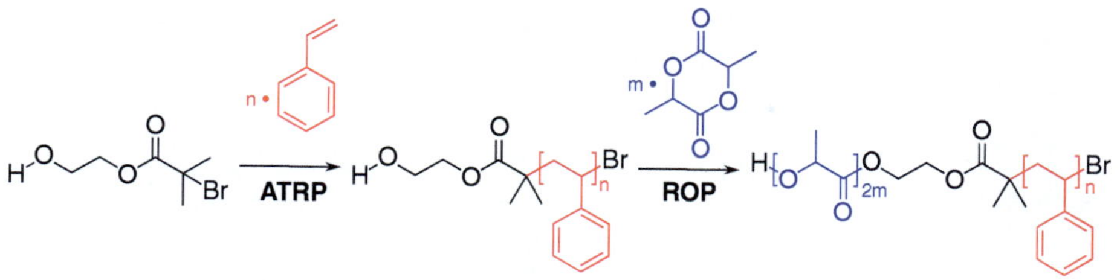

Figure 11.6 Two-step synthesis of PS-*b*-PLA via the ATRP of styrene and organocatalytic ROP of lactide employing a bifunctional initiator. (Reproduced with permission from [39])

organocatalytic ring-opening polymerisation, initiated by a bifunctional initiator (Figure 11.6) [59, 60]. Importantly, both polymerisation techniques offer precise and robust control over the copolymer composition, which is an essential requirement to reliably target the DG's narrow location in phase space [11]. A detailed synthesis description can be found in our earlier publications [37, 39, 40]. The DG-forming PS-*b*-PLA ($M_W = 21.6$ kg mol^{-1}, PDI = 1.09) has a 39.8%$_{wt}$(37.1%$_{vol}$) PLA fraction [61]. Similarly, the polymer containing 30%$_{mol}$ 4-fluorostyrene ($M_W = 23.8$ kg mol^{-1}, PDI = 1.1) exhibits a 39.9%$_{wt}$(37.9%$_{vol}$) PLA fraction.

Next, micrometre-thick polymer films were spun on glass with a transparent, conductive fluorine-doped tin oxide (FTO) coating; see Figure 11.2. Due to fast solvent evaporation during spin coating, the polymer films are typically trapped in a non-equilibrium state possessing a worm-like structure that lacks any order. Self-assembly into the DG morphology occurs during brief annealing at 173 °C for 20 min. Selective degradation of the PLA minority phase in a mild alkaline solution (pH $\approx$ 13) yields the desired voided bicontinuous DG polystyrene film with 38%$_{vol}$ porosity, see the scanning electron microscopy (SEM) images presented in Figure 11.7 [62]. The voided films exhibit a pore diameter of about 11 nm and a specific surface-to-bulk volume ratio of 161.4 μm^{-1}. The temperature-induced phase transition was confirmed by synchrotron small-angle X-ray scattering (SAXS) of bulk samples (Figure 11.7d). The peaks observed for the annealed samples are consistent with a DG phase exhibiting a cubic unit cell length $a = 2\pi q^{-1}(h^2 + k^2 + l^2)^{0.5} = 43.2$ nm, where q is the reciprocal lattice vector [63].

A crucial requirement for the successful replication of the porous polymer scaffold by electroplating is the continuity of the mesopores to both film interfaces to enable electrolyte infiltration. Thus, prevention of preferential wetting of the substrate by either of the two polymers blocks is essential. A neutral FTO surface was achieved by cleaning with a Piranha solution, followed by the deposition of a partial octyltrichlorosilane (OTS) monolayer, resulting in a water contact angle on the modified substrate of 59 ± 5°. The FTO surface was further modified using photolithography to create an isolating SU-8 pattern, which locally inhibited electroplating, thereby forming visible design patterns with electrochromic contrast.

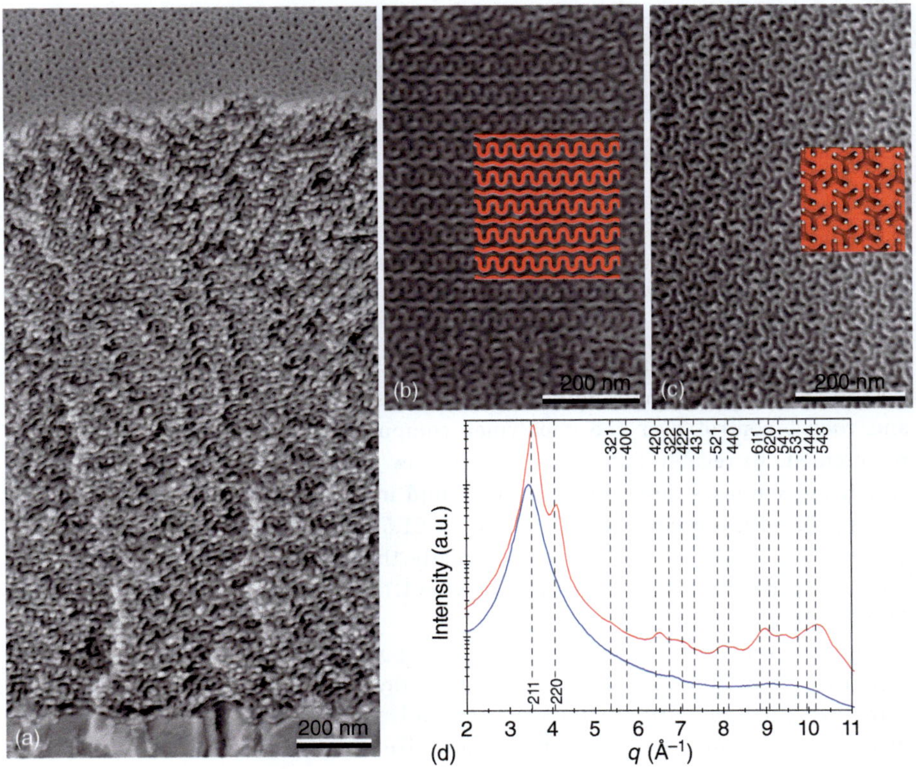

Figure 11.7 (a) Cross-sectional SEM image of a representative voided DG template. (b, c) SEM images of the free surface showing the characteristic double-wave pattern of the (211) plane (b) and the (111) plane (c). (d) SAXS spectra of bulk samples of PS-*b*-PLA with a disordered worm-like morphology (blue curve) and the DG structure after annealing at 173°C (red curve).

11.3 Gyroid-Structured Vanadium Pentoxide

Orthorhombic crystalline vanadium pentoxide (V_2O_5) is a typical intercalation compound as a result of its layered structure [64]. It offers the essential advantages of low cost, abundant availability, easy synthesis and high intercalation densities [65, 66] and, therefore, finds widespread use in lithium ion intercalation applications such as electrochromic cells [67], high energy density batteries [68], supercapacitors [69] and sensors [70]. However, V_2O_5 suffers from an inherently low ion diffusion coefficient ($D \leq 10^{-12}$ cm^2 s^{-1}) and a moderate electrical conductivity ($\sigma = 10^{-2}-10^{-3}$ S cm^{-1}) [66].

Lithium ion insertion is accompanied by several structural modifications that occur in consecutive stages [71, 72]

$$V_2O_5 + x\ Li^+ + x\ e^- \rightleftharpoons Li_xV_2O_5. \quad (11.2)$$

The α-, ϵ- and δ-phase exist for $x < 0.01, 0.35 < x < 0.7$ and $x = 1$, respectively, and the transitions are fully reversible. The ϵ-phase is characterised by an increasing puckering of the layers, while δ-phase is formed by gliding of one layer out of the surrounding two [73]. Further lithium intercalation $x > 1$ results in an irreversible transformation to the γ-phase, which itself can reversibly be cycled in the stoichiometric range $0 < x < 2$ without causing a structural change. Upon further lithiation $x = 3$, the irreversibly rock-salt ω-phase is formed. In the bulk, this bronze can be reversibly cycled as the ω-phase, but a large fraction of the lithium cannot be removed, causing a decrease in energy capacity [74].

The refilling of the DG-structured PS templates on FTO glass with V_2O_5 via electrodeposition is fairly straightforward. Anodic deposition [16, 75] was performed at a constant potential of 1.5 V vs Ag/AgCl from a 1:1 mixture of deionised water and ethanol containing $VOSO_4$. The low surface tension of ethanol enabled the infiltration of the hydrophobic PS pores by the electrolyte. A post-deposition annealing in air for 2 hours at a temperature of 275 °C was found to be optimal for the preparation of electrochromic electrodes [76, 77]. Figure 11.8a shows the highly periodic DG V_2O_5 replica with a strut radius of ≈5.5 nm formed upon removal of the styrenic template by dissolution. The density of electrodeposited nontemplated vanadia is assumed to be ≈ 2.87 g cm^{-3}; thus, the average density of DG-structured vanadia with a volume fraction of 37.9% is ≈ 1.09 g cm^{-3} with a specific surface area of ≈ 1.48 m^2 g^{-1} [66]. Inverse opal (IO) and nontemplated, compact V_2O_5 electrodes were prepared under identical conditions in order to investigate the implications of nanostructuring as well as the structure size. The IOs were synthesised using an array of 400 nm polystyrene spheres as template (Figure 11.8b) [78].

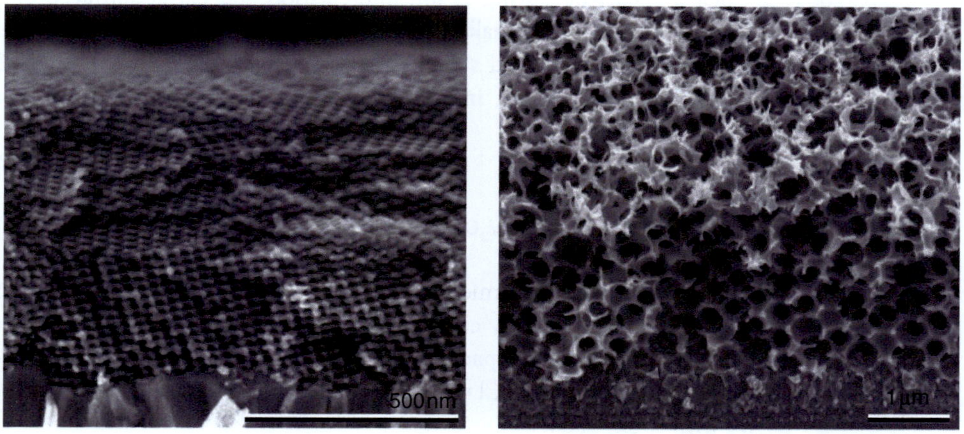

(a) (b)

Figure 11.8 Cross-sectional SEM images of templated V_2O_5 electrodes highlighting the differences in structural dimensions. (a) Mesoporous DG electrode. (b) Macroporous IO structure. Note the different scale bars. (Reproduced with permission from [39].)

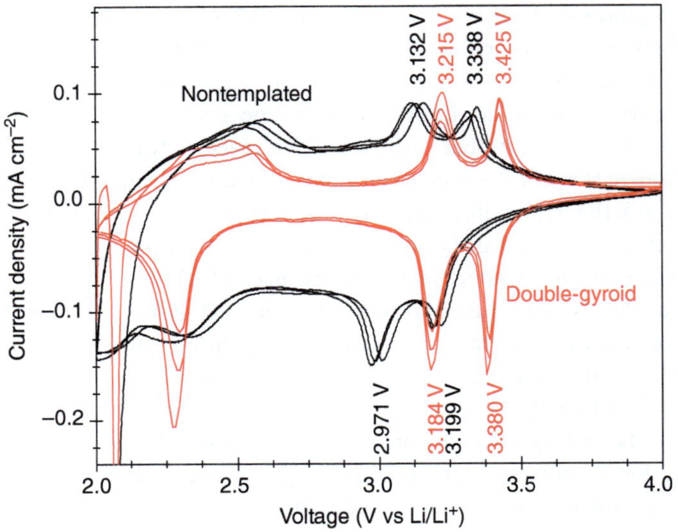

Figure 11.9 Cyclic voltammograms of nontemplated and DG V_2O_5 electrodes recorded in 1 M $LiClO_4$ propylene carbonate using lithium foils as counter and reference electrodes (1 mV s^{-1}). (Reproduced with permission from [38].)

11.3.1
Electrochemical Characterisation of V_2O_5 Electrodes

The electrochemical Li intercalation behaviour of the DG and nontemplated V_2O_5 electrodes in 1 M $LiClO_4$ propylene carbonate was investigated by cyclic voltammetry, see Figure 11.9. The CV curves show the two well-resolved sets of redox peaks occurring during Li ion intercalation and extraction according to Eq. (11.2) [67]. The redox peaks of the DG electrode are higher and sharper than those of the nontemplated sample. This is a first indication that the DG nanostructure enhances Li ion intercalation and consequently, improves the electrochemical properties.

11.3.2
Electrochromic Displays Based on V_2O_5 Electrodes

The assembly of the electrochromic cells from 1.1 μm thick DG and 1.8 μm thick IO electrodes was completed by capping with a FTO counter electrode using a precut thermoplastic gasket as spacer, infiltration with 1 M $LiClO_4$ in propylene carbonate, insertion of a Ag/AgCl wire as reference electrode and, finally, sealing the device with epoxy glue.

Cycling the DG device between −1 V and +1 V gave rise to a colour change from blue-grey to green-yellow as shown in Figure 11.10a. Figure 11.10b presents the optical transmittance spectra of the DG and IO device that exhibit approximately the same transmission values at $\lambda = 430$ nm in the coloured

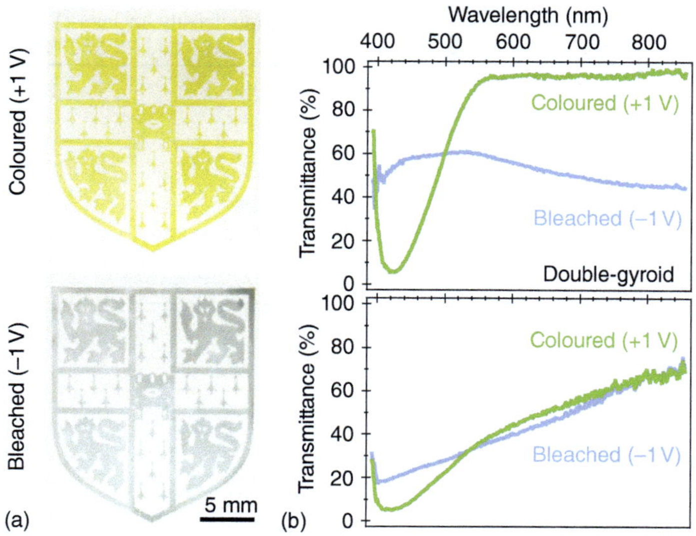

Figure 11.10 (a), Photographs of DG-structured V_2O_5 device in the bluish-grey reduced (−1 V) and the yellow-green oxidised state (1 V). The pattern shows the crest of the University of Cambridge. (b), Optical transmittance spectra of DG-nanopatterned (top row) and macroporous IO devices (bottom row). (Reproduced with permission from [37].)

oxidised state. The colour contrast of the DG devices benefits from a pronounced variation across the entire wavelength spectrum, whereas the colour change in the IO structure stems only from the 400–500 nm spectral range. In the visible wavelength range, the DG device shows a maximal transmission variation at $\lambda = 430$ nm of $\Delta T = 49.8\%$ where the IO only reaches a value of $\Delta T = 14.1\%$. The IO's low contrast stems from the limited Li^+ ion penetration depth and the resulting low number of accessible intercalation sites. The difference in chromatic contrast highlights the distinct effect the nanostructure's feature size has on the material's intercalation properties.

A second important aspect of electrochromism is the temporal response under alternating potential steps (Figure 11.11a). In contrast to the IO device, the DG electrodes showed sharp and distinct transitions between the coloured/oxidised and bleached/reduced state. The characteristic response time τ was determined by fitting the exponential function $T(t) = T_0 + \Delta T{\cdot}e^{-t/\tau}$ to the switching curves. The DG devices showed short switching times below 82.5 ± 2.9 ms for the bleaching step and 86.7 ± 2.3 ms for the reverse process. The IOs exhibited a much slower response with switching times above 88 and 205 ms for a considerably smaller change in transmittance. The DG enables response times below 100 ms because the strut radius is comparable to the corresponding lithium diffusion distance $x = \sqrt{2Dt} \approx 5$ nm.

Finally, the charge per area q that was required to induce a gradient in optical density Δ OD = ln (T/T_0) was measured, where T_0 is the initial transmittance.

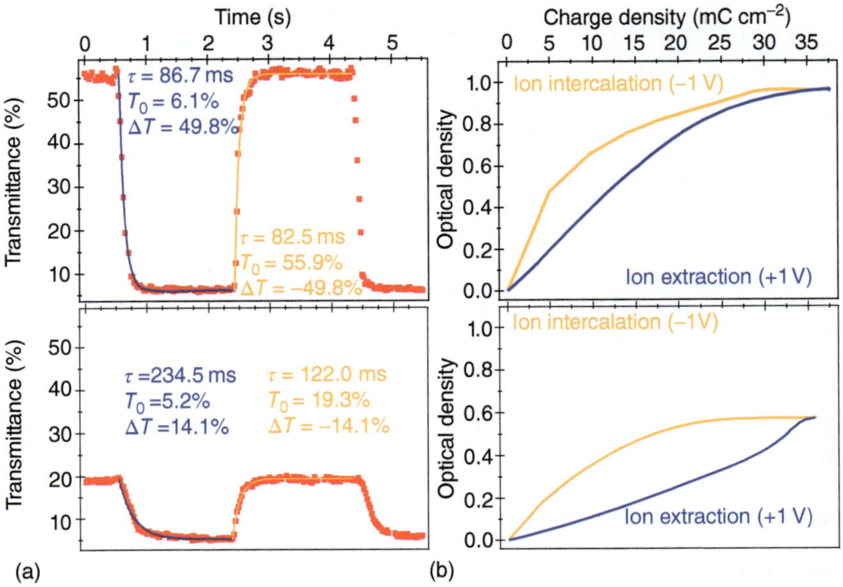

Figure 11.11 Electrochromic performance of DG (top row) and IO devices (bottom row) at $\lambda = 430$ nm. (a), Switching behaviour under alternating 2 s potential steps (± 1 V). (b), Optical density $|\Delta$ OD$|$ versus the charge density $|q|$ during ion intercalation (orange) and extraction (blue). Significantly higher changes in optical density were observed for the gyroid device compared to the IO for equal amounts of charge density passed through the cells. (Reproduced with permission from [38].)

In (Figure 11.11b), the optical density $|\Delta$ OD$|$ at 430 nm is plotted against charge density $|q|$ during ion intercalation (orange) and extraction (blue). Significantly higher changes in optical density were observed for the DG device compared to the IO for equal amounts of charge density passed through the cells. The ratio of these two quantities defines the coloration efficiency (CCE). The ideal electrochromic material would display a large transmittance change with a small amount of charge, giving rise to a high CCE [2, 79, 80]. For the DG device, CCE values of −33.9 and −35.8 $cm^2\ C^{-1}$ were required to achieve a 95% switch at 430 nm, for ion insertion and extraction, respectively. This corresponds to respective CCE values of −24.0 and −15.6 $cm^2\ C^{-1}$ of the IO electrode. The DG's excellent CCE values are a direct consequence of the highly interconnected network morphology, which combines a very high specific surface area providing good ion access with high electric conductivities.

11.3.3 Electrochromic V_2O_5 Supercapacitors

Similarly to the electrochromic devices, supercapacitor cells were fabricated by sandwiching a precut gasket between a pair of laterally offset V_2O_5 electrodes.

Instead of propylene carbonate, the highly conducting room-temperature ionic liquid triethylsulfonium bis(trifluoromethylsulfonyl) imide ([SET3][TFSI]) in combination with the corresponding lithium salt, LiTFSI, was chosen as electrolyte and the integration of Ag/AgCl reference electrode was omitted.

The electrochemical properties of such supercapacitors with 1 μm thick DG and nontemplated electrodes were studied by CV. The capacitive current density of the DG V_2O_5 was found to be much higher compared to the significantly thicker and denser nontemplated V_2O_5 film.

The specific capacitance of the DG-structured V_2O_5 pseudocapacitor during the initial charge/discharge cycles was 422 F g^{-1}, which equilibrated to the steady state of 155 F g^{-1} after 10 switching cycles at a high current density of > 10 A g^{-1}. The nontemplated layer material showed a much more gradual variation, dropping from an initial 20 F g^{-1} to 14 F g^{-1}. The overall capacitance of the devices comprises faradaic and double-layer capacitance, with a predominating faradaic contribution. The involved processes are kinetically facile and thus considered pseudocapacitive, but phase transitions may occur. Although it is difficult to distinguish between redox and intercalation pseudocapacitance, the latter is likely to be present in DG bicontinuous materials. The charge storage ability of supercapacitors is directly linked to the geometric confinement of ions in extremely small pores and the availability of a large surface area. The much higher specific surface area of the DG-structured V_2O_5 compared to the nontemplated material therefore partly explains the substantial improvement in capacitive current.

The overall performance of the V_2O_5-based supercapacitors is summarised in the Ragone plot describing the relation between energy density and power density (Figure 11.12) [81]. The highest obtained energy density of nontemplated vanadia supercapacitors is 1 W h kg^{-1} at a power density of 50 W kg^{-1}, which drops to one-half at a power density of 0.4 kW kg^{-1}. Such low values arise from the limited ion diffusion into the thick nontemplated V_2O_5 electrode, resulting also in a lack of responsiveness of the device. The maximisation of the response time in nontemplated layer devices limits the thickness of most pseudocapacitive electrodes to sub-micrometre values [82–84].

Devices made from DG V_2O_5, on the other hand, exhibit energy densities as high as 52 W h kg^{-1} at power densities above 1 kW kg^{-1}. Even at power densities as high as 16.7 kW kg^{-1}, the best performing device still had an energy density of 13.9 W h kg^{-1}, placing them among the top performing V_2O_5 devices [81]. Combined with superb electrochemical properties of [SET3][TFSI], the introduction of interconnected mesoporosity in DG V_2O_5 significantly enhances the supercapacitor performance, stemming from a combination of electrolyte access, short Li^+ penetration lengths, good electronic transport and large surface area of nanostructure enabling Li^+ insertion/extraction [83]. This leads to the observed conjunction of high specific energy and power densities of DG V_2O_5 supercapacitors, with energy densities approaching the lower end of lithium batteries.

Nontemplated V_2O_5 devices exhibit an electrochromic switching time of several minutes, and the colour change is therefore not sufficiently rapid for observation when charging and discharging the device. The improvement of ion

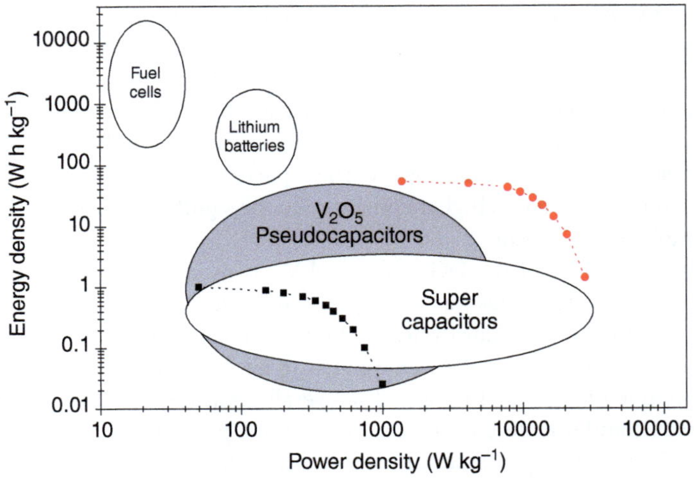

Figure 11.12 Ragone plot of electrochemical energy storage devices including the DG (•) and nontemplated layer V_2O_5(■) supercapacitors. (Reproduced with permission from [44].)

intercalation in the DG V_2O_5 supercapacitor gives rise to rapid colour changes and high chromatic contrast. This, the intrinsic electrochromic response, allows visualisation of the device charge state, since the degree of colour saturation indicates the energy storage level of the device, see Figure 11.13a. The dynamic optical response during the charge and discharge process is shown in Figure 11.13b. The *in situ* UV-Vis-NIR measurements were carried out during chronopotentiometry with a current density of ± 2 mA cm^{-2}. The anode attained the fully oxidised state, revealing a vivid yellow colour, after 20 s when the potential reached 3.5 V, whereas the cathode exhibited the green/grey reduced state. The coloration change during oxidation commenced at a potential of -1 V and reached the fully oxidised state at $+3.5$ V, while the coloration response started at $+1$ V and ended at -3.5 V during reduction.

11.4 Gyroid-Structured Nickel Oxide

As naturally abundant and low-cost semiconductor, nickel oxide (NiO) is widely used in electrochromic windows [85], batteries [86], supercapacitors [87] and sensors [88]. However, the templated electrodeposition of NiO is very complicated and time-consuming. In contrast, electroplating of nickel into voided DG templates is far less demanding. A two-step synthesis route involving a post-nanostructuring metal oxidation was therefore chosen rather than the direct synthesis of the desired DG NiO nanostructure, see Figure 11.14a. In fact, more porous and sophisticated structures, such as hollow nanospheres, nanotubes and nanopeapods, can be synthesised based on the nanoscale Kirkendall effect occurring during thermal oxidation of a nanostructured metal [89–92].

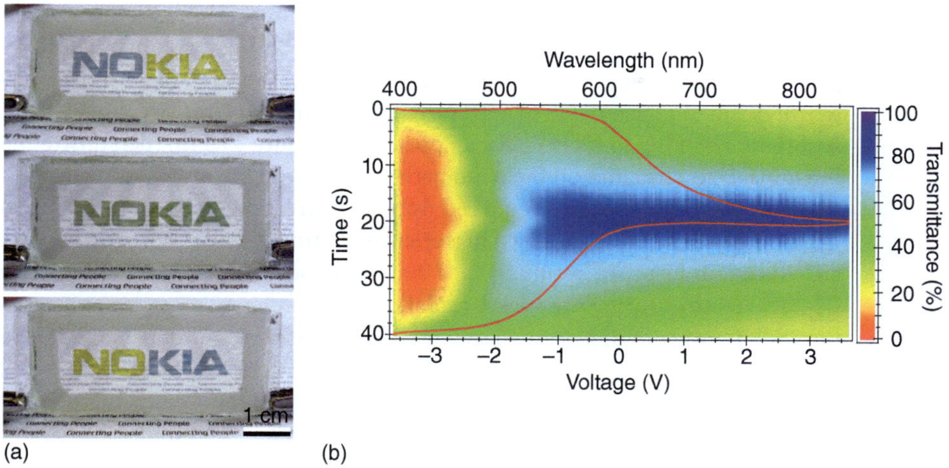

Figure 11.13 (a) Photographs of a DG-structured electrochromic supercapacitor with the top and bottom electrodes patterned as 'NO' and 'KIA', respectively. Discharged green state at 0 V (middle row) and charged state at ±3.5 V (top/bottom row). (b), Transmittance variation during chronopotentiometry with alternating current polarity (+2 mA cm^{-2} for the initial 20 s and −2 mA cm^{-2} for the remaining time). The red line is the potentiometric response. (Reproduced with permission from [44].)

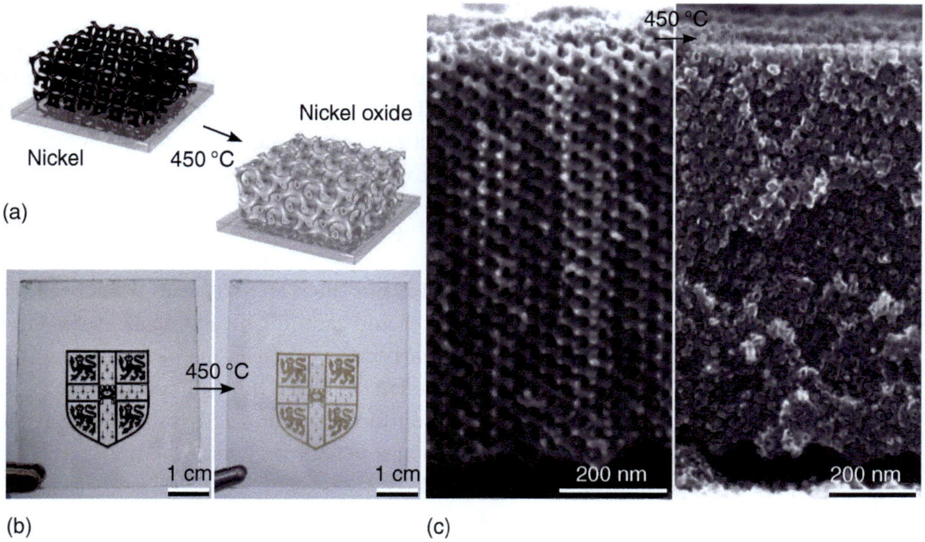

Figure 11.14 Preparation of NiO electrodes with a nanotubular DG morphology. (a) Schematic illustration of the thermal oxidation of an electrodeposited nickel DG into a porous NiO nanotubular array with DG morphology via the nanoscale Kirkendall effect. (b) Photograph of a pitch-black as-deposited DG Ni film that turns semi-transparent on oxidation at 450°C. (c) Cross-sectional SEM images of the Ni network after template removal (left) and the NiO nanotube array (right). (Reproduced with permission from [40].)

Cathodic templated and nontemplated nickel deposition was performed at a constant potential of −1.05 V vs Ag/ AgCl at 50 °C using a commercial nickel plating solution (bright finish, Alfa Aesar). In contrast to the bright, mirror-like finish of nontemplated Ni deposits, gyroid-structured films are highly light absorbing resulting in a pitch-black appearance (Figure 11.14b), demonstrating the impact the structure on the nanometre scale has on the optical properties of the material. Ni films were then freed from the styrenic template by dissolution in toluene and thermally oxidised in a preheated furnace at 450 °C for 10 h under an oxygen atmosphere resulting in a brownish, semi-transparent appearance (Figure 11.14b). During the nickel oxidation, the compact DG struts are transformed into a nanotubular DG array with hollow struts, without sacrificing its connectivity. Responsible for this transformation is the nanoscale Kirkendall effect, which is based on the fast outward diffusion of metal cations through the forming metal oxide/chalcogenide layer, balanced not by a considerably slower inward diffusion of chalcogen anions but rather by an influx of vacancies [90, 93, 94]. These vacancies gradually accumulate giving rise to porosity and finally void formation, see Figure 11.14c.

11.4.1
Electrochromic Displays Based on NiO Electrodes

To investigate the electrochromic performance, transparent cells were fabricated similar to the V_2O_5 device assembly and filled with a 1 M KOH(aq) electrolyte. On infiltration of the alkaline electrolyte, a spontaneous chemical conversion of NiO into a hydrous metal oxide phase $Ni(OH)_2$ takes place [95]. The Ni(II)/Ni(III) couple responsible for the reversible colour change on electrochemical cycling has been identified as the nickel hydroxide/oxy-hydroxide phases [95, 96]

$$\mathrm{Ni(OH)_2} \rightleftharpoons \mathrm{NiOOH} + \mathrm{H^+} + \mathrm{e^-}. \tag{11.3}$$

Figure 11.15a shows photographs of a 460 nm thick DG-structured NiO sample that exhibits anodic coloration at +0.55 V and cathodic bleaching at −0.55 V (vs Ag/AgCl). The corresponding spectral response of the DG in comparison with a 70 nm thick nontemplated film is displayed in Figure 11.15b. The colour contrast of the DG benefits from a pronounced and uniform variation across the entire visible wavelength spectrum with $\Delta T = 79.6\%$ at $\lambda = 630$ nm where the nontemplated sample only reaches a value of $\Delta T = 6.3\%$.

A second important aspect of electrochromism is the temporal response under alternating potentials (±0.55 V). The DG showed sharp and distinct transitions between the coloured/oxidised and bleached/reduced state across the entire visible spectrum (Figure 11.16a). This time-resolved switching behaviour was analysed in more detail at $\lambda = 630$ nm (Figure 11.16b). The DG device showed characteristic response times of $t_b = 53$ ms for the bleaching step and $t_c = 63$ ms for the reverse process, determined by fitting exponential functions to the switching curves. These short response times which are close to video rate

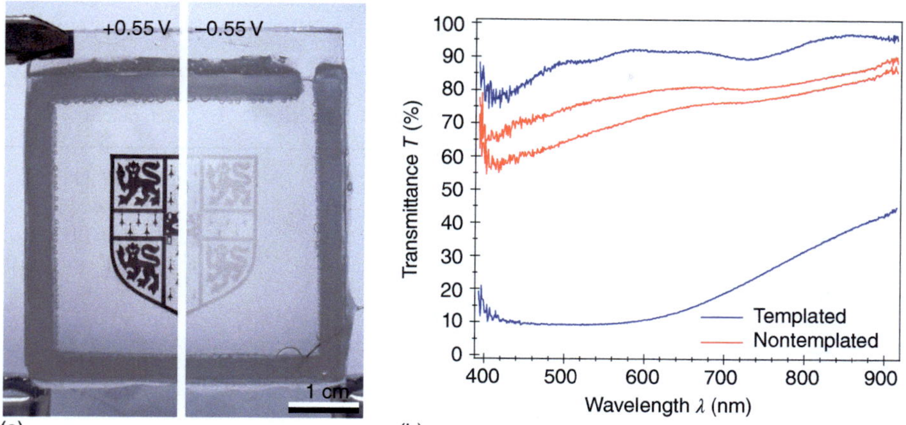

Figure 11.15 (a) DG-structured NiO electrochromic device showing the coloured oxidised (NiOOH) and the bleached reduced ($Ni(OH)_2$) state. (b) Optical transmittance spectra of a 460 nm thick DG-nanopatterned NiO film in the coloured (+0.55 V) and bleached (−0.55 V) state after 750 full switching cycles juxtaposed with a 70 nm thick nontemplated film. (Reproduced with permission from [40].)

(24 frames per second) are enabled by the short ion diffusion distance across the NiO $\approx$ 5 nm thick hollow NiO strut walls.

Electrochromic window applications, for example, may require an adjustable transmittance control which can be achieved by applying a cell potential between −0.55 and +0.55 V, as shown in Figure 11.15c. As expected, a steep change in transmittance occurs at the two redox peak potentials giving rise to hysteresis. However, the transmittance change in the range from 60 to 90 % is more gradual with a very low current passing through the cell.

For the DG NiO morphology, average coloration efficiencies at $\lambda = 630$ nm of −47 and −40 $cm^2\ C^{-1}$ were measured for the 10th and 750th switching cycle (Figure 11.15d) [97]. These η values in combination with a high Δ OD = 2.1 are a direct consequence of the highly interconnected gyroid morphology, which combines a very high specific surface area providing good ion access with a high electric conductivity.

Overall, the DG-nanopatterned sample not only outperforms the nontemplated counterpart but also stacks up well against NiO prepared by electrodeposition, chemical bath and chemical vapour deposition (see Table 11.1) [19, 24, 98–100]. Importantly, compared to the listed earlier studies, the DG device performs exceptionally well in all relevant electrochromic parameters (Figure 11.16).

11.5 Concluding Remarks

In summary, nanostructuring the electrodes of electrochromic and electrochemical energy storage devices can lead to significant performance enhancements

Table 11.1 Comparison of performances between the 460 nm thick DG-nanopatterned and other selected NiO-based electrochromic devices.

λ(nm)	$\|\Delta T\|$(%)	t_c(s)	t_b(s)	$\|\Delta$ OD$\|$	$\|$CE$\|$(cm^2 C^{-1})	References
550	66.2	3.5	4.0	—	25.5	Lin *et al.* [98]
550	76	3.0	5.0	—	41	Yuan *et al.* [24]
550	82	8.0	10	—	42	Xia *et al.* [99]
630	45	1.7	2.0	0.71	107	Sonavane *et al.* [100]
630	58	—	—	1.16	54	Dalavi *et al.* [19]
630	79.6	0.063	0.053	2.1	40—47	This study

(Reproduced with permission from [44].)

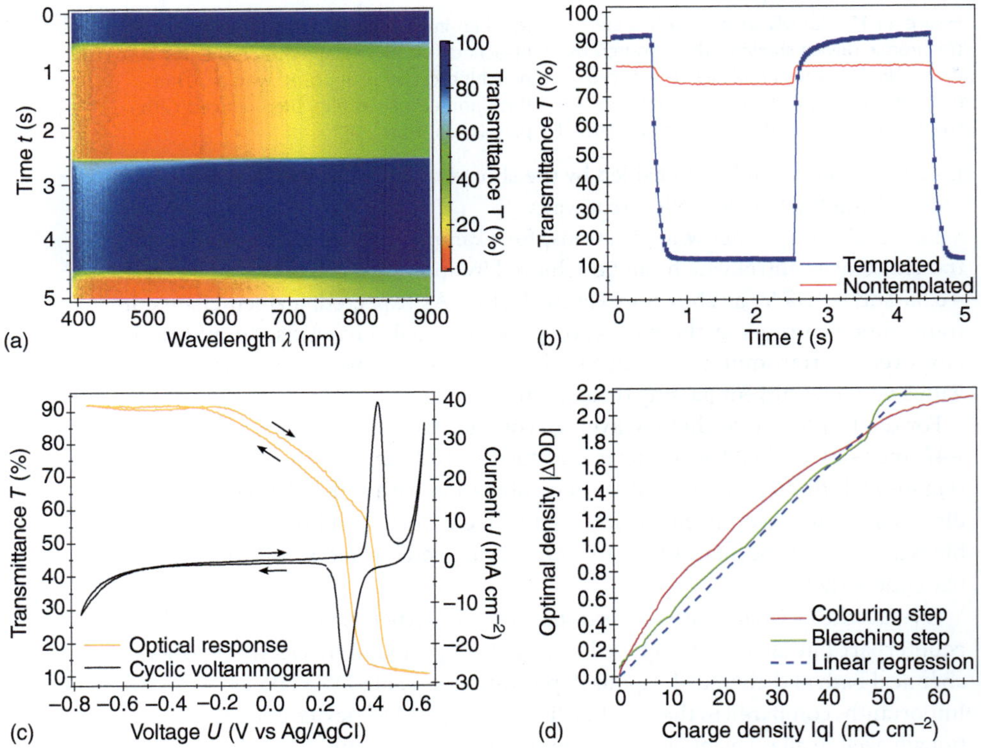

Figure 11.16 (a) Transmittance variation upon alternating 2 s potential steps ($\pm$0.55 V) recorded with a 25 ms integration time visualising a consistently fast switching response over the entire visible spectrum. (b) Corresponding switching behaviour at λ = 630 nm. (c) Optical transmittance response at λ = 630 nm as function of the applied cell potential and corresponding CV recorded at a 50 mV s^{-1} scan rate. (d) Optical density $|\Delta$ OD$|$ at λ = 630 nm versus the charge density $|q|$. (Reproduced with permission from [37])

compared to unstructured, compact electrodes. The results presented here illustrate the benefits of the mesoporous DG morphology with a strut diameter of around 11 nm. Further reduction of the structural feature size is probably counterproductive because of the lowered conductivity, arising quantum effects and lack of structural stability of the network.

References

1. Rosseinsky, D.R. and Mortimer, R.J. (2001) Electrochromic systems and the prospects for devices. *Adv. Mater.*, **13** (11), 783–793, doi: 10.1002/1521-4095(200106)13:11<783::AID-ADMA783>3.0.CO;2-D.
2. Monk, P.M.S., Mortimer, R.J. and Rosseinsky, D.R. (2007) *Electrochromism and Electrochromic Devices*, Cambridge University Press.
3. Granqvist, C.G. (2006) Electrochromic materials: out of a niche. *Nat. Mater.*, **5**, 89–90, doi: 10.1038/nmat1577.
4. Deb, S.K. (1969) A novel electrophotographic system. *Appl. Opt.*, **8** (Suppl. 1), 192–195, *http://ao.osa.org/abstract.cfm?URI=ao-8-101-192.*
5. Vidotti, M. and Cordoba de Torresi, S.I. (2008) Nanochromics: old materials, new structures and architectures for high performance devices. *J. Braz. Chem. Soc.*, **19** (7), 1248–1257, doi: 10.1590/S0103-50532008000700004.
6. Cho, S.I., Kwon, W.J., Choi, S.J., Kim, P., Park, S.A., Kim, J., Son, S.J., Xiao, R., Kim, S.H. and Lee, S.B. (2005) Nanotube-based ultrafast electrochromic display. *Adv. Mater.*, **17** (2), 171–175, doi: 10.1002/adma.200400499.
7. Macedo, M., Dallantonia, L., Valla, B. and Aegerter, M. (1992) Electrochromic smart windows. *J. Non-Cryst. Solids*, **147**, 792–798, doi: 10.1016/S0022-3093(05)80718-5, WOS:A1992JW92000136.
8. Cinnsealach, R., Boschloo, G., Nagaraja Rao, S. and Fitzmaurice, D. (1998) Electrochromic windows based on viologen-modified nanostructured TiO2 films. *Sol. Energy Mater. Sol. Cells*, **55** (3), 215–223, doi: 10.1016/S0927-0248(98)00096-8.
9. Cummins, D., Boschloo, G., Ryan, M., Corr, D., Rao, S.N. and Fitzmaurice, D. (2000) Ultrafast electrochromic windows based on redox-chromophore modified nanostructured semiconducting and conducting films. *J. Phys. Chem. B*, **104** (48), 11 449–11 459, doi: 10.1021/jp001763b.
10. Scherer, M.R.J., Muresan, N.M., Steiner, U. and Reisner, E. (2013) RYB tricolour electrochromism based on a molecular cobaloxime. *Chem. Commun.*, **49** (89), 10 453–10 455, doi: 10.1039/C3CC45881D.
11. Meuler, A.J., Ellison, C.J., Qin, J., Evans, C.M., Hillmyer, M.A. and Bates, F.S. (2009) Polydispersity effects in poly(isoprene-b-styrene-b-ethylene oxide) triblock terpolymers. *J. Chem. Phys.*, **130** (23), 234 903–234 917, doi: 10.1063/1.3140205.
12. Gates, B.D., Xu, Q., Stewart, M., Ryan, D., Willson, C.G. and Whitesides, G.M. (2005) New approaches to nanofabrication: molding, printing, and other techniques. *Chem. Rev.*, **105** (4), 1171–1196, doi: 10.1021/cr030076o.
13. Li, M. and Ober, C.K. (2006) Block copolymer patterns and templates. *Mater. Today*, **9** (9), 30–39, doi: 10.1016/S1369-7021(06)71620-0.
14. Bonhôte, P., Gogniat, E., Campus, F., Walder, L. and Grätzel, M. (1999) Nanocrystalline electrochromic displays. *Displays*, **20** (3), 137–144, doi: 10.1016/S0141-9382(99)00015-3.
15. Lee, S.H., Deshpande, R., Parilla, P.A., Jones, K.M., To, B., Mahan, A.H. and Dillon, A.C. (2006) Crystalline WO_3 nanoparticles for highly improved electrochromic applications. *Adv. Mater.*, **18** (6), 763–766, doi: 10.1002/adma.200501953.

16. Takahashi, K., Wang, Y. and Cao, G. (2005) Growth and electrochromic properties of single-crystal v_2O_5 nanorod arrays. *Appl. Phys. Lett.*, **86** (5), 053 102, doi: 10.1063/1.1857087.
17. Wang, J., Khoo, E., Lee, P.S. and Ma, J. (2009) Controlled synthesis of WO_3 nanorods and their electrochromic properties in H_2SO_4 electrolyte. *J. Phys. Chem. C*, **113** (22), 9655–9658, doi: 10.1021/jp901650v.
18. Zhang, H., Yu, X. and Braun, P.V. (2011) Three-dimensional bicontinuous ultrafast-charge and -discharge bulk battery electrodes. *Nat. Nanotechnol.*, **6** (5), 277–281, doi: 10.1038/nnano.2011.38.
19. Dalavi, D., Suryavanshi, M., Patil, D., Mali, S., Moholkar, A., Kalagi, S., Vanalkar, S., Kang, S., Kim, J. and Patil, P. (2011) Nanoporous nickel oxide thin films and its improved electrochromic performance: effect of thickness. *Appl. Surf. Sci.*, **257** (7), 2647–2656, doi: 10.1016/j.apsusc.2010.10.037.
20. Dalavi, D.S., Suryavanshi, M.J., Mali, S.S., Patil, D.S. and Patil, P.S. (2011) Efficient maximization of coloration by modification in morphology of electrodeposited NiO thin films prepared with different surfactants. *J. Solid State Electrochem.*, **16** (1), 253–263, doi: 10.1007/s10008-011-1314-y.
21. Brezesinski, T., Fattakhova-Rohlfing, D., Sallard, S., Antonietti, M. and Smarsly, B.M. (2006) Highly crystalline WO_3 thin films with ordered 3D mesoporosity and improved electrochromic performance. *Small*, **2** (10), 1203–1211, doi: 10.1002/smll.200600176.
22. Deepa, M., Srivastava, A.K., Sharma, S.N., Govind and Shivaprasad, S.M. (2008) Microstructural and electrochromic properties of tungsten oxide thin films produced by surfactant mediated electrodeposition. *Appl. Surf. Sci.*, **254** (8), 2342–2352, doi: 10.1016/j.apsusc.2007.09.035.
23. Kuai, S.L., Bader, G. and Ashrit, P.V. (2005) Tunable electrochromic photonic crystals. *Appl. Phys. Lett.*, **86** (22), 221 110, doi: 10.1063/1.1929079.
24. Yuan, Y., Xia, X., Wu, J., Chen, Y., Yang, J. and Guo, S. (2011) Enhanced electrochromic properties of ordered porous nickel oxide thin film prepared by self-assembled colloidal crystal template-assisted electrodeposition. *Electrochim. Acta*, **56** (3), 1208–1212, doi: 10.1016/j.electacta.2010.10.097.
25. Li, L., Steiner, U. and Mahajan, S. (2010) Improved electrochromic performance in inverse opal vanadium oxide films. *J. Mater. Chem.*, **20** (34), 7131, doi: 10.1039/c0jm00558d.
26. Liu, F., Liu, Y., Huang, L., Hu, X., Dong, B., Shi, W., Xie, Y. and Ye, X. (2011) Replication of homologous optical and hydrophobic features by templating wings of butterflies Morpho menelaus. *Opt. Commun.*, **284** (9), 2376–2381, doi: 10.1016/j.optcom.2011.01.017.
27. Ge, D., Yang, L., Tong, Z., Ding, Y., Xin, W., Zhao, J. and Li, Y. (2013) Ion diffusion and optical switching performance of 3D ordered nanostructured polyaniline films for advanced electrochemical/electrochromic devices. *Electrochim. Acta*, **104**, 191–197, doi: 10.1016/j.electacta.2013.04.102.
28. Cheng, L., Zhang, X., Liu, B., Wang, H., Li, Y., Huang, Y. and Du, Z. (2005) Template synthesis and characterization of WO_3/TiO_2 composite nanotubes. *Nanotechnology*, **16** (8), 1341, doi: 10.1088/0957-4484/16/8/060.
29. Cho, H., Kim, J., Patil, P., Kim, J.Y. and Kim, T.H. (2007) Synthesis of succinylated poly(4-hydroxystyrene) and its application for negative-tone photoresist. *J. Appl. Polym. Sci.*, **103** (6), 3560–3566, doi: 10.1002/app.24979.
30. Byun, J., Kim, Y., Jeon, G. and Kim, J.K. (2011) Ultrahigh density array of free-standing poly(3-hexylthiophene) nanotubes on conducting substrates via solution wetting. *Macromolecules*, **44** (21), 8558–8562, doi: 10.1021/ma202018m.
31. Nguyen, T.M., Cho, S., Varongchayakul, N., Yoon, D., Seog, J., Zong, K. and Lee, S.B. (2012) Electrochemical synthesis and one step modification of PMProDot nanotubes and their enhanced electrochemical properties. *Chem. Commun.*, **48** (21), 2725–2727, doi: 10.1039/C2CC17152J.

32. Crossland, E.J.W. (2008) Block Copolymer Patterning of Functional Materials. PhD thesis. University of Cambridge.
33. Hsueh, H.Y., Huang, Y.C., Ho, R.M., Lai, C.H., Makida, T. and Hasegawa, H. (2011) Nanoporous gyroid nickel from block copolymer templates via electroless plating. *Adv. Mater.*, **23** (27), 3041–3046, doi: 10.1002/adma.201100883.
34. Hsueh, H.Y. and Ho, R.M. (2012) Bicontinuous ceramics with high surface area from block copolymer templates. *Langmuir*, **28** (22), 8518–8529, doi: 10.1021/la3009706.
35. Wagner, T., Sauerwald, T., Kohl, C.D., Waitz, T., Weidmann, C. and Tiemann, M. (2009) Gas sensor based on ordered mesoporous In_2O_3. *Thin Solid Films*, **517** (22), 6170–6175, doi: 10.1016/j.tsf.2009.04.013.
36. Crossland, E.J.W., Ludwigs, S., Hillmyer, M.A. and Steiner, U. (2007) Freestanding nanowire arrays from soft-etch block copolymer templates. *Soft Matter*, **3** (1), 94–98, doi: 10.1039/b609780d.
37. Scherer, M.R.J., Li, L., Cunha, P.M.S., Scherman, O.A. and Steiner, U. (2012) Enhanced electrochromism in Gyroid-Structured vanadium pentoxide. *Adv. Mater.*, **24** (9), 1217–1221, doi: 10.1002/adma.201104272.
38. Wei, D., Scherer, M.R.J., Bower, C., Andrew, P., Ryhänen, T. and Steiner, U. (2012) A nanostructured electrochromic supercapacitor. *Nano Lett.*, **12** (4), 1857–1862, doi: 10.1021/nl2042112.
39. Scherer, M.R.J. (2013) *Double-Gyroid-Structured Functional Materials - Synthesis and Applications*, Springer-Verlag, *http://www.springer.com/materials/structural+materials/book/978-3-319-00353-5*.
40. Scherer, M.R.J. and Steiner, U. (2013) Efficient electrochromic devices made from 3D nanotubular gyroid networks. *Nano Lett.*, **13** (7), 3005–3010, doi: 10.1021/nl303833h.
41. Segalman, R.A. (2005) Patterning with block copolymer thin films. *Mater. Sci. Eng., R*, **48** (6), 191–226, doi: 10.1016/j.mser.2004.12.003.
42. Flory, P.J. (1942) Thermodynamics of high polymer solutions. *J. Chem. Phys.*, **10** (1), 51–61, doi: 10.1063/1.1723621.
43. Bates, F.S. and Fredrickson, G.H. (1999) Block Copolymers-Designer soft materials. *Phys. Today*, **52** (2), 32, doi: 10.1063/1.882522.
44. Khandpur, A.K., Foerster, S., Bates, F.S., Hamley, I.W., Ryan, A.J., Bras, W., Almdal, K. and Mortensen, K. (1995) Polyisoprene-polystyrene diblock copolymer phase diagram near the order-disorder transition. *Macromolecules*, **28** (26), 8796–8806, doi: 10.1021/ma00130a012.
45. Kim, M.I., Wakada, T., Akasaka, S., Nishitsuji, S., Saijo, K., Hasegawa, H., Ito, K. and Takenaka, M. (2009) Determination of the Fddd phase boundary in polystyrene-block-polyisoprene diblock copolymer melts. *Macromolecules*, **42** (14), 5266–5271, doi: 10.1021/ma900205s.
46. Thomas, E.L., Alward, D.B., Kinning, D.J., Martin, D.C., Handlin, D.L. and Fetters, L.J. (1986) Ordered bicontinuous double-diamond structure of star block copolymers: a new equilibrium microdomain morphology. *Macromolecules*, **19** (8), 2197–2202, doi: 10.1021/ma00162a016.
47. Hajduk, D.A., Harper, P.E., Gruner, S.M., Honeker, C.C., Kim, G., Thomas, E.L. and Fetters, L.J. (1994) The gyroid: a new equilibrium morphology in weakly segregated diblock copolymers. *Macromolecules*, **27** (15), 4063–4075, doi: 10.1021/ma00093a006.
48. Schulz, M.F., Bates, F.S., Almdal, K. and Mortensen, K. (1994) Epitaxial relationship for hexagonal-to-cubic phase transition in a book copolymer mixture. *Phys. Rev. Lett.*, **73** (1), 86, doi: 10.1103/PhysRevLett.73.86.
49. Hajduk, D.A., Harper, P.E., Gruner, S.M., Honeker, C.C., Thomas, E.L. and Fetters, L.J. (1995) A reevaluation of bicontinuous cubic phases in starblock copolymers. *Macromolecules*, **28** (7), 2570–2573, doi: 10.1021/ma00111a061.
50. Floudas, G., Ulrich, R. and Wiesner, U. (1999) Microphase separation in poly(isoprene-b-ethylene oxide) diblock copolymer melts. I. Phase state and

kinetics of the order-to-order transitions. *J. Chem. Phys.*, **110** (1), 652–663, doi: 10.1063/1.478122.

51. Ndoni, S., Vigild, M.E. and Berg, R.H. (2003) Nanoporous materials with spherical and gyroid cavities created by quantitative etching of polydimethylsiloxane in Polystyrene-Polydimethylsiloxane block copolymers. *J. Am. Chem. Soc.*, **125** (44), 13 366–13 367, doi: 10.1021/ja0360034.
52. Urade, V.N., Wei, T.C., Tate, M.P., Kowalski, J.D. and Hillhouse, H.W. (2007) Nanofabrication of double-gyroid thin films. *Chem. Mater.*, **19** (4), 768–777, doi: 10.1021/cm062136n.
53. Dair, B.J., Honeker, C.C., Alward, D.B., Avgeropoulos, A., Hadjichristidis, N., Fetters, L.J., Capel, M. and Thomas, E.L. (1999) Mechanical properties and deformation behavior of the double gyroid phase in unoriented thermoplastic elastomers. *Macromolecules*, **32** (24), 8145–8152, doi: 10.1021/ma990666h.
54. Nykänen, A., Nuopponen, M., Laukkanen, A., Hirvonen, S.P., Rytelä, M., Turunen, O., Tenhu, H., Mezzenga, R., Ikkala, O. and Ruokolainen, J. (2007) Phase behavior and temperature-responsive molecular filters based on self-assembly of polystyrene-block-poly(N-isopropylacrylamide)-block-polystyrene. *Macromolecules*, **40** (16), 5827–5834, doi: 10.1021/ma070378i.
55. Epps, T.H., Cochran, E.W., Bailey, T.S., Waletzko, R.S., Hardy, C.M. and Bates, F.S. (2004) Ordered network phases in linear poly(isoprene-b-styrene-b-ethylene oxide) triblock copolymers. *Macromolecules*, **37** (22), 8325–8341, doi: 10.1021/ma048762s.
56. Michielsen, K. and Kole, J.S. (2003) Photonic band gaps in materials with triply periodic surfaces and related tubular structures. *Phys. Rev. B*, **68** (11), 115 107, doi: 10.1103/PhysRevB.68.115107.
57. Wohlgemuth, M., Yufa, N., Hoffman, J. and Thomas, E.L. (2001) Triply periodic bicontinuous cubic microdomain morphologies by symmetries. *Macromolecules*, **34** (17), 6083–6089, doi: 10.1021/ma0019499.
58. Schoen, A.H. (1970) Infinite Periodic Minimal Surfaces Without Self-Intersections, NASA Technical Note D-5541.
59. Matyjaszewski, K. and Xia, J. (2001) Atom transfer radical polymerization. *Chem. Rev.*, **101** (9), 2921–2990, doi: 10.1021/cr940534g.
60. Kamber, N.E., Jeong, W., Waymouth, R.M., Pratt, R.C., Lohmeijer, B.G.G. and Hedrick, J.L. (2007) Organocatalytic ring-opening polymerization. *Chem. Rev.*, **107** (12), 5813–5840, doi: 10.1021/cr068415b.
61. Zalusky, A.S., Olayo-Valles, R., Wolf, J.H. and Hillmyer, M.A. (2002) Ordered nanoporous polymers from Polystyrene-Polylactide block copolymers. *J. Am. Chem. Soc.*, **124** (43), 12 761–12 773, doi: 10.1021/ja0278584.
62. Hashimoto, T., Nishikawa, Y. and Tsutsumi, K. (2007) Identification of the "Voided double-gyroid-channel": A new morphology in block copolymers. *Macromolecules*, **40** (4), 1066–1072, doi: 10.1021/ma061739h.
63. Leal, C., Bouxsein, N.F., Ewert, K.K. and Safinya, C.R. (2010) Highly efficient gene silencing activity of siRNA embedded in a nanostructured gyroid cubic lipid matrix. *J. Am. Chem. Soc.*, **132** (47), 16 841–16 847, doi: 10.1021/ja1059763.
64. Whittingham, M.S. (1976) Electrical energy storage and intercalation chemistry. *Science*, **192** (4244), 1126–1127, doi: 10.1126/science.192.4244.1126.
65. Winter, M., Besenhard, J.O., Spahr, M.E. and Novák, P. (1998) Insertion electrode materials for rechargeable lithium batteries. *Adv. Mater.*, **10** (10), 725–763, doi: 10.1002/(SICI)1521-4095(199807)10:10<725::AID-ADMA725>3.0.CO;2-Z.
66. Wang, Y., Takahashi, K., Lee, K.H. and Cao, G.Z. (2006) Nanostructured vanadium oxide electrodes for enhanced lithium-ion intercalation. *Adv. Funct. Mater.*, **16** (9), 1133–1144, doi: 10.1002/adfm.200500662.
67. Talledo, A. and Granqvist, C.G. (1995) Electrochromic vanadium-pentoxide-based films: structural, electrochemical,

and optical properties. *J. Appl. Phys.*, **77** (9), 4655, doi: 10.1063/1.359433.

68. West, K., Zachau-Christiansen, B., Jacobsen, T. and Skaarup, S. (1993) Vanadium oxide xerogels as electrodes for lithium batteries. *Electrochim. Acta*, **38** (9), 1215–1220, doi: 10.1016/0013-4686(93)80052-2.
69. Kudo, T., Ikeda, Y., Watanabe, T., Hibino, M., Miyayama, M., Abe, H. and Kajita, K. (2002) Amorphous V_2O_5/carbon composites as electrochemical supercapacitor electrodes. *Solid State Ionics*, **152-153**, 833–841, doi: 10.1016/S0167-2738(02)00383-1.
70. Grigorieva, A.V., Badalyan, S.M., Goodilin, E.A., Rumyantseva, M.N., Gaskov, A.M., Birkner, A. and Tretyakov, Y.D. (2010) Synthesis, structure, and sensor properties of vanadium pentoxide nanorods. *Eur. J. Inorg. Chem.*, **2010** (33), 5247–5253, doi: 10.1002/ejic.201000372.
71. Lu, Z., Levi, M., Salitra, G., Gofer, Y., Levi, E. and Aurbach, D. (2000) Basic electroanalytical characterization of lithium insertion into thin, well-crystallized V_2O_5 films. *J. Electroanal. Chem.*, **491** (1-2), 211–221, doi: 10.1016/S0022-0728(00)00184-4.
72. Odani, A., Pol, V.G., Pol, S.V., Koltypin, M., Gedanken, A. and Aurbach, D. (2006) Testing Carbon-Coated VO_x prepared via reaction under autogenic pressure at elevated temperature as Li-Insertion materials. *Adv. Mater.*, **18** (11), 1431–1436, doi: 10.1002/adma.200501611.
73. Murphy, D., Christian, P., DiSalvo, F. and Waszczak, J. (1979) Lithium incorporation by vanadium pentoxide. *Inorg. Chem.*, **18** (10), 2800–2803.
74. Chan, C.K., Peng, H., Twesten, R.D., Jarausch, K., Zhang, X.F. and Cui, Y. (2007) Fast, completely reversible Li insertion in vanadium pentoxide nanoribbons. *Nano Lett.*, **7** (2), 490–495, doi: 10.1021/nl062883j.
75. Potiron, E. (1999) Electrochemically synthesized vanadium oxides as lithium insertion hosts. *Electrochim. Acta*, **45** (1-2), 197–214, *http://resolver.scholarsportal.info/resolve/00134686/v45i1-2/197_esvoalih.xml.*
76. Wang, W., Jia, F., Huang, Q. and Zhang, J. (2005) A new type of low power thermoelectric microgenerator fabricated by nanowire array thermoelectric material. *Microelectron. Eng.*, **77** (3-4), 223–229, doi: 10.1016/j.mee.2004.11.005.
77. Wang, Y., Shang, H., Chou, T. and Cao, G. (2005) Effects of thermal annealing on the Li^+ intercalation properties of $V_2O_5 \cdot nH_2O$ xerogel films. *J. Phys. Chem. B*, **109** (22), 11 361–11 366, doi: 10.1021/jp051275+.
78. Li, J.M., Chang, K.H. and Hu, C.C. (2010) A novel vanadium oxide deposit for the cathode of asymmetric lithium-ion supercapacitors. *Electrochem. Commun.*, **12** (12), 1800–1803, doi: 10.1016/j.elecom.2010.10.029.
79. Granqvist, C.G. (1995) *Handbook of Inorganic Electrochromic Materials*, Elsevier, Amsterdam, The Netherlands.
80. Gaupp, C.L., Welsh, D.M., Rauh, R.D. and Reynolds, J.R. (2002) Composite coloration efficiency measurements of electrochromic polymers based on 3,4-alkylenedioxythiophenes. *Chem. Mater.*, **14** (9), 3964–3970, doi: 10.1021/cm020433w.
81. Chen, Z., Augustyn, V., Wen, J., Zhang, Y., Shen, M., Dunn, B. and Lu, Y. (2011) High-Performance supercapacitors based on intertwined CNT/V_2O_5 nanowire nanocomposites. *Adv. Mater.*, **23** (6), 791–795, doi: 10.1002/adma.201003658.
82. Toupin, M., Brousse, T. and Bélanger, D. (2002) Influence of microstucture on the charge storage properties of chemically synthesized manganese dioxide. *Chem. Mater.*, **14** (9), 3946–3952, doi: 10.1021/cm020408q.
83. Brezesinski, T., Wang, J., Tolbert, S.H. and Dunn, B. (2010) Ordered mesoporous [alpha]-MoO_3 with iso-oriented nanocrystalline walls for thin-film pseudocapacitors. *Nat. Mater.*, **9** (2), 146–151, doi: 10.1038/nmat2612.
84. Lee, S.W., Yabuuchi, N., Gallant, B.M., Chen, S., Kim, B.S., Hammond, P.T. and Shao-Horn, Y. (2010) High-power lithium batteries from functionalized carbon-nanotube electrodes. *Nat.*

Nanotechnol., **5** (7), 531–537, doi: 10.1038/nnano.2010.116.

85. Niklasson, G.A. and Granqvist, C.G. (2007) Electrochromics for smart windows: thin films of tungsten oxide and nickel oxide, and devices based on these. *J. Mater. Chem.*, **17** (2), 127, doi: 10.1039/b612174h.
86. Shukla, A., Venugopalan, S. and Hariprakash, B. (2001) Nickel-based rechargeable batteries. *J. Power Sources*, **100** (1-2), 125–148, doi: 10.1016/S0378-7753(01)00890-4.
87. Nam, K.W. and Kim, K.B. (2002) A study of the preparation of NiO_x electrode via electrochemical route for supercapacitor applications and their charge storage mechanism. *J. Electrochem. Soc.*, **149** (3), A346, doi: 10.1149/1.1449951.
88. Dirksen, J.A., Duval, K. and Ring, T.A. (2001) NiO thin-film formaldehyde gas sensor. *Sens. Actuators, B*, **80** (2), 106–115, doi: 10.1016/S0925-4005(01)00898-X.
89. Yin, Y., Rioux, R.M., Erdonmez, C.K., Hughes, S., Somorjai, G.A. and Alivisatos, A.P. (2004) Formation of hollow nanocrystals through the nanoscale Kirkendall effect. *Science*, **304** (5671), 711–714, doi: 10.1126/science.1096566.
90. Nakamura, R., Lee, J.G., Mori, H. and Nakajima, H. (2008) Oxidation behaviour of Ni nanoparticles and formation process of hollow NiO. *Philos. Mag.*, **88** (2), 257–264, doi: 10.1080/14786430701819203.
91. Ren, Y., Chim, W.K., Chiam, S.Y., Huang, J.Q., Pi, C. and Pan, J.S. (2010) Formation of nickel oxide nanotubes with uniform wall thickness by Low-Temperature thermal oxidation through understanding the limiting effect of vacancy diffusion and the Kirkendall phenomenon. *Adv. Funct. Mater.*, **20** (19), 3336–3342, doi: 10.1002/adfm.201000498.
92. Yang, Y., Liu, L., Güder, F., Berger, A., Scholz, R., Albrecht, O. and Zacharias, M. (2011) Regulated oxidation of nickel in multisegmented Nickel-Platinum nanowires: An entry to wavy nanopeapods. *Angew. Chem. Int. Ed.*, **50** (46), 10 855–10 858, doi: 10.1002/anie.201102740.
93. Atkinson, A. and Taylor, R.I. (1979) The diffusion of Ni in the bulk and along dislocations in NiO single crystals. *Philos. Mag. A*, **39** (5), 581–595, doi: 10.1080/01418617908239293.
94. O'Keeffe, M. and Moore, W.J. (1961) Diffusion of oxygen in single crystals of nickel oxide. *J. Phys. Chem.*, **65** (8), 1438–1439, doi: 10.1021/j100826a039.
95. Bouessay, I., Rougier, A., Poizot, P., Moscovici, J., Michalowicz, A. and Tarascon, J.M. (2005) Electrochromic degradation in nickel oxide thin film: a self-discharge and dissolution phenomenon. *Electrochim. Acta*, **50** (18), 3737–3745, doi: 10.1016/j.electacta.2005.01.020.
96. Liu, H., Yan, G., Liu, F., Zhong, Y. and Feng, B. (2009) Structural, electrochemical and optical properties of NiO_xH_y thin films prepared by electrochemical deposition. *J. Alloys Compd.*, **481** (1-2), 385–389, doi: 10.1016/j.jallcom.2009.03.015.
97. Maruyama, T. and Arai, S. (1993) The electrochromic properties of nickel oxide thin films prepared by chemical vapor deposition. *Sol. Energy Mater. Sol. Cells*, **30** (3), 257–262, doi: 10.1016/0927-0248(93)90145-S.
98. Lin, S.H., Chen, F.R. and Kai, J.J. (2008) Electrochromic properties of nanocomposite nickel oxide film. *Appl. Surf. Sci.*, **254** (11), 3357–3363, doi: 10.1016/j.apsusc.2007.11.022.
99. Xia, X., Tu, J., Zhang, J., Huang, X., Wang, X., Zhang, W. and Huang, H. (2008) Enhanced electrochromics of nanoporous cobalt oxide thin film prepared by a facile chemical bath deposition. *Electrochem. Commun.*, **10** (11), 1815–1818, doi: 10.1016/j.elecom.2008.09.025.
100. Sonavane, A.C., Inamdar, A.I., Deshmukh, H.P. and Patil, P.S. (2010) Multicoloured electrochromic thin films of NiO/PANI. *J. Phys. D: Appl. Phys.*, **43** (31), 315 102, doi: 10.1088/0022-3727/43/31/315102.

12
Layer-by-Layer Assembly of Electrochromic Materials: On the Efficient Method for Immobilisation of Nanomaterials

Susana I. Córdoba de Torresi, Jose R. Martins Neto, Marcio Vidotti, and Fritz Huguenin

12.1
Introduction to the Layer-by-Layer Deposition Technique

The electrostatic layer-by-layer (LbL) deposition technique is a well-known method that comprises sequential dipping of the electrode in dilute solutions containing charged species. The major advantages of LbL adsorption from solution are that many different nanomaterials can be incorporated in individual multilayer films and that the film architectures are completely determined by the deposition sequence. Another difference between bulk systems and surface-confined multilayers is that bulk nanocomposites are often turbid materials, whereas LbL assembled films can be applied as wavelength-thick transparent coatings on, for example, optical devices.

Since the report of Decher [1] regarding the LbL technique, many different reports in the literature can be found where this methodology is applied using different materials beyond the polymeric structures initially proposed [2]. A very interesting contribution was provided by Lvov *et al.*, who studied the adsorption of SiO_2 nanoparticles [3]. In this paper, many experimental parameters were tested, such as size and concentration of SiO_2, ionic strength, nature of polycation (polydiallyldimethylamonium (PDDA) and polyethyleneimine (PEI) were used) and adsorption time. This is one of the few contributions that corroborate the layer growth by using the quartz crystal microbalance as seen in Figure 12.1. In this result, the effect of the deposition of PDDA and SiO_2 on the frequency change is observed by different slopes, clearly indicating the multilayer growth apart from the concentration of SiO_2. The same authors also studied the adsorption of other materials such as proteins by employing similar methodology [4].

12.2
Layer-by-Layer Assembly in Electrochromic Materials

The simplicity of the LbL technique was also extended to other materials, including the electrochromic ones. One important feature achieved by this deposition

Electrochromic Materials and Devices, First Edition.
Edited by Roger J. Mortimer, David R. Rosseinsky and Paul M. S. Monk.

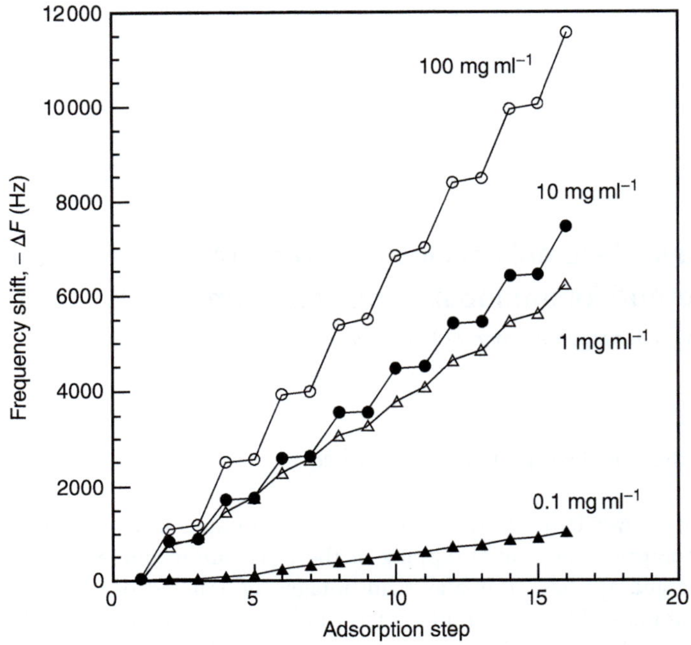

Figure 12.1 Influence of the particle concentration on the alternate adsorption of SiO_2 and PDDA at 22 °C and an adsorption time of 15 min; even-number steps, SiO_2 adsorption; odd-number steps, PDDA adsorption, SiO_2; particle diameter, 45 nm; dispersed at pH 10. PDDA; 3 mg ml^{-1}. (Figure reproduced from Ref. [3], with permission from The American Chemical Society.)

relies on the film homogeneity achieved which is very difficult to obtain by using classical electrode modifications such as casting or spin coating.

12.2.1
Layer-by-Layer Assembly of Conjugated Conducting Polymers

Among intrinsic (conjugated) conducting polymers with electrochromic properties, poly(3,4-ethylenedioxythiophene) (PEDOT) has been extensively studied over the years due to characteristics such as high transmittance changes from deep blue to transparent, besides its remarkable film stability. PEDOT can be commercially obtained in an aqueous dispersion known as Baytron®, with PSS (polystyrenesulfonate) as counterion and stabiliser, and the construction of electrochromic electrodes based on this suspension can be easily found in the literature, due to its simplicity and availability [5]. Apart from poly (3,4-ethylenedioxythiophene):poly(styrene sulfonate) (PEDOT:PSS) dispersions, Cutler *et al.* have demonstrated the LbL electrode modification by employing poly(4-(2,3-dihydrothieno [3,4-b]-[1,4] dioxin-2-yl-methoxyl)-1-butanesulfonic acid) (PEDOT-S), a PEDOT derivative highly soluble in water due to the presence of a hydrophilic chain attached to the EDOT structure, the electrode was modified

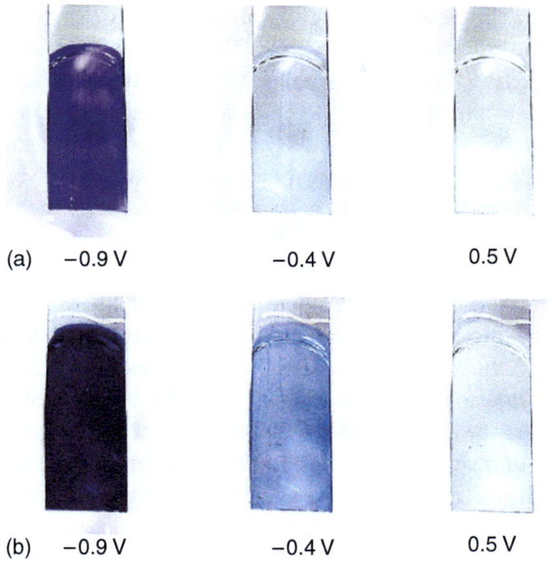

Figure 12.2 Digital photographs of the coloration and bleaching of (a) $(PXV/PEDOT:SPS)_{40}$ and (b) $(PXV/PEDOT:SPS)_{50}$. The films as shown are immersed in 0.1 M NH_4Cl within an electrochemical cell. Shadows and lighting gradients are responsible for the variations in brightness on the film surface as the films are smooth and uniform. (Figure reproduced from Ref. [8], with permission from The American Chemical Society.)

by alternating layers of PEDOT-S and poly(hydrochloride diallylamine) (PAH) [6]. Many other electrochromic polymers and their physical chemical properties can be found in Ref. [7].

Another example of using the LbL technique and electrochromic polymers can be found in the work published by DeLongchamp *et al.* [8] where alternate layers of both poly (hexyl viologen) (PXV) and PEDOT:PSS were used. Figure 12.2 shows the assembled electrodes which presented a 'dual electrochrome' character due to the effect of both PXV and PEDOT. According to the authors, the thicknesses of the modified electrodes grow linearly with the number of bilayers presenting approximately 1 μm for an electrode modified by 50 bilayers. As expected, by increasing the number of bilayers, the electrochromic response times increase, reaching values of 2–3 s for thicker films, but with higher chromatic contrast.

The incorporation of nanostructures in electrochromic electrodes has been of great importance in electrochromic electrodes. With their unique morphology, nanomaterials allow fast ionic intercalation leading to high-performance electrochromic electrodes with high efficiency and lower response times [9]. The synthesis of nanoparticles based on conducting polymers is widespread in the literature [10]. As mentioned before, PEDOT has many advantages in electrochromic electrodes and its utilisation in nanometric morphology can lead to an enhancement of all chromogenic properties.

Müller *et al.* have developed a simple and straightforward methodology to synthesise PEDOT nanoparticles in aqueous dispersion [11]. The nanoparticles were formed from the oxidation of EDOT by Fe^{3+} and stabilised by the cationic surfactant decyltrimethylammonium bromide (DETAB). The TEM images have shown mainly particles of about 25 nm with high conductivity. The electrode modification was performed by LbL using PSS as anionic polymer, each layer was grown by immersing gold modified electrodes for 15 min in each solution. The change in the optical properties of the modified electrode was measured by surface plasmon resonance (SPR) monitoring of the change in reflectivity upon oxidation and reduction, leading to swelling and shrinking of the nanoparticles accomplished with ionic diffusion in and out of the modified electrode. In order to assure the effect of the nanometric morphology, the authors have also shown the results for LbL modified electrodes using polymeric PEDOT (Baytron®) and PEI. By using SPR and chronoamperometry, it was shown that the electrode modified by nanoparticles presented a faster ionic diffusion rate, leading to response times about 10 times shorter than those of the electrodes modified by PEDOT in polymeric traditional form. In Figure 12.3, the images of the PEDOT nanoparticles and the change in reflectivity for electrodes modified by nanoparticles and Baytron® are compared.

12.2.2
Layer-by-Layer Assembly of Intervalence Charge Transfer Coloration Materials

Many different inorganic electrochromic nanoparticles have been employed for LbL modified electrodes. Iron hexacyanoferrate, mostly known as Prussian blue (PB), is one of the most versatile inorganic materials, its utilisation covering pigments, electrocatalytic systems and, obviously, electrochromic electrodes.

The PB synthetic procedure is very simple consisting basically of chemical precipitation of hexacyanoferrate and iron cations. For the growth of nanoparticles, it is very important to control the ratio of the ions as described by Hammond and co-worker [12], who synthesised PB nanoparticles of about 5 nm using a very rapid and straightforward methodology. These nanoparticles were assembled on ITO-coated electrodes by alternating layers of the linear polycation PEI and presented multiple colorations (bleached, cyan, green and yellow) depending on the voltage applied to the electrode. The PB modified electrode showed a decrease in its transmittance signal after cycling to high potentials where Prussian yellow was formed; this fact can be due either to the irreversible electrochemical reaction or to the loss of the negative net charge alongside the inorganic layer leading to a bleaching effect.

In another publication from the same research group [13], PB nanoparticles were intercalated with poly(aniline) (PANI) on the same electrode. This kind of assembly is particularly interesting because, differently from classical cationic polycations (such as PAH, PDDA, PEI), PANI is an electroactive polymer; by this way, the electrical contact between PB layers is enhanced. Besides, PANI also has electrochromic properties, so the modified electrode presented an additional

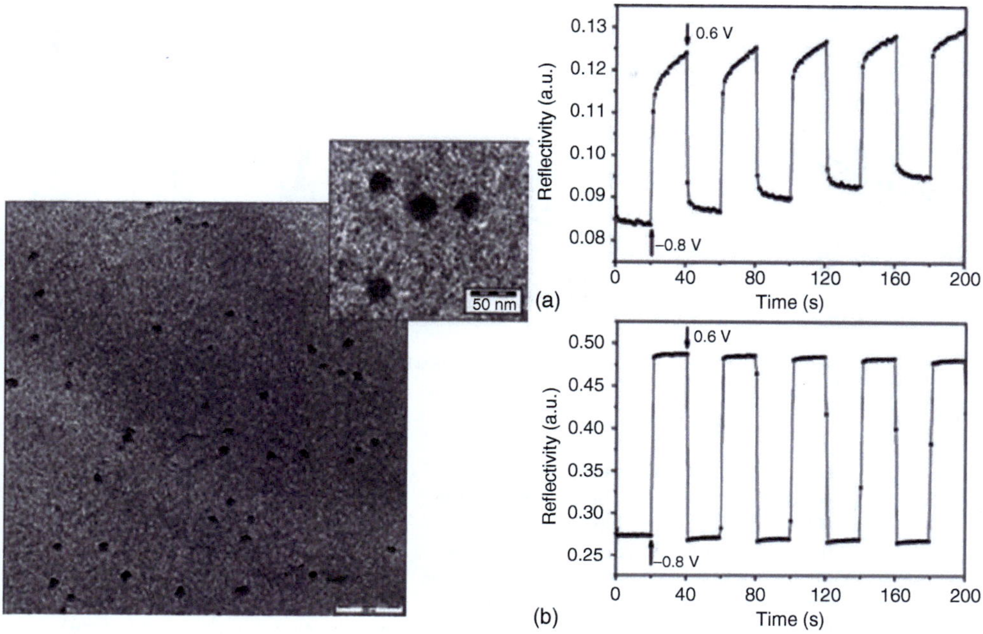

Figure 12.3 (a) TEM image of PEDOT nanoparticles. (b) Time-dependent SPR reflectivity changes of (a) (Baytron® P/PEI)$_3$ multi-layer on gold electrodes at a fixed incidence angle (61°) and (b) (PEDOT nanoparticles/PSS)$_4$ multi-layer at a fixed incidence angle (58°) upon application of potential square waves between −0.8 and 0.6 V (vs. Ag/AgCl) with a time step of 20 s. All experiments were measured in 50×10^{-3} M (pH = 7.40) aqueous solution. (Reproduced from Ref. [11], with permission from John Wiley and Sons.)

chromogenic change. The electrochemical properties were analysed by cyclic voltammetry, and the electric interconnection between the layers where both peaks, of PB and PANI, were found was clearly observed. By increasing the number of PB–PANI (Prussian blue–poly(aniline)) bilayers, the transmittance increased, reaching a value of 61.1% (750 nm), higher than that of PB electrode modified with the same number of bilayers (30). It is important to note that the response times (for bleaching and coloration) are diminished by intercalating PB electrodes with PANI indicating that the ionic diffusion through the electroactive layers has also suffered an enhancement. Figure 12.4 shows the voltammetric responses of PANI, PB and PANI–PB modified electrodes and the differences in the chromogenic behaviour of the modified electrodes.

As mentioned earlier, the range of applications of PB is quite broad; nevertheless, there are some electrochemical limitations, the most described being the lack of electroactivity of PB electrodes in electrolytes containing large hydrated cations such as Na^+ and NH_4^+ because they cannot intercalate within the inorganic structure of PB, leading to a pronounced irreversible behaviour [14]. In order to overcome this fact, many attempts have been made consisting of a replacement

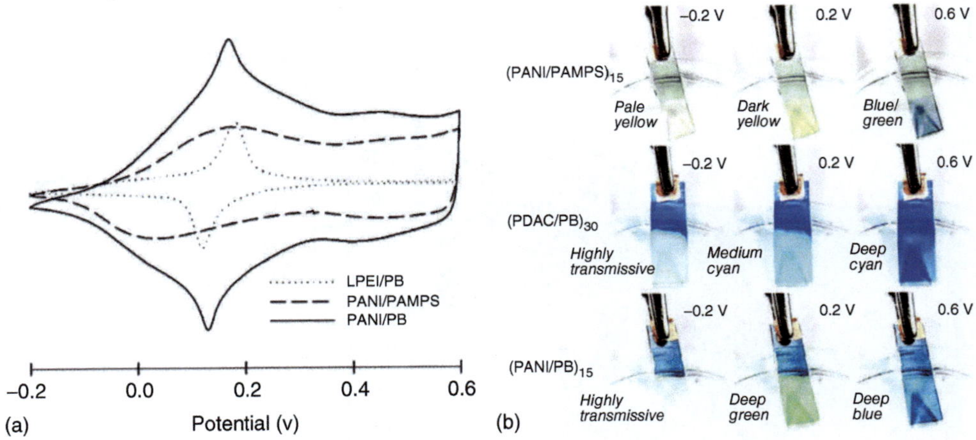

Figure 12.4 (a) Comparison of cyclic voltammetry of several LBL composites: the LPEI/PB composite exhibits only PB electrochemistry, the PANI/poly(2-acrylamido-2-methyl-1-propanesulfonic acid) (PANI/PAMPS) composite exhibits only PANI electrochemistry and the PANI/PB composite exhibits both. Scan rate of $25\,mV\,s^{-1}$, and potential versus K-SCE. (b) Photographs of the electrochromism of PANI/PB and the origins of multiple coloured electrochromism in this dual electrochromic composite. (Figure reproduced from Ref. [13], with permission from The American Chemical Society.)

of the Fe^{3+} by other metallic cations, such as those of copper, nickel and cobalt. Interestingly, all of the so-called PB analogues present electrochromic properties.

Baioni *et al.* [15] described the ultrasonic synthesis of copper hexacyanoferrate nanoparticles, with characterisation by spectroscopic and microscopic techniques. The electrochromic electrodes were built up using LbL, with poly(hydrochloride diallylamine)(PAH) as the cationic polyelectrolyte. Differently from PB, the copper hexacyanoferrate modified electrode presented a unique chromogenic process changing its coloration from deep brown to orange and achieving an electrochromic efficiency of about $3.8\,cm^2\,C^{-1}$ at 540 nm. This value is low when compared with PB modified electrodes but the electrochromic reversibility and the durability of the copper hexacyanoferrate nanoparticles were very impressive being cycled for over 5 h without a significant change in the chromatic contrast. In Figure 12.5, the voltammetric behaviour and the respective change in absorbance signal at 540 nm of electrodes modified by 5, 10 and 20 bilayers are shown, where, differently from PB electrodes, a unique redox process is found to occur.

12.3
Layer-by-Layer Assembly of Metal Oxides

Metal oxides as electrochromic films are widely studied in the literature. Indeed, LbL techniques have been applied as methods for film formation.

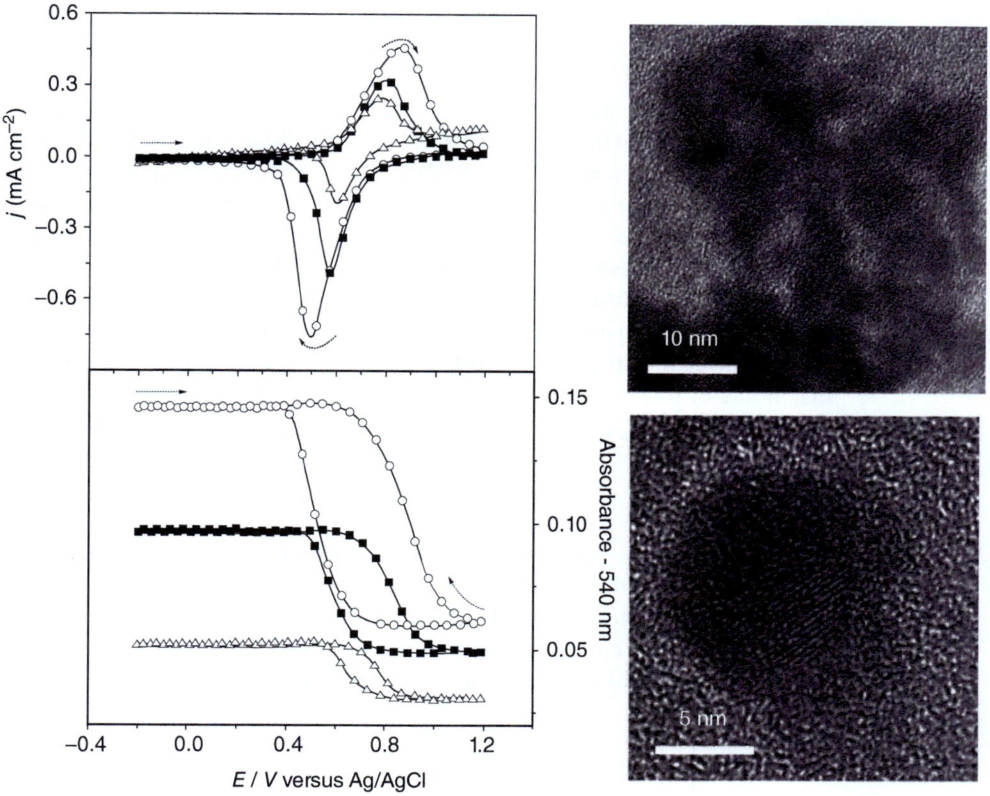

Figure 12.5 In situ *j/E* and absorbance/*E* potentiodynamic profiles recorded at 540 nm for nanostructured copper hexacyanoferrate electrodes modified with different amounts of bilayers: (-Δ-) 5, (-■-) 10 and (-o-) 20 bilayers. Electrolyte: 0.1 M KCl (pH 3). Scan rate = 0.01 V s^{-1}. HRTEM images obtained from a colloidal solution of copper hexacyanoferrate nanoparticles. (Figure reproduced from Ref. [15], with permission from The American Chemical Society.)

A spectroelectrochemical method was proposed for the examination of the chromogenic properties and the mass transport rate associated with the electroinserted lithium ions, which accompanies the coloration wave in the LbL nanocomposites formed from metal oxides [16, 17]. From this investigation, it is possible to predict the important factors for the design of high-performance electrodes for electrochromic devices (ECDs), particularly because of the possibility of manipulation at the nanometre scale in these self-assembled materials. On the basis of this spectroelectrochemical model, the lithium ion diffusion coefficient can be determined for distinct chromogenic components in composite electrodes that absorb at different wavelengths. Moreover, the influence of the amount of electroactive sites and of the diffusion process on the charge capacity can be investigated separately.

12.3.1
Tungsten Oxide

The electrochromic properties of WO_3 have been widely studied [18–22]. Its colour switches from transparent to blue during the lithium ion electro-insertion, when W(VI) ions are reduced to W(V) ions. The chemical reversibility and the electrochromic efficiency of electrodes formed from this metal oxide make them suitable for utilisation in display devices and smart windows [23]. However, the time needed for these electrodes to change their colour is long, because of the slow lithium ion diffusion into the host matrix.

The site saturation model (SSM) has been employed to describe the absorbance change during the lithium ion insertion into self-assembled WO_3 films, once the magnitude of this absorbance, associated with the intervalence charge transfer between the W(V) and W(VI) ions, depends on the amount of injected charge (q).

$$\Delta A = rq\left(1 - \frac{q}{K}\right) \tag{12.1}$$

where the r constant corresponds to the electrochromic efficiency ($\eta = \Delta A/q$) for low values of q and the K constant indicates the maximum charge injected in the electrode. Solving Fick's second law, with the corresponding initial and boundary conditions [24], the optical diffusion coefficient (D_{op}) can be determined, according to Eq. (12.2), which is associated with the lithium ion diffusion and the coloration wave front,

$$D_{op} = \left[\frac{iLr}{\sqrt{t\pi}} \frac{\left(1 - \frac{2q}{K}\right)}{\left(\frac{d\Delta A}{dt}\right)}\right]^2, \tag{12.2}$$

where i is the applied current, L is the thickness of the electrode and t is the time. On the basis of the fitting of the spectroelectrochemical data, the terms r and K can be obtained using the SSM. Meanwhile, the $d\Delta A/dt$ term can be obtained from the absorbance change as a function of the time during the current pulses.

Figure 12.6a shows the absorbance change as a function of injected charge for an LbL film formed from WO_3, chitosan and poly(ethylene oxide) (PEO). Using SSM to fit this curve, the K and r values obtained were 1.42 mC cm^{-2} and 109.0 cm^2 C^{-1}, respectively. These data allowed the determination of the amount of electroactive sites (about 14.71 nmol cm^{-2}) in the self-assembled material and the electrochromic efficiency at low values of injected charge. Figure 12.6b shows the values of absorbance change (at 660 nm) as a function of time for the WO_3/chitosan/PEO film, during a current pulse of 10 μA cm^{-2} by 10 s, and later at open circuit potential. We can note that the values of absorbance change decrease towards the initial value after the current pulse. This behaviour suggests dependence between the absorptivity and the lithium ion concentration inside the host matrix, possibly related with changes in the chemical environment. Due to low ionic diffusion rate, there is lithium ion accumulation close to the electrode/electrolytic solution interface at the non-equilibrium state, contributing

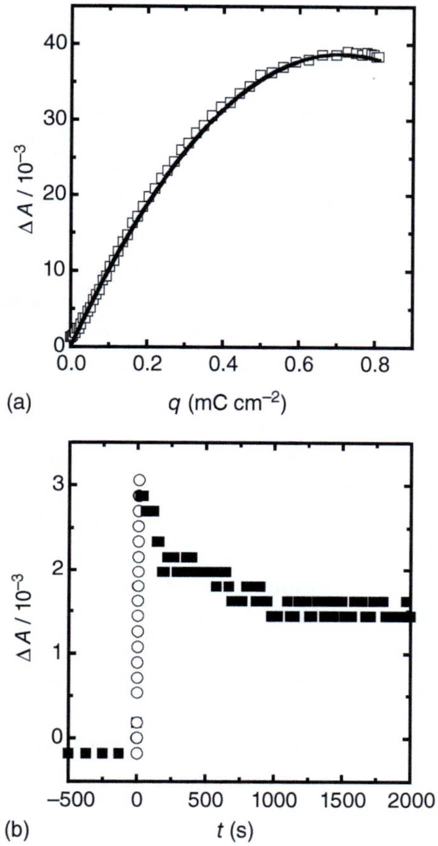

Figure 12.6 (a) Absorbance change, ΔA versus charge, at 660 nm for the 20-bilayer WO_3/Chitosan/PEO LbL film. (b) The absorbance change at 660 nm as a function of time (○) at a constant current pulse of 10 μA cm^{-2} for 10 s and (■) at open circuit potential. (Figure reproduced from Ref. [17], with permission from Elsevier.)

to enhancement in the absorptivity in comparison with that in the equilibrium state. Thus, after the current pulse, the absorptivity decreases towards its initial value, once the lithium ion distribution in the self-assembled matrix tends to be uniform at the novel equilibrium state.

On the basis of the r and K values and of the $d\Delta A/dt$ slopes of each of the current pulses, the D_{op} values can be determined according to Eq. (12.2). Thus, these data help to explain the values obtained for ion storage capacity and response time of electrochromic electrodes. For instance, Galiote *et al.* showed that the D_{op} and K values are higher for the WO_3/chitosan/PEO film than those in case of the WO_3/chitosan film, indicating that the presence of chitosan in the self-assembled structure increased both the rate of coloration wave front and the amount of electro-inserted lithium ions.

12.3.2
Hexaniobate

In the search for novel materials to produce nanostructured electrodes, some metal oxides have received much attention because of the possibility of delamination into nanosheets. These nanosheets correspond to a new class of nanoparticles in consequence of their strong morphological anisotropy described by a thickness at nanometric scale and lateral dimensions at micrometric scale. Moreover, these nanoparticles can be easily produced and used for the formation of LbL nanocomposites. Colloidal dispersion of hexaniobate single sheets has been obtained from the exfoliation of the acidic form of hexaniobate ($H_2K_2Nb_6O_{17}$) in aqueous solution, which has been used with a delaminating agent such as n-butylamine [25]. The plate-like niobate nanoparticles can be reassembled by the LbL method [26, 27], dip coating, Langmuir–Blodgett and electrostatic self-assembly deposition (ESD) techniques, [28, 29] forming multilayer films [30].

LbL films formed from $K_2Nb_6O_{17}{}^{2-}$ and poly(allylamine hydrochloride) (PAH) have been grown because of the electrostatic interactions between these components, and Rezende *et al.* have prepared these LbL films of hexaniobate to be used as a lithium ion insertion electrode for ECDs. Although the self-assembled and dip-coated electrochromic materials composed from this oxide have displayed electrochromic efficiency lower than that in the case of WO_3 electrodes, these materials show high electrochromic efficiency ($\approx$69.5 cm^2 C^{-1}). Moreover, the absorbance changes during the lithium ion electro-insertion occur in other regions of the visible spectrum (wavelength lower than 600 nm), which make these electrodes a possible alternative for ECDs.

12.3.3
Vanadium Oxide

V_2O_5 has also been widely studied due to its ionic electro-insertion and chromogenic properties, which makes this oxide a suitable host matrix for use as counter electrode in ECDs. This insertion electrode exhibits a specific capacity of about 250 A h kg^{-1} in the presence of lithium salts dissolved in organic solvent [31, 32]. This charge storage can still be increased depending on electrode mass, electro-insertion rate and synthesis and deposition methods [33]. During the lithium ion insertion in this electrode, there is a change of colour from yellow to blue. The light absorption (with low molar absorptivity) is attributed to the electronic transition between 2p O and 3d V bands at lower wavelengths as well the intervalence charge transfer from V(IV) to V(V) at higher wavelengths [34].

A significant reason for using self-assembled nanocomposites from V_2O_5 is the effect of intimate contact between their components on lithium ion diffusion, once the ionic mobility into this host matrix is low. The chemical diffusion coefficient (D_c) varies between 10^{-9} and 10^{-17} cm^2 s^{-1}, which depends on composition, nature of the precursor, preparation method and ageing time of the precursor gel [35, 36]. One of the reasons for low values of D_c is the interaction between lithium ions and the oxygen atoms, mainly those of the vanadyl groups that have the highest negative charge density.

Nanocomposites formed from V_2O_5, PAH and silver nanoparticles (AgNPs) have been assembly and characterised [37]. In this case, the PAH protonated amine group can specifically and strategically interact with the oxygen atom of the vanadyl group, allowing the spontaneous growth of these host matrices and giving rise to an electrostatic shield between the electro-inserted lithium ions and the sites with the highest electronic charge, increasing the ionic diffusion rate. However, the presence of non-conducting polyelectrolyte isolates some V(V) sites in the host matrix, due to its low electronic conductivity. Considering the versatility of the LbL method to build up insertion electrodes under manipulation on nanometric scale, AgNPs were stabilised in polymeric dispersion composed from PAH, allowing their incorporation into the self-assembled V_2O_5 matrices. Thus, conducting pathways were created to increase the number of electroactive sites under higher lithium ion mobility as compared to V_2O_5. Electrical and spectroelectrochemical experiments showed that the self-assembled nanoarchitectures obtained from the LbL method can lead to a better performance of ECDs, such as shorter response time and enhanced charge capacity.

Multilayer electrodes were produced from PANI and V_2O_5 with the LbL method, where control of thickness, electrochromic properties and charge capacity could be achieved. Using this self-assembled method, manipulations on the nanometric scale allowed the chromogenic properties of PANI to be maintained while the electrochemical profile of V_2O_5 xerogel was observed. Another advantage of this self-assembled film is the participation of lithium ions in the charge compensation mechanism. The deposition of a self-assembled PANI/V_2O_5 film on a PANI film changes the charge compensation process from anion to cation, which can decrease the mass and volume of the ECDs and may increase the ionic diffusion rate [38].

PEO is among the several examples of polymer insertion into V_2O_5 xerogels [39–41], which can be exploited to increase the mobility of lithium ions. Polymeric complexes from alkali metal salts and this polymer are known for their high ionic conductivity in comparison to other solid-state ionic conductors [42–44]. Huguenin *et al.* have verified that LbL films formed from PEO and V_2O_5 possess high charge storage and electrochromic capabilities, according to cyclic voltammetric, chronopotentiometric and chronoamperometric studies. Because the formation of multilayers of PEO and V_2O_5 is not efficient due to weak interactions between them, a mixture of PEO and chitosan was used to prepare V_2O_5/chitosan/PEO alternated LbL films [45]. The choice of chitosan was based on its ability to complex with metal oxides in solutions [46]. As compared with the V_2O_5/chitosan film, the V_2O_5/chitosan/PEO film shows better performance due to lithium ion mobility, as determined from electrochemical impedance spectroscopy. The values of the diffusion coefficients are 1×10^{-13} and 3×10^{-16} $cm^2\,s^{-1}$ for the V_2O_5/chitosan/PEO and V_2O_5/chitosan films, respectively.

12.3.4 Titanium Oxide

Among semiconductor materials, TiO_2 has been significantly investigated due to its high chemical stability, low cost and non-toxicity. Electrodes formed from TiO_2 have been used for electrochromic, photocatalytic and photoelectrochromic devices, photoelectrochemical solar cells, rocking-chair lithium batteries and so on [47–52].

Much attention has been given to nanostructured TiO_2 and its use for several applications. These nanostructured electrodes comprise interconnected colloidal particles that increase the surface area in comparison with semiconductor electrodes consisting of macroscopic particles. This fact has a substantial influence on the TiO_2 physical-chemistry properties, once reactions take place in the electrode bulk. An abundant amount of molecules can be adsorbed on surfaces of TiO_2 nanoparticles, which is advantageous in ECDs constituted from redox chromophores. The lithium ion diffusion rate within these nanoparticulated materials can also be increased, once the diffusion pathway is shortened as compared with conventional electrodes, thus decreasing the response time of electrochemical windows.

The details of composition, preparation and deposition methods influence the properties of anatase (TiO_2). There are several methods for deposition of TiO_2 on substrates, including the LbL method. Numerous works have reported on these self-assembled films of TiO_2 nanoparticles, which have been employed in photocatalytic, electronic and electrochemical systems [53–58]. Some works have shown that self-assembled films are promising for their use in ECDs because of the film uniformity and of the possibility of controlling its nanoarchitecture and thickness [59–61]. Moreover, highly wettable electrodes can be achieved, which depends on the nature and composition of these nanocomposites. This can increase the electrolytic connection between the nanoparticles in the host matrices, which guarantees the access of lithium ions to the electroactive sites, as well as an increase in the ionic diffusion rate [62, 63].

Gomes *et al.* used a polymeric dispersion of N,O-carboxymethylchitosan (NOCMCh) and PEO and a colloidal dispersion of TiO_2 to obtain self-assembled nanocomposites for the storage of lithium ions. The electrostatic interactions between the negative charge on the carboxyl group and the positively charged TiO_2 nanoparticles promote the growth of visually uniform LbL films with high thickness and nanoarchitecture control. Moreover, there was the transport of the polyether into the LbL structure during the self-assembly, which can still be plasticised with organic solvent normally used in electrolytic solutions [64, 65]. The contact angles after the spreading of 0.5 M $LiClO_4$/propylene carbonate (PC) solution drop on the TiO_2 and TiO_2/NOCMCh/PEO films are 23.0° and 2.3°, respectively. These data show lower surface energy for the electrolytic solution in the LbL nanocomposite compared with the TiO_2 electrode, which is associated with higher solvophilicity of (NOCMCh) and PEO. This favours higher accessibility of the lithium ions to the electroactive sites and higher charge capacity.

Although the presence of these polymers tends to hamper the charge (electron and lithium ion) transport at the electrode surface and bulk, NOCMCh and PEO do not interfere with these properties significantly. Indeed, the TiO_2 and TiO_2/NOCMCh/PEO electrodes have demonstrated similar charge transfer resistance ($R_{ct} \approx 80\,\Omega\,cm^2$) and electron diffusion coefficient ($D_e \approx 2.0 \times 10^{-8}\,cm^2\,s^{-1}$) values. The polymeric chains act mainly on the lithium ion accessibility to the electroactive sites and the lithium ion mobility. In fact, these D_e values indicate that the electron transport in these nanocomposites does not hinder the enhancement in the ionic motion associated with the internal electric field. On basis of these mass transport parameters, the inclusion of electronic conducting components in these LbL materials can further increase the lithium ion diffusion rate and, consequently, the charge capacity.

Galiote *et al.* analysed LbL electrodes formed from TiO_2 nanoparticles and poly(vinyl sulfonic acid) (PVS) in an electrolytic solution of 0.5 M $LiClO_4$/PC [66]. The electrostatic interaction between these components allowed the growth of visually uniform nanocomposites, with well-controlled thickness, which allowed a better understanding of the electrochemical and chromogenic properties of these nanoparticles in the self-assembled structures, as shown in Figure 12.7. The presence of a voltammetric peak and a shoulder was noted during the positive potential scan during the cyclic voltammetry (Figure 12.8), which were attributed to different mobilities of lithium ions in the host matrices associated with the trapping of charge carriers in the TiO_2 sites. Due to the control of thickness and of its relation with the nanoparticle number in these self-assembled matrices, it was possible to attribute the trapping effects to the colloidal particles, on the basis of the enhancement in the ratio between the charge associated with trapping sites (a shoulder was observed on the voltammetric peak) and that associated with non-trapping sites (the main voltammetric peak) as a function of the number of bilayers.

The total number of trapping sites was determined by monitoring the light absorption: the profile of absorbance change (ΔA) as a function of injected charge was evaluated with the SSM. On the basis of these spectroelectrochemical experiments, it has been suggested that trapping sites are formed after phase transition in the colloidal particles, when a stoichiometric ratio (x in Li_xTiO_2) equal to 0.55 is reached [67].

12.3.5 Nickel Hydroxide

Among inorganic nanoparticles, nickel hydroxide ('hydrated nickel oxide') is one of the most studied in electrochemical sciences due to its range of applications covering electrocatalysis, energy storage and electrochromism. Nickel hydroxide is transparent in its reduced form and becomes deep brown when oxidised. Electrochromic electrodes can be modified either by direct electrochemical precipitation of thin films or by the incorporation of nanoparticles. Vidotti *et al.* [68] compared these two techniques and the changes produced

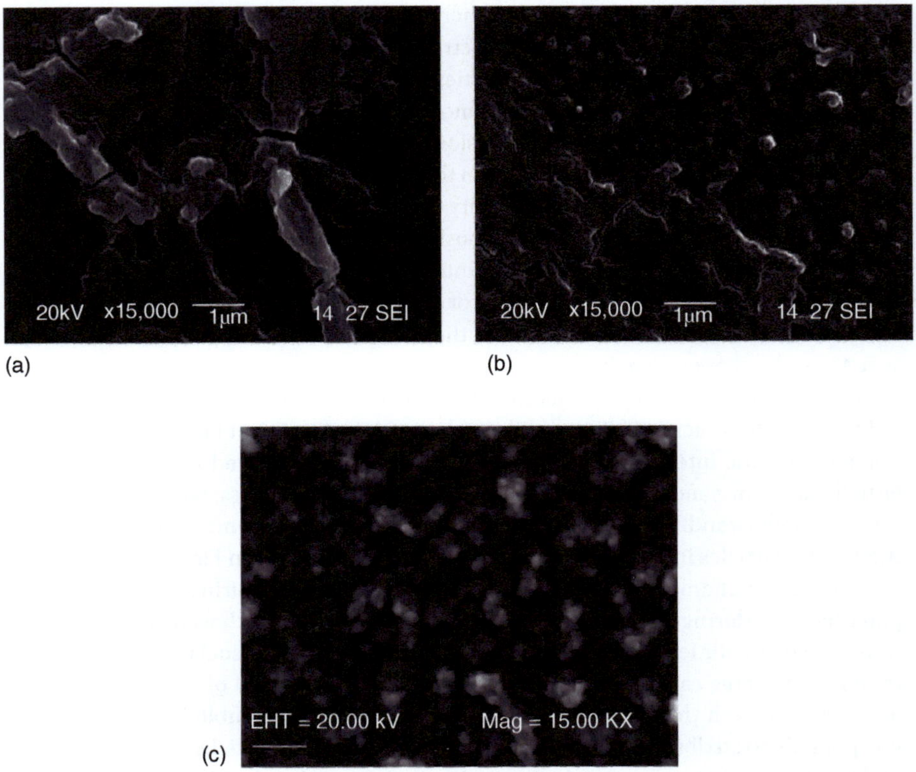

Figure 12.7 Scanning electron micrographs for (a) 5-, (b) 15- and (c) 20-bilayer TiO_2/PVS LbL films. Scale bar = 1 μm. (Reproduced from Ref. [66], with permission from The American Chemical Society.)

in the electrochromic properties. The LbL electrodes were grown by using sonochemically synthesised $Ni(OH)_2$ nanoparticles and PAH, and the nanoarchitecture has shown improvement in the response time and the efficiency (about 20% higher than the electrodeposited thin film). This work has also shown the formation of tetralayers by alternating $Ni(OH)_2$ and $Co(OH)_2$ nanoparticles on the same electrode with the respective change of optical properties. One intriguing point raised by this work is about the electrical connectivity between $Ni(OH)_2$ layers due to the presence of insulating PAH. The amount of electroactive material on electrodes modified by different number of bilayers was compared by calculating the voltammetric charge of each electrode and the respective amount of Ni^{2+} determined by ICP-OES. According to the authors, the amount of Ni^{2+} grows linearly with the number of bilayers; on the other hand, the electric charge (and the contrast $\Delta\%T$) increases until the 5th bilayer and maintains its value until the 10th bilayer, indicating the lack of electrical connectivity in thicker films.

The examples cited demonstrate that physical-chemistry characterisation can be performed on the basis of the electrochromic properties of LbL films. Moreover, these self-assembled electrodes can be manipulated at the

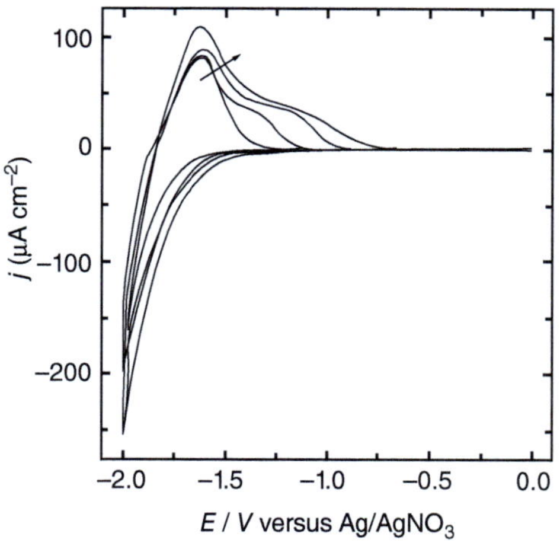

Figure 12.8 Cyclic voltammograms for 5, 10, 15 and 20-bilayer TiO_2/PVS LbL films, obtained at a sweep rate of 50 mV·s^{-1}. The arrow points to increasing number of bilayers, from 5 to 20. (Reproduced from Ref. [66], with permission from The American Chemical Society.)

nanometric scale, allowing the construction of suitable architecture, thickness and composition to be used in ECDs of high performance.

12.4 Layer-by-Layer and Electrophoretic Deposition for Nanoparticles Immobilisation

The superficial electrical charge presented by nanomaterials is also the basis for electrophoretic deposition (EPD), which is a traditional processing method in the ceramic industry [69]. EPD is achieved by the motion of charged particles towards an oppositely charged surface due to an external applied electric field (as schematically shown in Figure 12.9). Parameters such as deposition time, distance between electrodes and applied potential can be easily controlled in order to obtain different morphologies and thicknesses [70, 71].

12.4.1 Comparing Layer-by-Layer and Electrophoretic Deposition

A comparison between LbL and EPD methodologies has been performed by Vidotti and Córdoba de Torresi [72] by studying the electrochromic response of doped nickel hydroxide nanoparticles immobilised onto ITO-coated electrodes. The nanoparticles were synthesised by means of ultrasonic radiation reaching particles of 5 nm in diameter. The LbL electrodes were modified by intercalating PDDA between nanoparticle layers; in this case, the use of PDDA is very important as this polymer keeps its electrical charge even in basic media (necessary for the electrochromic behaviour of the nickel hydroxide). The EPD

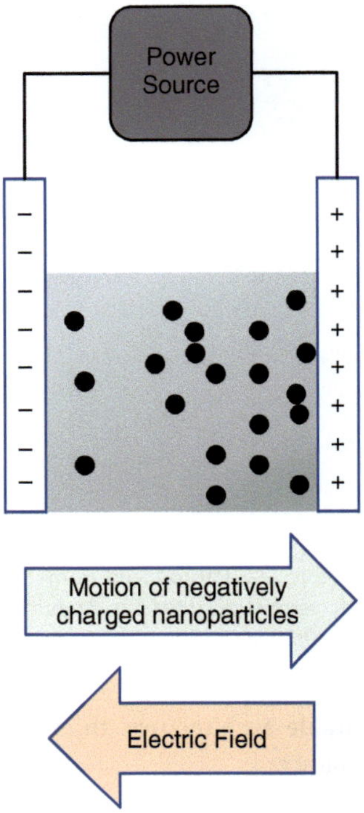

Figure 12.9 Schematically representation of electrophoretic deposition.

was achieved by applying an electric field of 1.15 V cm^{-1} for different times. It was verified that all electrochromic parameters were enhanced, especially the electrode durability, which was extremely high, with the electrochromic electrode keeping its contrast when cycled continuously for over 20 h. The comparison between the studied parameters can be viewed in Table 12.1. The authors attributed the remarkable electrochromic properties of the EPD electrodes over LbL electrodes due to the absence of a non-conductive polymeric layer between the inorganic nanoparticle layers besides being allied to the high roughness obtained by the EPD.

Many electrochromic electrodes are based on the traditional Grätzel electrode modified by nanostructured TiO_2 superficially decorated with organic dyes [73, 74]. In this case, the control of the inorganic morphology is very important for both the adsorption of large amounts of chromophores and to provide an effective contact between the electrolyte and the conductive layer. Bar *et al.* [75] employed EPD for the deposition of titania layers followed by compression and heat treatment in order to obtain a highly porous surface for the adsorption of red, green and blue viologen derivatives. The closed electrochromic electrode was assembled using $LiClO_4$ in PC and showed high contrast. On application of different values of potential (−5, −10 and −6 V), the colours displayed were red, green and blue, respectively.

EPD can be also employed for the deposition of other structures, such as nanotubes or nanorods [76]. Khoo *et al.* [77] synthesised WO_3 nanorods of

Table 12.1 Comparison between $Ni_{0.74}Co_{0.22}Cd_{0.04}(OH)_2$ and $Ni(OH)_2$ nanoparticles modified electrodes prepared by different methods.

	$Ni(OH)_2$ nanoparticles – LbL (six bilayers)	$Ni_{0.74}Co_{0.22}Cd_{0.04}(OH)_2$ nanoparticles – LbL (six bilayers)	$Ni_{0.74}Co_{0.22}Cd_{0.04}(OH)_2$ nanoparticles – EPD (15 min)
Chromatic contrast at 457 nm (%)	≈20	≈20	≈20
Charge (mC)	1.45	1.36	1.21
Electrochromic efficiency (457 nm) ($cm^2\,C^{-1}$)	75	80	90
Switching time (s)	1	0.65	0.3
Loss of absorbance after 1 h cycling (%)	45	30	<1
Time of electrode preparation (min)	75	75	15

Table from Ref. [72], with permission from Elsevier.

about 100 nm in diameter and 2 μm in length by hydrothermal methodology. The nanorods were dispersed in water and deposited onto ITO-coated electrodes under different electric fields reaching the best deposition at $6\,V\,cm^{-1}$ and creating a uniform distribution on the electrode surface. The electrochromic studies showed a high contrast, nearly 60% at 700 nm (Figure 12.10).

An interesting application of EPD in electrochromic WO_3 films has been developed by our group. Improvements in the electrochromic properties, including stability, have been achieved by using a protic ionic liquid (PIL) as electrolyte. Ionic liquids are defined as fused salts which have a melting point below 100 °C and can be divided into two broad categories: aprotic ionic liquids and PILs. PILs are produced through proton transfer from a Brønsted acid to a Brønsted base. They can be used as electrolytes without any addition of salt since they are intrinsically conductive. Indeed, they have attracted attention as electrolytes for their suitable electrochemical properties, including high electrochemical and dimensional stability, good ionic conductivity and long-term durability. For the electrolyte application in ECDs, the aforementioned properties mentioned have to include also the excellent optical transparency and high photostability of the electrolyte. PIL systems based on pyrrolidinium meet these requirements. Moreover, they are environmentally friendly and exhibit only a low toxicity when comparable to other electrolytes such as PC.

WO_3 nanoparticles with 50 nm average size were synthesised in non-aqueous solvent using an ultrasonic irradiation process, deposited by EPD process onto ITO-coated substrates using acetonitrile as solvent. The PIL *N*-methyl-pyrrolidinium tetrafluoroborate was chosen as electrolyte replacing traditional aqueous solvent. Coloration efficiencies ($52\,cm^2\,C^{-1}$) of films are in good agreement with other nanostructured WO_3 films. The proton intercalation/deintercalation reaction was satisfactorily achieved, resulting in fast

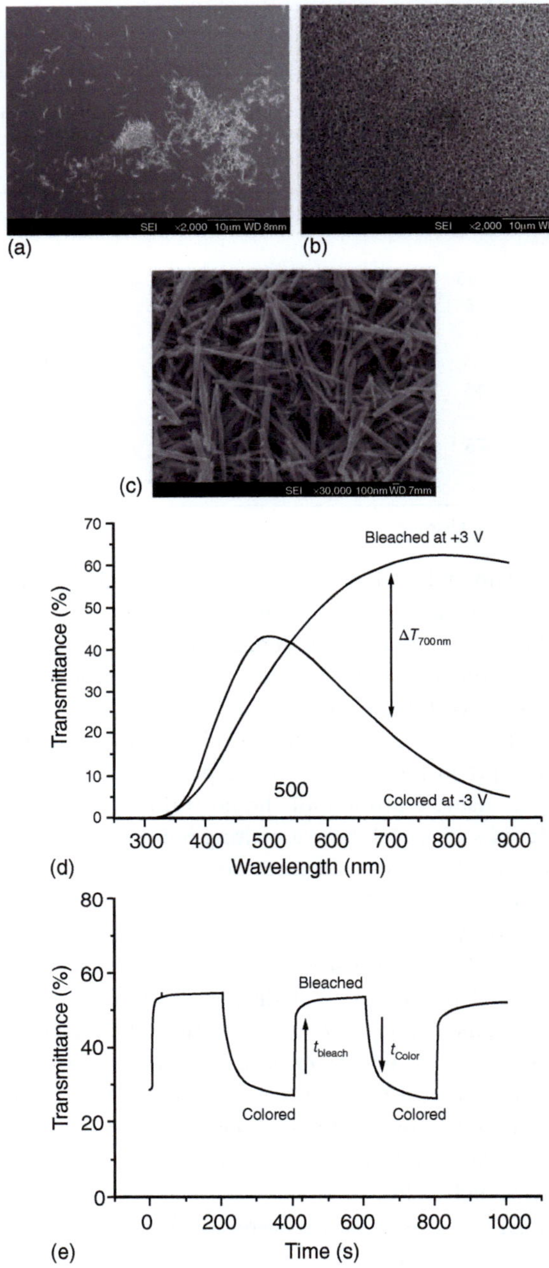

Figure 12.10 FESEM images of the deposited WO_3 nanorods on ITO glass under (a) $4\,V\,cm^{-1}$ and (b) $6\,V\,cm^{-1}$ and (c) higher magnification of $6\,V\,cm^{-1}$ deposited WO_3 nanorods. (d) UV–vis transmittance of WO_3 nanorods at bleaching (+3 V) and colouring (−3 V) and (e) switching study by applying a square wave signal ±3 V at 632.8 nm. All electrochemical studies were carried out in 1 M $LiClO_4$/PC electrolyte. (Figure reproduced from Ref. [77], with permission from Elsevier.)

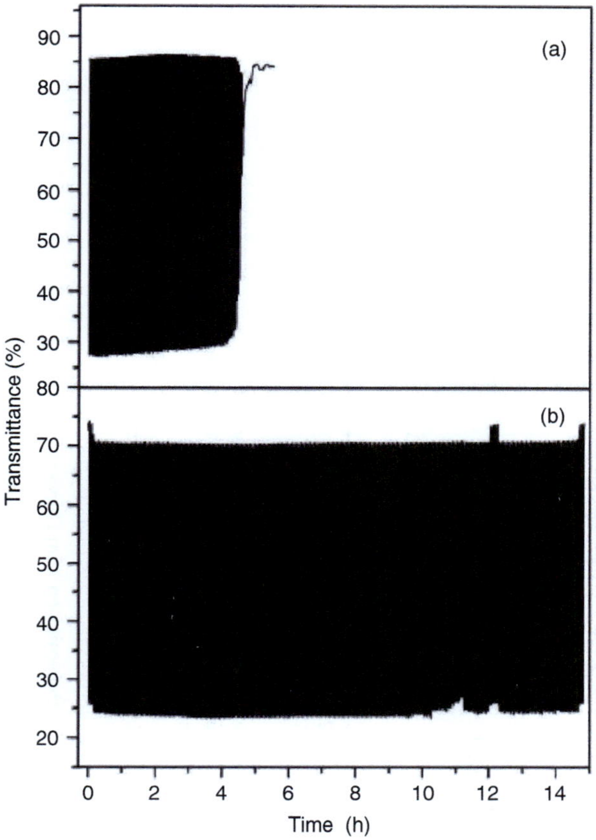

Figure 12.11 Electrochromic response at 633 nm of the tungsten oxide nanostructured films in (a) acid electrolyte 1.0 mol l^{-1} H_2SO_4 and (b) PIL electrolyte $[C_3mpyr][BF_4]$.

coloration and bleaching. The results using PIL as electrolyte, instead of H_2SO_4, show that all electrochromic parameters were improved, including the cycling durability, which was higher using PIL (Figure 12.11). Therefore, the use of a PIL enables proton intercalation, suggesting its use as a good electrolyte for ECDs.

Augusto *et al.* [78] synthesised core–shell Au@PEDOT nanoparticles of about 2–8 nm by one-pot synthesis. The composite materials are especially interesting because with the incorporation of a metallic core in the PEDOT structure, the electronic transfer and ionic diffusion through the polymeric matrix is enhanced. In this work in particular, Au@PEDOT nanocomposites were deposited onto ITO-coated electrodes by EPD on application of a 1.15 V cm^{-1} electric field for different times. The electrochromic behaviour of the nanocomposite modified electrode was compared with the traditional electrochemical deposition of PEDOT. In Figure 12.12, the dynamic transmittance changes are shown, recorded simultaneously with the cyclic voltammograms, for a 2 h EPD modified Au@PEDOT on electrodeposited PEDOT/PSS^- thin films. The transmittance

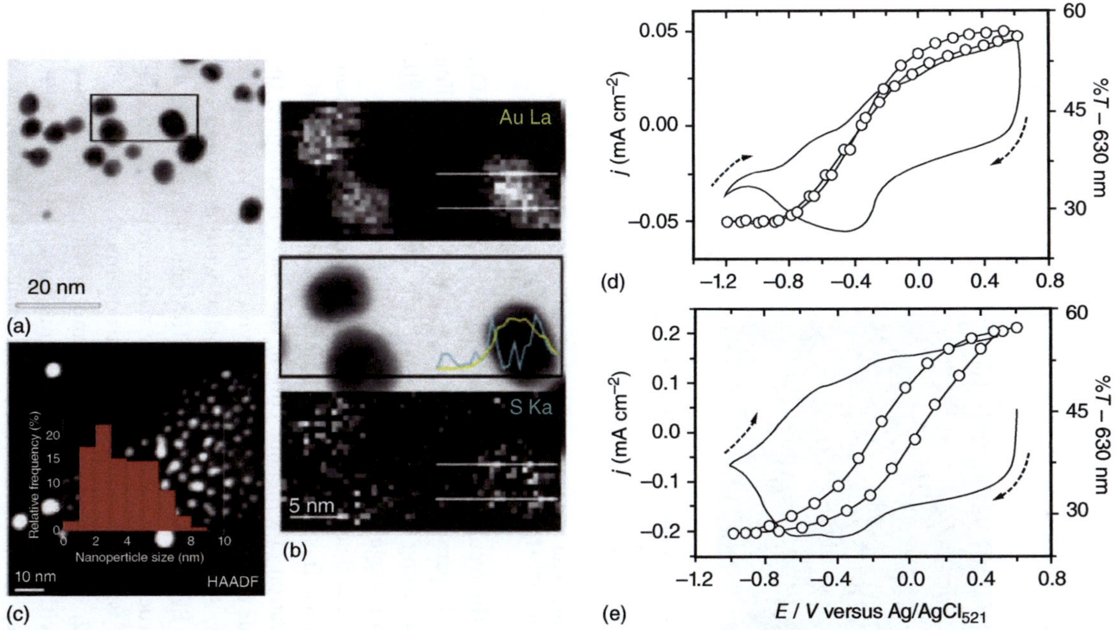

Figure 12.12 (a) TEM and (b) BF-STEM image of the Au/PEDOT nanocomposite and the XEDS-SI compositional maps of the area marked by a rectangle in the BF. The coloured line scans drawn in the BF are from the areas delimited by the parallel white lines depicted in the respective compositional maps. The hemispherical distribution of the XEDS counts measured along the particle in the Au map contrasts with the crown shape distribution measured in the S map. (c) Histogram of the Au nanoparticles size distribution within the PEDOT/PSS/AuNPs nanocomposites film superimposed on a HAADF image of the material. Cyclic voltammograms (__) and the percent transmittance at 630 nm (-o-) in a 0.1 mol dm^{-3} NaCl aqueous solution at 10 mV s^{-1} for (d) nanocomposite EPD modified electrode by 2 h and (e) 40 mC cm^{-2} electrodeposited PEDOT/PSS. (Figure reproduced from Ref. [78], with permission from Elsevier.)

changes at 630 nm were very close in the two modified electrodes and presented good electrochromic reversibility. The current densities for the electrodeposited PEDOT/PSS^- thin films were higher when compared to the EPD modified electrode, indicating that higher amounts of electrochromic material were needed to achieve similar contrast. The shape of the voltammogram obtained for PEDOT film showed a traditional capacitive behaviour with no pronounced waves. In contrast, the EPD modified electrode showed an intense reduction wave, and according to the authors, this fact can be attributed to the increase in the intrinsic PEDOT electroactivity in the presence of gold nanoparticles. The very low hysteresis viewed in the transmittance signal for the EPD modified electrode is related to the effective ionic diffusion into/out of the solid material, and this fact can be also corroborated by the very low response time presented, in the order of 60 ms. It can be concluded that the LbL technique allows an easy combination of different materials, and a direct result of this approach is that it could provide multiple coloured electrochromic films to support full-colour displays or cheap and flexible colour-adaptive clothing.

The combination of nanoparticles, chromophoric organic molecules, transition metal complexes and conjugated conducting polymers, together with the simplicity of the LbL method for preparing electrodes with defined architectures controlled at nanoscale level, is a novel direction in electrochromics. The LbL deposition is a low-cost and easily handled technique where nanoparticles, organic molecules and polymers can be alternately deposited. Depending on their electrochromic behaviour, many colours can be achieved, without counter-electrode modification, and by this means cheaper devices are found. The number of publications in this field is increasing sharply indicating that patents and large-scale commercial applications are becoming real in the not so far future. Superficially charged nanostructures can also be employed for EPD, which is a traditional processing method already used in the ceramic industry. This method can be used alternatively to LBL when the electrical connection of nanomaterials is limited. On the other hand, compact thicker films can be prepared easily with the concomitant increase of chromatic contrast.

Acknowledgements

The authors thank Brazilian agencies FAPESP, CAPES and CNPq for financial support. All the authors belong to the Smart Materials Center Research at the University of São Paulo.

References

1. Decher, G. (1997) Fuzzy nanoassemblies: toward layered polymeric multicomposities. *Science*, **277**, 1232–1237.
2. Hammond, P.T. (2004) Form and function in multilayer assembly: new applications at nanoscale. *Adv. Mater.*, **16**, 1271–1293.
3. Lvov, Y., Ariga, K., Onda, M., Ichinose, I., and Kunitake, T. (1997) Alternate assembly of ordered multilayers of SiO_2

and other nanoparticles and polyions. *Langmuir*, **13**, 6195–6203.
4. Lvov, Y., Ariga, K., Ichinose, I., and Kunitake, T. (1995) Assembly of multicomponent protein films by means of electrostatic layer-by-layer adsorption. *J. Am. Chem. Soc.*, **117**, 6117–6123.
5. Heuer, H.W., Wehrmann, R., and Kirchmeyer, S. (2002) Electrochromic window based on conducting poly (3,4-ethylenedioxythiophene)poly(styrene sulfonate). *Adv. Funct. Mater.*, **12**, 89–94.
6. Cutler, C.A., Bouguettaya, M., and Reynolds, J.R. (2002) PEDOT polyelectrolyte based electrochromic films via electrostatic adsorption. *Adv. Mater.*, **14**, 684–688.
7. Beaujuge, P.M. and Reynolds, J.R. (2010) Color control in pi-conjugated organic polymers for use in electrochromic devices. *Chem. Rev.*, **110**, 268–320.
8. DeLongchamp, D.M., Kastantin, M., and Hammond, P.T. (2003) High-contrast electrochromism from layer-by-layer polymer films. *Chem. Mater.*, **15**, 1575–1586.
9. Vidotti, M. and Torresi, S.I.C. (2008) Nanochromics: old materials, new structures and architectures for high performance devices. *J. Braz. Chem. Soc.*, **19**, 1248–1257.
10. Pecher, J. and Mecking, S. (2010) Nanoparticles of conjugated polymers. *Chem. Rev.*, **110**, 6260–6279.
11. Müller, K., Park, M.-K., Klapper, M., Knoll, W., and Müllen, K. (2007) Synthesis and layer-by-layer deposition of spherical poly(3,4-ethylenedioxythiophene) nanoparticles – toward fast switching times between reduced and oxidized states. *Macromol. Chem. Phys.*, **208**, 1394–1401.
12. Longchamps, D.M. and Hammond, P.T. (2004) High contrast electrochromism and controllable dissolution of assembled Prussian blue/polymer nanocomposites. *Adv. Funct. Mater.*, **14**, 224–232.
13. Longchamps, D.M. and Hammond, P.T. (2004) Multiple-color electrochromism from layer-by-layer-assembled polyaniline/Prussian blue nanocomposite thin films. *Chem. Mater.*, **16**, 4799–4805.
14. Karyakin, A.A. (2001) Prussian blue and its analogues: electrochemistry and analytical applications. *Electroanalysis*, **13**, 813–819.
15. Baioni, A.P., Vidotti, M., Fiorito, P.A., Ponzio, E.A., and Torresi, S.I.C. (2007) Synthesis and characterization of copper hexacyanoferrate nanoparticles for building up long-term stability electrochromic electrodes. *Langmuir*, **23**, 6796–6800.
16. Rezende, A.R., Bizeto, M.A., Constantino, V.R.L., and Huguenin, F. (2009) Lithium ion electro-insertion and spectroelectrochemical properties of films from hexaniobate. *J. Phys. Chem. C*, **113**, 10868–10876.
17. Galiote, N.A., Parreira, R.L.T., Rosolen, J.M., and Huguenin, F. (2010) Self-assembled films from WO_3: electrochromism and lithium ion diffusion. *Electrochem. Commun.*, **12**, 733–736.
18. Deb, S.K. (1973) Optical and photoelectric properties and colour centres in thin films of tungsten oxide. *Philos. Mag.*, **27**, 801–822.
19. Baetens, R., Jelle, B.P., and Gustavsen, A. (2010) Properties, requirements and possibilities of smart windows for dynamic daylight and solar energy control in buildings: a state-of-the-art review. *Sol. Energy Mater. Sol. Cells*, **94**, 87–105.
20. Nah, Y.-C., Ghicov, A., Kim, D., and Schmuki, P. (2008) Enhanced electrochromic properties of self-organized nanoporous WO_3. *Electrochem. Commun.*, **10**, 1777–1780.
21. Green, S., Backholm, J., Georén, P., Granqvist, C.G., and Niklasson, G.A. (2009) Electrochromism in nickel oxide and tungsten oxide thin films: ion intercalation from different electrolytes. *Sol. Energy Mater. Sol. Cells*, **93**, 2050–2055.
22. Granqvist, C.G. (2000) Electrochromic tungsten oxide films: review of progress 1993–1998. *Sol. Energy Mater. Sol. Cells*, **60**, 201–262.
23. Niklasson, G.A. and Granqvist, C.G. (2007) Electrochromics for smart windows: thin films of tungsten oxide and nickel oxide, and devices based on these. *J. Mater. Chem.*, **17**, 127–156.
24. Weppner, W. and Huggins, R.A. (1997) Determination of the kinetic parameters of mixed-conducting electrodes

and application to the system Li_3Sb. *J. Electrochem. Soc.*, **124**, 1569–1578.

25. Bizeto, M.A. and Constantino, V.R.L. (2004) Layered $H_2K_2Nb_6O_{17}$ exfoliation promoted by n-butylamine. *Mater. Res. Bull.*, **39**, 1811–1820.
26. Hoertz, P.G. and Mallouk, T.E. (2005) Light-to-chemical energy conversion in lamellar solids and thin films. *Inorg. Chem.*, **44**, 6828–6840.
27. Izawa, K., Yamada, T., Unal, U., Ida, S., Altuntasoglu, O., Koinuma, M., and Matsumoto, Y. (2006) Photoelectrochemical oxidation of methanol on oxide nanosheets. *J. Phys. Chem. B*, **110**, 4645–4650.
28. Unal, U., Matsumoto, Y., Tamoto, N., Koinuma, M., Machida, M., and Izawa, K. (2006) Visible light photoelectrochemical activity of $K_4Nb_6O_{17}$ intercalated with photoactive complexes by electrostatic self-assembly deposition. *J. Solid State Chem.*, **179**, 33–40.
29. Ida, S., Unal, U., Izawa, K., Ogata, C., Inoue, T., and Matsumoto, Y. (2007) Synthesis and photoluminescence properties of niobate layered oxides intercalated with rare earth ions by electrostatic self-assembly methods. *Mol. Cryst. Liq. Cryst.*, **470**, 393–402.
30. Saruwatari, K., Sato, H., Kameda, J., Yamagishi, A., and Domen, K. (2005) Evidence for the role of organic layers in photoconductivity of organic/inorganic hybrid nanosheets as prepared by Langmuir–Blodgett methods. *Chem. Commun.*, **15**, 1999–2001.
31. Livage, J. (1991) Vanadium pentoxide gels. *Chem. Mater.*, **3**, 578–593.
32. Baddour, R., Pereira-Ramos, J.P., Messina, R., and Perichon, J. (1991) A thermodynamic, structural and kinetic study of the electrochemical lithium intercalation into the xerogel V_2O_5.1.6 H_2O in a propylene. *J. Electroanal. Chem.*, **314**, 81–101.
33. Chernova, N.A., Roppolo, M., Dillon, A.C., and Whittingham, M.S. (2009) Layered vanadium and molybdenum oxides: batteries and electrochromics. *J. Mater. Chem.*, **19**, 2526–2552.
34. Bullot, J., Cordier, P., Gallais, O., Gauthier, M., and Babonneau, F. (1984) Thin layers deposited from V_2O_5 gels: II. An optical absorption study. *J. Non-Cryst. Solids*, **68**, 135–146.
35. Holland, G.P., Huguenin, F., Torresi, R.M., and Buttry, D.A. (2003) Comparison of V_2O_5 xerogels prepared by the vanadate and alkoxide routes using X-ray absorption and other methods. *J. Electrochem. Soc.*, **150**, A721–A725.
36. Mège, S., Levieux, Y., Ansart, F., Savariault, J.M., and Rousset, A. (2000) Electrochemical properties of a new V_2O_5 xerogel. *J. Appl. Electrochem.*, **30**, 657–664.
37. Galiote, N.A., Camargo, M.N.L., Iost, R.M., Crespilho, F., and Huguenin, F. (2011) Effects of self-assembled materials prepared from V_2O_5 for lithium ion electroinsertion. *Langmuir*, **27**, 12209–12217.
38. Huguenin, F., Ferreira, M., Zucolotto, V., Nart, F.C., Torresi, R.M., and Oliveira, O.N. (2004) Molecular-level manipulation of V_2O_5/polyaniline layer-by-layer films to control electrochromogenic and electrochemical properties. *Chem. Mater.*, **16**, 2293–2299.
39. Liu, Y.-J., DeGroot, D.C., Schindler, J.L., Kannewurf, C.R., and Kanatzidis, M.G. (1991) Intercalation of poly(ethylene oxide) in vanadium pentoxide (V_2O_5) xerogel. *Chem. Mater.*, **3**, 992–994.
40. Liu, Y.-J., Cowen, J.A., Kaplan, T.A., DeGroot, D.C., Schindler, J., Kannewurf, C.R., and Kanatzidis, M.G. (1995) Investigation of the alkali-metal vanadium oxide xerogel bronzes: $A_xV_2O_5.nH_2O$ (A = K and Cs). *Chem. Mater.*, **7**, 1616–1624.
41. Liu, Y.-J., Schindler, J., DeGroot, D.C., Kannewurf, C.R., Hirpo, W., and Kanatzidis, M.G. (1996) Synthesis, structure, and reactions of poly(ethylene oxide)/V_2O_5 intercalative nanocomposites. *Chem. Mater.*, **8**, 525–534.
42. Armand, M. (1983) Polymer solid electrolytes – an overview. *Solid State Ionics*, **9**, 745–754.
43. Croce, F., Appetecchi, G.B., Persi, L., and Scrosati, B. (1998) Nanocomposite polymer electrolytes for lithium batteries. *Nature*, **394**, 456–458.
44. Huguenin, F., Cavalcante, M.G., and Torresi, R.M. (1999) Impedance spectroscopy study of poly(ethylene oxide)

sodium hexafluortitanate complex. *Solid State Ionics*, **126**, 259–267.

45. Alexeev, V.L., Kelberg, E.A., Evmenenko, G.A., and Bronnikov, S.V. (2000) Improvement of the mechanical properties of chitosan films by the addition of poly(ethylene oxide). *Polym. Eng. Sci.*, **40**, 1211–1215.
46. Guzmán, J., Saucedo, I., Navarro, R., Revilla, J., and Guibal, E. (2002) Vanadium interactions with Chitosan: influence of polymer protonation and metal speciation. *Langmuir*, **18**, 1567–1573.
47. O'Regan, B. and Grätzel, M. (1991) A low-cost, high-efficiency solar cell based on dye-sensitized colloidal TiO_2 films. *Nature*, **353**, 737–740.
48. Ghicov, A., Tsuchiya, H., Hahn, R., Macak, J.M., Munõz, A.G., and Schmuki, P. (2006) TiO_2 nanotubes: H^+ insertion and strong electrochromic effects. *Electrochem. Commun.*, **8**, 528–532.
49. Georg, A., Georg, A., and Opara Krašovec, U. (2006) Photoelectrochromic window with Pt catalyst. *Thin Solid Films*, **502**, 246–251.
50. Exnar, E., Kavan, L., Huang, S.Y., and Grätzel, M. (1997) Novel 2 V rocking-chair lithium battery based on nano-crystalline titanium dioxide. *J. Power Sources*, **68**, 720–722.
51. Kiyonaga, T., Mitsui, T., Torikoshi, M., Takekawa, M., Soejima, T., and Tada, H. (2006) Ultrafast photosynthetic reduction of elemental sulfur by Au nanoparticle-loaded TiO_2. *J. Phys. Chem. B*, **110**, 10771–10778.
52. O'Regan, B.C. and Durrant, J.R. (2006) Calculation of activation energies for transport and recombination in mesoporous TiO_2/dye/electrolyte films – taking into account surface charge shifts with temperature. *J. Phys. Chem. B*, **110**, 8544–8547.
53. Kovtyukhova, N., Ollivier, P.J., Chizhik, S., Dubravin, A., Buzaneva, E., Gorchinskiy, A., Marchenko, A., and Smirnova, N. (1999) Self-assembly of ultrathin composite TiO_2/polymer films. *Thin Solid Films*, **337**, 166–170.
54. Cassagneau, T., Fendler, J.H., and Mallouk, T.E. (2000) Optical and electrical characterizations of ultrathin films self-assembled from 11-aminoundecanoic acid capped TiO_2 nanoparticles and polyallylamine hydrochloride. *Langmuir*, **16**, 241–246.
55. Kim, T.-H. and Sohn, B.-H. (2002) Photocatalytic thin films containing TiO_2 nanoparticles by the layer-by-layer self-assembling method. *Appl. Surf. Sci.*, **201**, 109–114.
56. Kim, J.H., Kim, S.H., and Shiratori, S. (2004) Fabrication of nanoporous and hetero structure thin film via a layer-by-layer self-assembly method for a gas sensor. *Sens. Actuators, B*, **102**, 241–247.
57. He, J.-A., Mosurkal, R., Samuelson, L.A., Li, L., and Kumar, J. (2003) Dye-sensitized solar cell fabricated by electrostatic layer-by-layer assembly of amphoteric TiO_2 nanoparticles. *Langmuir*, **19**, 2169–2174.
58. Paddon, C.A. and Marken, F. (2004) Hemoglobin adsorption into TiO_2 phytate multi-layer films: particle size and conductivity effects. *Electrochem. Commun.*, **6**, 1249–1253.
59. DeLongchamp, D.M. and Hammond, P.T. (2006) Multiple-color electrochromism from layer-by-layer-assembled polyaniline/Prussian blue nanocomposite thin films. *Chem. Mater.*, **16**, 4799–4805.
60. Huguenin, F., Gonzalez, E.R., and Oliveira, O.N. Jr., (2005) Electrochemical and electrochromic properties of layer-by-layer films from WO_3 and Chitosan. *J. Phys. Chem. B*, **109**, 12837–12844.
61. Huguenin, F., Zucolotto, V., Carvalho, A.J.F., Gonzalez, E.R., and Oliveira, O.N. Jr., (2005) Layer-by-layer hybrid films incorporating WO_3, TiO_2, and chitosan. *Chem. Mater.*, **17**, 6739–6745.
62. Su, X., Wu, Q., Zhan, X., Wu, J., Wei, S., and Guo, Z. (2012) Advanced titania nanostructures and composites for lithium ion battery. *J. Mater. Sci.*, **47**, 2519–2534.
63. Guo, Y.G., Hu, Y.S., Sigle, W., and Maier, J. (2007) Superior electrode performance of nanostructured mesoporous TiO_2 (Anatase) through efficient hierarchical mixed conducting networks. *Adv. Mater.*, **19**, 2087–2091.
64. Kim, Y.T. and Smotkin, E.S. (2002) The effect of plasticizers on transport and

electrochemical properties of PEO-based electrolytes for lithium rechargeable batteries. *Solid State Ionics*, **149**, 29–37.

65. Gomes, W.J.A.S., Araújo, D.M., Carvalho, A.J.F., Campana-Filho, S.P., and Huguenin, F. (2013) High lithium ion electroinsertion rate into self-assembled films formed from TiO_2. *J. Phys. Chem. C*, **117**, 16774–16782.
66. Galiote, N.A., Carvalho, A.J.F., and Huguenin, F. (2006) Trapping of charge carriers in colloidal particles of self-assembled films from TiO_2 and poly(vinyl sulfonic acid). *J. Phys. Chem. B*, **110**, 24612–24620.
67. Kavan, L., Grätzel, M., Gilbert, S.E., Klemenz, C., and Scheel, H.J. (1996) Electrochemical and photoelectrochemical investigation of single-crystal anatase. *J. Am. Chem. Soc.*, **118**, 6716–6723.
68. Vidotti, M., Greco, C.V., Ponzio, E.A., and Torresi, S.I.C. (2006) Sonochemically synthesized $Ni(OH)_2$ and $Co(OH)_2$ nanoparticles and their application in electrochromic electrodes. *Electrochem. Commun.*, **8**, 554–560.
69. Boccaccini, A. and Zhitomirsky, I. (2002) Application of electrophoretic and electrolytic deposition techniques in ceramics processing. *Curr. Opin. Solid State Mater. Sci.*, **6**, 251–260.
70. Corni, I., Ryan, M.P., and Boccaccini, A.R. (2008) Electrophoretic deposition: from traditional ceramics to nanotechnology. *J. Eur. Ceram. Soc.*, **28**, 1353–1367.
71. Ammam, M. (2012) Electrophoretic deposition under modulated electric fields: a review. *RCS Adv.*, **2**, 7633–7646.
72. Vidotti, M. and Córdoba de Torresi, S.I. (2009) Electrostatic layer-by-layer and electrophoretic depositions as methods for electrochromic nanoparticle immobilization. *Electrochim. Acta*, **54** (10), 2800–2804.
73. Campus, F., Bonhote, P., Grätzel, M., Heinen, S., and Walder, L. (1999) Electrochromic devices based on surface modified nanocrystalline TiO_2 thin film electrodes. *Sol. Energy Mater. Sol. Cells*, **56**, 281–297.
74. Cummins, D., Boschloo, G., Ryan, M., Corr, D., Rao, S.N., and Fitzmaurice, D. (2000) Ultrafast electrochromic windows based on redox chromophore modified nanostructured semiconducting and conducting films. *J. Phys. Chem. B*, **104**, 11449–11459.
75. Bar, G., Larina, N., Grinis, L., Lokshin, V., Gvishi, R., Kiryuschev, I., Zaban, A., and Khodorkovsky, V. (2012) RGB organic electrochromic cells. *Sol. Energy Mater. Sol. Cells*, **99**, 123–128.
76. Boccaccini, A.R., Cho, J., Roether, J.A., Thomas, B.J.C., Minay, E.J., and Shaffer, M.S.P. (2006) Electrophoretic deposition of carbon nanotubes. *Carbon*, **44**, 3149–3160.
77. Khoo, E., Lee, P.S., and Ma, J. (2010) Electrophoretic depositions of (WO_3) nanorods for electrochromic application. *J. Eur. Ceram. Soc.*, **30**, 1139–1144.
78. Augusto, T., Neto, E.T., Neto, A.A.T., Vichessi, R., Vidotti, M., and Torresi, S.I.C. (2013) Electrophoretic deposition of Au@PEDOT nanoparticles toward the construction of high-performance electrochromic electrodes. *Sol. Energy Mater. Sol. Cells*, **118**, 72–80.

13
Plasmonic Electrochromism of Metal Oxide Nanocrystals

Anna Llordes, Evan L. Runnerstrom, Sebastien D. Lounis, and Delia J. Milliron

13.1
Introduction to Plasmonic Electrochromic Nanocrystals

After more than 35 years of active research in the field of electrochromics, there are still significant challenges in order for smart windows to become an energy-efficient and cost-effective technology. Besides optimal performance parameters such as long durability, fast switching times and low manufacturing costs, an ideal smart window should be applicable across different building types and climate zones. The latter functionality requires the electrochromic material to be able to modulate the transmittance of visible sunlight and solar heat independently. Modulating visible sunlight and solar heat separately would directly affect the energy use in a building, as well as influence the occupant comfort. Therefore, achieving such optical functionality would contribute to optimisation of energy efficiency across a building's heating, cooling and artificial lighting systems.

Unfortunately, none of the commonly investigated electrochromic materials, such as the ubiquitous WO_3, show the required 'dual-band' functionality, that is, independent control over the visible and near-infrared (NIR) regions of the solar spectrum. In addition, given that the coloration mechanism of conventional materials is driven by Faradaic processes, that is, ion intercalation/deintercalation upon reduction/oxidation, these materials typically show slow switching kinetics and limited cycle life. Decreasing the diffusion length to the nanometre scale through nanostructuring approaches has proven to enhance electrochromic performances of conventional materials [1]. However, they do not offer the spectral selectivity required to modulate visible and NIR light separately. Thus, investigation of new materials with non-Faradaic mechanisms as well as with spectrally-selective optical responses is essential to improve the limited performances of current electrochromics (Scheme 13.1).

In this chapter, we describe a new type of electrochromic phenomenon, plasmonic electrochromism, which enables not only faster switching times and improved durability but also the required spectral selectivity. This new paradigm has been realised thanks to the development of colloidal plasmonic nanocrystals. Due to their small size (3–20 nm) and high surface area, they can

Electrochromic Materials and Devices, First Edition.
Edited by Roger J. Mortimer, David R. Rosseinsky and Paul M. S. Monk.

Materials	Spectral selectivity
WO_3, NiO, Nb_2O_5, V_2O_5, etc.	Limited
ITO, AZO, ATO	NIR
Nb_2O_5-ITO	Vis-NIR

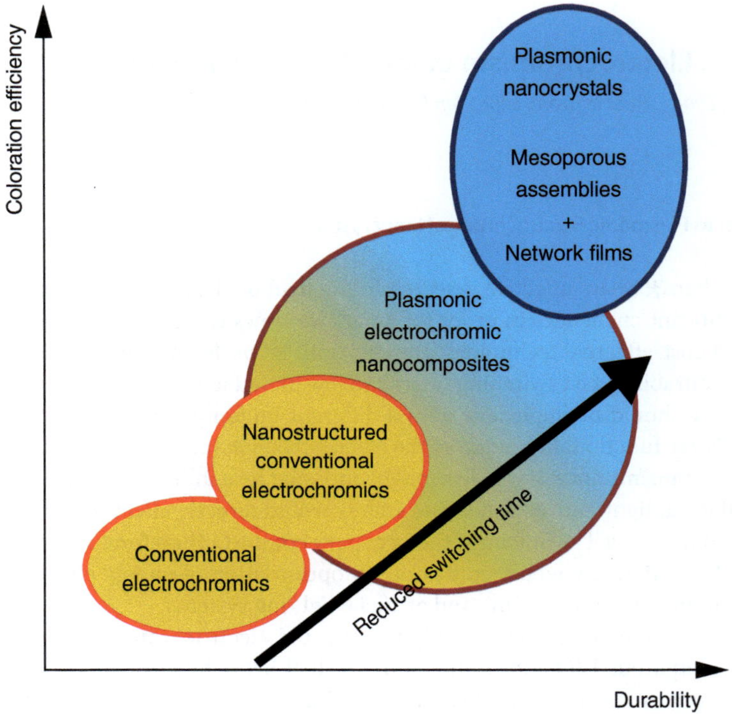

Scheme 13.1 Performance parameters for different electrochromic materials and electrodes. Devices with improved coloration efficiencies, faster switching times, and longer cycle lives can be achieved by nanostructuring conventional metal oxides (e.g. WO_3), which reduces diffusion length for ion intercalation. Realisation of a new type of electrochromic device that goes beyond the capabilities of commonly used electrochromic materials is achieved by using plasmonic transparent conductive oxide (TCO) nanocrystals, which operate under a capacitive mechanism. Electrochemical injection of electrons into plasmonic TCO nanocrystal films induces a shift in the plasmon frequency and gives rise to the new functionality of selective optical modulation in the near-infrared region of the solar spectrum, a functionality that is missing in conventional materials. Plasmonic nanocrystals can be used as building blocks to enable creation of advanced electrochromic electrodes, such as ordered mesoporous nanocrystals assemblies and nanocrystal-in-glass composites, the latter achieving the ideal smart window functionality: independent control over visible and NIR transmittance. Reprinted from Ref. [1], reproduced by permission of the Royal Society of Chemistry. dx.doi.org/10.1039/C4CC03109A.

easily be electrochemically charged in a capacitive manner. Furthermore, these nanocrystals absorb or reflect light through localised surface plasmon resonance (LSPR), which can be tuned across the entire solar spectrum by material selection and by doping. We will also show that electrochemical injection or extraction of electrons can dynamically change the wavelength and intensity of LSPR absorption. This discovery gave rise to a new type of electrochromic device that has NIR selectivity, fast switching kinetics and excellent durability, bringing us closer to the ideal smart window. In addition, integrating plasmonic colloidal nanocrystals into heterogeneous nanocomposite electrodes allowed, for the first time, dual-band functionality in an electrochromic electrode.

Besides the unique electrochromic performance of plasmonic colloidal nanocrystals, a key advantage is the capability to use them as building blocks and finely control their arrangement, which offers a unique platform to perform systematic studies and understand the role of structure and function at different length scales. A summary of the performance improvements for different electrochromic materials and electrodes is depicted in Scheme 13.1.

Before explaining the plasmonic electrochromic effect and the materials that made this new paradigm possible, we will briefly introduce the fundamentals of LSPR in nanocrystals.

Plasmonic nanocrystals are those showing light-induced collective oscillations of free charge carriers (Figure 13.1). These excitations are related to surface plasmon resonances (SPRs), which oscillate and propagate within a thin surface layer defined by the skin depth of the material (typically a metal) [2]. LSPRs, by contrast, involve the collective oscillation of all of the conduction electrons in a nanocrystal and emerge when the critical requirements for free carrier

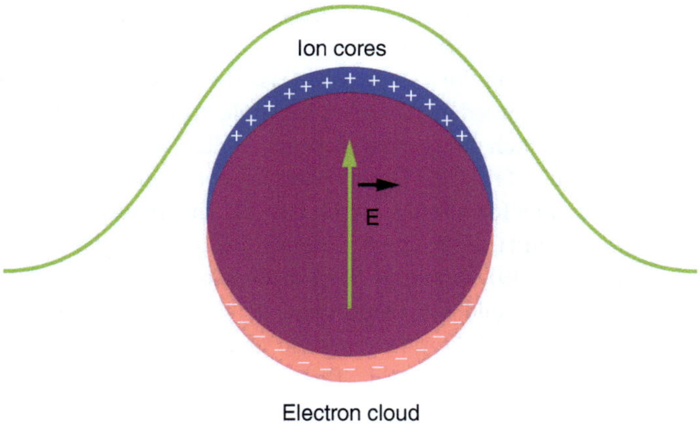

Figure 13.1 Illustration of a localised surface plasmon (LSP) in a conductive metal or semiconductor nanocrystal. Incident electromagnetic radiation (in green) displaces the free electron cloud, which then experiences a restoring force exerted by the positively charged ion cores. At the resonance frequency, scattering and absorption are strongly enhanced. Reprinted from Ref. [1], reproduced by permission of the Royal Society of Chemistry.

concentration and particle size are satisfied [2–4]. Since semiconductors can be doped to the point of degeneracy, both doped semiconductors and classical metal nanoparticles have been observed to have LSPRs. A doped semiconductor must have a negative real part of its dielectric function in order to display metallic optical properties. This condition is achieved above a critical free carrier concentration that depends on the electronic structure of the material, including the effect of defect states. Several wide-band-gap oxide semiconductors can be routinely doped above the threshold for metallic optical behaviour.

The advancement of colloidal synthesis of highly doped semiconductor nanocrystals [5–7] has enabled the production of a host of oxide materials demonstrating LSPRs. These collective resonant oscillations of free electrons or holes lead to enhanced optical absorption and scattering, as well as the possibility of strong enhancement of the electric field near the nanoparticle surface, and are sensitive to the particle size, shape and dielectric environment surrounding the particle. In addition, semiconductor LSPRs are inherently tunable by varying the doping level and electrochemical potential, allowing observations of plasmonic effects across visible, NIR and mid-infrared wavelengths and enabling new classes of optoelectronic devices with functionalities in these regions of the spectrum.

To support a dipolar LSPR, the diameter of a nanoparticle whose composition has the proper dielectric properties must be larger than a few nanometres but less than approximately one-fifth of the wavelength of the exciting radiation. Thus, particles less than 100 nm in diameter can exhibit dipolar LSPRs in the visible-to-infrared range, leading to strong optical extinction in these frequency ranges. In this regime, both the internal dielectric properties of the nanocrystal and the external electric field felt by the particle under optical excitation can be reasonably approximated as uniform across its entire volume, a condition known as the quasistatic limit. The cross terms in Maxwell's equations can thus be neglected and the LSPR can thus be understood as arising from purely electrostatic interaction between the exciting radiation and the free electron gas. The optical linewidth of the plasmon resonance is determined by the scattering of electrons in the bulk and at the surface of the crystal, which damps the harmonic oscillation.

The free electron properties of metallic nanocrystals are conventionally treated with the classical Drude model for a free electron gas [2]. This theory has been effectively applied to understand the plasmonic properties of both metal [3, 8] and semiconductor nanostructures [9], including metal oxide nanocrystals [10–12].

The optical properties of a solid are determined by its complex dielectric function, which describes the polarisation response of charge in the solid to an externally applied electromagnetic field. The Drude contribution to the frequency-dependent complex dielectric constant, $\varepsilon(\omega)$, takes the form:

$$\varepsilon(\omega) = \varepsilon_\infty - \frac{\omega_\mathrm{p}^2}{\omega^2 + i\omega\Gamma} \tag{13.1}$$

where the constant ε_∞ represents the high-frequency dielectric response of the ion cores, Γ is a damping parameter due to the scattering of electrons, related to the average scattering time (τ) by $\Gamma = 1/\tau$, and ω_p is the plasma frequency given by

$$\omega_p^2 = \frac{ne^2}{\varepsilon_0 m^*} \tag{13.2}$$

where n is the free electron concentration, e is the elementary electronic charge, ε_0 is the permittivity of free space and m^* is the effective mass of the electron within the crystal. A simple heuristic for understanding the plasma frequency is that of an n oscillator defined by the negatively charged sea of electrons oscillating against the positively charged lattice of ions in a bulk crystal. The plasma frequency is simply the resonance frequency of such an oscillator.

From the aforementioned equations, it is clear that the optical properties of the free electron gas will depend strongly on both the free carrier concentration and the damping of electron motion in this model. Semiconductors like the metal oxides differ from metals in important ways related to both of these parameters. While metals generally have a fixed n due to near-perfect screening of external electric fields, this value can be tuned widely in semiconductors by synthetic or electrochemical means, allowing variation of the absorption and scattering of light by changing the bulk plasma frequency. In addition, the damping parameter, which determines the width and line shape of the LSPR, is generally frequency-independent in metals, resulting from electron-phonon scattering, electron-electron scattering, radiative energy loss and surface scattering for sufficiently small nanocrystals, [13] though some metals, like gold, have frequency-dependent damping due to interband electron–electron scattering [14, 15]. Here, again, semiconductor nanocrystals must be treated differently due to the effect of ionised impurity scattering arising from dopant ions or ionised vacancies. Because the charge on ionised impurities is screened by the electron cloud, scattering takes on fundamentally different character below the bulk plasma frequency, where electrons can respond efficiently to the changing electric field and above the bulk plasma frequency, where the field changes too quickly for the electrons to respond. Thus, in addition to frequency-independent scattering due to the mechanisms listed, frequency-dependent scattering from ionised impurities must be considered in understanding the optical response of semiconductors, including metal oxides.

The dielectric properties of the medium surrounding the nanoparticles also play a crucial role in the overall optical response. In the quasistatic limit, Maxwell's equations can be solved to obtain the Mie–Gans cross section for optical absorbance:

$$\sigma_A(\omega) = 4\pi k R^3 Im \left\{ \frac{\varepsilon_P(\omega) - \varepsilon_H}{\varepsilon_P(\omega) + 2\varepsilon_H} \right\} \tag{13.3}$$

where $k = (2\pi(\varepsilon_H)^{1/2}\omega)/c$, ε_P is the particle dielectric function, including the Drude term for the free electron gas, ε_H is the dielectric constant of the host medium, R is the particle radius and c is the speed of light. The total absorbance and scattering can then be calculated using the Beer–Lambert law, taking into account the optical path length and particle volume fraction.

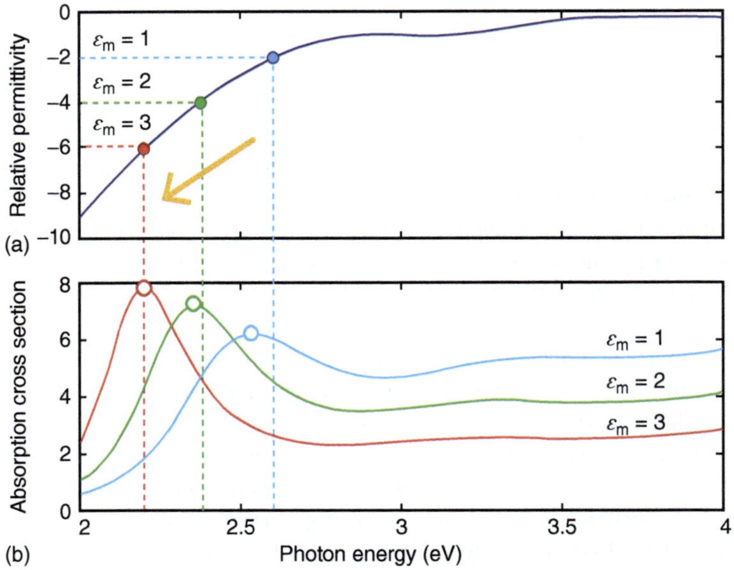

Figure 13.2 (a) Real part of the dielectric function for gold nanocrystals and (b) the corresponding absorption cross section calculated based on the Mie theory. The position in energy of the resonance condition is defined by setting $\varepsilon_P = -2\varepsilon_H$, corresponding to the intersection of the horizontal dotted lines with the curve in (a). Reproduced from Ref. [4] with permission from the Royal Society of Chemistry.

While it is important to consider the full complex dielectric function in calculating the Mie absorbance [10], one can visualise the influence of the medium by setting $\varepsilon_P = -2\varepsilon_H$, assuming that the imaginary parts of the dielectric function of both the particle and the medium are negligible [4]. As the dielectric constant of the medium increases, the peak of the LSPR absorbance progressively shifts to lower energy (Figure 13.2).

13.2 History of Electrochromism in Metal and Semiconductor Nanocrystals

Henglein, Mulvaney and Linnert were among the first to demonstrate chemical manipulation of LSPR in colloidal metal nanocrystals in 1991 [16]. Aqueous suspensions of silver clusters or nanoparticles were exposed to high-energy electrons under a nitrous oxide atmosphere. The electrons, nitrous oxide and water reacted to produce reactive hydroxide radicals. In the presence of 2-propanol, the radicals reacted to form 2-hydroxy-2-propanyl radicals, which, in turn, diffused to the Ag particle surfaces and injected electrons. This reduction was confirmed by an increase in the DC conductivity of the suspension (Figure 13.3b). When the experiment was repeated in the absence of 2-propanol, the hydroxide radicals directly reacted with the particles, injecting holes.

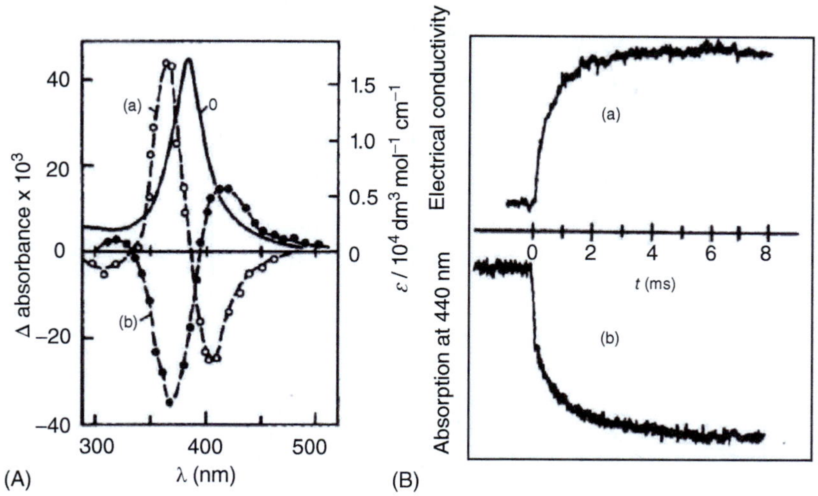

Figure 13.3 Chemically induced changes in LSPR in colloidal silver. (A) Chemical control of electron concentration with hydroxide radicals. The solid curve represents the original LSPR absorption peak, while the open circles and closed circles represent the change in absorption upon reduction and oxidation, respectively. (B) (a) Increase in electrical conductivity with time of Ag colloidal dispersion upon reduction. (b) Decrease in absorption at $\lambda = 440$ nm with time upon reduction. Reproduced from Ref. [16] with permission from the Royal Society of Chemistry.

Upon electron injection into the Ag colloids, the LSPR absorption peak blue-shifted, narrowed and increased in intensity (Figure 13.3a). Conversely, the absorption peak red-shifted, broadened and decreased in intensity when the colloids were oxidised either by hole injection or by adsorption of silver ions added to the suspension (Figure 13.4). Oxidation and reduction change the free electron density and the Fermi level within the particles, shifting the LSPR frequency according to Eq. (13.1). Furthermore, Mie theory can be used to show that increasing the free carrier density increases the particle conductivity, which decreases the LSPR bandwidth and also enhances the absorption coefficient at the LSPR wavelength, explaining the spectral changes that accompany electron or hole injection in Ag colloids [16, 17]. Despite the fact that these optical changes were not electrochromic in nature, these results were very important in demonstrating that electron injection or extraction into or out of metal nanocrystals shifts the LSPR wavelength and changes the absorption peak. This important result is the foundation for the following two decades of research and current studies of chemical, photochemical and electrochemical manipulation of LSPR features in nanocrystals.

The Mulvaney group expanded these results to electrochemical doping of colloidal silver nanocrystals in 1997 [18]. In this experiment, the authors used poly(acrylic acid) to stabilise Ag colloids in an aqueous sodium perchlorate electrolyte solution. In a spectroelectrochemical cell, the colloids were cathodically charged at a gold mesh working electrode while the absorption spectrum of

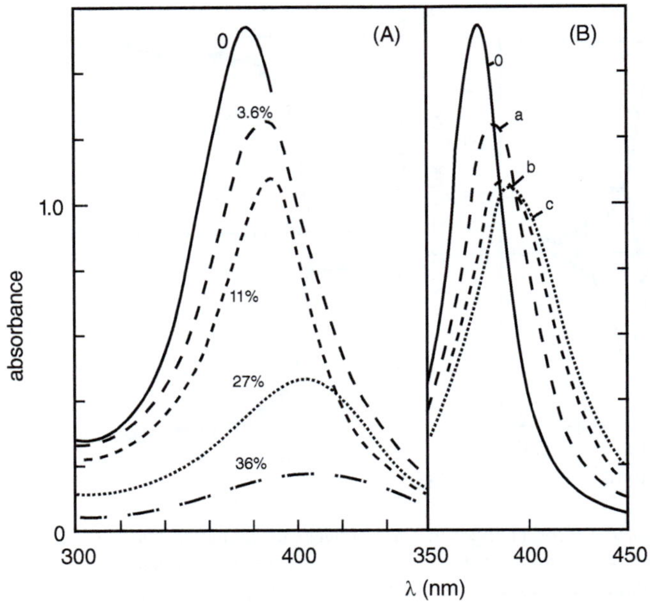

Figure 13.4 Chemically induced changes in LSPR in colloidal silver. (A) LSPR absorption spectra of colloidal Ag before and after oxidation by hydroxide radicals. The percentages next to each curve indicate the number of Ag atoms oxidised. (B) LSPR absorption of Ag colloids red-shifts as increasing numbers of Ag^{+} ions (a–c) are added to the dispersion. Reproduced from Ref. [16] with permission from the Royal Society of Chemistry.

the dispersion was monitored *in situ*. A potential range of −0.8 to 0.4 V versus Ag/AgCl applied to the solution produced a total shift in the LSPR wavelength of about 12 nm, between 392 and 404 nm (Figure 13.5). As previously observed, the blue shift of the LSPR peak was accompanied by a decrease in bandwidth and an increase in intensity. The authors determined that these optical changes resulted from between 1600 and 1790 electrons tunnelling between each nanocrystal (9.8 nm diameter) and the working electrode; the injected electrons were compensated by sodium ions in the electrolyte through a double layer interaction (Figure 13.6). The capacitance of this double layer was $80 \pm 10\,\mu F\,cm^{-2}$. The Mulvaney group has since examined electrochemical charging and shifting plasmon resonances in single gold nanorods using dark field optical microscopy [19]. In this study, the authors were able to blue-shift the LSPR absorption peak of Au nanorods by up to 11 nm, corresponding to injecting 85 000 electrons and a surface capacitance of $150\,\mu F\,cm^{-2}$.

While the 11 nm shift in the LSPR absorption peak was modest, due to the high intrinsic carrier densities of gold and silver, these seminal reports demonstrated the feasibility of electrochemically induced manipulation of plasmon absorption. In doing so, the authors discovered a new type of electrochromism based on shifting LSPR absorption peaks electrochemically. As we will see later, these

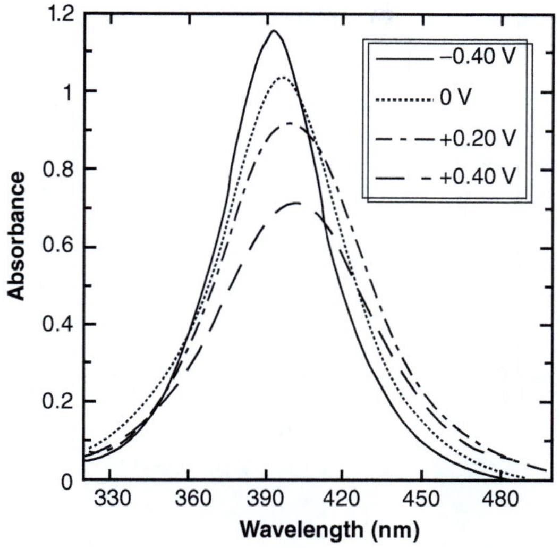

Figure 13.5 Absorbance of colloidal Ag nanocrystals at various potentials versus Ag/AgCl, demonstrating electrochemical manipulation of LSPR. Reprinted with permission from Ref. [18]. Copyright 1997 American Chemical Society.

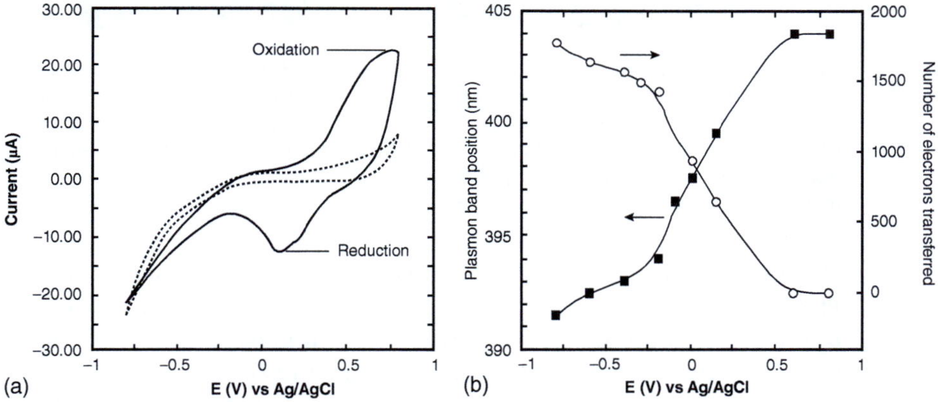

Figure 13.6 (a) Cyclic voltammogram (CV) of a colloidal dispersion of Ag nanocrystals. The dashed line is the CV of the blank electrolyte solution. (b) Ag LSPR peak position and number of electrons injected into each Ag nanocrystal as a function of applied potential. Reprinted with permission from Ref. [18]. Copyright 1997 American Chemical Society.

principles, when applied to the proper material, enable large LSPR shifts and high-performance NIR-selective electrochromic devices.

The practice of using electrochemical charging of nanostructures to affect optical changes was extended to semiconductor nanocrystals between 1999 and 2002. In 2001, Wang, Shim and Guyot-Sionnest discovered electrochromism in colloidal dispersions of cadmium selenide (CdSe) quantum dots stabilised in a tetrahydrofuran and tetrabutylammonium perchlorate (TBAP) electrolyte [20].

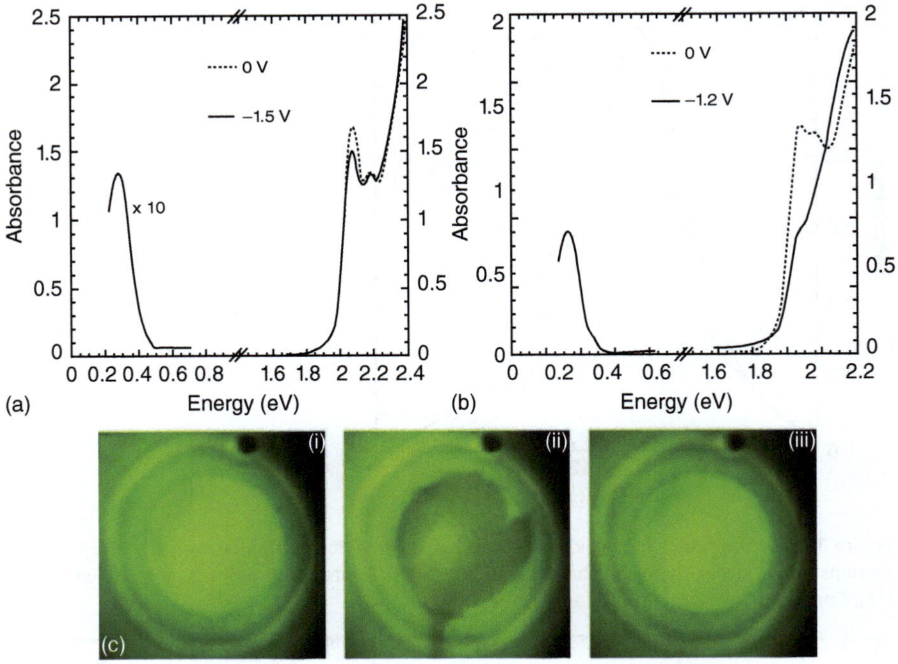

Figure 13.7 Electrochromism in CdSe quantum dots. (a) Electrochromic response of 5.4 nm colloidal CdSe upon applying negative potential. (b) Electrochromic response of 7.0 nm colloidal CdSe. (c) Photograph displaying the photoluminescence of colloidal CdSe/ZnS core/shell nanocrystals (4 nm CdSe core) at 0 V (i), quenching of the PL at −1.5 V (ii) and recovery of PL at 0 V (iii). Reprinted from Ref. [20] with permission from AAAS.

Using a spectroelectrochemical cell with a CaF_2 window and an optically polished Pt disk as a working electrode, the authors monitored the optical properties of the CdSe nanocrystals with Fourier transform infrared (FTIR), UV-visible and photoluminescence (PL) spectroscopy (Figure 13.7). At negative potentials, electrons were transferred into the nanocrystals from the working electrode, initiating a new absorption peak in the mid-IR, bleaching of the visible absorption and quenching the PL. Reduction of the CdSe and the optical changes were fully reversible. The optical changes were indicative of injected electrons completely filling the 1S state of the nanocrystals, creating a new optical 1S to 1P intraband transition in the mid-IR and preventing both the lowest energy 1S interband transitions and PL. Nanocrystal size had a strong effect on the intensity of the optical transitions, as smaller, 5.4 nm nanocrystals exhibited only a 10% visible bleach, corresponding to reduction of only 20% of nanocrystals within the dispersion, while 7.0 nm nanocrystals displayed a 50% bleach, with 100% of nanocrystals being reduced. The authors attributed this difference to a kinetic balance between reduction at the working electrode and oxidative processes at the nanocrystal surfaces, which are faster at smaller sizes. Differences in nanocrystal size also

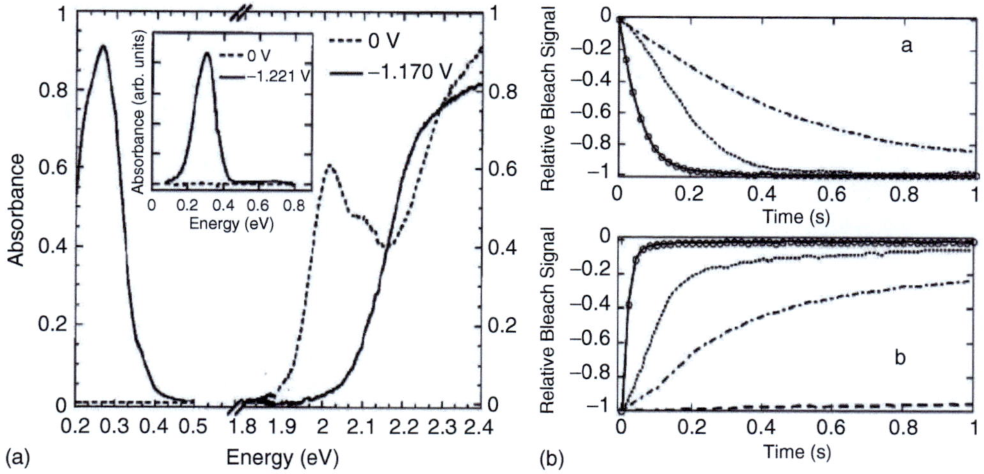

Figure 13.8 (A) Electrochromic response of a CdSe nanocrystal film (5.8 nm diameter) at negative potential. Panel (A) inset: IR response of a film of 5.0 nm CdSe nanocrystals. (B) (a) kinetics of electrochromic bleaching at 639.5 nm of 6.2 nm CdSe nanocrystal films exposed to various chemical treatments. (b) Recovery of absorption peak after potential set to zero. Dot-dashed line: TOPO-capped nanocrystals treated with 4,4′-biphenyldithiol. Dotted line: Pyridine-capped nanocrystals treated with 4,4′-biphenyldithiol. Solid line with circles: Pyridine-capped nanocrystals treated with 1,6-hexanedithiol. Dashed line in (b): pyridine-capped nanocrystals treated with 1,6-hexanedithiol show good electrochromic memory when the electrode is simply disconnected after charging. Left panel reprinted with permission from Ref. [21]. Copyright 2002 American Institute of Physics. Right panel reprinted with permission from Ref. [22]. Copyright 2003 American Chemical Society.

affected the energy of the optical transitions, as well as the reduction potential of the colloids (defined as the voltage at which electrons occupy the 1S state and create the new optical transitions), owing to the size-dependent quantum confinement effect in CdSe nanocrystals. Smaller nanocrystals therefore had a more negative reduction potential, requiring more energy to reduce and had higher energy absorption features.

The Guyot-Sionnest group continued to develop this discovery of electrochromism in CdSe by processing quantum dots into thin films for spectroelectrochemical experiments (Figure 13.8) [21, 22]. By eliminating particle diffusion towards the working electrode as a rate-limiting step, the electrochromic response was considerably quickened and enhanced, and a 100% bleach of the first exciton peak was achieved. Furthermore, chemical procedures to manipulate the nanocrystal surface, such as exchanging the trioctylphosphine oxide (TOPO) ligands for pyridine ligands or cross-linking the nanocrystals in the films, dramatically enhanced electron transport and film stability, providing a switching speed on the order of 40 ms for over 10 000 cycles in a 100 nm thick film of 6.2 nm nanocrystals. The surface capacitance of the reduced CdSe nanocrystal films was calculated to be 19 μFcm^{-2}. These experiments highlight the importance of

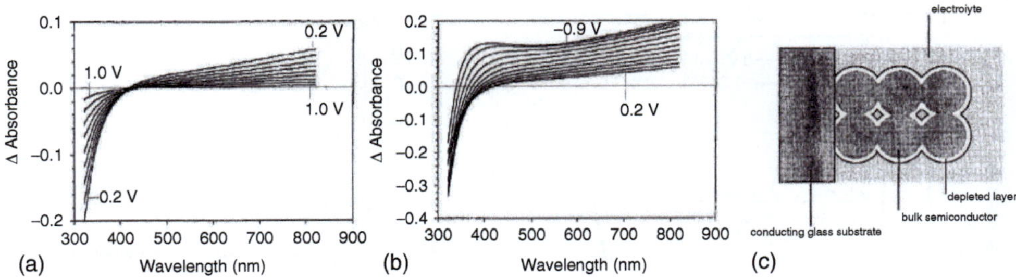

Figure 13.9 Electrochromic response of an ATO nanocrystal film. (a) Change in visible absorbance upon applying reducing potentials. Each curve is separated by a 0.1 V interval. (b) Increasing changes in absorbance at more strongly reducing potentials (0.1 V interval between curves). (c) Depiction of a depletion condition in a highly doped nanocrystal film in contact with an electrolyte. Reprinted with permission from Ref. [23]. Copyright 1999 American Chemical Society.

nanocrystal size on electrochromic response, as changing size not only affects the 'ease' of electrochemical reduction but also affords the opportunity to tune spectral range and intensity.

In addition to electrochromic chalcogenide nanocrystals, highly doped metal oxide semiconductor nanocrystals were also shown to display electrochromic properties. In 1999, Boschloo and Fitzmaurice showed that antimony-doped tin oxide ($Sb:SnO_2$ or ATO) nanocrystal thin films could be electrochemically charged and discharged in a $LiClO_4$ aqueous electrolyte to achieve a fully reversible electrochromic effect with a coloration efficiency of $25\,cm^2\,C^{-1}$ at 700 nm [23]. Upon electron injection into the nanocrystals, the visible absorption of the film increased, growing stronger with increasing negative potential (Figure 13.9a,b). The decrease in absorption at short wavelengths was due to an increase in the optical band-gap attributed to the Burstein–Moss effect, while at longer wavelengths, the absorption increase was ascribed to electron 'Drude absorption'. It should be noted that the spectral results in this experiment do not extend sufficiently far enough into the infrared to observe a blue-shifting absorption peak that would be characteristic of plasmonic electrochromism. The authors did, however, investigate the electrochemical characteristics of this system in depth by performing Mott–Schottky analysis and cyclic voltammetry along with spectroelectrochemistry. Mott–Schottky analysis revealed that the ATO nanocrystals had a pH-dependent flat band potential; at voltages below the flat band potential, an electron accumulation layer formed at the periphery of the nanocrystals, while at voltages positive of the flat band potential, a depletion layer formed as electrons were removed from the film (Figure 13.9c). These accumulation and depletion layers coincided with the observed changes in absorption, and the optical changes at different voltages above the flat band potential indicated that the nanocrystals were not fully depleted at these voltages, rather, the width of the depletion region was changing and causing the optical changes. The magnitude of changes in free carrier absorption at positive potentials supports

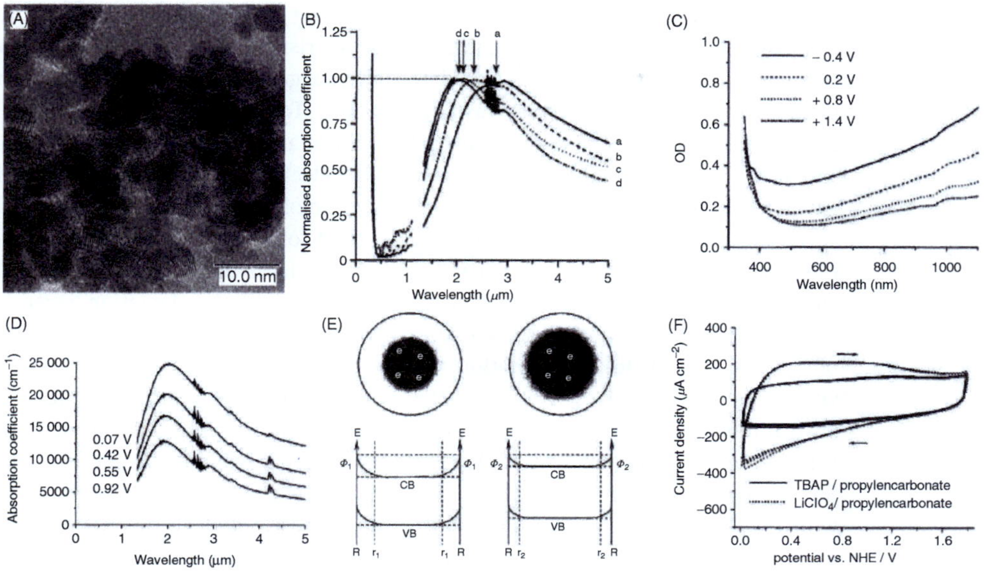

Figure 13.10 Electrochromic Sb-doped tin oxide nanocrystal films. (A) Transmission electron microscopy (TEM) image of synthesised ATO nanocrystals. (B) Normalised absorption spectra of ATO nanocrystal thin films with varying Sb content. (a) 4.8%, (b) 9.1%, (c) 16.7% and (d) 20.6%. (C) Visible absorption of an ATO nanocrystal electrode at various potentials versus Ag/AgCl. (D) NIR electrochromic response of an ATO nanocrystal electrode (16.7% Sb). (E) Depiction of depletion layer and band bending in ATO nanocrystals. When electrons are injected, the depletion layer width decreases, but the total carrier concentration does not increase. (F) Cyclic voltammogram of an ATO nanocrystal electrode (16.7% Sb) showing mostly capacitive behaviour. Using a bulky TBAP electrolyte limits ion insertion and reduction of Sb^{5+} to Sb^{3+} at low potentials. Reprinted with permission from Ref. [24]. Copyright 2000 American Chemical Society.

this assertion. The authors also demonstrated that the charge was injected capacitively, without significant contribution from Faradaic or ion insertion reactions.

The next year, zum Felde, Haase and Weller demonstrated a drastic improvement in the electrochromic response of semiconductor nanocrystals, also using thin films of ATO nanocrystals in an organic $LiClO_4$ electrolyte [24]. Much of the improvement was a result of the authors synthesising their own nanocrystals to better control size, Sb content and film morphology and porosity (Figure 13.10a). In doing so, they were able to achieve strong changes in absorption intensity in both the visible and NIR regions, with a coloration efficiency of $18\,cm^2\,C^{-1}$ at 650 nm, fast switching speeds of 150 ms or less and excellent stability over 1 000 000 cycles. In this spectroelectrochemical experiment, the spectral window did extend out into the near- and mid-IR, revealing a nanocrystal absorption peak centred between 2000 and 3000 nm, depending on the Sb doping level (Figure 13.10b). Because the peak position depended on the Sb content and therefore the free carrier concentration, it was determined to be an LSPR absorption feature. Upon electrochemical charging and electron injection, the absorption peak doubled in magnitude, giving rise to NIR electrochromism (Figure 13.10d). However, the peak did not shift, suggesting that the total electron density within the nanocrystals was not changing. To explain this behaviour, the authors invoked the argument made earlier [23] of a depletion region present at the ATO nanocrystal surface in contact with the electrolyte (Figure 13.10e). Such a depletion region can result from partial reduction of Sb^{5+} to Sb^{3+} at nanocrystal surfaces, other charged surface states, adsorbed ions or other electron traps. Electron injection would then cause this depletion layer to shrink, but the total electron concentration remains constant, at least until the depletion layer is completely eliminated and an accumulation region is formed (this was not observed). The shrinking depletion width corresponds to an increase in the volume occupied by free carriers in the nanocrystals, explaining the enhanced absorption. At the most negative potentials, a strong visible blue absorption feature appeared (Figure 13.10c), likely stemming from the reduction of Sb^{5+} to Sb^{3+} along with cation insertion at the nanocrystal surfaces. Using an electrolyte with a bulky cation, such as tetrabutylammonium, avoided ion insertion reactions and the associated blue edge absorption (Figure 13.10f). Pflughoefft and Weller further examined this system in 2002 by developing a new synthesis for alcohol-soluble ATO nanocrystals, enhancing their compatibility with processing techniques like dip coating and allowing for the use of Sb dopant precursors with different ligands and oxidation states [25]. The authors again achieved strong NIR electrochromism and were able to attribute the visible absorption to $Sb^{3+}Cl^{-}Sb^{5+}$ and $Sb^{3+}O^{2-}Sb^{5+}$ charge transfer complexes.

The experiments described here were foundational in elucidating a new type of electrochromic mechanism based on the electrochemical modulation of free electrons in plasmonic nanocrystals *via* capacitive charging. In such nanocrystals, a change in electron concentration should result in a shift in the LSPR absorption frequency and a change in absorption intensity, giving rise to a plasmonic electrochromic response. However, the materials used in these early experiments fell short for smart windows applications. For instance, in metal nanocrystals, the high carrier concentration ($\sim 10^{23}$ cm^{-3}) severely limited the spectral range over which the plasmon resonance was modulated. ATO nanocrystals showed much stronger plasmon modulation than metal nanocrystals, but the spectral response was located in the NIR and mid-IR, where there is not much solar insolation. Likewise, the optical contrast and spectral response of electrochromic CdSe nanocrystals were not suitable for optimal solar modulation. Therefore, to realise a functional plasmonic electrochromic, new materials needed to be developed.

Transparent conducting oxides (TCOs) are a class of doped metal oxide materials that have the potential to avoid the aforementioned shortcomings. ATO is one example, while other material compositions can achieve improved overlap of their spectral response with the solar spectrum. Bulk Sn-doped In_2O_3 and Al-doped ZnO, in particular, have carrier concentrations intermediate between semiconductors and metals ($\sim 10^{21}$ cm^{-3}), placing their plasma frequencies in the NIR [26].

13.3 Doped Metal Oxide Colloidal Nanocrystals as Plasmonic Electrochromic Materials

Doped metal oxides, like Sn-doped In_2O_3 (ITO) or Al-doped ZnO (AZO), are good candidates for NIR-selective plasmonic electrochromics owing to their wide optical band gap and intermediate carrier concentrations, which place their plasma frequencies in the NIR, while remaining transparent in the visible region [26]. Thanks to the efforts of many researchers over the past 5 years, both ITO and AZO materials can now be synthesised as monodisperse, colloidal nanocrystals, with tunable size, doping concentration and optical properties [27–30]. The synthetic control achieved over these nanocrystals has been crucial to realise the phenomenon of plasmonic electrochromism in doped metal oxide colloidal nanocrystals.

13.3.1 Colloidal Synthesis of Doped Metal Oxide Nanocrystals

The well-established surfactant-mediated method for colloidal synthesis allows one to grow nanocrystals from molecular precursors in solution phase at moderate temperatures, between 150 and 400 °C (Figure 13.11). Surfactant-assisted colloidal synthesis [31–34] offers excellent control over nanocrystal size and shape, as well as preventing their aggregation and precipitation from solution, which enables stable colloidal dispersions (Figure 13.11c), a key difference from other synthesis methods. Such dispersions can serve as inks for solution deposition of nanocrystal films using potentially low-cost and industrially scalable techniques such as spray coating or slot-die coating, as well as laboratory-scale techniques like spin coating.

While colloidal synthesis is well developed to achieve small-diameter nanocrystals (1–20 nm), achieving high doping levels has been more elusive and challenging. Growing colloidal nanocrystals from solution phase is a kinetically controlled process; introducing dopants within the lattice (Figure 13.11d) therefore requires a fine balance of the host nanocrystal growth rate and the dopant incorporation rate. A variety of doping strategies have been reported for several nanocrystal compositions, and they have been recently reviewed and discussed in depth [7]. For the particular case of binary metal oxides, the most popular doping strategy is based on tuning the relative chemical reactivity of the molecular precursors for the host nanocrystal and for the dopant atom. Typical precursors are metal-organic salts comprising the metal of interest and a ligand (e.g. an alkyl carboxylate). By selecting a ligand with the appropriate steric hindrance and negative charge delocalisation properties, it is possible to tune the ligand–metal bond strength to balance the reactivity of both precursors. By developing this strategy, our group was able to demonstrate the first synthesis of plasmonic, Al-doped ZnO nanocrystals with tunable dopant levels (0–7 at%) and well-controlled size distributions (from 5 to 20 nm with <15% std. deviation in diameter) [30]. Substitutional

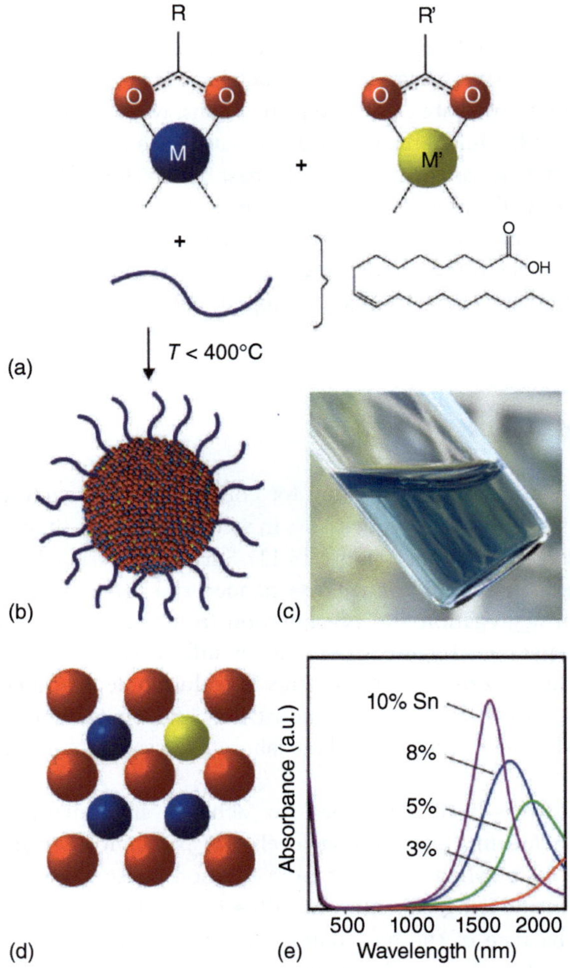

Figure 13.11 Colloidal synthesis of doped metal oxide nanocrystals. (a) Molecular precursors (depicted here as metal carboxylates) are combined with organic ligands (oleic acid depicted) in a non-polar solvent and heated to moderate temperatures. (b) After the precursors decompose and react, inorganic colloidal nanocrystals are formed, which are stabilised in solution by the ligands coordinated to the surface. (c) Optically clear dispersion of colloidal ITO nanocrystals in hexane. (d) Substitutional doping mechanism whereby a donor atom (i.e. Sn) occupies a metal (i.e. In) site in the nanocrystal, thereby introducing electrons into the conduction band. (e) Absorbance spectra of ITO nanocrystals, showing a blue shift, narrowing and enhanced intensity of the plasmon peak as doping increases. Part (e) adapted from Ref. [27]. Copyright 2009 American Chemical Society. Figure reprinted from Ref. [1], reproduced by permission of the Royal Society of Chemistry.

doping strategies have been also proven successful for other TCO nanocrystals, such as Sn-doped In_2O_3 [27–29], Nb-doped TiO_2 [35] and In-doped CdO [36].

The emergence of plasmonic semiconductor nanocrystals has done much to motivate research on doping strategies. Fortunately, today, chemists have achieved far better control over the incorporation of dopants into colloidal nanocrystals, and can now do so with a high degree of control over the dopant content [7], consequently tuning the LSPR frequency (Figure 13.11e). These advances have been critical for the development of LSPR-based nanocrystal electrochromics.

13.3.2 Plasmonic Electrochromic Electrodes Based on Colloidal ITO and AZO Nanocrystals

Our research on developing plasmonic nanocrystals as electrochromic smart window materials was first reported in 2011. Inspired by previous works that demonstrated capacitive electrochemical charging of colloidal nanocrystals [18–25] (described in Section 13.2), we applied similar concepts to develop a new class of electrochromic materials suitable for smart window applications. We first reported the unprecedented magnitude of plasmonic electrochromic response in ITO nanocrystals in 2011 [37]. After synthesising colloidal ITO nanocrystals of various sizes and doping levels (Figure 13.12a–d), they were processed into continuous, electrically connected microporous nanocrystal thin films (Figure 13.12e). This was accomplished by spin-coating the native nanocrystal dispersion (with organic surface ligands) onto ITO-coated glass substrates. The films were then post-treated in a formic acid solution to displace the organic ligands by mass action. Formic acid molecules at the nanocrystal surfaces were easily removed by annealing at low temperature (200 °C) in an inert atmosphere, providing electrical conductivity throughout the entire film while leaving the micropores between nanocrystals intact, which allowed for electrolyte penetration throughout the film. Placing the ITO nanocrystal film as a working electrode in a spectroelectrochemical cell, with $LiClO_4$ in propylene carbonate as the electrolyte, and applying a reducing potential, a large LSPR blue shift was observed, causing up to unity change in optical density at NIR wavelengths for 150 nm thick films (Figure 13.12g). Even greater modulation was observed for thicker films. The shift in LSPR absorption frequency indicated that the electron density was indeed changing with electrochemical potential, indicating that the ITO nanocrystals were operating in an electrochemically stable accumulation regime. The large shift also indicated that, unlike ATO nanocrystal films [24], the ITO nanocrystals were relatively free of surface defects and the associated depletion layer modulation. Applying a positive potential easily reversed the optical changes, and the plasmon could even be red-shifted from its equilibrium position, causing a bleach in absorption and indicating a transition to a depletion regime. Drude modelling confirmed that the carrier concentration increased over threefold from about $4.5 \times 10^{20}\,cm^{-3}$ at the most positive bias to over $1.4 \times 10^{21}\,cm^{-3}$ at the most negative bias.

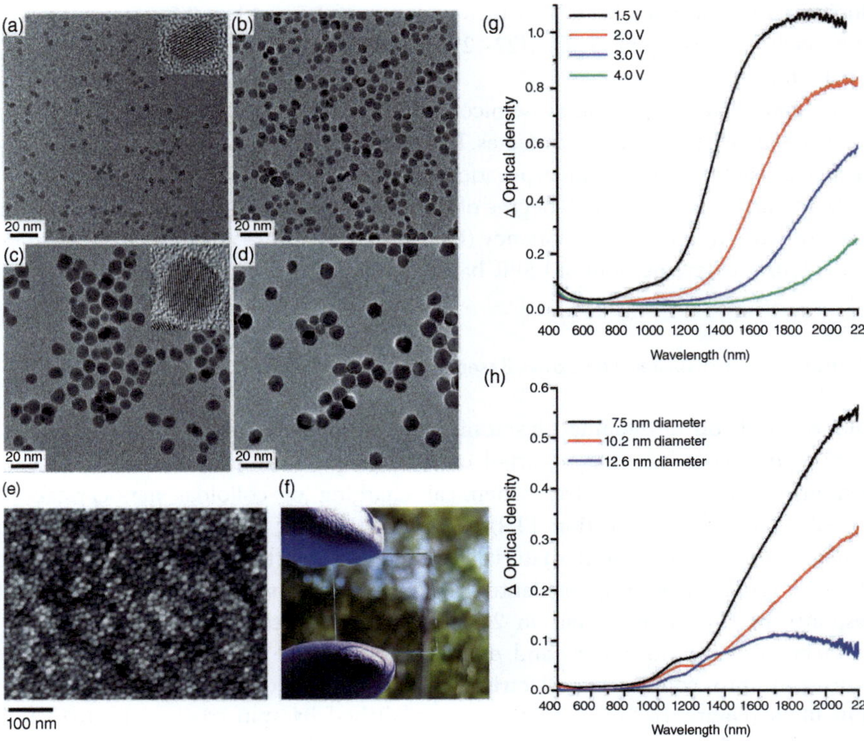

Figure 13.12 (a–d) Transmission electron microscopy images of ITO nanocrystals with sizes (nm): (a) 4.1 ± 0.6, (b) 7.4 ± 1.4, (c) 10.2 ± 1.7 and (d) 12.1 ± 1.5. (e) Scanning electron image (top view) and (f) photograph of an ITO NC network film on glass substrate. (g) Optical density at different applied voltages of an ITO nanocrystal network film with 4.1 nm diameter, 16.8% Sn nanocrystals. (h) Change in optical density between 1.5 and 4 V (vs Li/Li^+) for various nanocrystal sizes, each with 4.65 ± 0.25% Sn. Parts (a)–(d), (g), (h) reprinted from Ref. [37]. Copyright 2011 American Chemical Society. Figure reprinted from Ref. [1], reproduced by permission of the Royal Society of Chemistry.

Ultimately, smaller and highly doped nanocrystals exhibited the strongest modulation of solar transmittance (Figure 13.12h) by allowing accumulation and depletion layers to extend farther into their interior volume and increasing the blue shift of the LSPR peak into the range of the NIR with higher insolation. These nanocrystals modulated the integrated NIR solar transmittance by 35% with only a 6% change in integrated visible transmittance. The wavelength-dependent coloration efficiency ranged from 25 $cm^2\ C^{-1}$ at 1200 nm to nearly 200 $cm^2\ C^{-1}$ at 1700–1800 nm. The electrochromic response was the same in $LiClO_4$ electrolyte, as in tetrabutylammonium ion–containing electrolytes, where insertion was sterically prohibitive, indicating that capacitive charging is the dominant electrochromic mechanism. This mechanism enabled fast coloration and bleaching processes, and one should expect that capacitive operation should also promote

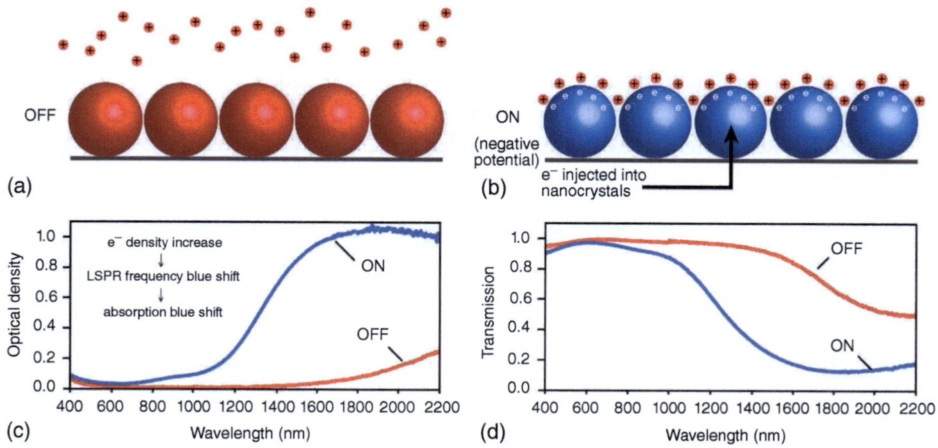

Figure 13.13 Depiction of the microscopic operation of a nanocrystal-based plasmonic electrochromic film, along with the associated optical changes. (a) In the OFF state, positive potential is applied to the nanocrystals, which are depleted of electrons and lithium ions are repelled. (b) In the ON state, a negative potential is applied to the nanocrystals, which injects electrons. Lithium ions are attracted to the nanocrystal surface to compensate the injected charge capacitively. (c) Optical density changes resulting from electron injection. The increase in carrier density causes a blue shift in the LSPR and absorption. (d) Corresponding changes in transmission of the film. Figure reprinted from Ref. [1], reproduced by permission of the Royal Society of Chemistry.

long cycle life. The capacitive nature of this electrochromic effect, and the associated optical changes in the ITO nanocrystals, is depicted in Figure 13.13.

Garcia *et al.* next examined the electrochromism of AZO nanocrystal films and compared them to ITO films [38], utilising a newly developed 'Meerwein's salt' ligand stripping procedure to avoid dissolution or etching of the AZO nanocrystals in formic acid [39]. The NIR modulation of AZO and particularly its durability improved with respect to that of ITO, which was ascribed to the enhanced electrochemical stability of Al^{3+} dopants in AZO compared to Sn^{4+} in ITO. The better electrochemical stability of Al^{3+} also allowed for more negative potentials to be applied to AZO, increasing the number of injected electrons without irreversibly reducing the metal ions. Overall, with changes in optical density of up to 2.5, the best AZO films achieved a 39% modulation in solar NIR transmittance, while the transmittance of visible sunlight changed by only 4%. This corresponded to coloration efficiencies of about 50 $cm^2\ C^{-1}$ at 1100 nm and up to over 400 $cm^2\ C^{-1}$ at 2000 nm. Coloration efficiencies in the ITO films ranged from about 50 $cm^2\ C^{-1}$ at 1200 nm to a maximum of about 375 $cm^2\ C^{-1}$ at 1700–1800 nm. These values are extremely high compared to conventional transition metal oxide electrochromic materials, which are typically in the range of 20–100 $cm^2\ C^{-1}$. In addition, the capacitive coloration mechanism enabled fast switching speeds, ranging between 0.06 and 0.9 s for AZO and 0.05–3.4 s as well as good durability after 20 000 electrochemical cycles [38]. The AZO film lost only 11% of its charge capacity with little change in optical performance. On the other hand, the ITO film suffered a 45% reduction in charge capacity. The loss in capacity moderately reduced the optical modulation range of ITO and was attributed to free carrier loss due to irreversible reduction of Sn^{4+} to Sn^{2+}.

These studies have demonstrated that the LSPR in colloidal TCO nanocrystals, such as ITO and AZO, can be electrochemically modulated within the NIR region, showing great promise as a NIR-selective smart window coating with fast and deep coloration as well as good durability. AZO also has the added benefits of low cost and earth abundance compared to ITO.

13.3.3
Design Principles for Nanocrystal-Based Plasmonic Electrochromics

Maximising the shift in the LSPR peak and the associated modulation of NIR radiation is still a key challenge to improving the functionality and the energy saving capabilities of plasmonic smart windows. As observed in ITO and other semiconductor nanocrystals, higher doping shifts the equilibrium position of the LSPR peak to higher frequency, moving it closer to a region of appreciable solar energy. In addition, decreasing nanocrystal size yields enhanced optical modulation given the larger surface-to-volume ratio, which allows for higher capacitance and thus more electrons to be injected [37]. As a general design principle for NIR-selective plasmonic electrochromics, small and highly doped plasmonic metal oxide nanocrystals should be used to maximise optical modulation in the NIR range of the solar spectrum. In addition, materials that weakly screen external electric fields will allow the electron accumulation region to extend as far as possible into the nanocrystal core, consequently maximising the NIR modulation. As this field progresses, we expect that the material that combines small nanocrystal size, high doping level and low screening of electric fields will provide the best electrochromic performance.

Given that most of the NIR solar energy is located at wavelengths between 750 and 1250 nm, it would be optimal to use nanocrystals with LSPRs near that range. Unfortunately, ITO nanocrystals, which so far display the 'bluest' plasmons of TCO nanocrystals, have a minimum LSPR wavelength of about 1600–1700 nm [27, 37, 38]. This limitation is due to the interplay and competition between doping and defect compensation mechanisms. In metal oxides, aliovalent dopants can be compensated by ionic defects like lattice vacancies and interstitials to maintain charge neutrality [40–42]. Consequently, optimum doping levels have been observed for ITO nanocrystals, above which the LSPR peak begins to red-shift with increasing doping level [27].

Development of new plasmonic colloidal metal oxide nanocrystals, with wavelengths closer to the 750–1250 nm range, will depend on selecting materials that overcome the limitations of defect compensation mechanisms and/or exploit new doping mechanisms, such as interstitial doping [43]. Vacancy- and caesium-doped WO_3 are two examples of recently developed plasmonic metal oxide nanocrystals with LSPR features within this region of interest [44, 45].

13.4 Advanced Electrochromic Electrodes Constructed from Colloidal Plasmonic NCs

The colloidal nature of plasmonic nanocrystals together with their unique NIR selectivity *via* capacitive charging offers new possibilities to construct advanced electrochromic electrodes with greater functionalities [46–50]. The surfaces of the nanocrystals, when stripped of their native organic ligands, offer an exceptional chemical platform to rationally attach other active components through interfacial bonding (Figure 13.14). These chemical components ultimately can direct the assembly of nanocrystals into ordered mesoporous architectures [47] or form a glassy matrix of electrochromic material surrounding the nanocrystals, thus adding new optical functionalities [49]. In this section, we describe two types of advanced electrochromic electrodes based on plasmonic nanocrystals: mesoporous NIR-selective nanocrystal assemblies with ordered, tunable pore sizes and enhanced charge capacity and switching kinetics [50], and dual-band 'nanocrystal-in-glass' composites with enhanced optical contrast and durability [49]. We discuss the relationships between composition, structure and observed enhanced functionalities and identify the design rules as a future perspective. At

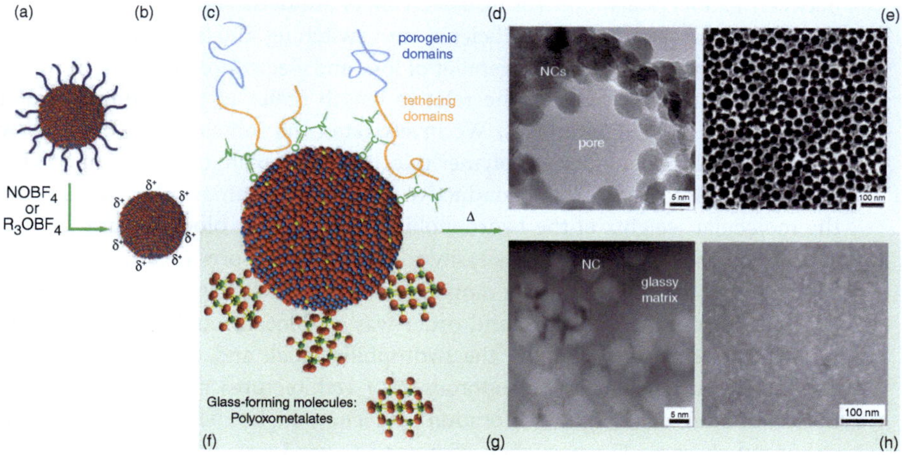

Figure 13.14 Colloidal nanocrystals as building blocks for advanced electrochromic electrodes. (a) Starting with the as-synthesised nanocrystals, a ligand-stripping agent is used to produce (b) nanocrystals with bare, positively charged surfaces. (c) The nanocrystal surfaces can be decorated with amphiphilic block copolymers and made to form (d) micelles decorated with nanocrystals (TEM image of a micelle). (e) Upon depositing the micelles and decomposing the polymer, an ordered mesoporous structure is obtained (scanning electron microscopy (SEM) image of film, top view). (f) Nanocrystal surfaces functionalised with polyoxometalate (POM) clusters. Upon deposition and annealing, the POMs decompose to form a glassy matrix surrounding the nanocrystals. (g) TEM cross-sectional image of a composite film. (h) SEM top-view image of a composite film. Parts (d) and (e) reprinted with permission from Ref. [47] Copyright 2012 American Chemical Society. Parts (g) and (h) reprinted with permission from Ref. [49]. Copyright 2013 Nature Publishing Group. Figure reprinted from Ref. [1], reproduced by permission of the Royal Society of Chemistry.

the end of this section, we also present other types of advanced electrochromic electrodes (e.g. nanocomposites) that do not necessarily employ colloidal approaches or plasmonic nanocrystals.

13.4.1 NIR-Selective Mesoporous Architectured Electrodes Based on Plasmonic Colloidal Nanocrystals

As presented in the previous section, the simplest nanostructured plasmonic electrochromic electrode is constructed from randomly packed nanocrystals forming an interconnected open network that yields 34% porosity. Given the capacitive operating mechanism, these electrodes must have high surface areas that are easily accessed by the electrolyte. An ITO nanocrystal electrode with 34% porosity functions well enough in liquid electrolyte to achieve dynamic changes of 40% in solar NIR transmittance [49]. Nevertheless, the accessible surface area of active material is still limited in these films owing to insufficient nanocrystal–electrolyte interfacial contact or interpenetration, inefficient electron conduction within the nanocrystal network or restricted electrolyte or ion motion within micropores between nanocrystals [50]. All of these effects are detrimental to the dynamic modulation range, coloration efficiency and switching kinetics of the electrode and are dependent on the arrangement of ion- and electron-conducting domains in the nanocrystal electrode, the relative length scales of those domains and the topology of their interfaces. We investigated the influence of pore size by utilising amphiphilic block copolymer micelles to template ligand-stripped ITO nanocrystals into ordered and conductive mesoporous frameworks. By changing the molecular weights of the hydrophobic or hydrophilic block (Figure 13.14c), the mesoporous architecture was easily controlled, with pores ranging from 29 to 37 nm and pore wall thicknesses ranging from 9 to 13 nm (Figure 13.14d,e). Such exceptional mesoscale control and order was enabled through tailored, strong enthalpic interactions between the hydrophilic block and the ligand-stripped nanocrystal surfaces [47]. Mesoporous ITO architectures showed comparable optical modulation, higher coloration efficiency, higher charge capacity and enhanced charging kinetics when compared to randomly packed, microporous nanocrystal networks. The enhancement in switching speed scaled with increasing pore size, indicating that larger pores provided for faster ion transport to the nanocrystal surfaces [50].

In addition to improved coloration efficiency, charge capacity and switching speed, mesostructured nanocrystal electrodes can be combined with polymer gel electrolytes to fabricate full-cell solid-state electrochromic devices. We created full-cell devices by using a mesostructured ITO nanocrystal electrochromic electrode, a CeO_2 nanocrystal counter electrode (which was optically passive) and a plasticised poly(vinyl butyral) (PVB) gel electrolyte containing a lithium salt [50]. Our PVB electrolyte was easily deposited on each electrode, and penetrated into the mesopores of the ITO electrochromic electrode, allowing the electrodes to be laminated together to produce a robust solid-state device.

The ionic conductivity of the electrolyte was $\sim 10^{-7}\ \mathrm{S\,cm^{-1}}$, which was within acceptable limits for typical electrochromic devices and provided good switching speed (<10 s) for our devices.

When compared to an unarchitectured solid-state device using the same PVB-based electrolyte, mesoporous ITO nanocrystal films displayed greater than twofold enhancements in optical modulation, coloration efficiency and charge capacity. The 60% porosity of the mesoporous film, compared to the 34% porosity in the unarchitectured film, provided the solid electrolyte with improved access to and intimate contact with the nanocrystal surfaces. This is a key advance, as the continued development of nanocrystal-based electrochromics will likely depend on the ability to integrate nanocrystals into solid-state devices and manufacturing methods. The capability to produce mesostructured nanocrystal electrodes is therefore incredibly important for the continued development of these devices, as is improved understanding of the interactions between polymer electrolytes, lithium ions within such electrolytes and nanocrystal surfaces.

13.4.2
Dual-Band Nanocrystal-in-Glass Composite Electrodes Based on Plasmonic Colloidal Nanocrystals and Conventional Electrochromic Materials

Precise control over nanocrystal surface chemistry also provides an opportunity to embed plasmonic NCs within a matrix of conventional electrochromic material. Such a device could approach the ideal functionality for a smart window, as the nanocrystals and conventional materials could be activated independently to modulate both the NIR and visible regions of the solar spectrum, respectively. 'Dual-band' functionality would allow the window to dynamically respond to changing weather conditions and comfort needs, as well as offering glare control, ultimately pushing past the current performance limits of either conventional or plasmonic nanocrystal electrochromics.

In 2013, we demonstrated that this elusive functionality could be achieved by creating a composite electrochromic material containing NIR-selective plasmonic nanocrystals (ITO) which are embedded within a matrix of visible-active amorphous niobium oxide (NbO_x) [49]. The realisation of such an exotic 'nanocrystal-in-glass' composite material would have not been possible without the key contributions of colloidal and surface chemistry. Colloidal approaches allow structural and compositional control at multiple length scales from the atomic (doping, bonding and crystal structure) to the nano- and mesoscale (crystallite sizes, shapes, interface densities). Non-colloidal methods to prepare glass-ceramic composites, instead, rely on thermally induced phase separation and *in situ* crystallisation processes, which lack the required control [51]. The time-temperature parameter space for selective nanocrystallisation is typically narrow and strongly influenced by the glass composition and structure. Consequently, to selectively crystallise the desired composition and phase with control over the interface area is a challenging task, especially for complex multi-component

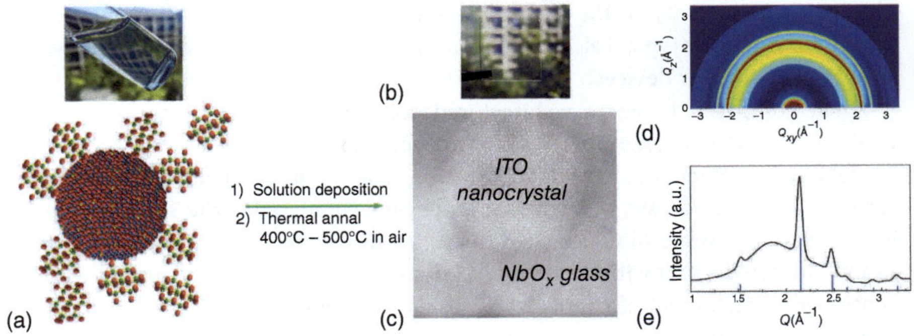

Figure 13.15 Nanocrystal-in-glass film preparation and structural characterisation. (a) Aqueous colloidal dispersion containing ITO plasmonic nanocrystals surface-stabilised with Nb polyoxometalate clusters (POMs), POM; $[N(CH_3)_4]_6$ $Nb_{10}O_{28}$. (b) NbO_x-ITO film obtained after solution deposition on glass substrate and thermal annealing processing. Upon annealing, the $[N(CH_3)_4]^+$counter ions thermally decompose and the anionic $[Nb_{10}O_{28}]^{6-}$ clusters condense to form an amorphous NbO_x matrix [46]. (c) High resolution cross-sectional STEM Z-contrast image showing the crystalline nature of NC and amorphous matrix. (d) 2D-XRD image of an NbO_x-ITO nanocrystal-in-glass film and (e) XRD pattern obtained after integrating along the out-of-plane scattering direction. The indexed diffraction lines result from randomly oriented ITO nanocrystals (cubic In_2O_3 bixbyite structure) while the broad peak at $Q = 1.85\,Å^{-1}$ is assigned to amorphous NbO_x. Adapted with permission from Ref. [49] Copyright 2013 Nature Publishing Group.

stoichiometries like doped metal oxide nanocrystals. Such compositions can, however, be easily accessed by using well-defined colloidal nanocrystals that are post-synthetically embedded in and chemically linked to glass.

In our colloidal approach, anionic polyniobate polyoxometalate (POM) clusters were chemically linked to the surfaces of ligand-stripped colloidal ITO nanocrystals in solution *via* bridging M−O−M′ bonds. This process results in POM-stabilised nanocrystal aqueous dispersions (Figure 13.15a), from which films are deposited and then thermally annealed in air, at $T = 400\,°C$ (Figure 13.15b), yielding a glassy metal oxide network that is covalently bonded to the nanocrystal surfaces. Scanning transmission electron microscopy (STEM) cross-sectional images and grazing incidence x–ray diffraction (XRD) patterns of films annealed at 400 °C confirmed the amorphous nature of the NbO_x matrix, which remains bonded to the crystalline ITO nanocrystals (Figure 13.15c–e).

Given that the reduction potential of NbO_x is lower than the potential range at which ITO nanocrystals are capacitively charged, the two optical responses can be independently controlled by simply modifying the applied voltage (Figure 13.16). At 4 V (vs Li/Li^+), ITO nanocrystals are depleted of free electrons and therefore, the film is transparent to solar radiation, allowing NIR and visible light to transmit through. Reducing the voltage to an intermediate value (2.3 V vs Li/Li^+) results in the capacitive charging of the ITO nanocrystals and consequent increase in the free electron concentration and appearance of LSPR in the NIR region. Lower potentials are needed (1.5 V vs Li/Li^+) to activate the visible response of amorphous NbO_x matrix *via* small polaron transitions caused by the Faradaic

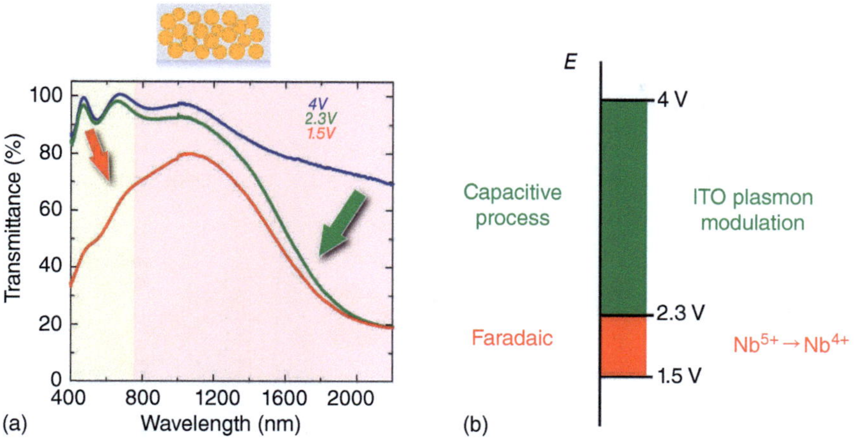

Figure 13.16 Dual-band solar control in nanocrystal-in-glass electrochromic electrodes. (a) Transmittance spectra under applied electrochemical voltage; 4 V in blue, 2.3 V in green and 1.5 V in red, versus Li/Li^+ of ITO-*in*-NbO_x with 40 vol% of ITO nanocrystal. (b) Schematics showing the different potential ranges where Faradaic and capacitive process occur.

reaction. Conceptually, these potential-dependent and spectrally-selective optical responses yield a three-way operating window suitable for regulating solar heat gain and lighting under widely variable climate and weather conditions. These three modes of operation are sketched in Scheme 13.2.

By systematically increasing the volume fraction of nanocrystals, the contribution of NIR modulation varies monotonically with the ITO nanocrystal content, as expected. However, a striking deviation from linearity is observed for the NbO_x matrix (Figure 13.17a,b). The modest visible-light modulation of pure amorphous NbO_x was greatly exceeded by that of the nanocrystal-in-glass composites, indicating that the NbO_x glass is far more electrochemically active when linked to the nanocrystals. As shown in (Figure 13.17b), the optical contrast in the visible range monotonically increased with nanocrystal content up to ~43% ITO, reaching a value that is three times greater than that of pure NbO_x (five times greater when normalised by NbO_x volume fraction). At higher nanocrystal content, the contrast decreased, indicating that the enhancement is maximised when the average nanocrystal spacing is around 0.7 nm. This thin matrix material between closely spaced nanocrystals can provide effective ion transport channels, and the overall electrochromic behaviour is strongly affected by the percolation of nanocrystal-glass interfaces. It is thus apparent that the density of glass–crystal interfaces plays a dominant role in the electrochromic properties of the glass.

Enhanced ion transport by percolating glass/crystal interfaces has been previously reported in other material systems [52–54]. For example, we recently showed that the overall ionic transport in mixed conductor Ag_2S-in-GeS_2 composites was enhanced compared to either single component [54]. Interestingly,

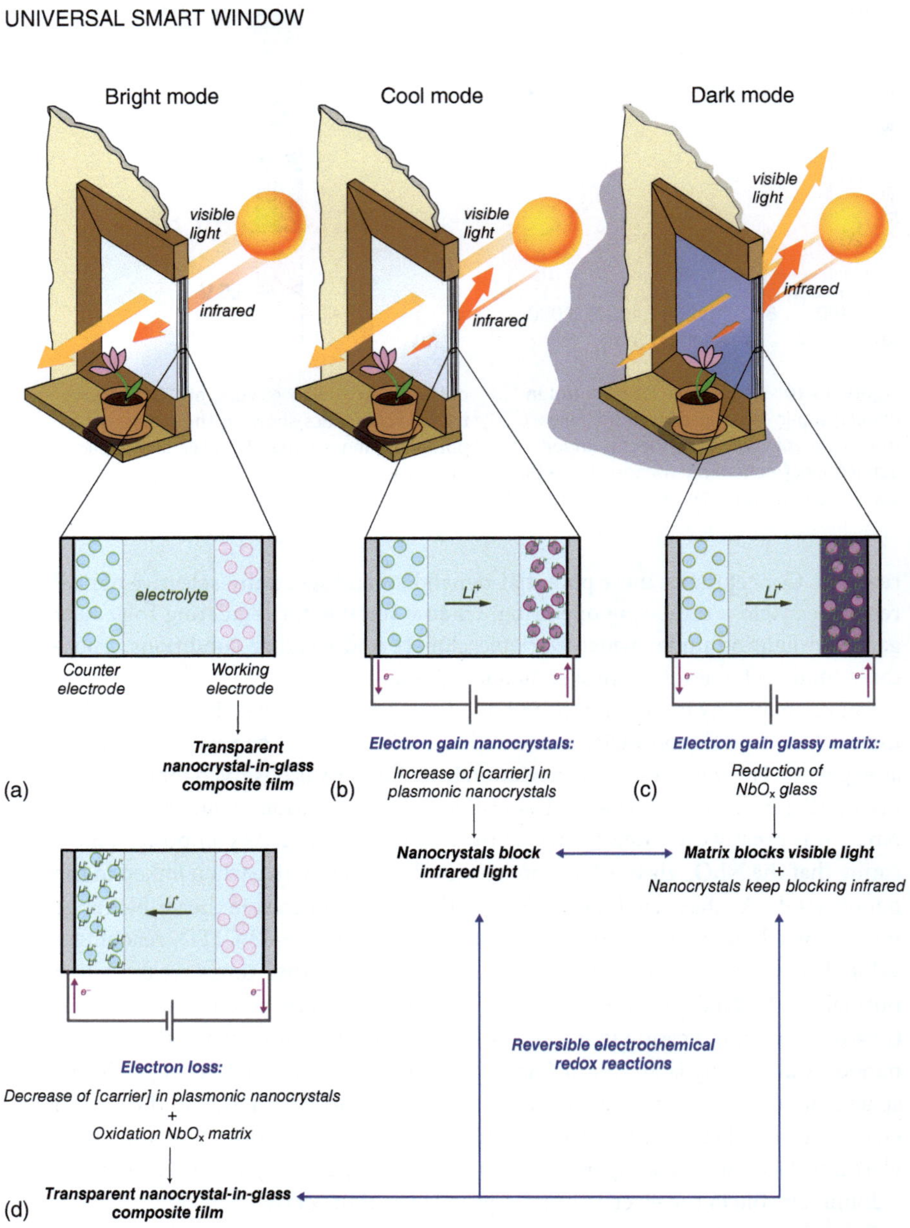
UNIVERSAL SMART WINDOW
Bright mode
Cool mode
Dark mode
visible light
infrared
electrolyte
Counter electrode
Working electrode
Li+
e−
Transparent nanocrystal-in-glass composite film
(a)
Electron gain nanocrystals:
Increase of [carrier] in plasmonic nanocrystals
(b)
Nanocrystals block infrared light
Electron gain glassy matrix:
Reduction of NbOx glass
(c)
Matrix blocks visible light
+
Nanocrystals keep blocking infrared
Reversible electrochemical redox reactions
Electron loss:
Decrease of [carrier] in plasmonic nanocrystals
+
Oxidation NbOx matrix
Transparent nanocrystal-in-glass composite film
(d)

Scheme 13.2 'Universal smart window' based on dual-band electrochromic concept: dynamic independent modulation of visible and near-infrared light transmitted through the window. The nanocrystal-in-glass electrochromic film is the working electrode in an electrochemical cell, in which an optically passive film acts as a counter electrode. Both electrodes are deposited on conducting glass panes, separated by a solid electrolyte layer. At open circuit voltage, the window is in bright mode, that is, both nanocrystals and matrix are transparent to solar radiation, allowing heat and natural light enter the room. Reducing the voltage to an intermediate value, the window switches to cool mode, blocking heat while allowing natural light enter the room. At lower potentials, the window switches to dark mode, limiting the amount of heat and natural light that enter the room. These three switching modes enable the window to operate at variable weather conditions, thus improving energy savings and comfort. Reprinted with permission from Ref. [49]. Copyright 2013 Nature Publishing Group.

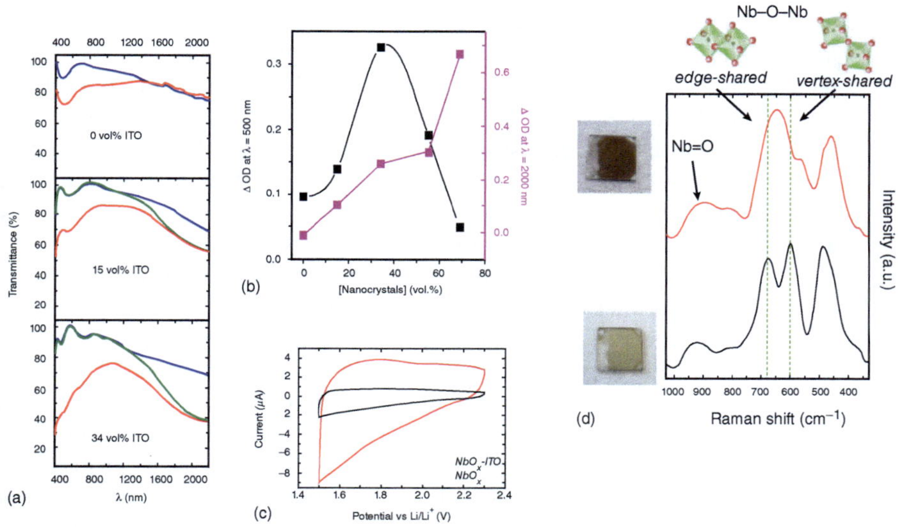

Figure 13.17 Optical contrast enhancement and interface reconstruction in nanocrystal-in-glass films. (a) Transmittance spectra under applied electrochemical voltage (4 V in blue, 2.3 V in green and 1.5 V in red, vs Li/Li$^+$), of ITO-*in*-NbO_x with different ITO nanocrystal content. The ITO nanocrystal vol% between 0% and 69% at constant size of 4.8 ± 0.8 nm, size chosen to maximise their electrochromic response. Sn at% in ITO was 12.2 ± 0.1% and film thickness = 150 ± 35 nm. (b) Change in optical density (ΔOD), at $\lambda = 500$ nm between 1.5 and 4 V, at different ITO NCs volume fraction. (c) Cyclic voltammetry curves for a pure NbO_x film and a ITO-in NbO_x composite (34 vol% ITO). (d) Raman spectra of pure NbO_x (black) and a composite (red) probing the structural change of the glass matrix when linked to the nanocrystals. Photographs of charged films (0 and 34% ITO) are also shown. Adapted with permission from Ref. [49]. Copyright 2013 Nature Publishing Group.

the optimal value for this enhancement corresponded to a spacing of 0.7 nm of matrix material between the closely spaced Ag_2S nanocrystals, which is the same value we found for ITO-in-NbO_x composites. It is thus very likely that percolating interfaces in ITO-in-NbO_x can provide effective ion transport channels as in the case of Ag_2S-in-GeS_2. Despite the several examples in the literature reporting enhancement of ionic conductivity in glass/crystal interfaces, a detailed understanding of this phenomenon and its correlation with structure is far from achieved. It is well-accepted, though, that the density of the amorphous structure (i.e. degree of connectivity in the bonding network) has a significant effect on the ionic transport properties [55, 56].

To identify structural changes in the NbO_x glass induced by the covalent bonding with the nanocrystals, Raman spectroscopy was carried out on pure amorphous NbO_x and ITO-in-NbO_x with different densities of crystal–glass interfaces (Figure 13.17d) [49]. Vibrational bands of the distorted $[NbO_6]$ octahedra that dominate the structure of NbO_x were readily observed in the range from 500 to $1000\,cm^{-1}$, and changes in their peak positions and relative intensities tracked with the interfacial area. By deconvoluting the bridging oxygen bands ($550–750\,cm^{-1}$), we found that highly distorted edge-sharing $[NbO_6]$ was prevalent in the nanocrystal-in-glass films, while pure NbO_x contained more regular, vertex sharing arrangements (Figure 13.17d). In addition, the relative intensity of Raman peaks at higher frequencies ($750–950\,cm^{-1}$) increased at higher nanocrystal loading, suggesting higher amount of terminal oxygens and thus reduced connectivity of the glassy network. The ratio between these terminal bonds and bridging ones was greater in the composites, which we attributed to internal interruption of the polymerised network. Generally, the degree of connectivity in oxide glasses can be reduced by addition of network modifiers or glass intermediates [57]. In our case, we suggested that indium and tin from ITO nanocrystals can diffuse locally into the NbO_x glass during thermal processing and disrupt the glassy network, inducing chain breaking and thus reducing the network connectivity. Indeed, X-ray diffraction patterns of films processed at increasing temperatures (400–550 °C) showed broader ITO peaks, pointing to progressive interfacial dissolution of ITO in the NbO_x matrix [49]. Hence, the structure of the NbO_x glass is profoundly altered by the covalent linking to ITO nanocrystals, becoming highly distorted and less interconnected. This lower density allows more charge to be inserted in tandem with increased optical contrast (Figure 13.17b,c). Consequently, the coloration efficiency in these composites remained constant in the range of $17–23\,cm^2\,C^{-1}$, which is in agreement with typical values for pure NbO_x.

Besides the increase in optical contrast and charge capacity, the stability upon electrochemical cycling was also improved. ITO-in-NbO_x composites showed only 4% of charge capacity loss after 2000 cycles, while pure NbO_x and especially ITO nanocrystal films degraded substantially (20% and 85% capacity loss, respectively) [49]. The high capacity retention of NbO_x can be also explained by the more open bonding network, which could facilitate ion insertion and extraction and help relax the stress induced by these electrochemical processes.

In summary, the colloidal 'nanocrystal-in-glass' approach not only enables adding optical functionalities through material selection but also provides a method to deliberately manipulate the structure of conventional electrochromics and enhance their performance limits. We anticipate that this material design concept will open the door to a new generation of smart window technologies, with better optical functionalities, better durability and faster switching kinetics. In addition, as new and 'better' plasmonic electrochromic NCs are being developed [44, 45] with LSPR absorption peaks located between 750 and 1250 nm, we anticipate that they will be readily integrated into dual-band composites, providing NIR modulation in the optimal spectral range. Other POM compositions could be also investigated as glassy matrix precursors; tungstates, niobates, vanadates, and their mixtures or doped versions could offer a wide selectron, from which heterogeneous matrices with distinct electrochromic properties could be prepared. From a practical viewpoint, combinations of conventional materials within the matrix phase could be utilized to reach a neutral colour in the dark state of the composite, a property desired for application in buildings.

13.4.3 Other Advanced Composite Electrochromic Electrodes Obtained from Non-Colloidal Approaches

Combining dissimilar materials in a single heterogeneous composite, *via* colloidal approaches, allows exceptional control of structure, composition and function. Other preparation techniques, however, have also been reported to yield conventional electrochromic nanocomposites (i.e. non-plasmonic) with enhanced performance [58–61]. This observation is relevant since it validates the generality of the concept of property enhancement *via* crystal/glass interfaces. In this section, we present some recently reported composites containing conventional metal oxide nanostructures embedded within an amorphous organic or inorganic matrix. The role of crystal/glass interfaces in the enhancement mechanism is still not well understood in these examples, mainly due to experimental limitations in controlling interface density in a systematic way as well as poor characterisation. We present these examples, though, to give a broader view of advanced composite electrodes and, in particular, as an inspiration for using colloidal approaches, which would allow the required structural and compositional control as well as adding the missing NIR-selective functionality.

13.4.3.1 Hybrid Electrochromic Nanocomposites

It is well known that Li^+ transport in polymer electrolytes, such as poly(ethylene oxide) (PEO), is greatly improved when glass/crystal interfaces are introduced [62]. Therefore, a promising hybrid composite would be one containing an electrochromic nanocrystal embedded in a matrix of ion conducting polymer, which would act as solid electrolyte. Such a nanocrystal-in-polymer composite has not yet been realised, although a similar concept was reported using layer-by-layer deposition of PEO combined with V_2O_5 [63] and WO_3 [64]. In both cases, the composites showed enhanced Li^+ transport, which contributed to an increase in the charge capacity and optical switching rate.

Combining organic electrochromes, such as conjugated polymers, and conventional metal oxides can also yield hybrid composites with enhanced performance [65]. Very recently, Cai *et al.* [60] showed that, when crystalline WO_3 nanowires were coated with electropolymerised polyaniline (PANI), the performance parameters were enhanced with respect to the single components. The hybrid WO_3-PANI composite showed a coloration efficiency of 86.3 $cm^2\,C^{-1}$ at 700 nm, significantly higher than the coloration efficiency of PANI and WO_3 (32.5 and 53.2 $cm^2\,C^{-1}$, respectively). The durability of the composites was also greatly improved, especially for the PANI component. These enhancements were attributed to the formation of a donor–acceptor interaction, which promotes charge transfer from PANI to WO_3, as well as stabilising semiquinone radicals during the redox process of PANI.

13.4.3.2 **Inorganic Nanocomposites for Advanced Counter Electrodes**

Nanocomposites show great promise to push the limited performances of often overlooked counter electrode materials. NiO is the preferred choice as an anodic material for complementary electrochromic devices. However, many challenges remain unsolved, including slow switching kinetics, weak optical modulation and poor bleached state transparency. Lin *et al.* [59] as well as Feng *et al.* [61] recently showed that these performances could be enhanced by embedding doped NiO nanocrystals in a Li_2O amorphous matrix. Using co-sputtering methods, the authors were able to tune the Li stoichiometry in the film, which had a direct influence in the nickel oxidation state, hole concentration as well as formation of the Li_2O glassy matrix. The most favourable electrochromic performance (close to the state of the art of WO_3) was observed for films with overall composition $Li_{1.97}NiZr_{0.23}O_x$, which were found to contain 5 nm nanocrystallites of Zr- and Li-doped NiO in a Li_2O amorphous matrix. Unfortunately, the *in situ* nature of the film growth process did not allow independent control of nanocrystal size, doping and volume fraction, which limited the possibility to optimise these parameters. This limitation also restricted the interpretation of the observed enhanced performance.

The examples just described demonstrate the generality of the interface-induced electrochromic performance enhancement. However, a detailed understanding of the origin of such enhancement is still far from being achieved. Most of the recent examples in literature did not show any systematic study varying the interface density or characterising the structural or electronic changes. This lack of information is due, in part, to experimental limitations of conventional *in situ* film growth methodologies. There is thus plenty of room for colloidal methodologies to contribute in the development of advanced electrochromic electrodes, not only by pushing the performance of conventional materials by integrating plasmonic nanocrystals, but also by enabling systematic studies that would help elucidate the role of heterogeneous interfaces in property enhancement.

13.5
Conclusions and Outlook

Plasmonic nanocrystals show great promise as new electrochromic materials for next-generation smart windows [37, 38]. Their NIR-selective response allows dynamic control of solar heat while allowing visible light to pass through a window. Energy savings coming from this functionality are related to cooling and heating systems as well as reducing the use of electric lighting [66]. In order to modulate the largest amount of solar heat and thus optimise the energy savings, plasmonic nanocrystals need to be responsive within the specific NIR ranges where most of the solar radiation is located. Given that the spectral response is determined by the concentration of free electrons in the nanocrystals, developing doping strategies that allow high concentrations of free electrons and thus a blue-shift of the LSPR peak will be crucial. Furthermore, the role of the nanocrystal dielectric constant in screening charges should influence the modulation range. The potential to exploit this effect has yet to be explored.

Besides facilitating the construction of well-controlled nanocrystal-in-matrix composite, different types of nanocrystals can be easily mixed to yield compound materials that integrate multiple optical response functions. A simple example would be mixing two or three compositions of plasmonic metal oxide nanocrystals to create the most efficient broad-band spectral response in modulating NIR light. This would be favourable since solar NIR radiation spans a broader range of wavelengths than a typical plasmon absorption peak.

The energy saving implications of independently modulating daylight and solar heat must be explored and control strategies developed to make the best use of such dual-band electrochromics [66]. The versatility of colloidal approaches allowed linking plasmonic nanocrystals and conventional electrochromics that are visibly active. The combination of these two dissimilar materials enabled dual-band behaviour in a composite film. In future developments, these components can be replaced or combined modularly to optimise the voltage-dependent spectral response for energy management in buildings or transportation vehicles. Not only can the nanocrystals be exchanged to tune properties, but the modularity of the preparation method can be also applied to the matrix material by using POMs as precursors to conventional metal oxides. Since these metal-oxo clusters can be prepared with many early transition metals, like Ti, V, Nb, Ta, Cr, Mo, W, Fe, and Co, they presumably can be used to form many different electrochromic metal oxide glasses. Furthermore, POMs can be metal-substituted (analogous to doping in a crystal), increasing even more the compositional flexibility of this approach [67, 68]. Finally, the terminal oxygens in these POMs can readily bond to open coordination sites on the nanocrystal surface, allowing for chemical linkage and providing new routes to combine materials in heterogeneous composite electrochromic films with tunable properties.

As a counter point, realisation of commercial technologies is made easier by simplicity, at least as an initial incarnation. Hence, we raise the question: is it possible to obtain dual-band (or more complex) functionality in one single material? In the context of this chapter, this would imply a nanocrystal with LSPR features (i.e. free electrons in the conduction band) as well as having the capability to undergo Faradaic reactions as in a conventional electrochromic transition metal oxide. Therefore, we expect that this functionality could be achieved by doped metal oxides in which the host crystals have a redox-active metal and layered structure to allow intercalation. Alternative strategies for decoupling NIR and visible response could include materials with two stimuli response mechanisms (e.g. combinations of materials) or materials with dynamic structural colour and dynamic absorptive colour that can be switched independently. Many exciting possibilities lie on the horizon, yet to be explored.

References

1. Runnerstrom, E.L., Llordes, A., Lounis, S.D., and Milliron, D.J. (2014) Nanostructured electrochromic smart windows: traditional materials and NIR-selective plasmonic nanocrystals. *Chem. Commun.*, **50**, 10555–10572
2. Maier, S.A. (2007) *Plasmonics: Fundamentals and Applications: Fundamentals and Applications*, Springer, New York.
3. Ghosh, S.K. and Pal, T. (2007) Interparticle coupling effect on the surface plasmon resonance of gold nanoparticles: from theory to applications. *Chem. Rev.*, **107**, 4797–4862.
4. Comin, A. and Manna, L. (2014) New materials for tunable plasmonic colloidal nanocrystals. *Chem. Soc. Rev.*, **43**, 3957–3975.
5. Erwin, S.C., Zu, L.J., Haftel, M.I., Efros, A.L., Kennedy, T.A., and Norris, D.J. (2005) Doping semiconductor nanocrystals. *Nature*, **436**, 91–94.
6. Norris, D.J., Efros, A.L., and Erwin, S.C. (2008) Doped nanocrystals. *Science*, **319**, 1776–1779.
7. Buonsanti, R. and Milliron, D.J. (2013) Chemistry of doped colloidal nanocrystals. *Chem. Mater.*, **25**, 1305–1317.
8. Moores, A. and Goettmann, F. (2006) The plasmon band in noble metal nanoparticles: an introduction to theory and applications. *New J. Chem.*, **30**, 1121–1132.
9. Scotognella, F., Della Valle, G., Kandada, A.R.S., Dorfs, D., Zavelani-Rossi, M., Conforti, M., Miszta, K., Comin, A., Korobcheyskaya, K., Lanzani, G., Manna, L., and Tassone, F. (2011) Plasmon dynamics in colloidal $Cu_{2-x}SE$ nanocrystals. *Nano Lett.*, **11**, 4711–4717.
10. Mendelsberg, R.J., Garcia, G., Li, H., Manna, L., and Milliron, D.J. (2012) Understanding the plasmon resonance in ensembles of degenerately doped semiconductor nanocrystals. *J. Phys. Chem. C*, **116**, 12226–12231.
11. Mendelsberg, R.J., Garcia, G., and Milliron, D.J. (2012) Extracting reliable electronic properties from transmission spectra of indium tin oxide thin films and nanocrystal films by careful application of the drude theory. *J. Appl. Phys.*, **111**, 063515.
12. Mendelsberg, R.J., Zhu, Y., and Anders, A. (2012) Determining the nonparabolicity factor of the CdO conduction band using indium doping and the Drude theory. *J. Phys. D: Appl. Phys.*, **45**, 425302.
13. Liu, M., Pelton, M., and Guyot-Sionnest, P. (2009) Reduced damping of surface plasmons at low temperatures. *Phys. Rev. B*, **79**, 035418.
14. Sonnichsen, C., Franzl, T., Wilk, T., von Plessen, G., Feldmann, J., Wilson, O., and Mulvaney, P. (2002) Drastic reduction of plasmon damping in gold nanorods. *Phys. Rev. Lett.*, **88**, 077402.
15. Henkel, A., Jakab, A., Brunklaus, G., and Soennichsen, C. (2009) Tuning plasmonic properties by alloying copper into gold nanorods. *J. Phys. Chem. C*, **113**, 2200–2204.
16. Henglein, A., Mulvaney, P., and Linnert, T. (1991) Chemistry of Ag_n aggregates in aqueous-solution – nonmetallic

oligomeric clusters and metallic particles. *Faraday Discuss.*, **92**, 31–44.
17. Doyle, W.T. and Agarwal, A. (1965) Optical extinction of metal spheres. *J. Opt. Soc. Am.*, **55**, 305–308.
18. Ung, T., Giersig, M., Dunstan, D., and Mulvaney, P. (1997) Spectroelectrochemistry of colloidal silver. *Langmuir*, **13**, 1773–1782.
19. Novo, C., Funston, A.M., Gooding, A.K., and Mulvaney, P. (2009) Electrochemical charging of single gold nanorods. *J. Am. Chem. Soc.*, **131**, 14664–14666.
20. Wang, C.J., Shim, M., and Guyot-Sionnest, P. (2001) Electrochromic nanocrystal quantum dots. *Science*, **291**, 2390–2392.
21. Wang, C.J., Shim, M., and Guyot-Sionnest, P. (2002) Electrochromic semiconductor nanocrystal films. *Appl. Phys. Lett.*, **80**, 4–6.
22. Guyot-Sionnest, P. and Wang, C. (2003) Fast voltammetric and electrochromic response of semiconductor nanocrystal thin films. *J. Phys. Chem. B*, **107**, 7355–7359.
23. Boschloo, G. and Fitzmaurice, D. (1999) Spectroelectrochemistry of highly doped nanostructured tin dioxide electrodes. *J. Phys. Chem. B*, **103**, 3093–3098.
24. zum Felde, U., Haase, M., and Weller, H. (2000) Electrochromism of highly doped nanocrystalline SnO_2: Sb. *J. Phys. Chem. B*, **104**, 9388–9395.
25. Pflughoefft, M. and Weller, H. (2002) Spectroelectrochemical analysis of the electrochromism of antimony-doped nanoparticulate tin-dioxide electrodes. *J. Phys. Chem. B*, **106**, 10530–10534.
26. Ellmer, K. (2012) Past achievements and future challenges in the development of optically transparent electrodes. *Nat. Photon.*, **6**, 809–817.
27. Kanehara, M., Koike, H., Yoshinaga, T., and Teranishi, T. (2009) Indium tin oxide nanoparticles with compositionally tunable surface plasmon resonance frequencies in the near-IR region. *J. Am. Chem. Soc.*, **131**, 17736–17737.
28. Choi, S.I., Nam, K.M., Park, B.K., Seo, W.S., and Park, J.T. (2008) Preparation and optical properties of colloidal, monodisperse, and highly crystalline ITO nanoparticles. *Chem. Mater.*, **20**, 2609–2611.
29. Gilstrap, R.A. Jr.,, Capozzi, C.J., Carson, C.G., Gerhardt, R.A., and Summers, C.J. (2008) Synthesis of a nonagglomerated indium tin oxide nanoparticle dispersion. *Adv. Mater.*, **20**, 4163–4166.
30. Buonsanti, R., Llordes, A., Aloni, S., Helms, B.A., and Milliron, D.J. (2011) Tunable infrared absorption and visible transparency of colloidal aluminum-doped zinc oxide nanocrystals. *Nano Lett.*, **11**, 4706–4710.
31. Yin, Y. and Alivisatos, A.P. (2005) Colloidal nanocrystal synthesis and the organic inorganic interface. *Nature*, **437**, 664–670.
32. Buonsanti, R., Grillo, V., Carlino, E., Giannini, C., Curri, M.L., Innocenti, C., Sangregorio, C., Achterhold, K., Parak, F.G., Agostiano, A., and Cozzoli, P.D. (2006) Seeded growth of asymmetric binary nanocrystals made of a semiconductor TiO_2 rodlike section and a magnetic gamma-Fe_2O_3 spherical domain. *J. Am. Chem. Soc.*, **128**, 16953–16970.
33. Jun, Y.W., Choi, J.S., and Cheon, J. (2006) Shape control of semiconductor and metal oxide nanocrystals through nonhydrolytic colloidal routes. *Angew. Chem. Int. Ed.*, **45**, 3414–3439.
34. Talapin, D.V., Lee, J.S., Kovalenko, M.V., and Shevchenko, E.V. (2010) Prospects of colloidal nanocrystals for electronic and optoelectronic applications. *Chem. Rev.*, **110**, 389–458.
35. De Trizio, L., Buonsanti, R., Schimpf, A.M., Llordes, A., Gamelin, D.R., Simonutti, R., and Milliron, D.J. (2013) Nb-doped colloidal TiO_2 nanocrystals with tunable infrared absorption. *Chem. Mater.*, **25**, 3383–3390.
36. Gordon, T.R., Paik, T., Klein, D.R., Naik, G.V., Caglayan, H., Boltasseva, A., and Murray, C.B. (2013) Shape-dependent plasmonic response and directed self-assembly in a new semiconductor building block, indium-doped cadmium oxide (ICO). *Nano Lett.*, **13**, 2857–2863.
37. Garcia, G., Buonsanti, R., Runnerstrom, E.L., Mendelsberg, R.J., Llordes, A., Anders, A., Richardson, T.J., and

Milliron, D.J. (2011) Dynamically modulating the surface plasmon resonance of doped semiconductor nanocrystals. *Nano Lett.*, **11**, 4415–4420.

38. Garcia, G., Buonsanti, R., Llordes, A., Runnerstrom, E.L., Bergerud, A., and Milliron, D.J. (2013) Near-infrared spectrally selective plasmonic electrochromic thin films. *Adv. Opt. Mater.*, **1**, 215–220.
39. Rosen, E.L., Buonsanti, R., Llordes, A., Sawvel, A.M., Milliron, D.J., and Helms, B.A. (2012) Exceptionally mild reactive stripping of native ligands from nanocrystal surfaces by using Meerwein's salt. *Angew. Chem. Int. Ed.*, **51**, 684–689.
40. Frank, G. and Köstlin, H. (1982) Electrical properties and defect model of tin-doped indium oxide layers. *Appl. Phys. A*, **27**, 197–206.
41. Hwang, J.H., Edwards, D.D., Kammler, D.R., and Mason, T.O. (2000) Point defects and electrical properties of Sn-doped In-based transparent conducting oxides. *Solid State Ionics*, **129**, 135–144.
42. Gonzalez, G.B., Mason, T.O., Quintana, J.P., Warschkow, O., Ellis, D.E., Hwang, J.-H., Hodges, J.P., and Jorgensen, J.D. (2004) Defect structure studies of bulk and nano-indium-tin oxide. *J. Appl. Phys.*, **96**, 3912–3920.
43. Lounis, S.D., Runnerstrorm, E.L., Llordes, A., and Milliron, D.J. (2014) Defect chemistry and plasmon physics of colloidal metal oxide nanocrystals. *J. Phys. Chem. Lett.*, **5**, 1564–1574.
44. Manthiram, K. and Alivisatos, A.P. (2012) Tunable localized surface plasmon resonances in tungsten oxide nanocrystals. *J. Am. Chem. Soc.*, **134**, 3995–3998.
45. Mattox, T.M., Bergerud, A., Agrawal, A., and Milliron, D.J. (2014) Influence of shape on the surface plasmon resonance of tungsten bronze nanocrystals. *Chem. Mater.*, **26**, 1779–1784.
46. Llordes, A., Hammack, A.T., Buonsanti, R., Tangirala, R., Aloni, S., Helms, B.A., and Milliron, D.J. (2011) Polyoxometalates and colloidal nanocrystals as building blocks for metal oxide nanocomposite films. *J. Mater. Chem.*, **21**, 11631–11638.
47. Buonsanti, R., Pick, T.E., Krins, N., Richardson, T.J., Helms, B.A., and Milliron, D.J. (2012) Assembly of ligand-stripped nanocrystals into precisely controlled mesoporous architectures. *Nano Lett.*, **12**, 3872–3877.
48. Milliron, D.J., Buonsanti, R., Llordes, A., and Helms, B.A. (2014) Constructing functional mesostructured materials from colloidal nanocrystal building blocks. *Acc. Chem. Res.*, **47**, 236–246.
49. Llordes, A., Garcia, G., Gazquez, J., and Milliron, D.J. (2013) Tunable near-infrared and visible-light transmittance in nanocrystal-in-glass composites. *Nature*, **500**, 323–326.
50. Williams, T.E., Chang, C.M., Rosen, E.L., Garcia, G., Runnerstrom, E.L., Williams, B.L., Koo, B., Buonsanti, R., Milliron, D.J., and Helms, B.A. (2014) NIR-selective electrochromic heteromaterial frameworks: a platform to understand mesoscale transport phenomena in solid-state electrochemical devices. *J. Mater. Chem. C*, **2**, 3328–3335.
51. Sakamoto, A. and Yamamoto, S. (2010) Glass–ceramics: engineering principles and applications. *Int. J. Appl. Glass Sci.*, **1**, 237–247.
52. Schirmeisen, A., Taskiran, A., Fuchs, H., Bracht, H., Murugavel, S., and Roling, B. (2007) Fast interfacial ionic conduction in nanostructured glass ceramics. *Phys. Rev. Lett.*, **98**, 225901.
53. Li, C., Gu, L., and Maier, J. (2012) Enhancement of the Li conductivity in LiF by introducing glass/crystal interfaces. *Adv. Funct. Mater.*, **22**, 1145–1149.
54. Wang, R.Y., Tangirala, R., Raoux, S., Jordan-Sweet, J.L., and Milliron, D.J. (2012) Ionic and electronic transport in Ag_2S nanocrystal–GeS_2 matrix composites with size-controlled Ag_2S nanocrystals. *Adv. Mater.*, **24**, 99–103.
55. Peters, A., Korte, C., Hesse, D., Zakharov, N., and Janek, J. (2007) Ionic conductivity and activation energy for oxygen ion transport in superlattices - the multilayer system CSZ $(ZrO_2 + CaO)/Al_2O_3$. *Solid State Ionics*, **178**, 67–76.
56. Kosacki, I., Rouleau, C.M., Becher, P.F., Bentley, J., and Lowndes, D.H. (2005)

Nanoscale effects on the ionic conductivity in highly textured YSZ thin films. *Solid State Ionics*, **176**, 1319–1326.

57. Shelby, J.E. (2005) *Introduction to Glass Science and Technology*, The Royal Society of Chemistry.

58. Gillaspie, D., Norman, A., Tracy, C.E., Pitts, J.R., Lee, S.-H., and Dillon, A. (2010) Nanocomposite counter electrode materials for electrochromic windows. *J. Electrochem. Soc.*, **157**, H328–H331.

59. Lin, F., Nordlund, D., Weng, T.-C., Sokaras, D., Jones, K.M., Reed, R.B., Gillaspie, D.T., Weir, D.G.J., Moore, R.G., Dillon, A.C., Richards, R.M., and Engtrakul, C. (2013) Origin of electrochromism in high-performing nanocomposite nickel oxide. *ACS Appl. Mater. Interfaces*, **5**, 3643–3649.

60. Cai, G.F., Tu, J.P., Zhou, D., Zhang, J.H., Wang, X.L., and Gu, C.D. (2014) Dual electrochromic film based on WO_3/polyaniline core/shell nanowire array. *Sol. Energy Mater. Sol. Cells*, **122**, 51–58.

61. Feng, L., Montano, M., Chixia, T., Yazhou, J., Nordlund, D., Tsu-Chien, W., Moore, R.G., Gillaspie, D.T., Jones, K.M., Dillon, A.C., Richards, R.M., and Engtrakul, C. (2014) Electrochromic performance of nanocomposite nickel oxide counter electrodes containing lithium and zirconium. *Sol. Energy Mater. Sol. Cells*, **126**, 206–212.

62. Croce, F., Appetecchi, G.B., Persi, L., and Scrosati, B. (1998) Nanocomposite polymer electrolytes for lithium batteries. *Nature*, **394**, 456–458.

63. Huguenin, F., dos Santos, D.S., Bassi, A., Nart, F.C., and Oliveira, O.N. (2004) Charge storage capability in nanoarchitectures of V_2O_5/chitosan/poly (ethylene oxide) produced using the layer-by-layer technique. *Adv. Funct. Mater.*, **14**, 985–991.

64. Galiote, N.A., Parreira, R.L.T., Rosolen, J.M., and Huguenin, F. (2010) Self-assembled films from WO_3: electrochromism and lithium ion diffusion. *Electrochem. Commun.*, **12**, 733–736.

65. Thakur, V.K., Ding, G.Q., Ma, J., Lee, P.S., and Lu, X.H. (2012) Hybrid materials and polymer electrolytes for electrochromic device applications. *Adv. Mater.*, **24**, 4071–4096.

66. DeForest, N., Shehabi, A., Garcia, G., Greenblatt, J., Masanet, E., Lee, E.S., Selkowitz, S., and Milliron, D.J. (2013) Regional performance targets for transparent near-infrared switching electrochromic window glazings. *Build. Environ.*, **61**, 160–168.

67. Ammam, M. (2013) Polyoxometalates: formation, structures, principal properties, main deposition methods and application in sensing. *J. Mater. Chem. A*, **1**, 6291–6312.

68. Long, D.-L., Tsunashima, R., and Cronin, L. (2010) Polyoxometalates: building blocks for functional nanoscale systems. *Angew. Chem. Int. Ed.*, **49**, 1736–1758.

Part III
Applications of Electrochromic Materials

Electrochromic Materials and Devices, First Edition.
Edited by Roger J. Mortimer, David R. Rosseinsky and Paul M. S. Monk.

14
Solution-Phase Electrochromic Devices and Systems

Harlan J. Byker

14.1
Introduction

Electrochromic (EC) materials, devices and systems often involve electrochemical redox processes. Solution-phase EC materials are distinguished from surface-confined and electrodeposition-type EC materials by the characteristic that the functional EC materials are dissolved and remain dissolved in a liquid solution prior to and after electrochemical reaction. This distinction is illustrated in Table 14.1, which characterises some EC materials by the type of chemistry involved and their location in EC devices. While a significant number of EC materials are listed in this table, a great many others are known as well. It may be somewhat surprising to note from Table 14.1 how widespread are EC phenomena until one considers how likely it is that a change in oxidation state by electrochemical or other means will populate or depopulate orbitals and lead to a change in light absorption character.

The first column of Table 14.1 contains surface-confined EC materials, which are layers on an electrode surface that remain in direct electronic contact with the electrode during device operation. These include extensively investigated materials like WO_3 and NiO, where electrons cross the interface between these EC materials and the electrode, while the WO_3 and NiO layers remain in contact with the electrode during the reduction–oxidation or oxidation–reduction cycles.

The centre column of Table 14.1 cites various metals, metal oxides and certain viologen–solvent combinations in which the electrochromic material starts in solution, but then forms a solid deposit on the electrode surface as a result of electrochemical reaction. The solid layer may be removed by reversing the electrochemical process, which causes the solid (coloured) electrochromic material to re-enter the solution.

The right-hand column of Table 14.1 cites the simplest and probably the oldest type of EC materials, all of which are solution-phase. As shown, these may be inorganic, metal–organic or organic.

Electrochromic Materials and Devices, First Edition.
Edited by Roger J. Mortimer, David R. Rosseinsky and Paul M. S. Monk.

Table 14.1 Various types of electrochromic materials and devices.

EC systems	Surface-confined solid state	Electrodeposition	Solution-phase
Inorganic	WO_3 MoO_3 V_2O_5 TiO_2 CeO_2 MnO_2 Nb_2O_5 $Ir(OH)_3$ NiO CoO Metal hydrides Prussian blue	Metals (e.g. Bi and Ag) Metal oxides (e.g. PbO_2 and MnO_2)	$Na_3PW_{12}O_{40}$ I^-/I_2
Metal–organic	Phthalocyanines (pc) [pc Lu pc H]$^{2+}$ red [pc Lu pc H]$^{+}$ yellow [pc Lu pc H] green [pc Lu pc H]$^{-}$ light blue [pc Lu pc H]$^{2-}$ blue		Ferrocene Ferrocene derivatives Other metallocenes
Organic	Redox polymers such as polythiophene or polyaniline High surface area electrode or polymer with attached redox groups	Certain viologen – solvent combinations	pH indicators Viologens Quinones Phenazines Tertiary amines

In principle, an anodic EC material from any location in Table 14.1 may be combined with a cathodic material from any other location in Table 14.1 to form a 'complementary' EC system. However, the remainder of this chapter reviews and discusses only systems which operate completely in the solution phase.

Inorganic materials and even metal–organic materials are often considered more robust than organic materials because they are likely to be more stable. Also, surface-confined EC systems and systems relying on electrodeposition have been the subject of extensive commercialisation efforts over the last 40 years. As a consequence, it is interesting to consider that the only significant commercial success for EC devices to date has come from the lower right-hand corner of Table 14.1 with the use of organic, solution-phase EC materials. This success has been largely related to the specific application of electrochromic devices in automatic dimming rear-view mirrors for which well over 250 million devices have been sold, starting in 1987 [1]. Today, over 15% of all the cars and light trucks produced in the world come with one or more of these EC devices.

14.2 Early History of Solution-Phase EC

A review of older EC technology which is still quite relevant today was given by Chang [2]. Chang suggested the term 'electrochemichromic' for systems that changed colour in conjunction with an electrochemical reaction. This nomenclature was his means of distinguishing electrochemical EC systems from the systems that were originally designated EC systems yet based on purely electric-field effects, without change in redox state. Eventually, EC systems which involve electrochemical reactions became so dominant that the term 'electrochromic' is now generally understood to mainly refer to systems involving electrochemical redox reactions.

What appears to be a solution-phase EC device is found in an early US Patent which discloses an 'electrographic display apparatus' [3]. That patent was filed on 30 October 1905 and describes the use of an electrochemical or electrolytic process using a 'cathode reaction to produce a precipitate or solution, preferably of striking colour'. The inventor appears to electrolytically induce a pH change resulting in the pink-coloured basic form of phenolphthalein being formed near the cathode surface.

Another early description involving oxidation of iodide to iodine was given by Smith in a 1929 British Patent [4]. The yellow-brown colour formed in this EC device was intended for use in fog-penetrating head lamps.

The electrolytically induced change in colour of an aqueous solution of phenolphthalein was again suggested for device usage in 1953 by Sziklai of the RCA Corporation [5]; the same chemistry was subsequently suggested for use in a dimming rear-view mirror by Donnelly and Cooper of Donnelly Mirrors Inc. in 1966 [6].

While the pH-type devices were occasionally suggested to involve coloration due to an electric-field effect, this is deemed unlikely in the devices referred to here. Those devices use electrolyte solutions with high ionic conductivity, and thus, most of the high electric-field drop occurs across the thin double layer adjacent to the electrode surface, and this layer is so thin that its path length is too small for significant optical absorbance. More likely those devices involved electrolysis of dissolved oxygen which, during reduction to form water, reacted with protons in solution and thereby raised the pH in the volume immediately adjacent to the cathode. That rise in pH would cause the phenolphthalein indicator to deprotonate, open its lactone ring and thereby turn pink.

Solution-phase pH indicators were revisited many years later in a collaboration between the Dow Chemical Company and Aveso in an attempt to commercialise an inexpensive display [7]. Starting in the mid-1960s, research activity increased on solution-phase electrochromic systems and a significant number of large companies were awarded patents during the subsequent 15 years.

The Dow Chemical Company patented several solution-phase electrochromic systems. They involved Cu(I) oxidised to form Cu(II), thereby forming a blue colour [8]; the reduction of what they termed oxyacid anions of vanadium, tungsten or molybdenum [9]; Mn(II) oxidised to Mn(III) in the presence of thiocyanate [10] and the reduction of Fe(III) to Fe(II) in the presence of various complexing agents [11]. The Polaroid Corporation patented solution-phase electrochromic systems involving the reduction of viologens to form various colours [12, 13]. While Schwarz *et al.* [14] give an early description of the electrochemistry of viologens, the Polaroid patent of 1972 [12] was one of the first suggestions of using viologens in EC devices. The IBM Corporation patented systems with triaryl pyrazolines as anodic materials and nitro-compounds and quinones as cathodic materials [15]; thienylidene pyrazoline was used instead of triaryl pyrazolines as an anodic material [16] and alkoxy-substituted fluorenes as anodic materials with nitro-compounds and quinones as cathodic materials [17]. The Xerox Corporation patented electrochromic systems involving various inorganic redox systems such as Fe(II)/Fe(III) in combination with leuco dyes and pH indicators [18].

The DuPont Corporation patented solution-phase EC systems with anodic materials such as ferrous ions, leuco dyes, hydroquinones, *N,N,N′,N′*-tetramethylphenylenediamine and apparently novel redox materials based on hydroxyaryl imidazoles [19]. Cathodic materials included anthraquinone, indigo, thioindigo, indophenol, indoaniline, diphenoquinone and oxo-arylidene-imidazole dyes. The cathodic material occasionally involved electrodeposition of a metal, in which case, the devices were no longer solution-phase. The inventor Manos [19] coined the term 'self-erasing' for single-compartment solution-phase EC devices in which materials oxidise at the anode and other, different, materials reduce at the cathode. The complementary electrochromic materials are free to diffuse into the bulk of the solution layer and meet and thence undergo a homogeneous electron transfer; thus they then return to their original, often colourless, oxidation states. Self-erasing electrochromism spontaneously bleaches the device after the current ceases flowing. The Manos devices optionally included an 'opacifier' (an additive used to achieve opacity) to ensure that the coloration at only one electrode was observed.

Eliminating this self-erasing effect is necessary in order to create an electrochromic 'memory'. To this end, the DuPont Corporation patented a solution-phase device with a selectively permeable membrane between the anode and cathode compartments [20].

The Philips Corporation and ICI in England did much work with viologen compounds, focusing on electrodepositing reduced forms of the compound as highly coloured films on solid electrode surfaces. Philips also patented solution-phase viologen systems which combine anodic materials (like hydroquinones, ferrous salts and 1,4-di(dialkylamino)benzenes) with the viologens [21]. The combination of a viologen and a 1,4-di(dialkylamino) benzene was studied in detail later on by Leventis *et al.* [22a] and Ho *et al.* [22b,c].

Yet another approach to the use of viologens was patented by Dai Nippon Toryo Company [23]. In their system, the device was filled with a viologen in the blue state formed by one-electron reduction of the colourless dication. At the anode, the blue colour was bleached by oxidation back to the colourless dication during device operation. In the presence of a white opacifier, a transparent conducting anode appears white during activation and returns to its blue appearance when the device is switched off.

A rather simple solution-phase system was patented by the Bell Labs in which quinone was the cathodically colouring electrochromic material and iodide or bromide ion was the anodically colouring material [24].

In the 1970s, a group in Russia also started publishing work on solution-phase EC systems. They appear to be the first to suggest using 5,10-dimethyl-5,10-dihydrophenazine in an EC device [25–28].

14.3 The World's Most Widely Used Electrochromic Material

Many believe that WO_3 is the world's most commonly used electrochromic material. In fact, the organic species 5,10-dimethyl-5,10-dihydrophenazine has been used much more often.

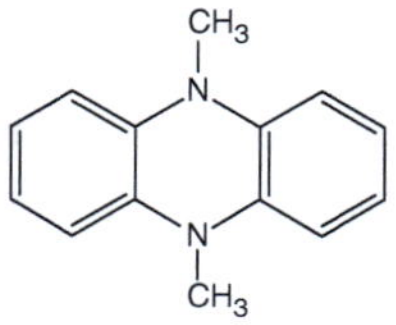

By 2014, 5,10-dimethyl-5,10-dihydrophenazine had been utilised in excess of 200 million electrochromic devices. While the IUPAC name is 5,10-dimethyl-5,10-dihydrophenazine, its common names include 5,10-dihydro-5,10-dimethylphenazine, dimethylphenazine, 9,10-dimethylphenazine, DMPhen, DMPZ or DMP.

Early descriptions and syntheses of DMP originate in the 1930s [29, 30]. A more direct synthesis was reported by Gilman and Dietrich [31] and that remained as the most widely used approach until the dramatically improved synthesis of Giri *et al.* [32] was disclosed in 2001.

Nelson *et al.* have given a detailed discussion of the highly reversible electrochemistry of DMP [33].They showed that not only is the DMP oxidised to form a highly stable cation radical, but at higher anodic potentials, it is oxidised further to form a stable dication. On at least the timescale of a cyclic voltammogram, the stability of those oxidised species is such that there appears to be complete chemically reversibility for both oxidised species.

Kuwana and Winograd and Gruver and Kuwana were among the very first to study the spectroelectrochemisty of organic electrochromic materials such as DMP [34, 35]. The spectroelectrochemistry of 1,1′-dimethyl-4,4′-bipyridinium dichloride (methyl viologen) was investigated in the same device, although there is no evidence that those two materials were combined in the same solution. The rather crude spectrum of the DMP cation radical shown in Figure 4 of Ref. [35] has a moderately strong absorption band between 400 and 500 nm and a further weak absorption peak around 680 nm. Those absorptions result in a pleasant lime-green colour for the cation radical when its concentration is relatively low.

While the Kuwana group studied the separate spectroelectrochemistry of DMP and of methyl viologen, these two electrochromic materials were combined in the late 1970s by Shelepin *et al.* [26] and Ushakov *et al.* [27]. Unfortunately, those investigators chose *N*-dimethylformamide (DMF) as their solvent for device fabrication, which led to nucleophilic attack on the dication of the twice-oxidised anodic materials compromising the stability of the EC system as discussed later.

14.4 Commercialisation of EC Devices

In 1983, the Michigan-based Gentex Corporation sponsored Battelle Columbus Laboratories to investigate electrochromic materials. The research was directed towards an automatic dimming rear-view mirror application for which there had already been significant research efforts by others throughout the world. That research project initially focused on EC materials and devices relying on surface-confined metal oxides. At that time, Gentex believed that transparent conducting electrodes were too expensive for a commercially successful device, so the research involved metallic reflector electrodes that were either inter-digitated (microfabricated with spaces between adjacent pairs) or used in conjunction with a hidden counter electrode.

In September 1984, the research at Battelle lead to the combination of heptyl viologen and DMP, both dissolved in propylene carbonate as an inert solvent and under an inert atmosphere. A potential of about 1 V was applied between two platinum electrodes and a deep blue colour formed around the cathode while a lime-green colour formed around the anode. Stirring the solution removed all traces of colour. This simple system was cycled repeatedly for several days with no deterioration.

This system proved so promising that transparent electrodes (even if expensive) were employed in device prototypes, with the EC solution between two electrodes, the front being an indium-tin oxide (ITO)-coated sheet of glass while the rear was a reflector comprising thin-film chromium on glass as both reflector and electrode. Even though the cell spacing was thin, the activated electrochromic materials were able to absorb significant amounts of light and thereby resulted in a significant decrease in reflectance from the chromium mirror. Open-circuiting the device caused it to clear and the reflectance to increase over about 30 s. Short-circuiting the device caused complete colour removal after about 15 s. While relatively slow in reflectance changes the glare-sensing electronics and the automatic control of the EC device ensured that these speeds were adequate to relieve night-time glare from headlights of following vehicles, as discussed by Bechtel and Byker [36].

Three years after the first prototype, auto-dimming mirrors based on this electrochromic combination were being sold in small volumes. That volume increased rapidly such that, today, tens of thousands of these devices are produced each working day. During the 3-year period of development (1984–1987), many key features were advanced, as discussed next.

The syntheses of electrochromic materials were improved. All electrochromic materials were purified rigorously. For example, DMP was zone-refined to an extreme purity to minimize impurity-based side reactions that might compromise device durability.

The Applied Films Corporation successfully improved the quality of transparent electrodes. Once optimised, Gentex used glass coated with ITO with a sheet resistance of 15 $\Omega\ sq^{-1}$ or less for a half-wave thickness coating with 80% or more visible light transmission and acceptable colour. This ITO coated glass was produced economically for the very first time.

A good perimeter seal was crucial to ensure device longevity. In some respects, this aspect of the device required more intensive development than the electrochromic system itself. Eventually, Gentex developed a rapid-cure epoxy resin with outstanding solvent resistance, low gas permeability, excellent adhesion to glass and long-term resistance to humidity.

Cell spacing was optimised by making cells having a variety of different pound-test fishing lines as the spacer. Only later did Gentex use precision-sized glass beads in this role. Gentex then optimised the cell spacing and concentration of electrochromic material for speed, uniformity and depth of coloration and for the colour-switch response time. Interestingly, when these variables are considered together, there was essentially no scope for altering solution viscosity or the diffusion coefficients of the electrochromic material to improve all performance variables at once.

Gentex next addressed the flatness of the float glass. At that time, the glass used in these mirrors was usually selected from particular portions of the float-line ribbon during the float-glass production. Optimising flatness was critical because a double image appears at intermediate reflectance levels if the rear reflector is not properly aligned with the front surface.

The success of the Gentex electrochromic mirror caused market competition. Several attempts were made to find variations on this single-compartment, solution-phase and self-erasing device.

One of the first competitors was the Donnelly Corporation, which had formerly partnered with Gentex to develop the glare-sensing technology for auto-dimming mirrors. Donnelly started developing their electrochromic mirror before Gentex, but initially focused on surface-confined layers such as metal oxides. As the Gentex device was nearing commercialisation, Donnelly changed direction and started looking at solution-phase EC devices. They dubbed one of their early attempts 'superchromic' [37] owing to the high molar absorption coefficients ε of the dyes formed at the anode. The activated state of this solution-phase system comprised the oxidised form of a leuco dye and a cathodically coloured material such as a viologen. Just prior to filling the device with their electrochromic solution, the dye containing solution was past by reducing agent such as zinc metal which reduced the oxidised form of the dye to its leuco form. A weak organic acid (such as acetic acid) supplied the protons required for this transformation. The conjugate base of the weak acid was presumably then available to accept the protons released during the electrochemical oxidation of the leuco dye during device activation.

Despite many years of research and hundreds of patents issued to the Donnelly Corporation, today their EC mirror is virtually identical to the Gentex system. (This similarity resulted from protracted patent ligation and a settlement with a 4-year moratorium period in which only Gentex was allowed to make EC devices of the type with a viologen and dimethylphenazine.)

With the commercial success of a solution-phase EC device, Dow Chemicals patented an electrochromic device based on polyoxometalates such as phosphotungstic acid as a cathodic material and metallocenes (or a transition-metal complex) as the anodic material [38] and polyoxometalates that were surface-substituted with a transition metal ion to form a cathodically colouring electrochromic material with an enhanced photochemical stability [39].

Siegfried Hünig of the Univeristät Würburg was a key consultant on organic redox systems for Gentex Corporation for several years. Hünig and his former student, Horst Berneth (who went to work for the Bayer Aktiengesellschaft), co-authored an extensive review on organic redox materials and their electrochemistry [40].

Following the success of the Gentex electrochromic mirror, Berneth helped Bayer obtain a large number of patents in the United States describing electrochromic devices using solution-phase materials and systems: Ref. [41]. No commercial products are known to have come from that effort. The Russian group of Shelepin *et al.* helped develop two of these Bayer patents. One of these revealed the benefits of using mediated electron transfer in a solution-phase EC device [42], and the other used anodic electrochromic materials tethered to a cathodic electrochromic material [43].

After Gentex, the second company to commercialise an electrochromic device was Murikami Kaimeido of Japan. Their product was a solid-state mirror for cars [1]. Murikami Kaimeido also tried to develop solution-phase EC systems based on metallocenes and some special viologens [44, 45].

Further research at Gentex sought to improve the performance and stability of electrochromic systems and devices. Many of the organic redox systems they considered for EC systems form coloured radicals. For this reason, Gentex deliberately sought other compounds with a perceived superior stability. A notable development occurred when they joined together two DMP moieties as in Scheme 14.1.

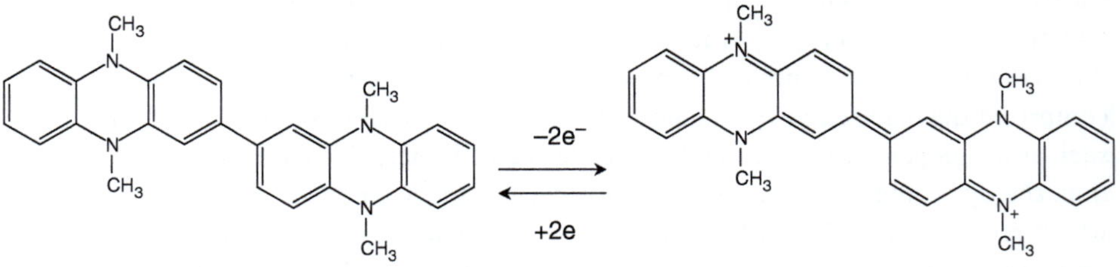

Scheme 14.1 Redox reactions of a bi-phenazine.

Oxidation of this dimeric phenazine to its coloured form yields a non-radical with a closed outer electron shell [46, 47]. While this dimer could yield an enhanced durability, the additional complications and hence preparation costs are difficult to justify.

A further electrochromic development occurred when it was appreciated how substantial changes in absorbance could be achieved by electrochromic modulation in the near-infrared region of the spectrum [48].

Furthermore, while most organic redox electrochromic materials have two donor or acceptor sites per molecule, Hünig *et al.* showed that materials having *four* electron donor or acceptor sites in a single molecule [49–52] have the possibility of higher molar absorption coefficients ε. In principle, therefore, significant improvements in ε are possible as compared to the molar absorption coefficients of about $10\,000\,l\,mol^{-1}\,cm^{-1}$ (at 600 nm) for a typical viologen radical cation and about $10\,000\,l\,mol^{-1}\,cm^{-1}$ (at about 460 nm) for the DMP cation radical.

In another approach to solution-phase electrochromic operation, Hünig *et al.* described electrochemical processes followed by intramolecular radical–radical dimer formation (which in turn causes these coloured species to disappear [53] or coloured species to appear [54]).

14.5
Reversibility and Stability in Solution-Phase EC Systems

Reversibility in electrochemistry may be separated into thermodynamic reversibility and chemical reversibility. While poor chemical reversibility makes a determination of thermodynamic reversibility difficult, these are separate concepts. Unfortunately, they are often intertwined without distinction in discussions on reversibility. Thermodynamic reversibility in electrochemical systems assumes that the electrode processes are at equilibrium and a small shift in potential in either direction rapidly establishes a new equilibrium condition near the electrode surface. If electron transfer kinetics at the electrode surface are slow then equilibrium may take some time to reestablish and the electrochemical reaction is termed thermodynamically quasi-reversible or irreversible [55]. In principle, this has nothing to do with chemical stability or chemical reversibility of the different oxidation states involved. A reaction which has very poor electrode kinetics may be considered irreversible thermodynamically and yet it may be totally chemically reversible if the material involved is chemically stable in both oxidation states.

Thermodynamic reversibility may be determined by the difference in peak potential between an oxidation on an initial scan and the peak potential of the reduction back to the original species on the reverse scan of a cyclic voltammogram. Conversely, it may involve the peak potential difference between a reduction on an initial scan and the peak potential of the oxidation back the original species on the reverse scan. This difference in peak potential is hard to measure if there are fast follow up reactions even if the kinetics of electron transfer would have been fast for electron transfer in both directions and the system would have been thermodynamically reversible. On the other hand if the difference in peak potentials for the forward scan and the reverse scan in a voltammogram is relatively large the redox system may be termed thermodynamically irreversible even though the redox species involved is highly chemically stable in both oxidation states.

Chemical reversibility is paramount when it comes to stability of solution-phase EC systems and thermodynamic reversibility is generally only of passing interest. Chemical reversibility or chemical stability of the redox material in its various oxidation states is related to the peak current for an initial change in oxidation state compared to the peak current for the reverse change in oxidation state in a cyclic voltammogram. A quintessential case for both chemical and thermodynamic reversibility is illustrated in Figure 1 of Ref. [33] in a cyclic voltammogram for DMPhen in acetonitrile.

Gruver and Kuwana [35], have pointed out that stable organic compounds have paired electrons and that adding or removing electrons generally requires two electrons to return to a closed shell system. These electrons are often added or removed electrochemically at different electrode potentials. The first oxidation or reduction typically forms a radical. But the fact that a second oxidation or reduction might also form a chemically stable species implies that the radical might be a long lived species despite its radical nature.

A key patent, that was related to the commercially successful EC system [56], required that the anodic EC material display two chemically reversible oxidation waves in the solvent of the system and that the cathodic EC material also display two chemically reversible waves in the solvent of the system. During the extensive litigation between Gentex and Donnelly on that patent it was legally determined in a Federal District Court in the United States that a chemically reversible wave for an anodic material has a peak current ratio $i_a/i_c \geq 0.9$ and likewise a cathodic material has a ratio $i_c/i_a \geq 0.9$ for both waves of a voltammogram run at 100 mV s^{-1}. It was also determined that it was legitimate to extend the base line due to non-faradaic charging of the double layer at the electrode surface by hand, via eyeballing, to allow peak heights to be determined. Fortunately, the Gentex attorneys were able to enlist Allen Bard of the University of Texas at Austin as an expert witness for this case and he was able to convincingly demonstrate that a Donnelly system in use at the time infringed that patent.

For many organic redox systems this requirement of two such chemically reversible waves in a voltammogram is a minimum requirement, that is, it is necessary but not always sufficient to give a highly stable device, depending on the application of course. Excellent inherent stability is essential along with exceptional purity of all materials and outstanding seals for the device. Excellent stability along with low cost and high performance are generally the keys to commercial viability and these features were exemplified in the several different viologens and the DMPhen used commercially in EC devices over the years. In fact this combination has not been surpassed in more than twenty six years of commercial use and is even somewhat similar to that currently used in the windows of the Boeing 787 airliner. It is fascinating to consider that the electrochemistry of the key materials for this commercially successful EC system, viologens and DMPhen, are described only 4 pages apart in a book published in 1970 and coauthored by Mann and Barnes [57].

14.6 Thickened and Gelled Solution-Phase Systems

While solution-phase electrochromic materials are dissolved and remain dissolved in a liquid solution prior to and after electrochemical activation, the liquid may be thickened to the point of being highly viscous or gelled to the point where it forms a free-standing film. There are at least two motivations for thickening or gelling the solution. The first is simply to help prevent a low-viscosity liquid leaking out from a device should it break.

The second use of a thickener is to minimise the so-called segregation [19]. In a solution-phase device, colour segregation is due to natural convection at a cell's electrodes and is seen as coloured bands that are slow to dissipate after a device is darkened for prolonged periods of time. This segregation effect is believed to occur in solution-phase devices when an organic electrochromic material is dissolved in a non-aqueous solvent because the coloured solution at the anode becomes denser and thence sinks (owing to gravity) while the coloured solution at the cathode becomes less dense and rises. This effect is illustrated schematically in Figure 14.1, which is a pictorial representation of a single-compartment, solution-phase and self-erasing EC device.

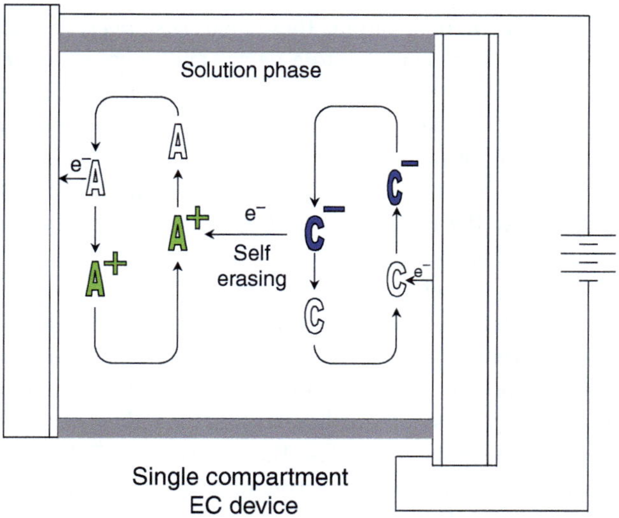

Figure 14.1 A solution-phase electrochromic device with illustration of segregation phenomenon.

A simple explanation of the segregation phenomena could be based on the way cations are often better solvated than anions by non-aqueous solvents and thus an increase in cationic or decrease in anionic character generated at the anode causes solvent molecules to bind more tightly resulting in a density increase. Likewise, the decreased cationic or increased anionic character generated at the cathode causes solvent molecules to bind less tightly and thence decrease solution density relative to the bulk solution. Increased solution viscosity slows the natural convection and colour separation or segregation process. It is possible to form free-standing gels in which the polymer matrix represents as little as 5–10 w/w% of the total system.

Gels for electrochromic systems may be physical gels or chemical gels [58] as follows. A physical gel is merely a highly viscous solution that is free-standing. By contrast, a chemical gel comprises two phases, with a liquid phase interspersed within a solid phase, and in which the internal structure has formed a three-dimensional inorganic or organic polymer matrix. The formation of such a matrix is likely to be synergistic with the polymer supporting the liquid phase such that it eliminates hydrostatic pressure from the liquid, and the solvent can solvate the polymer chains such that it stops agglomeration and/or light scattering. A carefully designed gel helps avoid the problems of syneresis, that is, the problem whereby the gel exudes free liquid. The electrochromic material can still diffuse to and from an electrode in the solution phase of the gel even if the solution is highly gelled.

In order to avoid the solution-phase limitation of several of Gentex devices (as described in their patents), the Donnelly Corporation made early attempts to develop other electrochromic products in which a solution-phase system also comprised electrochromic materials with polymerisable groups to yield a so-called electrochromic monomer composition. These materials, together with other polymerisable monomers, were loaded into a device and polymerised *in situ*, presumably to tie up some of the electrochromic 'monomer' within the polymer chains. Unfortunately, Donnelly termed these devices 'polychromic' in an apparent allusion to the polymeric nature of the system. This usage should not to be confused with the more widely used definition of polychromic for systems that exhibit a variety of colours. The Donnelly 'polychromic' EC technology of Varaprasad *et al.* is described in Ref. [59].

Substantial anodic and cathodic electrochromic materials that were not monomeric were also included in these 'polychromic' solutions so that essentially the operation of the devices remained solution-phase. The concentrations of electrochromic 'monomers' reported were substantially below those required for direct electrochemical activity by, say, redox hopping via the redox materials that were possibly confined to the polymer chains. The diffusion coefficients of solution-based redox materials immersed in a gelled solution (as determined at a micro electrode) are nearly as high as those in un-gelled solutions. They were nevertheless legally defined as 'solid electrochromic systems' by several US courts.

The group of Shelepin has recently addressed the thickening issues that minimise segregation (although they call it 'gravitation-induced stratification') [60]. Since it is difficult to fill devices with a thickened viscous solution, they suggested that fine polymer particles be added to the EC solution just prior to filling and that this slurry be used to quickly fill the device. After filling, the polymer particles dissolve in the solvent to increase its viscosity after the device is filled and sealed. This approach also helps minimise segregation.

Yet another approach to address segregation was suggested by a group at Bayer. Viologen-based redox materials were studied in which a substituted phenazine was tethered to a viologen with an alkyl chain [61], as below:

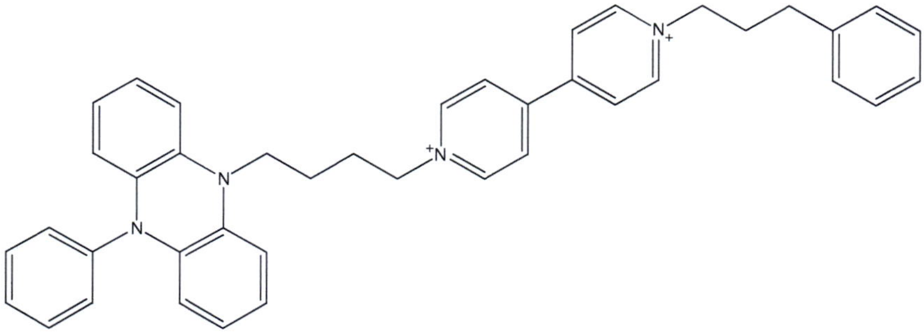

Density gradients formed near the electrodes were nearly eliminated in electrochromic devices when dodecylphenylsulfonate was chosen as the counter ion for the viologen dication; indeed, in some situations, the gradients were even reversed. This case where the anodic and cathodic species are tethered together presents an interesting example of the driving force provided by diffusion through a concentration gradient. Even when bound to a dication, the neutral anodic material diffuses to the positive electrode and is electrochemically oxidised at rates almost as if the cationic character was not there. Also interesting is the spectra reported by this group in Figure 5a in Ref. [61] as it closely resembles the spectra of most commercial EC mirrors.

14.7 Nernst Equilibrium, Disproportionation and Stability

The following equilibrium is rapidly established when two thermodynamically reversible redox materials are placed in solution:

$$A + C \rightleftharpoons A^+ + C^- \qquad (14.1)$$

If the standard oxidation potentials $E^\ominus$ of A and C are known, the equilibrium concentrations (with no applied voltage) of A^+ and C^- may be calculated from

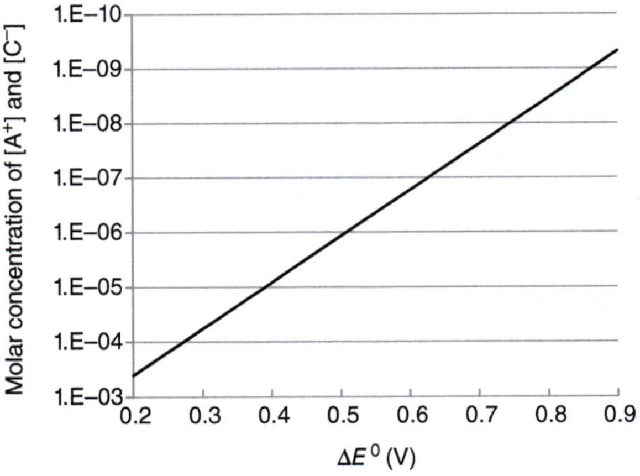

Figure 14.2 Equilibrium concentration of oxidised and reduced species with no applied voltage based on the Nernst equation.

the Nernst Equation:

$$\Delta E^{\ominus} = -\frac{RT}{nF} \ln \frac{[A^+][C^-]}{[A][C]} \tag{14.2}$$

Assuming a temperature of 25 °C, no applied voltage, activity coefficients equal to 1 and an arbitrarily chosen initial concentration of [A] = [C] = 0.02 M, the equilibrium concentrations of A^+ and C^- based on the Nernst equation may be plotted as shown in Figure 14.2.

The well-known disproportionation reactions for A^+ and C^- are as follows:

$$2A^+ \rightleftharpoons A + A^{2+} \text{ or } 2C^- \rightleftharpoons C + C^{2-} \tag{14.3}$$

Essentially, the Nernst equation is used again to determine the equilibrium constants for these reactions but now based on the difference in redox potential between the first and second steps in the oxidation or in the reduction of species A and C; see, for example, Ref. [40].

$$\Delta E^{\ominus} = -\frac{RT}{nF} \ln \frac{[A][A^{2+}]}{[A^+]^2} \text{ or } \Delta E^{\ominus} = -\frac{RT}{nF} \ln \frac{[C][C^{2-}]}{[C^-]^2} \tag{14.4}$$

In the case of some viologens, the reduction potential for the first reduction is separated from the reduction potential for the second reduction in non-aqueous solvents by about 0.4 V; see Chapter 9 of the book by Monk on viologens [62]. If 0.02 M viologen is placed in a solution with 0.02 M of an anodic redox material which has a first oxidation potential 0.5 V higher than the first reduction potential of the viologen, Figure 14.2 or Eq. (14.2) shows that the equilibrium concentration of once reduced C^- is 1.19×10^{-6} M. If the disproportionation reaction is valid at these concentrations, then one can even calculate an equilibrium concentration $[C^{2-}] = 1.22 \times 10^{-17}$ M. (*These calculations assume that the speciation is simple, as mentioned, and that there is no precipitation of products.*) While this is truly an infinitesimal concentration, reactions of this species over 20+ years of EC device life might even be a concern depending on the reaction rate for decomposition of C^{2-}. This concern becomes many times greater when one considers device activation. For simplicity, one may consider having a 0.005 M solution of a viologen cation radical in the presence of 0.015 M viologen dication. With a 0.4 V difference between reduction waves, Eq. (14.4) suggests an equilibrium concentration of zero-valent, doubly reduced viologen of 2.9×10^{-10}. The voltage difference of 0.72 V [33] between two oxidation waves for DMP is a serious advantage in this regard. For a solution of 0.005 M DMP cation radical in the presence of 0.015 M unoxidised neutral DMP, the concentration of the twice-oxidised DMP is only 1.12×10^{-15}, which is more than 5 orders of magnitude less that the concentration with a ΔE^{o} of 0.4 V. Even without an applied voltage and especially during device activation, the chemical stabilities of the once-reduced or once oxidized and the twice-reduced or twice oxidized species are both of interest for long-term durability and a large ΔE^{o} between waves is generally advantageous.

14.8 Closing Remarks

For many years a key concern for device development has been enhancing the cycle life. In the case of solid-state metal-oxide systems, the cycling itself contributes to the degradation as a result of intercalation and de-intercalation of counter ions causing repeated expansion and contraction of the metal oxide lattice. Even in the 1970s, groups were claiming cycle lives in excess of 10^7 for WO_3-based devices, yet there has been no commercial success in terms of returns from investment in well over 40 years. By contrast, redox cycling of solution-phase devices does not have the same impact on device durability. Leaving a solution-phase device in its activated state at elevated temperatures might even be a better accelerated test of durability. Conversely, properly prepared solution-phase electrochromic devices with organic redox materials may be cycled at 70 °C for many hundreds of thousands of full-darkening and full-clearing cycles without any observable loss in performance.

When anodic and cathodic organic redox materials are placed in a solution, there is often light absorption or colour formation that exceeds that which would be expected from the Nernst equilibrium concentrations, as discussed earlier. This coloration is likely to be due to charge-transfer complexation processes. Such charge-transfer complexation is extremely common for solution-phase viologens according to Chapter 5 of Ref. [62].

While the stability discussions centre on thermal or chemical stability, the photochemical stability of solution-phase systems is also of great importance.

References

1. Byker, H.J. (1994) in *Proceeding of the Symposium on Electrochromic Materials II* (eds K.-C. Ho and D.A. MacArthur), The Electrochemical Society Inc., pp. 3–13.
2. Chang, I.F. (1976) in *Non-Emissive Electrooptic Displays* (eds A.R. Kmetz and F.K. von Willisen), Pergamon Press, New York, pp. 155–196.
3. Hutchison, M.R. (1913) Electrographic Display Apparatus and Method. US Patent 1,068,774, Publication date July 29, 1913.
4. Smith, F.H. (1929) British Patent 328,017.
5. Sziklai, G.C. (1953) Electrochemical Color Filter. US Patent 2,632,045, Mar. 17, 1953.
6. Donnelly, J.F. and Cooper, R.C. (1966) Optically Variable One-Way Mirror. US Patent 3,280,701, Oct. 25, 1966.
7. Vincent, J.B., Babinec, S.J., Dermody, D.L. and Chen, Y. (2005) Electrochromic display device and compositions useful in making such devices. US Patent 6,879,424, filed Mar. 19, 2002 and issued Apr. 12, 2005, Assignee, Aveso, Inc.
8. Friedrich, R.E. and Jones, G.D. (1966) Colour reversible electrochemical light filter utilizing electrolytic solution. US Patent 3,282,158, Dow Chemical Company, Nov. 1, 1966.
9. Friedrich, R.E. and Jones, G.D. (1966) Colour reversible electrochemical light filter utilizing electrolytic solution. US Patent 3,283,656, Dow Chemical Company, Nov. 8, 1966.
10. Friedrich, R.E. and Jones, G.D. (1966) Colour reversible electrochemical light filter utilizing electrolytic solution. US Patent 3,282,157, Dow Chemical Company, Nov. 1, 1966.
11. Friedrich, R.E. and Jones, G.D. (1966) Colour reversible electrochemical light filter utilizing electrolytic solution. US Patent 3,282,160, Dow Chemical Company, Nov. 1, 1966.
12. Rogers, H.G. (1972) Variable light-filtering device with a redox compound which functions as its own electrolyte. US Patent 3,652,149, Polaroid Corporation, Mar. 28, 1972.
13. Rogers, H.G. (1975) Variable light-filtering device. US Patent 3,873,185, Polaroid Corporation, Mar. 25, 1975.
14. Schwarz, W.M., Kosower, E.M., and Shain, I. (1961) Electroreduction of N-alkylpyridinium ions. *J. Am. Chem. Soc.*, **83**, 3164–3165.
15. Shattuck, M.D. and Sincerbox, G.T. (1978) High efficiency electrochromic display device. US Patent 4,093,358, IBM Corporation, June 6, 1978.
16. Bredfeldt, K.E., Champ, R.B. and Fowler, K.J. (1978) Electrochromic display devices comprising thienylidene pyrazoline compounds. US Patent 4,090,782, IBM Corporation, May 23, 1978.
17. Clecak, N.J., Grant, B.D., Jaffe, A.B. and Keller, G.S. (1979) Electrochromic display devices comprising substituted fluorene compounds. US Patent 4,139,276, IBM Corporation, Feb. 13, 1979.
18. Findl, E. (1976) Reversible radiant energy filter and process of using same. US Patent 3,951,521, Xerox Corporation, Apr. 20, 1976.
19. Manos, P. (1969) Electrochromic device. US Patent 3,451,741, E. I. du Pont de Nemours and Company, June 24, 1969.
20. Kissa, E., Manos, P. and Wahlig, C.F. (1969) Compartmented electrochromic device. US Patent 3,453,038, E. I. du Pont de Nemours and Company, Jul. 1, 1969.
21. Schoot, C.J. and Ponjée, J.J. (1974) Image display apparatus. US Patent 3,806,229, U.S. Philips Corporation, Apr. 23, 1974.
22. (a) Leventis, N., Chen, M., Liapis, A.I., Johnson, J.W., and Jain, A. (1998) Characterization of 3 × 3 matrix arrays of solution-phase electrochromic cells. *J. Electrochem. Soc.*, **145**, L55–L58; (b) Ho, K.-C., Fang, Y.-W., Hsu, Y.-C., and Chen, L.-C. (2003) The influences of operating voltage and cell gap on the performance of a solution-phase electrochromic device containing HV and TMPD. *Solid State Ionics*, **165** (1-4), 279–287; (c) Ho, K.-C., Fang, Y.-W., Hsu, Y.-C., and Yu, F.-C. (2003) *Proc. Electrochem. Soc.* (Electrochromic Materials and Applications), **2003–17**, 266–278.

23. Yaguchi, M. and Haruhiro, K. (1980) Electrochromic display device. US Patent 4192581, Dai Nippon Toryo Co, Ltd., Mar. 11, 1980.
24. Camlibel, I., Singh, S., Stocker, H.J., Van Ultert, L.G. and Zydzik, G.J. (1982) Display devices. US Patent 4,309,081, Bell Telephone Laboratories, Incorporated Jan. 5, 1982.
25. Shelepin, I.V. *et al.* (1977) Electrochromism of organic compounds I. Electrochemical and spectral properties of a system based on methylviologen and 3-ethyl-2-benzothiazolone. *Electrokhima*, **13** (1), 32–37.
26. Shelepin, I.V. *et al.* (1977) Electrochromism of organic compounds II. Spectral and electrochemical examination of a system based on methylviologen and 5,10-dihydro-5,10-dimethylphenazine. *Electrokhimya*, **13** (3), 404–408.
27. Ushakov, O.A. *et al.* (1978) Electrochromism of organic compounds. Some properties of two-electrode cells. *Electrokhimya*, **14** (2), 319–322.
28. Ushakov, O.A. *et al.* (1985) A calculation of steady-state electrocoloration parameters of electrochromic systems. *Elektrokhimiya*, **21**, 918–922.
29. Clemo, G.R. and Mcllwain, H. (1935) Syntheses and molecular complexes in the phenazine series. *J. Chem. Soc.*, 738–741.
30. Campbell, I.G.M., Le Fevre, C.G., Le Fevre, R.J.W., and Turner, E.E. (1938) Configuration of heterocyclic compounds. VIII. Configuration of anthracene, 9,10-dihydroanthracene, phenazine, 9,10-dimethyl-9,10-dihydrophenazine, thianthrene and selenanthrene. *J. Chem. Soc.*, 404–409.
31. Gilman, H. and Dietrich, J.J. (1957) Direct preparation of some dihydro and other phenazine derivatives. *J. Am. Chem. Soc.*, **79**, 6178–6179.
32. Giri, P., Byker, H.J. and Baumann, K.L. (2001) One pot synthesis of 5,10-dihydrophenazine compounds and 5,10-substituted dihydrophenazines. US Patent 6,242,602.
33. Nelson, R.F., Leedy, D.W., Seo, E.T., and Adams, R.N. (1967) Anodic oxidation of 5,10-dihydro-5,10-dimethyl-phenazine. *Z. Anal. Chem.*, **224** (1), 184–196.
34. Kuwana, T. and Winograd, N. (1971) Homogeneous electron-transfer reactions studied by internal reflection spectroelectrochemistry. *J. Am. Chem. Soc.*, **93** (18), 4343–4350.
35. Gruver, G.A. and Kuwana, T. (1972) Spectroelectrochemical studies of E.E. and E.E.C. mechanisms. *J. Electroanal. Chem. Interfacial Electrochem.*, **36** (1), 85–99.
36. Bechtel, J.H. and Byker, H.J. Automatic rearview mirror system for automotive vehicles. US Patent 4,917,477, filed Apr. 6, 1987 and issued Apr. 17, 1990, Gentex Corporation, Zeeland, MI.
37. Varaprasad, D.V., Looman, S.D., Zhao, M., Habibi, H.R. and Lynam, N.R. Electrochemichromic solutions, processes for preparing and using the same, and devices manufactured with the same. US Patent 5,239,405, filed Sep. 6, 1991 and issued Aug. 24, 1993, Assigned to Donnelly Corporation.
38. Babinec, S.J. (1995) Photostable self-erasing electrochromic device. US Patent 5,471,337, filed Aug. 25, 1994 and issued Nov. 28, 1995, Assignee: The Dow Chemical Company.
39. Babinec, S.J. (1997) Single compartment self-erasing electrochromic device. US Patent 5,671,082, filed, Aug. 25, 1994 and issued Sept. 23, 1997, Assignee: The Dow Chemical Company.
40. Hünig, S. and Berneth, H. (1980) Two step reversible redox systems of the weitz type. *Top. Curr. Chem.*, **92**, 1–44.
41. US Patents 6,183,878, 6,207,292, 6,241,916, 6,277,307, 6,344,918, 6,372,159, 6,388,796, 6,404,532, 6,417,951, 6,433,912, 6,466,355, 6,509,999, 6,545,793, 6,569,361, 6,600,589, 6,631,023, 6,767,481, 6,785,036, 6,806,989, 6,930,815, 7,106,489 and 7,190,506.
42. Shelepin, I., Gavrilov, V., Berneth, H., Jacobsen, W. and Neigl, R. (2003) Electrochromic device having improved switching behavior. US Patent 6,509,999, assigned to Bayer Aktiengesellschaft, Jan. 21, 2003.

43. Berneth, H., Neigl, R., Shelepin, I. and Gavrilov, V. Electrochromic device. US Patent 7,106,489, PCT Published Nov. 22, 2001 and issued Sept. 12, 2006, Assignee: Bayer Aktiengesellschaft
44. Mochizuka, T. (1999) Solution-phase electrochromic device. Jpn Kokai Tokkyo Koho JP 11271814 A 19991008.
45. Mochizuka, T., Uchida, S., Okamura, S., Yamada, Y. and Maesawa, A. (2001) Solution-phase electrochromic devices uses CH2-3-pyrindine substituted viologen with DMP. Jpn. Kokai Tokkyo Koho, JP 2001021926 A 20010126.
46. Theiste, D.A., Baumann, K.L., Giri, P. and Guarr, T.F. (2001) Coupled electrochromic compounds with photostable dication oxidation states. US Patent 6,249,369, filed Jul. 9, 1999 and issued Jun. 19, 2001, Assigned Gentex Corporation, Zeeland, MI.
47. Theiste, D.A., Baumann, K.L., Giri, P. and Guarr, T.F. (2003) Coupled electrochromic compounds with photostable dication oxidation states. US Patent 6,560,004, filed Jun. 19, 2001 and issued May 6, 2003, Assigned Gentex Corporation, Zeeland, MI.
48. Thieste, D., Byker, H.J., Baumann, K. and Srinivasa, R. (2001) US Patent 6,193,912, filed Mar. 3, 1998 and issued Feb. 27, 2001, Assignee Gentex Corporation, Zeeland, MI.
49. Hünig, S., Kemmer, M., Wenner, H., Perepichka, I.F., Bäuerle, P., Emge, A., and Gescheidt, G. (1999) Violene/cyanine hybrids: a general structure for electrochromic systems. *Chem. Eur. J.*, **5** (7), 1969–1973.
50. Hünig, S., Kemmer, M., Wenner, H., Barbosa, F., Gescheidt, G., Perepichka, I.F., Bäuerle, P., Emge, A., and Peters, K. (2000) Violene/cyanine hybrids as electrochromics part 2: tetrakis(4-dimethylaminophenyl)ethene and its derivatives. *Chem. Eur. J.*, **6** (14), 2618–2632.
51. Hünig, S., Perepichka, I.F., Kemmer, M., Wenner, H., BaÈuerle, P., and Emge, A. (2000) *Tetrahedron*, **56**, 4203–4211.
52. Hünig, S., Langels, A., Schmittel, M., Wenner, H., Perepichka, I.F., and Peters, K. (2001) Violene/cyanine hybrids as electrochromic systems: a new variation of the general structure. *Eur. J. Org. Chem.*, **2001** (7), 1393–1399.
53. Hünig, S., Briehn, C.A., Bäuerle, P., and Emge, A. (2001) Electrochromics by intramolecular redox switching of single bonds. *Chem. Eur. J.*, **7** (13), 2745–2757.
54. Hünig, S., Aldenkortt, S., Bäuerle, P., Briehn, C.A., Schäferling, M., Perepichka, I.F., Stalke, D., and Walfort, B. (2002) Electrochromics by intramolecular redox switching of single bonds, Part II: type OS/CL. *Eur. J. Org. Chem.*, **2002**, 1603–1613.
55. Bard, A.J. and Faulkner, L.R. (2001) *Electrochemical Methods Fundamentals and Applications*, 2nd edn, John Wiley & Sons, Inc., New York, pp. 195–242.
56. Byker, H.J. (1990) Single-compartment, self-erasing, solution-phase electrochromic devices, solutions for use therein, and uses thereof. US Patent 4,902,108, filed Mar. 31, 1986 and issued Feb. 20, 1990.
57. Mann, C.K. and Barnes, K.K. (1970) *Electrochemical Reactions in Nonaqueous Systems*, Marcel Dekker, see pp. 314 and 318.
58. Byker, H.J. (2001) Electrochromics and polymers. *Electrochim. Acta*, **46**, 2015–2022.
59. US Patents 5,910,854, 6,002,511, 6,154,306, 6,245,262, 6,420,036, 6,855,431, 6,954,300 and 7,202,987.
60. Gavrilov, V.I. and Shelepin, I.V. (2007) Method for producing an electrochromic device and said electrochromic device. US Patent 7,295,361, Close Joint Stock Company 'Technoglass Engineering', Nov. 13, 2007.
61. Michaelis, A., Berneth, H., Haarer, D., Kostromine, S., Neigl, R., and Schmidt, R. (2001) Electrochromic dye system for smart window applications. *Adv. Mater.*, **13** (23), 1825–1828.
62. Monk, P.M.S. (1998) *The Viologens, Physicochemical Properties, Synthesis and Applications of the Salts of 4,4′-Bipyridine*, John Wiley & Sons, Ltd, Chichester.

15
Electrochromic Smart Windows for Dynamic Daylight and Solar Energy Control in Buildings

Bjørn Petter Jelle

15.1
Introduction

The application of large areas of window panes and glass structures in buildings has in recent years become more widespread and extensive, thus the construction and design of glass material properties have become more important. Glass can have material additives and different surface coatings tailor-made to fulfil specific requirements for each building. The energy transfer in windows consists of solar radiation, thermal (infrared) radiation, thermal conduction in solids and gases, and gas convection. The energy from solar radiation will diminish the need for heating, but it needs to be balanced against possible cooling costs. The measurement and calculation of quantities such as solar transmittance and reflectance values are important in this respect.

Smart windows are able to dynamically control the solar radiation throughput. Of the available technologies, promising electrochromic window (ECW) devices have been developed into commercial products [1, 2]. ECWs are able to control the solar radiation passage by varying the applied voltage, hence offering an elegant and dynamic way to regulate the solar transmittance. Various aspects and information on electrochromic materials and windows are widely referred to in the available literature: Alamri [3], Azens *et al.* [4], Baetens *et al.* [1], Bhadra *et al.* [5], Bouessay *et al.* [6], Carpenter *et al.* [7], Carpenter and Conell [8], Chen *et al.* [9], Chen *et al.* [10], Fantini and Gorenstein [11], Gao and Jelle [12], Geniès *et al.* [13], Goldner *et al.* [14–16], Granqvist [17–23], Granqvist *et al.* [24–26], Green *et al.* [27, 28], Jelle *et al.* [29–34], Jelle and Hagen [35–37], Jelle [38], Kalagi *et al.* [39], Karuppasamy and Subrahmanyam [40], Lampert [41–46], Lampert and Ma [47], Lampert *et al.* [48], Lee and DiBartolomeo [49], Lee *et al.* [50, 51], Leventis and Chung [52], Makimura *et al.* [53], Monk *et al.* [54, 55], Mortimer [56], Mortimer *et al.* [57, 58], Moulki *et al.* [59], Penin *et al.* [60], Piccolo [61], Rougier *et al.* [62, 63], Rougier and Blyr [64], Sauvet *et al.* [65, 66], Stilwell *et al.* [67], Syed and Dinesan [68], Yu *et al.* [69], Yu and Lampert [70] and Zhang *et al.* [71]. Solar regulation may also be achieved by smart windows other than electrochromic ones, that is, photochromic windows, thermochromic

Electrochromic Materials and Devices, First Edition.
Edited by Roger J. Mortimer, David R. Rosseinsky and Paul M. S. Monk.

windows, gasochromic windows, liquid crystal windows and suspended particle (electrophoretic) windows; see, for example, the studies by Adami *et al.* [72], Anders *et al.* [73], Baetens *et al.* [1], Chen and Lo [74], Cupelli *et al.* [75], Gao *et al.* [76], Gao and Jelle [77], Gardiner *et al.* [78], Georg *et al.* [79], Huang *et al.* [80], Lampert [44, 81, 45], Li *et al.* [82], Mennig *et al.* [83], Mlyuka *et al.* [84], Park and Hong [85], Vergaz *et al.* [86], Wittwer *et al.* [87], Ye *et al.* [88] and Yoshimura *et al.* [89]. Energy performance and life cycle cost of smart windows have been studied by Sekhar and Toon [90], whereas an eco-efficiency evaluation of a smart-window prototype has been performed by Syrrakou *et al.* [91]. Furthermore, an evaluation of control strategies for different smart-window combinations using computer simulations has been carried out by Jonsson and Roos [92]. In order to be able to achieve as large solar regulation as possible in smart windows, dirt accumulation should be avoided as for all window panes; thus, self-cleaning windows and glazing products [93, 94] may play an important role.

Several commercial ECWs are already available [1, 2], and applications in combination with other technologies like solar cells, solar cell glazing and building integrated photovoltaics (BIPV) may also be possible [2, 81, 95–101]. The BIPV also have to fulfil requirements of the building envelope [99], e.g. wind-driven rain tightness [102]. Snow and ice related friction and adhesion issues and self-cleaning properties are important for ECWs, BIPV, and solar cells and glazing products in general [94]. Translucent windows when applying e.g. silica aerogel granules may also be an option [103], where a combination with an ECW will thus control translucent daylight and solar energy. Concerning high-performance window frames and window spacers, see Gustavsen *et al.* [104–106], Jelle *et al.* [2] and Van Den Bergh *et al.* [107]. Grynning *et al.* [108] studied the energy balance of a window addressing both heat losses and heat gains. The solar radiation throughput in glass and other materials (e.g. transparent polymers) is obviously also highly interesting for solar cell and ECW applications in order to be able to utilize as much as possible of the solar radiation. Furthermore, recent investigations on thermal insulation materials, which may also be applied in future high-performance window frames, have been conducted by Baetens *et al.* [109], Baetens *et al.* [110], Gao *et al.* [103, 111–114], Jelle *et al.* [115–117], Jelle [118], Kalnæs and Jelle [119], Sandberg *et al.* [120] and Schlanbusch *et al.* [121]. Regarding ECW ageing and durability issues see Czanderna *et al.* [122], Lampert [43], Lampert *et al.* [48], Nagai *et al.* [123], Tajima *et al.* [124] and Tracy *et al.* [125].

It is also important to emphasize the degradation of building materials by solar radiation [34, 126–129]. A substantial part of the ultraviolet (UV) radiation is blocked by the glass itself, but nevertheless a significant amount of UV radiation passes through the glass and into the buildings. This transmitted UV radiation affects both materials and living species inside the buildings (e.g. degradation and discoloration of paintings and furniture).

The most important solar radiation glazing factors are, in general:

1) Ultraviolet Solar Transmittance, T_{uv}
2) Visible Solar Transmittance, T_{vis}

3) Solar Transmittance, T_{sol}
4) Solar Material Protection Factor, SMPF
5) Solar Skin Protection Factor, SSPF
6) External Visible Solar Reflectance, $R_{vis,ext}$
7) Internal Visible Solar Reflectance, $R_{vis,int}$
8) Solar Reflectance, R_{sol}
9) Solar Absorbance, A_{sol}
10) Emissivity, ε
11) Solar Factor, SF (from T_{sol}, R_{sol} and ε)
12) Colour Rendering Factor, CRF.

To quantify and compare solar characteristics of different glass materials with and without coatings, with a special focus on electrochromic materials, and in general other materials transmitting solar radiation (e.g. various polymers), this study presents a comprehensive review of measurement and calculation of the most important solar radiation glazing factors, including three different ECW devices at various coloration levels and for comparison reasons several examples for various glass materials and several two-layer and three-layer window pane configurations. The work presented herein is based on a journal article by Jelle [38].

15.2
Solar Radiation

Solar radiation at the earth's surface is roughly located between 300 and 3000 nm (0.3 and 3 μm, respectively), where the visible (VIS) radiation (light) lies between 380 and 780 nm. Ultraviolet (UV) and near infrared (NIR) radiation are located below and above the VIS region, respectively. Figure 15.1 depicts the solar radiation in outer space and at the earth's surface, both with and without molecular absorption in the atmosphere. Above 3000 nm, and not part of the direct solar radiation, lies the thermal, that is, infrared (IR) radiation, which all materials radiate above 0 K (peaking at around 10 μm at room temperature).

The UV radiation is further divided into three sub-regions: UVA (320–400 nm), UVB (280–320 nm) and UVC (100–280 nm) (wavelength borders can vary slightly in the literature, e.g. 315 nm instead of 320 nm). Photodegradation examples are given in Croll and Skaja [131], Fufa *et al.* [132], Gerlock *et al.* [133], Jelle *et al.* [34], Jelle [128], Rånby and Rabek [134], Rabek [135, 136] and Tylli *et al.* [137].

15.3
Solar Radiation through Window Panes and Glass Structures

The solar radiation falling onto a material will be transmitted, absorbed or reflected depending upon the wavelength (λ) of the radiation, the incident angle and the optical properties of the material, subject to the material's transmittance

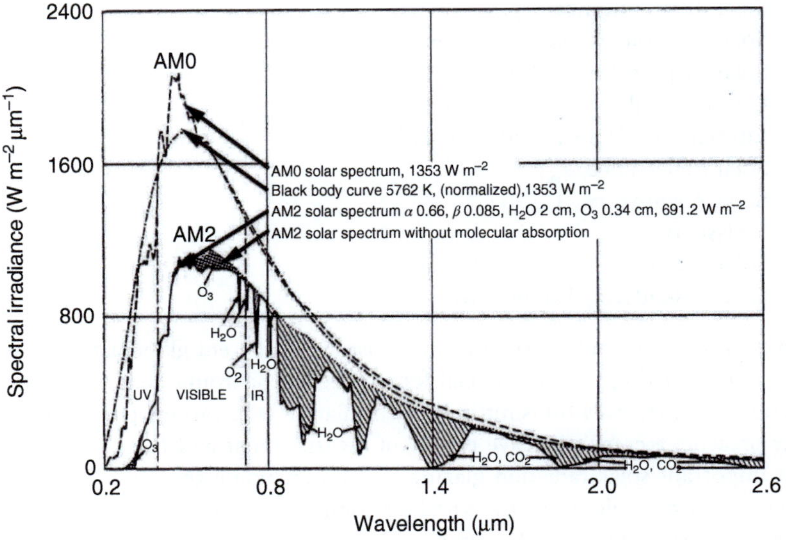

Figure 15.1 Solar radiation, comparing the AM0 (outer space) and AM2 (at the earth's surface, the sun 30° above the horizon) spectra. The AM2 spectrum is shown both with and without molecular absorption (in O_2, O_3, H_2O and CO_2). Somewhat redrawn from Fahrenbruch and Bube [130].

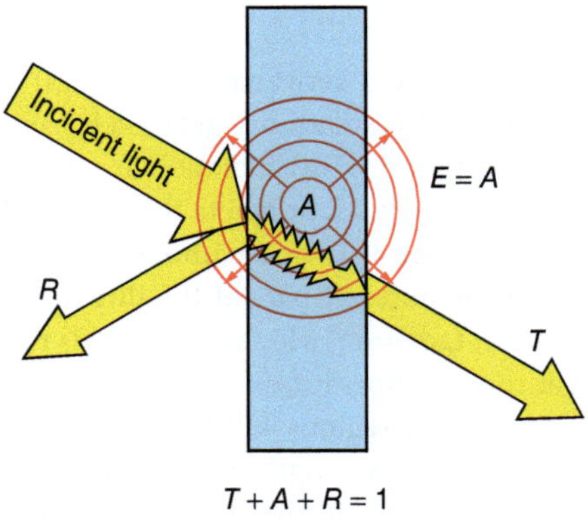

Figure 15.2 The relationship between transmittance (*T*), absorbance (*A*) and reflectance (*R*), in addition to emittance (*E*), in a single glass pane. Diffuse transmittance, diffuse reflectance and retroreflectance are not depicted, nor is reflectance from the interface glass/air within the glass material.

(*T*), absorbance (*A*) and reflectance (*R*). The fraction of incident radiation intensity transmitted, absorbed or reflected by the material is depicted in Figure 15.2. Conservation of the total energy in the solar radiation beam requires that the sum of *T*, *A* and *R* is given by [17, 18, 34, 38, 138–144]:

$$T(\lambda) + A(\lambda) + R(\lambda) = 1\,(100\%) \tag{15.1}$$

For a body in thermodynamic equilibrium with its surroundings, the energy absorbed in the material must be equal to the emitted energy (Kirchhoff's law of thermal radiation), that is [17, 18, 34, 38, 139, 140, 143]:

$$E(\lambda) = A(\lambda) \tag{15.2}$$

where E denotes emittance.

The diffuse transmittance, diffuse reflectance and retroreflectance have been omitted in Figure 15.2, that is, only the specular (mirror) reflectance and regular transmittance are depicted. Regular reflectance may be used to refer to specular reflectance (e.g. [145]). A solar radiation beam which is reflected back in the same path as the incident beam is called a retroreflected radiation beam (e.g. used in reflector discs to be seen by car headlights), whereas a refracted radiation beam refers to a beam transmitted into a second medium.

While Figure 15.2 depicts a single glass pane with a simple relationship between the spectroscopic quantities, the situation is more complex as in Figure 15.3 with multiple transmittance, absorbance and reflectance in a two-layer glass pane. Both Figures 15.2 and 15.3 are simplified with respect to actual refracted radiation beam paths, for example, parallel displacement due to different media with unequal refraction indices. Contrary to real situations, this is avoided in the calculations by assuming a radiation beam with normal incidence. The real transmittance values including the total solar energy transmittance are then somewhat lower. Also, the reflectance from the interface glass/air(gas) within the glass material is not depicted either. The calculations for a three-layer window pane will clearly be even more complex.

In a two-layer window pane with the purpose of acting as a blocking screen towards solar radiation, a coating is placed on the inside of the outer glass. In a

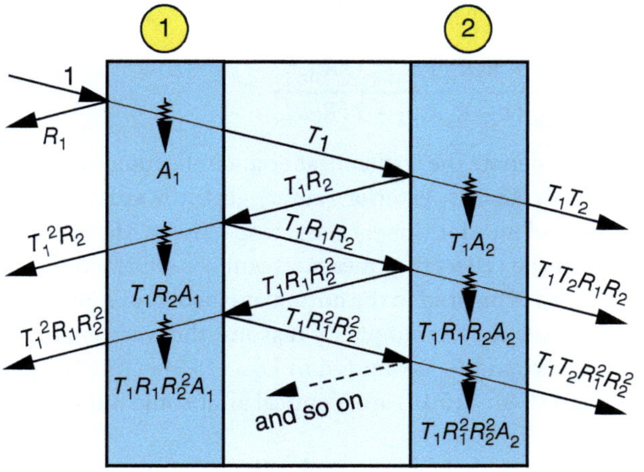

Figure 15.3 Multiple transmittance (T), absorbance (A) and reflectance (R), in a two-layer glass pane. See, for example, Davies [146] and Kimura [147]. Diffuse transmittance, diffuse reflectance and retroreflectance are not depicted, nor is reflectance from the interface glass/air(gas) within the glass material.

two-layer window pane, intended to reduce the heat loss from the inside, a coating is placed on the outside of the inner glass. In both cases, the coating is facing the window pane cavity (which is to be protected).

Transmittance T and reflectance R as functions of wavelength λ for a ***single glass pane*** given by

$$T(\lambda) = T_1 \tag{15.3}$$

$$R(\lambda) = R_1 \tag{15.4}$$

may be calculated for a ***two-layer window pane*** by applying Figure 15.3 and infinite series expansion to give the following expressions [146–149]:

$$T(\lambda) = \frac{T_1 T_2}{1 - R_{1b}R_2} \tag{15.5}$$

$$R_{ext}(\lambda) = R_1 + \frac{T_1^2 R_2}{1 - R_{1b}R_2} \tag{15.6}$$

$$R_{int}(\lambda) = R_{2b} + \frac{T_2^2 R_{1b}}{1 - R_2 R_{1b}} \tag{15.7}$$

and furthermore for a ***three-layer window pane*** the following formulas [148, 149]:

$$T(\lambda) = \frac{T_1 T_2 T_3}{[1 - R_{1b}R_2][1 - R_{2b}R_3] - T_2^2 R_{1b} R_3} \tag{15.8}$$

$$R_{ext}(\lambda) = R_1 + \frac{T_1^2 R_2 [1 - R_{2b}R_3] + T_1^2 T_2^2 R_3}{[1 - R_{1b}R_2][1 - R_{2b}R_3] - T_2^2 R_{1b} R_3} \tag{15.9}$$

$$R_{int}(\lambda) = R_{3b} + \frac{T_3^2 R_{2b} [1 - R_2 R_{1b}] + T_3^2 T_2^2 R_{1b}}{[1 - R_3 R_{2b}][1 - R_2 R_{1b}] - T_2^2 R_3 R_{1b}} \tag{15.10}$$

where T_1, T_2, T_3, R_1, R_2 and R_3 denote the transmittance and reflectance for glass number 1, 2 and 3, respectively, that is, exterior (outer) glass towards incident radiation beam, middle glass and interior (inner) glass, respectively. The index 'b' for R_{1b} and R_{2b} designates that the reflectance measurement is performed on the back (reverse) side of the glass as compared to the normal incident radiation beam direction, for example, R_{1b} versus R_1. For simplicity reasons, the wavelength (λ) dependence of T_1, T_2, T_3, R_1, R_2 and R_3 is omitted in Eqs. (15.3)–(15.10). Note that the denominators in Eqs. (15.8)–(15.10) are identical after some minor rearrangements in Eq. (15.10).

$R_{ext}(\lambda)$ and $R_{int}(\lambda)$ denote the external and internal solar radiation reflectance, that is, outdoor solar radiation reflected back to the outside and indoor solar radiation reflected back to the inside, respectively. Note that the denominators in the last term for R_{ext} and R_{int} are equal for both a two-layer and a three-layer window pane, as the apparent distinction is due to the chosen symmetrical notation.

All the calculated solar radiation glazing factors for multilayer window panes are based on transmittance and reflectance measurements carried out on single glass panes, with subsequent calculations applying Eqs. (15.3)–(15.10). The absorbance is then calculated by applying the expression in Eq. (15.1).

15.4 Solar Radiation Modulation by Electrochromic Windows

As explained, electrochromic windows (ECW) control the solar radiation transmittance by an applied external voltage, which is achieved by regulation of either the absorbance or the reflectance as shown in Figures 15.4 and 15.5, respectively. The former one (Figure 15.4) represents a broadband (i.e. over a specified broad wavelength range) transmittance modulation, while the latter (Figure 15.5) represents a transmittance modulation by a movable transmittance edge. (See also, e.g. the early studies by Goldner and Rauh [150], Goldner *et al.* [151] and Lampert [42].) Figure 15.6 illustrates transmittance modulation by a combined regulation of both the absorbance and the reflectance. The graphs in Figures 15.4–15.6 depict an ideal or maximum modulation, whereas real ECWs will impose more limited modulation.

Most of the studies on ECWs for both fundamental research and commercial applications have been performed with solar radiation absorbing materials. See, for example, the studies by Baetens *et al.* [1] and Jelle *et al.* [2] on the available state-of-the-art commercial ECWs.

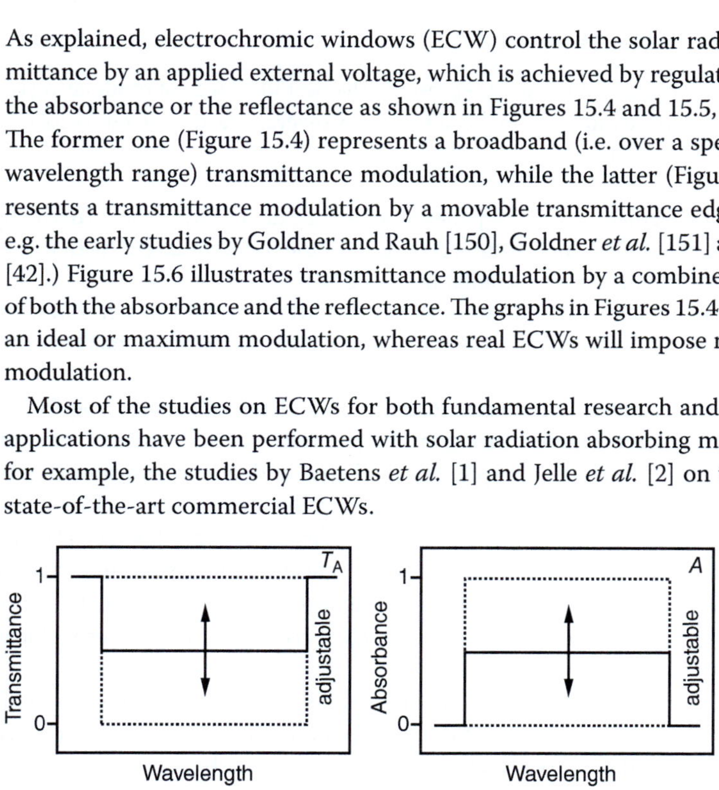

Figure 15.4 Transmittance modulation(T_A) by regulation of the absorbance (A).

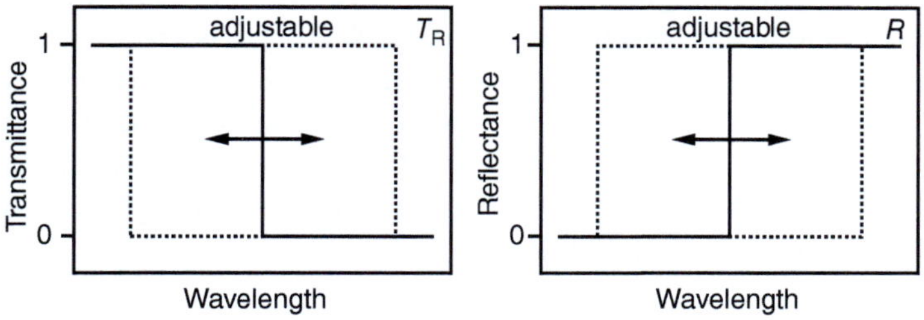

Figure 15.5 Transmittance modulation (T_R) by regulation of the reflectance (R).

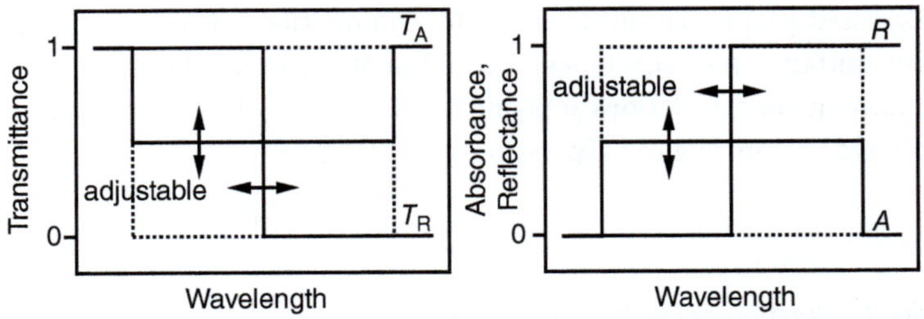

Figure 15.6 Transmittance modulation (T_A and T_R) by combined regulation of both the absorbance (*A*) and the reflectance (*R*).

Reflectance regulation possibly represents the best ideal way of modulating the solar radiation transmittance in ECWs, due to several reasons given as follows [15, 16, 143, 150–154]:

- Absorbance regulating window panes may become unacceptably hot. Thermally induced strain may then lead to degradation of the different layers (with different thermal expansion coefficients) in the ECWs.
- Absorbance regulating window panes have a lower thermal transfer efficiency since, of the energy absorbed, 50% is reradiated into the building, thus resulting in uncontrolled heat gains.
- Absorbance regulation requires thicker films than reflectance regulation, leading to higher manufacturing and operation costs, from the amount of materials needed and the electrical charge and energy consumed. The absorbance depends on the film thickness and the absorption coefficient (via the Beer–Lambert law), while reflectance depends upon the density and dynamics of the free charge carriers in the reflector. Apart from interference effects, reflectance should be relatively independent of the film thickness for thicknesses larger than about 10–50 nm. Furthermore, thinner films will lead to shorter switching times.
- Spectrally selective modulation or 'filtering' is believed to be more readily achieved by reflectance regulation than by bound electron absorbance regulation, as illustrated by the difference in Figure 15.4 (absorbance regulation) and Figure 15.5 (reflectance regulation).

To exemplify, reflectance regulation may be obtained with crystalline tungsten oxide (c-WO_3), while amorphous tungsten oxide (a-WO_3) exerts predominantly only absorbance regulation. In c-WO_3 free electrons govern the reflectance regulation, while the absorbance in a-WO_3 is due to excitation of bound or highly localized electrons. That is, the reflectance may be changed by variation of the free carrier (electron) density, which alters the plasma wavelength λ_p given by the following expression [15, 16, 143, 152, 153]:

$$\lambda_p = \frac{2\pi c}{\omega_p} = \left(\frac{2\pi c}{q_e}\right)\left(\frac{m_e \varepsilon_0}{n_e}\right)^{1/2} \tag{15.11}$$

where ω_p, c, q_e, m_e, n_e and ε_0 denote the plasma frequency, the velocity of light, the electron charge, the free electron effective mass, the free electron density and the dielectric coefficient of vacuum, respectively. Materials with a high free electron density (short λ_p) will be highly reflecting materials with a low emissivity, whereas a low free electron density (long λ_p) corresponds to a low reflectivity and thus a high emissivity. By changing λ_p, that is, changing the free electron density, one may regulate the reflectance with an adjustable vertical reflectance edge as, for example, depicted in Figure 15.5, where λ_p is located at just this vertical reflectance edge. Despite these considerations, most studies of WO_3 have involved amorphous or microcrystalline materials, which are easier to make and handle.

Hence, in order to be able to regulate the reflectance, the electrochromic materials should be crystalline so that the injected or extracted electrons can be free. Changing the free electron density changes the plasma wavelength λ_p, thereby making it possible to regulate the reflectance. Increasing the electron density decreases λ_p and moves the reflectance edge from higher to lower wavelengths (Figure 15.5), thus, a material with a high free electron density is a highly reflecting material. And vice versa, decreasing the electron density increases λ_p and moves the reflectance edge from lower to higher wavelengths, that is, a material with a low free electron density has a low reflectivity.

Various studies on tungsten oxide have been performed regarding its crystallinity and reflectance regulating possibilities. For further details and elaborations on these and related aspects, including more general information about the plasma wavelength and frequency, see for example, the investigations by Ashrit *et al.* [155], Beni and Shay [156], Cogan *et al.* [157], Deb [158], Faughnan and Crandall [159], Goldner and Rauh [150], Goldner *et al.* [15, 16, 151–154, 160], Granqvist [23, 161, 162], Hamberg and Granqvist [163, 164], Hecht [141], Jelle [143], Lim [165], Mendelsohn and Goldner [166], Nagai *et al.* [167], Schirmer *et al.* [168], Schuster *et al.* [169], Svensson and Granqvist [170–172] and Yamanaka [173].

15.5 Experimental

15.5.1 Glass Samples and Window Pane Configurations

In order to illustrate various transmittance, absorbance and reflectance levels in the solar spectrum, one float glass, one glass with low emittance coating, one dark silver coated glass, several two-layer and three-layer window pane combinations and three different electrochromic window (ECW) devices at various coloration levels were selected as examples and for comparisons. Based on the spectroscopic measurements, the solar radiation glazing factors were calculated. The detailed fabrication and miscellaneous testing and characterization of the ECWs are described elsewhere [34–36, 174–176].

15.5.2
UV-VIS-NIR Spectrophotometry

A Cary 5 UV-VIS-NIR spectrophotometer, with an absolute reflectance accessory (Strong-type, VW principle), was used to measure the transmittance and reflectance of the glass samples in the UV, VIS and NIR regions, from 290 to 3300 nm. The absorbance was calculated from the expression given in Eq. (15.1). However, during the period of fabrication and characterization of the ECW devices, no laboratory resources for determining the absolute reflectance of the ECWs were available. Nevertheless, as the three ECWs consist of solar absorbing electrochromic materials, and not reflecting modulating electrochromics, the measured (low) reflectance values for float glass are applied in the calculation of the various reflectance-based solar radiation glazing factors.

15.5.3
Emissivity Determination by Specular IR Reflectance

The standard ISO 9050:2003(E) [148] refers to ISO 10292:1994 [177] for emissivity determinations, which according to ISO 10292:1994 [177] are to be carried out with an infrared spectrometer, measuring the near normal reflectance ($\leq 10°$) at a temperature of 283 K. More details of the emissivity determinations and measurements are found in EN 12898:2001 [178]. In order to minimize polarization effects, the angle of incidence with respect to the normal of the sample must be 10° or less [179]. For ambient temperatures other than 283 K ($\approx 10\,°C$), the emissivity is not strongly dependent on the mean temperature [177].

15.5.4
Emissivity Determination by Heat Flow Meter

The emissivity may be determined by applying a heat flow meter apparatus according to the standard EN 1946-3:1999 E [180]. For theoretical considerations, see EN 1946-2:1999 E [181] and EN 1946-3:1999 E [180]. Note that in general, glass with a low emittance coating is assumed to have an emissivity value below 0.1.

In order to validate the method and to achieve a satisfactory accuracy level, five experiments have been carried out with varying thickness of the air gap between two sample glass plates facing each other. The air gap thickness was controlled and varied by use of three columns consisting of 1, 2, 3, 4 or 5 small squares of glass located at the outer glass-sample edges as depicted in Figures 15.7 and 15.8. The low emittance coatings on each glass sample are facing each other towards the air gap. Each column height, that is, from 1 to 5 glass squares, was measured and adjusted to be of the same height with an accuracy of 0.1 mm. For glass dimensions see legends to Figures 15.7 and 15.8.

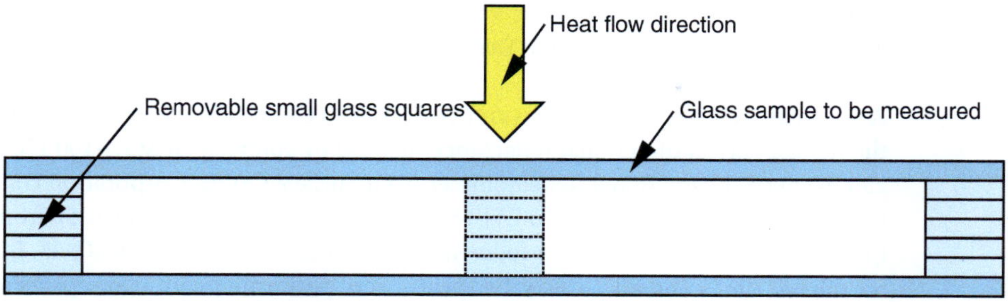

Figure 15.7 Air gap thickness control between two glass plates by application of three glass columns, each column with 1, 2, 3, 4 or 5 small glass squares. The glass sample and removable small glass squares have dimensions 500 mm × 500 mm × 4 mm and approximately 10 mm × 10 mm × 4 mm, respectively. The low emittance coatings on each glass sample are facing each other towards the air gap. Vertical cross-section.

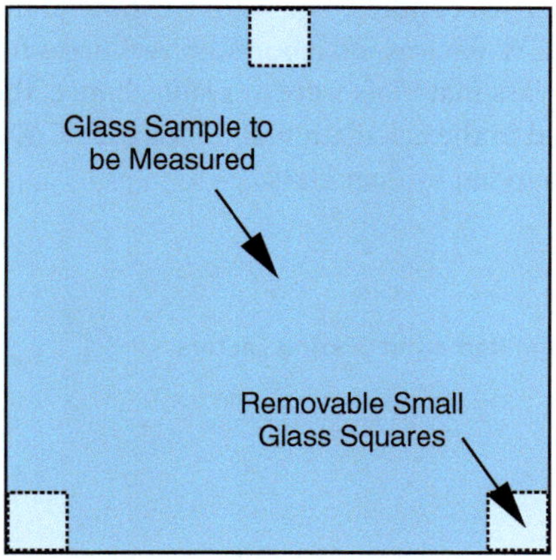

Figure 15.8 Air gap thickness control between two glass plates by application of three glass columns, each column with 1, 2, 3, 4 or 5 small glass squares. The glass sample and removable small glass squares have dimensions 500 mm × 500 mm × 4 mm and approximately 10 mm × 10 mm × 4 mm, respectively. The low emittance coatings on each glass sample are facing each other towards the air gap. Horizontal cross-section.

15.5.5 Emissivity Determination by Hemispherical Reflectance

The emissivity may also be determined by measuring the directional hemispherical reflectance (DHR, direct mode) or the hemispherical directional reflectance (HDR, reciprocal mode). In the DHR method, the sample is illuminated from a single direction and all the reflected radiation into the hemisphere surrounding the sample is measured. In the HDR method, the sample is uniformly illuminated from all directions by use of a hemisphere and the radiation reflected into a single direction is measured. For both the DHR and HDR methods, the single direction may be varied and controlled for miscellaneous instruments, that is, illuminating or detecting at varying angles, respectively.

15.5.6 Actual Emissivity Determinations in This Study

A float glass, a low emittance glass and a dark silver coated glass were measured by the hemispherical directional reflectance method by applying an SOC-100 HDR Hemispherical Directional Reflectometer from Surface Optics Corporation connected to a Thermo Nicolet 8700 FTIR Spectrometer. The reflected radiation from the sample was detected at the following incident angles: 10, 20, 30, 40, 45, 50, 55, 60, 65, 70, 75 and 80°. Thirty-two scans were performed with two repeats at a resolution of 16 cm^{-1} in the wavelength range 2–25 μm. The IR source temperature was 704 °C (maximum). The results were $\varepsilon_{float} = 0.836$, $\varepsilon_{lowe} = 0.071$ and $\varepsilon_{silver} = 0.543$, for the float glass, low emittance glass and dark silver coated glass, respectively. The ε_{float} value was directly applied in the calculation of the solar factor (SF) as the ε value in the SF calculations is with respect to the inside facing surface (of a building) of the innermost glass pane, that is, normally a float glass, whereas the ε_{lowe} and ε_{silver} values were indirectly applied in the SF calculations through the calculations of the thermal conductance Λ. At the time of the fabrication and characterization of the ECW devices, no laboratory resources for determining the emissivity of the ECW glass materials were available, hence, the value $\varepsilon_{float} = 0.836$ for float glass is applied in the calculation of SF for the ECWs, as the electrochromic materials are surrounded by float glass.

15.6 Measurement and Calculation Method of Solar Radiation Glazing Factors

15.6.1 Ultraviolet Solar Transmittance

The *Ultraviolet Solar Transmittance* (T_{uv}) is given by the following expression [148]:

$$T_{uv} = \frac{\sum_{\lambda=300\,nm}^{380\,nm} T(\lambda) S_\lambda \Delta\lambda}{\sum_{\lambda=300\,nm}^{380\,nm} S_\lambda \Delta\lambda} \tag{15.12}$$

where

S_λ	=	relative spectral distribution of ultraviolet solar radiation [148, 182]
$T(\lambda)$	=	spectral transmittance of the glass
λ	=	wavelength
$\Delta\lambda$	=	wavelength interval
$S_\lambda \Delta\lambda$	=	values at different wavelengths given in Table A.1 in Appendix 15.A [148]

The T_{uv} value will thus be a number between 0 and 1, calculated for the ultraviolet part of the solar spectrum, that is, 300–380 nm. A low number indicates a low transmission of ultraviolet solar radiation, and a high number represents a high ultraviolet solar radiation transmission. The T_{uv} values may commonly be given as a percentage, which can range between 0% and 100%.

Note that the whole ultraviolet solar spectrum is not covered in the calculation of T_{uv}, and in future versions of ISO 9050:2003(E) [148], the wavelength range may favourably be extended to cover an even larger part of the ultraviolet solar radiation, for example, from 290 to 380 nm. Note also that T_{uv} is not directly correlated to solar radiation damage of materials and human skin (given in the following chapters on SMPF and SSPF).

15.6.2
Visible Solar Transmittance

The *Visible Solar Transmittance* (T_{vis}), often denoted as *Light Transmittance*, is given by the following expression [148]:

$$T_{vis} = \frac{\sum_{\lambda=380\,\text{nm}}^{780\,\text{nm}} T(\lambda) D_\lambda V(\lambda) \Delta\lambda}{\sum_{\lambda=380\,\text{nm}}^{780\,\text{nm}} D_\lambda V(\lambda) \Delta\lambda} \tag{15.13}$$

where

D_λ	=	relative spectral distribution of illuminant D65 [148, 183]
$V(\lambda)$	=	spectral luminous efficiency for photopic vision defining the standard observer for photometry [148, 184]
$T(\lambda)$	=	spectral transmittance of the glass
λ	=	wavelength
$\Delta\lambda$	=	wavelength interval
$D_\lambda V(\lambda)\Delta\lambda$	=	values at different wavelengths given in Table A.2 in Appendix 15.A [148]

The T_{vis} value will thus be a number between 0 and 1 for the visible part of the solar spectrum, 380–780 nm. A low T_{vis} indicates a low transmission of visible light, and high T_{vis} represents high visible light transmission. Commonly, percentages between 0% and 100% may be given.

15.6.3
Solar Transmittance

The *Solar Transmittance* (T_{sol}) is given by the following expression [148]:

$$T_{sol} = \frac{\sum_{\lambda=300\,nm}^{2500\,nm} T(\lambda)S_\lambda \Delta\lambda}{\sum_{\lambda=300\,nm}^{2500\,nm} S_\lambda \Delta\lambda} \tag{15.14}$$

where

S_λ	=	relative spectral distribution of solar radiation [148, 182]
$T(\lambda)$	=	spectral transmittance of the glass
λ	=	wavelength
$\Delta\lambda$	=	wavelength interval
$S_\lambda \Delta\lambda$	=	values at different wavelengths given in Table A.3 in Appendix 15.A [148, 182]

The T_{sol} value will thus be a number between 0 and 1 for the main part of the solar spectrum, 300–2500 nm. A low T_{sol} indicates a low transmission of solar radiation, and a high T_{sol} represents a high solar radiation transmission. Commonly, percentages between 0% and 100% may be given.

The whole solar spectrum is not covered in this calculation of T_{sol}, and in future versions of ISO 9050:2003(E) [148], an extended wavelength range may cover a larger part of the solar radiation, for example, 290–3000 nm.

15.6.4 Solar Material Protection Factor (SMPF)

The *Solar Material Protection Factor* (SMPF) is given by the following expression [34]:

$$\text{SMPF} = 1 - \tau_{df} = 1 - \frac{\sum_{\lambda=300\,nm}^{600\,nm} T(\lambda)C_\lambda S_\lambda \Delta\lambda}{\sum_{\lambda=300\,nm}^{600\,nm} C_\lambda S_\lambda \Delta\lambda} \tag{15.15}$$

where

τ_{df}	=	CIE damage factor [148, 185]
C_λ	=	$e^{-0.012\lambda}$ (λ given in nm)
S_λ	=	relative spectral distribution of solar radiation [148, 182]
$T(\lambda)$	=	spectral transmittance of the glass
λ	=	wavelength
$\Delta\lambda$	=	wavelength interval
$C_\lambda S_\lambda \Delta\lambda$	=	values at different wavelengths given in Table A.4 in Appendix 15.A [148]

The SMPF value will thus be a number between 0 and 1, similar to and consistent with related values like solar transmittance, emissivity, solar factor and so on. A low number indicates a low material protection, and a high number represents a high degree of material protection. One should also note that both SMPF and SSPF (next section) are protection factors. The word ‘protection’ evokes positive associations, while ‘damage’, for example, in the CIE damage factor, may arouse negative associations. It may be easier and better for a potential window customer to look for a window with the highest protection factor and not the lowest damage factor (probably selling far more windows). Commonly, SMPF values may be given as percentages between 0% and 100%.

The calculation of SMPF has recently been extended from the earlier 500 nm upper limit to today’s 600 nm [148], because a much larger part of the visible solar spectrum is now known to contribute to the degradation of materials. The earlier Krochmann damage factor for materials was calculated for 300 to 500 nm limits [186].

Some of the short-wavelength part of the ultraviolet solar spectrum is not covered in the aforementioned calculation of SMPF; thus, future versions ISO 9050:2003(E) [148] may extend the wavelength range to a larger part of the ultraviolet and visible solar radiation, for example, from 290 to 600 nm.

15.6.5 Solar Skin Protection Factor (SSPF)

The *Solar Skin Protection Factor* (SSPF) is given by the following expression [34]:

$$\mathrm{SSPF} = 1 - F_{\mathrm{sd}} = 1 - \frac{\sum_{\lambda=300\,\mathrm{nm}}^{400\,\mathrm{nm}} T(\lambda) E_\lambda S_\lambda \Delta\lambda}{\sum_{\lambda=300\,\mathrm{nm}}^{400\,\mathrm{nm}} E_\lambda S_\lambda \Delta\lambda} \tag{15.16}$$

where

F_{sd}	=	skin damage factor [148, 187]
E_λ	=	CIE erythemal effectiveness spectrum
S_λ	=	relative spectral distribution of solar radiation [148, 182]
$T(\lambda)$	=	spectral transmittance of the glass
λ	=	wavelength
$\Delta\lambda$	=	wavelength interval
$E_\lambda S_\lambda \Delta\lambda$	=	values at different wavelengths given in Table A.5 in Appendix 15.A [148]

The SSPF value will thus be a number between 0 and 1, similar to and consistent with related values like solar transmittance, emissivity, solar factor and so on. A low number indicates a low skin protection, and a high number represents a high degree of skin protection. One should also note that both SSPF and SMPF (previous section) are protection factors. The word ‘protection’ evokes positive associations, while ‘damage’, for example, in the skin damage factor, may arouse negative associations. It may be easier and better for a potential window customer to look for a window with the highest protection factor and not the lowest damage factor (probably selling far more windows). Commonly, SSPF values may be given as percentages between 0% and 100%.

The calculation of the SSPF extends over the ultraviolet spectrum (at earth's surface) and the low wavelength part of the visible spectrum, which may contribute to the solar radiation damage to the human skin. Earlier there was another definition of a skin protection factor, denoted as SPF [186], with the following correlation between the different terms: SSPF = 1 – (1/SPF) = 1 – F_{sd} = (SPF – 1)/SPF.

Notably, SSPF does not cover all of the short-wavelength part of the ultraviolet solar spectrum, and in future versions of ISO 9050:2003(E) [148], the wavelength range may be extended to cover a larger part of the ultraviolet and visible solar radiation, for example, from 290 to 400 nm.

15.6.6 External Visible Solar Reflectance

The *External Visible Solar Reflectance* ($R_{vis,ext}$), often denoted as *External Light Reflectance*, is given by the following expression [148]:

$$R_{vis,ext} = \frac{\sum_{\lambda=380\,nm}^{780\,nm} R_{ext}(\lambda) D_\lambda V(\lambda) \Delta\lambda}{\sum_{\lambda=380\,nm}^{780\,nm} D_\lambda V(\lambda) \Delta\lambda} \quad (15.17)$$

where

D_λ	=	relative spectral distribution of illuminant D65 [148, 183]
$V(\lambda)$	=	spectral luminous efficiency for photopic vision defining the standard observer for photometry [148, 184]
$R_{ext}(\lambda)$	=	external spectral reflectance of the glass
λ	=	wavelength
$\Delta\lambda$	=	wavelength interval
$D_\lambda V(\lambda)\Delta\lambda$	=	values at different wavelengths given in Table A.2 in Appendix 15.A [148]

The $R_{vis,ext}$ value will thus be a number between 0 and 1 (or as corresponding percentages), calculated for the visible part of the solar spectrum, that is, 380–780 nm. A low number indicates a low external reflection of visible light, whereas a high number represents a high external visible light reflection.

15.6.7 Internal Visible Solar Reflectance

The *Internal Visible Solar Reflectance* ($R_{vis,int}$), often denoted as *Internal Light Reflectance*, is given by the following expression [148]:

$$R_{\mathrm{vis},int} = \frac{\sum_{\lambda=380\,\mathrm{nm}}^{780\,\mathrm{nm}} R_{\mathrm{int}}(\lambda) D_\lambda V(\lambda) \Delta\lambda}{\sum_{\lambda=380\,\mathrm{nm}}^{780\,\mathrm{nm}} D_\lambda V(\lambda) \Delta\lambda} \tag{15.18}$$

where

D_λ	=	relative spectral distribution of illuminant D65 [148, 183]
$V(\lambda)$	=	spectral luminous efficiency for photopic vision defining the standard observer for photometry [148, 184]
$R_{\mathrm{int}}(\lambda)$	=	internal spectral reflectance of the glass
λ	=	wavelength
$\Delta\lambda$	=	wavelength interval
$D_\lambda V(\lambda)\Delta\lambda$	=	values at different wavelengths given in Table A.2 in Appendix 15.A [148]

The $R_{\mathrm{vis,int}}$ value will thus be a number between 0 and 1 (or as corresponding percentages), calculated for the visible part of the solar spectrum, that is, 380–780 nm. A low number indicates a low internal reflection of visible light, whereas a high number represents a high internal visible light reflection.

15.6.8
Solar Reflectance

The *Solar Reflectance* (R_{sol}), implicitly external solar reflectance, is given by the following expression [148]:

$$R_{\mathrm{sol}} = \frac{\sum_{\lambda=300\,\mathrm{nm}}^{2500\,\mathrm{nm}} R_{\mathrm{ext}}(\lambda) S_\lambda \Delta\lambda}{\sum_{\lambda=300\,\mathrm{nm}}^{2500\,\mathrm{nm}} S_\lambda \Delta\lambda} \tag{15.19}$$

where

S_λ	=	relative spectral distribution of solar radiation [148, 182]
$R_{\mathrm{ext}}(\lambda)$	=	external spectral reflectance of the glass
λ	=	wavelength
$\Delta\lambda$	=	wavelength interval
$S_\lambda\Delta\lambda$	=	values at different wavelengths given in Table A.3 in Appendix 15.A (ISO 9050:2003(E) [148])

The R_{sol} value will thus be a number between 0 and 1 (or as corresponding percentages), calculated in the main part of the solar spectrum, that is, 300–2500 nm. A low number indicates a low reflection of solar radiation, whereas a high number represents a high solar radiation reflection. The whole solar spectrum is not covered in the calculation of R_{sol}, and in future versions of ISO 9050:2003(E) [148] the wavelength range may be extended to cover an even larger part of the solar radiation, for example, from 290 nm to 3000 nm.

15.6.9
Solar Absorbance

The *Solar Absorbance* (A_{sol}) is calculated from the expression in Eq. (15.1), which is transformed into the following:

$$A_{sol} = 1 - T_{sol} - R_{sol} \tag{15.20}$$

where insertion of T_{sol} and R_{sol} from Eqs. (15.14) and (15.19) gives:

$$A_{sol} = 1 - \frac{\sum_{\lambda=300\,\text{nm}}^{2500\,\text{nm}} T(\lambda) S_\lambda \Delta\lambda}{\sum_{\lambda=300\,\text{nm}}^{2500\,\text{nm}} S_\lambda \Delta\lambda} - \frac{\sum_{\lambda=300\,\text{nm}}^{2500\,\text{nm}} R_{ext}(\lambda) S_\lambda \Delta\lambda}{\sum_{\lambda=300\,\text{nm}}^{2500\,\text{nm}} S_\lambda \Delta\lambda} \tag{15.21}$$

where

S_λ	=	relative spectral distribution of solar radiation [148, 182]
$T(\lambda)$	=	spectral transmittance of the glass
$R_{ext}(\lambda)$	=	external spectral reflectance of the glass
λ	=	wavelength
$\Delta\lambda$	=	wavelength interval
$S_\lambda \Delta\lambda$	=	values at different wavelengths given in Table A.3 in Appendix 15.A [148].

The A_{sol} value will thus be a number between 0 and 1 (or as corresponding percentages), calculated in the main part of the solar spectrum, that is, 300–2500 nm. A low number indicates a low absorption of solar radiation, whereas a high number represents a high solar radiation absorption. The whole solar spectrum is not covered in the calculation of A_{sol}, and in future versions of ISO 9050:2003(E) [148] the wavelength range may favourably be extended to cover an even larger part of the solar radiation, for example, from 290 nm to 3000 nm. It should also be noted that A_{sol} is a calculated value from measured transmittance and reflectance spectra, that is, no direct measurements of absorbance spectra are performed.

15.6.10
Emissivity

15.6.10.1 Emissivity in General

The emissivity (ε), having a value between 0 and 1, is a measure of a material's radiative, that is, emission, properties of infrared radiation. Highly reflective materials of infrared radiation have low emissivity values, for example, polished surfaces of gold, silver, aluminium or copper.

Oxidation of metallic surfaces will increase the emissivity substantially, for example, polished aluminium with $\varepsilon = 0.05$ (reflectance 0.95) and oxidised aluminium with $\varepsilon = 0.30$ (reflectance 0.70). Confer Eqs. (15.1) and (15.2) with zero transmittance and the emittance equal to the absorbance.

Determination of the emissivity is required in order to further determine the solar factor (SF) and the thermal transmittance (U-value).

15.6.10.2 Emissivity by Specular IR Reflectance Measurements

According to ISO 9050:2003(E) [148], by referring to ISO 10292:1994 [177], the emissivity may be determined with an infrared spectrometer, measuring the near normal reflectance (≤10°) at a temperature of 283 K. Further details of the emissivity determinations and measurements are found in EN 12898:2001 E [178]. The normal emissivity (ε_n) is given by (ISO 10292:1994 [177], EN 12898:2001 E [178]):

$$\varepsilon_n = 1 - R_n = 1 - \frac{1}{30}\sum_{i=1}^{30} R_n(\lambda_i) \tag{15.22}$$

where

R_n	=	average spectral reflectance calculated by summation of spectral reflectance values at 30 distinct wavelengths and divided by 30 as shown in Eq. (15.22)
λ_i	=	wavelength
λ_i	=	values for the 30 wavelengths given in Table A.6 in Appendix 15.A [177, 178]

The corrected emissivity (ε) is determined from:

$$\varepsilon = c_{corr} \cdot \varepsilon_n = \frac{\varepsilon}{\varepsilon_n}\varepsilon_n \tag{15.23}$$

where

c_{corr}	=	$\varepsilon/\varepsilon_n$ = correction coefficient found in Table A.7 in Appendix 15.A [177, 178] for corresponding ε_n values
ε_n	=	normal emissivity from Eq. (15.22).

In order to minimize polarization effects, the angle of incidence with respect to the normal of the sample must be 10° or less [179]. For other ambient temperatures than 283 K (≈10 °C), the emissivity is not strongly dependent on the mean temperature [177].

15.6.10.3 Emissivity by Heat Flow Meter Apparatus

The emissivity may be determined by applying a heat flow meter apparatus according to the standard EN 1946-3:1999 E [180]. For theoretical considerations, referral is made to EN 1946-2:1999 E [181] and EN 1946-3:1999 E [180]. A short excerpt is given in the following.

The total heat flow density q_{tot} (in W m^{-2}) between two parallel, flat infinite isothermal surfaces may be written as [180, 181]:

$$q_{tot} = q_{rad} + \frac{\kappa}{d}\Delta T \tag{15.24}$$

where

$$q_{rad} = \frac{4\sigma T_m^3 \Delta T}{\frac{1}{\varepsilon_1} + \frac{1}{\varepsilon_2} - 1} = \text{radiation flow density } (\text{W m}^{-2}) \tag{15.25}$$

κ	=	thermal conductivity of the medium separating the two surfaces (W m^{-1} K^{-1})
d	=	distance between the two surfaces (m)
ΔT	=	temperature difference between the two surfaces (K)
T_m	=	mean temperature of the two surfaces (K)

$$\sigma = \pi^2 k^4/(60\hbar^3 c^2) \approx 5.67 \cdot 10^{-8}\,\text{W m}^{-2}\,\text{K}^{-4} = \text{Stefan–Boltzmann's constant} \tag{15.26}$$

k	=	Boltzmann's constant ≈ 1.38·10^{-23} J K^{-1}
c	=	velocity of light ≈ 3.00·10^{8} m s^{-1}
$\hbar$	=	$h/(2\pi)$ ≈ 1.05·10^{-34} J s = reduced Planck's constant
h	=	Planck's constant ≈ 6.63·10^{-34} J s
ε_1	=	emissivity of surface 1
ε_2	=	emissivity of surface 2

Equation 15.25 is an approximation of the Stefan-Boltzmann law, which describes the radiation flow density q_{rad} as proportional to the fourth power of the temperature T of the radiant object with emissivity ε:

$$q_{rad} = \sigma\varepsilon T^4 \tag{15.27}$$

or the radiation flow density between a radiation emitting object at temperature T_1 and surroundings at temperature T_2:

$$q_{rad} = \sigma\varepsilon(T_1^4 - T_2^4) \tag{15.28}$$

or the radiation flow density between two radiant parallel, flat isothermal surfaces with temperatures T_1 and T_2, and emissivities ε_1 and ε_2, respectively:

$$q_{rad,12} = \frac{\sigma}{\frac{1}{\varepsilon_1} + \frac{1}{\varepsilon_2} - 1}(T_1^4 - T_2^4) \tag{15.29}$$

With air between the plane surfaces, κ is κ_{air} given by the following [180, 181]:

$$\kappa_{air} = 0.0242396(1 + 0.0030528\theta - 1.282 \cdot 10^{-6}\theta^2)\ (\text{W m}^{-1}\,\text{K}^{-1}) \tag{15.30}$$

where θ is given in °C and is calculated from

$$\theta = (T_m - 273.15\,\text{K})°\text{C/K}\ (°\text{C}) \tag{15.31}$$

and values are accurate to 0.6% between $\theta = 10$ and 70 °C.

With two identical glass plates, it follows that

$$\varepsilon_1 = \varepsilon_2 = \varepsilon \tag{15.32}$$

Inserting Eq. (15.32) in Eq. (15.25) yields the following expression:

$$q_{\text{rad}} = \frac{4\sigma T_{\text{m}}^3 \Delta T}{\frac{2}{\varepsilon} - 1} = \frac{4\varepsilon\sigma T_{\text{m}}^3 \Delta T}{2 - \varepsilon} \tag{15.33}$$

Furthermore, inserting Eq. (15.33) in Eq. (15.24) gives

$$q_{\text{tot}} = \frac{4\varepsilon\sigma T_{\text{m}}^3 \Delta T}{2 - \varepsilon} + \frac{\kappa}{d}\Delta T \tag{15.34}$$

Solving Eq. (15.34) with respect to the emissivity ε yields

$$\varepsilon = \frac{2\left(q_{\text{tot}} - \frac{\kappa}{d}\Delta T\right)}{4\sigma T_{\text{m}}^3 \Delta T + q_{\text{tot}} - \frac{\kappa}{d}\Delta T} \tag{15.35}$$

which may be used for calculation of the emissivity from heat flow meter measurements.

The uncertainty in the emissivity value, $\Delta\varepsilon$, is calculated as follows:

$$\Delta\varepsilon = \sqrt{\left(\frac{\partial\varepsilon}{\partial q_{\text{tot}}}\Delta q_{\text{tot}}\right)^2 + \left(\frac{\partial\varepsilon}{\partial\kappa}\Delta\kappa\right)^2 + \left(\frac{\partial\varepsilon}{\partial d}\Delta d\right)^2 + \left(\frac{\partial\varepsilon}{\partial(\Delta T)}\Delta(\Delta T)\right)^2 + \left(\frac{\partial\varepsilon}{\partial\sigma}\Delta\sigma\right)^2 + \left(\frac{\partial\varepsilon}{\partial T_{\text{m}}}\Delta T_{\text{m}}\right)^2} \tag{15.36}$$

where

$$\frac{\partial\varepsilon}{\partial q_{\text{tot}}} = \frac{8\sigma T_{\text{m}}^3 \Delta T}{\left(4\sigma T_{\text{m}}^3 \Delta T + q_{\text{tot}} - \frac{\kappa}{d}\Delta T\right)^2} \tag{15.37}$$

$$\frac{\partial\varepsilon}{\partial\kappa} = \frac{-8\sigma T_{\text{m}}^3 (\Delta T)^2 / d}{\left(4\sigma T_{\text{m}}^3 \Delta T + q_{\text{tot}} - \frac{\kappa}{d}\Delta T\right)^2} \tag{15.38}$$

$$\frac{\partial\varepsilon}{\partial d} = \frac{8\sigma T_{\text{m}}^3 (\Delta T)^2 / d^2}{\left(4\sigma T_{\text{m}}^3 \Delta T + q_{\text{tot}} - \frac{\kappa}{d}\Delta T\right)^2} \tag{15.39}$$

$$\frac{\partial\varepsilon}{\partial(\Delta T)} = \frac{-8\sigma T_{\text{m}}^3 q_{\text{tot}}}{\left(4\sigma T_{\text{m}}^3 \Delta T + q_{\text{tot}} - \frac{\kappa}{d}\Delta T\right)^2} \tag{15.40}$$

$$\frac{\partial\varepsilon}{\partial\sigma} = \frac{-8\left(q_{\text{tot}} - \frac{\kappa}{d}\Delta T\right) T_{\text{m}}^3 \Delta T}{\left(4\sigma T_{\text{m}}^3 \Delta T + q_{\text{tot}} - \frac{\kappa}{d}\Delta T\right)^2} \tag{15.41}$$

$$\frac{\partial \varepsilon}{\partial T_{\mathrm{m}}} = \frac{-24\left(q_{\mathrm{tot}} - \frac{\kappa}{d}\Delta T\right)\sigma T_{\mathrm{m}}^{2}\Delta T}{\left(4\sigma T_{\mathrm{m}}^{3}\Delta T + q_{\mathrm{tot}} - \frac{\kappa}{d}\Delta T\right)^{2}} \tag{15.42}$$

The uncertainties for the different parameters, that is, Δq_{tot}, $\Delta\kappa$, Δd, $\Delta(\Delta T)$, $\Delta\sigma$ and $\Delta \mathrm{T}_{m}$, may be estimated, and hence, numerical values for the emissivity uncertainties may then be calculated.

15.6.10.4 Emissivity by Hemispherical Reflectance

The emissivity may be determined as the total hemispherical emissivity by applying a hemispherical reflectometer and integrating over the hemisphere by [188]:

$$\varepsilon = 2\int_{0}^{\pi/2} \varepsilon_{t}(\theta) \sin\theta \cos\theta \, d\theta \tag{15.43}$$

where

$$\varepsilon_{t}(\theta,\phi,\lambda) = 1 - \frac{\int_{0}^{\infty} R(\lambda)P(\lambda,T)\,d\lambda}{\int_{0}^{\infty} P(\lambda,T)\,d\lambda} \tag{15.44}$$

$$P(\lambda,T) = \frac{8\pi hc}{\lambda^{5}(e^{hc/(\lambda kT)} - 1)} = \text{Planck's function} \tag{15.45}$$

R	=	hemispherical reflectance
λ	=	wavelength (m)
T	=	temperature (K)
θ and ϕ	=	integrating angles over the hemisphere
h	=	Planck's constant $\approx 6.63 \cdot 10^{-34}$ J s
k	=	Boltzmann's constant $\approx 1.38 \cdot 10^{-23}$ J K^{-1}
c	=	velocity of light $\approx 3.00 \cdot 10^{8}$ m s^{-1}

15.6.11 Solar Factor (SF)

15.6.11.1 Solar Factor in General

The *Solar Factor* (SF) is the *Total Solar Energy Transmittance,* also known as *solar heat gain coefficient* (SHGC) and *g-value,* which is given by the following [148]:

$$\mathrm{SF} = T_{\mathrm{sol}} + q_{\mathrm{i}} \tag{15.46}$$

T_{sol}	=	solar transmittance (Eq. 15.14)
q_{i}	=	secondary heat transfer factor towards the inside

The secondary heat transfer factor towards the inside results from heat transfer by convection and long-wave infrared radiation of the incident solar radiation part which has been absorbed by the glazing.

The solar absorbance A_{sol} may be divided into two parts:

$$A_{sol} = q_i + q_e \qquad (15.47)$$

where

q_i	=	secondary heat transfer factor towards the inside
q_e	=	secondary heat transfer factor towards the outside

That is, the absorbed energy in the glazing is transferred to the inside (q_i) and to the outside (q_e).

15.6.11.2 Heat Transfer Coefficients of Glazing towards the Outside and Inside

In order to calculate the secondary heat transfer factor towards the inside (q_i), the heat transfer coefficients of the glazing towards the outside (h_e) and towards the inside (h_i), are required. These coefficients depend mainly on the glazing position, wind velocity, inside and outside temperatures and the temperature of the two external glazing surfaces.

Under the following assumptions:

- vertical glazing position
- wind velocity of approximately $4\,m\,s^{-1}$ and corrected emissivity of 0.837 for the outside surface
- natural convection and optional emissivity for the inside surface
- unventilated air spaces

the heat transfer coefficients of the glazing towards the outside (h_e) and towards the inside (h_i) are given by the following:

$$h_e = 23\,W\,m^{-2}K^{-1} \qquad (15.48)$$

$$h_i = \left(3.6 + \frac{4.4\varepsilon}{0.837}\right)\,W\,m^{-2}\,K^{-1} \qquad (15.49)$$

where ε = corrected emissivity of the inside surface.

For soda lime glass with $\varepsilon = 0.837$, it follows from Eq. (15.49) that $h_i = 8\,W\,m^{-2}\,K^{-1}$. Note that $1/h_e \approx 0.04\,m^2\,KW^{-1}$ and $1/h_i \approx 0.13\,m^2\,K\,W^{-1}$ express the external and internal thermal surface resistance, respectively, for soda lime glass and horizontal heat flow direction (±30° from the horizontal plane). See ISO 10292:1994 [177] and EN-ISO 6946:1996 [189] for further details.

15.6.11.3 Secondary Heat Transfer Factor towards the Inside for Multiple Glazing

The secondary heat transfer factor towards the inside (q_i) of a multiple glazing is given by ISO 9050:2003(E) [148]:

$$q_i = \frac{\frac{A_{sol,1}+A_{sol,2}+A_{sol,3}+\cdots+A_{sol,n}}{h_e} + \frac{A_{sol,2}+A_{sol,3}+\cdots+A_{sol,n}}{\Lambda_{12}} + \frac{A_{sol,3}+\cdots+A_{sol,n}}{\Lambda_{23}} + \cdots + \frac{A_{sol,n}}{\Lambda_{(n-1)n}}}{\frac{1}{h_i} + \frac{1}{h_e} + \frac{1}{\Lambda_{12}} + \frac{1}{\Lambda_{23}} + \cdots + \frac{1}{\Lambda_{(n-1)n}}} \quad (15.50)$$

where

$A_{sol,1}$	=	solar absorbance of the outer (first) pane within the n-fold glazing
$A_{sol,2}$	=	solar absorbance of the second pane within the n-fold glazing
$A_{sol,n}$	=	solar absorbance of the n-th (inner) pane within the n-fold glazing
h_e	=	heat transfer coefficient of the glazing towards the outside
h_i	=	heat transfer coefficient of the glazing towards the inside
Λ_{12}	=	thermal conductance between the outer surface of the outer (first) pane and the centre of the second pane
Λ_{23}	=	thermal conductance between the centre of the second pane and the centre of the third pane
$\Lambda_{(n-1)n}$	=	thermal conductance between the centre of the $(n-1)$-th pane and the outer surface of the n-th (inner) pane

15.6.11.4 Thermal Conductance

As seen the earlier discussion, the thermal conductance (Λ) is required in order to calculate the secondary heat transfer factor towards the inside (q_i) (Eq. 15.50) of a multiple glazing, which is further applied in the calculation of the SF (Eq. 15.46).

The thermal conductance (Λ) is defined in the expression for the thermal transmittance (U-value) as given by:

$$\frac{1}{U} = \frac{1}{h_e} + \frac{1}{\Lambda} + \frac{1}{h_i} \quad (15.51)$$

where the thermal conductance (Λ) then may be expressed as follows:

$$\Lambda = \left(\sum^{N} \frac{1}{h_s} + \sum^{M} d_m r_m \right)^{-1} \quad (15.52)$$

where

$$h_s = h_g + h_r = \text{gas space conductance} \quad (15.53)$$

where

h_g	=	gas conductance (conduction and convection)
h_r	=	radiation conductance
h_e	=	heat transfer coefficient of the glazing towards the outside
h_i	=	heat transfer coefficient of the glazing towards the inside
d_m	=	total thickness of each material (glass)
r_m	=	thermal resistivity of each material, 1 mK W^{-1} for glass
N	=	number of spaces
M	=	number of materials (glass panes)

Note that the last term in Eq. (15.52), containing the material thickness and thermal resistivity, may be viewed as a material U-value.

The gas conductance (h_g) is given by:

$$h_g = Nu\frac{\kappa}{s} \tag{15.54}$$

and the radiation conductance (h_r) is given by:

$$h_r = \frac{4\sigma T_m^3}{\frac{1}{\varepsilon_1} + \frac{1}{\varepsilon_2} - 1} \tag{15.55}$$

where

κ	=	gas thermal conductivity ($W\,m^{-1}\,K^{-1}$) as given in Table A.8 in Appendix 15.A
s	=	space width
σ	≈	$5.67\cdot10^{-8}\,W\,m^{-2}\,K^{-4}$ = Stefan-Boltzmann's constant (Eq. 15.26)
T_m	=	gas mean temperature, where $T_m = 283\,K$ is a principal reference value [177]
ε_1	=	corrected emissivity for material 1 (glass 1) at the mean absolute temperature T_m of the gas space
ε_2	=	corrected emissivity for material 2 (glass 2) at the mean absolute temperature T_m of the gas space

and where

$$\mathrm{Nu} = A(\mathrm{Gr}\cdot \mathrm{Pr})^n = \text{Nusselt number for Nu} > 1 \text{ (heat flow enhanced by convection)} \tag{15.56}$$

$$\mathrm{Nu} = 1 = \text{Nusselt number for Nu} \leq 1 \text{ (heat flow by conduction only, limit value of 1)} \tag{15.57}$$

where

$$\mathrm{Gr} = \frac{9.81 s^3 \rho^2 \Delta T}{\mu^2 T_m} = \text{Grashof number} \tag{15.58}$$

$$\mathrm{Pr} = \frac{\mu c_m}{\kappa} = \text{Prandtl number} \tag{15.59}$$

where

A	=	constant depending on space inclination as given in Table A.9 in Appendix 15.A
n	=	constant depending on space inclination as given in Table A.9 in Appendix 15.A
ρ	=	mass density of gas ($kg\,m^{-3}$) as given in Table A.8 in Appendix 15.A
μ	=	gas dynamic viscosity ($kg\,m^{-1}\,s^{-1}$) as given in Table A.8 in Appendix 15.A
c_m	=	specific heat (heat capacity per unit mass) of gas ($J\,kg^{-1}\,K^{-1}$) as given in Table A.8 in Appendix 15.A
ΔT	=	temperature difference on either side of the glazing, where $\Delta T = 15\,K$ is a principal reference value [177]

The corrected emissivities (ε) of the surfaces bounding the enclosed gas spaces are required to calculate the radiation conductance (h_r) in Eq. (15.55). A corrected emissivity of 0.837 is to be applied for glass surfaces. The normal emissivity (ε_n) and corrected emissivity (ε) for coated surfaces are found from the procedures described in Sections 15.6.10.2 and 15.6.10.3. The mean temperature (T_m) of the gas space is fixed at 283 K (≈10 °C) for comparison purposes.

The gas properties required for calculation of the Nusselt number (Nu) (Eq. 15.56), which is part of the calculation of the thermal conductance (Λ) (Eq. 15.52), are found in Table A.8 in Appendix 15.A. For gas mixtures, the gas properties are proportioned in volume ratios as the following [177]:

$$P = R_1P_1 + R_2P_2 + \cdots + R_mP_m \tag{15.60}$$

where

P	=	relevant property, that is mass density, dynamic viscosity, thermal conductivity or specific heat
R_1	=	volume ratio of gas 1
R_2	=	volume ratio of gas 2
⋮		
R_m	=	volume ratio of gas m

The heat transfer coefficient of the glazing towards the outside, that is, the external heat transfer coefficient (h_e), is dependent on the wind velocity close to the glazing according to the following approximate formula [177]:

$$h_e = (10.0 + 4.1\text{v})\,\text{W}\,\text{m}^{-2}\text{K}^{-1} \tag{15.61}$$

where

v	=	wind velocity given in m s^{-1}
h_e	=	23 W m^{-2} K^{-1}) (Eq. 15.48)

The $h_e = 23$ W m^{-2} K^{-1} value is used for comparing U-values of window panes. The reciprocal

$$R_e = 1/h_e = 0.04\,\text{m}^2\text{K}\,\text{W}^{-1} \tag{15.62}$$

expresses the external thermal surface resistance.

Equation 15.61 shows that h_e increases with increasing outdoor wind velocity, that is, the external thermal surface resistance (R_e) will decrease with increasing wind velocity on the external window surface.

The heat transfer coefficient of the glazing towards the inside, that is, the internal heat transfer coefficient (h_i), is given by the following expression; see Eq. (15.49) [177]:

$$h_i = h_c + h_r = \left(3.6 + \frac{4.4\varepsilon}{0.837}\right)\ \text{W}\,\text{m}^{-2}\,\text{K}^{-1} \tag{15.63}$$

where

$$h_c = 3.6\,\mathrm{W\,m^{-2}\,K^{-1}} \quad \text{for free convection} \tag{15.64}$$

and

$$h_r = \frac{4.4\varepsilon}{0.837}\ \mathrm{W\,m^{-2}\,K^{-1}} \tag{15.65}$$

The expression for h_r in Eq. (15.65) is only valid if there is no condensation on the coated surface. For a clear, uncoated glass, the corrected emissivity is 0.837, and Eq. (15.65) reduces to:

$$h_r = 4.4\,\mathrm{W\,m^{-2}\,K^{-1}} \quad \text{for clear, uncoated glass} \tag{15.66}$$

Furthermore, Eq. (15.63) becomes:

$$h_i = 8.0\,\mathrm{W\,m^{-2}\,K^{-1}} \quad \text{for clear, uncoated glass and free convection} \tag{15.67}$$

The $h_i = 8.0\ \mathrm{W\,m^{-2}\,K^{-1}}$ is used as a standardized value for comparing U-values of window panes.

The reciprocal

$$R_i = 1/h_i = 0.13\,\mathrm{m^2K\,W^{-1}} \tag{15.68}$$

expresses the internal thermal surface resistance.

The value of h_c, and thereby h_i, will be larger if a current of air is blown over the window from a fan-blown heater or similar situated below or above the window. That is, the internal thermal surface resistance (R_i) will decrease with increasing air current over the window surface.

Actual values applied in the calculations of the thermal conductance Λ for different cases are given in Tables B.1 and B.2 in Appendix 15.B for two-layer and three-layer window pane configurations, respectively, which are thereby applied in the calculations of the solar factors. For the three-layer window pane, each gas (air) space is treated in each column in Table B.2, where a temperature difference of 7.5 K (i.e. half of 15 K) is assumed over both gas spaces and with gas mean temperatures of 279.25 and 286.75 K for the exterior and interior gas space, respectively, which are simplifications as the different material configurations around one gas space will influence the temperatures around the other gas space. Nevertheless, these simplifications are assumed to not influence the thermal conductance calculations significantly. Furthermore, for the three-layer window pane, the glass thickness is stated as 3 mm (2 × 3 mm = 6 mm) in Table B.2 as to add up to a full glass pane (4 mm, exterior or interior side) plus half a glass pane (2 mm, to the centre of the middle pane), that is, 6 mm in total.

15.6.11.5 Solar Factor for Single Glazing

The *Solar Factor* (SF) is the *Total Solar Energy Transmittance*, also known as *solar heat gain coefficient* (SHGC) and *g-value*, which is given by the following for *single glazing*:

$$\mathrm{SF} = T_{sol} + q_i \tag{15.69}$$

where

$$q_{\mathrm{i}} = A_{\mathrm{sol}} \frac{h_{\mathrm{i}}}{h_{\mathrm{e}} + h_{\mathrm{i}}} \tag{15.70}$$

and as before

T_{sol}	=	solar transmittance (Eq. 15.14)
R_{sol}	=	solar reflectance (Eq. 15.19)
A_{sol}	=	$1 - T_{\mathrm{sol}} - R_{\mathrm{sol}}$ = solar absorbance (Eqs. 15.20 and 15.21)
h_{e}	=	23 W m^{-2} K^{-1} (Eq. 15.48, see applicable assumptions)
h_i	=	$\left(3.6 + \frac{4.4\varepsilon}{0.837}\right)$ W m^{-2}K^{-1} (Eq. 15.49, see applicable assumptions)
ε	=	corrected emissivity of the inside surface

15.6.11.6 Solar Factor for Double Glazing

The *Solar Factor* (SF) is the *Total Solar Energy Transmittance,* also known as *solar heat gain coefficient* (SHGC) and *g-value,* which is given by the following for *double glazing*:

$$\mathrm{SF} = T_{\mathrm{sol}} + q_{\mathrm{i}} \tag{15.71}$$

where

$$q_i = \frac{\dfrac{A_{\mathrm{sol},1} + A_{\mathrm{sol},2}}{h_{\mathrm{e}}} + \dfrac{A_{\mathrm{sol},2}}{\Lambda}}{\dfrac{1}{h_{\mathrm{i}}} + \dfrac{1}{h_{\mathrm{e}}} + \dfrac{1}{\Lambda}} \tag{15.72}$$

$A_{\mathrm{sol},1}$ = solar absorbance of the outer (first) pane within the double glazing given by:

$$A_{\mathrm{sol},1} = \frac{\sum_{\lambda=300\ \mathrm{nm}}^{2500\ \mathrm{nm}} \left\{ A_1 + \dfrac{A_{1\mathrm{b}} T_1 R_2}{1 - R_{1\mathrm{b}} R_2} \right\} S_\lambda \Delta\lambda}{\sum_{\lambda=300\ \mathrm{nm}}^{2500\ \mathrm{nm}} S_\lambda \Delta\lambda} \tag{15.73}$$

$A_{\mathrm{sol},2}$ = solar absorbance of the inner (second) pane within the double glazing given by:

$$A_{\mathrm{sol},2} = \frac{\sum_{\lambda=300\ \mathrm{nm}}^{2500\ \mathrm{nm}} \left\{ \dfrac{A_2 T_1}{1 - R_{1\mathrm{b}} R_2} \right\} S_\lambda \Delta\lambda}{\sum_{\lambda=300\ \mathrm{nm}}^{2500\ \mathrm{nm}} S_\lambda \Delta\lambda} \tag{15.74}$$

Λ = thermal conductance between the outer surface of the outer (first) pane and the innermost surface of the inner (second) pane.

T_1, T_2, A_1, A_2, R_1 and R_2 denote the transmittance, absorbance and reflectance for glass numbers 1 and 2, respectively, that is, exterior (outer) glass towards incident radiation beam, and interior (inner) glass, respectively. The index 'b' for $R_{1\mathrm{b}}$

designates that the reflectance measurement is performed on the back (reverse) side of the glass as compared to the normal incident radiation beam direction, that is, R_{1b} versus R_1. For simplicity, the wavelength (λ) dependence of T_1, T_2, A_1, A_2, R_1 and R_2 is omitted in Eqs. (15.73) and (15.74).

and where

$$A_1 = 1 - T_1 - R_1 \tag{15.75}$$

$$A_{1b} = 1 - T_1 - R_{1b} \tag{15.76}$$

$$A_2 = 1 - T_2 - R_2 \tag{15.77}$$

and as before

T_{sol}	=	solar transmittance (Eq. 15.14)
R_{sol}	=	solar reflectance (Eq. 15.19)
A_{sol}	=	$1 - T_{sol} - R_{sol}$ = solar absorbance (Eqs. 15.20 and 15.21)
S_λ	=	relative spectral distribution of solar radiation [148, 182]
λ	=	wavelength
$\Delta\lambda$	=	wavelength interval
$S_\lambda \Delta\lambda$	=	values at different wavelengths given in Table A.3 in Appendix 15.A [148]
h_e	=	$23\ \mathrm{W\,m^{-2}\,K^{-1}}$ (Eq. 15.48, see applicable assumptions)
h_i	=	$\left(3.6 + \frac{4.4\varepsilon}{0.837}\right)\ \mathrm{W\,m^{-2}K^{-1}}$ (Eq. 15.49, see applicable assumptions)

15.6.11.7 **Solar Factor for Triple Glazing**

The *Solar Factor* (SF) is the *Total Solar Energy Transmittance,* also known as *solar heat gain coefficient* (SHGC) and *g-value*, which is given by the following for *triple glazing*:

$$\mathrm{SF} = T_{sol} + q_i \tag{15.78}$$

where

$$q_i = \frac{\dfrac{A_{sol,1} + A_{sol,2} + A_{sol,3}}{h_e} + \dfrac{A_{sol,2} + A_{sol,3}}{\Lambda_{12}} + \dfrac{A_{sol,3}}{\Lambda_{23}}}{\dfrac{1}{h_i} + \dfrac{1}{h_e} + \dfrac{1}{\Lambda_{12}} + \dfrac{1}{\Lambda_{23}}} \tag{15.79}$$

$A_{sol,1}$ = solar absorbance of the outer (first) pane within the triple glazing given by:

$$A_{sol,1} = \frac{\displaystyle\sum_{\lambda=300\ \mathrm{nm}}^{2500\ \mathrm{nm}} \left\{ A_1 + \frac{T_1 A_{1b} R_2 \left(1 - R_{2b} R_3\right) + T_1 T_2^2 A_{1b} R_3}{(1 - R_{1b} R_2)(1 - R_{2b} R_3) - T_2^2 R_{1b} R_3} \right\} S_\lambda \Delta\lambda}{\displaystyle\sum_{\lambda=300\ \mathrm{nm}}^{2500\ \mathrm{nm}} S_\lambda \Delta\lambda} \tag{15.80}$$

$A_{sol,2}$ = solar absorbance of the middle (second) pane within the triple glazing given by:

$$A_{sol,2} = \frac{\sum_{\lambda=300\text{ nm}}^{2500\text{ nm}} \left\{ \frac{T_1A_2\left(1-R_{2b}R_3\right)+T_1T_2A_{2b}R_3}{(1-R_{1b}R_2)(1-R_{2b}R_3)-T_2^2R_{1b}R_3} \right\} S_\lambda \Delta\lambda}{\sum_{\lambda=300\text{ nm}}^{2500\text{ nm}} S_\lambda \Delta\lambda} \tag{15.81}$$

$A_{sol,3}$ = solar absorbance of the inner (third) pane within the triple glazing given by:

$$A_{sol,3} = \frac{\sum_{\lambda=300\text{ nm}}^{2500\text{ nm}} \left\{ \frac{T_1T_2A_3}{\left(1-R_{1b}R_2\right)\left(1-R_{2b}R_3\right)-T_2^2R_{1b}R_3} \right\} S_\lambda \Delta\lambda}{\sum_{\lambda=300\text{ nm}}^{2500\text{ nm}} S_\lambda \Delta\lambda} \tag{15.82}$$

Λ_{12}	=	thermal conductance between the outer surface of the outer (first) pane and the centre of the middle (second) pane
Λ_{23}	=	thermal conductance between the centre of the middle (second) pane and the innermost surface of the inner (third) pane

$T_1, T_2, T_3, A_1, A_2, A_3, R_1, R_2$ and R_3 denote the transmittance, absorbance and reflectance for glass numbers 1, 2 and 3, respectively, that is, exterior (outer) glass towards incident radiation beam, middle glass and interior (inner) glass, respectively. The index 'b' for R_{1b} and R_{2b} designates that the reflectance measurement is performed on the back (reverse) side of the glass as compared to the normal incident radiation beam direction, for example, R_{1b} versus R_1. For simplicity, the wavelength (λ) dependence of $T_1, T_2, T_3, A_1, A_2, A_3, R_1, R_2$ and R_3 is omitted in Eqs. (15.80)-(15.82).

and where

A_1	=	$1-T_1-R_1$ (Eq. 15.75)
A_{1b}	=	$1-T_1-R_{1b}$ (Eq. 15.76)
A_2	=	$1-T_2-R_2$ (Eq. 15.77)

$$A_{2b} = 1 - T_2 - R_{2b} \tag{15.83}$$

$$A_3 = 1 - T_3 - R_3 \tag{15.84}$$

and as before

T_{sol}	=	solar transmittance (Eq. 15.14)
R_{sol}	=	solar reflectance (Eq. 15.19)
A_{sol}	=	$1 - T_{sol} - R_{sol}$ = solar absorbance (Eqs. 15.20 and 15.21)
S_λ	=	relative spectral distribution of solar radiation [148, 182]
λ	=	wavelength
$\Delta\lambda$	=	wavelength interval
$S_\lambda \Delta\lambda$	=	values at different wavelengths given in Table A.3 in Appendix 15.A [148]
h_e	=	$23\,\mathrm{W\,m^{-2}K^{-1}}$ (Eq. 15.48, see applicable assumptions)
h_i	=	$\left(3.6 + \frac{4.4\epsilon}{0.837}\right)\ \mathrm{W\,m^{-2}\,K^{-1}}$ (Eq. 15.49, see applicable assumptions)

15.6.12 Colour Rendering Factor (CRF)

The *Colour Rendering Factor* (CRF) is given by the following expression [190]:

$$\mathrm{CRF} = \frac{R_a}{100} = \frac{1}{800}\sum_{i=1}^{8} R_i \tag{15.85}$$

where

$$R_a = \frac{1}{8}\sum_{i=1}^{8} R_i = \text{general colour rendering index (EN 410:1998 E[190])} \tag{15.86}$$

$$R_i = 100 - 4.6\Delta E_i = \text{specific colour rendering index} \tag{15.87}$$

$$\Delta E_i = \sqrt{(U^*_{t,i} - U^*_{r,i})^2 + (V^*_{t,i} - V^*_{r,i})^2 + (W^*_{t,i} - W^*_{r,i})^2} = \text{total distortion of colour } i \tag{15.88}$$

$U^*_{r,i}$, $V^*_{r,i}$ and $W^*_{r,i}$ are given in Table A.10 in Appendix 15.A [190]

where conversion into the CIE 1964 uniform colour space system is applied [190]:

$$W^*_{t,i} = 25\left(\frac{100Y_{t,i}}{Y_t}\right)^{1/3} - 17 \tag{15.89}$$

$$U^*_{t,i} = 13W^*_{t,i}(u^*_{t,i} - 0.1978) \tag{15.90}$$

$$V^*_{t,i} = 13W^*_{t,i}(v^*_{t,i} - 0.3122) \tag{15.91}$$

where the trichromatic coordinates corrected in terms of distortion by chromatic adaptation, for the eight test colours illuminated by the transmitted light, are given by:

$$u^*_{t,i} = \frac{10.872 + 0.8802\frac{c_{t,i}}{c_t} - 8.2544\frac{d_{t,i}}{d_t}}{16.518 + 3.2267\frac{c_{t,i}}{c_t} - 2.0636\frac{d_{t,i}}{d_t}} \quad (15.92)$$

$$v^*_{t,i} = \frac{5.520}{16.518 + 3.2267\frac{c_{t,i}}{c_t} - 2.0636\frac{d_{t,i}}{d_t}} \quad (15.93)$$

where for transmitted light:

$$c_t = \frac{1}{v_t}(4 - u_t - 10v_t) \quad (15.94)$$

$$d_t = \frac{1}{v_t}(1.708v_t + 0.404 - 1.481u_t) \quad (15.95)$$

where for transmitted light, then reflected by the test colour i:

$$c_{t,i} = \frac{1}{v_{t,i}}(4 - u_{t,i} - 10v_{t,i}) \quad (15.96)$$

$$d_{t,i} = \frac{1}{v_{t,i}}(1.708v_{t,i} + 0.404 - 1.481u_{t,i}) \quad (15.97)$$

where the trichromatic coordinates in the CIE 1960 uniform chromaticity diagram are calculated as follows[190]:
for transmitted light:

$$u_t = \frac{4X_t}{X_t + 15Y_t + 3Z_t} \quad (15.98)$$

$$v_t = \frac{6Y_t}{X_t + 15Y_t + 3Z_t} \quad (15.99)$$

for transmitted light, then reflected by the test colour i:

$$u_{t,i} = \frac{4X_{t,i}}{X_{t,i} + 15Y_{t,i} + 3Z_{t,i}} \quad (15.100)$$

$$v_{t,i} = \frac{6Y_{t,i}}{X_{t,i} + 15Y_{t,i} + 3Z_{t,i}} \quad (15.101)$$

where the tristimulus values:
for transmitted light:

$$X_t = \sum_{\lambda=380\,\text{nm}}^{780\,\text{nm}} \frac{d\Phi}{d\lambda} T(\lambda)\overline{x}(\lambda)\Delta\lambda \quad (15.102)$$

$$Y_t = \sum_{\lambda=380\,\text{nm}}^{780\,\text{nm}} \frac{d\Phi}{d\lambda} T(\lambda)\overline{y}(\lambda)\Delta\lambda \quad (15.103)$$

$$Z_t = \sum_{\lambda=380\,\text{nm}}^{780\,\text{nm}} \frac{d\Phi}{d\lambda} T(\lambda)\overline{z}(\lambda)\Delta\lambda \quad (15.104)$$

for transmitted light, then reflected by each of the eight test colours:

$$X_{t,i} = \sum_{\lambda=380\,\text{nm}}^{780\,\text{nm}} \frac{d\Phi}{d\lambda} T(\lambda)\beta_i(\lambda)\bar{x}(\lambda)\Delta\lambda \tag{15.105}$$

$$Y_{t,i} = \sum_{\lambda=380\,\text{nm}}^{780\,\text{nm}} \frac{d\Phi}{d\lambda} T(\lambda)\beta_i(\lambda)\bar{y}(\lambda)\Delta\lambda \tag{15.106}$$

$$Z_{t,i} = \sum_{\lambda=380\,\text{nm}}^{780\,\text{nm}} \frac{d\Phi}{d\lambda} T(\lambda)\beta_i(\lambda)\bar{z}(\lambda)\Delta\lambda \tag{15.107}$$

where

λ	=	wavelength
$\Delta\lambda$	=	wavelength interval
$(d\Phi/d\lambda)\Delta\lambda$	=	the relative spectral energy distribution of illuminant D65 as given in Table A.11 in Appendix 15.A [190]
$\beta_i(\lambda)$	=	the spectral reflectance of each test colour (i from 1 to 8) as given in Table A.12 in Appendix 15.A [190]
$\bar{x}(\lambda), \bar{y}(\lambda)$ and $\bar{z}(\lambda)$	=	the spectral tristimulus values for the CIE 1931 colorimetric standard observer as given in Table A.13 in Appendix 15.A [190]
$T(\lambda)$	=	spectral transmittance of the glazing

The CRF value expresses a quantitative evaluation of the colour differences between eight test colours lighted directly by the reference illuminant D65 and by the same illuminant transmitted through the glazing.

Thus, the CRF value will be a number between 0 and 1 (often given as a percentage), calculated for the visible part of the solar spectrum 380–780 nm. A high number indicates a good colour rendering. Ideally, the maximum value of 1 will be obtained by glazing whose spectral transmittance is completely constant in the whole visible spectral range, that is, no variation of transmittance with wavelength. CRF > 0.9 characterizes a very good colour rendering, while CRF > 0.8 represents a good colour rendering [148, 190]. The recent studies by Mortimer and Varley [191, 192] on colour measurements of electrochromic materials are notable.

15.6.13 Additional Heat Transfer

In addition to the heat transfer directly related to the solar radiation, an additional heat transfer q_u occurs if the room (interior) temperature θ_i differs from the outside (exterior) temperature θ_e, and may be calculated as follows:

$$q_u = U(\theta_i - \theta_e) \tag{15.108}$$

where U denotes the U-value (W m^{-2} K^{-1}), that is, the thermal transmittance, of the glazing, determined according to ISO 10291:1994(E) [193], ISO 10292:1994 [177] or ISO 10293:1997(E) [194]. Also see EN 1946-3:1999 E [180], EN 1946-2:1999 E [181], EN 1946-4:2000 E [195], EN 673:1997 E [196], EN 674:1997 E [197], EN 675:1997-E [198], ISO 8301:1991(E) [199] and ISO 8302:1991(E) [200].

15.6.14 Number of Glass Layers in a Window Pane

The solar radiation glazing factors are dependent upon the number of glass layers in the actual window pane. Transmittance and reflectance calculations for single glass panes, two-layer window panes and three-layer window panes are given in Eqs. (15.3) – (15.10). Furthermore, the corresponding absorbance calculations are carried out with Eq. (15.1).

15.6.15 General Calculation Procedures

The solar radiation glazing factors may readily be calculated via simple computer programs. However, with today's sophisticated data collection tools in computers and instruments, even higher resolutions are accessible, that is, narrower (wavelength) data intervals, should be listed in future versions of Tables A.1 – A.5 in Appendix 15.A. To simplify the programming of the calculations, *one* wavelength interval should be chosen for the whole table. For easy implementation of the table values in calculations, a reference in the standard ISO 9050:2003(E) [148] should be given to an internet website where the table values may be downloaded as ASCII files.

15.7 Spectroscopic Measurement and Calculation of Solar Radiation Glazing Factors

Spectrophotometric measurements along with calculations of the solar radiation glazing factors will be shown in the following for one float glass, one glass with a low emittance coating, one dark silver coated glass, several two-layer and three-layer window pane combinations and three different electrochromic window (ECW) devices at various coloration levels. The ECWs for building applications will both replace and be used together with other glass panes and coatings; thus, it is very important to be able to compare the different ECWs with the various glass pane types and coatings. The absorbance spectra depicted in the following sections comply with the absorbance definition in Eq. (15.1), a number between 0 and 1, that is, the absorbance is not written on the often used/measured logarithmic scale called optical density defined in Eq. (15.114) in a following section [37].

15.7.1
Spectroscopic Data for Float Glass and Low Emittance Glass

For comparison, the transmittance, absorbance and reflectance in the whole solar spectrum were measured (from 290 to 3300 nm) for one float glass and one glass with low emittance coating, depicted in Figures 15.9 and 15.10, respectively. The upper border of 3300 nm represents the spectrophotometer's long wavelength limit, while below 290 nm, the absorption by glass becomes very large. The most noticeable difference is the large reflectance values, and thereby small transmittance values, in the NIR region for the glass with the low emittance coating. Details of the UV and VIS region (Figures 15.11 and 15.12) show that the low emittance glass absorbs more solar radiation at these lower wavelengths than

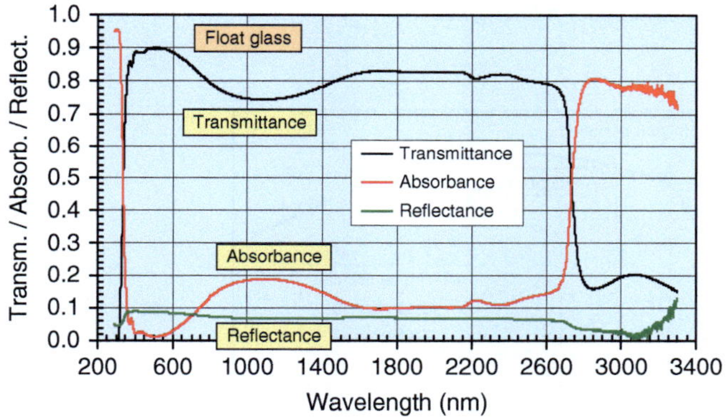

Figure 15.9 Transmittance, absorbance and reflectance versus wavelength for the whole solar spectrum of a float glass.

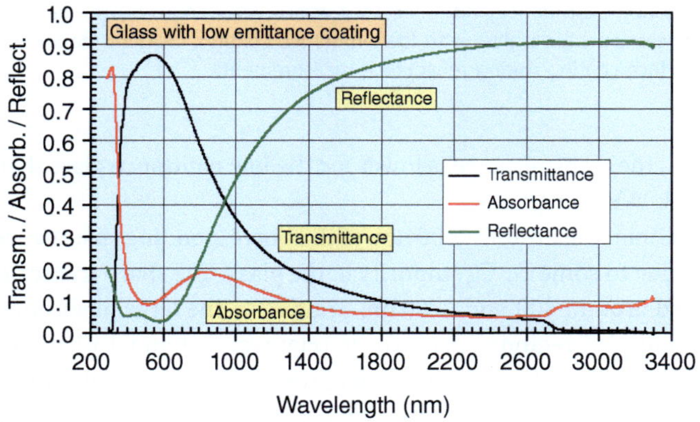

Figure 15.10 Transmittance, absorbance and reflectance versus wavelength for the whole solar spectrum measured for a glass with low emittance coating. Incident radiation beam is towards the surface coating during reflectance measurements.

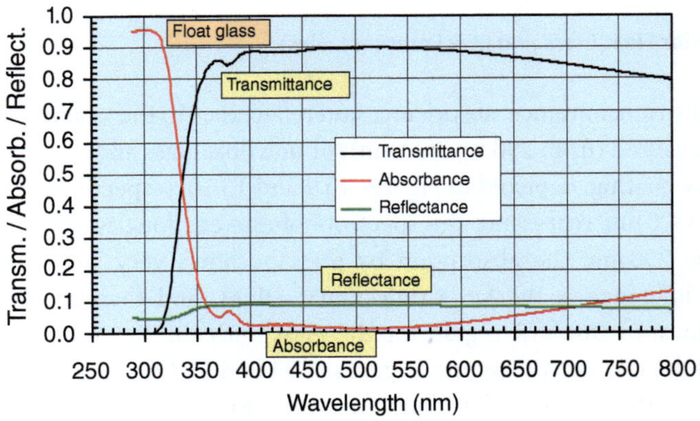

Figure 15.11 Transmittance, absorbance and reflectance versus wavelength for the ultraviolet and visible region measured for a float glass.

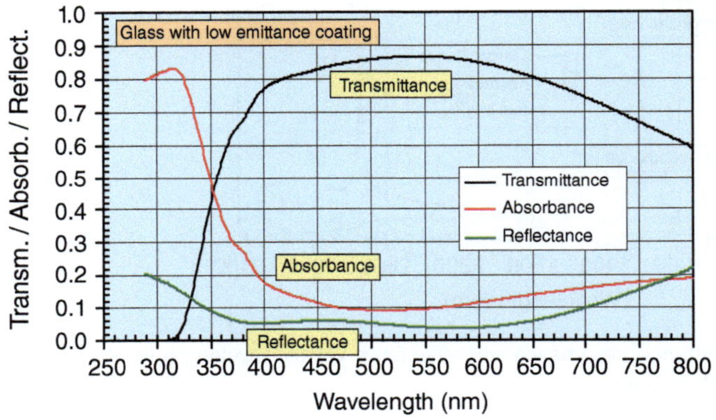

Figure 15.12 Transmittance, absorbance and reflectance versus wavelength for the ultraviolet and visible region measured for a glass with low emittance coating. Incident radiation beam is towards the surface coating during reflectance measurements.

the float glass, that is, the transmittance is lower for the low emittance glass than the float glass in the UV-VIS region.

The drop in transmittance at around 1000 nm, with a corresponding absorbance peak, Figure 15.9, is due to some Fe_2O_3 impurity in the glass. The sharp transmittance cut-offs located around 400 and 2700 nm are due to the large absorption in glass in the UV and NIR regions, respectively (Figures 15.9 and 15.10). The sharp transmittance cut-off, with corresponding absorbance increase, at around 2700 nm, is also observed in Figure 15.10 for the low emittance glass, but much smaller in value since the largest part of the incoming radiation in this wavelength region is reflected from the low emittance coating (the coating surface faces the incident radiation beam in the spectrophotometer).

15.7.2
Spectroscopic Data for Dark Silver Coated Glass

For comparison, the transmittance, absorbance and reflectance in the whole solar spectrum were measured for one dark silver coated glass, depicted in Figure 15.13. The ultraviolet and visible regions are shown in Figure 15.14. The transmittance is found to be rather low, between 0 and 0.2, over the whole UV-VIS-NIR region. Both the reflectance and absorbance are quite high in the whole UV-VIS-NIR region, with the absorbance (0.45–0.7) dominating in the UV-VIS region while still retaining a substantial reflectance (0.3–0.4), thus giving rise to the dark absorbing and reflecting colour. As will be seen later (Table 15.1), this coating will result in high SMPF and SSPF values (high material and skin protection.

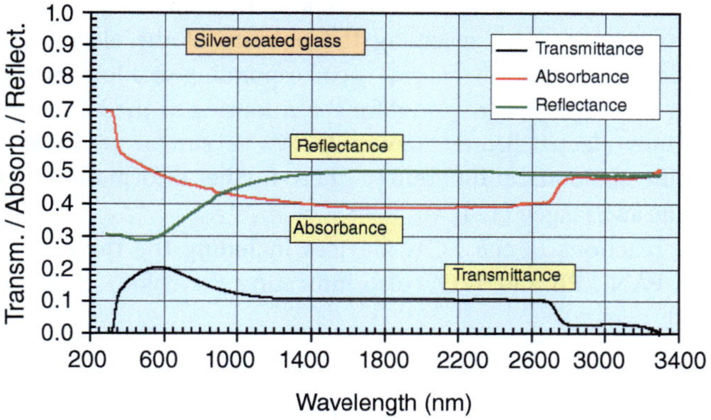

Figure 15.13 Transmittance, absorbance and reflectance versus wavelength for the whole solar spectrum measured for a dark silver coated glass. Incident radiation beam is towards surface coating during reflectance measurements.

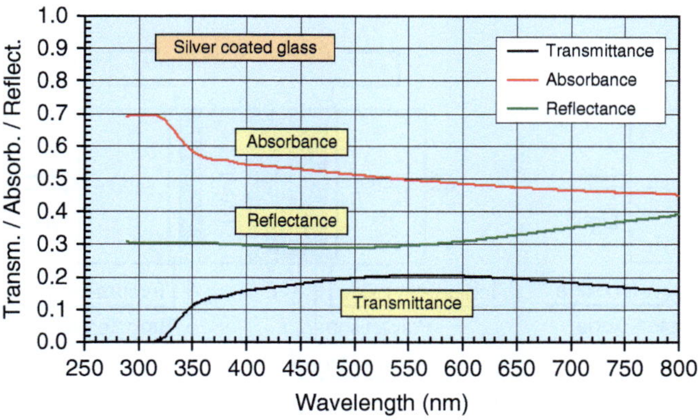

Figure 15.14 Transmittance, absorbance and reflectance versus wavelength for the ultraviolet and visible region measured for a dark silver coated glass. Incident radiation beam is towards surface coating during reflectance measurements.

15.7.3 Spectroscopic Data for Electrochromic Windows

To recapitulate, electrochromic windows (ECW) control the colour of the window, thereby also the solar radiation throughput, by variation of the applied electrical potential. Two ECWs (ECW1 and ECW2) are shown schematically in Figure 15.15, constructed in a sandwich form from the electrochromic materials polyaniline (PANI), prussian blue (PB) and tungsten oxide (WO_3), transparent conducting glass plates with an indium-tin oxide coating (ITO) with a typical surface resistivity of 90 Ω/□, and the solid-state polymer electrolyte poly(2-acrylamido-2-methyl-propane-sulphonic acid) (PAMPS) as an ionic conductor. The PANI, PB and WO_3 coating thicknesses were less than 1 μm, while the PAMPS layer thickness was about 0.1 mm. Applying a positive potential to the PANI/PB electrode, PANI, PB and WO_3 turn to a blue colour, while the window is bleached (almost transparent) by reversing the polarity of the electrodes. Only a small charge density of about 3 mC cm^{-2}, corresponding to a low energy consumption of about 5 mWh m^{-2}, is required for the colouring or the bleaching process [175]. The third electrochromic device ECW3 is similar to ECW1; however, with a specific geometrical difference, where further information and details are given by Jelle and Hagen [174].

The schematic cell reactions in the ECW devices including the three electrochromic materials PANI, PB and WO_3, also indicating the colour changes, may be written as [36]:

$$\begin{aligned}&WO_3 + (\text{PANI})(y\text{H}^+\text{A}^-) + (x-y)\text{K}_2\text{Fe}^{\text{II}}[\text{Fe(CN)}_6] + (x-y)\text{H}^+(\text{transparent}) \leftrightarrow \\ &\text{H}_x\text{WO}_3 + (\text{PANI}^{y+})(\text{A}^-)_y + (x-y)\text{KFe}^{\text{III}}[\text{Fe(CN)}_6] + (x-y)\text{K}^+(\text{blue})\end{aligned} \quad (15.109)$$

where x denotes the number of protons (H^+) exchanged at the WO_3 electrode, and y and $(x-y)$ are the numbers of cations, H^+ and K^+, transferred in the PANI and

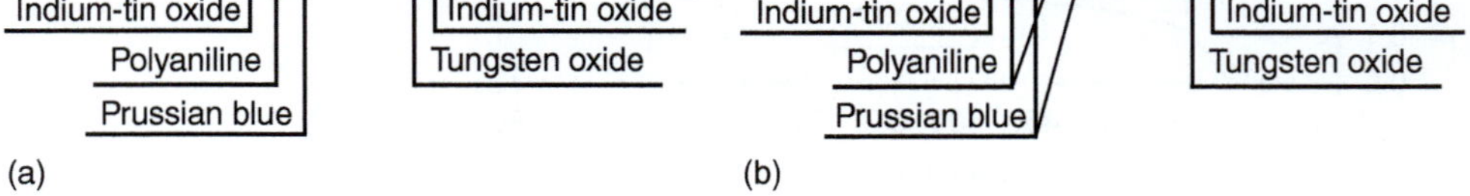

Figure 15.15 Schematic drawings of electrochromic window configurations ECW1 (a) and ECW2 (b, PANI-PB multilayer) based on the electrochromic materials polyaniline (PANI), prussian blue (PB) and tungsten oxide (WO_3). From Jelle and Hagen [36].

PB matrix, respectively. Setting $y = x$ in Eq. (15.109), we obtain the cell reaction in an ECW device including only PANI and WO_3 as electrochromic materials [37]:

$$\underset{\text{(transparent)}}{WO_3 + (PANI)(xH^+A^-)} \leftrightarrow \underset{\text{(blue)}}{H_xWO_3 + (PANI^{x+})(A^-)_x} \tag{15.110}$$

Experiments conducted by the hole method by characterization of individual electrochromic layers in solid-state ECWs based on PANI, PB and WO_3 [175] revealed that (i) PANI regulates the transmittance in the whole VIS and NIR region, with a characteristic shift in modulation from the NIR to the VIS region by application of high positive potentials, while (ii) PB modulates the transmittance only in the VIS and at the start of the NIR region (400–1300 nm) and (iii) WO_3 regulates the transmittance in the whole VIS and NIR region where the modulation in the VIS region is largest at the longest wavelengths.

A high transmission regulation and solar modulation (solar regulation $\Delta T_{sol} = 53\%$, T_{sol} calculated based on the solar spectral irradiance given in *CRC Handbook of Chemistry and Physics* [201]) [175] have been achieved with this type of ECW (ECW1, Figure 15.15a), which is depicted in Figure 15.16 (whole solar spectrum) and Figure 15.17 (UV-VIS region). Note that applying the solar spectral irradiance given in ISO 9050:2003(E) [148] gives a slightly higher $\Delta T_{sol} = 57\%$ for ECW1 (Table 15.3).

To include PB in PANI enhances the coloration (wavelength-dependent absorption), while the adhesion of PB is improved by PANI, that is, in this respect, there exists a symbiotic relationship between PANI and PB [36, 175].

Transmittance curves for a second ECW (ECW2, Figure 15.15b) of the same construction, though with PANI-PB multilayers and a very dark colour in the coloured state, are shown in Figures 15.18 and 15.19. The solar regulation is $\Delta T_{sol} = 49\%$ [176] with solar spectral irradiance from *CRC Handbook of Chemistry and Physics* [201] and slightly higher $\Delta T_{sol} = 51\%$ (Table 15.3) with solar spectral irradiance given in ISO 9050:2003(E) [148] for ECW2.

Transmittance curves for a third ECW (ECW3) of a similar construction (not depicted here) as ECW1, though with a specific geometrical difference and a very dark colour in the coloured state (measured at a location with thick electrochromic coatings, see [174]), are shown in Figures 15.20 and 15.21. The solar regulation is $\Delta T_{sol} = 56\%$ [174] with solar spectral irradiance from *CRC Handbook of Chemistry and Physics* [201] and somewhat higher $\Delta T_{sol} = 59\%$ (Table 15.3) with solar spectral irradiance given in ISO 9050:2003(E) [148] for ECW3. The $\Delta T_{sol} = 59\%$ for ECW3 is notably high.

It is observed that both ECW2 and ECW3 are considerably darker than ECW1 in the coloured state as depicted by Figures 15.16–15.21. Comparing ECW2 and ECW3, it is seen that ECW2 absorbs very much of the solar radiation around 700 nm, whereas ECW3 absorbs substantially more in the NIR region than ECW2. In that respect, as the solar irradiance decreases rapidly into the NIR region (Figure 15.1), the further out into the NIR region the less solar radiation energy there is to be modulated by the ECWs.

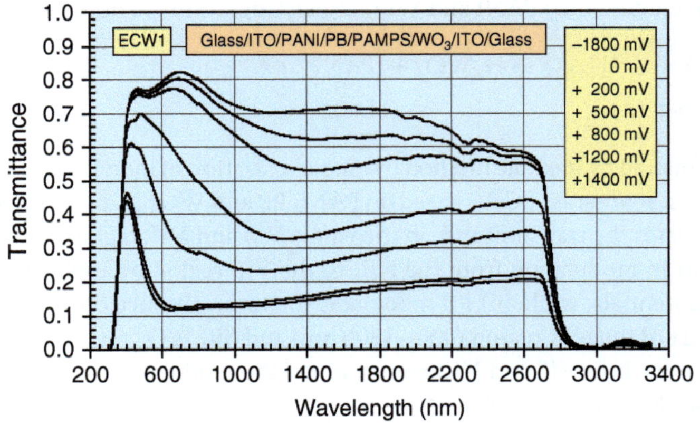

Figure 15.16 Transmittance versus wavelength for the whole solar spectrum measured for an electrochromic window ECW1 at different applied potentials. Highest coloration level is at +1400 mV. Redrawn from Jelle *et al.* [175].

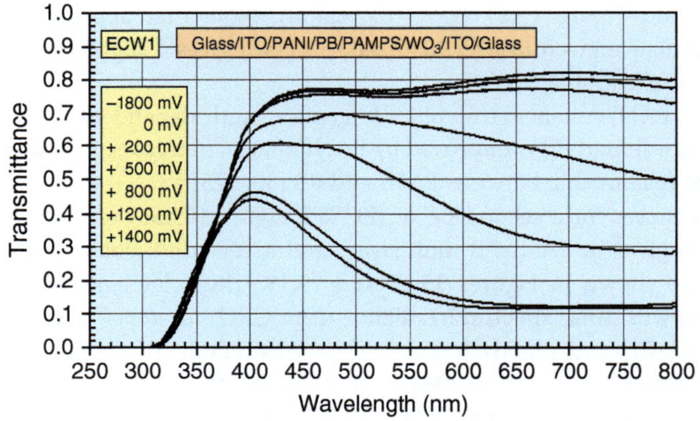

Figure 15.17 Transmittance versus wavelength for the ultraviolet and visible region measured for an electrochromic window ECW1 at different applied potentials. Highest coloration level is at +1400 mV.

The doping mechanisms in PANI include both redox processes and proton doping [33, 202, 204], and to show the characteristic absorbance shift from the NIR to the VIS region for PANI, we plot the absorbance versus both wavelength and applied electrical potential in three dimensions in order to more clearly demonstrate the absorbance changes. Hence, two- and three-dimensional graphical plots of light absorbance versus wavelength for PANI on ITO glass in an aqueous solution of 0.1 M $Na_2SO_4 \cdot 10H_2O$ and 0.001 M H_2SO_4 (pH 3.7) at different applied potentials versus Ag/AgCl (3.5 M KCl) are depicted in Figure 15.22 [33, 36].

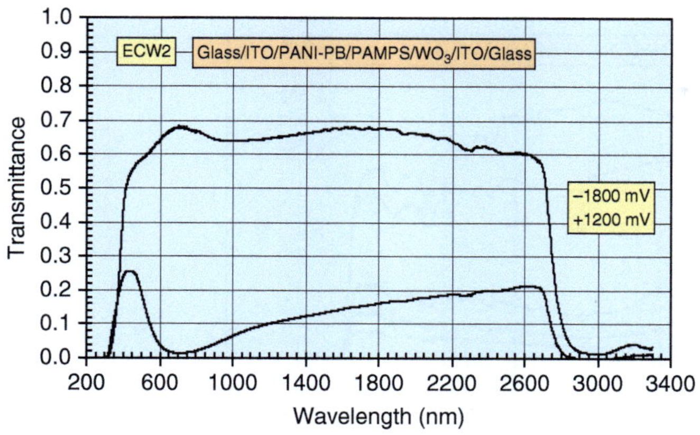

Figure 15.18 Transmittance versus wavelength for the whole solar spectrum measured for a second electrochromic window ECW2 (PANI-PB multilayer) at different applied potentials. Highest coloration level is at +1200 mV. Redrawn from Jelle and Hagen [176].

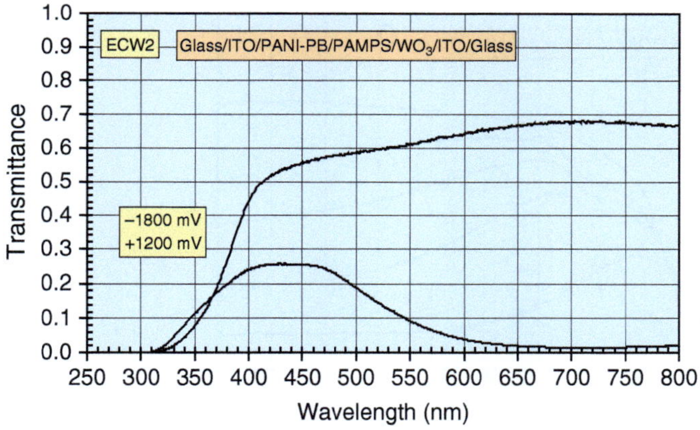

Figure 15.19 Transmittance versus wavelength for the ultraviolet and visible region measured for a second electrochromic window ECW2 (PANI-PB multilayer) at different applied potentials. Highest coloration level is at +1200 mV.

The dynamic characteristics of ECWs may allow *diffuse* daylight through the window panes in the required amount in order to obtain a satisfactory room illumination, whereas under *direct* solar radiation, the SMPF and SSPF values can be increased to a sufficiently high protection level.

Several experimental investigations have been performed on these ECWs with PANI, PB and WO_3. Transmission properties for individual electrochromic layers in solid-state ECWs based on PANI, PB and WO_3 have been studied by applying the described hole method by Jelle *et al.* [175]. Correlation between light absorption and electric charge in solid-state ECWs based on

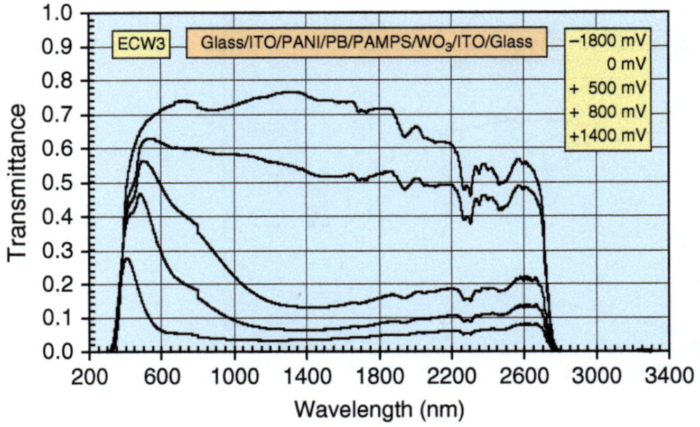

Figure 15.20 Transmittance versus wavelength for the whole solar spectrum measured for a third electrochromic window ECW3 at different applied potentials. Highest coloration level is at +1400 mV. Redrawn from Jelle and Hagen [174].

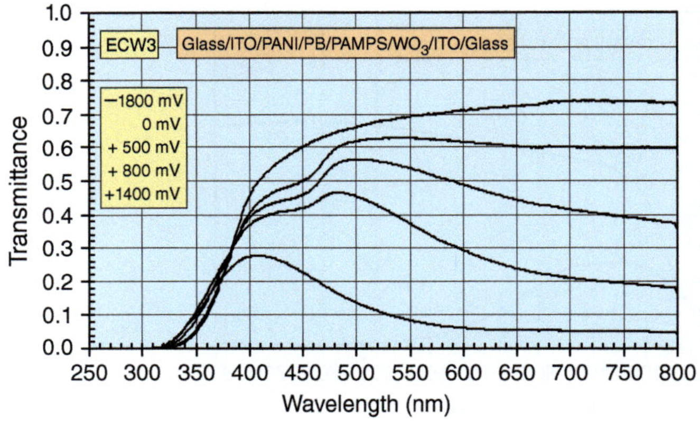

Figure 15.21 Transmittance versus wavelength for the ultraviolet and visible region measured for a third electrochromic window ECW3 at different applied potentials. Highest coloration level is at +1400 mV.

PANI, PB and WO_3 has been explored by Jelle and Hagen [37], including memory effect, electrochromic efficiency and impedance spectra investigations. The reduction factor for PANI films has been investigated and determined from both cyclic voltammetry and visible absorption spectra, showing in general good agreement, by Jelle *et al.* [33]. Furthermore, the solar material and skin protection factors SMPF and SSPF have been defined and studied for solid-state ECWs based on PANI, PB and WO_3 by Jelle *et al.* [34], where also various examples of photodegradation processes have been shown.

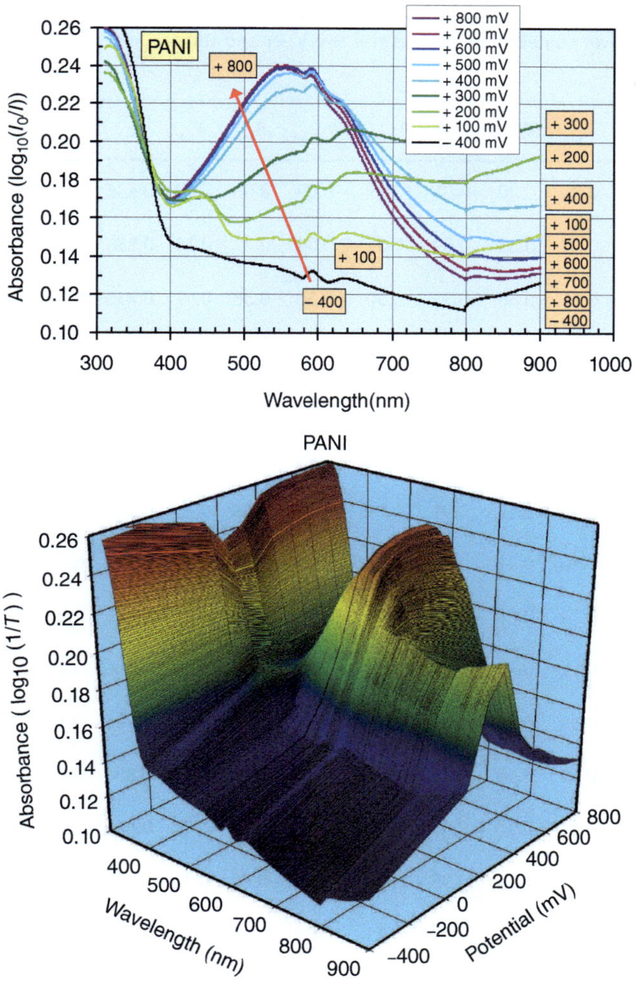

Figure 15.22 Two- and three-dimensional graphical plots of light absorbance versus wavelength for PANI on ITO glass in an aqueous solution of 0.1 M $Na_2SO_4 \cdot 10H_2O$ and 0.001 M H_2SO_4 (pH 3.7) at different applied potentials versus Ag/AgCl (3.5 M KCl). The absorbance in the ITO glass, electrolyte and cell has been subtracted by the use of a reference cell in the double beam spectrophotometer. Replotted from Jelle and Hagen [36].

15.7.4
Solar Radiation Glazing Factors for Float Glass, Low Emittance Glass, Dark Silver Coated Glass and Two-Layer and Three-Layer Window Pane Configurations

Table 15.1 gives calculated solar radiation glazing factor values for one float glass, one low emittance glass, one dark silver coated glass and several two-layer and three-layer window pane combinations.

Table 15.1 Calculated solar radiation glazing factor values for one float glass, one glass with low emittance coating, one dark silver coated glass and several two-layer and three-layer window pane configurations.

Glass configuration	n	T_{uv}	T_{vis}	T_{sol}	SMPF	SSPF	$R_{vis,ext}$	$R_{vis,int}$	R_{sol}	A_{sol}	ε	SF	CRF
Float Glass G	1	0.65	0.89	0.83	0.20	0.81	0.09	0.09	0.08	0.10	0.836	0.85	0.99
Low emittance glass LE/G	1	0.41	0.86	0.59	0.32	0.89	0.04	0.04	0.27	0.14	0.836	0.62	0.98
Dark silver glass S/G	1	0.10	0.20	0.16	0.85	0.97	0.30	0.30	0.38	0.47	0.836	0.28	0.97
Float/float G/A/G	2	0.50	0.80	0.69	0.31	0.87	0.16	0.04	0.13	0.17	0.836	0.76	0.97
Float/LowE G/A/LE/G	2	0.32	0.77	0.50	0.41	0.92	0.12	0.03	0.26	0.24	0.836	0.60	0.96
Float/Silver G/A/S/G	2	0.08	0.19	0.14	0.87	0.98	0.33	0.00	0.34	0.52	0.836	0.42	0.97
LowE/LowE G/LE/A/LE/G	2	0.22	0.74	0.43	0.48	0.94	0.09	0.01	0.26	0.31	0.836	0.52	0.95
LowE/Float G/LE/A/G	2	0.32	0.77	0.50	0.41	0.92	0.12	0.02	0.24	0.25	0.836	0.55	0.96
Silver/Float G/S/A/G	2	0.08	0.19	0.14	0.87	0.98	0.26	0.15	0.24	0.62	0.836	0.21	0.97
Silver/LowE G/S/A/LE/G	2	0.05	0.18	0.11	0.88	0.99	0.26	0.04	0.24	0.65	0.836	0.18	0.96
Silver/Silver G/S/A/S/G	2	0.01	0.05	0.03	0.97	1.00	0.27	0.00	0.25	0.72	0.836	0.16	0.95
Float/Float/Float G/A/G/A/G	3	0.40	0.73	0.59	0.40	0.90	0.21	0.17	0.17	0.24	0.836	0.68	0.96
Float/Float/LowE G/A/G/A/LE/G	3	0.26	0.69	0.44	0.48	0.93	0.18	0.10	0.25	0.31	0.836	0.55	0.95
Float/LowE/LowE G/A/LE/G/A/LE/G	3	0.18	0.66	0.37	0.54	0.95	0.15	0.09	0.29	0.34	0.836	0.50	0.94
LowE/LowE/LowE G/LE/A/LE/G/A/LE/G	3	0.12	0.63	0.33	0.59	0.97	0.11	0.08	0.28	0.39	0.836	0.45	0.93
Silver/Float/Float G/S/A/G/A/G	3	0.06	0.17	0.12	0.88	0.98	0.26	0.26	0.24	0.64	0.836	0.18	0.96
Silver/Float/LowE G/S/A/G/A/LE/G	3	0.04	0.16	0.09	0.89	0.99	0.26	0.15	0.24	0.66	0.836	0.15	0.95
Silver/LowE/LowE G/S/A/LE/G/A/LE/G	3	0.03	0.15	0.08	0.91	0.99	0.26	0.14	0.25	0.67	0.836	0.14	0.94
Silver/Silver/Silver G/S/A/S/G/A/S/G	3	0.00	0.01	0.01	0.99	1.00	0.27	0.26	0.25	0.74	0.836	0.11	0.93

All coatings on each glass are facing towards the outside, except the low emittance coating on the outermost glass pane, which is facing towards the inside. In the table, the glass pane sequence from left to right corresponds to going from outside to inside. Corresponding transmittance spectra are given in Figures 15.9–15.14. The emissivity of the float glass was determined to be $\varepsilon_{float} = 0.836$ by hemispherical directional reflectance measurements and was directly applied in the calculation of SF as the emissivity value in the SF calculations is with respect to the inside facing surface (of the building) of the innermost glass pane, that is, normally a float glass, whereas $\varepsilon_{lowe} = 0.071$ and $\varepsilon_{silver} = 0.543$ were indirectly applied in the SF calculations through the calculations of the thermal conductance Λ. G = glass, LE = low emittance coating, S = dark silver coating, A = air cavity.

From Table 15.1 it is seen that the float glass and the low emittance glass have rather large and similar T_{vis} values, that is, 0.89 and 0.86, respectively. However, the T_{sol} value for float glass is substantially higher than the T_{sol} value for the low emittance glass, that is, 0.83 versus 0.59, as shown in the transmittance spectra in Figures 15.9 and 15.10. Furthermore, float glass has substantially lower R_{sol} value than the low emittance glass, that is, 0.08 versus 0.27, confirmed by the reflectance spectra in Figures 15.9 and 15.10. The float glass has a larger SF value than the low emittance glass, that is, 0.85 versus 0.62.

The A_{sol} values for float glass and low emittance glass are 0.10 and 0.14, respectively, which are considerably lower values than the A_{sol} value of 0.47 for dark silver coated glass. The dark silver glass has furthermore a higher R_{sol} (0.38) value, and lower T_{vis} (0.20), T_{sol} (0.16) and SF (0.28) values than both the float glass and the low emittance glass.

The glass with the low emittance coating has SMPF and SSPF values of 0.32 and 0.89, respectively, which gives better protection than the float glass with SMPF and SSPF values of 0.20 and 0.81, respectively. The dark silver coated glass has SMPF and SSPF values of 0.85 and 0.97, respectively, which gives a far better material protection and also substantially better skin protection than both the float glass and the low emittance glass.

The observed large differences in transmittance, absorbance and reflectance spectra for the float glass, low emittance glass and dark silver coated glass are also expressed through the solar radiation glazing factors, for example, compare $T_{sol} = 0.83$, $A_{sol} = 0.10$ and $R_{sol} = 0.08$ for float glass with $T_{sol} = 0.59$, $A_{sol} = 0.14$ and $R_{sol} = 0.27$ for low emittance glass and $T_{sol} = 0.16$, $A_{sol} = 0.47$ and $R_{sol} = 0.38$ for dark silver coated glass. For further comparisons of these three single glass panes, see Table 15.1, including comparison of transmittance spectra in Figures 15.9, 15.10 and 15.13 for the whole solar region and Figures 15.11, 15.12 and 15.14 for close-ups of the ultraviolet and visible region.

Obviously, the two-layer and three-layer window panes transmit less solar radiation than their single glass pane counterparts. Thus (Table 15.1), a single float glass pane has $T_{vis} = 0.89$ and $T_{sol} = 0.83$, whereas the two-layer float/float window pane has $T_{vis} = 0.80$ and $T_{sol} = 0.69$ and the three-layer float/float/float window pane has $T_{vis} = 0.73$ and $T_{sol} = 0.59$. Furthermore, a single low emittance glass pane has $T_{vis} = 0.86$ and $T_{sol} = 0.59$, whereas the two-layer LowE/LowE window pane has $T_{vis} = 0.74$ and $T_{sol} = 0.43$ and the three-layer LowE/LowE/LowE window pane has $T_{vis} = 0.63$ and $T_{sol} = 0.33$. On the more extreme dark side, a single dark silver coated glass pane has $T_{vis} = 0.20$ and $T_{sol} = 0.16$, whereas the two-layer dark silver/silver window pane has $T_{vis} = 0.05$ and $T_{sol} = 0.03$ and the three-layer dark silver/silver/silver window pane has $T_{vis} = 0.01$ and $T_{sol} = 0.01$. Normally, only one or two low emittance coated glass panes and only a maximum of one dark silver coated glass pane are applied in a two-layer or three-layer window pane unit. Several configurations are calculated as in Table 15.1. Note that all the calculated configurations here have glass (i.e. not a coating) as the outermost surface both at the exterior and interior side of the window pane, that is, outside and inside the building, respectively; thus, all coatings are protected from the exterior and interior environments by glass.

The two-layer window panes give even better material and skin protection compared with their single pane counterparts, with SMPF values of 0.31, 0.41 and 0.87, and with SSPF values of 0.87, 0.92 and 0.98, for float/float, float/LowE and float/silver, respectively. Note that the SMPF is as high as 0.87 for the two-layer window pane consisting of one float glass and one dark silver coated glass. The three-layer window pane counterparts have even higher SMPF and SSPF values, exemplified by the very high SMPF and SSPF values of 0.88 and 0.98 for the three-layer dark silver/float/float window pane, and 0.99 and 1.00 for the three-layer dark silver/silver/silver window pane, respectively. Clearly, materials inside buildings are far better protected with the two-layer or three-layer window pane with one dark silver coated surface than with the float or low emittance coating alternatives. However, the dark silver coated window pane alternatives admit far less visible solar radiation and also total solar radiation as demonstrated by the much lower T_{vis} and T_{sol} values.

Comparison of the relatively low SMPF (with dark silver coated glass as an exception) and relatively high SSPF values explains why some materials, for example, books in book shelves and various wood and paint materials, become discoloured and bleached, whereas we human beings usually do not get tanned or sunburnt behind glass inside buildings. Note that with the earlier upper wavelength limit of 500 nm (compared with 600 nm today) in the calculation of SMPF, the material protection values were apparently larger, 0.34, 0.52 and 0.89 for the single float glass, the low emittance glass and the dark silver coated glass, respectively. Thus, a much larger part of the visible solar radiation is considered today to contribute to the solar deterioration of materials. For further details and visual comparison of SMPF and SSPF values with T_{vis} values, see Jelle *et al.* [34]. Notably, all glass configurations in Table 15.1 have very high CRF values, including float glass, low emittance glass and dark silver coated glass in single, two-layer and three-layer window pane combinations, with CRF values ranging from 0.93 to 0.99, that is, a very good colour rendering, even for the various dark silver coated glass combinations.

Figure 15.23 depicts an example of applying measured transmittance and reflectance spectra for single float glass panes (Eqs. 15.3 and 15.4) for calculating the transmittance spectra through two-layer and three-layer window panes according to Eqs. (15.5) and (15.8), respectively. Thus, the addition of extra glass panes decreases the transmittance in the whole solar spectral range. That is, adding more glass panes to a window or glass structure decreases the available amount of daylight and solar radiation to be exploited and utilized inside a building. The two-layer and three-layer window pane transmittance spectra have been calculated with the full spectral resolution and wavelength range applied in the spectrophotometric measurements, that is, avoiding the considerably lower resolution and narrower wavelength range applied when calculating the solar radiation glazing factors like e.g. T_{sol} (Table A.3).

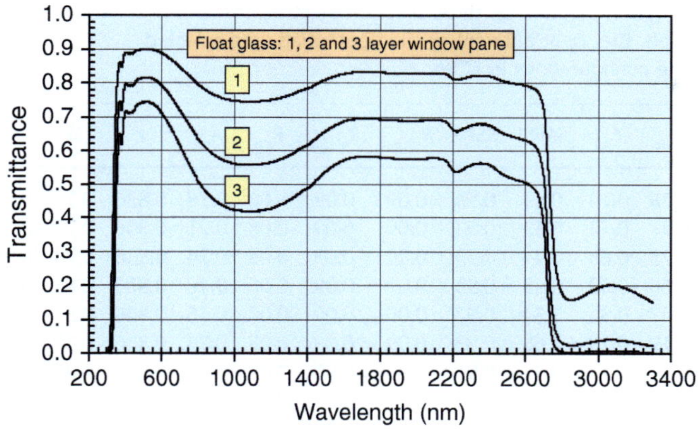

Figure 15.23 Transmittance versus wavelength for the whole solar spectrum measured for a float glass (1) and calculated for two-layer (2) and three-layer (3) float glass window panes (Eqs. 15.5 and 15.8).

Designing and choosing the appropriate window glass or glass structure require many considerations and evaluations to be made, with respect to solar material protection, solar skin protection, solar energy aspects, thermal radiation aspects, daylight, visual appearance and so on, where many of these parameters influence each other. Thus, measurement and calculation of the solar radiation glazing factors represent the most adequate tool for making quantitative evaluations, and hence the most appropriate decisions in this regard.

15.7.5 Solar Radiation Glazing Factors for Electrochromic Windows

Table 15.2 gives the solar radiation glazing factors for different coloration levels, that is, at different applied electrical potentials, for three ECWs (ECW1, ECW2 and ECW3), in addition to selected two-layer and three-layer window pane configurations with incorporated electrochromic materials.

From Table 15.2, it is observed that various solar radiation glazing factors may obtain both high and low values depending on the applied electrical potential in the ECWs, for example, changing the T_{vis} value from 0.78 to 0.17 for the ECW1 device and from 0.62 to 0.10 for the darker ECW2 device. It is also noted that these ECWs contain solar radiation *absorbing* electrochromic materials, that is, not *reflecting* materials, as the changes with applied potential occur in the transmittance (e.g. T_{sol}) and absorbance (e.g. A_{sol}) values and not in the reflectance (e.g. R_{sol}) values. The ECW3 device changes T_{vis} from 0.69 in the transparent state to 0.09 in the coloured state.

As expected, the highest coloration level gives the largest SMPF values, that is, the best protection of materials is achieved with the darkest ECW, for example, compare an SMPF value of 0.71 for ECW1, 0.82 for ECW2 and 0.83 for ECW3

Table 15.2 Calculated solar radiation glazing factors for three different electrochromic windows (ECW) at different coloration levels, that is, at different applied potentials, and selected two-layer and three-layer window pane configurations in ECWs.

Glass configuration	n	T_{uv}	T_{vis}	T_{sol}	SMPF	SSPF	$R_{vis,ext}$	$R_{vis,int}$	R_{sol}	A_{sol}	ε	SF	CRF
ECW1 (−1800 mV)	1	0.23	0.78	0.74	0.43	0.93	0.09	0.09	0.08	0.18	0.836	0.79	0.98
ECW1 (0 mV)	1	0.23	0.77	0.72	0.43	0.93	0.09	0.09	0.08	0.21	0.836	0.77	0.98
ECW1 (+200 mV)	1	0.24	0.75	0.68	0.44	0.93	0.09	0.09	0.08	0.24	0.836	0.74	0.99
ECW1 (+500 mV)	1	0.25	0.66	0.52	0.48	0.93	0.09	0.09	0.08	0.40	0.836	0.62	0.95
ECW1 (+800 mV)	1	0.26	0.47	0.36	0.54	0.92	0.09	0.09	0.08	0.56	0.836	0.51	0.82
ECW1 (+1200 mV)	1	0.24	0.19	0.19	0.68	0.93	0.09	0.09	0.08	0.73	0.836	0.38	0.68
ECW1 (+1400 mV)	1	0.23	0.17	0.17	0.71	0.93	0.09	0.09	0.08	0.75	0.836	0.37	0.68
ECW2 (−1800 mV)	1	0.10	0.62	0.61	0.61	0.97	0.09	0.09	0.08	0.31	0.836	0.69	0.95
ECW2 (+1200 mV)	1	0.12	0.10	0.10	0.82	0.97	0.09	0.09	0.08	0.82	0.836	0.31	0.31
ECW3 (−1800 mV)	1	0.08	0.69	0.67	0.59	0.97	0.09	0.09	0.08	0.25	0.836	0.74	0.96
ECW3 (+1400 mV)	1	0.12	0.09	0.08	0.83	0.97	0.09	0.09	0.08	0.84	0.836	0.30	0.59
EC1/Float (−) EC1T/A/G	2	0.18	0.70	0.62	0.50	0.95	0.14	0.04	0.12	0.26	0.836	0.67	0.99
EC1/Float (+) EC1C/A/G	2	0.18	0.15	0.15	0.75	0.95	0.09	0.04	0.08	0.77	0.836	0.25	0.67
EC1/LowE (−) EC1T/A/LE/G	2	0.12	0.67	0.45	0.55	0.97	0.11	0.03	0.23	0.32	0.836	0.55	0.99
EC1/LowE (+) EC1C/A/LE/G	2	0.12	0.14	0.11	0.78	0.97	0.09	0.03	0.09	0.80	0.836	0.19	0.68
EC1/Float/Float (−) EC1T/A/G/A/G	3	0.15	0.63	0.53	0.55	0.96	0.18	0.17	0.15	0.32	0.836	0.60	0.99
EC1/Float/Float (+) EC1C/A/G/A/G	3	0.14	0.14	0.13	0.78	0.96	0.09	0.17	0.08	0.79	0.836	0.20	0.67
EC1/Float/LowE (−) EC1T/A/G/A/LE/G	3	0.10	0.60	0.39	0.60	0.97	0.16	0.10	0.22	0.39	0.836	0.50	0.98
EC1/Float/LowE (+) EC1C/A/G/A/LE/G	3	0.10	0.13	0.10	0.81	0.97	0.09	0.10	0.09	0.82	0.836	0.16	0.67
EC1/LowE/LowE (−) EC1T/A/LE/G/A/LE/G	3	0.07	0.58	0.33	0.64	0.98	0.13	0.09	0.25	0.41	0.836	0.45	0.97
EC1/LowE/LowE (+) EC1C/A/LE/G/A/LE/G	3	0.07	0.12	0.08	0.83	0.98	0.09	0.09	0.09	0.83	0.836	0.14	0.67
EC2/LowE/LowE (−) EC2T/A/LE/G/A/LE/G	3	0.03	0.46	0.26	0.74	0.99	0.11	0.09	0.22	0.52	0.836	0.37	0.96
EC2/LowE/LowE (+) EC2C/A/LE/G/A/LE/G	3	0.03	0.07	0.04	0.89	0.99	0.09	0.09	0.08	0.87	0.836	0.10	0.30

Highest coloration level is at +1400 mV (ECW1 and ECW3) and +1200 mV (ECW2, PANI-PB multilayer). Corresponding transmittance spectra are given in Figures 15.16–15.21. Reflectance values of the ECWs have not been measured, but as the (absorbing) electrochromic coatings are located between two glass plates, the (low) reflectance values will be close to the values for float glass, and these are hence employed in the current calculations. The emissivity of the float glass was determined as $\varepsilon_{float} = 0.836$ by hemispherical directional reflectance measurements and was directly applied in the calculation of SF as the emissivity value in the SF calculations is with respect to the inside-facing surface (of the building) of the innermost glass pane, that is, normally a float glass, whereas $\varepsilon_{lowe} = 0.071$ and $\varepsilon_{silver} = 0.543$ were indirectly applied in the SF calculations through the calculations of the thermal conductance Λ. G = glass, LE = low emittance coating, A = air cavity, EC1 = ECW1 device between two glass plates, EC2 = ECW2 device between two glass plates (PANI-PB multilayer), T or C behind EC1 and EC2 denote transparent or coloured state, respectively.

in the coloured state, with 0.43 (ECW1), 0.61 (ECW2) and 0.59 (ECW3) in the transparent (bleached) state, respectively.

Incorporating the ECWs into two-layer and three-layer window pane configurations reduces the total solar energy throughput in the windows, for example, as seen in the T_{sol} and SF values, as several layers of glass and coatings will increase the total reflectance and absorbance. Note that some of the reflectance values $R_{vis,ext}$, $R_{vis,int}$ and R_{sol} may have errors due to parallel displacement of solar radiation through glass causing parts of the radiation to not enter the spectrophotometer detector during the measurements.

The CRF values are very high, that is, a very good colour rendering, for all the ECW configurations in the transparent (bleached) state, with CRF values ranging from 0.95 to 0.99, also including two-layer and three-layer window pane combinations. ECW2 has the lowest CRF value in the transparent state, that is, a CRF = 0.95. However, in the coloured state, the CRF values are substantially reduced for all the three different ECWs. ECW1 has CRF = 0.68, ECW2 has CRF = 0.31 and ECW3 has CRF = 0.59 in the coloured state. Note that especially ECW2 has a very low CRF (0.31) value. The large differences between the CRF values for the three ECWs in the coloured state may be deduced from comparing the transmittance spectra in the visible region between 380 and 780 nm (Figures 15.16–15.21).

Furthermore, based on values in Table 15.2, solar radiation glazing factor modulations are calculated for the three different ECWs and selected two-layer and three-layer window pane configurations with ECWs as given in Table 15.3. The modulation level is calculated by subtracting the solar radiation glazing factors for the same ECW at the high and low potentials given in Table 15.2, for example, as for ΔT_{sol}:

$$\Delta T_{sol} = T_{sol}(\text{bleached}) - T_{sol}(\text{coloured}) \qquad (15.111)$$

where the T_{sol} values in Table 15.2 are calculated from Eq. (15.14) and likewise for the other solar radiation glazing factors.

The ECW1, ECW2 and ECW3 devices have rather large solar radiation modulation abilities, for example, $\Delta T_{vis} = 0.61$ and $\Delta T_{sol} = 0.57$ for ECW1, $\Delta T_{vis} = 0.52$ and $\Delta T_{sol} = 0.51$ for ECW2 and $\Delta T_{vis} = 0.60$ and $\Delta T_{sol} = 0.59$ for ECW3, where the transmittance modulation is assumed to be due to absorbance regulation, that is, $\Delta A_{sol} = -0.57$ for ECW1, $\Delta A_{sol} = -0.51$ for ECW2 and $\Delta A_{sol} = -0.59$ for ECW3. Note that reflectance values of the ECWs have not been measured, but as the (absorbing) electrochromic coatings are located between two glass plates, the (low) reflectance values will be close to the values for float glass, and these are hence employed in the current calculations.

Although the SF modulations are lower than their solar transmission counterparts for the ECWs, the SF modulations are still quite high, that is, ΔSF = 0.42, ΔSF = 0.38 and ΔSF = 0.44 for ECW1, ECW2 and ECW3, respectively.

As mentioned earlier, applying the ECWs into two-layer and three-layer window pane configurations reduces the total solar energy throughput modulation in the windows, for example, as also seen in the ΔT_{sol} and ΔSF values, as several layers

Table 15.3 Calculated solar radiation glazing factor modulations for three different electrochromic windows (ECW) and selected two-layer and three-layer window pane configurations with ECWs.

Glass configuration	n	ΔT_{uv}	ΔT_{vis}	ΔT_{sol}	ΔSMPF	ΔSSPF	$\Delta R_{vis,ext}$	$\Delta R_{vis,int}$	ΔR_{sol}	ΔA_{sol}	ε	ΔSF	ΔCRF
ECW1 (−1800 mV) ECW1 (+1400 mV)	1	0.00	0.61	0.57	−0.28	0.00	0.00	0.00	0.00	−0.57	–	0.42	0.30
ECW2 (−1800 mV) ECW2 (+1200 mV)	1	−0.02	0.52	0.51	−0.21	0.00	0.00	0.00	0.00	−0.51	–	0.38	0.64
ECW3 (−1800 mV) ECW3 (+1400 mV)	1	−0.04	0.60	0.59	−0.24	0.00	0.00	0.00	0.00	−0.59	–	0.44	0.37
EC1/Float (−) EC1T/A/G EC1/Float (+) EC1C/A/G	2	0.00	0.55	0.47	−0.25	0.00	0.05	0.00	0.04	−0.51	–	0.42	0.32
EC1/LowE (−) EC1T/A/LE/G EC1/LowE (+) EC1C/A/LE/G	2	0.00	0.53	0.34	−0.23	0.00	0.02	0.00	0.14	−0.48	–	0.36	0.31
EC1/Float/Float (−) EC1T/A/G/A/G EC1/Float/Float (+) EC1C/A/G/A/G	3	0.01	0.49	0.40	−0.23	0.00	0.09	0.00	0.07	−0.47	–	0.40	0.32
EC1/Float/LowE (−) EC1T/A/G/A/LE/G EC1/Float/LowE (+) EC1C/A/G/A/LE/G	3	0.00	0.47	0.29	−0.21	0.00	0.07	0.00	0.13	−0.43	–	0.34	0.31
EC1/LowE/LowE (−) EC1T/A/LE/G/A/LE/G EC1/LowE/LowE (+) EC1C/A/LE/G/A/LE/G	3	0.00	0.46	0.25	−0.19	0.00	0.04	0.00	0.16	−0.42	–	0.31	0.30
EC2/LowE/LowE (−) EC2T/A/LE/G/A/LE/G EC2/LowE/LowE (+) EC2C/A/LE/G/A/LE/G	3	0.00	0.39	0.22	−0.15	0.00	0.02	0.00	0.14	−0.35	–	0.27	0.66

The modulation level is calculated by subtracting the solar radiation glazing factors for the same ECW at the high and low potentials given in Table 15.2 (e.g. Eq. 15.111).

of glass and coatings will increase the total reflectance and absorbance, that is, less solar radiation is left for the ECWs to modulate (regulate). That is, the solar radiation modulation by an ECW will decrease with the number of glass panes and low emittance coatings added to the total window configuration. For example, the single glass ECW1 values $\Delta T_{vis} = 0.61$ and $\Delta T_{sol} = 0.57$ are decreased to the two-layer EC1/float window pane values $\Delta T_{vis} = 0.55$ and $\Delta T_{sol} = 0.47$ and three-layer EC1/float/float window pane values $\Delta T_{vis} = 0.49$ and $\Delta T_{sol} = 0.40$. Nevertheless, even if these values are lower, they still represent a large solar modulation by the ECW.

Employing one or two low emittance coatings in the window configurations will decrease, for example, ΔT_{vis} and ΔT_{sol} even further, where especially the ΔT_{sol} value is decreased. For example, the three-layer EC1/LowE/LowE window pane has $\Delta T_{vis} = 0.46$ and $\Delta T_{sol} = 0.25$. For further comparisons of solar radiation glazing factors for various ECW pane configurations, refer to Table 15.3.

It is observed that the ΔSSPF modulation is more or less insignificant for the ECW glass configurations given in Table 15.3, as the change in ECW coloration state at low wavelengths is almost negligible due to the highly increasing absorption in the glass system from 400 nm and below (see Figures 15.16–15.21). It should be noted that the ECWs referred to were constructed in order to achieve the highest possible solar energy regulation, with no optimization with respect to SMPF and SSPF values. Direct investigation in this area may therefore improve the SMPF regulation in ECWs substantially (but not SSPF due to the already high absorption in glass as already mentioned). For further details and visual comparison of SMPF and SSPF values with T_{vis} values for ECWs, see Jelle *et al.* [34].

The application of glass panes with low emittance coatings together with ECWs decreases the solar modulation abilities of the ECWs, as there is much less solar radiation to be regulated due to the large reflectance in the NIR region caused by the low emittance coating (Figure 15.10). Nevertheless, the combined ECW low-emittance window pane alternatives represent appropriate and useful window configurations as the low emittance coating reduces the heat loss (infrared radiation) through windows, and hence examples are given in Tables 15.2 and 15.3. Applying glass panes with dark silver coatings together with ECWs decreases the solar modulation abilities of the ECWs as well, as there is much less solar radiation to be regulated due to the large absorbance and reflectance in both the visible and the NIR region caused by the dark silver coating (Figure 15.13). However, as the combined ECW dark silver coated window pane alternatives normally do not represent appropriate and useful window configurations, no such examples are given in Tables 15.2 and 15.3.

Figure 15.24 shows an example of applying measured transmittance and reflectance spectra for the ECW1 single glass pane structure (Eqs. 15.3 and 15.4) for calculating the transmittance spectra through two-layer and three-layer ECW1 window panes according to Eqs. (15.5) and (15.8), respectively. It is observed that the addition of extra glass panes decreases the transmittance in the whole solar spectral range. Thus, more glass panes in an ECW decrease the amount of daylight and solar radiation to be controlled and regulated. The two-layer and three-layer window pane transmittance spectra have been calculated with the full spectral resolution and wavelength range as applied in the spectrophotometric measurements, in contrast with the considerably lower resolution and narrower wavelength range applied in calculating the solar radiation glazing factors like e.g. T_{sol} (Table A.3).

Thus, the ECWs may contribute to elegant, flexible glazing systems with dynamical control of the solar radiation with regard to daylight, solar energy aspects and

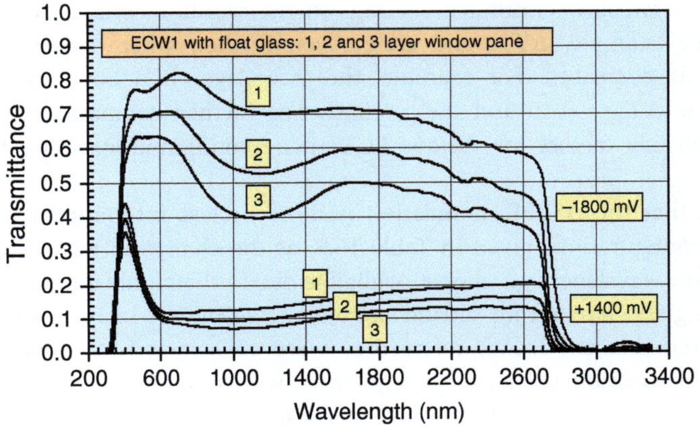

Figure 15.24 Transmittance versus wavelength for the whole solar spectrum measured for an electrochromic window ECW1 and calculated for two-layer and three-layer ECW1 float glass window pane combinations (Eqs. 15.5 and 15.8) at two different applied potentials. Highest coloration level is at +1400 mV.

protection of materials inside buildings. The ECWs may readily be characterized by spectroscopic measurements and subsequent calculations of the solar radiation glazing factors.

15.7.6 Miscellaneous Other Electrochromic Properties

15.7.6.1 General

The solar radiation glazing factors represent some of the most important properties for ECWs. In addition, there are also miscellaneous other electrochromic properties which are crucial for the performance of ECWs. Clearly, the production, transport, installation, operation, maintenance and demolition of ECWs should not pose any hazard to human beings, nearby surroundings and the environment in general.

15.7.6.2 Colour Coordinates

Colour measurements involve the determination of among others the colour coordinates L, a and b in a three-dimensional colour space, where higher positive values denote a more light colour (on a white–black scale, i.e. higher luminance), a more reddish colour (on a red–green scale) and a more yellowish colour (on a yellow–blue scale), respectively [128]. Furthermore, a higher gloss value denotes a more glossy (shiny) material. One may also calculate a colour difference ΔE as:

$$\Delta E = \sqrt{(\Delta L)^2 + (\Delta a)^2 + (\Delta b)^2} \tag{15.112}$$

where $\Delta L = L_2 - L_1$, $\Delta a = a_2 - a_1$ and $\Delta b = b_2 - b_1$, where (L_2, a_2, b_2) and (L_1, a_1, b_1) represent two colours or colour (chromaticity) coordinates in the three-dimensional CIELab colour space. Thus, ΔE represents the 'distance' between two colours, for example, between a non-aged and an aged condition, or alternatively a colour difference for an ECW changing colour between two states. However, these colour coordinates have not been measured for the treated ECWs here, as the solar radiation glazing factors including the colour rendering factor (CRF) are the important factors for window applications, that is, also requiring a transparent state with as low colour distortion as possible. For colour displays, the L, a and b coordinates would be interesting, though.

15.7.6.3 Electrochromic Efficiency

One characteristic quantity for ECWs is the bleaching and colouring efficiency at a specific wavelength λ, the electrochromic efficiency η defined by [37, 143, 156, 174, 203, 205–209]:

$$\eta(\lambda) = \Delta\text{OD}(\lambda)/Q = \log_{10}(T_{\text{bleach}}(\lambda)/T_{\text{col}}(\lambda))/Q \tag{15.113}$$

where ΔOD is the change in optical density and Q is the consumed electrical charge density stepping from a bleached state (with transmittance T_{bleach}) to a coloured state (with transmittance T_{col}) or vice versa. The optical density (OD) is defined as the absorbance A' written on a logarithmic form by [37, 143]:

$$\text{OD} = A' = \log_{10}\left(\frac{1}{T}\right) = \log_{10}\left(\frac{I_0}{I}\right) = \alpha\log_{10}(e)x = \alpha' x \tag{15.114}$$

which is deduced from the well-known Beer–Lambert law given by [141–143, 210–213]:

$$I = I_0 e^{-\alpha x} \tag{15.115}$$

where the transmittance T is given by:

$$T = I/I_0 \tag{15.116}$$

where the transmitted radiation intensity I decreases exponentially with the penetration length or depth x, I_0 is the incident radiation intensity and α and α' denote absorption coefficients depending on what form is used.

In order to account for the whole visible solar radiation spectral region and the whole solar radiation spectral region, visible and solar spectral electrochromic efficiencies η_{vis} and η_{sol} may be respectively defined by:

$$\eta_{\text{vis}} = \Delta\text{OD}_{\text{vis}}/Q = \log_{10}(T_{\text{vis}}(\text{bleached})/T_{\text{vis}}(\text{coloured}))/Q \tag{15.117}$$

$$\eta_{\text{sol}} = \log_{10}(T_{\text{sol}}(\text{bleached})/T_{\text{sol}}(\text{coloured}))/Q \tag{15.118}$$

where T_{vis} and T_{sol} denote the visible solar transmittance (Eq. 15.13) and the solar transmittance (Eq. 15.14), respectively. The term 'optical density' is not used in the expression for η_{sol} as 'optical' normally refers to the visible region.

Electrochromic efficiencies have been calculated from measurements on the ECW3 configuration (transmittance spectra in Figure 15.20), yielding the following bleaching and colouring efficiencies [174]: $\eta_{\text{bleach}}(1000\,\text{nm}) = 200\,\text{cm}^2\,\text{C}^{-1}$, $\eta_{\text{col}}(1000\,\text{nm}) = 120\,\text{cm}^2\,\text{C}^{-1}$, $\eta_{\text{bleach}}(550\,\text{nm}) = 120\,\text{cm}^2\,\text{C}^{-1}$ and

$\eta_{col}(550\,nm) = 70\,cm^2\,C^{-1}$. The observed longer colouring time, compared to the bleaching time, is reflected in the lower colouring efficiency versus the bleaching efficiency [174]. Assuming approximately the same consumed electrical charge densities as utilized in the aforementioned calculations (see [174] for details), and T_{vis}(bleached) = 0.69, T_{vis}(coloured) = 0.09, T_{sol}(bleached) = 0.67 and T_{sol}(coloured) = 0.08 for ECW3 given in Table 15.2, the following approximate electrochromic efficiencies are obtained: η_{vis}(bleached) = $150\,cm^2\,C^{-1}$, η_{vis}(coloured) = $90\,cm^2\,C^{-1}$, η_{sol}(bleached) = $160\,cm^2\,C^{-1}$ and η_{sol}(coloured) = $90\,cm^2\,C^{-1}$.

15.7.6.4 **Energy Consumption, Memory and Switching Time**

The energy consumption in ECWs represents an important value with respect to the energy savings and overall performance of ECWs. Typically for the ECWs discussed in this work, only a small charge density of about $3\,mC\,cm^{-2}$, corresponding to a low energy consumption of about $5\,mWh\,m^{-2}$, is required for either the colouring or the bleaching process, that is, about $0.01\,Wh\,m^{-2}$ for one whole cycle including both bleaching and colouring [35, 174, 175].

The memory of ECWs, that is, the ability to maintain the bleached and coloured states without any applied voltage, is another important property. However, the lower the required energy consumption for intended bleaching and colouring, the less important is the memory of the ECWs. Jelle and Hagen [37, 174] present memory experiments for ECWs based on PANI and WO_3 and based on PANI, PB and WO_3.

Switching times, that is, bleaching and colouring times, are also crucial properties for ECWs. The shorter switching times, the better. Nevertheless, for ECW applications, switching times of some minutes may be acceptable. Typical switching times for the ECWs discussed in this work are in the order of half a minute, although repeated cycling may increase the switching time to several minutes [32, 35, 174]. However, for display applications, switching times considerably shorter than a second may be required.

15.7.6.5 **Durability**

A satisfactory durability of ECWs is crucial. The durability of ECWs comprises the following aspects:

- Climate exposure durability
- Cycling durability
- Elapsed time durability.

Climate exposure durability covers resistance towards various climate exposure factors the ECWs may be subjected to during their lifetime, including the production, transport, installation and operational period until planned demolition. In general, these climate exposure factors are solar radiation (i.e. UV, VIS and NIR radiation), ambient IR heat radiation (the resulting elevated temperature increases the rate of chemical degradation reactions and also the rate of growth of rot and fungus up to limiting temperatures), high and low temperatures, temperature changes/cycles (relative temperature movements between different materials, number of freezing point passes during freezing/thawing), water (e.g. moisture, relative air humidity, rain (precipitation) and wind-driven rain), physical strains (e.g. snow loads), wind, erosion (also from the aforementioned factors), pollutions (e.g. gases and particles in air), microorganisms, oxygen and time (determining the effect for all the aforementioned factors to work). For a general overview of these climate strains and the role of accelerated climate ageing of building materials, components and structures in the laboratory, including new materials and solutions, see Jelle *et al.* [127] and Jelle [128, 129]. Furthermore, performing a robustness assessment of these materials and components may also be found to be beneficial [214].

Cycling durability covers the ability of the ECWs to be cycled (or stepped) back and forth between various coloration levels, that is, between highly transparent and very dark coloured states during bleaching and colouring of the ECWs, without any significant degradation. The highest number of complete cycles (from transparent to coloured and then back to transparent state) for typical ECWs based on PANI, PB and WO_3 as reported herein was 3745 cycles (52 days), then with a significantly increased switching time, especially the colouring time [35]. Note that in the available literature, there are also reported much longer cycling lifetimes, for example, 10^7 cycles at 22 °C and 5×10^6 cycles at 50 °C without any failure detection for an ECW based on WO_3 and a solid polymer electrolyte (PAMPS) [43].

Elapsed time durability means the durability performance of ECWs with respect to elapsed time independent of any external strains, that is, including the durability during no climate exposure and no applied voltage (no cycling), often termed the shelf-life of ECWs and other devices.

The studies by Czanderna *et al.* [122], Lampert [43], Lampert *et al.* [48], Nagai *et al.* [123], Tajima *et al.* [124] and Tracy *et al.* [125] investigate miscellaneous ageing and durability issues concerning ECWs, where also general considerations and several specific ECW configurations are treated.

15.7.6.6 **Electrochromic Window Configuration**

To recapitulate, an ECW normally consists in principle of an electrochromic (EC) material or coating, an ionic conductor (IC) and a counter electrode (CE) interposed between two glass panes each with a transparent conductor (TC). By applying a voltage over the two transparent conductors, ions will go through the ionic conductor between the electrochromic coating and the counter electrode and thereby change the colour of the electrochromic layer. The counter electrode acts as an ion storage, supplying the necessary ions for the electrochromic reaction, which, via the ionic conductor, are injected into or withdrawn from the electrochromic layer depending on the polarity of the applied voltage. The counter electrode may also have electrochromic properties complementary to the electrochromic coating (CE = EC*), thus enhancing the colour changes [143].

Such a synergetic complementary electrochromic material pair is sometimes called a *rocking chair* configuration [215–217].

Naturally, the properties of all these layers and their interactions have to be optimized. The ionic conductor should have a low electrical conductivity and a high ionic conductivity in order to prevent loss of electrons (short circuit) and to obtain satisfactory fast bleaching and colouring times. Transparent ionic conductors are required for use in ECWs, and solid-state ionic conductors are preferred before liquid ones. As the name indicates, the transparent conductor has to provide both a high transparency and a high electrical conductivity, that is, a low electrical surface resistance. A lower electrical surface resistance for a specific material is obtained by increasing the thickness of the transparent conductor, which unfortunately in turn decreases the transparency. An obvious requirement for window applications is that the electrochromic material has to have a clear and transparent state, which is not required in display devices. In addition to the solar energy regulation of ECWs, one may also exploit the visible colour changes in an architectural way, that is, the so-called *fancy windows* as introduced by Jelle and Hagen [35, 174].

15.7.6.7 **Reflectance-Induced Limitations**

Reflection of solar radiation from each air/glass interface places limitations to how transparent a glass and window pane configuration may become and hence also limits the total solar radiation modulation potential of ECWs. The solar radiation is reflected at every boundary and is dependent upon the refractive indices n_1 and n_2 for the ambient and substrate medium, respectively. For normal incidence and in a spectral range where the absorbance is low, the reflectance is given by the following expression [141, 142, 211]:

$$R \approx \frac{(n_2 - n_1)^2}{(n_2 + n_1)^2} \tag{15.119}$$

Setting $n_1 = 1$ for air and $n_2 = 1.5$ for glass (typical), we obtain $R = 0.04$, that is, 4% of the radiation is reflected at each air/glass and glass/air interface. Hence, the maximum transmittance through a single glass pane (two air/glass interfaces) becomes 92%, and in fact as low as about 84% for a two-layer window pane configuration (with air between the two glass panes). For normal incidence and low absorbance, the transmittance may be written as [141]:

$$T \approx \frac{4n_2 n_1}{(n_2 + n_1)^2} \tag{15.120}$$

Applying $n_1 = 1$ for air and $n_2 = 1.5$ for glass again and inserting in Eq. (15.120), a transmittance of 96% is obtained, which is in agreement with the results from Eq. (15.119) (assuming normal incidence and low absorbance). These results clearly demonstrate the limitations to achieve as large transmittance (and transparency) as possible in windows in general and in ECWs in particular, especially when applying several glass panes. In addition, the several different material layers with somewhat different refractive indices, and the several back and forth reflections from each material interface (especially the air/glass interfaces), will result in deviations from the simple calculated examples in the above.

15.8 Commercial Electrochromic Windows and the Path Ahead

Commercial ECWs are already available from several manufacturers, comprehensively reviewed by Baetens *et al.* [1] and Jelle *et al.* [2], with various properties among which several important solar radiation glazing factors are being noted. Materials exhibiting both photovoltaic and electrochromic properties may be envisioned. However, a window still needs a transparent (in some cases translucent) state, and when in the transparent state, a photovoltaic window cannot produce electricity from the visible part of the solar spectrum, as one single photon obviously cannot be used simultaneously for both electricity generation and providing daylight. Nevertheless, for the UV and NIR parts, the solar radiation may still in principle be exploited in the transparent state. In the coloured state, all the solar radiation including the visible part may be utilized.

It is unfortunately often hard to obtain all the desired information concerning the commercial product properties from all the manufacturers [2, 99]. In general, many property values are often not available at the manufacturers' websites or other open information channels, including the solar radiation glazing factors. Hopefully, the addressing of this fact in these studies could act as an incentive for the manufacturers to state all the important properties of their products at their websites and other information channels, and also as an incentive and reminder for the contractors, consumers and users to demand these values from the respective manufacturers.

In addition to the solar radiation glazing factors, and with regard to increased application of ECWs in the years ahead, it should be stressed that the durability of commercial ECWs is essential. The durability of ECWs should therefore be rigorously tested and characterized with respect to (i) climate exposure durability, (ii) cycling durability and (iii) elapsed time durability. Hence, the durability of ECWs should also be stated by the manufacturers.

As discussed earlier herein (see e.g. Figures 15.4–15.6) and noted by Jelle *et al.* [2], as today's commercial and almost all research ECWs are solar radiation absorbing ECWs, research carried out on reflecting ECWs may pay off as these have a large potential in solar energy control. Reflecting ECWs avoid any heating problems that absorbing ECWs may be subjected to. Furthermore, reflecting ECWs may in principle regulate more of the solar energy as these windows block off the solar radiation by reflecting it back towards the outside, while absorbing ECWs reemit the absorbed solar radiation in all directions both towards the outside and towards the inside, the latter process hence decreasing the overall solar energy regulation for absorbing ECWs.

Finally, a lowered ECW cost and increased awareness of potential savings and enhanced indoor comfort would contribute to facilitate the application of ECWs in the buildings of tomorrow, also with respect to retrofitting of existing buildings.

15.9 Increased Application of Solar Radiation Glazing Factors

The solar radiation glazing factors represent a comprehensive, versatile and valuable characterization method of glass products, also including various two-layer and three-layer window pane structures, giving a set of specific and crucial parameters for the glass products with regard to transmittance, absorbance and reflectance of solar radiation in selected wavelength regions and in addition to the protection level of materials and human skin. Hence, in order to be able to evaluate and choose the most appropriate glass material or product for a specific building application, the desire and demand for available solar radiation glazing factors are believed to increase. This is important with respect to daylight, solar energy aspects and protection of materials inside buildings. The dynamic solar radiation control enabled by electrochromic windows or other smart windows may readily and beneficially be characterized by spectroscopic measurements and subsequent calculations of the solar radiation glazing factors.

15.10 Conclusions

Daylight and solar energy in buildings may be dynamically controlled in an efficient way by applying electrochromic windows. It is of major importance to be able to quantify this solar radiation control, which may readily be obtained through characterization by a number of solar radiation glazing factors, that is, ultraviolet solar transmittance, visible solar transmittance, solar transmittance, solar material protection factor, solar skin protection factor, external visible solar reflectance, internal visible solar reflectance, solar reflectance, solar absorbance, emissivity, solar factor and colour rendering factor. These factors may readily be compared for different glass fabrications in order to select the most appropriate glass material for the specific building application. Spectroscopic measurements and corresponding calculations of the solar radiation glazing factors were performed on three different electrochromic window devices at various coloration levels, and for comparison reasons various glass materials and several two-layer and three-layer window pane configurations were also treated.

Acknowledgements

This work has been supported by the Research Council of Norway and several partners through the SINTEF and NTNU research projects *'Environmentally Favourable Energy Use in Buildings'*, *'Robust Envelope Construction Details for Buildings of the 21st Century'* (ROBUST), *'Improved Window Technologies for Energy Efficient Buildings'* (EffWin) and *'The Research Centre on Zero Emission Buildings'* (ZEB).

15.A
Appendix: Tables for Calculation of Solar Radiation Glazing Factors

Table A.1 Normalized relative spectral distribution of the ultraviolet part of global solar radiation for the calculation of the ultraviolet solar transmittance (T_{uv}).

Wavelength, λ (nm)	$S_\lambda \Delta\lambda$
300	0
305	0.001859
310	0.007665
315	0.017961
320	0.029732
325	0.042466
330	0.062108
335	0.065462
340	0.071020
345	0.073326
350	0.079330
355	0.082894
360	0.087039
365	0.097963
370	0.108987
375	0.113837
380	0.058351

The table has been drawn up with relative values so that $\Sigma S_\lambda \Delta\lambda = 1$ for the wavelength range 300–380 nm. Table values are given in ISO 9050:2003(E) [148], based upon air mass 1.5 values [182].

Table A.2 Normalized relative spectral distribution $D_\lambda V(\lambda)\Delta\lambda$ for the calculation of the visible solar transmittance (T_{vis}).

Wavelength, λ (nm)	$D_\lambda V(\lambda)\Delta\lambda$
380	0
390	0.000005
400	0.000030
410	0.000103
420	0.000352
430	0.000948
440	0.002274
450	0.004192
460	0.006663
470	0.009850
480	0.015189
490	0.021336
500	0.033491
510	0.051393
520	0.070523
530	0.087990
540	0.094427
550	0.098077
560	0.094306
570	0.086891
580	0.078994
590	0.063306
600	0.053542
610	0.042491
620	0.031502
630	0.020812
640	0.013810
650	0.008070
660	0.004612
670	0.002485
680	0.001255
690	0.000536
700	0.000276
710	0.000146
720	0.000057
730	0.000035
740	0.000021
750	0.000008
760	0.000001
770	0.000000
780	0.000000

The table has been drawn up with relative values so that $\Sigma D_\lambda V(\lambda)\Delta\lambda = 1$ for the wavelength range 380–780 nm. Table values are given in ISO 9050:2003(E) [148].

Table A.3 Normalized relative spectral distribution of global solar radiation for the calculation of the solar transmittance (T_{sol}).

Wavelength, λ (nm)	$S_\lambda \Delta\lambda$
300	0
305	0.000057
310	0.000236
315	0.000554
320	0.000916
325	0.001309
330	0.001914
335	0.002018
340	0.002189
345	0.002260
350	0.002445
355	0.002555
360	0.002683
365	0.003020
370	0.003359
375	0.003509
380	0.003600
385	0.003529
390	0.003551
395	0.004294
400	0.007812
410	0.011638
420	0.011877
430	0.011347
440	0.013246
450	0.015343
460	0.016166
470	0.016178
480	0.016402
490	0.015794
500	0.015801
510	0.015973
520	0.015357
530	0.015867
540	0.015827
550	0.015844
560	0.015590
570	0.015256
580	0.014745
590	0.014330
600	0.014663
610	0.015030
620	0.014859
630	0.014622
640	0.014526
650	0.014445
660	0.014313
670	0.014023

Table A.3 (*Continued*)

Wavelength, λ (nm)	$S_\lambda \Delta\lambda$
680	0.012838
690	0.011788
700	0.012453
710	0.012798
720	0.010589
730	0.011233
740	0.012175
750	0.012181
760	0.009515
770	0.010479
780	0.011381
790	0.011262
800	0.028718
850	0.048240
900	0.040297
950	0.021384
1000	0.036097
1050	0.034110
1100	0.018861
1150	0.013228
1200	0.022551
1250	0.023376
1300	0.017756
1350	0.003743
1400	0.000741
1450	0.003792
1500	0.009693
1550	0.013693
1600	0.012203
1650	0.010615
1700	0.007256
1750	0.007183
1800	0.002157
1850	0.000398
1900	0.000082
1950	0.001087
2000	0.003024
2050	0.003988
2100	0.004229
2150	0.004142
2200	0.003690
2250	0.003592
2300	0.003436
2350	0.003163
2400	0.002233
2450	0.001202
2500	0.000475

The table has been drawn up with relative values so that $\Sigma S_\lambda \Delta\lambda = 1$ for the wavelength range 300–2500 nm. Table values are given in ISO 9050:2003(E) [148], based upon air mass 1.5 values [182].

Table A.4 Normalized relative spectral distribution factors for the calculation of the solar material protection factor (SMPF).

Wavelength, λ (nm)	$C_\lambda S_\lambda \Delta\lambda$
300	0
305	0.001003
310	0.003896
315	0.008597
320	0.013402
325	0.018028
330	0.024831
335	0.024648
340	0.025183
345	0.024487
350	0.024949
355	0.024551
360	0.024278
365	0.025734
370	0.026962
375	0.026522
380	0.025624
385	0.023656
390	0.022418
395	0.025529
400	0.043742
410	0.057799
420	0.052317
430	0.044328
440	0.045896
450	0.047150
460	0.044062
470	0.039108
480	0.035167
490	0.030034
500	0.026650
510	0.023893
520	0.020373
530	0.018671
540	0.016517
550	0.014665
560	0.012799
570	0.011108
580	0.009522
590	0.008208
600	0.003695

The table has been drawn up with relative values so that $\Sigma C_\lambda S_\lambda \Delta\lambda = 1$ for the wavelength range 300–600 nm. Table values are given in ISO 9050:2003(E) [148], based upon air mass 1.5 values [182].

Table A.5 Normalized relative spectral distribution factors for the calculation of the solar skin protection factor (SSPF).

Wavelength, λ (nm)	$E_\lambda S_\lambda \Delta\lambda$
300	0
305	0.168176
310	0.230555
315	0.187429
320	0.102699
325	0.050895
330	0.034134
335	0.030432
340	0.027729
345	0.024094
350	0.021930
355	0.019298
360	0.017028
365	0.016157
370	0.015108
375	0.013298
380	0.011471
385	0.009440
390	0.008009
395	0.008165
400	0.003953

The table has been drawn up with relative values so that $\Sigma E_\lambda S_\lambda \Delta\lambda = 1$ for the wavelength range 300–400 nm. Table values are given in ISO 9050:2003(E) [148], based upon air mass 1.5 values [182].

Table A.6 Wavelengths for determining normal reflectance (R_n) for the emissivity (ε) determination in Eq. (15.22) at 283 K [177, 178].

No. *i*	Wavelength, λ_i (μm)
1	5.5
2	6.7
3	7.4
4	8.1
5	8.6
6	9.2
7	9.7
8	10.2
9	10.7
10	11.3
11	11.8
12	12.4
13	12.9
14	13.5
15	14.2
16	14.8
17	15.6
18	16.3
19	17.2
20	18.1
21	19.2
22	20.3
23	21.7
24	23.3
25	25.2
26	27.7
27	30.9
28	35.7
29	43.9
30	50.0

Table A.7 Corresponding correction coefficient ($\varepsilon/\varepsilon_n$) and normal emissivity (ε_n) values, for determination of corrected emissivity (ε) values according to Eq. (15.23), that is, $\varepsilon = c_{corr} \cdot \varepsilon_n = (\varepsilon/\varepsilon_n) \cdot \varepsilon_n$.

Normal emissivity, ε_n	0.03	0.05	0.1	0.2	0.3	0.4	0.5	0.6	0.7	0.8	0.89
Correction coefficient, $\varepsilon/\varepsilon_n$	1.22	1.18	1.14	1.10	1.06	1.03	1.00	0.98	0.96	0.95	0.94

Other values may be obtained with sufficient accuracy by linear interpolation or extrapolation [177, 178].

Table A.8 Gas properties required for calculation of the Nusselt number (Nu) (Eq. 15.56), which is part of the calculation of the thermal conductance (Λ) (Eq. 15.52) [177].

Gas	Temperature, θ (°C)	Mass density, ρ (kg m^{-3})	Dynamic viscosity, μ (10^{-5} kg m^{-1} s^{-1})	Thermal conductivity, κ (10^{-2} W m^{-1} K^{-1})	Specific heat, c_m (10^3 J kg^{-1} K^{-1})
Air	−10	1.326	1.661	2.336	1.008
	0	1.277	1.711	2.416	
	+10	1.232	1.761	2.496	
	+20	1.189	1.811	2.576	
Argon (Ar)	−10	1.829	2.038	1.584	0.519
	0	1.762	2.101	1.634	
	+10	1.699	2.164	1.684	
	+20	1.640	2.228	1.734	
SF_6	−10	6.844	1.383	1.119	0.614
	0	6.602	1.421	1.197	
	+10	6.360	1.459	1.275	
	+20	6.118	1.497	1.354	
Krypton (Kr)	−10	3.832	2.260	0.842	0.245
	0	3.690	2.330	0.870	
	+10	3.560	2.400	0.900	
	+20	3.430	2.470	0.926	

Equation 15.31 gives the relationship between the absolute mean temperature T_m in K and the mean temperature θ in °C: $\theta = (T_m - 273.15\,\text{K})\,°\text{C/K}$.

Table A.9 Values of the constants *A* and *n*, depending on the space inclination, for calculation of the Nusselt number (Nu), Eq. (15.56) [177].

Space inclination	*A*	*n*
Vertical	0.035	0.38
45°	0.10	0.31
Horizontal	0.16	0.28

Table A.10 Values of $U^*_{r,i}$, $V^*_{r,i}$ and $W^*_{r,i}$ for the test colours lighted by the standard illuminant D65, applicable in calculation of the colour rendering factor (CRF) [190].

Test colour number, *i*	$U^*_{r,i}$	$V^*_{r,i}$	$W^*_{r,i}$
1	31.92	8.41	60.48
2	15.22	23.76	59.73
3	−8.34	36.29	61.08
4	−33.29	18.64	60.25
5	−26.82	−6.55	61.41
6	−18.80	−28.80	60.52
7	9.77	−26.50	60.14
8	28.78	−16.24	61.83

Table A.11 Relative spectral energy distribution of illuminant D65 for wavelengths between 380 and 780 nm normalized to the value of 100 at 560 nm, applicable in calculation of the colour rendering factor (CRF) [190].

Wavelength, λ (nm)	$(d\Phi/d\lambda)\Delta\lambda$
380	50.0
390	54.6
400	82.8
410	91.5
420	93.4
430	86.7
440	104.9
450	117.0
460	117.8
470	114.9
480	115.9
490	108.8
500	109.4
510	107.8
520	104.8
530	107.7
540	104.4
550	104.0
560	100.0
570	96.3
580	95.8
590	88.7
600	90.0
610	89.6
620	87.7
630	83.3
640	83.7
650	80.0
660	80.2
670	82.3
680	78.3
690	69.7
700	71.6
710	74.3
720	61.6
730	69.9
740	75.1
750	63.6
760	46.4
770	66.8
780	63.4

Table A.12 Spectral reflectance $\beta_i(\lambda)$ of the eight test colours (i from 1 to 8) for calculation of the colour rendering factor (CRF) [190].

λ (nm)	Test colour number, i							
	1	2	3	4	5	6	7	8
380	0.219	0.070	0.065	0.074	0.295	0.151	0.378	0.104
390	0.252	0.089	0.070	0.093	0.310	0.265	0.524	0.170
400	0.256	0.111	0.073	0.116	0.313	0.410	0.551	0.319
410	0.252	0.118	0.074	0.124	0.319	0.492	0.559	0.462
420	0.244	0.121	0.074	0.128	0.326	0.517	0.561	0.490
430	0.237	0.122	0.073	0.135	0.334	0.531	0.556	0.482
440	0.230	0.123	0.073	0.144	0.346	0.544	0.544	0.462
450	0.225	0.127	0.074	0.161	0.360	0.556	0.522	0.439
460	0.220	0.131	0.077	0.186	0.381	0.554	0.488	0.413
470	0.216	0.138	0.085	0.229	0.403	0.541	0.448	0.382
480	0.214	0.150	0.109	0.281	0.415	0.519	0.408	0.352
490	0.216	0.174	0.148	0.332	0.419	0.488	0.363	0.325
500	0.223	0.207	0.198	0.370	0.413	0.450	0.324	0.299
510	0.226	0.242	0.241	0.390	0.403	0.414	0.301	0.283
520	0.225	0.260	0.278	0.395	0.389	0.377	0.283	0.270
530	0.227	0.267	0.339	0.385	0.372	0.341	0.265	0.256
540	0.236	0.272	0.392	0.367	0.353	0.309	0.257	0.250
550	0.253	0.282	0.400	0.341	0.331	0.279	0.259	0.254
560	0.272	0.299	0.380	0.312	0.308	0.253	0.260	0.264
570	0.298	0.322	0.349	0.280	0.284	0.234	0.256	0.272
580	0.341	0.335	0.315	0.247	0.260	0.225	0.254	0.278
590	0.390	0.341	0.285	0.214	0.232	0.221	0.270	0.295
600	0.424	0.342	0.264	0.185	0.210	0.220	0.302	0.348
610	0.442	0.342	0.252	0.169	0.194	0.220	0.344	0.434
620	0.450	0.341	0.241	0.160	0.185	0.223	0.377	0.528
630	0.451	0.339	0.229	0.154	0.180	0.233	0.400	0.604
640	0.451	0.338	0.220	0.151	0.176	0.244	0.420	0.648
650	0.450	0.336	0.216	0.148	0.175	0.258	0.438	0.676
660	0.451	0.334	0.219	0.148	0.175	0.268	0.452	0.693
670	0.453	0.332	0.230	0.151	0.180	0.278	0.462	0.705
680	0.455	0.331	0.251	0.158	0.186	0.283	0.468	0.712
690	0.458	0.329	0.288	0.165	0.192	0.291	0.473	0.717
700	0.462	0.328	0.340	0.170	0.199	0.302	0.483	0.721
710	0.464	0.326	0.390	0.170	0.199	0.325	0.496	0.719
720	0.466	0.324	0.431	0.166	0.196	0.351	0.511	0.725
730	0.466	0.324	0.460	0.164	0.195	0.376	0.525	0.729
740	0.467	0.322	0.481	0.168	0.197	0.401	0.539	0.730
750	0.467	0.320	0.493	0.177	0.203	0.425	0.553	0.730
760	0.467	0.316	0.500	0.185	0.208	0.447	0.565	0.730
770	0.467	0.315	0.505	0.192	0.215	0.469	0.575	0.730
780	0.467	0.314	0.516	0.197	0.219	0.485	0.581	0.730

Table A.13 Abridged set of spectral tristimulus values $\bar{x}(\lambda), \bar{y}(\lambda)$ and $\bar{z}(\lambda)$ for the CIE 1931 standard colorimetric observer, for wavelengths between 380 and 780 nm at 10 nm intervals, applicable in calculation of the colour rendering factor (CRF) [190].

Wavelength, λ (nm)	$\bar{x}(\lambda)$	$\bar{y}(\lambda)$	$\bar{z}(\lambda)$
380	0.0014	0.0000	0.0065
390	0.0042	0.0001	0.0201
400	0.0143	0.0004	0.0679
410	0.0435	0.0012	0.2074
420	0.1344	0.0040	0.6456
430	0.2839	0.0116	1.3856
440	0.3483	0.0230	1.7471
450	0.3362	0.0380	1.7721
460	0.2908	0.0600	1.6692
470	0.1954	0.0910	1.2876
480	0.0956	0.1390	0.8130
490	0.0320	0.2080	0.4652
500	0.0049	0.3230	0.2720
510	0.0093	0.5030	0.1582
520	0.0633	0.7100	0.0782
530	0.1655	0.8620	0.0422
540	0.2904	0.9540	0.0203
550	0.4334	0.9950	0.0087
560	0.5945	0.9950	0.0039
570	0.7621	0.9520	0.0021
580	0.9163	0.8700	0.0017
590	1.0263	0.7570	0.0011
600	1.0622	0.6310	0.0008
610	1.0026	0.5030	0.0003
620	0.8544	0.3810	0.0002
630	0.6424	0.2650	0.0000
640	0.4479	0.1750	0.0000
650	0.2835	0.1070	0.0000
660	0.1649	0.0610	0.0000
670	0.0874	0.0320	0.0000
680	0.0468	0.0170	0.0000
690	0.0227	0.0082	0.0000
700	0.0114	0.0041	0.0000
710	0.0058	0.0021	0.0000
720	0.0029	0.0010	0.0000
730	0.0014	0.0005	0.0000
740	0.0007	0.0002	0.0000
750	0.0003	0.0001	0.0000
760	0.0002	0.0001	0.0000
770	0.0001	0.0000	0.0000
780	0.0000	0.0000	0.0000

15.B
Appendix: Tables for Calculation of Thermal Conductance

Table B.1 Actual given and calculated values applied in the calculations of the thermal conductance Λ for different cases, which are hence applied in the calculations of the solar factor (SF).

Properties – two-layer window with air	**G/A/G**	**G/A/LE/G**	**G/A/S/G**	**G/LE/A/LE/G**	**G/LE/A/G**	**G/S/A/G**	**G/S/A/LE/G**	**G/S/A/S/G**
Stefan–Boltzmann's constant (10^{-8} W m^{-2} K^{-4})	5.6696	5.6696	5.6696	5.6696	5.6696	5.6696	5.6696	5.6696
External heat transfer coefficient (W m^{-2} K^{-1})	23	23	23	23	23	23	23	23
Internal heat transfer coefficient (W m^{-2} K^{-1})	8	8	8	8	8	8	8	8
Thickness of each material (glass) (m)	0.004	0.004	0.004	0.004	0.004	0.004	0.004	0.004
Thermal resistivity of each material (glass) (mK W^{-1})	1	1	1	1	1	1	1	1
Number of spaces	1	1	1	1	1	1	1	1
Number of materials (glass panes)	2	2	2	2	2	2	2	2
Gas thermal conductivity at +10 °C (W m^{-1} K^{-1})	0.02496	0.02496	0.02496	0.02496	0.02496	0.02496	0.02496	0.02496
Space width of gas (m)	0.012	0.012	0.012	0.012	0.012	0.012	0.012	0.012
Gas mean temperature (K)	283	283	283	283	283	283	283	283
Corrected emissivity for material 1 (glass 1)	0.836	0.836	0.836	0.071	0.071	0.543	0.543	0.543
Corrected emissivity for material 2 (glass 2)	0.836	0.071	0.543	0.071	0.836	0.836	0.071	0.543
Space inclination constant	0.035	0.035	0.035	0.035	0.035	0.035	0.035	0.035
Space inclination exponent	0.38	0.38	0.38	0.38	0.38	0.38	0.38	0.38
Gas mass density at +10 °C (kg m^{-3})	1.232	1.232	1.232	1.232	1.232	1.232	1.232	1.232

Table B.1 (*Continued*)

Properties – two-layer window with air	G/A/G	G/A/LE/G	G/A/S/G	G/LE/A/LE/G	G/LE/A/G	G/S/A/G	G/S/A/LE/G	G/S/A/S/G
Gas dynamic viscosity at +10 °C (10^{-5} kg m^{-1} s^{-1})	1.761	1.761	1.761	1.761	1.761	1.761	1.761	1.761
Gas specific heat capacity per unit mass (J kg^{-1} K^{-1})	1008	1008	1008	1008	1008	1008	1008	1008
Temperature difference on either side of glazing (K)	15	15	15	15	15	15	15	15
Prandtl number	0.71117308	0.71117308	0.71117308	0.71117308	0.71117308	0.71117308	0.71117308	0.71117308
Grashof number	4397.64606	4397.64606	4397.64606	4397.64606	4397.64606	4397.64606	4397.64606	4397.64606
Nusselt number	0.7451477	0.7451477	0.7451477	0.7451477	0.7451477	0.7451477	0.7451477	0.7451477
Corrected Nusselt number	1	1	1	1	1	1	1	1
Gas conductance (W m^{-2} K^{-1})	2.08	2.08	2.08	2.08	2.08	2.08	2.08	2.08
Radiation conductance (W m^{-2} K^{-1})	3.69168821	0.35993398	2.52238676	0.18918985	0.35993398	2.52238676	0.34436941	1.91563161
Gas space conductance (W m^{-2} K^{-1})	5.77168821	2.43993398	4.60238676	2.26918985	2.43993398	4.60238676	2.42436941	3.99563161
Total gas space resistance (m^2 K W^{-1})	0.17325953	0.40984715	0.21727857	0.44068591	0.40984715	0.21727857	0.41247839	0.25027332
Material resistance (m^2 K W^{-1})	0.004	0.004	0.004	0.004	0.004	0.004	0.004	0.004
Total material resistance (m^2 K W^{-1})	0.008	0.008	0.008	0.008	0.008	0.008	0.008	0.008
Thermal conductance (W m^{-2} K^{-1})	5.51695123	2.3932196	4.43894869	2.22873055	2.3932196	4.43894869	2.37824348	3.87186717
Thermal transmittance – U-value (W m^{-2} K^{-1})	2.85928493	1.70553753	2.5396385	1.62031441	1.70553753	2.5396385	1.69791782	2.34328363

The emissivity values $\varepsilon_{float} = 0.836$, $\varepsilon_{lowe} = 0.071$ and $\varepsilon_{silver} = 0.543$, for the float glass, low emittance glass and dark silver coated glass, respectively, were measured by hemispherical directional reflectance. Two-layer window pane with air in the gas space. G = glass, LE = low emittance coating, S = dark silver coating and A = air cavity.

Table B.2 Actual given and calculated values applied in the calculations of the thermal conductance Λ for different cases, which are hence applied in the calculations of the solar factor (SF).

Properties – three-layer window with air	G/A/G e	G/A/G i	G/A/LE/G i	G/A/LE/G e	G/LE/A/LE/G e	G/S/A/G e	G/S/A/LE/G e	G/S/A/S/G e	G/A/S/G i
Stefan-Boltzmann's constant (10^{-8} W m^{-2} K^{-4})	5.6696	5.6696	5.6696	5.6696	5.6696	5.6696	5.6696	5.6696	5.6696
External heat transfer coefficient (W m^{-2} K^{-1})	23	23	23	23	23	23	23	23	23
Internal heat transfer coefficient (W m^{-2} K^{-1})	8	8	8	8	8	8	8	8	8
Thickness of each material (glass) (m)	0.003	0.003	0.003	0.003	0.003	0.003	0.003	0.003	0.003
Thermal resistivity of each material (glass) (mK W^{-1})	1	1	1	1	1	1	1	1	1
Number of spaces	1	1	1	1	1	1	1	1	1
Number of materials (glass panes)	2	2	2	2	2	2	2	2	2
Gas thermal conductivity at +10 °C (W m^{-1} K^{-1})	0.02496	0.02496	0.02496	0.02496	0.02496	0.02496	0.02496	0.02496	0.02496
Space width of gas (m)	0.012	0.012	0.012	0.012	0.012	0.012	0.012	0.012	0.012
Gas mean temperature (K)	279.25	286.75	286.75	279.25	279.25	279.25	279.25	279.25	286.75
Corrected emissivity for material 1 (glass 1)	0.836	0.836	0.836	0.836	0.071	0.543	0.543	0.543	0.836
Corrected emissivity for material 2 (glass 2)	0.836	0.836	0.071	0.071	0.071	0.836	0.071	0.543	0.543
Space inclination constant	0.035	0.035	0.035	0.035	0.035	0.035	0.035	0.035	0.035
Space inclination exponent	0.38	0.38	0.38	0.38	0.38	0.38	0.38	0.38	0.38
Gas mass density at +10 °C (kg m^{-3})	1.232	1.232	1.232	1.232	1.232	1.232	1.232	1.232	1.232

Table B.2 (*Continued*)

Properties – three-layer window with air	G/A/G e	G/A/G i	G/A/LE/G i	G/A/LE/G e	G/LE/A/LE/G e	G/S/A/G e	G/S/A/LE/G e	G/S/A/S/G e	G/A/S/G i
Gas dynamic viscosity at +10 °C (10^{-5} kg m^{-1} s^{-1})	1.761	1.761	1.761	1.761	1.761	1.761	1.761	1.761	1.761
Gas specific heat capacity per unit mass (J kg^{-1} K^{-1})	1008	1008	1008	1008	1008	1008	1008	1008	1008
Temperature difference on either side of glazing (K)	7.5	7.5	7.5	7.5	7.5	7.5	7.5	7.5	7.5
Prandtl number	0.71117308	0.71117308	0.71117308	0.71117308	0.71117308	0.71117308	0.71117308	0.71117308	0.71117308
Grashof number	2228.35064	2170.06772	2170.06772	2228.35064	2228.35064	2228.35064	2228.35064	2228.35064	2170.06772
Nusselt number	0.57550938	0.56974236	0.56974236	0.57550938	0.57550938	0.57550938	0.57550938	0.57550938	0.56974236
Corrected Nusselt number	1	1	1	1	1	1	1	1	1
Gas conductance (W m^{-2} K^{-1})	2.08	2.08	2.08	2.08	2.08	2.08	2.08	2.08	2.08
Radiation conductance (W m^{-2} K^{-1})	3.54686986	3.84039582	0.37443275	0.34581441	0.18176827	2.42343802	0.33086041	1.8404848	2.62399287
Gas space conductance (W m^{-2} K^{-1})	5.62686986	5.92039582	2.45443275	2.42581441	2.26176827	4.50343802	2.41086041	3.9204848	4.70399287
Total gas space resistance (m^2 K W^{-1})	0.1777187	0.16890763	0.40742612	0.41223269	0.44213194	0.22205257	0.41478967	0.2550705	0.21258536
Material resistance (m^2 K W^{-1})	0.003	0.003	0.003	0.003	0.003	0.003	0.003	0.003	0.003
Total material resistance (m^2 K W^{-1})	0.006	0.006	0.006	0.006	0.006	0.006	0.006	0.006	0.006
Thermal conductance (W m^{-2} K^{-1})	5.44310403	5.71730361	2.41881188	2.39101348	2.23148565	4.38495381	2.37648418	3.83038305	4.57487189
Thermal transmittance – *U*-value (W m^{-2} K^{-1})	2.83932035	2.9121756	1.71849541	1.7044168	1.62177011	2.52187198	1.6970209	2.32802443	2.58355463

The emissivity values $\varepsilon_{float} = 0.836$, $\varepsilon_{lowe} = 0.071$ and $\varepsilon_{silver} = 0.543$, for the float glass, low emittance glass and dark silver coated glass, respectively, were measured by hemispherical directional reflectance. Three-layer window pane with air in the two gas spaces, where each gas space is treated in each column. G = glass, LE = low emittance coating, S = dark silver coating, A = air cavity, e = exterior side (gas space) and i = interior side (gas space).

References

1. Baetens, R., Jelle, B.P., and Gustavsen, A. (2010) Properties, requirements and possibilities of smart windows for dynamic daylight and solar energy control in buildings: a state-of-the-art review. *Sol. Energy Mater. Sol. Cells*, **94**, 87–105.
2. Jelle, B.P., Hynd, A., Gustavsen, A., Arasteh, D., Goudey, H., and Hart, R. (2012) Fenestration of today and tomorrow: a state-of-the-art review and future research opportunities. *Sol. Energy Mater. Sol. Cells*, **96**, 1–28.
3. Alamri, S.N. (2009) The temperature behavior of smart windows under direct solar radiation. *Sol. Energy Mater. Sol. Cells*, **93**, 1657–1662.
4. Azens, A., Avendaño, E., Backholm, J., Berggren, L., Gustavsson, G., Karmhag, R., Niklasson, G.A., Roos, A., and Granqvist, C.G. (2005) Flexible foils with electrochromic coatings: science, technology and applications. *Mater. Sci. Eng., B*, **119**, 214–223.
5. Bhadra, S., Khastgir, D., Singha, N.K., and Lee, J.H. (2009) Progress in preparation, processing and applications of polyaniline. *Prog. Polym. Sci.*, **34**, 783–810.
6. Bouessay, I., Rougier, A., Beaudoin, B., and Leriche, J.B. (2002) Pulsed laser-deposited nickel oxide thin films as electrochromic anodic materials. *Appl. Surf. Sci.*, **186**, 490–495.
7. Carpenter, M.K., Conell, R.S., and Corrigan, D.A. (1987) The electrochromic properties of hydrous nickel oxide. *Sol. Energy Mater.*, **16**, 333–346.
8. Carpenter, M.K. and Conell, R.S. (1990) A single-film electrochromic device. *J. Electrochem. Soc.*, **137**, 2464–2467.
9. Chen, K.-C., Hsu, C.-Y., Hu, C.-W., and Ho, K.-C. (2011) A complementary electrochromic device based on prussian blue and poly(ProDOT-Et_2) with high contrast and high coloration efficiency. *Sol. Energy Mater. Sol. Cells*, **95**, 2238–2245.
10. Chen, Z., Gao, Y., Kang, L., Du, J., Zhang, Z., Luo, H., Miao, H., and Tan, G. (2011) VO_2-based double-layered films for smart windows: optical design, all-solution preparation and improved properties. *Sol. Energy Mater. Sol. Cells*, **95**, 2677–2684.
11. Fantini, M. and Gorenstein, A. (1987) Electrochromic nickel hydroxide films on transparent/conducting substrates. *Sol. Energy Mater.*, **16**, 487–500.
12. Gao, T. and Jelle, B.P. (2013) Paraotwayite-type α-$Ni(OH)_2$ nanowires: structural, optical and electrochemical properties. *J. Phys. Chem. C*, **117**, 17294–17302.
13. Geniès, E.M., Boule, A., Lapkowski, M., and Tsintavis, C. (1990) Polyaniline: a historical survey. *Synth. Met.*, **36**, 139–182.
14. Goldner, R.B., Chapman, R.I., Foley, G., Goldner, E.L., Haas, T., Norton, P., Seward, G., and Wong, K.K. (1986) Recent research related to the development of electrochromic windows. *Sol. Energy Mater.*, **14**, 195–203.
15. Goldner, R.B., Wong, K., Foley, G., Norton, P., Wamboldt, L., Seward, G., Haas, T., and Chapman, R. (1987) Thin films for practical electrochromic windows. *Sol. Energy Mater.*, **16**, 365–370.
16. Goldner, R.B., Haas, T.E., Seward, G., Wong, K.K., Norton, P., Foley, G., Berera, G., Wei, G., Schulz, S., and Chapman, R. (1988) Thin film solid state ionic materials for electrochromic smart window glass. *Solid State Ionics*, **28-30**, 1715–1721.
17. Granqvist, C.G. (1989) in *Electricity. Efficient End-Use and New Generation Technologies, and their Planning Implications* (eds T.B. Johansson, B. Bodlund, and R.H. Williams), Lund University Press, pp. 89–123.
18. Granqvist, C.G. (1991) in *Materials Science for Solar Energy Conversion Systems* (ed. C.G. Granqvist), Pergamon Press, pp. 106–167.
19. Granqvist, C.G. (1995) *Handbook of Inorganic Electrochromic Materials*, Elsevier, Amsterdam.
20. Granqvist, C.G. (2000) Electrochromic tungsten oxide films: review of progress 1993–1998. *Sol. Energy Mater. Sol. Cells*, **60**, 201–262.

21. Granqvist, C.G. (2005) Electrochromic devices. *J. Eur. Ceram. Soc.*, **25**, 2907–2912.
22. Granqvist, C.G. (2008) Oxide electrochromics: why, how, and whither. *Sol. Energy Mater. Sol. Cells*, **92**, 203–208.
23. Granqvist, C.G. (2012) Oxide electrochromics: an introduction to devices and materials. *Sol. Energy Mater. Sol. Cells*, **99**, 1–13.
24. Granqvist, C.G., Azens, A., Isidorsson, J., Kharrazi, M., Kullman, L., Lindström, T., Niklasson, G.A., Ribbing, C.-G., Rönnow, D., Mattsson, M.S., and Veszelei, M. (1997) Towards the smart window: progress in electrochromics. *J. Non-Cryst. Solids*, **218**, 273–279.
25. Granqvist, C.G., Avendaño, E., and Azens, A. (2003) Electrochromic coatings and devices: survey of some recent advances. *Thin Solid Films*, **442**, 201–211.
26. Granqvist, C.G., Green, S., Niklasson, G.A., Mlyuka, N.R., von Kræmer, S., and Georén, P. (2010) Advances in chromogenic materials and devices. *Thin Solid Films*, **518**, 3046–3053.
27. Green, S.V., Pehlivan, E., Granqvist, C.G., and Niklasson, G.A. (2012) Electrochromism in sputter deposited nickel-containing tungsten oxide films. *Sol. Energy Mater. Sol. Cells*, **99**, 339–344.
28. Green, S.V., Watanabe, M., Oka, N., Niklasson, G.A., Granqvist, C.G., and Shigesato, Y. (2012) Electrochromic properties of nickel oxide based thin films sputter deposited in the presence of water vapor. *Thin Solid Films*, **520**, 3839–3842.
29. Jelle, B.P., Hagen, G., Hesjevik, S.M., and Ødegård, R. (1992) Transmission through an electrochromic window based on polyaniline, tungsten oxide and a solid polymer electrolyte. *Mater. Sci. Eng., B*, **13**, 239–241.
30. Jelle, B.P., Hagen, G., and Ødegård, R. (1992) Transmission spectra of an electrochromic window based on polyaniline, tungsten oxide and a solid polymer electrolyte. *Electrochim. Acta*, **37**, 1377–1380.
31. Jelle, B.P., Hagen, G., Sunde, S., and Ødegård, R. (1993) Dynamic light modulation in an electrochromic window consisting of polyaniline, tungsten oxide and a solid polymer electrolyte. *Synth. Met.*, **54**, 315–320.
32. Jelle, B.P., Hagen, G., and Nødland, S. (1993) Transmission spectra of an electrochromic window consisting of polyaniline, prussian blue and tungsten oxide. *Electrochim. Acta*, **38**, 1497–1500.
33. Jelle, B.P., Hagen, G., Hesjevik, S.M., and Ødegård, R. (1993) Reduction factor for polyaniline films on ITO from cyclic voltammetry and visible absorption spectra. *Electrochim. Acta*, **38**, 1643–1647.
34. Jelle, B.P., Gustavsen, A., Nilsen, T.-N., and Jacobsen, T. (2007) Solar material protection factor (SMPF) and solar skin protection factor (SSPF) for window panes and other glass structures in buildings. *Sol. Energy Mater. Sol. Cells*, **91**, 342–354.
35. Jelle, B.P. and Hagen, G. (1993) Transmission spectra of an electrochromic window based on polyaniline, prussian blue and tungsten oxide. *J. Electrochem. Soc.*, **140**, 3560–3564.
36. Jelle, B.P. and Hagen, G. (1999) Performance of an electrochromic window based on polyaniline, prussian blue and tungsten oxide. *Sol. Energy Mater. Sol. Cells*, **58**, 277–286.
37. Jelle, B.P. and Hagen, G. (1999) Correlation between light absorption and electric charge in solid state electrochromic windows. *J. Appl. Electrochem.*, **29**, 1103–1110.
38. Jelle, B.P. (2013) Solar radiation glazing factors for window panes, glass structures and electrochromic windows in buildings – Measurement and calculation. *Sol. Energy Mater. Sol. Cells*, **116**, 291–323.
39. Kalagi, S.S., Mali, S.S., Dalavi, D.S., Inamdar, A.I., Im, H., and Patil, P.S. (2011) Limitations of dual and complementary inorganic-organic electrochromic device for smart window application and its colorimetric analysis. *Synth. Met.*, **161**, 1105–1112.

40. Karuppasamy, A. and Subrahmanyam, A. (2007) Studies on electrochromic smart windows based on titanium doped WO_3 thin films. *Thin Solid Films*, **516**, 175–178.
41. Lampert, C.M. (1980) *Thin Film Electrochromic Materials for Energy Efficient Windows*, Lawrence Berkeley Laboratory, Berkeley, CA, LBL-10862, October 1980.
42. Lampert, C.M. (1984) Electrochromic materials and devices for energy efficient windows. *Sol. Energy Mater.*, **11**, 1–27.
43. Lampert, C.M. (ed.) (1989) *Failure and Degradation Modes in Selected Solar Materials: A Review*, The International Energy Agency, Solar Heating and Cooling Program, Task 10: Solar Materials R&D, May 1989.
44. Lampert, C.M. (1998) Smart switchable glazing for solar energy and daylight control. *Sol. Energy Mater. Sol. Cells*, **52**, 207–221.
45. Lampert, C.M. (2004) Chromogenic smart materials. *Mater. Today*, **7**, 28–35.
46. Lampert, C.M., Omstead, T.R., and Yu, P.C. (1986) Chemical and optical properties of electrochromic nickel oxide films. *Sol. Energy Mater.*, **14**, 161–174.
47. Lampert, C.M. and Ma, Y.-P. (1992) *Advanced Glazing Technology: Fenestration 2000 Project- Phase III: Glazing Materials*, Lawrence Berkeley Laboratory, Berkeley, CA, LBL-31616, June 1992.
48. Lampert, C.M., Agrawal, A., Baertlien, C., and Nagai, J. (1999) Durability evaluation of electrochromic devices - an industry perspective. *Sol. Energy Mater. Sol. Cells*, **56**, 449–463.
49. Lee, E.S. and DiBartolomeo, D.L. (2002) Application issues for large-area electrochromic windows in commercial buildings. *Sol. Energy Mater. Sol. Cells*, **71**, 465–491.
50. Lee, E.S., DiBartolomeo, D.L., and Selkowitz, S.E. (2006) Daylighting control performance of a thin-film ceramic electrochromic window: field study results. *Energy Build.*, **38**, 30–44.
51. Lee, E.S., Selkowitz, S.E., Clear, R.D., DiBartolomeo, D.L., Klems, J.H., Fernandes, L.L., Ward, G.J., Inkarojrit, V., and Yazdanian, M. (2006) *Advancement of Electrochromic Windows*, Lawrence Berkeley National Laboratory, California Energy Commission, PIER, Publication number CEC-500-2006-052.
52. Leventis, N. and Chung, Y.C. (1990) Polyaniline-prussian blue novel composite material for electrochromic applications. *J. Electrochem. Soc.*, **137**, 3321–3322.
53. Makimura, Y., Rougier, A., and Tarascon, J.-M. (2006) Cobalt and tantalum additions for enhanced electrochromic performances of nickel-based oxide thin films grown by pulsed laser deposition. *Appl. Surf. Sci.*, **252**, 4593–4598.
54. Monk, P.M.S., Mortimer, R.J., and Rosseinsky, D.R. (1995) *Electrochromism: Fundamentals and Applications*, VCH.
55. Monk, P.M.S., Akhtar, S.P., Boutevin, J., and Duffield, J.R. (2001) Toward the tailoring of electrochromic bands of metal-oxide mixtures. *Electrochim. Acta*, **46**, 2091–2096.
56. Mortimer, R.J. (1999) Organic electrochromic materials. *Electrochim. Acta*, **44**, 2971–2981.
57. Mortimer, R.J., Rosseinsky, D.R., and Glidle, A. (1992) Polyelectrochromic prussian blue: a chronoamperometric study of the electrodeposition. *Sol. Energy Mater. Sol. Cells*, **25**, 211–223.
58. Mortimer, R.J., Dyer, A.L., and Reynolds, J.R. (2006) Electrochromic organic and polymeric materials for display applications. *Displays*, **27**, 2–18.
59. Moulki, H., Park, D.H., Min, B.-K., Kwon, H., Hwang, S.-J., Choy, J.-H., Toupance, T., Campet, G., and Rougier, A. (2012) Improved electrochromic performances of NiO based thin films by lithium addition: from single layers to devices. *Electrochim. Acta*, **74**, 46–52.
60. Penin, N., Rougier, A., Laffont, L., Poizot, P., and Tarascon, J.-M. (2006) Improved cyclability by tungsten addition in electrochromic NiO thin films. *Sol. Energy Mater. Sol. Cells*, **90**, 422–433.

61. Piccolo, A. (2010) Thermal performance of an electrochromic smart window tested in an environmental test cell. *Energy Build.*, **42**, 1409–1417.
62. Rougier, A., Portemer, F., Quédé, A., and El Marssi, M. (1999) Characterization of pulsed laser deposited WO_3 thin films for electrochromic devices. *Appl. Surf. Sci.*, **153**, 1–9.
63. Rougier, A., Blyr, A., Garcia, J., Zhang, Q., and Impey, S.A. (2002) Electrochromic W-M-O (M=V, Nb) sol-gel thin films: a way to neutral colour. *Sol. Energy Mater. Sol. Cells*, **71**, 343–357.
64. Rougier, A. and Blyr, A. (2001) Electrochromic properties of vanadium tungsten oxide thin films grown by pulsed laser deposition. *Electrochim. Acta*, **46**, 1945–1950.
65. Sauvet, K., Rougier, A., and Sauques, L. (2008) Electrochromic WO_3 thin films active in the IR region. *Sol. Energy Mater. Sol. Cells*, **92**, 209–215.
66. Sauvet, K., Sauques, L., and Rougier, A. (2009) IR electrochromic WO_3 thin films: from optimization to devices. *Sol. Energy Mater. Sol. Cells*, **93**, 2045–2049.
67. Stilwell, D.E., Park, K.H., and Miles, M.H. (1992) Electrochemical studies of the factors influencing the cycle stability of prussian blue films. *J. Appl. Electrochem.*, **22**, 325–331.
68. Syed, A.A. and Dinesan, M.K. (1991) Review: polyaniline – a novel polymeric material. *Talanta*, **38**, 815–837.
69. Yu, P.C., Nazri, G., and Lampert, C.M. (1987) Spectroscopic and electrochemical studies of electrochromic hydrated nickel oxide films. *Sol. Energy Mater.*, **16**, 1–17.
70. Yu, P.C. and Lampert, C.M. (1989) In-situ spectroscopic studies of electrochromic hydrated nickel oxide films. *Sol. Energy Mater.*, **19**, 1–16.
71. Zhang, J., Wang, X.L., Xia, X.H., Gu, C.D., and Tu, J.P. (2011) Electrochromic behavior of WO_3 nanotree films prepared by hydrothermal oxidation. *Sol. Energy Mater. Sol. Cells*, **95**, 2107–2112.
72. Adami, N., Fazzi, D., Bianco, A., Bertarelli, C., and Castiglioni, C. (2010) Enhancing the light driven modulation of the refractive index in organic photochromic materials: a quantum chemical strategy. *J. Photochem. Photobiol., A*, **214**, 61–68.
73. Anders, A., Slack, J.L., and Richardson, T.J. (2008) Electrochromically switched, gas-reservoir metal hydride devices with application to energy-efficient windows. *Thin Solid Films*, **517**, 1021–1026.
74. Chen, C.-Y. and Lo, Y.-L. (2009) Integration of a-Si:H solar cell with novel twist nematic liquid crystal cell for adjustable brightness and enhanced power characteristics. *Sol. Energy Mater. Sol. Cells*, **93**, 1268–1275.
75. Cupelli, D., Nicoletta, F.P., Manfredi, S., De Filpo, G., and Chidichimo, G. (2009) Electrically switchable chromogenic materials for external glazing. *Sol. Energy Mater. Sol. Cells*, **93**, 329–333.
76. Gao, L.-P., Ding, G.-J., Wang, Y.-C., and Yang, Y.-L. (2011) Preparation of UV curing crosslinked polyviologen film and its photochromic and electrochromic performances. *Appl. Surf. Sci.*, **258**, 1184–1191.
77. Gao, T. and Jelle, B.P. (2013) Visible-light-driven photochromism of hexagonal sodium tungsten bronze nanorods. *J. Phys. Chem. C*, **117**, 13753–13761.
78. Gardiner, D.J., Morris, S.M., and Coles, H.J. (2009) High-efficiency multistable switchable glazing using smectic a liquid crystals. *Sol. Energy Mater. Sol. Cells*, **93**, 301–306.
79. Georg, A., Graf, W., Schweiger, D., Wittwer, V., Nitz, P., and Wilson, H.R. (1998) Switchable glazing with a large dynamic range in total solar energy transmittance (TSET). *Sol. Energy*, **62**, 215–228.
80. Huang, R., Shen, Y., Zhao, L., and Yan, M. (2012) Effect of hydrothermal temperature on structure and photochromic properties of WO_3 powder. *Adv. Powder Technol.*, **23**, 211–214.
81. Lampert, C.M. (2003) Large-area smart glass and integrated photovoltaics. *Sol. Energy Mater. Sol. Cells*, **76**, 489–499.
82. Li, S.-Y., Niklasson, G.A., and Granqvist, C.G. (2012) Thermochromic

fenestration with VO_2-based materials: three challenges and how they can be met. *Thin Solid Films*, **520**, 3823–3828.

83. Mennig, M., Fries, K., Lindenstruth, M., and Schmidt, H. (1999) Development of fast switching photochromic coatings on transparent plastics and glass. *Thin Solid Films*, **351**, 230–234.
84. Mlyuka, N.R., Niklasson, G.A., and Granqvist, C.G. (2009) Mg doping of thermochromic VO_2 films enhances the optical transmittance and decreases the metal-insulator transition temperature. *Appl. Phys. Lett.*, **95**, 171909-1–171909-3.
85. Park, S. and Hong, J.W. (2009) Polymer dispersed liquid crystal film for variable-transparency glazing. *Thin Solid Films*, **517**, 3183–3186.
86. Vergaz, R., Sánchez-Pena, J.-M., Barrios, D., Vázquez, C., and Contreras-Lallana, P. (2008) Modelling and electro-optical testing of suspended particle devices. *Sol. Energy Mater. Sol. Cells*, **92**, 1483–1487.
87. Wittwer, V., Datz, M., Ell, J., Georg, A., Graf, W., and Walze, G. (2004) Gasochromic windows. *Sol. Energy Mater. Sol. Cells*, **84**, 305–314.
88. Ye, H., Meng, X., and Xu, B. (2012) Theoretical discussions of perfect window, ideal near infrared solar spectrum regulating window and current thermochromic window. *Energy Build.*, **49**, 164–172.
89. Yoshimura, K., Yamada, Y., Bao, S., Tajima, K., and Okada, M. (2009) Preparation and characterization of gasochromic switchable-mirror window with practical size. *Sol. Energy Mater. Sol. Cells*, **93**, 2138–2142.
90. Sekhar, S.C. and Toon, K.L.C. (1998) On the study of energy performance and life cycle cost of smart window. *Energy Build.*, **28**, 307–316.
91. Syrrakou, E., Papaefthimiou, S., and Yianoulis, P. (2006) Eco-efficiency evaluation of a smart window prototype. *Sci. Total Environ.*, **359**, 267–282.
92. Jonsson, A. and Roos, A. (2010) Evaluation of control strategies for different smart window combinations using computer simulations. *Sol. Energy*, **84**, 1–9.
93. Midtdal, K. and Jelle, B.P. (2013) Self-cleaning glazing products: a state-of-the-art review and future research pathways. *Sol. Energy Mater. Sol. Cells*, **109**, 126–141.
94. Jelle, B.P. (2013) The challenge of removing snow downfall on photovoltaic solar cell roofs in order to maximize solar energy efficiency – Research opportunities for the future. *Energy Build.*, **67**, 334–351.
95. Ahn, K.-S., Yoo, S.J., Kang, M.-S., Lee, J.-W., and Sung, Y.-E. (2007) Tandem dye-sensitized solar cell-powered electrochromic devices for the photovoltaic-powered smart window. *J. Power Sources*, **168**, 533–536.
96. Deb, S.K., Lee, S.-H., Tracy, C.E., Pitts, J.R., Gregg, B.A., and Branz, H.M. (2001) Stand-alone photovoltaic-powered electrochromic smart window. *Electrochim. Acta*, **46**, 2125–2130.
97. Bullock, J.N., Bechinger, C., Benson, D.K., and Branz, H.M. (1996) Semi-transparent a-SiC: H solar cells for self-powered photovoltaic-electrochromic devices. *J. Non-Cryst. Solids*, **198-200**, 1163–1167.
98. Gao, W., Lee, S.H., Bullock, J., Xu, Y., Benson, D.K., Morrison, S., and Branz, H.M. (1999) First a-SiC: H photovoltaic-powered monolithic tandem electrochromic smart window device. *Sol. Energy Mater. Sol. Cells*, **59**, 243–254.
99. Jelle, B.P., Breivik, C., and Røkenes, H.D. (2012) Building integrated photovoltaic products: a state-of-the-art review and future research opportunities. *Sol. Energy Mater. Sol. Cells*, **100**, 69–96.
100. Jelle, B.P. and Breivik, C. (2012) State-of-the-art building integrated photovoltaics. *Energy Procedia*, **20**, 68–77.
101. Jelle, B.P. and Breivik, C. (2012) The path to the building integrated photovoltaics of tomorrow. *Energy Procedia*, **20**, 78–87.
102. Breivik, C., Jelle, B.P., Time, B., Holmberget, Ø., Nygård, J., Bergheim, E., and Dalehaug, A. (2013) Large-scale experimental wind-driven rain exposure

investigations of building integrated photovoltaics. *Sol. Energy*, **90**, 179–187.
103. Gao, T., Jelle, B.P., Ihara, T., and Gustavsen, A. (2014) Insulating glazing units with silica aerogel granules: the impact of particle size. *Appl. Energy*, **128**, 27–34.
104. Gustavsen, A., Jelle, B.P., Arasteh, D. and Kohler, C. (2007) State-of-the-Art Highly Insulating Window Frames – Research and Market Review. Project Report 6, SINTEF Building and Infrastructure.
105. Gustavsen, A., Arasteh, D., Jelle, B.P., Curcija, C., and Kohler, C. (2008) Developing low-conductance window frames: capabilities and limitations of current window heat-transfer design tools. *J. Building Phys.*, **32**, 131–153.
106. Gustavsen, A., Grynning, S., Arasteh, D., Jelle, B.P., and Goudey, H. (2011) Key elements of and materials performance targets for highly insulating window frames. *Energy Build.*, **43**, 2583–2594.
107. Van Den Bergh, S., Hart, R., Jelle, B.P., and Gustavsen, A. (2013) Window spacers and edge seals in insulating glass units: a state-of-the-art review and future perspectives. *Energy Build.*, **58**, 263–280.
108. Grynning, S., Gustavsen, A., Time, B., and Jelle, B.P. (2013) Windows in the buildings of tomorrow: energy losers or energy gainers? *Energy Build.*, **61**, 185–192.
109. Baetens, R., Jelle, B.P., Thue, J.V., Tenpierik, M.J., Grynning, S., Uvsløkk, S., and Gustavsen, A. (2010) Vacuum insulation panels for building applications: a review and beyond. *Energy Build.*, **42**, 147–172.
110. Baetens, R., Jelle, B.P., and Gustavsen, A. (2011) Aerogel insulation for building applications: a state-of-the-art review. *Energy Build.*, **43**, 761–769.
111. Gao, T., Sandberg, L.I.C., Jelle, B.P., and Gustavsen, A. (2012) in *Fuelling the Future: Advances in Science and Technologies for Energy Generation, Transmission and Storage* (ed. A. Mendez-Vilas), BrownWalker Press, pp. 535–539.
112. Gao, T., Jelle, B.P., Sandberg, L.I.C., and Gustavsen, A. (2013) Monodisperse hollow silica nanospheres for nano insulation materials: synthesis, characterization, and life cycle assessment. *ACS Appl. Mater. Interfaces*, **5**, 761–767.
113. Gao, T., Jelle, B.P., Gustavsen, A., and Jacobsen, S. (2014) Aerogel-incorporated concrete: an experimental study. *Constr. Build. Mater.*, **52**, 130–136.
114. Gao, T., Sandberg, L.I.C., and Jelle, B.P. (2014) Nano insulation materials: synthesis and life cycle assessment. *Procedia CIRP*, **15**, 490–495.
115. Jelle, B.P., Gustavsen, A. and Baetens, R. (2009) Beyond vacuum insulation panels – how may it be achieved? Proceedings of the 9th International Vacuum Insulation Symposium (IVIS 2009), London, UK, September 17–18, 2009.
116. Jelle, B.P., Gustavsen, A., and Baetens, R. (2010) The path to the high performance thermal building insulation materials and solutions of tomorrow. *J. Building Phys.*, **34**, 99–123.
117. Jelle, B.P., Gao, T., Sandberg, L.I.C., Tilset, B.G., Grandcolas, M., and Gustavsen, A. (2014) Thermal superinsulation for building applications - From concepts to experimental investigations. *Int. J. Struct. Anal. Des.*, **1**, 43–50.
118. Jelle, B.P. (2011) Traditional, state-of-the-art and future thermal building insulation materials and solutions - properties, requirements and possibilities. *Energy Build.*, **43**, 2549–2563.
119. Kalnæs, S.E. and Jelle, B.P. (2014) Vacuum insulation panel products: a state-of-the-art review and future research pathways. *Appl. Energy*, **116**, 355–375.
120. L.I.C. Sandberg, T. Gao, B.P. Jelle and A. Gustavsen, Synthesis of hollow silica nanospheres by sacrificial polystyrene templates for thermal insulation applications, *Adv. Mater. Sci. Eng.*, 2013, 6p, Article ID 483651, 2013.
121. Schlanbusch, R.D., Jelle, B.P., Sandberg, L.I.C., Fufa, S.M., and Gao, T. (2014) Integration of life cycle assessment in

the design of hollow silica nanospheres for thermal insulation applications. *Build. Environ.*, **80**, 115–124.

122. Czanderna, A.W., Benson, D.K., Jorgensen, G.J., Zhang, J.-G., Tracy, C.E., and Deb, S.K. (1999) Durability issues and service lifetime prediction of electrochromic windows for buildings applications. *Sol. Energy Mater. Sol. Cells*, **56**, 419–436.
123. Nagai, J., McMeeking, G.D., and Saitoh, Y. (1999) Durability of electrochromic glazing. *Sol. Energy Mater. Sol. Cells*, **56**, 309–319.
124. Tajima, K., Hotta, H., Yamada, Y., Okada, M., and Yoshimura, K. (2012) Environmental durability of electrochromic switchable mirror glass at sub-zero temperature. *Sol. Energy Mater. Sol. Cells*, **104**, 146–151.
125. Tracy, C.E., Zhang, J.-G., Benson, D.K., Czanderna, A.W., and Deb, S.K. (1999) Accelerated durability testing of electrochromic windows. *Electrochim. Acta*, **44**, 3195–3202.
126. Jelle, B.P. and Nilsen, T.-N. (2011) Comparison of accelerated climate ageing methods of polymer building materials by attenuated total reflectance Fourier transform infrared radiation spectroscopy. *Constr. Build. Mater.*, **25**, 2122–2132.
127. Jelle, B.P., Nilsen, T.-N., Hovde, P.J., and Gustavsen, A. (2012) Accelerated climate aging of building materials and their characterization by Fourier transform infrared radiation analysis. *J. Building Phys.*, **36**, 99–112.
128. Jelle, B.P. (2012) Accelerated climate ageing of building materials, components and structures in the laboratory. *J. Mater. Sci.*, **47**, 6475–6496.
129. Jelle, B.P. (2013) The role of accelerated climate ageing of building materials, components and structures in the laboratory. Proceedings of the 7th Nordic Conference on Construction Economics and Organisation 2013, Trondheim, Norway, June 12–14, 2013, pp. 111–122.
130. Fahrenbruch, A.L. and Bube, R.H. (1983) *Fundamentals of Solar Cells. Photovoltaic Solar Energy Conversion*, Academic Press, pp. 26–31.
131. Croll, S.G. and Skaja, A.D. (2003) Quantitative spectroscopy to determine the effects of photodegradation on a model polyester-urethane coating. *J. Coat. Technol.*, **75**, 85–94.
132. Fufa, S.M., Jelle, B.P., Hovde, P.J., and Rørvik, P.M. (2012) Coated wooden claddings and the influence of nanoparticles on the weathering performance. *Prog. Org. Coat.*, **75**, 72–78.
133. Gerlock, J.L., Smith, C.A., Cooper, V.A., Dusbiber, T.G., and Weber, W.H. (1998) On the use of Fourier transform infrared spectroscopy and ultraviolet spectroscopy to assess the weathering performance of isolated clearcoats from different chemical families. *Polym. Degrad. Stab.*, **62**, 225–234.
134. Rånby, B. and Rabek, J.F. (1975) *Photodegradation, Photo-oxidation and Photostabilization of Polymers, Principles and Applications*, John Wiley & Sons, Inc., New York.
135. Rabek, J.F. (1995) *Polymer Photodegradation. Mechanisms and Experimental Methods*, Chapman & Hall.
136. Rabek, J.F. (1996) *Photodegradation of Polymers. Physical Characteristics and Applications*, Springer-Verlag.
137. Tylli, H., Olkkonen, C., and Forsskåhl, I. (1989) A sensitivity analysis of the kinetic mechanism for the photodegradation of a model system of relevance to lignin yellowing. *J. Photochem. Photobiol., A*, **49**, 397–408.
138. Andersson, A.M., Talledo, A., Granqvist, C.G., and Stevens, J.R. (1990) in *Electrochromic Materials*, Proceedings Volume 90–2, Proceedings of the Symposium on Electrochromic Materials, The 176th Meeting of The Electrochemical Society, Hollywood, FL, 16–18 October, 1989 (eds M.K. Carpenter and D.A. Corrigan), The Electrochemical Society, Pennington, NJ, pp. 261–273.
139. Granqvist, C.G. (1981) Radiative heating and cooling with spectrally selective surfaces. *Appl. Opt.*, **20**, 2606–2615.
140. Granqvist, C.G. (1984) Spectrally selective surface coatings for energy efficiency and solar applications. *Phys. Teach.*, **22**, 372–383.

141. Hecht, E. (1987) *Optics*, 2nd edn, Addison-Wesley, p. 112.
142. Jastrzebski, Z.D. (1987) *The Nature and Properties of Engineering Materials*, 3rd edn, John Wiley & Sons, Inc., New York, pp. 464–468.
143. Jelle, B.P. (1993) Electrochemical and spectroscopic studies of electrochromic materials. PhD thesis, 1993:131, Department of Applied Electrochemistry, The Norwegian Institute of Technology, Trondheim, Norway.
144. Stjerna, B. and Granqvist, C.G. (1990) Electrical conductivity and optical transmittance of sputter-deposited SnO_x thin films. *Sol. Energy Mater.*, **20**, 225–233.
145. ASTM E179-81. (1981) *Standard Practice for Selection of Geometric Conditions for Measurement of Reflectance and Transmittance.*
146. Davies, M.G. (2004) *Building Heat Transfer*, John Wiley & Sons, Inc, New York.
147. Kimura, K.I. (1977) *Scientific Basis of Air Conditioning*, Applied Science Publishers, London, pp. 99–105.
148. International Standards Organization ISO 9050:2003(E). (2003) *Glass in Building – Determination of Light Transmittance, Solar Direct Transmittance, Total Solar Energy Transmittance, Ultraviolet Transmittance and Related Glazing Factors*, International Standards Organization.
149. Rubin, M., von Rottkay, K., and Powles, R. (1998) Window optics. *Sol. Energy*, **62**, 149–161.
150. Goldner, R.B. and Rauh, R.D. (1984) Electrochromic materials for controlled radiant energy transfer in buildings. *Sol. Energy Mater.*, **11**, 177–185.
151. Goldner, R.B., Brofos, A., Foley, G., Goldner, E.L., Haas, T.E., Henderson, W., Norton, P., Ratnam, B.A., Weis, N., and Wong, K.K. (1985) Optical frequencies free electron scattering studies on electrochromic materials for variable reflectivity windows. *Sol. Energy Mater.*, **12**, 403–410.
152. Goldner, R.B., Seward, G., Wong, K., Haas, T., Foley, G.H., Chapman, R., and Schulz, S. (1989) Completely solid lithiated smart windows. *Sol. Energy Mater.*, **19**, 17–26.
153. Goldner, R.B., Berera, G., Arntz, F.O., Haas, T.E., Morel, B., and Wong, K.K. (1990) in *Electrochromic Materials*, Proceedings Volume 90–2, Proceedings of the Symposium on Electrochromic Materials, The 176th Meeting of The Electrochemical Society, Hollywood, FL, 16-18 October, 1989 (eds M.K. Carpenter and D.A. Corrigan), The Electrochemical Society, Pennington, NJ, pp. 14–22.
154. Goldner, R.B., Arntz, F.O., Berera, G., Haas, T.E., Wei, G., Wong, K.K., and Yu, P.C. (1992) A monolithic thin-film electrochromic window. *Solid State Ionics*, **53–56**, 617–627.
155. Ashrit, P.V., Benaissa, K., Bader, G., Girouard, F.E., and Truong, V.V. (1992) in *Proceedings of SPIE – Optical Materials Technology for Energy Efficiency and Solar Energy Conversion XI*, vol. 1728, Toulouse-Labege, France, 19 May, 1992 (eds A. Hugot-Le Goff, C.G. Granqvist, and C.M. Lampert), The Society of Photo-Optical Instrumentation Engineers, Bellingham, WA, pp. 232–240.
156. Beni, G. and Shay, J.L. (1982) in *Image Pickup and Display*, vol. 5 (ed. B. Kazan), Academic Press, pp. 83–136.
157. Cogan, S.F., Plante, T.D., Parker, M.A., and Rauh, R.D. (1986) Electrochromic solar attenuation in crystalline and amorphous Li_xWO_3. *Sol. Energy Mater.*, **14**, 185–193.
158. Deb, S.K. (2008) Opportunities and challenges in science and technology of WO_3 for electrochromic and related applications. *Sol. Energy Mater. Sol. Cells*, **92**, 245–258.
159. Faughnan, B.W. and Crandall, R.S. (1980) in *Display Devices*, Topics in Applied Physics, vol. 40 (ed. J.I. Pankove), Springer-Verlag, pp. 181–211.
160. Goldner, R.B., Mendelsohn, D.H., Alexander, J., Henderson, W.R., Fitzpatrick, D., Haas, T.E., Sample, H.H., Rauh, R.D., Parker, M.A., and Rose, T.L. (1983) High near-infrared

reflectivity modulation with polycrystalline electrochromic WO_3 films. *Appl. Phys. Lett.*, **43**, 1093–1095.

161. Granqvist, C.G. (1993) Electrochromics and smart windows. *Solid State Ionics*, **60**, 213–214.

162. Granqvist, C.G. (2007) Transparent conductors as solar energy materials: a panoramic review. *Sol. Energy Mater. Sol. Cells*, **91**, 1529–1598.

163. Hamberg, I. and Granqvist, C.G. (1984) Optical properties of transparent and heat-reflecting indium tin oxide films: the role of ionized impurity scattering. *Appl. Phys. Lett.*, **44**, 721–723.

164. Hamberg, I. and Granqvist, C.G. (1984) Optical properties of transparent and heat-reflecting indium-tin-oxide films: experimental data and theoretical analysis. *Sol. Energy Mater.*, **11**, 239–248.

165. Lim, Y.K. (1986) *Introduction to Classical Electrodynamics*, World Scientific, p. 242.

166. Mendelsohn, D.H. and Goldner, R.B. (1984) Ellipsometry measurements as direct evidence of the Drude model for polycrystalline electrochromic WO_3 films. *J. Electrochem. Soc.*, **131**, 857–860.

167. Nagai, J., Kamimori, T., and Mizuhashi, M. (1986) Transmissive electrochromic device. *Sol. Energy Mater.*, **14**, 175–184.

168. Schirmer, O.F., Wittver, V., Baur, G., and Brandt, G. (1977) Dependence of WO_3 electrochromic absorption on crystallinity. *J. Electrochem. Soc.*, **124**, 749–753.

169. Schuster, A.P., Nguyen, D., and Caporaletti, O. (1986) Solid state electrochromic infrared switchable windows. *Sol. Energy Mater.*, **13**, 153–160.

170. Svensson, J.S.E.M. and Granqvist, C.G. (1984) Modulated transmittance and reflectance in crystalline electrochromic WO_3 films: theoretical limits. *Appl. Phys. Lett.*, **45**, 828–830.

171. Svensson, J.S.E.M. and Granqvist, C.G. (1984) in *Proceedings of SPIE - Optical Materials Technology for Energy Efficiency and Solar Energy Conversion III*, vol. 502, San Diego, CA, August 21–23, 1984 (ed. C.M. Lampert), The Society of Photo-Optical Instrumentation Engineers, Bellingham, WA, pp. 30–37.

172. Svensson, J.S.E.M. and Granqvist, C.G. (1985) Electrochromic coatings for «smart windows». *Sol. Energy Mater.*, **12**, 391–402.

173. Yamanaka, K. (1987) Electrodeposited films from aqueous tungstic acid-hydrogen peroxide solutions for electrochromic display devices. *Jpn. J. Appl. Phys.*, **26**, 1884–1890.

174. Jelle, B.P. and Hagen, G. (1994) in *Electrochromic Materials II*, Proceedings Volume 94–2, Proceedings of The 184th Meeting of Symposium on Electrochromic Materials, The Electrochemical Society, New Orleans, LA, October 10–15, 1993 (eds K.C. Ho and D.A. MacArthur), The Electrochemical Society, Pennington, NJ, pp. 324–338.

175. Jelle, B.P., Hagen, G., and Birketveit, Ø. (1998) Transmission properties for individual electrochromic layers in solid state devices based on polyaniline, prussian blue and tungsten oxide. *J. Appl. Electrochem.*, **28**, 483–489.

176. Jelle, B.P. and Hagen, G. (1998) Electrochemical multilayer deposition of polyaniline and prussian blue and their application in solid state electrochromic windows. *J. Appl. Electrochem.*, **28**, 1061–1065.

177. International Standards Organization ISO 10292:1994(E). (1994) *Glass in Building – Calculation of Steady-State U values (Thermal Transmittance) of Multiple Glazing*.

178. EN 12898:2001 E. (2001) *Glass in Building – Determination of the Emissivity*.

179. ASTM E 1585-93 (1993) *Standard Test Method for Measuring and Calculating Emittance of Architectural Flat Glass Products Using Spectrometric Measurements*.

180. EN 1946-3:1999 E. (1999) *Thermal Performance of Building Products and Components – Specific Criteria for the Assessment of Laboratories Measuring Heat Transfer Properties – Part 3: Measurements by Heat Flow Meter Method*.

181. EN 1946-2:1999 E. (1999) *Thermal Performance of Building Products and*

Components – Specific Criteria for the Assessment of Laboratories Measuring Heat Transfer Properties – Part 2: Measurements by Guarded Hot Plate Method.

182. International Standards Organization ISO 9845-1:1992(E). (1992) *Solar Energy – Reference Solar Spectral Irradiance at the Ground at Different Receiving Conditions – Part 1: Direct Normal and Hemispherical Solar Irradiance for Air Mass 1.5,* International Standards Organization.

183. International Standards Organization ISO 10526:1999(E). (1999) *CIE Standard Illuminants for Colorimetry,* International Standards Organization.

184. International Standards Organization ISO/CIE 10527:1991(E). (1991) *CIE Standard Illuminants for Colorimetry,* International Standards Organization.

185. CIE No 89/3:1990. (1990) *On the Deterioration of Exhibited Museum Objects by Optical Radiation.*

186. International Standards Organization ISO/DIS 9050:2001. (2001) *Glass in Building - Determination of Light Transmittance, Solar Direct Transmittance, Total Solar Energy Transmittance, Ultraviolet Transmittance and Related Glazing Factors,* International Standards Organization.

187. McKinlay, A.F. and Diffey, B.L. (1987) A reference action spectrum for ultraviolet induced erythema in human skin. *CIE J.*, **6**, 17–22.

188. Surface Optics Corporation (2009) *SOC-100 User's Manual,* San Diego, CA.

189. EN-ISO 6946:1996. (1996) *Building Components and Building Elements. Thermal Resistance and Thermal Transmittance. Calculation Method.*

190. EN 410:1998 E. (1998) *Glass in Building – Determination of Luminous and Solar Characteristics of Glazing.*

191. Mortimer, R.J. and Varley, T.S. (2011) Quantification of colour stimuli through the calculation of CIE chromaticity coordinates and luminance data for application to in situ colorimetry studies of electrochromic materials. *Displays,* **32**, 35–44.

192. Mortimer, R.J. and Varley, T.S. (2012) In situ spectroelectrochemistry and colour measurement of a complementary electrochromic device based on surface-confined prussian blue and aqueous solution-phase methyl viologen. *Sol. Energy Mater. Sol. Cells,* **99**, 213–220.

193. International Standards Organization ISO 10291:1994(E). (1994) *Glass in Building – Determination of Steady-state U Values (Thermal Transmittance) of Multiple Glazing – Guarded Hot Plate Method.*

194. International Standards Organization ISO 10293:1997(E). (1997) *Glass in Building – Determination of Steady-State U Values (Thermal Transmittance) of Multiple Glazing – Heat Flow Meter Method.*

195. EN 1946-4:2000 E. (2000) *Thermal Performance of Building Products and Components – Specific Criteria for the Assessment of Laboratories Measuring Heat Transfer Properties – Part 4: Measurements by Hot Box Methods.*

196. EN 673:1997 E. (1997) *Glass in Building – Determination of Thermal Transmittance (U Value) – Calculation Method.*

197. EN 674:1997 E. (1997) *Glass in Building – Determination of Thermal Transmittance (U value) – Guarded Hot Plate Method.*

198. EN 675:1997-E. (1997) *Glass in Building – Determination of Thermal Transmittance (U Value) – Heat Flow Meter Method.*

199. International Standards Organization ISO 8301:1991(E). (1991) *Thermal Insulation – Determination of Steady-State Thermal Resistance and Related Properties – Heat Flow Meter Apparatus,* International Standards Organization.

200. International Standards Organization ISO 8302:1991(E). (1991) *Thermal Insulation – Determination of Steady-State Thermal Resistance and Related Properties – Guarded Hot Plate Apparatus,* International Standards Organization.

201. Weast, R.C., Lide, D.R., Astle, M.J., and Beyer, W.H. (eds) (1989–1990) *CRC Handbook of Chemistry and Physics*

1989-1990, 70th edn, CRC Press, Boca Raton, FL, p. F-167.

202. Chiang, J.-C. and MacDiarmid, A.G. (1986) 'Polyaniline': protonic acid doping of the emeraldine form to the metallic regime. *Synth. Met.*, **13**, 193–205.

203. Cogan, S.F., Rauh, R.D., Plante, T.D., Nguyen, N.M., and Westwood, J.D. (1990) in *Electrochromic Materials*, Proceedings Volume 94–2, Proceedings of the Symposium on Electrochromic Materials, The 176th Meeting of The Electrochemical Society, Hollywood, FL, 16–18 October, 1989 (eds M.K. Carpenter and D.A. Corrigan), The Electrochemical Society, Pennington, NJ, pp. 99–115.

204. Huang, W.-S., Humphrey, B.D., and MacDiarmid, A.G. (1986) Polyaniline, a novel conducting polymer. Morphology and chemistry of its oxidation and reduction in aqueous electrolytes. *J. Chem. Soc., Faraday Trans. 1*, **82**, 2385–2400.

205. Córdoba de Torresi, S.I., Gorenstein, A., Torresi, R.M., and Vázquez, M.V. (1991) Electrochromism of WO_3 in acid solutions. An electrochemical, optical and electrogravimetric study. *J. Electroanal. Chem.*, **318**, 131–144.

206. Córdoba de Torresi, S.I. and Gorenstein, A. (1992) Electrochromic behaviour of manganese dioxide electrodes in slightly alkaline solutions. *Electrochim. Acta*, **37**, 2015–2019.

207. Kitao, M., Yamada, S., Yoshida, S., Akram, H., and Urabe, K. (1992) Preparation conditions of sputtered electrochromic WO_3 films and their infrared absorption spectra. *Sol. Energy Mater. Sol. Cells*, **25**, 241–255.

208. Seike, T. and Nagai, J. (1991) Electrochromism of 3d transition metal oxides. *Sol. Energy Mater.*, **22**, 107–117.

209. Zhang, J.G., Tracy, C.E., Benson, D.K., and Deb, S.K. (1993) The influence of microstructure on the electrochromic properties of Li_xWO_3 thin films: part I. Ion diffusion and electrochromic properties. *J. Mater. Res.*, **8**, 2649–2656.

210. Batchelder, D.N. (1988) Colour and chromism of conjugated polymers. *Contemp. Phys.*, **29**, 3–31.

211. Gale, R.J. (1988) *Spectroelectrochemistry. Theory and Practice*, Plenum Press, pp. 193–205.

212. Loudon, R. (1990) *The Quantum Theory of Light*, 2nd edn, Oxford University Press, pp. 24–30.

213. Silverstein, R.M., Bassler, G.C., and Morrill, T.C. (1981) *Spectrometric Identification of Organic Compounds*, 4th edn, John Wiley & Sons, Inc., New York, p. 307.

214. Jelle, B.P., Sveipe, E., Wegger, E., Gustavsen, A., Grynning, S., Thue, J.V., Time, B., and Lisø, K.R. (2014) Robustness classification of materials, assemblies and buildings. *J. Building Phys.*, **37**, 213–245.

215. Armand, M.B. (1980) in *Materials for Advanced Batteries* (eds D.W. Murphy, J. Broadhead, and B.C.H. Steele), Plenum Press, New York, pp. 145–161.

216. Passerini, S., Scrosati, B., and Gorenstein, A. (1990) The intercalation of lithium in nickel oxide and its electrochromic properties. *J. Electrochem. Soc.*, **137**, 3297–3300.

217. Di Pietro, B., Patriarca, M., and Scrosati, B. (1982) On the use of rocking chair configurations for cyclable lithium organic electrolyte batteries. *J. Power Sources*, **8**, 289–299.

16
Fabric Electrochromic Displays for Adaptive Camouflage, Biomimicry, Wearable Displays and Fashion

Michael T. Otley, Michael A. Invernale, and Gregory A. Sotzing

16.1
Introduction

Chromatism is the ability to change colour through some type of stimulus, and a variety of chromatic examples exist, such as thermochromism, photochromism and electrochromism. Electrochromism is the ability to change colour in response to an electrical charge, and the term electrochromism was originally used in 1932 to describe the phenomena of Franz–Keldysh and Stark's effects [1, 2]. Since the 1970s, electrochromic materials have become popular due to the development of conjugated conducting polymers. Inorganic electrochromism [3–5], viologens [6–9] and small-molecule organics have transformed the field [10–12]. Electrochromic applications that are currently commercialised include rear-view auto-dimming mirrors, smart windows, automotive sunroofs and aeroplane windows [13]. Gentex and Donnelly make use of small molecules, viologens, to create the auto-dimming feature for a reflective mirror device [14]. Smart windows and sunroof applications make use of a variety of technologies; Sage makes windows and skylights based on an inorganic electrochromic material, tungsten trioxide, whereas a variety of other companies use LCDs (liquid crystal displays) or SPDs (suspended particle displays) [15]. These systems all have a variety of advantages and disadvantages, but when it comes to making a smart fabric, the major issue is that their architectures cannot be translated without the use of an optically transparent conductor. Practically, this means that another technology must be sought if wearable and flexible colour-changing devices are ever to become a commercial reality.

The attractiveness of ECDs (electrochromic devices) is the possibility of making flexible devices and displays with more potential applications than non-flexible displays. This is currently accomplished by using tin-doped indium oxide (ITO) coated onto poly(ethylene terephthalate) PET substrates. These flexible ECDs segue into the next generation of ECDs: electrochromic fabric devices (EFDs). EFDs are not only flexible but also stretchable. They are fully integrated fabrics that possess the ability to change colour, through using all-organic materials, without the limitation of a rigid substrate.

Electrochromic Materials and Devices, First Edition.
Edited by Roger J. Mortimer, David R. Rosseinsky and Paul M. S. Monk.

Figure 16.1 (a) A chameleon and (b) a cuttlefish showing their ability to use adaptive camouflage to conceal themselves from predators.

Adaptive camouflage has been around for thousands of years in nature. For example, the cuttlefish and the chameleon each have the ability to adaptively change their colour to camouflage into their surroundings for protection from predators, Figure 16.1. Also, their skin can change colour due to emotion, when frightened, during fraternisation and more.

Ideally, EFDs would provide high contrast, sharpness of colours and also a variety of colours in order to provide functions such as adaptive camouflage, fashion, wearable displays (advertising) and so on. The military would benefit from adaptive camouflage not only for soldier protection but also for stealth vehicles, aeroplanes, ships, drones and the like. However, EFDs could change fashion forever. The designs could be infinite and forever changing, and some companies may even trend towards human advertising (Figure 16.2).

16.1.1 Colour-Changing Technologies Background

There are two major classes of electrochromic materials: organic and inorganic. The organic class is subdivided into small molecules and polymeric systems. Other colour-changing technologies, like LCDs, light emitting diodes (LEDs) and SPDs, are not strictly 'electrochromic' as the voltage applied to effect a colour change does not induce redox changes. Beyond these, there are two passive colour-changing technologies: photochromic and thermochromic, using light and heat, respectively, to modulate colour. Less common systems that also involve colour exhibit stimuli such as pressure or changes in pH, as well as those that employ structural colour, mainly found in nature (iridescent butterfly wings, peacock feathers and innumerable insect shells). The focus of this chapter is on organic polymeric electrochromic materials, as this class represents the best technical approach to developing EFDs. Several examples of the other technologies, particularly those which have seen some measure of commercialisation, are also briefly discussed.

Figure 16.2 A conceptualised image of how adaptive camouflage technology could conceal a soldier.

ECDs are electrochemical cells consisting of two electrodes separated by an ion-containing electrolyte. The electrolyte can be a liquid or gel or solid that is sandwiched between two substrates. The most commonly used substrates are ITO-coated on glass or PET [16]. Organic polymer electrochromics are conjugated polymers whose defining characteristic is an alternating single–double-bond backbone. This backbone is amenable to either oxidation or reduction and thus delocalisation of charge resulting in electrical conductivity. Accompanying the oxidation and reduction processes is a change in the absorption characteristics of the material. Normally, conjugated conducting polymers are electrochemically deposited from a monomer/electrolyte bath onto the desired substrate. However, a variety of processing improvements have made this process more versatile, including the use of spray-castable precursors [17]. Figure 16.3 shows a schematic representation of how an ECD can be assembled using such a precursor.

16.1.2 Previous Work

The first published work on the subject of EFDs was in 2010 with the invention of Invernale *et al.* [18]. There had not been a fully functional electrochromic fabric

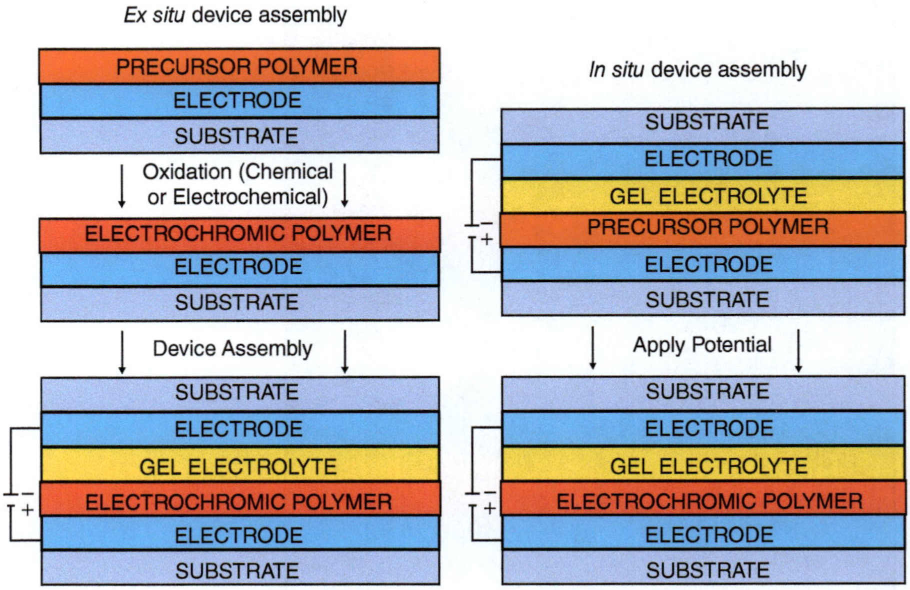

Figure 16.3 Schematic of the ECD assembly steps utilising spray-castable precursor polymers and an external oxidative conversion (a) or an internal electrochemical oxidative conversion (b).

previously, though attempts at colour-changing clothing had been made. These other types of colour-changing clothing include thermochromic, photochromic and LED/LCD technologies, and these are discussed briefly at the end of this chapter. The seminal work on EFDs, by Invernale *et al.*, described an ECD based on spandex (a polyester-polyurethane copolymer, alternatively known as Lycra or Elastane) electrodes and using a device architecture that has never before existed in the field. The fundamental advantages of this colour-changing fabric system are numerous, though there are three key elements that make it the most promising technology for the true commercial realisation of colour-changing garments, adaptive camouflage and wearable displays. The key benefits are the ability to use all-organic components, the architecture itself and the processability of the system. Figure 16.3 shows a schematic of the device stack (bottom) as well as a representative electrochromic polymeric material (top). The conducting material was poly(3,4-ethylenedioxythiophene)/poly(styrene sulfonate) (PEDOT–PSS), which is an aqueous dispersion of the conjugated conducting polymer PEDOT and the polymeric anionic dopant PSS. This material was soaked into the chosen substrate, in this case spandex, though we discuss other fabric substrates in the following section, as the follow-up paper on the seminal work closely examined this subject. The combination of substrate and conductor was then allowed to dry. The resulting fabric was now conductive, on the order of $0.1\ S\,cm^{-1}$, and it retained all of the characteristics of the base fabric, namely flexibility, stretchability, nap and feel.

The formerly white spandex was then a light-blue colour, due to the slight coloration of the PEDOT conductor. The impacts of colour on the EFDs are discussed based on the third publication, which considers a variety of factors affecting the perceived colours. It is important to note that the conductor is not a thin-film coating of any kind, but is rather an intercalated, imbedded system of polymeric components of sufficient proximity to achieve percolation. The entirety of the spandex is an electrode; it cannot be damaged or conduction broken by rubbing, bending or stretching the fabric. This is a unique advantage over other conjugated conducting polymer works performed on fabrics [19–29]. Two spandex electrodes were then cut to size, approximately 1 inch by 0.5 inch and were assembled into an ECD. The working electrode was spray-coated with a processable conjugated conducting polymer precursor [30, 31]. Once a sufficient thickness of this precursor was cast onto one of the spandex electrodes, it can be converted to a functional electrochromic polymer via chemical or electrochemical oxidation. Then, each electrode is coated with a UV-curable gel electrolyte composed of a cross-linkable polymer matrix, a lithium salt and a photoinitiator. Once coated, the electrolyte is cured using a UV chamber with a 365 nm light source. The electrodes are then stacked such that the counter electrode is *beneath* the working electrode. One of the key elements of this system is that it creates an ECD using this never-before-utilised architecture. The gel electrolyte provides a separation between electrodes, preventing electrical shorting, and its encasement of both electrodes also allows for an applied bias to accomplish electrochromic switching. The ion movement which accompanies the reversible oxidation/reduction reaction being experienced by the conjugated conducting polymer is able to be achieved using the 'reflective device' architecture developed by Invernale *et al.* This proof-of-concept laminar stack easily expands to the conception of interwoven devices using individual electrochromic threads, as well as printable, flat EFDs that achieve the necessary chemistry using coplanar electrodes. It is a powerful and versatile device conception that forms a very promising foundation for the development of an EFD garment.

Secondly, the electrode material, the fabric, the electrolyte and the colour-changing components are all composed of organic materials. No optically transparent conductor is needed to complete the device. Since this system does not require ITO (an optically transparent conductor that is ubiquitous to the display industry, as well as touch screen electronic devices), these devices have numerous properties concomitant with the desired specifications for a wearable ECD. The ability to avoid ITO, which is brittle and difficult to process, allows for a fabric material with the potential to bend, flex, stretch and conform the way a garment should. As mentioned previously, the spandex electrodes retain the properties of the base fabric after impregnation with PEDOT–PSS. The formation of this percolated network of conducting material is accomplished at very low loading and is thus not restricting to the material. Figure 16.4 shows an electrode coated with conjugated conducting polymer that has been stretched and switched. You can clearly see a black background through the holes in the threading. The system is still completely functional. With the appropriate

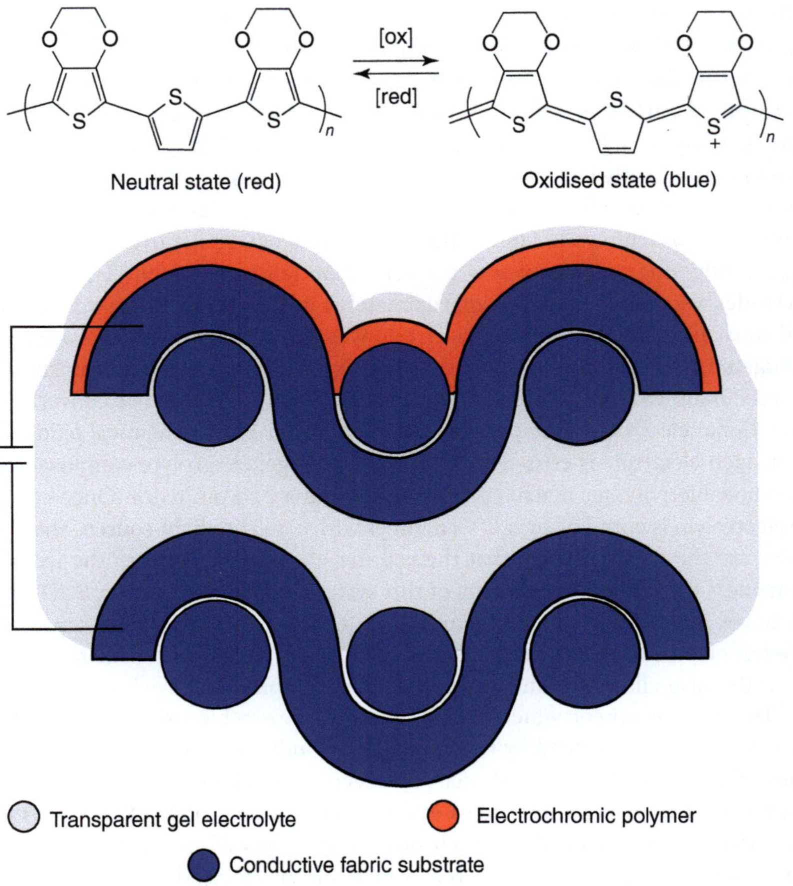

Figure 16.4 Schematic of the all-organic electrochromic spandex device developed by Invernale *et al.* (a) Chemical structure of the colour-changing material in its two extreme states. (b) Device architecture showing the unique stack for an EFD.

development of complementary electrolytes, this use of all-organic materials for the EFD's construction points towards the potential for the creation of a fully flexible, stretchable, breathable and wearable EFD (Figure 16.5).

The third key advantage is the processability of these devices. For any technology, it is critical that the system must be amenable to a cost-effective product from both a material cost perspective and the mechanical realities of assembly and manufacturing. Otherwise, its benefits will never outweigh its price. The use of an all-organic system allows for low raw material costs. The simplicity of the electrode design, coupled with the low loading percentage of conductor and the breadth of ways it could be impregnated (dip, soak, spray, inkjet, etc.), makes the fabric electrode components readily suited to fit into many current manufacturing processes. The processable conjugated conducting polymers utilising the precursor method also share these advantages. Figure 16.6 shows an image of an

Figure 16.5 Images of a spandex electrode coated with an electrochromic polymer. The electrode was stretched to its maximum recommended elongation of 20% and was then cycled between its coloured states. This system is fully functional in this manner due to the lack of ITO and the use of all-organic materials.

Figure 16.6 A simple phrase, UCONN, patterned in conjugated conducting polymer on a spandex EFD. The tremendous versatility in manufacturing choices affords not only innumerable processing options but also creative applications of the technology.

easily patterned simple word on a spandex EFD. This type of versatility allows any unforeseen obstacles in the assembly process to be overcome via compromises and alternative choices in processing. It also allows for creative designs and applications of the technology. For example, double-sided fabric ECDs are also possible using this architecture. They can be alternating-colour or same-colour, depending on the choice of stack. Figure 16.7 shows an example of an alternating-colour device with a mirror behind it. The only step that is not currently a common piece of equipment for a garment manufacturer would be the UV curing process. However, there are also thermally curable gel electrolyte compositions, and thus, it is entirely feasible that current manufacturers have everything they would need to implement this system. This sort of advantage can often be overlooked, but it is important to keep in mind if researchers wish to translate their ideas to real-world applications and have an impact on a global scale. Current projects in the Sotzing laboratory are focused on bringing newer and better electrolyte systems to this fabric device architecture, imparting durability, flexibility and stretchability, among other characteristics.

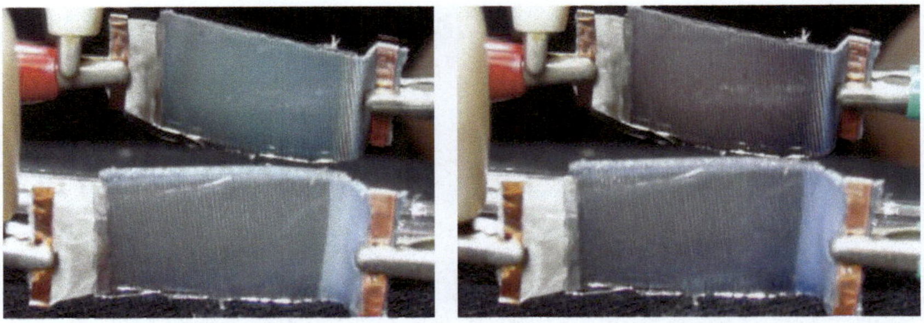

Figure 16.7 An example of a creative application of EFD technology: a double-sided fabric ECD. Both sides of the device have been coated with a conjugated conducting polymer in an alternating-colour device stack. The mirror shows the colour of the other side of the device.

16.1.3
Conductivity Trends of PEDOT–PSS Impregnated Fabric and the Effect of Conductivity on Electrochromic Textile

Following the invention of EFD architecture, a variety of related and continuing studies were carried out. Ding *et al.* sought to more fully understand the nature of the conductor within a fabric matrix. Ding *et al.* [32] showed the differences between fabric types with respect to uptake and conductivity, the effect of processing conditions, conductor loading optimisation, as well as an evaluation of the conductor's stability under mechanical stresses typically experienced by the fabric matrix [32].

The spandex studied was composed of 5% lycra and 95% cotton. To better understand the differences between fabric type and their function as electrodes for EFDs, other combinations and compositions were tested. 100% cotton, 100% polyester, 60% cotton/40% polyester, 58% polyester/40% rayon, 20% lycra/80% nylon and 50% nylon/50% polyurethane fabrics were all examined during the study. Figure 16.8 represents the data for conductivity observed as well as a table of water uptake for the various materials. There is a correlation between the ability of a given fabric to absorb water, which is the carrier fluid for PEDOT–PSS, and the resultant conductivity of the electrode. However, this, alone, is not enough to explain the differences. Factors such as thread size or additives to the fabric can also interfere with or enhance the conductivity of the base fabric. Significant ongoing work in the Sotzing laboratory continues to investigate these effects in an effort to maximise the conductivity of fabric for a variety of applications, including use as a component in EFDs.

While water uptake was found to play a role, the variance between these fabrics with respect to their conductivities was relatively small. There were no order of magnitude increases, and the highest value obtained using the standard processing procedure was only $0.25\,S\,cm^{-1}$. Traditional ECDs use electrodes,

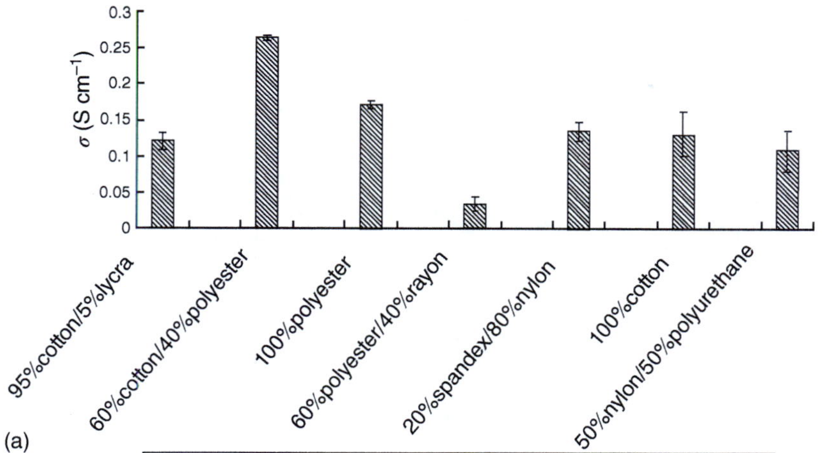

(a)

(b)

Textile type	Water absorbed (mg)
95%cotton/5%lycra	299.36
60%cotton/40%polyester	310.21
100%polyester	207.5
58%polyester/40%rayon	287.66
20%spandex/80%nylon	177.84
100%cotton	326.73
50%nylon/50%polyurethane	163.77

Figure 16.8 (a) Conductivity of various textiles soaked with PEDOT–PSS. (b) Water absorbed by different fabrics.

such as ITO-coated substrates, with conductivities of 1000–3000 S cm^{-1}. It is well worth noting the observation that a 10 000-fold lower conductivity can sufficiently supply the necessary power to switch these fabric devices. However, low conductivities will also result in slower diffusion rates for the electrolyte, as ion movement is dependent upon the strength of the applied bias that is driving their migration. Using the spandex fabric, it was found that by increasing the number of soaking cycles and soaking time, an optimum weight percent of PEDOT–PSS could be loaded into the fabric, resulting in a maximisation of its conductivity. It was found that fabric was saturated with conductor after 60 s, but that multiple cycles in a refreshed solution resulted in increased loading. After 10 cycles, the fabric was as loaded with PEDOT–PSS as possible. This translated into a conductivity enhancement from 0.1 to 1.5 S cm^{-1}. This order-of-magnitude increase in conductivity resulted in a reduction in switching time from 16 to 10 s. This result was not expected to be an order of magnitude change, however, because there are many other factors which effect ion diffusion in addition to conductivity.

Finally, hysteresis was studied for repeated stretching of the spandex electrode. This was done to ensure that the mechanical properties of the base fabric were

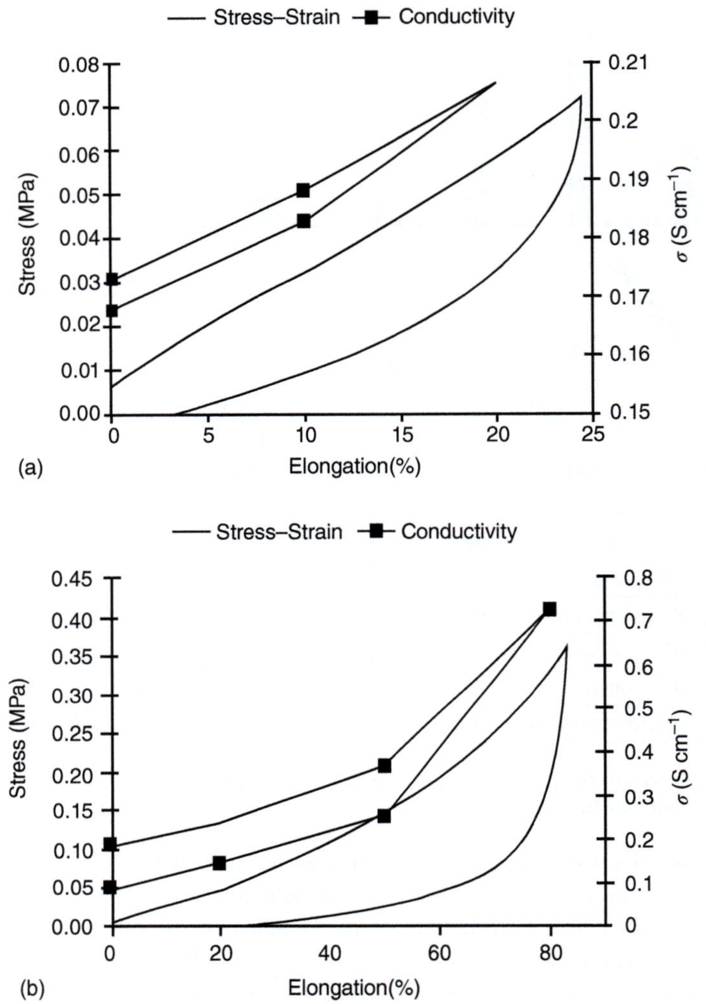

Figure 16.9 (a) Hysteresis for conductivity and stress-strain for spandex electrodes at an elongation of 20%. (b) Hysteresis for conductivity and stress–strain for spandex electrodes at an elongation of 80%.

not impeded by the addition of PEDOT–PSS and that conductivity does not suffer as a result. Figure 16.9 shows the hysteresis curves for the effect of conductivity under repeated stretching to different elongations. It was found that, under normal strain (for spandex, 20–25% elongation is the recommended maximum), the electrodes experienced slight losses in conductivity, changing by 12.5%; however, when stretched to 80% elongation, the conductivity was halved. This is expected, as elongation of the fibre beyond a certain point will cause loosening of the thread and fibre bundles that make up the material. This, then, causes a separation of conductive chains and a decrease in the substrate's ability to carry charges effectively.

16.1.4
The Effects of Coloured-Based Fabric on Electrochromic Textile

One of the most widely studied aspects of conjugated conducting polymers is the ability to tune their coloration to any desired hue and saturation. This work has been done towards the development of full-colour displays, among other applications. For EFDs, colour tuning is also a critical aspect and a crucial feature that must be understood and controlled if wearable displays, adaptive camouflage or other products are to be developed. There is, however, a fundamental difference between a traditional ECD and an EFD when it comes to the perceived colour. Namely, EFDs are reflective devices, and most ECDs and electrochromic polymers are evaluated via UV–Vis–NIR transmission spectra. Reflective devices must consider subtractive colour mixing during their design, or else the resulting colour of the material will not be as intended. For an electrochromic window, it is obvious that the spectrum of the colour-changing polymer will tell you the colour of the device very reliably, given the appropriate illuminant; however, using the same chromogenic polymer on a fabric device could yield a drastically different result. 'The effects of coloured-based fabric on electrochromic textile' was published in the *Journal of Coloration Technology* in 2011 by Invernale *et al.*, and it addressed this particular dilemma [33].

The architecture of these EFDs is such that the observed colour is actually an agglomeration of the colours from the fabric, the conductor and the colour-changing polymer. These colours are then further affected by the electrolyte which surrounds them, as it changes the way light gets in to and out of the reflective device. The end result for the observer is simple, they only have to look at it and their eyes will calculate the perceived colour for them, but for the researcher and the product development engineer, colour description is very complex. There is still a significant amount of work to be done regarding colour tuning for fabric electrochromics, but one of the pieces of the puzzle was explored and elucidated in this paper [33].

The study focused on a single chromophore, a red-to-blue electrochromic polymer of bisEDOT-thiophene, which was used in the previous two studies. The aim was to explore how this material's colour properties could be shifted without using a chemically distinct electrochromic material. Fifteen different coloured cotton fabrics were used in the production of EFDs, and the colours of each were monitored at each stage of the process. The thickness of the chromophore was kept constant, as were all other processing conditions. Figure 16.10 shows actual images of the fabric in each of its states from bare, electrode substrate, precursor coated, oxidised colour state and neutral colour state. The CIE chromaticity coordinates for the transformations are also plotted as in Figure 16.11. From this work, it is clear that the underlying substrate plays a critical role not only in its ability to perform the duties of an electrode but also as a chromatic component of the EFD.

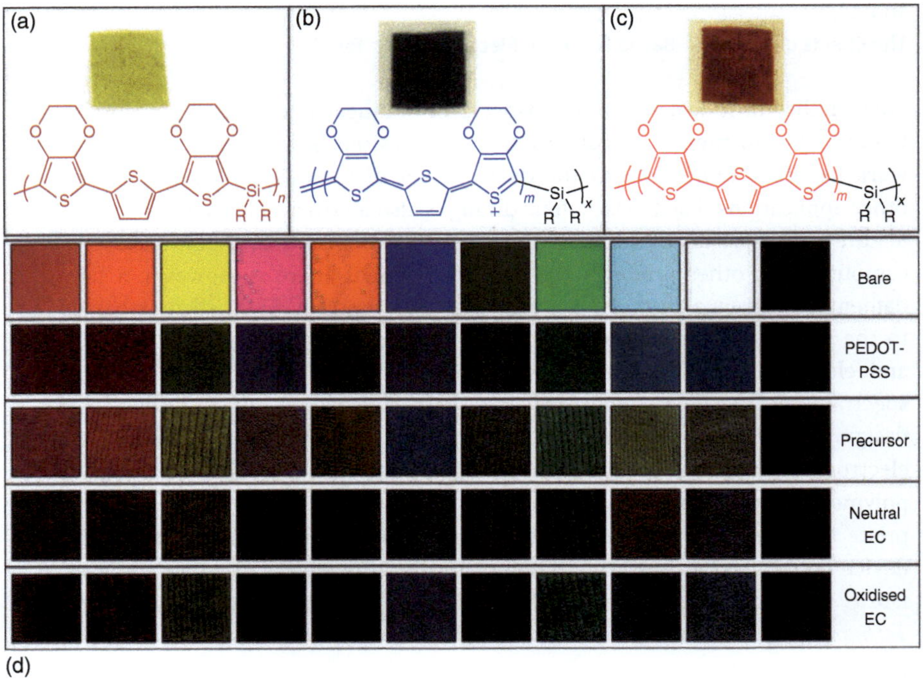

Figure 16.10 (a–c) Images of white cotton fabric without PEDOT–PSS soaking that were spray-cast with a processable precursor polymer and converted chemically to the electrochromic. The chemical structures of the polymer in each coloured stage are shown below. (d) Photographs of actual fabric showing the various colours obtained at each stage of EFD preparation procedure. Bare = base fabric; PEDOT–PSS = soaked fabric that is now an electrode; precursor = spray-coated electrode with the precursor polymer; neutral EC = converted electrochromic in the neutral state; oxidised EC = converted electrochromic in the oxidised state. The fabrics are always in the same order, left to right: red, orange, yellow, hot pink, tangerine, blue, army green, Irish green, light blue, white and black.

16.1.5
Other Electrochromic Fabric

There have been other types of EFDs reported in the literature, each of them exhibits significant architectural differences from the seminal EFDs discussed in the previous sections. One example, reported by Meunier *et al.*, illustrates the use of Prussian blue on a flexible substrate [34]. The device they fabricated exhibits the use of what they describe as a four-layer system, consisting of a textile substrate, a flexible electrode (silver or carbon black), an electrochromic ink contained in a spacer and a second electrode made of PET-ITO on top. The fabric used in the device consisted of PET coated with a layer of polyurethane. Then, on top of the PET and polyurethane, the first electrode, made of either carbon black or silver,

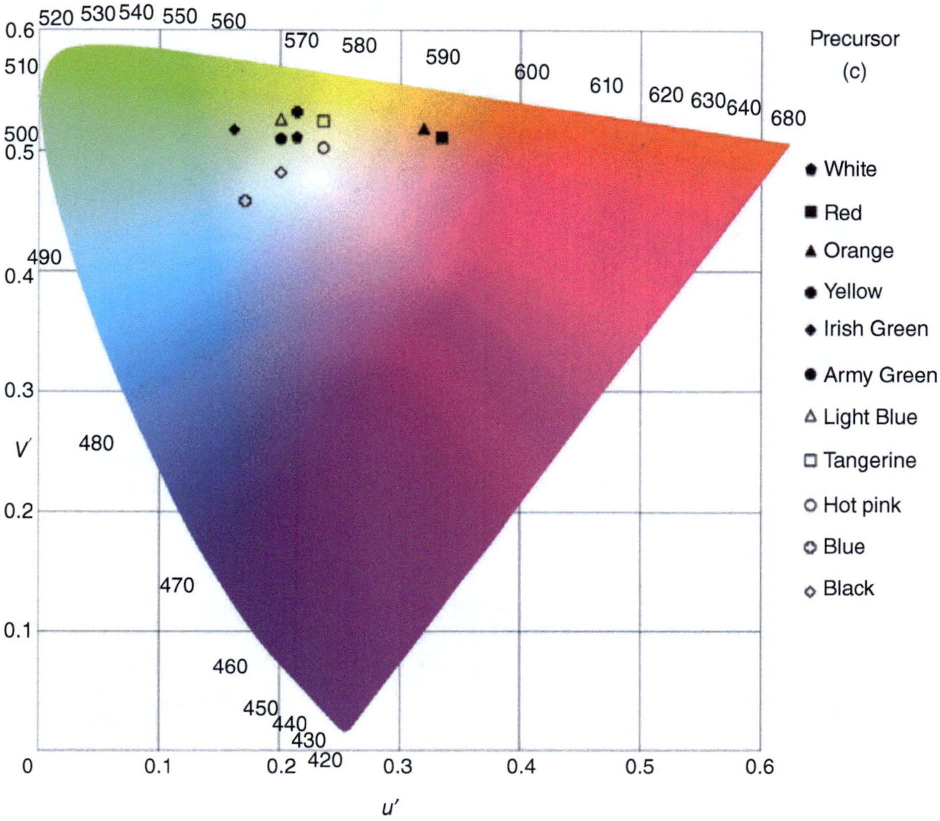

Figure 16.11 CIE Lu′v′ chromaticity coordinates for PEDOT–PSS loaded fabric that has been coated with the precursor polymer.

was coated. A suspension of Prussian blue was sandwiched between the ITO-coated PET and the fabric electrode assembly, the edges sealed with neoprene glue for flexibility. However, as can be seen in Figure 16.12, the neoprene glue is highly visible, and the applications for EFDs will be limited due to the use of a rigid top layer of PET-ITO. The device architecture is essentially a reproduction of a standard reflective-type ECD, using a modified fabric as the opaque electrode instead of a metallic mirror or titanium dioxide-based white surface.

Another EFD construction exhibited an electrochromic effect but its operational mechanism relied on more than just the voltage of a traditional ECD. Laforgue used PEDOT nanofibres as a conducting matrix to resistively heat a fabric that had thermochromic microcapsules imbedded in the PEDOT regions, thus producing a colour-changing effect [35]. Laforgue's work could be practically described as a thermoelectrochromic fabric system. This system could be manufactured with nanofibre non-wovens and then employing the printing

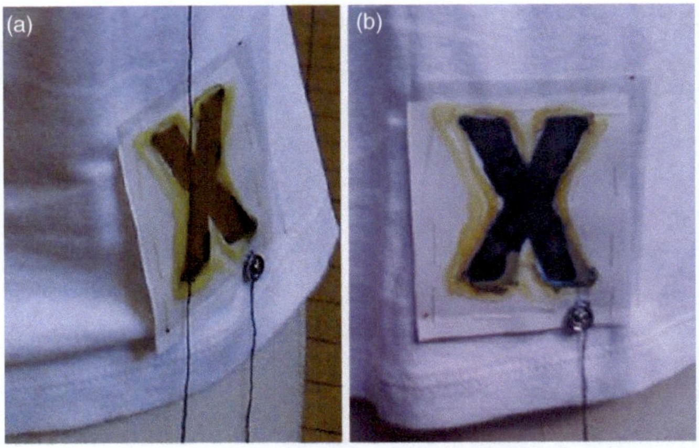

Figure 16.12 A flexible electrochromic fabric device using silver-coated fabric and PET-ITO as the electrodes and Prussian blue as the electrochromic material, shown in its (a) oxidised and (b) neutral states.

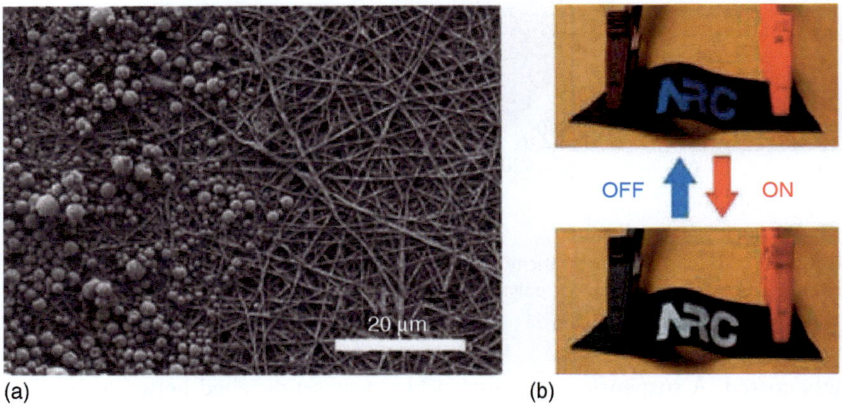

Figure 16.13 Laforgue *et al.* have produced a thermoelectrochromic system that uses resistive heating of a conductive, all-organic fabric to effect a colour change in thermochromic microcapsules. (a) SEM image of thermochromic microcapsules painted on the PEDOT nanofibres. (b) Illustration of the electrochromic effect by applying 0/120 mA to the conductive fabric.

of logos with a thermochromic ink to produce colour-changing fabric, as seen in Figure 16.13. This is a unique approach in that it does not require a second electrode or a sandwich, and it is entirely fabric-based, making it amenable to flexibility and wearability. The downside of such a system is the heavy impact of ambient temperature on thermochromic systems, as well as the energy requirements for sustained resistive heating.

16.2 Non-Electrochromic Colour-Changing Fabric

Colour-changing fabric is not limited to electrochromic materials. There have been several technologies besides electrochromism used for colour-changing fabric, including thermochromic [36, 37], photochromic [38–42], fibre optics [43], LCDs [44] and LEDs [45, 46]. Some examples are described in detail in this section, ranging from shoes to shirts to intimate apparel. Some of the technologies have been commercialised and could inform possible potential applications of ECDs.

16.2.1 Thermochromic Fabric

Thermochromism is the ability to change colour due to a change in temperature. Generally, thermochromic fabrics are manufactured with two separate dyes, where one dye stays constant despite a change in temperature and the other, leuco dye (thermochromic), which is enclosed in microcapsules, is attached to the fibres of the fabric [37]. The thermochromic dye is usually colourless in one state and coloured in the other, which is caused by a chemical reaction that occurs when it reaches a certain temperature range (either a rise in temperature or a drop in temperature, depending on the dye). The colour is dependent upon the dye of the fabric and the thermochromic dye in its coloured and/or bleached state. There are several products currently commercialised exhibiting thermochromic fabrics. For example, in 2011, Puma released thermochromic shoes in conjunction with the Freezy Freakies brand, whose thermochromic dyes undergo a colour change with a drop in temperature, as seen in Figure 16.14 [47].

16.2.2 Photochromic Fabric

Photochromism is the ability to reversibly change colour when exposed to light, and the phenomenon was originally discovered in the 1880s. Photochromic dyes are typically made of organic molecules and are in their clear state until they are

Figure 16.14 An example of a thermochromic fabric system, here in the form of a shoe marketed by Puma in 2011.

Figure 16.15 An example of photochromic fabric from Del Sol.

irradiated by ultraviolet light [40]. The λ_{max} (peak absorbance wavelength), which is 365 nm, is generally referred to as a 'black light', and ultraviolet light is mostly in the 320 and 400 nm range. The photochromic dyes undergo a molecular excitation transition when exposed to ultraviolet light causing the dye to shift its energy into the visible range [42]. When not in the presence of heat and ultraviolet light, the photochromic dye undergoes the reverse reaction and reverts to its clear state [42]. Thus, photochromic fabric has significant limitations to adaptive camouflage, displays and other applications because it is a passively controlled system and cannot be made user-controlled. An example of a photochromic fabric can be seen in Figure 16.15. Del Sol has been manufacturing photochromic clothing since 1994 and uses its own proprietary Spectrachrome® photochromic dyes [48].

16.2.3 LED and LCD Technology

LEDs have been around since the early 1960s, originally only emitting red light. Since then, they have been developed to cover nearly the entire spectral range and have found utility in many display and lighting applications. The use of LEDs in shoes has been around since the early 1990s, with LA Gear's L.A. Lights [49], which came out in 1992. However, the lights were only imbedded in the rubber portion sole of the shoe and not on the fabric itself. Philips, however, launched Lumalive in 2007, which uses LEDs sewn into clothing and fabrics, such as couches

Figure 16.16 An example of Philips Lumalive clothing.

and pillows, to display simple imagery as shown in Figure 16.16 [46]. These devices require a lot of power, however, and the battery constraints make portability and washability a concern.

LCDs have been around even longer since liquid crystals were originally discovered in 1888 and have been used for simple monochromatic displays in everything from calculators to gas pumps. Their use in apparel seems limited; however, more recently, Studio Roosegaarde has been developing colour-changing apparel that uses sensors to change the colour of an LCD-garment according to the mood [50]. The garment measures the person's heart rate, and an increase in pulse signals the LCD clothing to go from dark to light. As stated by the designer, it can be used on the red carpet or in sensual situations. Figure 16.17 shows a dress from the Intimacy Black clothing line released in 2010.

16.3 Conclusion

The realm of EFDs is still in its infancy, with much research and development still needed in order to realise the full potential of this technology. The groundwork laid by Invernale *et al.* with the invention of all-organic fabric electrochromics is a promising start for the field. It enables scientists to continue to work towards colour-changing clothing from fabric that is flexible with uses in adaptive camouflage, biomimicry, wearable displays and fashion.

Figure 16.17 The left side shows the dark state and on the clear is the dark state for Intimacy Black, an LCD-based apparel product line that links the colour change to the wearer's heart rate.

There remain many challenges to delivering on the promise of EFDs, however. The need for full colour displays necessitates the synthesis of a variety of electrochromic polymers that can cover the full visible range of colours. Enhancement of electrode conductivity and an understanding of substrate colour are needed for the longevity, speed and colour tuning of the final devices. The durability of the electronic system must also be optimised. These challenges are not insurmountable, however, and will require simply time and effort on the part of talented engineers.

EFDs also have the potential to serve as sensors, because the fabric is already conductive. Under Armour is envisioning clothing that monitors heart rate, body temperature and changes colour [51]. The example is shown in Figure 16.18 where the woman's jogging suit changes from grey to blue. The conductive fabric can double as a sensor to monitor these parameters as well as serve as the display for the information.

The fashion world can benefit from electrochromic fabric as well. The designs can be endless and ever-changing. A person could form their own design and download it to the clothing. Adaptive camouflage and military applications are logical extensions of this technology as well. The future of this field may appear to be in the realm of science fiction, but it is rapidly becoming science fact. The basic technology exists to make it a reality; it is only a matter of time before the challenges are met and EFDs become a part of our everyday lives.

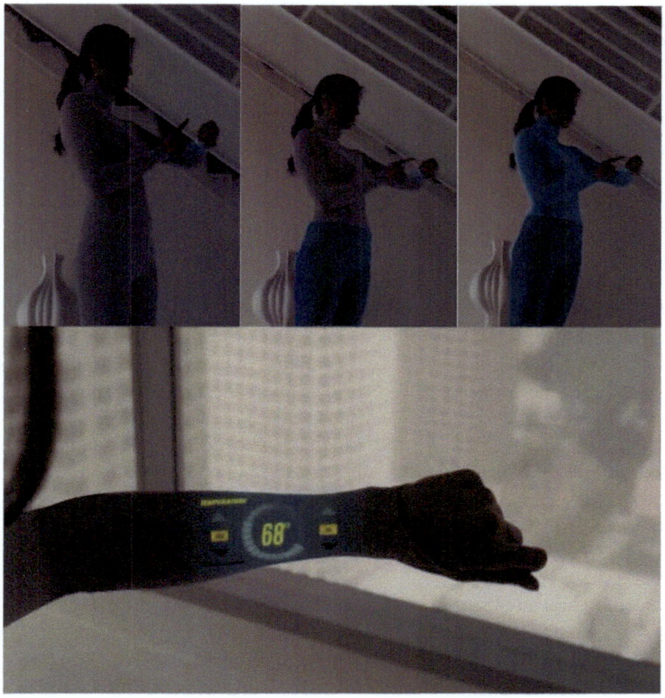

Figure 16.18 A concept of a future Under Armour product: a full-body running suit capable changing colour from grey to blue with an integrated arm-display and control mechanism.

References

1. Leventis, N. (1995) Polymers in Electrochromics. *Polym. News*, **20**, 5.
2. Mortimer, R.J. (2013) Switching colors with electricity. *Am. Sci.*, **101**, 38.
3. Granqvist, C.G. (2000) Electrochromic tungsten oxide films: review of progress 1993–1998. *Sol. Energy Mater. Sol. Cells*, **60** (3), 201–262.
4. Deb, S.K. (2008) Opportunities and challenges in science and technology of WO_3 for electrochromic and related applications. *Sol. Energy Mater. Sol. Cells*, **92** (2), 245–258.
5. Avendaño, E., Berggren, L., Niklasson, G.A., Granqvist, C.G., and Azens, A. (2006) Electrochromic materials and devices: brief survey and new data on optical absorption in tungsten oxide and nickel oxide films. *Thin Solid Films*, **496** (1), 30–36.
6. Monk, P.M.S. (1998) *The Viologens: Physicochemical Properties, Synthesis and Applications of the Salts of 4,4'-Bipyridine*, John Wiley & Sons, Ltd, Chichester.
7. Cinnsealach, R., Boschloo, G., Rao, S.N., and Fitzmaurice, D. (1998) Electrochromic windows based on viologen-modified nanostructured TiO_2 films. *Sol. Energy Mater. Sol. Cells*, **55** (3), 215–223.
8. Sun, X.W. and Wang, J.X. (2008) Fast switching electrochromic display using a viologen-modified ZnO nanowire array electrode. *Nano Lett.*, **8** (7), 1884–1889.
9. Kurth, D.G., López, J.P., and Dong, W. (2005) A new Co(II)-metalloviologen-based electrochromic material integrated in thin multilayer films. *Chem. Commun.*, **16**, 2119–2121.

10. Granqvist, C.G. (1995) *Handbook of Inorganic Electrochromic Materials*, Elsevier, p. 663.
11. Ohsedo, Y., Imae, I., and Shirota, Y. (2003) Synthesis and electrochromic properties of a new family of methacrylate polymers containing pendant oligothiophenes. *J. Polym. Sci., Part B: Polym. Phys.*, **41**, 2471–2484.
12. Imae, I., Nawa, K., Ohsedo, Y., Noma, N., and Shirota, Y. (1997) Synthesis of a novel family of electrochemically-doped vinyl polymers containing pendant oligothiophenes and their electrical and electrochromic properties. *Macromolecules*, **30**, 380–386.
13. Pawlicka, A. (2009) Development of electrochromic devices. *Recent Pat. Nanotechnol.*, **3**, 177.
14. Desmond, J.P. and Lawlor, P.J. (1994) Rearview mirror. US Patent D351,370, Oct. 11, 1994.
15. Lampert, C.M. (2003) Large-area smart glass and integrated photovoltaics. *Sol. Energy Mater. Sol. Cells*, **76**, 489–499.
16. Granqvist, C.G. and Hultåker, A. (2002) Transparent and conducting ITO films: new developments and applications. *Thin Solid Films*, **411**, 1–5.
17. Beaujuge, P.M. and Reynolds, J.R. (2010) Color control in π-conjugated organic polymers for use in electrochromic devices. *Chem. Rev.*, **110**, 268–320.
18. Invernale, M.A., Ding, Y., and Sotzing, G.A. (2010) All-organic electrochromic spandex. *ACS Appl. Mater. Interfaces*, **2**, 296–300.
19. Gregory, R.V., Kimbrell, W.C., and Kuhn, H.H. (1991) Electrically conductive nonmetallic textile coatings. *J. Coated Fabr.*, **20**, 167–175.
20. Beaupré, S., Dumas, J., and Leclerc, M. (2006) Toward the development of new textile/plastic electrochromic cells using triphenylamine-based copolymers. *Chem. Mater.*, **18** (17), 4011–4018.
21. Li, X., Zhao, G., Qian, J., and Fu, Z. (2009) Preparation, structure and electrochromic properties of poly(3,4-ethylenedioxythiophene) based conducting textile. *Gaodeng Xuexiao Huaxue Xuebao/Chem. J. Chin. Univ.*, **30** (5), 1052–1054.
22. Li, X., Qian, J., and Fu, Z. (2009) Preparation and electrochromic properties of polyaniline based on conducting textile. *J. Beijing Inst. Clothing Technol. (Nat. Sci. Ed)*, **29** (2), 12–17.
23. Takamatsu, S., Matsumoto, K. and Shimoyama, I. (2009) Stretchable yarn of display elements. Paper presented at the Proceedings of the IEEE International Conference on Micro Electro Mechanical Systems (MEMS), pp. 1023–1026
24. Zhang, Q., Xin, B., and Lin, L. (2013) Preparation and characterisation of electrochromic fabrics based on polyaniline. *Adv. Mater. Res.*, **651**, 77–82.
25. Molina, J., Esteves, M.F., Fernández, J., Bonastre, J., and Cases, F. (2011) Polyaniline coated conducting fabrics chemical and electrochemical characterization. *Eur. Polym. J.*, **47** (10), 2003–2015.
26. Mokhtari, J. and Nouri, M. (2012) Electrical conductivity and chromic behavior of poly (3-methylthiophene) – coated polyester fabrics. *Fibers Polym.*, **13** (2), 139–144.
27. Kelly, F.M., Meunier, L., Cochrane, C., and Koncar, V. (2013) Polyaniline: application as solid state electrochromic in a flexible textile display. *Displays*, **34** (1), 1–7.
28. Kelly, F.M., Meunier, L., Cochrane, C., and Koncar, V. (2013) Evaluation of solid or liquid phase conducting polymers within a flexible textile electrochromic device. *IEEE/OSA J. Disp. Technol.*, **9** (8), 626–631.
29. Kelly, F.M., Johnston, J.H., Borrmann, T., and Richardson, M.J. (2007) Functionalised hybrid materials of conducting polymers with individual fibres of cellulose. *Eur. J. Inorg. Chem.*, **2007**, 5571–5577.
30. Invernale, M.A., Ding, Y., Mamangun, D.M.D., Yavuz, M.S., and Sotzing, G.A. (2010) Preparation of conjugated polymers inside assembled solid-state devices. *Adv. Mater.*, **22** (12), 1379–1382.
31. Sotzing, G.A., Reynolds, J.R., and Steel, P.J. (1997) Poly (3, 4-ethylenedioxythiophene)(PEDOT)

prepared via electrochemical polymerization of EDOT, 2, 2′-Bis (3, 4-ethylenedioxythiophene)(BiEDOT), and their TMS derivatives. *Adv. Mater.*, **9** (10), 795–798.
32. Ding, Y., Invernale, M.A., and Sotzing, G.A. (2010) Conductivity trends of PEDOT–PSS impregnated fabric and the effect of conductivity on electrochromic textile. *ACS Appl. Mater. Interfaces*, **2** (6), 1588–1593.
33. Invernale, M.A., Ding, Y., and Sotzing, G.A. (2011) The effects of coloured base fabric on electrochromic textile. *Color. Technol.*, **127** (3), 167–172.
34. Meunier, L., Kelly, F.M., Cochrane, C., and Koncar, V. (2011) Flexible displays for smart clothing: part II-electrochromic displays. *Indian J. Fibre Text. Res.*, **36** (4), 429–435.
35. Laforgue, A. (2010) Electrically controlled colour-changing textiles using the resistive heating properties of PEDOT nanofibers. *J. Mater. Chem.*, **20** (38), 8233–8235.
36. Fujita, K., Inagaki, H., Ishimura, N., Kataoka, T., Kito, T., Matunami, N., Nakasuji, N., Ozaki, M. and Shibahashi, Y. (1987) Thermochromic textile material. US Patent 4,681,791, issued Jul. 21, 1987.
37. Christie, R.M. and Bryant, I.D. (2005) An evaluation of thermochromic prints based on microencapsulated liquid crystals using variable temperature colour measurement. *Color. Technol.*, **121** (4), 187–192.
38. Cheng, T., Lin, T., Fang, J., and Brady, R. (2007) Photochromic wool fabrics from a hybrid silica coating. *Text. Res. J.*, **77** (12), 923–928.
39. Cheng, T., Lin, T., Brady, R., and Wang, X. (2008) Photochromic fabrics with improved durability and photochromic performance. *Fibers Polym.*, **9** (5), 521–526.
40. Billah, S., Reduwan, M., Christie, R.M., and Shamey, R. (2008) Direct coloration of textiles with photochromic dyes. Part 1: application of spiroindolinonaphthoxazines as disperse dyes to polyester, nylon and acrylic fabrics. *Color. Technol.*, **124** (4), 223–228.
41. Billah, S., Reduwan, M., Christie, R.M., and Morgan, K.M. (2008) Direct coloration of textiles with photochromic dyes. Part 2: the effect of solvents on the colour change of photochromic textiles. *Color. Technol.*, **124** (4), 229–233.
42. Aldib, M. and Christie, R.M. (2011) Textile applications of photochromic dyes. Part 4: application of commercial photochromic dyes as disperse dyes to polyester by exhaust dyeing. *Color. Technol.*, **127** (5), 282–287.
43. Guy, J.K. (2004) Fiber optic fabric. US Patent 20,040,037,091, Feb. 26, 2004.
44. Swain, D. (2010) Display system in article of clothing. US Patent 7,787,240, Aug. 31, 2010.
45. Cheung, J. (2008) Clothing or footwear illumination system having electroluminescent and LED light sources. US Patent 7,329,019, Feb. 12, 2008.
46. Graham-Rowe, D. (2007) Photonic fabrics take shape. *Nat. Photonics*, **1** (1), 6–7.
47. Freeman, W. (2011) Puma x Freezy Freakies (Color Change) *www.kicksonfire.com* (accessed 14 February 2011).
48. Del Sol *https://www.delsol.com* (accessed 28 February 2015).
49. Bemis, J.L., Godinez, M. and Goldston, M.R. (1994) Footwear with flashing lights. US Patent 5,303,485, issued Apr. 19, 1994.
50. INTIMACY *http://www.studioroosegaarde.net/project/intimacy-black/* (accessed 25 February 2015).
51. Riggs, T. (2013) Under Armour Thinks Touchscreen Clothing is the Next Big Thing. Zaggs, March 25, 2013.

Part IV
Device Case Studies, Environmental Impact Issues and Elaborations

Electrochromic Materials and Devices, First Edition.
Edited by Roger J. Mortimer, David R. Rosseinsky and Paul M. S. Monk.

17
Electrochromic Foil: A Case Study

Claes-Göran Granqvist

17.1
Introduction

The electrochromism of thin films of tungsten oxide was first discussed during the late 1960s and early 1970s [1]. Such films colour/bleach under joint insertion/extraction of small ions and electrons and are comprised of a 'cathodic' electrochromic (EC) material. Other oxides in the same category are based on Mo, Nb and Ti [2]. Another class of EC materials includes 'anodic' ones, which colour/bleach under extraction/insertion of ions and electrons, that is, oppositely to the 'cathodic' ones; this class includes oxides based on Ir and Ni [2]. Electrochromism in Ni oxide was discovered in the mid-1980s [3]. It is obvious that a combination of 'cathodic' and 'anodic' films in a standard 'battery-type' EC device, discussed in Chapter 1, is advantageous and can lead to stronger colouring/bleaching than an EC device based on only one of the two types of films. If one further contemplates that the optical modulation should be as large as possible in the wavelength range for luminous and/or solar radiation, and that the cost of the films should not be excessive, one finds that devices with thin films based on W oxide and Ni oxide are very interesting. This combination of materials was first explored in 1989 for the case of EC devices with an intervening polymer electrolyte [4].

Numerous investigations on EC devices including W oxide and Ni oxide films have been performed over the years; they have considered both glass-based and hence rigid constructions [5–23] and polymer-based and flexible ones [24–31]. Figure 17.1 illustrates an EC device of the latter kind and its application as a lamination foil between two glass panes. This type of device was recently (2014) integrated in some full-scale EC 'smart' windows; it is discussed in this chapter.

Section 17.2 outlines the design and general optical properties of the foil, including effects of device size, antireflection coating and superposition of foils. Section 17.3 then discusses some aspects of durability and shows, in particular, that current noise can give information of degradation. Section 17.4 considers electrolyte functionalisation and demonstrates that the addition of nanoparticles can yield favourable effects on the device performance and, specifically, the

Electrochromic Materials and Devices, First Edition.
Edited by Roger J. Mortimer, David R. Rosseinsky and Paul M. S. Monk.

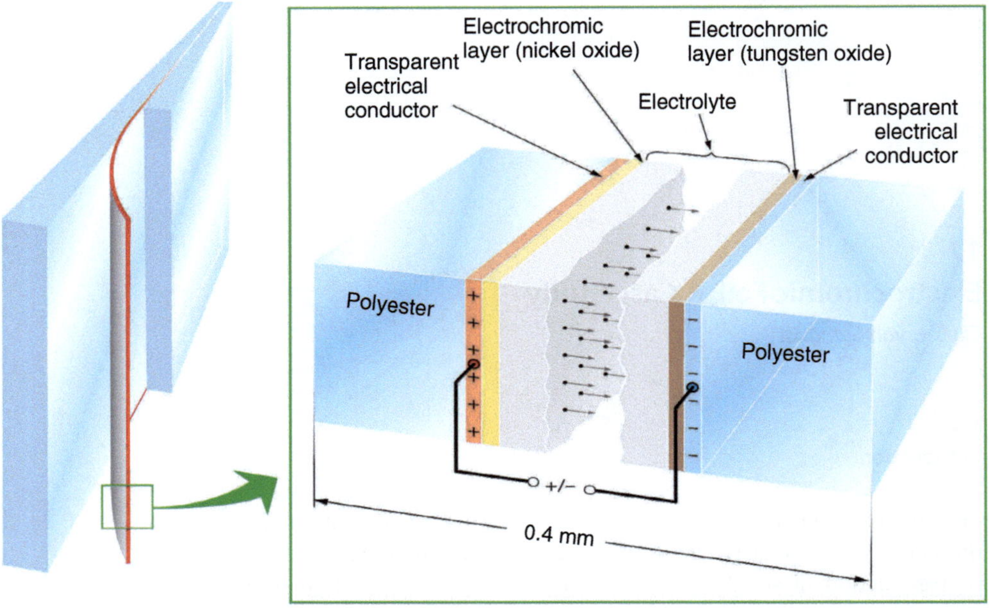

Figure 17.1 Principle construction of a foil-based EC device. The tiny arrows indicate ion movement when a voltage is applied between the transparent electrical conductors. The entire foil can be used to laminate glass panes, as illustrated in the left-hand part.

solar energy throughput can be diminished while the luminous performance is essentially unaltered. Finally, Section 17.5 provides some comments and conclusions. This chapter can be seen as an elaboration of Chapter 1 of this book and assumes that the reader has some familiarity with the general material presented therein.

17.2
Device Design and Optical Properties of Electrochromic Foil

The EC device in Figure 17.1 is constructed as follows: A 0.175-mm-thick foil of polyethylene terephthalate (PET) is first coated with a transparent electrical conductor and W oxide, a second foil of the same kind is coated with a transparent electrical conductor and Ni oxide and the two mentioned oxides are joined via a polymer electrolyte. The transparent conductors consist of In_2O_3:Sn, also known as indium tin oxide or ITO. The optical transmittance is modulated when a voltage is applied between the transparent electrical conductors and electrical charge is shuttled between the two EC films. The device maintains its optical properties during days or weeks under open circuit conditions, that is, energy is dissipated only when transmittance is changed.

The foil can be used in different ways: Firstly, it can be attached to the surface of an existing glass pane using known technology. Secondly, it can be suspended between two glass panes in an insulated glass unit (IGU), thus effectively changing it from a double-glazed to a triple-glazed construction without increasing its weight to any significant degree; the technology for maintaining a PET-based foil under tension inside an IGU – without causing visual defects – is well established. Thirdly, the foil can be used for lamination purposes, as illustrated in the left-hand part of Figure 17.1. With this latter construction, the EC device can also provide added functionality and impart spall shielding, burglar protection and more.

Figure 17.2 gives some background for understanding the optical properties of an as-prepared EC foil of the type illustrated in Figure 17.1 [30]. The upper panel shows spectral optical transmittance in the luminous spectral range (wavelength interval 400–800 nm) after charge exchange to the stated levels and demonstrates that the optical properties can be modulated within a wide range. The middle and lower panels of Figure 17.2 show the optical modulation of the individual W-oxide-based and Ni-oxide-based parts after disassembly of the device. The complementarity of the oxide films is apparent: the W-oxide-based film colours (cathodically) mainly in the long-wavelength part of the shown spectrum while the Ni-oxide-based film colours (anodically) predominantly in the short-wavelength part.

Optical modulation under extended colour/bleach cycling is shown in Figure 17.3 for a 240-cm^2-size EC foil device [30]. The upper panel illustrates mid-luminous transmittance (at a wavelength of 550 nm) for two consecutive colour/bleach cycles deliberately adjusted so that the transmittance modulation ΔT was 55%. Half of the transmittance range (from point 1 to 2) took ~10 s, 90% of this range (from point 1 to 3) took ~20 s, while the full range (from point 1 to 4) took ~30 s. Slower cycling could in principle have yielded a larger value of ΔT. The lower panel of Figure 17.2 illustrates the evolution of the maximum and minimum transmittance values during the first 2500 colour/bleach cycles. After some initial changes, the optical performance remained almost unchanged for several thousand cycles.

When EC foils are used for practical fenestration, they are likely to be exposed to ultraviolet irradiation. This radiation can give a photochromic effect that adds to the electrochromism [29]. Figure 17.4 shows data on charge insertion/extraction for a 200-cm^2-size EC foil both in the dark and as subjected to weak ultraviolet irradiation from an overcast sky. The increased charge exchange in the latter case translates to a more optically absorbing state. A similar effect has been reported also by others [14].

The colour/bleach dynamics is important for EC devices. For small constructions, the time for transitioning from dark to fully transparent states, or *vice versa*, can be as short as a few seconds, whereas larger devices are slower for the simple reason that greater amounts of electrical charge must be inserted/extracted via electrical contacts at one or more of the edges of the device. Devices show an 'iris effect' meaning that the colour change is more rapid at the edge than in the centre of a device. These features are illustrated in Figure 17.5, which reports recent data on a $5 \times 20\ cm^2$-size EC foil device, specifically on time evolution of

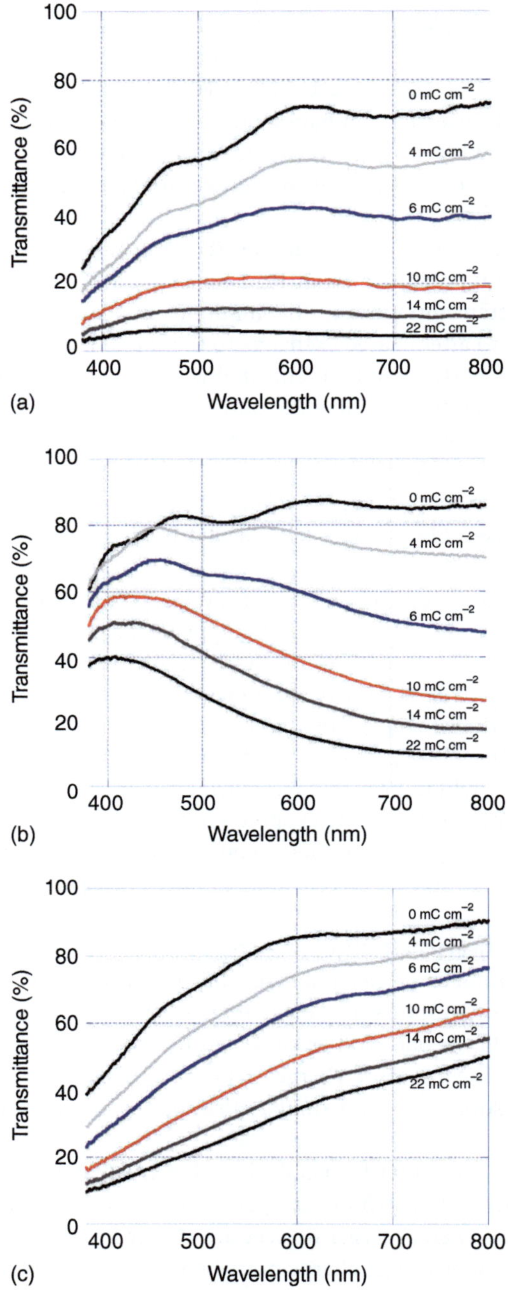

Figure 17.2 Spectral transmittance for an as-prepared EC foil device of the kind shown in Figure 17.1 (a), and similar data for the W-oxide-based (b) and Ni-oxide-based (c) parts of this device. (From Ref. [30].)

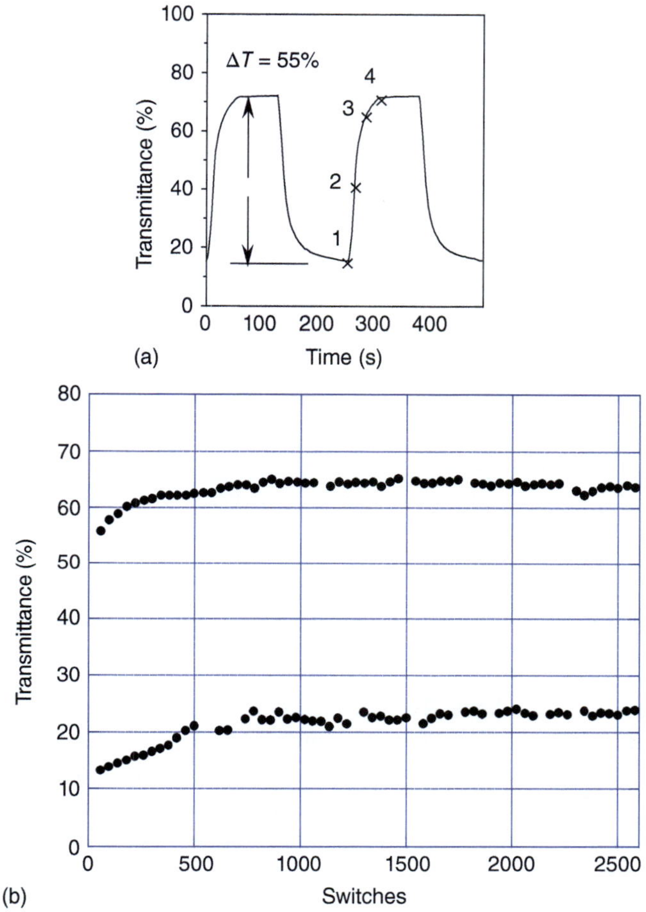

Figure 17.3 Mid-luminous transmittance for an EC foil device of the kind shown in Figure 17.1 adjusted so that the transmittance modulation ΔT was 55%. Panel (a) shows transmittance for two consecutive colour/bleach cycles (numbers 1–4 are explained in the main text) and panel (b) shows evolution of maximum and minimum transmittance during extended colour/bleach cycling. (From Ref. [30].)

transmittance at a wavelength of 532 nm for several distances from the current contact [32]. Clearly, the transmittance change progresses much faster close to the contact than far away from it. Also shown are results based on a circuit model of the current distribution, and one concludes that theory and experiment are in satisfactory agreement implying that the coloration dynamics can be modelled. Furthermore, electronic circuitry can be devised to minimise the unevenness of the coloration, if needed. Another modelling study appeared recently [33].

The desired modulation range depends on the intended application for the EC foil. In buildings, it may often be suitable to have glazings with large bleached-state transmittance, and antireflection coatings can be used for this purpose provided

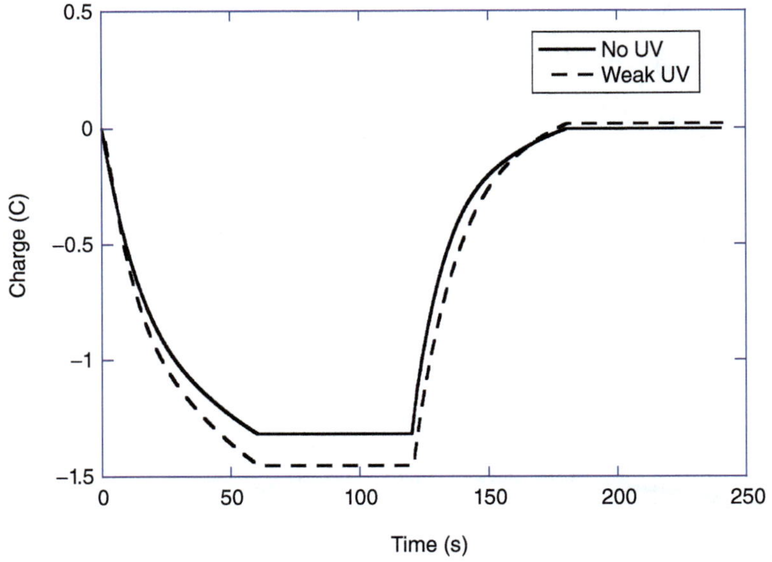

Figure 17.4 Time-dependent insertion/extraction of charge for an EC foil device of the kind shown in Figure 17.1 in the absence and presence of ultraviolet (UV) radiation. (From Ref. [29].)

they do not scatter too much light [29, 34]. Figure 17.6 shows the effect on the spectral transmittance when an EC foil is given a dip-coated silica-based antireflection coating on both sides. The mid-luminous transmittance was then enhanced from 73.1 to 79.1%, that is, by as much as 6% units. It is obvious that this enhancement holds for a freely suspended EC foil and does not apply if this foil is used for glass lamination.

If glare control is, instead, of prime importance, it is possible to diminish the coloured-state transmittance radically by use of superimposed foils. If, for example, the transmittance is 10% in the dark state for a certain EC device, then two superimposed devices yield a transmittance of only ~1%. Figure 17.7 reports transmittance measured during colouring and bleaching for this latter situation and demonstrates that a very low transmittance indeed can be achieved [29].

17.3 Comments on Lifetime and Durability

Long-term durability and reproducibility are multifaceted and critical aspects for EC devices; they are only rarely addressed more than superficially in scientific papers [35–37]. A detailed discussion of these matters is not feasible for the EC foil of present interest, but two aspects are discussed here.

The shelf-life of the EC foil was tested by storing it for extended periods of time and then testing it, before and after disassembly, in the same way as for the freshly

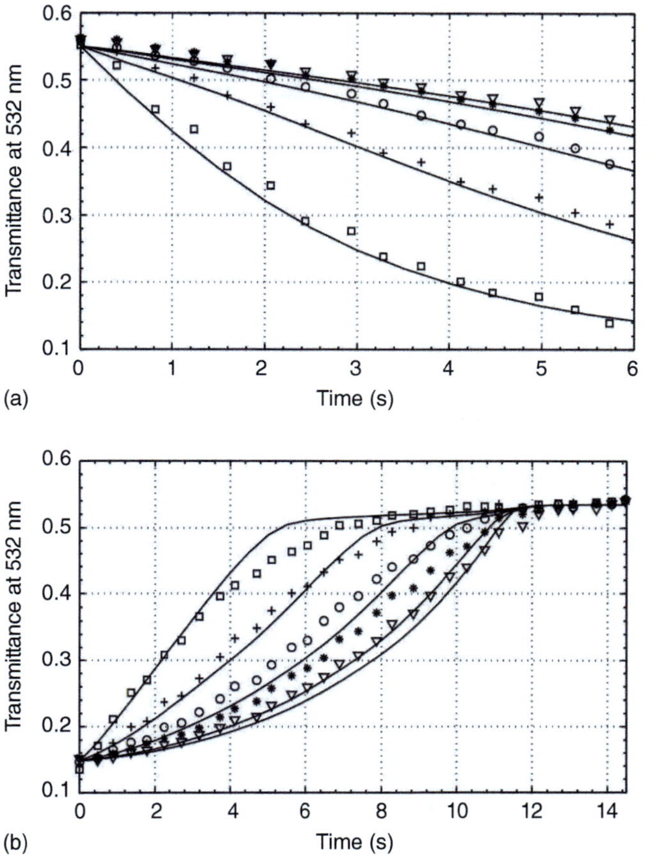

Figure 17.5 Transmittance for colouring (a) and bleaching (b) of an EC foil device of the kind shown in Figure 17.1. Squares, crosses, circles, asterisks and triangles represent data measured 1.5, 3.7, 6.0, 8.2 and 10.5 cm, respectively, from the current contact; curves signify calculated results. (From Ref. [32].)

prepared foil reported on in Figure 17.2. It is seen from Figure 17.8 that storing for 1.5 years has a very minor effect and that transmittance is only marginally increased.

There are many degradation modes for EC devices [38], and the decline of the properties can progress over time depending on how the device is operated. There is clearly a need for rapid techniques to assess degradation and to screen among devices in order to single out the ones that may be prone to show strong degradation during the course of their lifetime. Electromagnetic noise is such a technique and can be used for quality assessment of EC devices; it has been applied to the foil-type construction of present concern [29, 39]. A similar technique can be used to study corrosion processes [40], ageing of solar cells [42] and so on. Electrical noise is a general feature, which is well known to exist in many different types of materials and at interfaces between different media [42]. Fluctuations of current

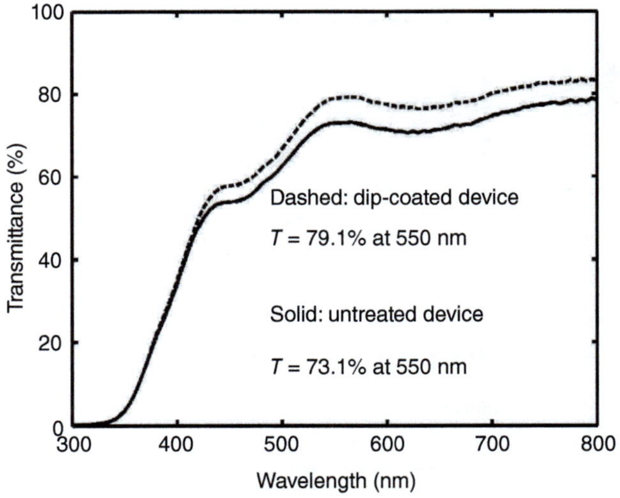

Figure 17.6 Spectral transmittance T of an EC foil device of the kind described in Figure 17.1 with and without a dip-coated antireflection treatment. (From Ref. [29].)

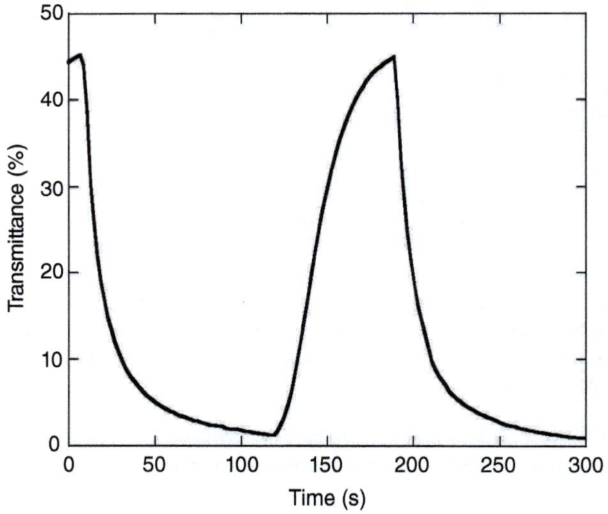

Figure 17.7 Mid-luminous transmittance dynamics for two superimposed EC foils of the kind described in Figure 17.1. (From Ref. [29].)

and voltage can ensue from thermal equilibrium processes as well as from non-equilibrium phenomena and lead to a variety of noise types known as thermal noise, shot noise, burst noise, generation–recombination noise, $1/f$ noise and $1/f^2$ noise [43]. Here we consider $1/f$ noise, where f denotes frequency, during the discharge of EC foils and with changes of this noise under device degradation.

Current noise was recorded on EC foil devices, as discussed in detail elsewhere [29, 39], under conditions that yielded easily measurable $1/f$ noise. Figure 17.9a

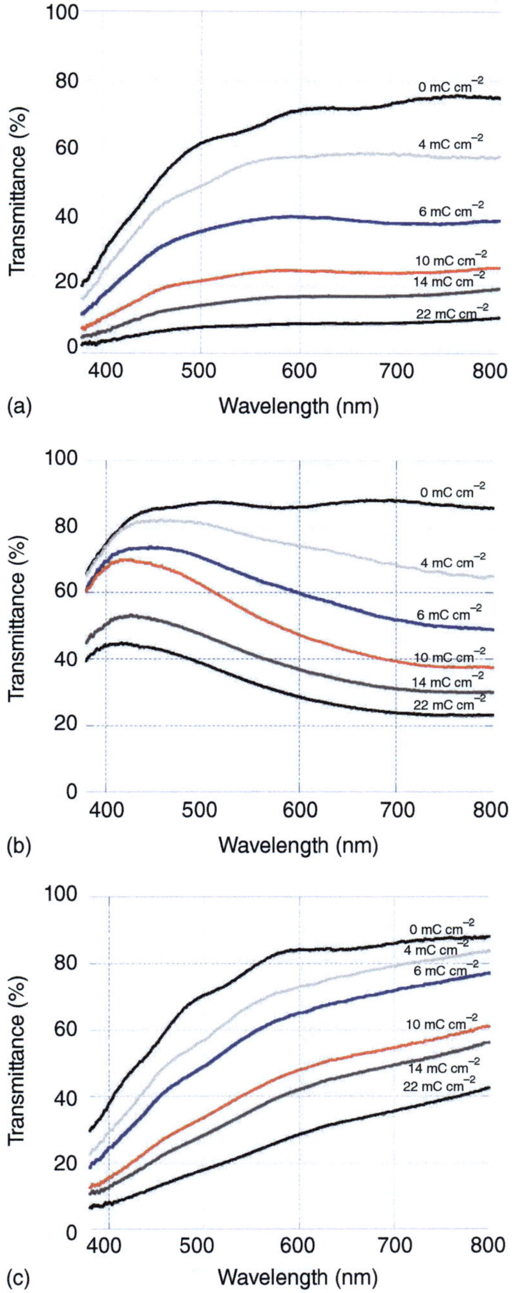

Figure 17.8 Spectral transmittance for an EC foil device of the kind shown in Figure 17.1 after storage for 1.5 years (a), and similar data for the W-oxide-based (b) and Ni-oxide-based (c) parts of this device. (From Ref. [30].)

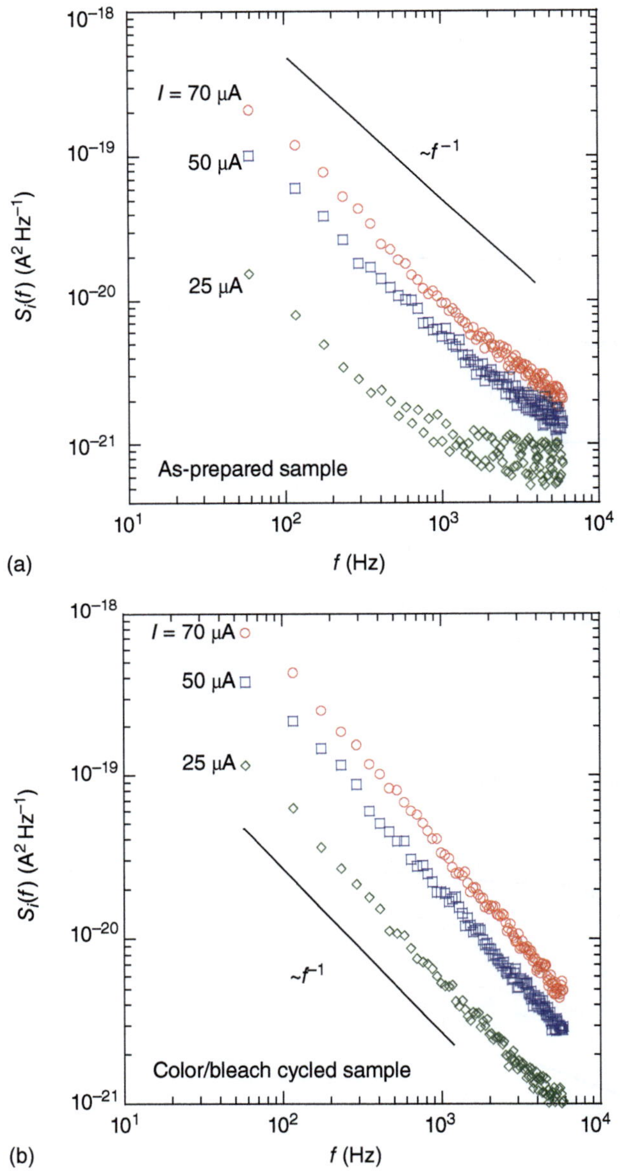

Figure 17.9 Noise spectra, denoted as $S_i(f)$ where f is frequency, of an EC device of the kind shown in Figure 17.1 before (a) and after (b) extensive colour/bleach cycling. Data were taken at the current I. The line indicates f^{-1}-dependence. (From Ref. [39].)

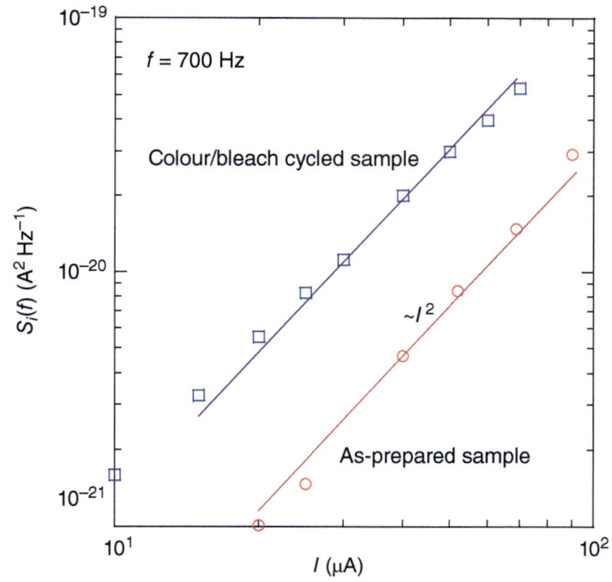

Figure 17.10 Current noise power spectra, denoted as $S_i(f)$ where $f = 700$ Hz, as a function of current I of an EC device of the kind shown in Figure 17.1 before and after extensive colour/bleach cycling. The lines denote I^2-dependence. (From Ref. [39].)

shows current power spectral density S_i for three values of the dc current for an as-prepared device $2.5 \times 2.5\ cm^2$ in size. Low-frequency noise is found to dominate at $f > 1$ kHz; its frequency dependence can be approximated by $S_i \sim 1/f^x$ with $1.0 < x < 1.2$. Figure 17.9b provides data on S_i for a device that has undergone severe colour/bleach cycling until it was visibly degraded. It is evident that the low-frequency noise has increased significantly and that $1/f$ noise now dominates in the full investigated frequency range even for the smallest current.

$S_i(f)$ as a function of current I was studied at $f = 700$ Hz. Figure 17.10 shows that $S_i(f) \sim I^2$ and that the cycling treatment has increased the magnitude of S_i roughly by a factor of 4 irrespective of I. This relationship suggests that the noise originates from resistance/conductance fluctuations and that their magnitude is a measure of device degradation.

The data in Figures 17.9 and 17.10 do not give any information on the origin of the noise. In order to shed some light on this issue, investigations were performed on $7 \times 7\ cm^2$-size 'symmetrical' devices (in principle agreement with samples studied in some prior work [44]) – comprising either two laminated W-oxide-based films or two laminated Ni-oxide-based films – and investigated $S_i(f)$ as a function of I at $f = 175$ Hz as before. Figure 17.11 shows that the relation $S_i(f) \sim I^2$ persists. The Ni-oxide-based EC foil displays noise that is about 4 orders of magnitude stronger than in the W-oxide-based foil. The reason for this difference is as yet unknown.

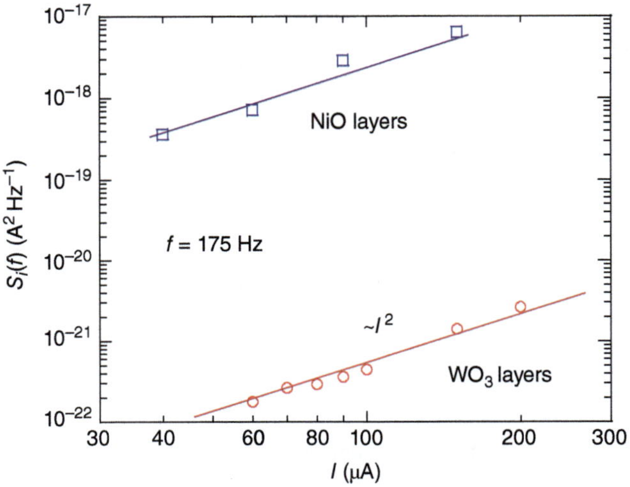

Figure 17.11 Current noise power spectra, denoted as $S_i(f)$ where $f = 175$ Hz, as a function of current I for symmetrical EC devices with laminated Ni-oxide-based and W-oxide-based films. The lines denote I^2-dependence. (From Ref. [39].)

17.4
Electrolyte Functionalisation by Nanoparticles

The foil-type EC devices have a centrally positioned layer of polymer electrolyte. This construction allows the electrolyte to be functionalised by nanoparticles. Explorative work on such functionalisation was performed recently; most of it was carried out by use of a 'model electrolyte' consisting of

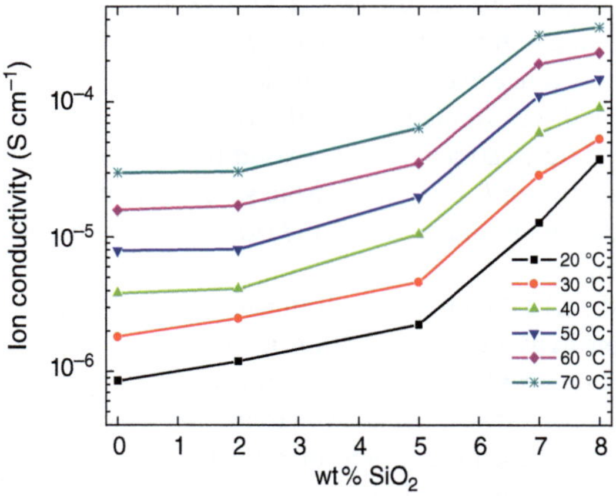

Figure 17.12 Ion conductivity at the shown temperatures as a function of the content (in wt%) of SiO_2 nanoparticles in electrolytes of (PEI–SiO_2):LiTFSI. Data points indicate measurements, which are joined by straight lines for convenience. (From Ref. [45].)

polyethyleneimine–lithium bis(trifluoromethylsulfonyl) (denoted PEI LiTFSI) [45, 46], but similar functionalisation can be accomplished for the electrolyte used in EC foil devices according to Figure 17.1 [47]. The work reported subsequently used glass substrates with ITO coatings rather than PET-based substrates.

We first consider the effect of electrolyte functionalisation with various amounts of 7-nm-diameter SiO_2 nanoparticles (fumed silica) [45]. Figure 17.12 shows ion conductivity – known to influence the colour/bleach rate of an EC device – which is found to rise monotonically as the amount of SiO_2 is increased, irrespective of temperature. Spectral total and diffuse transmittance are shown in Figure 17.13 where it can be seen that the luminous transmittance is high for all SiO_2 contents and that the diffuse component (haze) remains low for SiO_2 contents below ~6 wt%.

If nanoparticles of a transparent electrical conductor are added to the electrolyte, instead of silica, it is possible to obtain near-infrared plasmon-based

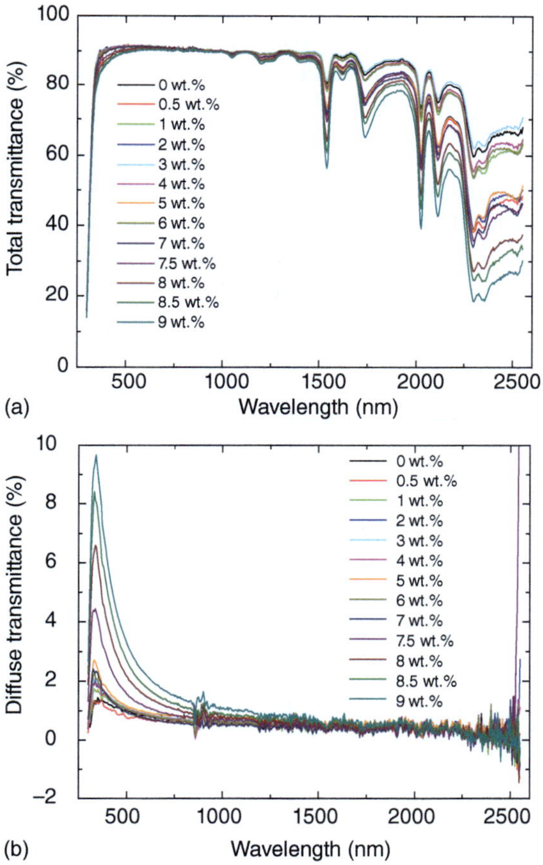

Figure 17.13 Total (a) and diffuse (b) spectral transmittance spectra for (PEI-SiO_2):LiTFSI electrolytes with different amounts of SiO_2. (From Ref. [45].)

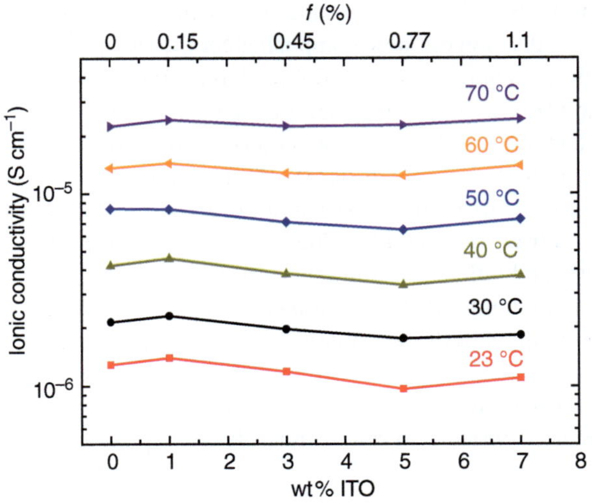

Figure 17.14 Ion conductivity at the shown temperatures as a function of the content (in wt% and in volume fraction *f*) of ITO nanoparticles in electrolytes of (PEI-ITO):LiTFSI. Data points indicate measurements, which are joined by straight lines for convenience. (From Ref. [46].)

absorption and diminish the transmittance of solar energy (in the 300–2500 nm wavelength range) without lowering the luminous transmittance to any significant degree. This property can be important for EC smart windows especially in warm climates. Figure 17.14 shows ion conductivity data for PEI LiTFSI electrolytes with different amounts of 13-nm-diameter ITO nanoparticles [46] and indicates

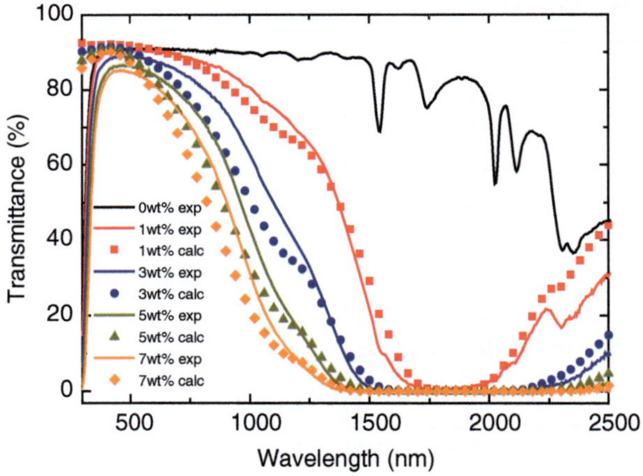

Figure 17.15 Spectral transmittance as a function of the amount of ITO nanoparticles in electrolytes of (PEI-ITO):LiTFSI. Experimental data (curves) and calculations (symbols) are reported. (From Ref. [46].)

that the nanoparticles play only a minor role in the ionic properties. Figure 17.15 reports spectral transmittance for the same samples and demonstrates that strong absorption occurs in the near-infrared (700–2500 nm range) as the quantity of nanoparticles is increased. With 7 wt% of ITO, to take one example, one obtains a wavelength-integrated luminous transmittance of 83.3% and solar energy transmittance of 56.3% while the electrolyte remains essentially free of haze. The measured data can be reconciled with accurate calculations, as indicated by the symbols in Figure 17.15, which is a consequence of the good theoretical understanding of the properties of ITO [48] and of well-established models for the optical properties of dilute suspensions of nanoparticles [49].

17.5 Comments and Conclusion

Cheap large-scale production is essential for any commodity product and must be combined with large-area manufacturability for EC devices intended for use on extended surfaces such as those of windows. The EC foil, shown in Figure 17.1, offers special possibilities and roll-to-roll coating – which is a low-cost technology [50, 51] – has been demonstrated [31]. This coating process can be combined with continuous lamination of the W-oxide-coated and Ni-oxide-coated PET foils by use of the polymer electrolyte, and with subsequent cross-linking of this electrolyte, thereby forming large flexible sheets ready for glass lamination. These sheets are rugged and easy to transport and store. Thus, manufacturing of the final product incorporating the EC foil, such as an EC-based window, can be separated from the actual foil production site. The large EC foils can be cut to any size and shape, and electrical contacts can then be applied reliably via proprietary

Figure 17.16 Office equipped with EC glazing. Two of the windows are in their fully dark state.

techniques. The latter procedures define a 'free-form' design concept that can be implemented at the site for the final product manufacturing.

Figure 17.16 shows an installation of full-size windows incorporating EC-based foil laminated between glass panes by use of standard procedures. Two of the panes are dark and display a strong contrast to the adjacent clear window. The time for switching between fully dark and coloured states is of the order of 10 min. No visible haze or other defects are apparent.

Acknowledgements

Useful comments on the manuscript were provided by G. Gregard. Financial support during the writing of this chapter was received from the European Research Council under the European Community's Seventh Framework Program (FP7/2007–2013)/ERC Grant Agreement No. 267234 (GRINDOOR).

References

1. Deb, S.K. (1973) Optical and photo-electric properties and colour centres in thin-films of tungsten oxide. *Philos. Mag.*, **27**, 801–822.
2. Granqvist, C.G. (1995) *Handbook of Inorganic Electrochromic Materials*, Elsevier, Amsterdam.
3. Svensson, J.S.E.M. and Granqvist, C.G. (1986) Electrochromic hydrated nickel oxide coatings for energy efficient windows: optical properties and coloration mechanism. *Appl. Phys. Lett.*, **49**, 1566–1568.
4. Passerini, S., Scrosati, B., Gorenstein, A., Andersson, A.M., and Granqvist, C.G. (1989) An electrochromic window based on $Li_xWO_3/(PEO)_8LiClO_4/NiO$. *J. Electrochem. Soc.*, **136**, 3394–3395.
5. Mathew, J.G.H., Sapers, S.P., Cumbo, M.J., O'Brien, N.A., Sargent, R.B., Rakhsha, V.P., Lahaderne, R.B., and Hichwa, B.P. (1997) Large-area electrochromics for architectural applications. *J. Non-Cryst. Solids*, **218**, 342–346.
6. Mathew, J.G.H., Sapers, S.P., Cumbo, M.J., O'Brien, N.A., Sargent, R.B., Rakhsha, V.P., Lahaderne, R.B., and Hichwa, B.P. (1997) in *Proceedings of the Third Symposium on Electrochromic Materials*, vol. 96-24 (eds K.-C. Ho, C.B. Greenberg, and D.M. MacArthur), The Electrochemical Society, Pennington, NJ, pp. 311–324.
7. Azens, A., Vaivars, G., Veszelei, M., Kullman, L., and Granqvist, C.G. (2001) Electrochromic devices embodying W oxide/Ni oxide tandem films. *J. Appl. Phys.*, **89**, 7885–7887.
8. Larsson, A.-L. and Niklasson, G.A. (2004) Optical properties of electrochromic all-solid-state devices. *Sol. Energy Mater. Sol. Cells*, **84**, 351–360.
9. Karuppasamy, A. and Subrahmanyam, A. (2007) Study on electrochromic smart windows based on titanium doped WO_3 thin films. *Thin Solid Films*, **516**, 175–178.
10. Subrahmanyam, A., Kumar, C.S., and Karuppasamy, K.M. (2007) A note on fast protonic solid state electrochromic device $NiO_x/Ta_2O_5/WO_{3-x}$. *Sol. Energy Mater. Sol. Cells*, **91**, 62–66.
11. Fang, Y.K., Chou, T.-H., Lin, C.-Y., Chiang, Y.T., Chen, S.-F., Yang, C.-Y., Chang, S.-H., and Lin, C.-S. (2008) A novel electrochromic device with high optical switching speed. *J. Phys. Chem. Solids*, **69**, 734–737.
12. Yang, H., Wang, C., Diao, X., Wang, H., Wang, T., and Zhu, K. (2008) A new all-thin-film electrochromic

device using LiBSO as the ion conducting layer. *J. Phys. D: Appl. Phys.*, **41**, 115301/1–115301/5.
13. Zelazowska, E. and Rysiakiewicz-Pasek, E. (2008) WO_3-based electrochromic system with organic–inorganic gel electrolyte. *J. Non-Cryst. Solids*, **354**, 4500–4505.
14. Zhang, Y., Lee, S.-H., Mascarenhas, A., and Deb, S.K. (2008) An UV-photochromic memory effect in proton-based WO_3 electrochromic devices. *Appl. Phys. Lett.*, **93**, 203508/1–203508/2.
15. Purushothaman, K.K. and Muralidharan, G. (2009) Nanoporous NiO based electrochromic window. *Funct. Mater. Lett.*, **2**, 143–145.
16. Sauvet, K., Sauques, L., and Rougier, A. (2009) IR electrochromic thin films: from optimization to device. *Sol. Energy Mater. Sol. Cells*, **93**, 2045–2049.
17. Zhang, J., Tu, J.P., Xia, X.H., Qiao, Y., and Lu, Y. (2009) An all-solid-state electrochromic device based on NiO/WO_3 complementary structure and solid hybrid polyelectrolyte. *Sol. Energy Mater. Sol. Cells*, **93**, 1840–1845.
18. Huang, H., Tian, J., Zhang, W.K., Gan, Y.P., Tao, X.Y., Xia, X.H., and Tu, J.P. (2011) Electrochromic properties of porous NiO thin film as a counter electrode for NiO/WO_3 complementary electrochromic window. *Electrochim. Acta*, **56**, 4281–4286.
19. Wang, S.-C., Liu, K.-Y., and Huang, J.-L. (2011) Tantalum oxide film prepared by reactive magnetron sputtering deposition for all-solid-state electrochromic device. *Thin Solid Films*, **520**, 1454–1459.
20. Chen, C.C. and Jheng, W.D. (2012) Effect of Ar/O_2ratio on double-sided electrochromic glass performance. *Ceram. Int.*, **38**, 4195–4200.
21. Chen, C.C. and Jheng, W.D. (2012) Logotype-selective electrochromic glass display. *Ceram. Int.*, **38**, 5835–5842.
22. Liao, C.-C. (2012) Lithium-driven electrochromic properties of electrodeposited nickel hydroxide electrodes. *Sol. Energy Mater. Sol. Cells*, **99**, 26–30.
23. Park, S.H., Lim, J.W., Yoo, S.J., Cha, I.Y., and Sung, Y.-E. (2012) The improving electrochromic performance of nickel oxide film using aqueous N,N-dimethylaminoethanol solution. *Sol. Energy Mater. Sol. Cells*, **99**, 31–37.
24. Azens, A., Gustavsson, G., Karmhag, R., and Granqvist, C.G. (2003) Electrochromic devices on polyester foil. *Solid State Ionics*, **165**, 1–5.
25. Azens, A., Avendaño, E., Backholm, J., Berggren, L., Gustavsson, G., Karmhag, R., Niklasson, G.A., Roos, A., and Granqvist, C.G. (2005) Flexible foils with electrochromic coatings: science, technology and applications. *Mater. Sci. Eng., B*, **119**, 214–223.
26. Niklasson, G.A. and Granqvist, C.G. (2007) Electrochromics for smart windows: thin films of tungsten oxide and nickel oxide, and devices based on these. *J. Mater. Chem.*, **17**, 127–156.
27. Granqvist, C.G., Niklasson, G.A., and Azens, A. (2007) Electrochromics: fundamentals and energy-related applications of oxide-based devices. *Appl. Phys. A*, **89**, 29–35.
28. Granqvist, C.G., Azens, A., Smulko, J., and Kish, L.B. (2007) Oxide-based electrochromics for energy efficient buildings: materials, technologies, testing, and perspectives. *J. Phys. Conf. Ser.*, **93**, 012021/1–012021/9.
29. Granqvist, C.G., Green, S., Jonson, E.K., Marsal, R., Niklasson, G.A., Roos, A., Topalian, Z., Azens, A., Georén, P., Gustavsson, G., Karmhag, R., Smulko, J., and Kish, L.B. (2008) Electrochromic foil-based devices: optical transmittance and modulation range, effect of ultraviolet irradiation, and quality assessment by $1/f$ current noise. *Thin Solid Films*, **516**, 5921–5926.
30. Granqvist, C.G., Green, S., Niklasson, G.A., Mlyuka, N.R., von Kræmer, S., and Georén, P. (2010) Advances in chromogenic materials and devices. *Thin Solid Films*, **518**, 3046–3053.
31. Granqvist, C.G. (2012) Progress in oxide-based electrochromics: towards roll-to-roll manufacturing. *Mater. Res. Soc. Symp. Proc.*, **1315**, 89–100.
32. Degerman Engfeldt, J., Georen, P., Lagergren, C., and Lindbergh, G. (2011) Methodology for measuring current distribution effects in electrochromic smart windows. *Appl. Opt.*, **50**, 5639–5646.

33. Liu, Y., Sun, L., Sikha, G., Isidorsson, J., Lim, S., Anders, A., Kwak, B.L., and Gordon, J.G. II, (2014) 2-D mathematical modeling for a large electrochromic window – part I. *Sol. Energy Mater. Sol. Cells*, **120**, 1–8.
34. Jonsson, A., Roos, A., and Jonson, E.K. (2010) The effect on transparency and light scattering of dip coated antireflection coatings on window glass and electrochromic foil. *Sol. Energy Mater. Sol. Cells*, **94**, 992–997.
35. Badding, M.E., Schulz, S.C., Michalski, L.A., and Budziak, R. (1997) in *Proceedings of the Third Symposium on Electrochromic Materials*, vol. 96-24 (eds K.-C. Ho, C.B. Greenberg, and D.M. MacArthur), The Electrochemical Society, Pennington, NJ, pp. 369–384.
36. Lampert, C.M., Agrawal, A., Baertlien, C., and Nagai, J. (1999) Durability evaluation of electrochromic devices: an industry perspective. *Sol. Energy Mater. Sol. Cells*, **56**, 449–463.
37. Tracy, C.E., Zhang, J.-G., Benson, D.K., Czanderna, A.W., and Deb, S.K. (1999) Accelerated durability testing of electrochromic windows. *Electrochim. Acta*, **44**, 3195–3202.
38. Bell, J.M. and Skryabin, I.L. (1999) Failure modes of sol–gel deposited electrochromic devices. *Sol. Energy Mater. Sol. Cells*, **56**, 437–448.
39. Smulko, J., Azens, A., Marsal, R., Kish, L.B., Green, S., and Granqvist, C.G. (2008) Application of $1/f$ current noise for quality and age monitoring of electrochromic devices. *Sol. Energy Mater. Sol. Cells*, **92**, 914–918.
40. Cottis, R.A. (2001) Interpretation of electrochemical noise data. *Corrosion*, **57**, 265–285.
41. Barone, C., Landi, G., De Sio, A., Neitzert, H.C., and Pagano, S. (2014) Thermal ageing of bulk heterojunction solar cells investigated by electric noise analysis. *Sol. Energy Mater. Sol. Cells*, **122**, 40–45.
42. Van Kampen, N.G. (1992) *Stochastic Processes in Physics and Chemistry*, Elsevier, Amsterdam.
43. Vandamme, L.K.J. (1994) Noise as a diagnostic tool for quality and reliability of electronic devices. *IEEE Trans. Electron Devices*, **41**, 2176–2187.
44. Lefrou, C., Gentilhomme, C., and Ast, M. (2006) Testing of electrochromic materials using symmetrical devices. *J. Appl. Electrochem.*, **36**, 1011–1019.
45. Bayrak Pehlivan, İ., Granqvist, C.G., Marsal, R., Georén, P., and Niklasson, G.A. (2012) [PEI–SiO_2]:[LiTFSI] nanocomposite polymer electrolytes: ion conduction and optical properties. *Sol. Energy Mater. Sol. Cells*, **98**, 465–471.
46. Bayrak Pehlivan, İ., Runnerstrom, E.L., Li, S.-Y., Niklasson, G.A., Milliron, D.J., and Granqvist, C.G. (2012) A polymer electrolyte with high luminous transmittance and low solar throughput: polyethyleneimine-lithium bis(trifluoromethylsulfonyl) imide with In_2O_3:Sn nanocrystals. *Appl. Phys. Lett.*, **100**, 241902/1–241902/4.
47. Bayrak Pehlivan, İ., Marsal, R., Pehlivan, E., Runnerstrom, E.L., Milliron, D.J., Granqvist, C.G., and Niklasson, G.A. (2014) Electrochromic devices with polymer electrolytes functionalized by SiO_2 and In_2O_3:Sn nanoparticles: rapid coloration/bleaching dynamics and strong near-infrared absorption. *Sol. Energy Mater. Sol. Cells*, **126**, 241–247.
48. Hamberg, I. and Granqvist, C.G. (1986) Evaporated Sn-doped In_2O_3 films: basic optical properties and applications to energy-efficient windows. *J. Appl. Phys.*, **60**, R123–R159.
49. Niklasson, G.A., Granqvist, C.G., and Hunderi, O. (1981) Effective medium models for the optical properties of inhomogeneous materials. *Appl. Opt.*, **20**, 26–30.
50. Bishop, C.A. (2010) *Roll-to-Roll Deposition of Barrier Coatings*, John Wiley & Sons, Inc, Hoboken, NJ.
51. Bishop, C.A. (2011) *Vacuum Deposition onto Webs, Films, and Foils*, 2nd edn, William Andrew, Waltham, MA.

18
Life Cycle Analysis (LCA) of Electrochromic Smart Windows

Uwe Posset and Matthias Harsch

18.1
Life Cycle Analysis

Life cycle assessment (LCA) is invaluable as it helps to optimise the environmental design and showcase the environmental performance of the components of a device and their environmental benefits, when in the operational phase. It allows for an insightful view on the causes of the major environmental burdens. Based on the LCA results, adaptations to more environmentally friendly alternatives can be made – at least if technically available and feasible [1]. LCA methodology has indeed continuously been applied during our research projects to provide steady feedback about optimisation possibilities from a life cycle perspective, that is, to support sustainable development. It enabled us, for instance, to improve the pilot-plant-scale production of novel but synthetically demanding plastic electrochromic (EC) electrodes. The LCA feedback helped to optimise the process, the new ecoprofile of the EC material being about 30% better, as will be detailed later.

Developing and implementing sustainable products is a major challenge in today's global business world. Due both to the general public's growing awareness of the environment and to more restrictive environmental legislation, product and production decisions are now based not only on economic and technological criteria but also on environmental aspects. Therefore, it is essential to know each process step across the entire life cycle of a product to reach an optimal solution. Industries need to know and understand the environmental impacts associated with their products and processes, and how to manage these negative effects in the best way possible, in ecological and economic terms. Especially with regard to products that comprise multiple coatings such as EC windows, the interaction between components' details, their influence on the application process and the resulting performance in the use phase are important yet challenging. The measurement of sustainability can be performed by the methods of life cycle assessment (ecology) and life cycle costing (LCC, economy). All inputs to (energy, resources) and outputs from (emissions, waste) the product system are analysed for each life cycle stage. LCA methodology hence allows production decisions

Electrochromic Materials and Devices, First Edition.
Edited by Roger J. Mortimer, David R. Rosseinsky and Paul M. S. Monk.

to be made from the point of view of environmental aspects, from raw material through production, use and recycling or disposal.

Multilayer coatings utilised in EC windows are complex multi-material systems of organic and inorganic compounds. The challenge is to

- obtain transparency of the life cycle of the coatings throughout the supply chain (life cycle perspective is a success factor);
- identify drivers influencing sustainability;
- find options to substitute unsustainable materials (e.g. change to renewable raw materials);
- improve cost, material and energy efficiency in each process step over the life cycle.

For this purpose, it is necessary to perform a compilation and evaluation of the inputs and outputs as well as potential environmental impacts associated with these, in each step of the production process and throughout the life cycle of the product. All the material and energy flows, as well as the relevant environmental releases involved in the complete life cycle of the product, are defined, characterised and quantified. In addition, the procedure of decision making for long-term investments is of essential importance. This awareness helps to expand the classical procedure with two additional steps, that is,

- a systematic market overview of promising technical alternatives and
- a holistic evaluation process by simulating the technical performance of the chosen alternatives to identify strengths, weaknesses and potentials.

The procedure of Life Cycle Assessment (Figure 18.1) has been established since 1996 and was revised in 2006 by applying ISO standards 14040 and 14044.

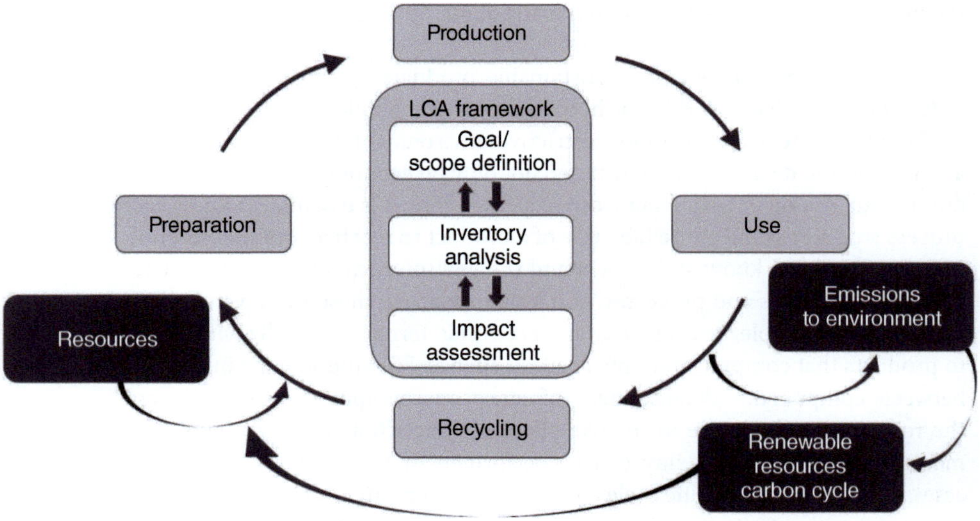

Figure 18.1 Life cycle assessment according to ISO Standards 14040 and 14044.

It is common consensus to measure with this tool the ecological sustainability of products, technologies and services [2, 3]. An LCA study comprises four major steps:

- The definition of goal and scope, depending on subject and intended use of the study.
- The Life Cycle Inventory (LCI), meaning the compilation of an inventory of any inputs and outputs of each step of the process.
- The Life Cycle Impact Assessment (LCIA), where the potential environmental impacts associated with the inputs and outputs are evaluated.
- The analysis and interpretation of the results.

A detailed description of the procedure and its application is available in the International Reference Life Cycle Data System (ILCD) handbook [4] and Life Cycle Assessment Handbook [5].

Economic sustainability can be measured by LCC, a best practice approach, which enhances the established cost calculation. LCC of EC smart windows is a topic that is just about to be investigated and shall thus not be addressed further in this book.

It is important to keep in mind that the results from an LCA depend on many different factors, such as

- the goal and scope definition of the process including the exact functional unit;
- the data available and their quality;
- the nature of choices, assumptions and definitions, which limit the model and can be subjective;
- the models or indicators used for the inventory analysis or in the impact assessment phase;
- the reasons why the study is carried out (purpose, kind of decisions to be supported) and the audience aimed at (company, general public, etc).

For definition of scope, the following parameters have to be unambiguously stated: the functional unit, the product system, the system boundaries, the allocation procedures and the data collection. Due to the importance of these parameters, each of them is defined and explained as follows.

- The EN ISO 14040 definition for a *functional unit* is: 'a quantified performance of a product system for use as a reference unit in a LCA study' [2]. In other words, the functional unit is the reference, which all inputs and outputs of the process are related to.
- According to EN ISO 14040, the *product system* is defined as 'the collection of materially and energetically connected unit-processes performing one or more defined functions'. As systems may have many different functions, the function under concern of the product system must be defined clearly.
- *System boundaries* are the 'interface between a product system and the environment or other product systems'. The system should be modelled in such a way that all inputs and outputs, included in the defined boundaries, are elementary

flows. An input elementary flow is a material or energy flow entering the system without human transformation (i.e. a resource like iron ore or surface water). An output elementary flow enters the environment without human transformation (i.e. emissions and waste set free into the environment after a possible emission or waste treatment). Since it is not possible to consider all production systems inter-linked worldwide, it is necessary to define applicable cut-off criteria, according to the goal and scope.

- *Allocation procedures* – Often, the processes of the targeted system generate by-products that may be input materials for processes in other systems and have hence to be treated as valuable products. Following ISO 14044 [2, 3], an allocation is always necessary where the allocation of a process with two or more outputs cannot be avoided, for example, by dividing the multi-output unit process into single-output processes or by extending the product system. In order to avoid that this allocation is carried out arbitrarily or subjectively, it is necessary to introduce allocation procedures to determine the contribution to the environmental impacts of all the (by-) products. At present, no internationally accepted standard exists for this purpose. The principal criteria followed to make an allocation are:
  - *Worst-case approach* – The main product ('the intention why the process is carried out') is considered responsible for all environmental effects and all the by-products are free of environmental harmfulness.
  - *Allocation according to economic criteria (price)* – Production and the demand of a product are strongly influenced by its market value. Hence, environmental effects can be assigned to the different products concerning their market value.
  - *Allocation according to the mass or energy content of the products* – The products with higher weight or calorific values are assigned higher environmental effects, according to their mass percentage and energy distribution.

It is important to keep in mind that almost every property can be used for allocation rules. Therefore, allocation criteria have to be documented and justified.

- The procedures employed for the *data collection* are dependent on the goal and the scope or the unit process. However, there are some data quality requirements, which have to be described and have to be taken into account, that is, the sources, level of detail, precision and reproducibility of the data, the consistency and reproducibility of the LCA methodology as well as the time-related geographical and technological coverage of the data. A good approach for collecting data is the description of the product system as a flow diagram, including all the unit processes and their interrelationships. Next, quantitative and/or qualitative data is collected for each unit process included within the system boundaries. It is important to keep in mind that the collection of data is an iterative and flexible process.

18.2 Application of LCA to Electrochromic Smart Windows

A lot of LCA data have been published on standard non-dynamic fenestration, which shall not be discussed further. A useful review on the Life Cycle Assessment and the Environmental Impact of Buildings (including windows) was published by Banfill *et al.* [6]. However, to date, there are only a limited number of studies on the environmental assessment of EC or other smart windows. P. Yianoulis *et al.* were probably the first to enter this interesting field by publishing a study on the environmental impact of EC glazing production in 2005 [7]. The scope of this early work is based on primary energy only and therefore rather an LCI than a full Life Cycle Assessment. The study gives detailed insights into the material production routes and provides details concerning energy and material depletion in the glass substrate production, including providing an emission analysis and a discussion on safety and environmental issues of EC glazing based on tungsten oxide and a PMMA/PC (Poly(methyl methacrylate)/Propylene carbonate) electrolyte with $LiClO_4$ as electrolyte. The contribution of materials in the EC device embodied energy is broken down, revealing that two-thirds of the embodied energy can be assigned to the K-glass substrate (FTO-glass: Fluorine doped tin oxide). A comparison of the total embodied energies for an EC smart window (including frame) and a state-of-the-art argon-filled insulating glass unit (IGU) revealed an approximately 10% higher value for the electrochromic device (ECD), which was considered promising with regard to the energy payback time. However, difficulties in finding reliable data sources and the lack of available inventory data have been denoted and no counter electrodes/ion storage layers have been modelled, which may account for differences in data reported there and in other studies. This particularly concerns the high energy demand of some usual counter electrode materials, as discussed later. In a follow-up study published in 2009 [8], a novel combined LCA and eco-efficiency approach was proposed for the rating of advanced glazing, aiming to add economic and environmental aspects to the existing evaluation and labelling systems established in the United Kingdom, the United States and Europe. The methodology was applied to a $40 \times 40\ cm^2$ sized sandwich-type EC window as a case study, with the window comprising a five-layer configuration including a lithium-doped vanadium pentoxide film deposited by thermal evaporation as the ion storage layer. Besides, it was found that a market price reduction for architectural EC glazing down to 200 € m^{-2} would be necessary for the operational energy benefits to exceed the initial cost.

In 2012, we published a comprehensive benchmark (of criteria) analysis performed on five electrically controlled state-of-the-art transmittance modulation devices including their production routes, from 'cradle-to-gate' [9]. (Cradle-to-gate is an assessment of a *partial* product life cycle from resource extraction (*cradle*) to the factory gate (i.e. before it is transported to the consumer).) The goal of this study was to generate an ecological LCA of prototypes envisaged for the electrically controlled modulation of visible light transmittance, that is, EC or electrotropic (ET) smart windows (the term 'electrotropic' refers to variable light *scattering* properties, as opposed to EC devices that change colour (i.e. light *absorption*) as a result of charge carrier exchange processes occurring between two electrode materials upon the application of a small voltage). In contrast to electrochromism, the ET effect is usually observed when liquid crystals are brought into an electric field (on: transparent; off: translucent).

A variety of structures and configurations of ECDs are comprehensively described earlier in this book. As current collectors and electronic conductors in a sandwich-type device, n-type Transparent Conductive Oxides (TCOs) on mineral glass or plastic film are by far the most prevalent systems [10, 11], such as In_2O_3:Sn (ITO: tin-doped indium oxide (sic!)) and SnO_2:F (FTO), produced via physical (PVD) and chemical vapour deposition (CVD), respectively. A large variety of substances can be employed as EC materials, including inorganic metal oxides (WO_3, MoO_3, NiO_x, IrO_x, mixed vanadium oxides, etc.), organic redox dyes, conducting polymers, as well as inorganic and organometallic complexes (metallopolymers, Prussian blue, etc.). The electrolyte in its role as pure ion conductor and electronic insulator can be either organic or inorganic and may be liquid or solid (e.g. solid polymer electrolytes based on polyethylene oxides or polymer gel electrolytes containing organic carbonates).

The LCA methodology we proposed for EC windows is illustrated in Figure 18.2. The LCA goal was to generate an environmental benchmark of electrically switchable shading devices according to ISO standard specifications 14040 and 14044 [2, 3] and thus identify environmental strengths, weak points and optimisation potential to develop an internal R&D-specific optimisation strategy for new technologies. Only commercially and technically viable products constituting clear technological benchmarks have been considered, that is, benchmarks 1a/b-4. The benchmark study was supposed to provide – from an LCA point of view – decision support for research and development on smart windows and anticipate possible environmental effects concerning the production, use and end

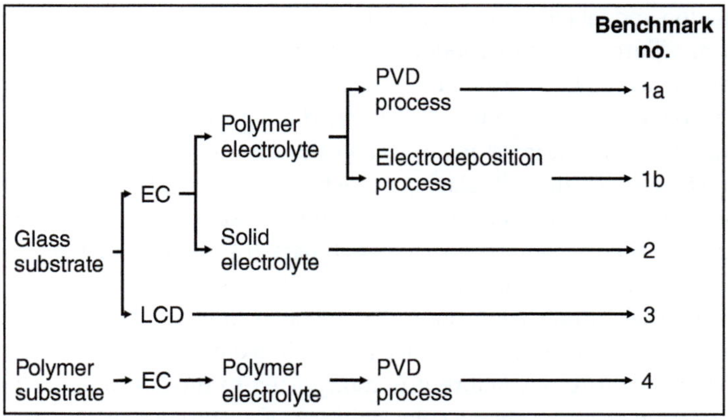

Figure 18.2 Overview of investigated, archetypical benchmark systems. Reproduced from Ref. [9] with the permission of The Royal Society of Chemistry.

of life of these products. It may also serve as a basis for the comparison of current smart shading technologies in terms of their ecological impact.

Usually, the assessment is based on a common functional unit for all technologies under concern, for example, a window (or segment thereof) of specified dimensions (the 'LCA scope'). In our study, we used as the main functional unit one piece of product, 20 cm × 30 cm in size; however, different sizes are viable as well. The system boundaries are cradle to gate, thus comprising production (including the whole value chain), the supply of all chemical precursors, operating supplies, auxiliary materials, energy sources and so on. It is planned to extend the system boundaries from cradle to grave at a later stage, then including the use phase and the end of life/recycling of the products as well. The LCI data basis used comprises literature and patent research about the chosen benchmark systems and their production routes, the use of the LCA software system and data basis GaBi [12], the European Life Cycle Database (ELCD) [13], the *ecoinvent* database [14] and internal calculations performed by the company LCS Life Cycle Simulation GmbH (modelling of surface-treatment processes PVD, CVD and electrodeposition with available industrial process technology information).

For performing an environmental assessment according to the LCA procedure, environmental indicators ('impact categories') need to be defined. The choice of impact categories should generally aim towards sustainable development, preserving resources, global protection of the eco-sphere, protecting human health and the stability of ecosystems [15, 16]. Many impact categories have been published, but only a few are generally and globally accepted. On a regular basis, the Institute of Environmental Sciences – Leiden University (CML – Centrum voor Milieuwetenschappen Leiden) publishes compilations of its own, externally developed characterisation and normalisation[1] factors [17] that may be used to show environmental impacts of several inventory categories. They constitute one of the two most used sets in LCA practice and are summarised in Table 18.1.

The European Parliament and Council published 'The Sixth Environment Action Programme of the European Community' [18], from which the impact categories used here are mainly derived. The environmental priorities within this programme are climate change, nature and biodiversity, environment and health and natural resources and wastes. For the quantified reduction goals for emissions in the European Union, see Refs. [19–22]. The choice of categories in Table 18.1 implements these priorities into a life cycle assessment. Especially due to the effect that the layers of EC devices also contain rare elements such as certain metals, the abiotic resource depletion potential (ADP) (see Table 18.1) is taken into account [23, 24].

Five archetypical benchmark systems have been chosen (Figure 18.2) and modelled by means of the LCA software GaBi [12]. The systems are further detailed in Figures 18.3–18.7. Benchmarks 1a/b–3 address glass-based variable

1) For normalisation, the yearly life cycle impacts of the production processes are set in relation to the yearly impacts of all EU 28 countries. The results are characteristic values without a specific unit, which thus can be compared to their relative value.

Table 18.1 Description of impact categories used for comparison of benchmark systems.

	Unit	Description	Examples
Parameter life cycle inventory			
Primary energy (PE)	MJ	Total of heating values of non-renewable and renewable energy used	Crude oil, natural gas, hydro power and so on
Parameter life cycle impact assessment			
Abiotic resource depletion potential (ADP)	kg Sb-equiv.	Use of non-renewable resources	Crude oil and metal ores
Greenhouse warming potential (GWP)	kg CO_2-equiv.	Emissions to air which influence the heat balance of the atmosphere	CH_4 and CO_2
Photochemical ozone creation potential (POCP)	kg ethene-equiv.	Emissions to air which act as ozone creators at ground level	Hydrocarbons
Acidification potential (AP)	kg SO_2-equiv.	Emissions to air which create acidification of rain water	NO_x and SO_2
Eutrophication potential (EP)	kg $[PO_4]^{3-}$-equiv.	Over-fertilisation of water and soil	P- and N-compounds

Reproduced from Ref. [9] with the permission of The Royal Society of Chemistry.

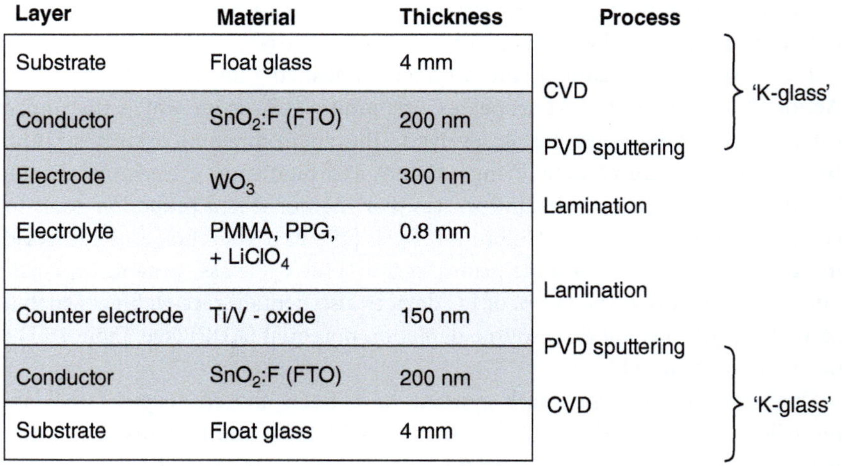

Figure 18.3 Benchmark 1a: FTO glass substrate, polymer electrolyte, EC films prepared via PVD [25, 26]. Reproduced from Ref. [9] with the permission of The Royal Society of Chemistry.

Layer	Material	Thickness	Process	
Substrate	Float glass	4 mm		
Conductor	SnO_2:F (FTO)	200 nm	CVD	'K-glass'
Electrode	WO_3	900 nm	Electro deposition	
Electrolyte	PVB + $LiClO_4$	0.8 mm	Lamination	
Counter electrode	$Fe^{(III)}_4[Fe^{(II)}(CN)_6]_3$	600 nm	Lamination	
Conductor	SnO_2:F (FTO)	200 nm	Electro deposition	'K-glass'
Substrate	Float glass	4 mm	CVD	

Figure 18.4 Benchmark 1b: FTO glass substrate, polymer electrolyte, EC films prepared via electrodeposition [27]. Reproduced from Ref. [9] with the permission of The Royal Society of Chemistry.

Layer	Material	Thickness	Process	
Substrate	Float glass	4 mm		
Conductor	SnO_2:F (FTO)	200 nm	CVD	'K-glass'
Electrode	WO_3	390 nm	PVD sputtering	
Electrolyte	$Ta_2O_5 \cdot nH_2O$	75 nm	PVD sputtering	
Counter electrode	IrO_x	55 nm	PVD sputtering	
Conductor	SnO_2:F (FTO)	200 nm	PVD sputtering	
Laminate	PVB	0.4 mm	Lamination	
Substrate	Float glass	4 mm	Lamination	

Figure 18.5 Benchmark 2: FTO glass substrate, EC films and electrolyte prepared via PVD (monolithic) [28].[2)] Reproduced from Ref. [9] with the permission of The Royal Society of Chemistry.

transmittance devices, whereas benchmark 4 describes a plastic-film-based EC device [30, 31]. Benchmark 3 concerns an ET polymer-dispersed liquid crystal (PDLC) technology. The liquid crystals are embedded in a polymer matrix

2) A lamination step is needed for connection of the substrate to the all-solid sputtered layers. No information about this step was available. Best guess: 0.4 mm PVB layer common for safety glass and used in benchmark 3 for the same purpose.

Layer	Material	Thickness	Process
Substrate	Float glass	4 mm	
			Lamination
Laminate	PVB	0.4 mm	
			Lamination
Carrier	PET	175 μm	
			PVD sputtering
Conductor	In_2O_3:Sn(ITO)	100 nm	
			Filling
Liquid crystal	Liquid crystal	5 μm	
			Filling
Conductor	In_2O_3:Sn(ITO)	100 nm	
			PVD sputtering
Carrier	PET	175 μm	
			Lamination
Laminate	PVB	0.4 mm	
			Lamination
Substrate	Float glass	4 mm	

'liquid crystal layer' (bracket spanning Carrier to Carrier)

Figure 18.6 Benchmark 3: active film laminated between glass panes, PDLC type [29]. Reproduced from Ref. [9] with the permission of The Royal Society of Chemistry.

the structure of which produces modulated light scattering, also known as the Christiansen effect [32]. Liquid Crystal Display (LCD) windows are translucent when in an unpowered state and become instantly clear (with a noticeable haze) when power is applied. Power must be applied continuously for the window to remain clear (24–100 V AC or 0.5 W ft^{-2} of glass) [33].

PDLC films are not used stand-alone, but integrated in glass doors, windows or room dividers (the so-called privacy glazing) [29], which has been taken into account for the assessment (Figure 18.6). Benchmarks 1a/b and 2 both concern EC systems comprising an ion conducting polymer electrolyte and a

Layer	Material	Thickness	Process
Substrate	PET	175 μm	
			PVD sputtering
Conductor	In_2O_3:Sn (ITO)	100 nm	
			PVD sputtering
Electrode	WO_3	300 nm	
			Lamination
Electrolyte	PMMA, PPG, + $LiClO_4$	50 μm	
			Lamination
Counter electrode	Ni/V -oxide	220 nm	
			PVD sputtering
Conductor	In_2O_3:Sn (ITO)	100 nm	
			PVD sputtering
Substrate	PET	175 μm	

Figure 18.7 Benchmark 4: ITO-coated polymer substrate, polymer electrolyte, EC films prepared via PVD [30, 31]. Reproduced from Ref. [9] with the permission of The Royal Society of Chemistry.

sputter-deposited solid proton conductor, respectively, located between the main EC film and the counter electrode layer. Benchmark 2 is sometimes also referred to as all-solid-state technology based on a monolithic stack of thin films deposited on a single substrate [28, 34, 35]. Benchmarks 1a and 1b further distinguish between two different well-established deposition processes for EC films, that is, PVD [25, 36] and electrodeposition [27]. Benchmarks 1a and 4 resemble each other in terms of EC materials (inorganic metal oxides), deposition processes (PVD) and type of electrolyte (polymeric ion conductor), but differ in the type of substrate (FTO glass vs ITO PET) [37].

The authors of Ref. [9] emphasise that the given device configurations do not necessarily fully comply with the situation in real existing products or prototypes. They comprise a number of assumptions made to accomplish the modelling tasks and are based on published data and information. Consequently, all models are open for variation of important parameters (toolbox) such as the materials used in substrates and functional layers, the thickness of layers and material deposition and lamination processes. Smart window half-cells based on tungsten or nickel oxides must usually be pre-treated ('charged' and 'pre-inserted') before they can be assembled to form a full device. Besides direct sputtering of lithiated compounds [38], pre-lithiation [39] and ozone exposure [40] have been proposed to accomplish that. It is emphasised that the energetic contribution of any of such activation steps has not been taken into account in our study [9].

The environmental assessment results shall be detailed with a focus on primary energy demand in the following. Figure 18.8 provides an overview of the relative energetic impact data obtained in Ref. [9]. To demonstrate the environmental

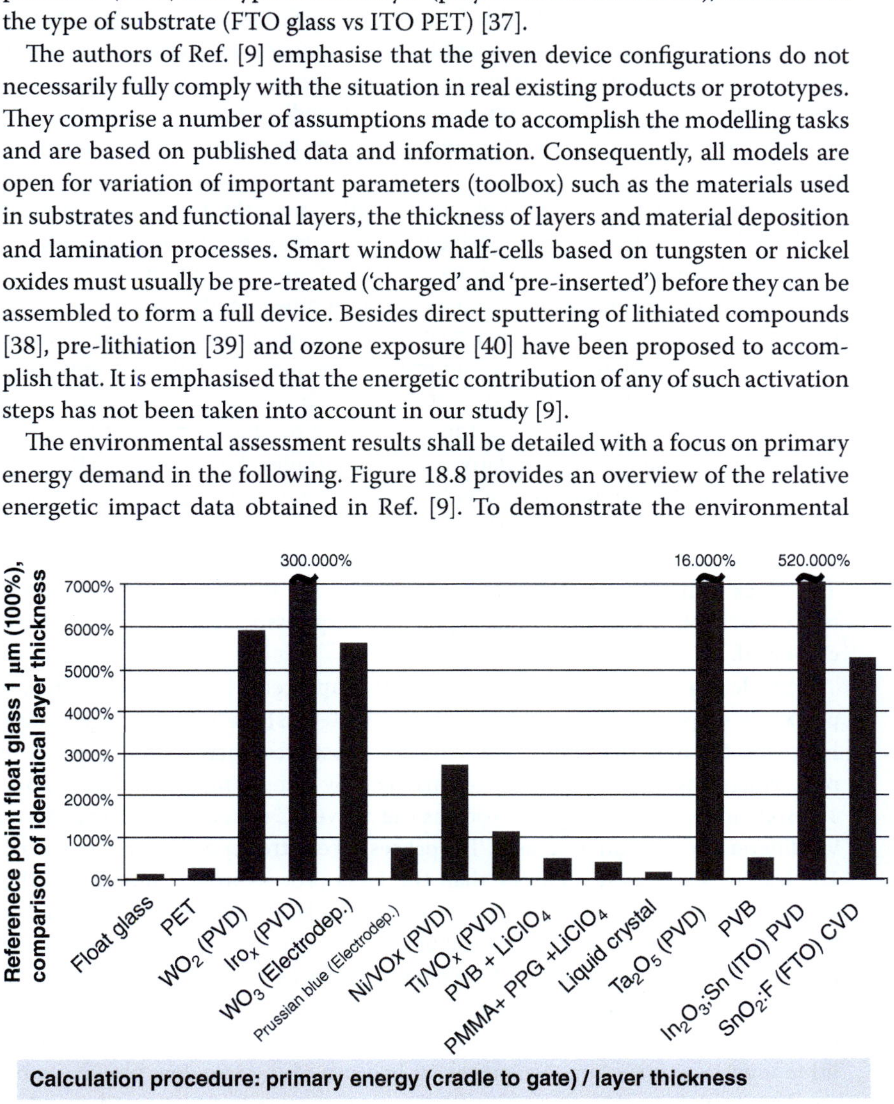

Figure 18.8 Specific importance of the different materials employed in the selected types of light-modulation devices. Reproduced from Ref. [9] with the permission of The Royal Society of Chemistry.

relevance, all investigated materials and their production processes have been normalised to the same layer thickness of 1 μm, the reference point being float glass and the normalised parameter being the primary energy demand.

It is obvious that iridium oxide and ITO clearly have the highest ecoprofiles among the materials studied due to extensive resource extraction and material processing steps, which is considered an important finding with respect to the future use of these oxides in smart windows (and perhaps other devices). An ITO film 1 μm in thickness has the 5200-fold energetic impact as compared to a 1-μm-thick sheet of float glass. This finding is remarkable, in particular, because float glass is a high-temperature (and hence high-energy) material. Moreover, it provides an important input for the discussion currently evolving on indium recycling. In this context, it shall be remarked that (doped) zinc oxides seem to emerge as viable and environmentally friendly candidates to replace ITO in electronic devices and displays [41, 42].

Sputtering (which is usually employed for oxides such as tungsten oxide, iridium oxide and ITO) is an energy-intensive deposition process. However, the usually rather low layer thickness these oxides are applied with may put into perspective their impact on the entire benchmark system. In other words, low consumption of a material with high embodied energy may result in lower overall resource depletion compared to high consumption of a material with low embodied energy, which must be considered carefully for each system under concern. WO_3 and Ta_2O_5 have high ecoprofiles, too, but can be produced in a more efficient way. Polymer materials have relatively low ecoprofiles due to higher efficiency factors during production. They are, however, usually applied with much higher film thickness. As mentioned earlier, the overall performance is important.

In Figure 18.9, the primary energy demand is summarised for all benchmarks. The black columns show the energy demand for the device manufacturing processes, while the grey columns depict the energy associated with the materials employed.

Considering the primary energy usages, it is apparent that the manufacturing processes employed (starting from the raw materials) are of some importance. However, the materials used (including the substrates!) play a more significant role, which may be demonstrated, for instance, by comparing benchmarks 1a and 1b. Both are based on tungsten oxide as the active EC material but are produced via different deposition processes. The process of electrodeposition in benchmark 1b clearly consumes less energy than the PVD process in benchmark 1a. This is, however, compensated for by the higher thickness of the electrodeposited layers, resulting in a virtually identical primary energy value for the wet-chemical approach 1b. In contrast to that, benchmark 2 shows a comparatively high energy demand for the all-solid-state process comprising the subsequent sputtering of three thin films (two EC films plus an inorganic ion conducting film) to a monolithic multilayer structure. Nevertheless, because of the very low film thickness employed in this technology, the total energy demand is almost identical to Benchmark 1a, that is, the system with sputter-deposited EC films and a polymer electrolyte. As mentioned before, the impact of 'high-energy materials' on the

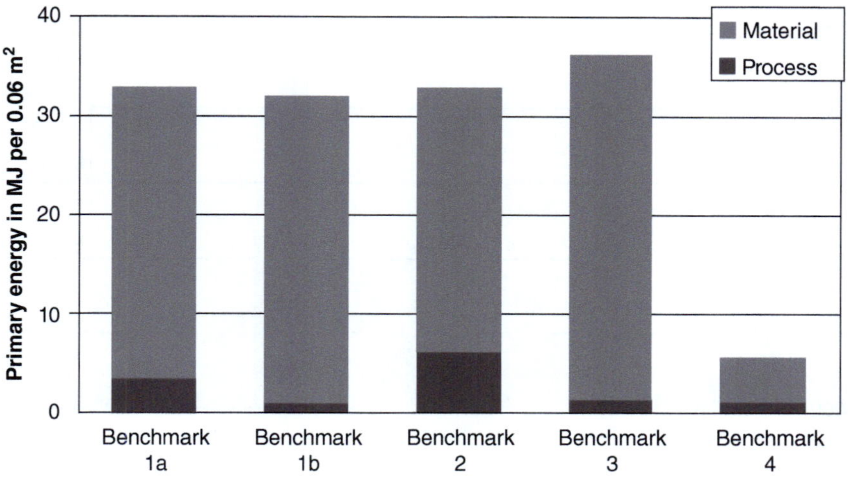

Figure 18.9 Primary energy demand per functional unit. Reproduced from Ref. [9] with the permission of The Royal Society of Chemistry.

entire system may be put into perspective when these are applied with low layer thickness and *vice versa*.

In benchmark 3, the energy requirements are observed to be high, which is caused by two construction steps involving lamination to glass and the use of ITO as a conductor in the PDLC film. Benchmark 4 is different from the other benchmarks, because it does not employ a glass substrate but, rather, a thin film of Poly(ethylene terephthalate) (PET). As the preparation of FTO glass is a highly energy- and resource-intensive process, the use of a plastic substrate results in much lower primary energy demand that would more than compensate for any other effects such as a change in the process. It should be noted, however, that in the course of this study, the choice not to laminate the benchmark 4 plastic EC film between two glass panes was merely made for reasons of clarity and comprehensibility, that is, to demonstrate the environmental impact of the type of substrates used. Realistically, one would rather assume that an EC film would not only be used stand-alone but also be laminated between glass panes or at least attached to an existing window (as a retrofit). Looking at Figure 18.9, the latter option would, however, result in a close to 50% reduction of the primary energy of a glass laminate, if single glazing is considered.

It should be noted that the glass demand of the functional unit in the studies of Yianoulis *et al.* [7] and Papaefthimiou *et al.* [8] is comparable to the benchmarks in the present study. However, in the latter, the sheet glass ecoprofile (meaning the main impact factor in the whole ecoprofile) is about 75% higher, which can be attributed to the availability of up-to-date LCA data and first-hand information from glass manufacturers confirming the higher values.

There was little difference between the tendencies found for the impact categories in comparison to the primary energy. Figure 18.10 provides an overview illustrating this. Benchmarks 1a, 1b, 2 and 3 generally have a higher impact than

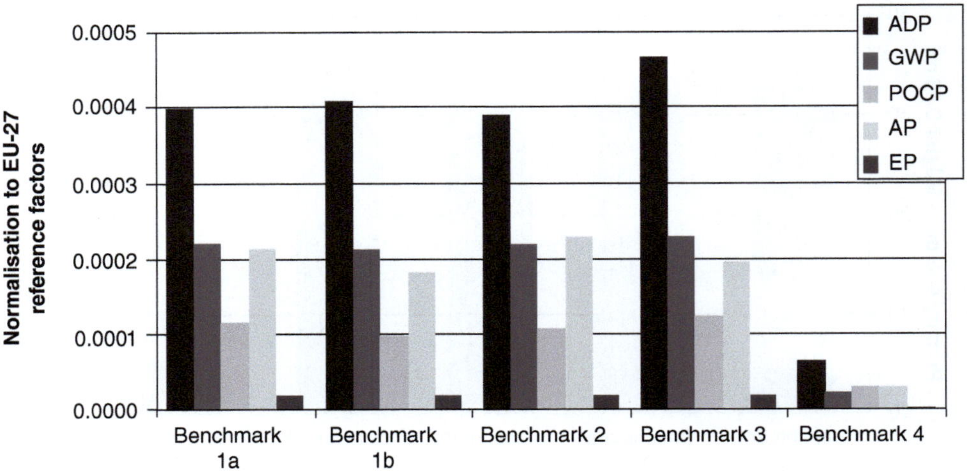

Figure 18.10 Environmental problem fields normalised to EU27 per functional unit (order from left to right within each benchmark): ADP: abiotic resource depletion potential; GWP: global warming potential, POCP: photochemical ozone creation potential, AP: acidification potential and EP: eutrophication potential.

benchmark 4. The ratios between the different categories are almost identical; there is only minor variation. The main impact category of the ADP (large black columns) is followed by the acidification potential (AP; light grey columns) and the global warming potential (GWP) – also known as carbon footprint – (GWP; dark grey columns), with this last category showing approximately half the level of person equivalents of the ADP. The photochemical ozone creation potential (POCP; medium grey columns) has a less marked impact. For the plastic-based benchmark 4, the situation was slightly different as the POCP was higher than the GWP at a level of around 50% of the ADP. However, since the absolute POCP value was still low compared to the other benchmarks, this is of no further relevance. The impact category eutrophication (EP; small black columns) may generally be considered of minor importance and thus omitted in the future discussion. EP becomes a relevant category if renewable materials are significantly involved, which is not the case here.

There are a number of important findings from Figures 18.9 and 18.10. The process of manufacturing from the raw materials to the final device has a marked environmental effect, which makes the choice of appropriate manufacturing technologies highly relevant. As opposed to the mineral-glass-based benchmarks, plastic-film-based systems (i.e. electrochromic films) are particularly favourable, implying that the substrate is a factor with a strong environmental impact in transmittance modulation devices. The layers of conductors and electrodes are generally of high specific environmental importance. They should be kept as thin as possible when it comes to designing devices. The most striking result was that the use of ITO as transparent conductor may more than compensate for any other effect from an environmental point of view. Hence, any discussion of the pros and cons of different deposition techniques in terms of energy depletion might be redundant as long as ITO is used as the transparent conductor. At least, this ought to be considered carefully.

The environmental impact of the EC layers in tungsten-oxide-based devices is much higher than the impact of the polymer electrolytes. This is remarkable keeping in mind that the EC and polymer electrolyte layers differ strongly in their layer thicknesses (usually 1–2 orders of magnitude). The use of an inorganic proton electrolyte (hydrated Ta_2O_5), interestingly, does not significantly change this situation, despite a high energy demand for the related sputter deposition process. This is due to the low layer thickness usually employed for such ion conductors. Among the EC layers, the working electrode material tungsten oxide (where applicable) has at least a twofold higher impact than usual counter electrode materials such as titanium, nickel or vanadium oxides. Tungsten is considered a 'specialty' metal, meaning a by-product or co-product of large mining or smelting operations, with a relatively small value mined, processed and used on a yearly basis. Specialty metals do not trade on the open markets, so pricing is negotiated via off-take contracts or spot sales [43]. Coming back to the LCA results shown earlier, the situation is different for iridium oxide, a much less abundant material, which has an extremely high environmental impact. Iridium is one of the rarest elements in the Earth's crust (40 and 10 times less abundant than gold and platinum, respectively), with annual production and consumption of only 3 tons. It is obtained commercially as a by-product from nickel and copper mining and processing [44]. Conversely, Prussian blue is by far the most favourable counter electrode material in terms of abundance[3] and environmental impact (of those materials modelled in this study).

In the following, we would like to make a short excursus and go a little deeper into the energy efficiency (and thus cost-effectiveness) discussion of EC layers produced by physical vapour deposition. For performing a life cycle assessment on EC windows, roughly, the balance is drawn of the energy needed to produce the window (i.e. the energy embodied within the entirety of the layers and substrates employed in the window) and the energy presumably saved during the use phase. As outlined earlier, some prominent sputter materials such as tungsten oxide may have a high environmental impact due to the energy needed to win the ore from a mine, then extract and purify the material needed from the ore, that is, tungsten metal. Further energy is needed to prepare the sputter targets from the metal and eventually produce the active films by sputtering. However, a closer look on the energy balance reveals that the energy embodied cannot in all cases be held responsible for the actual cost of smart windows based on sputtered tungsten oxide. Even high energy demands can be irrelevant for the process costs. Rather, the main cost drivers are investments in machinery, facilities and installations needed for the production, which

3) That is, from iron pre-products such as iron oxide and salts derived there from.

> can be elevated for sputter-deposited films (can amount to €1–20 million depending on size and throughput) (M. Vergoehl, 2013, Fraunhofer IST, personal communication). It is obvious that the eventual product costs will depend on the amount of product produced. Producing 10 million square metres on a €20 million machine would strongly relativise this investment cost factor, however, a production volume of 100 000 m^2 would clearly not.
>
> From the holistic perspective of life cycle assessment, energy is not expensive enough to account for high process costs in automated PVD coating. Hence, investment costs may account to as much as 50% of the full process costs for highly automated production.

Important parameters within the benchmark toolbox can be varied. Examples include the materials used as substrate and functional layers, the thickness of those layers and the type of material deposition and lamination processes. The work is currently focusing on stand-alone plastic-film-based chromogenic devices, for instance those employing conjugated conducting polymers and nanomaterials as active components [45–49]. These approaches have attracted increasing attention in the community in recent years as a way of addressing the demand for a cost-effective, lightweight and retrofittable technology for light and heat modulation, for example, in aircraft cabins. An exemplary novel prototype technology of such types shall further be discussed in the following.

18.3 LCA of Novel Plastic-Film-Based Electrochromic Devices

In the frame of the European collaborative research project *Innoshade* ('Innovative Shading Devices') Life Cycle Assessment has been employed to study the environmental impact of a new plastic-film-based technology and compare the results to the benchmarks described above. The technology under concern is mechanically flexible, low-cost EC films that could be put on to any surface and used in applications where glass windows are not viable for weight reasons, such as in aircraft [50]. The system was investigated over the whole life cycle (i.e. system boundaries: cradle to grave) in comparison to three different applications, that is, a household oven window (appliance), a car sunroof (automotive) and an aircraft cabin window

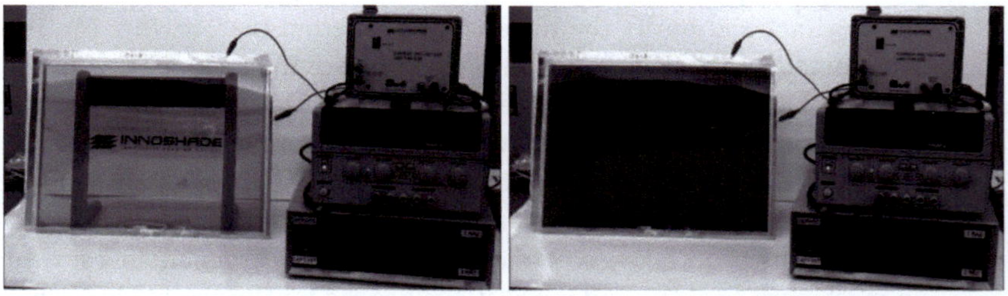

Figure 18.11 Visual appearance and EC contrast of an EC film device prepared with the *Innoshade* project. Effective size: 355 × 260 mm^2.

Layer	Material	Thickness	Process
Substrate	PET	130 μm	
			PVD sputtering
Conductor	In_2O_3:Sn (ITO)	100 nm	
			Coating
Counter electrode	Prussian blue	100 nm	
			Lamination
Electrolyte	PEO-acrylate LiTFSI	37 μm	
			Lamination
Electrode	Polythiopene	100 nm	
			Coating
Conductor	In_2O_3:Sn (ITO)	100 nm	
			PVD sputtering
Substrate	PET	130 μm	

Figure 18.12 *Innoshade* device configuration assumed for LCA.

(aviation). As an exemplar system, the 'automotive sunroof' application will be described in some detail. The study is about to be extended to architectural glazing in a follow-up project [51] (Figure 18.11).

One possible device system configuration is given in Figure 18.12. The technology, except for the transparent conductors, builds on materials that are favourable from an environmental point of view. The cathodically colouring material is a novel patented polythiophene nanocomposite showing fast EC response, adapted colour in the darkened state and a fully colourless bleached state [49]. Prussian blue was used as an anodically colouring material and was deposited from aqueous nanoparticle suspensions. A UV cross-linkable polyether acrylate composition was used as the polymeric electrolyte, with Lithium trifluoromethylsulfonyl imide (LiTFSI) as the salt [52].

Table 18.2 shows the general goal and scope definition of the assessment. For benchmark system configurations, LCA methodology, and chosen impact categories it is referred to the sections above.

Furthermore, viable production processes have been designed; see Figure 18.13. All deposition processes are capable of being implemented by roll-to-roll technology, as is the lamination of the pre-manufactured half-cells to full devices, thus enabling cost-effective and energy-efficient production. Process steps and parameters have been fixed in such a way to fulfil a state-of-the-art production process with regard to legal requirements, for example, Volatile organic compounds (VOC) incineration of the used solvents. If, due to the process design, not all parameters can be fixed for the time being, best and worst case assumptions can be made for energy consumption factors of aggregates (ventilators, pumps) and the scrap factor, which then provides a basis for energy and material flow simulations of the process steps, including air conditioning for production hall and clean room, VOC treatment and assembly processes.

Table 18.2 Goal and scope definition of LCA.

Goal	• Environmental assessment of electrically switchable shading devices according to ISO standard 14040 and 14044 • Identification of environmental strengths, weak points and optimisation potentials • Intended use: evaluation of a new development versus an existing reference systems for further optimisation in the industrial up-scaling process • Target audience: Project Partners, EU Commission, electrochromic window players (research, development, producers and suppliers)
Functional unit	• Product size: $20\,\text{cm} \times 30\,\text{cm} = 0.06\,\text{m}^2$
Devices for comparison	• Plastic film device, according to Figure 18.12 • Benchmarks 1–4, according to Figure 18.2
System boundary	• Cradle to gate (build-up of devices including sealing and contacting)
Data sources	• Scientific literature, patents and patent applications; adjustment by expert know-how of the authors
LCA databases	• GaBi professional database; *www.pe-international.com* • Ecoinvent database; *www.ecoinvent.ch* • GEMIS database; *www.oeko.de* • European Life Cycle database; *www.lca.jrc.ec.europa.eu* • Approved industrial database (LCS) • Calculations and simulations by LCS • Relevant literature

The most likely candidate materials for use in a possible pilot-plant-scale *Innoshade* production process were identified. The assessment of the EC polythiophene electrode showed a tremendous environmental impact for the monomer synthesis due to a relatively low yield alongside a high material demand for solvents. The subsequent process-optimisation driven by the LCA feedback resulted in an ecoprofile of the polymerogenic material being about 30% better in terms of primary energy. Due to the layer thickness, that is, the main material input to device in matters of mass, the PET substrate still had the highest contribution to the ecoprofile. Surprisingly, the very thin ITO layer is second with 33% of the ecoprofile of the whole process. The reasons have been discussed already (see Figure 18.8). The electrode production process itself (compare Figure 18.13) and the remaining materials have less relevance (Figure 18.14).

The results of the environmental comparison are exemplarily shown in Figure 18.15, again for the most interesting parameter Primary Energy that is comparable to the impact category of Fossil Resource Depletion. Benchmarks 1–4 are identical to those given in Figure 18.9. The plastic-film-based device described above is now included for both best case and worst case scenarios.

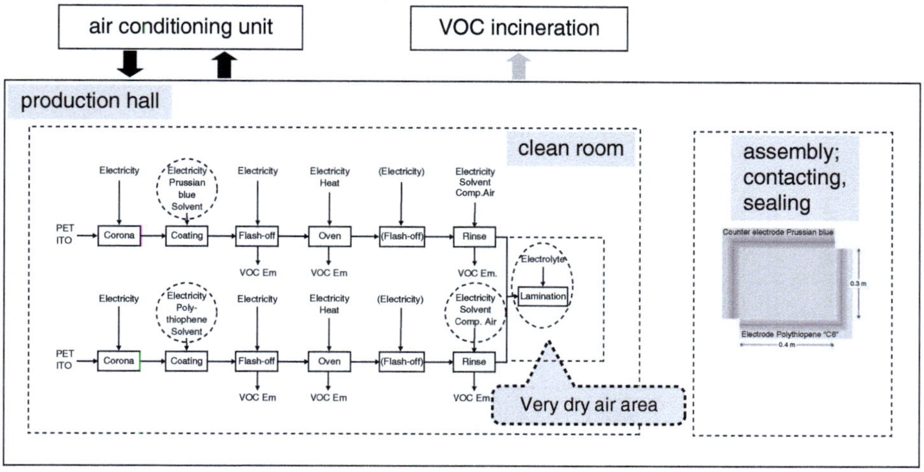

Figure 18.13 Proposed production process for an EC device built up, according to Figure 18.12.

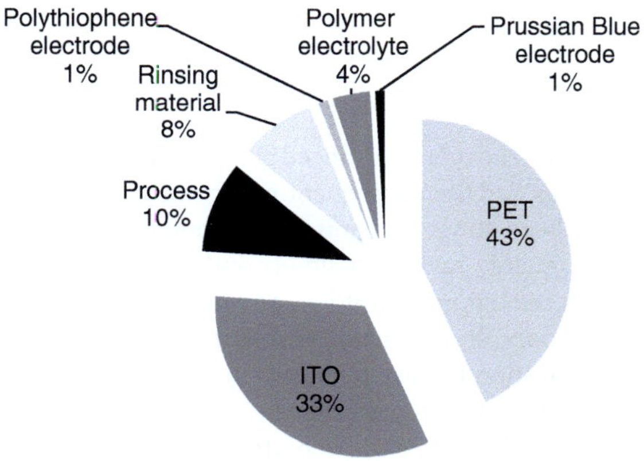

Figure 18.14 Plastic EC device based on Prussian blue and an *in-situ* polymerised polythiophene film – distribution of primary energy of materials and production process (cradle to gate).

It is apparent that benchmark 4 (= ITO-coated polymer substrate, polymer electrolyte, EC films prepared via PVD) and the *Innoshade* device (= ITO-coated polymer substrate, polymer electrolyte, EC films prepared wet-chemically) have a similar environmental performance, whereby the best and worst case scenario analysis for the latter's production process has only a minor effect to the overall results. Benchmarks 1–3 have a significantly worse environmental performance, which is clearly due to the glass substrate. A closer look on the energy efficiency of surface coating technologies or treatments revealed that much less electricity is needed for wet-chemical coating than for electrodeposition and a lot less than for sputtering.

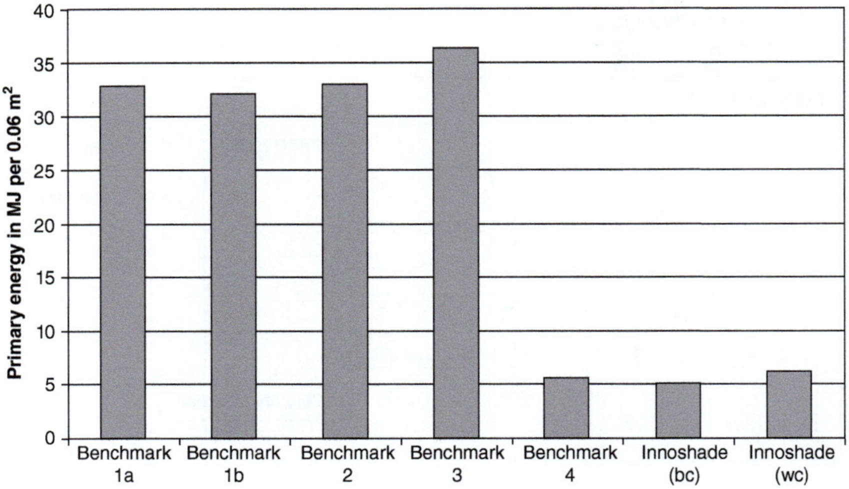

Figure 18.15 Comparison of devices, system boundaries: cradle to gate.

18.4
LCA for EC Target Applications

According to Papaefthimiou *et al.* [8], the role of advanced glazing in the reduction of energy consumption in buildings and the consequent reduction in the greenhouse gas (GHG) emissions has been so far recognised. We found in our LCA studies that there is considerable energy-saving possibility and other benefits in some other cases as well, such as reduction of heat loss and heat penetration in appliances (e.g. house-hold oven doors) and reduction of solar heat gain in vehicle cabins by EC sun roofs. Even in aviation the use of light-weight smart EC shading systems might make mechanical shades dispensable in aircraft cabins which would lead to high weight savings, while increasing comfort for the passengers at the same time. In the following, we summarise life cycle assessments for a few selected EC applications.

18.4.1
Automotive Sunroof Case

Table 18.3 exemplarily amends the general goal and scope definition of Table 18.2 with regard to the application of an EC film to a car sunroof.

From Figure 18.16, verifiable savings of energy needed for air conditioning in vehicles can be seen if the energy savings through solar energy reduction is equal to or exceeding 15%. The additional insulation effect of the EC devices has the dominating environmental impact of their life cycle. Interestingly, this holds for both the glass-based, energy-intensive benchmark 2 system and the less energy-demanding plastic film device. For benchmark 2, the glass panes

Table 18.3 LCA of an automotive sunroof, amendment to Table 18.2 (differences only).

Functional unit	• Functional unit: 42 cm × 29 cm = 0.122 m^2, plus switching equipment, that is, contacting and sealing materials. Power supply equipment is negligible • Lifetime: 5 years, 20 000 cycles (approximately 5 per day)
Devices for comparison	• Sunroof with plastic-film-based *Innoshade* device according to Figure 18.12 • Sunroof with glass-based benchmark 2 device according to Figure 18.5 • Sunroof without device (glass window only)
System boundary	• Production of devices (cradle to gate) • Use phase • End-of-Life of devices
Data sources	• Use phase: Energy savings (through solar energy reduction) of shading devices estimated to be approximately 15% (S.L. Mendaro, Microelectrónica MASER, Spain, 2012, personal communication) • Energy Mix: EU power mix 2011 (LCS database) • End-of-life: thermal incineration (LCS database)

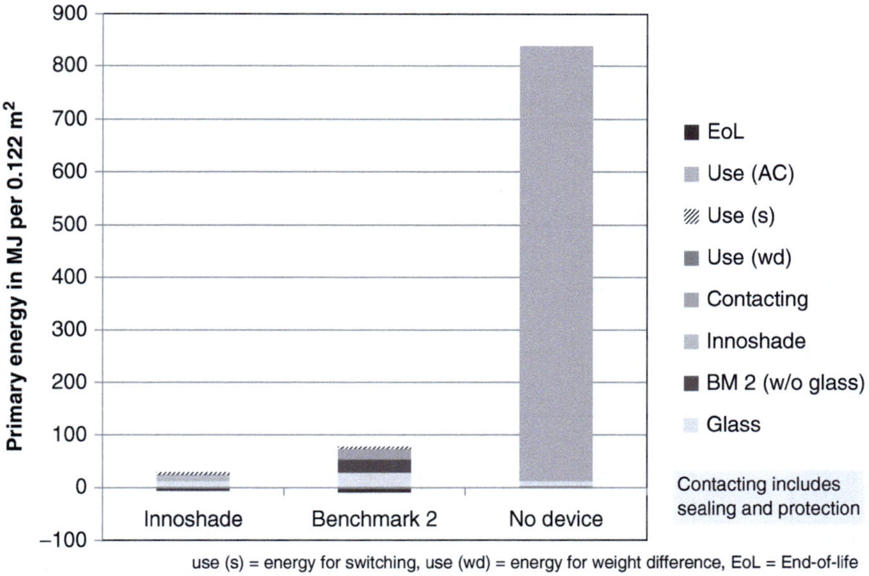

Figure 18.16 Primary energy demand (cradle-to-grave) for EC automotive sunroofs (shown are two selected configurations).

have been omitted in this particular case as the EC system was expected to be integrated in the roof's double glazing structure (no need for two additional panes). In any case, considerable amounts of energy can be saved by introducing EC windows in a car sunroof. A second important aspect is that all production processes (device, contacting, sealing, etc.) and the End-of-Life, in particular for the plastic device, are negligible for all analysed environmental parameters in reference to the energy savings, which means there is a clear net environmental benefit. The environmental impacts for the car sunroof case correlate with the primary energy demand in a very similar manner as has been demonstrated for the different smart window benchmarks in Figures 18.9 and 18.10. Again, the ADP is the most prominent impact category, followed by POCP, GWP and AP, with EP being virtually irrelevant (Figure 18.17). The totals of all categories correlate with the energy demand for the three cases 'Innoshade', 'Benchmark 2' and 'no device', the latter resulting in by far the highest impact.

18.4.2
Appliance Example: Window Case for a House-Hold Oven

A very cost-effective solution must be sought in order to economically endow household oven windows with an electrochromic function. This is due to the extremely high costs and low profit margins in the market of domestic appliances. We have therefore studied whether cooking oven windows can be equipped with

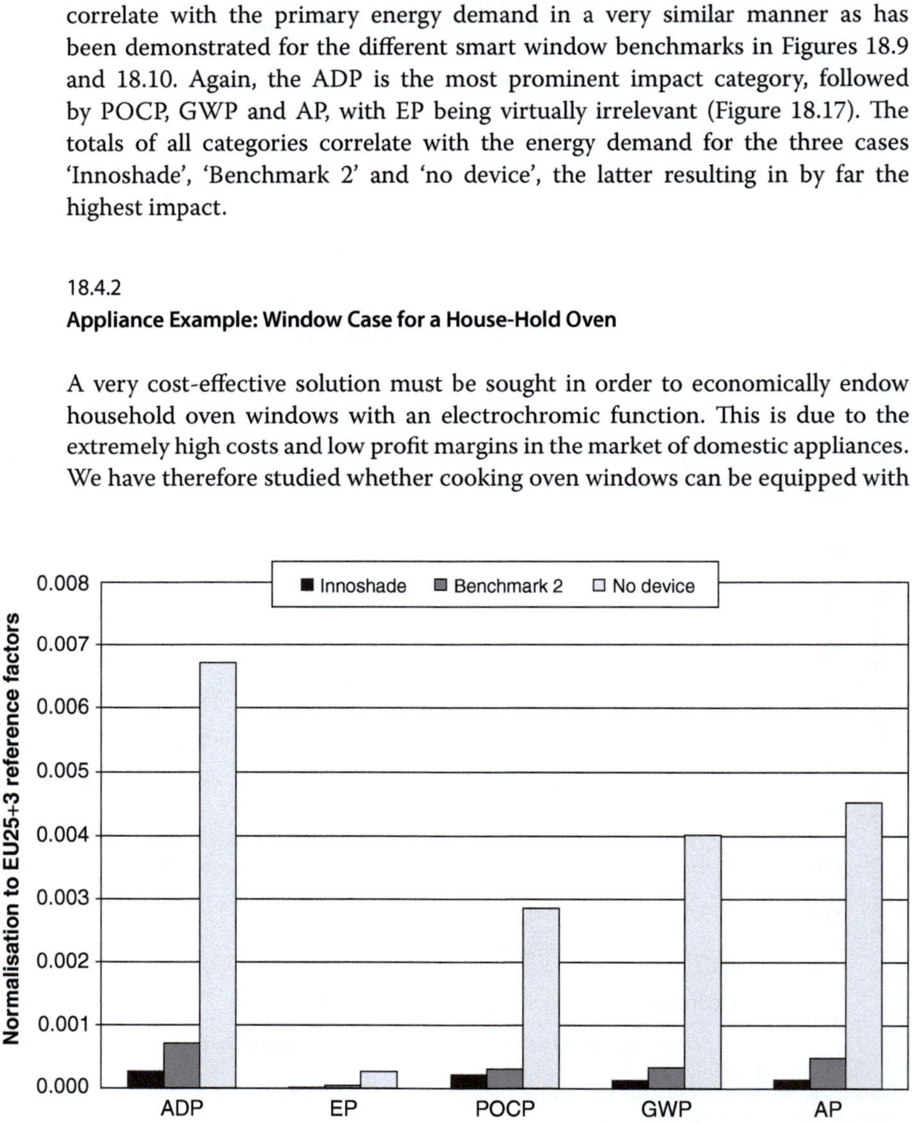

Figure 18.17 Normalisation of life cycle (cradle-to-grave) for EC automotive sunroof.

EC plastic films (for functional unit and use phase assumptions see[4]). A 'smart' EC oven door can surely be a fancy feature, but is generally expected to enable some heat energy saving as well by avoiding door opening during cooking or baking to inspect the contents. Some additional insulation effect was expected as two transparent electrodes are employed in the device known to reflect IR radiation. It could be confirmed by energy consumption modelling (according to EN 50304:2001) and cooking performance tests (according to EN 60350) that the energy consumption of a specially designed prototype oven endowed with an EC film device is reduced by up to 1.5% and 2% in the dark and clear state, respectively. This encouraging result has been fed into the life cycle assessment. Again, it turned out that the energy embodied in production processes and End-of-Life are negligible in reference to the projected energy saving. The additional insulation effect (up to 2% electrical energy saving) has the dominating environmental impact of the life cycle of the device. EC devices applied to oven windows thus generate a win-win situation: a new (and fancy) technical feature is combined with a reduced environmental impact.

18.4.3 Aircraft Cabin Window Case

The application of ECDs is also envisaged for the interior design of aeroplanes. Aircraft passengers could block light through cabin windows by using a dimmable window shade. Transferring the LCA approach to such aircraft cabin windows (for functional unit and use phase assumptions see[5]) reveals that any environmental benefit strongly depends on how the EC window is incorporated in the aircraft structure (U. Posset, M. Harsch, unpublished results). The main energy saving possibility is clearly attributed to the possible replacement of (heavy) mechanical shutters by (lighter) EC films. It is obvious that using a glass-based device is counter-productive in this sense, and finding the lightest alternative would mean the least fuel consumption during the use phase. The option of adding a mechanical shutter to an ECD leads to worse results compared to the conventional aircraft window plus shutter. This latter finding correlates to the expectation that EC windows would not significantly contribute to saving energy needed for air conditioning in an aeroplane. In this context, it should be noted that state-of-the-art electrically dimmable shades are said by passengers to not darken enough to block sunlight completely. So, by now, mechanical shutters are considered necessary.

4) Functional unit: 0.12 m^2 (prototype window dimensions), plus switching equipment. Lifetime: three times usage per week, five switching cycles per cooking and 20 years' appliance lifetime; approximately 20 000 cycles. Data sources: Use phase: energy consumption (electricity) 250–400 kWh (worst and best case, respectively) depending on working mode and temperature of the oven: Energy Mix: EU power mix 2011 [LCS database] and End-of-Life: thermal incineration [LCS database].
5) Functional unit: 0.07 m^2 (A320 cabin window dimensions) plus switching equipment. Eighty windows per aircraft life time of aircraft = 25 years and 55 000 000 km. Data sources: Use phase: EADS data for fuel savings due to weight differences. End-of-Life: thermal incineration [LCS database].

18.5 Conclusion

The environmental assessments performed on electrically controlled light-modulation devices ('smart windows') clearly highlight those EC materials and processes that are favourable from an environmental and energy demand point of view. Moreover, the findings provide strong support for major current development trends in the area such as the use of organic or polymeric EC materials, the quest for ITO replacement materials (e.g. zinc-oxide-based TCOs or polymeric conductors), and, in general, the replacement of energy- and resource-intensive processes (e.g. the sputter deposition of specialty or critical metal oxides) by less demanding methods. By life cycle analysis, light may also be shed on environmental impacts that are becoming more and more important against the background of the current lively discussion on greenhouse-gas emissions and the climate change associated with them.

There is a tremendous difference in terms of resource depletion and environmental impact as to whether mineral glass or plastic film is used as the substrate in EC devices. As regards resource depletion, the use of glass may even more than compensate for the perceived positive effect of wet-chemical processing. The same holds for ITO, the use of which may overcompensate for any environmental benefits attained with other components of EC (and probably other) devices. The substitution of ITO is thus considered an important issue for next-generation devices from both the economic and environmental point of view. It is hoped that, in the future, the environmental impact concerning the production, use and end-of-life of up-and-coming smart window technology can be traded off against their prospects, in particular, to lower the consumption of energy needed for air-conditioning in buildings. According to joint studies for the European Commission, nearly 60% of the dwellings in the European countries are still single glazed, which leads to excessive heat losses (and hence a high carbon footprint). EC technology can significantly contribute to improve this situation by lowering energy losses unless the energy balance is negative. This is where LCA comes into play.

References

1. Clear-up FP7 – EU Collaborative Project Clear-up (Clean and Resource-efficient Building for Real Life), *www.clear-up.eu* (accessed 25 February 2015).
2. International Organisation for Standardisation (ISO) ISO 14040. (2006) *Environmental Management – Life Cycle Assessment – Principles and Framework*, International Organisation for Standardisation (ISO), Geneva.
3. International Organisation for Standardisation (ISO) ISO 14044. (2006) *Environmental Management – Life Cycle Assessment – Requirements and Guidelines*, International Organisation for Standardisation (ISO), Geneve.
4. European Commission Joint Research Centre Institute for Environment and Sustainability: International Reference Life Cycle Data System (ILCD) (2010) *Handbook – General Guide for Life Cycle Assessment - Detailed Guidance*, 1st edn, Publications Office of the European Union, Luxembourg, EUR 24708 EN, March 2010.

5. Curran, M.A. (ed) (2012) *Life Cycle Assessment Handbook*, Scrivener Publishing LLC, Salem, MA.
6. Khasreen, M.M., Banfill, P.F.G., and Menzies, G.F. (2009) *Sustainability*, **1**, 674–701.
7. Syrrakou, E., Papaefthimiou, S., and Yianoulis, P. (2005) *Sol. Energy Mater. Sol. Cells*, **85**, 205.
8. Papaefthimiou, S., Syrrakou, E., and Yianoulis, P. (2009) *Energy Build.*, **41**, 17–26.
9. Posset, U., Harsch, M., Rougier, A., Herbig, B., Schottner, G., and Sextl, G. (2012) *RSC Adv.*, **2**, 5990–5996.
10. Granqvist, C.G. (2007) *Sol. Energy Mater. Sol. Cells*, **91**, 1529.
11. Exarkhos, G.J. and Zhou, X.-D. (2007) *Thin Solid Films*, **515**, 7025.
12. PE International AG (2011) GaBi Software and Data Base for Life Cycle Engineering, Echterdingen.
13. European Life Cycle Data Base *www.lca.jrc.ec.europa.eu* (accessed 25 February 2015).
14. Ecoinvent (2011) Ecoinvent Data v2.2, *www.ecoinvent.ch* (accessed 25 February 2015).
15. Kreißig, J., Baitz, M., Betz, M., Eyerer, P., Kümmel, J., and Reinhardt, H.-W. (1997) *Leitfaden zur Erstellung von Sachbilanzen in Betrieben der Steine-Erden-Industrie*, Bundesverband Steine + Erden, Frankfurt.
16. ENQUETE- Kommission (1993) Schutz des Menschen und der Umwelt des Bundestages (Hrsg.), in *Verantwortung für die Zukunft – Wege zum nachhaltigen Umgang mit Stoff- und Materialströmen*, Economica Verlag, Bonn.
17. Centre of Environmental Science – Leiden University (CML) (2001) Characterisation and Normalization Factors, Leiden.
18. EU (2002) EU: Decision No 1600/2002/EC Of The European Parliament and of the Council of 22 July 2002 Laying down the Sixth Community Environment Action Programme.
19. UNEP (2000) The Montreal Protocol on Substances that Deplete the Ozone Layer, published 2000 by Secretariat for The Vienna Convention for the Protection of the Ozone Layer & The Montreal Protocol on Substances that Deplete the Ozone Layer, United Nations Environment Program, Nairobi, Kenya
20. EU (2000) EU: REGULATION (EC) No 2037/2000 Of The European Parliament and of The Council of 29 June 2000 on Substances that Deplete the Ozone Layer.
21. UN ECE (1999) The 1999 Gothenburg Protocol to Abate Acidification, Eutrophication and Ground-Level Ozone. Convention on Long-Range Transboundary Air Pollution.
22. UNFCCC (1997) Kyoto Protocol to the United Nations Framework Convention on Climate Change.
23. Huppes, G. *et al.* (2002) *Abiotic Resource Depletion in LCA*, Ministerie van Verkeen en Waterstraat of the Netherlands, Amsterdam.
24. Guinee, J.B. (ed) (2002) *Handbook on Life Cycle Assessment*, Kluwer Academic Publisher, Dordrecht.
25. Wittkopf, H. (2010) *Vak. Forsch. Prax.*, **22**, 26.
26. Masetti, E., Varsano, F., Decker, F., and Krasilnikova, A. (2001) *Electrochim. Acta*, **46**, 2085.
27. Kraft, A., Rottmann, M., and Heckner, K.-H. (2006) *Sol. Energy Mater. Sol. Cells*, **90**, 469.
28. Taunier, S., Guery, C., and Tarascon, J.-M. (1999) *Electrochim. Acta*, **44**, 3219.
29. Lampert, C.M. (2003) *Sol. Energy Mater. Sol. Cells*, **76**, 489.
30. Widjaja, E.J., Delporte, G., Vandevelde, F., and Vanterwyngen, B. (2008) *Sol. Energy Mater. Sol. Cells*, **92**, 97.
31. Azens, A., Avendano, E., Backholm, J., Berggren, L., Gustavsson, G., Karmhag, R., Niklasson, G.A., Roos, A., and Granqvist, C.-G. (2005) *Mater. Sci. Eng.*, **B 119**, 214.
32. Lampert, C.M. (1995) Society of vacuum coaters. Annual TechCon Proceedings, 1995.
33. Eleanor, Lee (2006) A Design Guide for Early-Market Electrochromic Windows, Lawrence Berkeley National Laboratory.
34. Beteille, F., Boire, P., and Giron, J.-C. (1999) in *Proceeding of SPIE – The International Society for Optical Engineering, Switchable Materials and Flat Panel Displays* vol. 3788. (ed C.M. Lampert), Denver, CO.

35. Sone, Y., Kishimoto, A., and Kudo, T. (1994) *Solid State Ionics*, **70-71**, 316.
36. Granqvist, C.-G. (1999) *Electrochim. Acta*, **44**, 3005.
37. Baetens, R., Jelle, B.P., and Gustavsen, A. (2010) *Sol. Energy Mater. Sol. Cells*, **94**, 87.
38. Michalak, F., von Rottkay, K., Richardson, T., Slack, J., and Rubin, M. (1999) *Electrochim. Acta*, **44**, 3085.
39. Ye, Y., Zhang, J., Gu, P., and Tang, J. (1997) *Sol. Energy Mater. Sol. Cells*, **46**, 349.
40. Azens, A. and Granqvist, C.-G. (2003) *J. Solid State Electrochem.*, **7**, 64.
41. Fahland, M., Vogt, T., Schoenberger, A., and Mosch, S. (2010) *Adv. Sci. Technol.*, **75**, 9–15.
42. Fahland, M. (2011) *Vak. Forsch. Prax.*, **23**, 30–34.
43. Fulp, M. (2012) Murdock Capital Partners Critical Metals/Strategic Elements Symposium, New York City, Summary *www.kitco.com/ind/fulp/20120815.html* (accessed 25 February 2015).
44. Xiao, Z. and Laplante, A.R. (2004) *Miner. Eng.*, **17**, 961–979.
45. Ma, C., Taya, M., and Xu, C. (2008) *Electrochim. Acta*, **54**, 598.
46. Granqvist, C.-G., Lansåker, P.C., Mlyuka, N.R., Niklasson, G.A., and Avendaño, E. (2009) *Sol. Energy Mater. Sol. Cells*, **93**, 2032.
47. Wu, J., Potts, J.E., Waller, C.P., Jr., Weber, B.T., Wu, J.-S., and Lofgren, M.W. (2007) Electrochromic device. WIPO Patent WO 2008/133737 A2, 3M Innovative Properties Company.
48. INNOSHADE (2008) FP7 – EU Collaborative Project INNOSHADE (Innovative Switching Devices), *www.innoshade.eu* (accessed 25 February 2015).
49. Abbotto, A., Beverina, L., Cochet, A., Mari, C.-M., Pagani, G., Patriarca, G., Posset, U., Ruffo, R., Schottner, G., (2009) Highly transparent electrochromic coating material, method for producing the same and use of the material. EP Patent 1928000 B1, Fraunhofer Gesellschaft and Università degli Studi di Milano-Bicocca.
50. European Commission (2013) Thin-Film "Smart Glass" Puts Rivals in the Shade, *http://ec.europa.eu/research/infocentre/article_en.cfm?artid=29853* (accessed 25 February 2015).
51. EELICON FP7 – EU Collaborative Project EELICON (Enhanced Energy Efficiency and Contrast by Smart Light Transmittance Control), *www.eelicon.eu* (accessed 25 February 2015).
52. Zaghib, K., Perrier, M., Guerfi, A., Dupuis, E., Charest, P., Allaire, F., and Armand, M. (2011) Polymer Electrolyte with High Stability, its Use in Electrochemical Systems, US Patent 7897674 B2, HydroQuébec.

19
Electrochromic Glazing in Buildings: A Case Study

John Mardaljevic, Ruth Kelly Waskett, and Birgit Painter

19.1
Introduction

A major application area for electrochromic devices is architectural glazing, particularly that for office buildings which are often highly glazed. This chapter begins with an overview of daylighting in non-residential buildings and explains why the traditional control solutions, for example, blinds, often lead to the under-exploitation of the daylighting potential of the building. In a memorable statement made in 1998, Steve Selkowitz, the Leader of the Windows and Envelope Materials Group at Lawrence Berkeley Laboratory (CA, USA) called, the dynamic control of daylight the 'Holy Grail' of the fenestration industry [1]. Since then there has been considerable effort and many tens of millions of dollars (US) spent to achieve a viable product that is a practical alternative to ordinary or fixed-tint glass. Electrochromic (EC) glass is believed to be the leading contender in the race to manufacture a glazing technology that will achieve this 'big prize', resulting in the transformation of building facade design. With recent investment in the scaling-up of production capacity, EC glass is now set to become a mainstream glazing product. There are now a number of buildings throughout the world with production EC glass.

A review of the control of daylight in buildings is given together with reasons why the traditional methods, for example, venetian blinds, are habitually used in a sub-optimal fashion, often negating the daylight potential afforded by the glazing design. Various types of chromogenic and variable transmission glazing (VTG) are outlined, and the potential for practical use in buildings together with operational factors and performance issues is discussed. The chapter concludes with a description of and preliminary findings from a case study evaluation of a pair of offices spaces in the United Kingdom fitted with EC glazing. The installation of the EC glass and the post-occupancy monitoring of the space and its users are described.

Electrochromic Materials and Devices, First Edition.
Edited by Roger J. Mortimer, David R. Rosseinsky and Paul M. S. Monk.

19.1.1 Daylight in Buildings

The use of daylight in office buildings is generally considered to be a greatly under-exploited resource [2]. In large part, this is because of the highly variable nature of daylight illumination. The natural, large variability in daylight means that users will often need to use shades to moderate excessive ingress of daylight. Alternatively, the building may have fixed structures to block, redirect and/or attenuate the daylight, for example, light-shelves, brise-soleil, fixed-tint glazing.

With moveable shades, the users rarely make the effort to adjust them once the external condition has changed. Thus, the shades are often left deployed for much of the day resulting in the commonplace occurrence of 'blinds down, lights on', Figure 19.1. Using traditional approaches, there are essentially three ways to control daylight from facade windows:

(i) Block the light, for example, an opaque roller blind.
(ii) Redirect the light, for example, a reflective light-shelf, prismatic glazing and so on.
(iii) Attenuate the light, for example, a fixed-tint glazing.

Many shading devices—fixed and moveable—utilise two of the aforementioned mechanisms. For example, a light-shelf is usually designed to block direct sun near to the window and also redirect the (blocked) daylight to the ceiling where it can help to improve daylight penetration. Users can operate venetian blinds to both block and redirect light, though it is more common to see them used with the slats closed.

19.1.2 The Importance of View

Another key consideration is the mode of transmission: specular, diffuse or some mixture of the two. This is particularly important for primary view windows. A specular transmission is one where the light rays exhibit no noticeable deviation, and so the view always appears sharp. For example, even with heavily tinted glass, the view to the outside (Figure 19.2a) will appear sharp, albeit darkened, during daytime (Figure 19.2b). However, if the glazing is of the diffusing type (e.g. a privacy screen), then the light rays are scattered and there will be no discernible view.

Figure 19.1 Manual use of blinds is habitually sub-optimal: 'blinds down, lights on'. Fixed external structures (e.g. 'brise-soleil') can also used to provide shading.

(a) (b) (c) (d)

Figure 19.2 (a–d) Importance of view (simulated)—clear glass, fixed-tint glass, glass with some diffusion and clear glass with venetian blinds lowered.

Some transparent materials exhibit both specular and diffuse properties, and the quality of the view through them will depend on the relative proportions for the two transmission modes—see the (simulated) example of blur due to scattering of light in Figure 19.2c. Note that most occupants find that even a small degree of scattering by glazing is unacceptable for primary view windows. Slatted blinds (i.e. horizontal or vertical) can be adjusted to allow a partially obstructed view out, Figure 19.2d. However, depending on the particular configuration of the slats and the distance from the window, the occupant may find that their focus is drawn to the slats; thus, making it difficult to gain the beneficial relaxation of the eyes that is afforded when they are focused on the distant view beyond. This effective 'shortening' of the distant view can also occur with external shading structures such as brise-soleil.

A glazing with a transmissivity that varies continuously between clear and dark extremes could offer a much greater degree of control over the luminous environment and avoid many of the drawbacks of the traditional approaches to shading and solar protection [1]. In contrast to fixed-tint glass, a VTG would *modulate* the daylight in response to occurring conditions rather than simply attenuate it by a constant fraction.

19.2 Variable Transmission Glazing for Use in Buildings

The principle behind VTG is straightforward: the transmission properties of the glazing are varied to achieve an 'optimum' luminous and/or thermal environment. The various types of VTG can be grouped into three broad classes: chromogenic coatings, suspended particle device and micro-electro-mechanical systems (MEMS) (though there are also other types under development).

19.2.1 Chromogenic Glass

There are four distinct types of formulations for coatings that have variable transmission properties. These are: electrochromic, gasochromic, photochromic and thermochromic. The agents causing the change in transmission are:

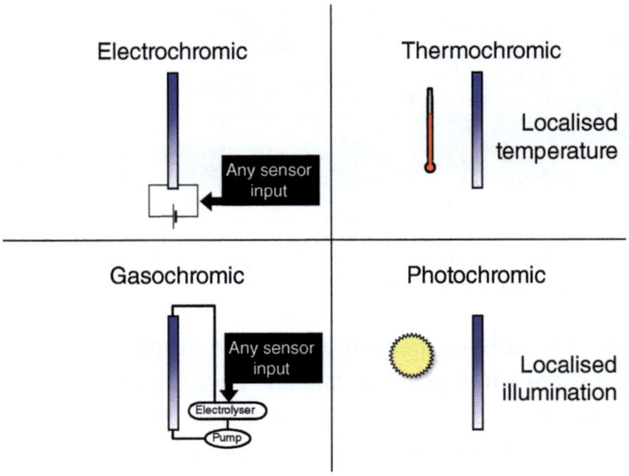

Figure 19.3 Chromogenic types.

voltage (electrochromic); concentration of pumped gas (gasochromic); localised illumination (photochromic); and localised temperature (thermochromic), Figure 19.3. Thermochromic and photochromic are essentially passive devices which respond to changes in the environment, whereas electrochromic and gasochromic are active devices that can be configured to respond to any sensor input, for example, illumination, temperature or some combination of the two.

Suspended particle device is a plastic-film-based technology. The plastic film is in fact a thin layer containing a suspension of rod-like particles in billions of liquid droplets. An applied voltage alters the orientation of the particles and therefore the transmission properties of the film. A VTG based on MEMS has tiny, micron-scale structures that move in response to the applied electrostatic field, thereby altering the transmission properties of the glazing.

19.2.2 VTG Performance Characteristics

The key to performance of a VTG is high (visible) transmission in the clear state and sufficiently low (visible) transmission in the darkened (or tinted) state. To be perceived as acceptable to the majority of occupants in non-residential buildings, the VTG in the clear state should appear like ordinary (un-tinted) double glazing, and so there is a visible transmission of 60% or greater. In the darkened state, the transmission should be low enough so that additional shading is required only very rarely or perhaps not at all. In practice, this means a minimum visible transmission of around 2% or less. Additionally, the building occupants should have some degree of control of the glazing, for example, to manually override an automated control setting. So, whilst a 'passive' VTG might seem attractive at first because it allows for autonomous operating behaviour, the corollary of this is a lack of control, for example, modulation of the glazing transmission by (localised)

Table 19.1 Glazing variable transmission type and visible transmission range where known.

VTG type	Visible transmittance	Solar heat gain coefficient
Electrochromic[a)]	1–60%	0.47–0.09
Thermochromic[b)]	13–60% or 6–30%	0.37–0.17 or 0.24–0.13
Photochromic	[N/A]	[N/A]
Gasochromic	13–60% [R]	[N/A]
Susp. particle device	≪ 1%–50% [R]	[N/A]
MEMS	[N/A]	[N/A]

a) Sage Glass.
b) Performance data for Sunintuitive products.

window temperature will not necessarily offer the luminous environment desired by the occupants.

The performance characteristics of various VTG types (where known) are summarised in Table 19.1. Visible solar transmittance (T_{vis}) is the fraction of visible radiation transmitted by a glazing system weighted according to the photopic response of the human eye. The spectral power distribution of incident radiation used for the evaluation of visible transmittance should be that of the CIE standard daylight illuminant D65. Visible transmittance may be denoted as a value between 0 and 1 or as a percentage between 0% and 100%. The visible solar transmittance is often referred to simply as the visible transmittance. The solar heat gain coefficient (SHGC) is the fraction of total solar radiation admitted through a transparent material, for example, a window with clear, tinted or diffusing glass. SHGC accounts for directly transmitted radiation and that part which is absorbed and subsequently released (e.g. re-radiated) inward. The range of SHGC is therefore never as large as that for the visible transmittance because of the release of absorbed short-wave (i.e. visible) and long-wave (i.e. thermal) radiation into the space as thermal radiation. Where no performance data was found to be available is indicated by 'N/A', and where the data was taken from research publications rather than a product specification is indicated by '[R]'.

There are examples of thermochromic glazing on the market, though the narrow visible transmission range (e.g. 13 – 60% or 6 – 30%) indicates that additional shading would be needed to control glare. Thermochromic glazing therefore seems better suited to offering a degree of moderation of the thermal rather than the luminous environment. Gasochromic has the potential advantage of rapid switching speeds. A gasochromic system requires that the glazing unit is literally 'plumbed-in'—connected to an electrolyser and pump by piping. The practicalities of a gasochromic installation are such that the technology is still considered the preserve of research. The situation regarding suspended particle device glazing for clear (i.e. view) glass is uncertain. There appear to be some products on the market, but limited examples of actual installations. MEMS window technology is still under development. Thus, of the technologies described, only EC glazing appears to have the necessary optical properties (i.e.

(a) (b)

Figure 19.4 Images showing electrochromic glazing in (a) clear and (b) darkened state (photos courtesy SAGE Electrochromics Inc.)

wide visible transmission range), is relatively straightforward to install, is already in the marketplace and is under large-scale production [3]. The two images in Figure 19.4 show a large electrochromic glazing installation using SageGlass. For the generation of the product, the visible transmittance in the clear state was 62% with a minimum of 2% when fully tinted. The performance of EC glazing between the years 2003 and 2013 in terms of the maximum and minimum visible transmittance values is summarised in Table 19.2. The goal of the manufacturers has been to achieve a very low minimum T_{vis} so that additional blinds/shades will not, in most instances, be required to moderate the internal luminous and/or thermal environment.

User acceptance for any daylight control technology depends on a number of performance and operational characteristics. For EC glazing, these include performance with respect to glazing transmission range (i.e. the values for the maximum and minimum visible transmittances), the switching time between the clear and tinted states and the effectiveness of the automated control to minimise user interventions (e.g. manual overrides). Another key factor for user acceptance is the *quality* of the luminous environment produced by EC glazing—meaning, the spectral composition of the daylight that is 'filtered' through tinted EC glass. This is because the spectral transmission properties of the EC coating vary as the glass changes state. This can be seen in Figure 19.4 showing a pair of photographs with EC glass in the clear state (a) and at full tint (b). As the glass darkens (i.e. 'tints'),

Table 19.2 Changes in the max/min range of visible transmittance for commercially available EC glazing.

Date	Range in T_{vis}
2003–2006	58–4%
2007–2009	62–4%
2010–2012	62–2%
2013–to date	60–1%

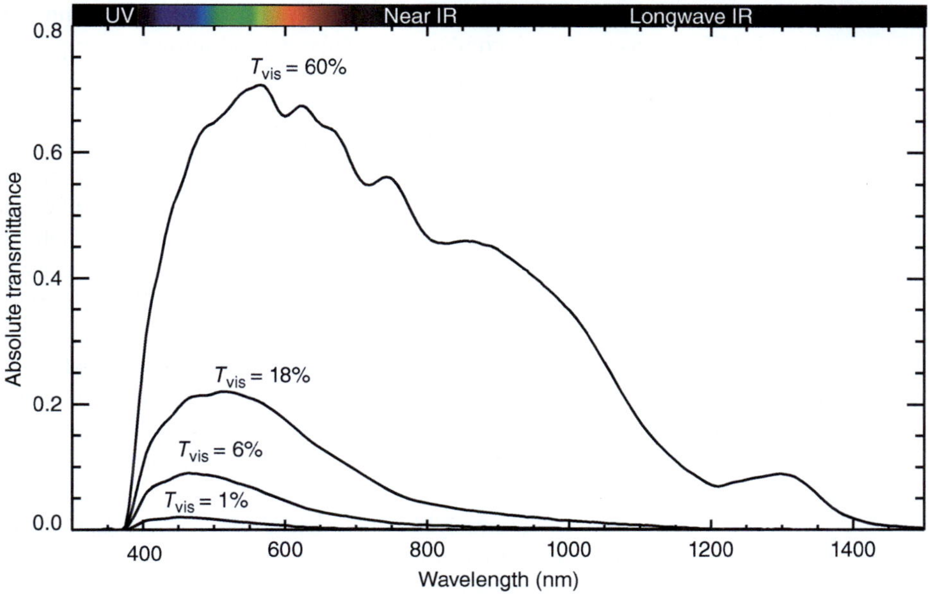

Figure 19.5 Absolute spectral transmission curves for SageGlass in clear (60%), fully tinted (1%) and two intermediate states. Note, the curves are for monolithic coated EC glass whilst the labels refer to the visible transmittance when used in conjunction with clear glass in an insulating glazed unit.

the longer wavelengths are diminished proportionally to a greater degree than the shorter wavelengths, providing the EC glazing good solar control properties to help prevent overheating. Optically, the consequence of this is to shift the peak in visible transmission to the blue end of the spectrum.

This can be seen in the transmission curves shown in Figure 19.5. In the clear state, the glazing exhibits a visible transmittance of 60% and appears effectively neutral to the eye. The SageGlass product shown in Figure 19.5 exhibits a minimum visible transmission of 1% when fully tinted, and it can be varied continuously between this and the clear state. However, a small number of intermediate states is considered adequate for most practical installations, for example, 'light-tint' (18%) and 'mid-tint' (6%).

19.2.3
EC Product Details and Practicalities

The data in this section was compiled from information on the SAGE Electrochromics Inc. website[1] and is specific to SageGlass EC glazing.

EC glass is manufactured by coating float glass with nanolayers of metal oxides using a sputter deposition system. It is a process similar to that used to make

1) *http://sageglass.com*

low-e glass. The coated glass is then fabricated into an insulating glass unit. The control of EC glazing can be any combination of manual and automatic, employing whatever sensor inputs are required, for example, illuminance levels, sun position, internal air temperature and so on.

EC glazing can, in principle, be used wherever daylight and/or solar control is needed. SHGC ranges between 0.47 in the clear state and 0.09 when fully tinted, giving a high degree of solar control to prevent overheating. Another application of EC glazing has been the control of 'light pollution' from the artificial lights in buildings. In addition to controlling the ingress of daylight/sun during the day, EC glazing is set to tint at night-time to prevent artificial light 'spillage' from the building.

Practical limitations are largely due to restrictions in the size of EC panels. Until recently, the maximum EC panel size was around 1 m by 1.5 m; however, panels of size 1.5 m by 3 m are now available. For the larger sizes, SAGE have introduced intra-pane zoning whereby EC glass in one frame can be 'zoned' to create two or three sections that tint independently.

EC glass panels are installed much like any other glazing system, the key difference being that the glass needs to be 'wired-in' by an electrician. However, a recent development is an integrated unit comprising EC glass, frame, photovoltaic strip, battery storage and WiFi controller. These units generate their own electricity and, since they have built-in WiFi, wiring for both power and control is no longer needed. SageGlass requires very little power—the equivalent of a 60 W lamp will operate 186 m^2 of EC glazing.

It will take on average 3 to 5 min for the EC pane to tint over 90 per cent of its range in warm weather and direct sun, 5 to 10 min over the same range in moderate weather and somewhat longer in colder weather in the absence of direct sun on the glass. Switching speed is also dependent upon the size of the pane.

To be most effective, the artificial lighting should allow for dimming and the lighting controls should be integrated with the EC controls to maintain optimum task lighting levels during normal operation. Where the EC glazing also serves to protect from overheating, the control should also be linked to building energy management systems. Effective integration of the EC controls with other building control systems is the key to realising the significant energy saving potential of EC glazing.

In 2013, an 'autonomous' EC window unit was demonstrated by Sage Electrochromics Inc. This window unit is powered by a small, integrated photovoltaic panel with a wireless connection to the external controller. Users can also control the glazing from a mobile WiFi device such as an iPad.

19.2.4 Operational Factors

Automatic operation of EC glazing is vital to maintain occupant comfort and also to achieve any hoped-for reductions in energy consumption over standard glazing. Depending on the building type and the prevailing climate, automatic control of the EC glazing state can be a function of any sensed environmental parameter for the building (singly or in combination) in addition to the time of day/year (i.e. to determine the sun position). However effective the automatic operation of the EC glazing, it is essential to provide the option for manual override by the occupants even if it will only be used infrequently. Experience has shown that occupants will often resort to sabotage if an automated building control system fails to do what they wish [4]. The sensors that provide data to the glazing controller could be any of the following:

- External facade-mounted illuminance sensor.
- Internal window-mounted illuminance sensor.
- Room illuminance or 'brightness' sensors, for example, on or above desks.
- Room-temperature sensors.
- Roof-mounted illuminance or irradiance sensors.

Care needs to be given to sensor placement since some environmental parameters, for example, illuminance, can vary dramatically across relatively small spatial scales.

The effectiveness of EC glazing to temper the indoor thermal environment has been demonstrated in a number of theoretical and empirical studies, and modelling its performance to control solar heat gain in a building simulation program is relatively straightforward (see Section 19.2.6). It is, however, the user acceptance of the visual environment produced by EC glazing that will be the key determinant for the success of this VTG technology. In particular, the ability of EC glazing to prevent visual discomfort *without* the need for an additional shading device, for example, venetian or roller blind, will be a major determinant for the long-term success of this technology. The most difficult visual discomfort condition for tinted glazing to control is direct sun in the field of view of the building occupant. Under clear sky conditions, the luminance of the sun can be as high as 1.5×10^9 cd m^{-2}. Studies have shown that 3% minimum visible transmittance is not sufficient to control glare, with one-quarter of participants resorting to occasional blinds use [5]. People vary in their tolerance to glare, and whilst 2% minimum visible transmittance will result in fewer occurrences of direct sun glare compared to 3%, it may not be sufficient to entirely eliminate the occasional need for blinds. It remains to be determined if a minimum visible transmittance of 1% is sufficient to control direct sun glare for those most susceptible to it.

Another operational factor is the time taken to switch, particularly important for occupants when a glare condition occurs abruptly, for example, direct sun suddenly appears from behind a cloud or obstruction. This condition invariably causes occupants to immediately lower a shading device to block the sun. Without that option, the few minutes taken for the EC glass to transition to a fully tinted state will be a period of visual and/or thermal discomfort, with the potential to disrupt normal work activities, for example, if the reflected 'sun dazzle' from a monitor renders it unusable until the EC glass has fully darkened. Thus, the automatic control system should attempt to pre-empt these occurrences as often as possible. A pre-emptive element to the control will account for the sun's position relative to the facade in addition to sensed parameters such as external (vertical) illuminance measured at the facade. The control system may also account for shading resulting from nearby buildings, thereby predicting the progression of shadow patterns across the facade and commencing the transition of the EC glass *before* it would occur under local sensing only.

The occupants of buildings are known to prefer daylight with a neutral spectrum over any other form of illumination. Fixed-tint glass has been shown to have lower approval rating than clear glass, particularly on overcast days when the tint will make a dull sky appear even duller, thereby exacerbating any sense of drabness regarding the illumination. Additionally, tints with a hue may be in discord with user expectations with regard to actual weather conditions. For example, on an overcast day, occupants would not expect the sky seen through the windows to appear blue. With electrochromic glazing under normal operation, it is possible to avoid the undesirable effects associated with fixed-tint glass. This is achieved through optimum control and zoning of the glass: (a) having the glass transition to full tint only when required and (b) ensuring that a proportion of the glass remains in the clear state. Zoning is described in the following section.

19.2.5 Zoning of EC Glazing

Zoning is the term used to describe the operation of electrochromic glazing when different tint settings are used on different panes or groups of panes at any one instant in a room or space. This is often desirable for a number of reasons:

a) to ensure that glare control can be achieved (needing $T_{vis} \leq 2\%$);
b) to provide sufficient daylight illumination in the space;
c) to ensure that the spectrum of illumination in the space remains neutral; and
d) to maximise the opportunity to save energy through photoelectric dimming of lights and/or reduction in cooling loads.

Consider the following example for EC glazing in horizontal rows. A window extending from the sill to the ceiling could be comprised of three glass panes (in three frames) or a single pane in one frame which has been 'zoned' to create three sections that tint independently—referred to as 'between-pane' and 'in-pane' zoning. In either case, it is possible to control the zones independently. A common zoning arrangement that works effectively for much of the year when skies are clear is shown in Figure 19.6. In this example, zones 2 and 3 are set to full tint to moderate the ingress of high-angle sun, whilst for zone 1, the glazing is set to be fully clear.

In addition to allowing for an (un-tinted) view to the outside, a zone pane set to fully clear ($T_{vis} = 60\%$) will allow in about 60 times more daylight than a zone pane of the same size set to full tint where $T_{vis} = 1\%$. This huge difference in absolute visible transmittance between clear and fully tinted means that, with just a relatively small proportion of the EC glazing set to clear (the rest at full tint), the daylight illumination in the space is dominated by the neutral light entering

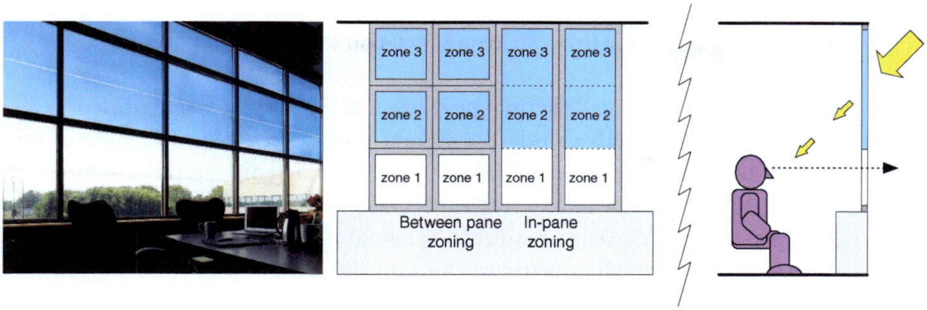

Figure 19.6 Zoning of electrochromic glazing by individual panes or within-pane zoning (photo courtesy SAGE Electrochromics Inc.)

through the clear glazing. For example, consider the case where there is just one pane set to clear ($T_{vis} = 60\%$) for every eight set to full tint ($T_{vis} = 1\%$). In this state, the clear glass occupies only 1/9 of the glazed area, but it accounts for ~ 9/10 of the overall visible transmittance of all nine panes of glass. Additionally, the overall visible transmittance for all nine panes is just 8%. This explains why the illumination in a space with EC glazing controlled in this manner generally appears neutral—as is evident in Figure 19.4 where less than one-third of the glass is clear (the remainder at full tint).

The design of the space and the zoning/operation of the EC glazing both should encourage the mixing of light. Allowing, for example, some (un-filtered) sunlight through clear EC to reflect off a white surface (sill or wall) can help to maintain a neutral spectrum of illumination in the space. Zoning can also be used to effect more subtle changes in the luminous environment. Tinting the top and bottom rows of EC glazing in the library space shown in Figure 19.7 reduces the contrast between the glazed facade and the internal surfaces, for example, the darker bookshelves.

Figure 19.7 Century College (photos courtesy: SAGE Electrochromics Inc.)

19.2.6
Performance Prediction Using Building Simulation Tools

Prediction of the overall thermal properties of EC glazing was reported in 1991 by Reilly *et al* [6]. In that study, the authors compared the 'performance of three prototypical electrochromic glazings' using simplified assumptions to model the overall energy balance of different window-to-wall areas. Modelling of the actual performance of EC glazing requires a dynamic simulation of the thermo-physical response of the building fabric to the external climate (e.g. solar gain through windows) in combination with the operation of any plant (e.g. heating, cooling, ventilation, etc).

Dynamic thermal simulation (DTS) of building performance became an established modelling technique in the 1980s [7]. By the mid-to-late 1990s, most of the large consulting engineering practices had sufficient in-house modelling skills to carry out DTS to evaluate and optimise building performance at the design stage. In cooling-dominated climates, a key building performance parameter is the required size and operation of the air-conditioning plant since this can have a significant impact on both the capital and the running costs of the building. Provided that the visible and thermal transmission properties are known, it is a relatively straightforward matter to model a building with EC glazing using DTS. The DTS program simply adjusts the transmission properties of the EC glazing on a time-step basis according to the prescribed control function, for example, predicted internal temperature. In the last two decades, there were a number of simulation studies on the energy performance of EC glazing using actual (i.e. based on production samples) and theoretical transmission values [8]. These studies invariably focus on the energy saving potential of EC glazing to reduce cooling load by moderating solar gain. Less numerous are field studies on scale-model or full-size spaces with EC glazing (see the following section).

Until recently, the evaluation of daylight in buildings was carried out using a highly idealised method called the daylight factor (DF) approach [9]. The basis of the DF was conceived over a century ago as a means of rating daylighting performance independently of the actually occurring, instantaneous sky conditions. Hence, it was defined as the ratio of the internal horizontal illuminance to the unobstructed (external) horizontal illuminance under a single, overcast sky condition. The ratio of internal to external illuminance was usually expressed as a percentage. The DF approach makes no account of building orientation because the overcast sky is rotationally symmetric about the zenith axis. Of course, there is no sun. Thus, for a given building design, the predicted DF is insensitive to either the building orientation (due to the symmetry of the sky) or the intended locale (since it is simply a ratio). Because the sun is not considered, any design strategies dependent on solar angle, solar intensity or redirection of sunlight have no influence on the DF. The DF approach is now widely considered to be a crude basis for the evaluation of buildings with standard glazing [2]. And, of course, woefully inadequate for any dynamic glazing system where realistic illumination conditions (i.e. sun and sky together) need to be modelled to make any meaningful evaluation of daylighting performance.

The dynamic simulation of daylight using sun and sky conditions derived from the same climate files used for DTS was only achieved in the late 1990s [10, 11]. Now called climate-based daylight modelling (CBDM), the technique has the potential to accurately model both illuminance and luminance in buildings of arbitrary complexity illuminated by realistic sun and sky conditions. The vast amount of simulated data from CBDM needs to be reduced to some readily intelligible measure, usually referred to as a 'metric', which serves as an indicator of predicted performance. Daylight metrics founded on CBDM include useful daylight illuminance (UDI) and daylight autonomy (DA) [2, 12]. The prediction of daylighting performance can only be reliably assessed using CBDM, regardless of the window type, though this is particularly so for EC glazing where the dynamic range in visible transmittance can be a factor of 60 (i.e. $1\% \leq T_{vis} \leq 60\%$). CBDM evaluation of EC glazing has been reported by a number of authors [13, 14].

The prediction of visual comfort may require the simulation of the predicted brightness (i.e. luminance) of the field of view as 'seen' by a (virtual) occupant. This could be important for the performance simulation of EC glazing because long-term visual comfort is a key consideration in the overall evaluation of any facade system designed to 'temper' the internal luminous environment [15].

19.2.7 Occupant-Based Studies

Whilst scale-model appraisals of EC performance offer some insights [16], the findings cannot be extrapolated to reliably predict human response and/or acceptance of conditions in an actual, say, office space. The research literature on buildings contains numerous examples of new technologies that, in reality, performed far less well than the simulations indicated [17]. In many ways, reliable prediction of the occupant's response to the luminous environment is far more challenging than, say, predicting the response to the thermal conditions. The luminous environment is generally far more complex and difficult to characterise than the thermal conditions in a space. Consequentially, many researchers believe that occupant-based studies in full-size spaces are needed to properly evaluate the user acceptance of new building devices/systems, not least this potentially 'game changing' glass technology.

One of the first significant studies using human subjects was reported in 2006 by researchers at the Lawrence Berkeley National Laboratory (LBNL), CA, USA. That study evaluated the responses of 43 subjects in a private office with switchable electrochromic windows, manually operated venetian blinds and dimmable fluorescent lights [18]. That study, using EC glazing with a minimum transmittance of 3%, found '*that the electrochromic windows reduced the incidence of glare compared to working under a fixed transmittance (60%) condition. Subjects used the Venetian blinds less often and preferred the variable transmittance condition*'.

A later study, also from researchers at LBNL, analysed the occurrence and causes of manual overrides in a conference room retrofitted with EC glazing with a range in visible transmittance of 3–50% [19]. The occurrence of a manual override was used to indicate dissatisfaction with the existing condition. The authors reported that: *'Of the 328 meetings that occurred over the six-month period, the automatic system was manually overridden on 14 or 4% of the meetings for reasons other than demonstration purposes. When overridden, occupants appeared to have switched the individual zones with deliberation, using a combination of clear and tinted zones and the interior Venetian blinds to produce the desired interior environment.'* Conference rooms are occupied intermittently and often by a varying group of users. Care should therefore be taken in extrapolating the findings from such studies to more general, continuously occupied office spaces.

19.3
Case Study: The De Montfort EC Office Installation

The final section of this chapter describes a case study evaluation of two office spaces fitted with electrochromic glazing in the United Kingdom. The user acceptance of the installation is being evaluated as part of a long-term study [20]. The study has been designed to assess the impact of the EC glazing on physical parameters in the rooms as well as the subjective impact of this new technology on the room occupants. The measurements, subjective and physical, are made at regular intervals (of varying length depending on the measure) throughout the monitoring period which will be 2 years in total. The study was in progress at the time of writing; however, some preliminary findings are discussed in the sections that follow.

19.3.1
Background

The first building in the United Kingdom to have production EC glazing installed and commissioned was the Clephan Building on the campus of De Montfort University (Leicester, UK). Two offices in the Clephan Building were fitted with SageGlass EC glazing in late 2012. The lighting in the offices was upgraded at the same time, but otherwise the offices and the occupants are as before. The EC product used has a range in visible transmittance of 2–62%. The external facade has a southerly aspect and offices on that elevation were observed to have lowered/closed blinds for much of the occupied period, Figure 19.8. Hence, they were identified as ideal settings to evaluate EC glazing. The approximate internal plan dimensions for each of the two rooms are 7 m deep by 5 m wide. The floor-to-false-ceiling height is approximately 2.7 m, with some additional height at one of the windows in Room 0.30. The facade normal has an orientation 30° east of south, with some overshadowing to the east. Midday and afternoon sun is more of a problem for the occupants of the rooms than early morning sun,

Figure 19.8 Offices with EC glazing installed (Clephan Building, De Montfort University, Leicester, UK)

especially in the winter when the sun angle is low and the rooms are shaded by the adjacent building.

Although limited in the number of subjects—a possible total of seven in both offices—a longitudinal study has the advantage of revealing any variation in glazing performance and/or user acceptance across the seasons. Importantly, the study also offers an insight into the practicalities of retrofit when an advanced facade technology is applied to an existing building. It can thus inform future retrofit projects, in terms of both technical issues and occupant perception of the change. It is estimated that 87% of the existing buildings will still be in use in 2050 [21], and so retrofit rather than new build could well be the largest market sector for facade technologies such as EC glazing.

19.3.2 Installation of the EC Glazing

The two offices (Rooms 0.30 and 0.29) comprise three large window bays, each with six panels (enclosed by the yellow box in Figure 19.8). However, the dividing wall between the two offices bisects the central bay. Additionally, the false ceiling in the offices meets the facade wall at the shared window and the window exclusive to Room 0.29. Thus, the upper panels for these two bays are either for ventilation or 'false' windows, that is, they do not provide any illumination to the offices. For the remaining bay in Room 0.30, the false ceiling is stepped back from the window and all six panels can illuminate the space. Thus, there are eight EC panels in Room 0.30 and six in Room 0.29—they were all set to full tint when the external photograph was taken and shown as 'black' windows (Figure 19.8). This was a retrofit of the previously remodelled spaces in what was originally a Victorian-era

Figure 19.9 Prior to (Room 0.29) and during installation (Office 0.30)

factory building. Photos of Room 0.29 prior to installation and Room 0.30 during installation are shown in Figure 19.9.

EC windows, of course, require electricity in order to operate. Thus, an electrician was required at the time of the installation of the windows in addition to a glazier. The artificial lighting was also upgraded with new luminaires and daylight dimming controls that were integrated with the overall EC control system, but otherwise the two offices and their use remained unchanged. For this installation, it was the visual environment of the spaces that was the key concern. Thus, automatic control of the EC glazing was based on measurements of illuminance recorded by sensors that are part of the EC control system. There were zone illuminance sensors on the inside of the window frames, and a single external mounted sensor that recorded the vertical illuminance in the plane of the facade. One of the aims of the study is to determine which sensors and what control algorithms worked best for the occupants, that is, which minimised the necessity to effect a manual override. The zoning arrangement for Room 0.30 is shown in Figure 19.10.

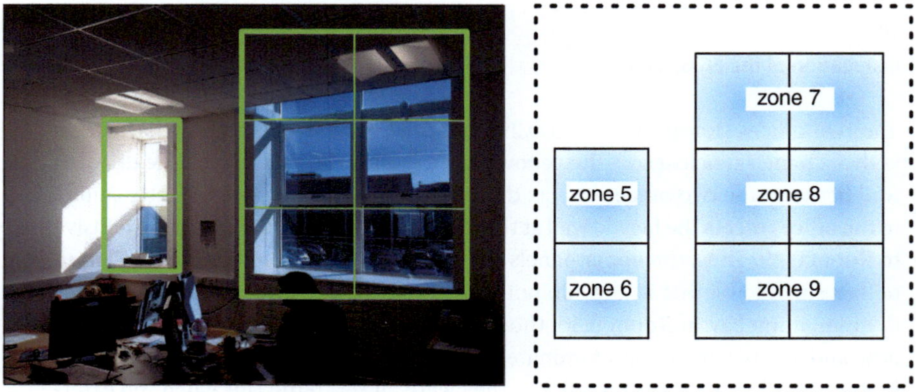

Figure 19.10 The eight EC panes in Room 0.30 are grouped into five zones (the zone numbers 1–4 were allocated to Room 0.29)

19.3.3 Subjective Data Collection

In order to understand the factors related to the long-term user acceptance of EC glazing by the occupants in the two offices, a variety of qualitative and quantitative approaches are employed. Across the approaches, the frequency of engagement with the participant is balanced with the level of detail of data obtained, that is, less frequent, more detailed versus more frequent, less detailed. This data collection procedure has been developed to reduce and carefully manage participant burden. The self-report and questionnaire data collection approaches are summarised as follows:

- **Daily experience form**—This allows users to record their experience in very general terms, using a traffic light system; 'good', 'bad' or 'neutral'. It provides a coarse data set of occupant perceptions that can be linked with the physical monitoring data.
- **Twice-monthly online questionnaire**—This goes into more detail about different aspects of the experience, for example, in terms of visual comfort and glare. The questionnaire has been designed to collect a good level of detail while still being short enough to be completed relatively quickly. This is a more traditional approach that will facilitate a deeper analysis of the links between physical conditions in the room and occupant's perception of the glazing performance.
- **Quarterly one-to-one semi-structured interview**—This allows deeper exploration of the subjective narrative. While this is necessarily more time intensive, it adds important qualitative data that can help put the measurement data from the other sources into context.
- **Ad-hoc participant feedback/researcher observation**—On an ad hoc basis, the occupants report their experiences verbally and via email. Participant feedback, along with pertinent observations by the researcher, is recorded in a diary by the researcher.
- **Blinds diary**—This is a simple pro-forma sheet mounted on the window wall that allows occupants to record instances when they needed to use the blinds. When an occupant needs to lower a roller blind, they note the time and date, the initials of the occupant who benefits and the reason (screen reflection, direct glare or other).

The control system also monitors the overrides made by the occupants. When an occupant selects to override the 'tint' set by the automatic control system, the user selected state is maintained for 2 h after which that zone reverts back to automatic control.

19.3.4 Measurement of Physical Quantities

One of the goals of the study is to understand the relation between the subjective response of the occupants to the visual environment and the measurable, physical parameters of the space. And, with that understanding, to use the knowledge gained to help codify recommendations for both the design/layout of spaces with EC glazing and the automatic control for the tinting of the glass. The physical measurements made include those that are part of the normal operation of EC glazing (e.g. facade vertical illuminance) and those carried out solely for the purpose of the case study (e.g. horizontal workplane illuminance).

In addition to the measurements/sensors already mentioned, the luminance of room surfaces and some of the windows was captured using a technique called high-dynamic-range (HDR) imaging. A consumer digital camera is used to capture a controlled sequence of exposures covering a wide range of exposure values, for example, from 1/1000 s to 1 s at a fixed aperture. The images are 'compiled' into a single HDR image, where each pixel now contains a (derived) measure of the luminance (in cd/m^2) of that point in the scene. With suitable calibration using a single spot measurement taken with a traditional photometer, the absolute accuracy of the luminance values in the HDR image is typically better than ±20% and often as good as ±10%.

Example HDR images taken in Room 0.30 are shown in Figure 19.11. Two cameras with fish-eye lenses were used for a period of the monitoring campaign. Set facing each other, together they record the luminance of most of the visible walls and the ceiling. The cameras also record the luminance of views through the windows. However, since the patterns of brightness through windows are invariably view dependent, the camera 'view' through the window will be similar to that for those occupants whose heads are fairly close to the camera lens. The database of HDR images will be used to determine the relation between the degree and magnitude of surface brightness in the offices and the tint setting of the various EC zones, under both automatic control and manual override.

As noted, the study was ongoing at the time of writing. However, one evaluation depending only on physical measurements has been completed—a summary of which is given in the following section.

19.3.5
The Daylight Illumination Spectrum with EC Glazing

One of the key illumination parameters for the case study is the *quality* of the daylight—that is, the daylight illumination spectrum—under normal operation of the EC glazing. There have been reports that 'blue' fixed-tinted glazing had lower approval ratings from test subjects than neutral or warm fixed-tinted glass [22]. Thus, the question regarding the neutrality of the illumination spectrum is an important one that needs to be addressed. It has been observed that, provided a small proportion of the EC glazing is left in the clear state, a space often has the subjective appearance of being illuminated by daylight that is devoid of any noticeable tint or hue, that is, the illumination is perceived as being 'neutral'. This observation refers to the *prevailing* illumination of the space rather than any localised illumination of surfaces by filtered direct sunlight (e.g. the 'blue' sun patches on the wall seen in Figure 19.10). This is illustrated in Figure 19.12 which

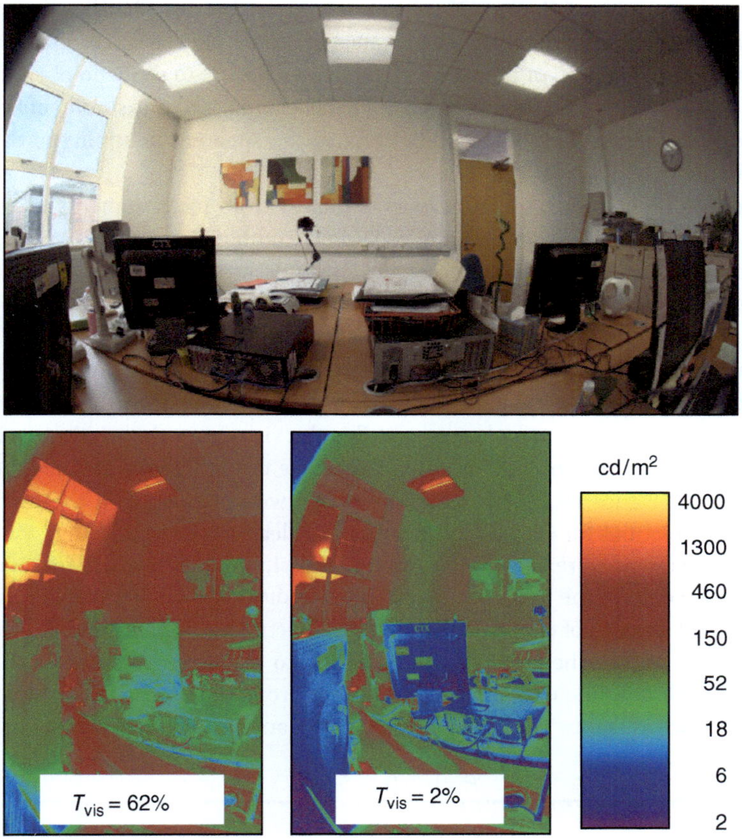

Figure 19.11 High-dynamic-range images illustrating how the environment luminance changes when the glazing tints (Room 0.30)

Figure 19.12 Photographs of the window and rear wall of Room 0.30 and a schematic of the office layout showing the occupant workstation positions and the six view directions (A–F) at which spectra were measured.

shows two photographs of Room 0.30 taken within moments of each other. Zones 6 and 9 (i.e. three panes) were set to clear state, the remaining zones (i.e. 5 panes) were at full tint. The photo of the rear wall certainly *appears* to show no obvious blue tint or hue. However, photographs can be misleading, and measurements of the illumination spectrum are needed to confirm the apparent neutrality of the daylight in the space.

The daylight illumination spectrum was evaluated from measurements taken in Room 0.30 using a spectrometer under various combinations of clear, tinted and partially tinted EC glass. Additionally, the measurements were used to validate a theoretical model formulated to predict the illumination spectrum for arbitrary combinations of clear and tinted EC glazing. The graphic in Figure 19.12 shows the approximate positions of the four occupants and their typical view directions (A–D) whilst working. View E is a general view from the back of the room looking towards the glazing, and F is the view from the window wall towards the back of the room, that is, similar to the camera views used for the photographs in Figure 19.12. Illumination spectra were measured at these six view points/directions for various combinations of clear and tinted glazing.

The measurements were taken under sunny, clear sky conditions on a weekend in order not to disrupt the usual occupants. The conditions were very stable with an almost total absence of clouds. The measured spectra were normalised to give the same illuminance—the important thing here is to compare the shapes of the curves, not their absolute magnitudes. Results are presented for the zoning configuration shown in Figure 19.12, that is, three clear panes and five panes set to full

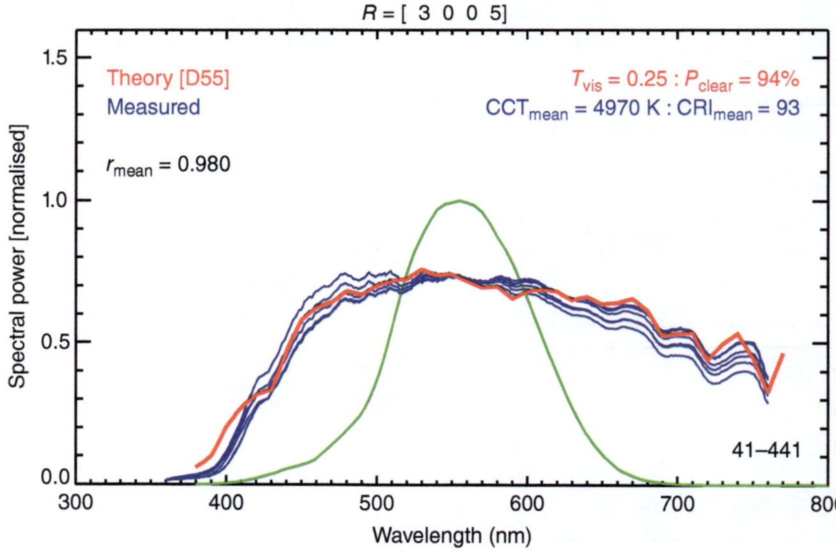

Figure 19.13 Six measured spectra (blue) for a combination of five fully tinted and three fully clear panels. Predicted spectrum (red) and the human eye's visual sensitivity curve (green).

tint. Spectra measured at the six view points/directions are plotted as blue curves in Figure 19.13. Also shown are a curve to predict the spectrum from a theoretical model (red) and the human eye's visual sensitivity curve (green).

Subjectively, at the time of measurement, the illumination in the space appeared effectively neutral for this case. The correlated colour temperature (CCT) and colour rendering index (CRI) were also measured by the spectrometer. These further support the qualitative appearance of neutrality. For this combination of clear and tinted glazing, the mean CCT was 4970 K and the mean CRI was 93. Objectively, the measured spectra show illumination that is neutral with no pronounced shift to the blue. Also, the spectra measured at the six points are very similar. This demonstrates that the daylight in the space is 'well mixed' and dominated by the illumination through the clear EC panes. The theoretical model (red curve) showed very good agreement with the measurements—the model is described in full in a paper by the authors of this chapter [23]. The model shows that it is possible to reliably predict the spectral performance of EC glazing at the design stage—particularly important for museums, art galleries and other critical applications that demand excellent colour rendering under daylight conditions.

19.4 Summary

Electrochromic glazing offers the potential to greatly reduce, or even eliminate, the need for moveable internal shading (e.g. venetian blinds) and fixed external blinds (e.g. brise-soleil)—thus transforming the way the glass is used in architecture. Highly glazed buildings tend to have poor energy performance, often due to uncontrolled solar gains and poor utilisation of daylight (i.e. blinds down, lights on). Glazing that changes its visible-optical and solar-thermal properties dynamically in response to external conditions should improve both the internal environment for occupants and the sustainability of glazed buildings in general. Recent experience has shown that the knowledge gained from occupant-based studies is vital to help ensure the successful design, integration and control of electrochromic glazing.

References

1. Selkowitz, S. and Lee, E.S. (1998) Advanced fenestration systems for improved daylight performance. *Daylighting '98 Conference Proceedings Ottawa, Ontario, Canada.*
2. Mardaljevic, J., Heschong, L. and Lee, E. (2009) Daylight metrics and energy savings. *Light. Res. Technol.*, **41** (3), 261–283.
3. Sanders, H. and Andreau, A. (2013) The next generation of dynamic glazing product performance. *Proceedings of Glass Performance Days, Tampere, Finland, June 13–15, 2013.*
4. Bordass, B., Cohen, R., Standeven, M. and Leaman, A. (2001) Assessing building performance in use 3: energy performance of the Probe buildings. *Build. Res. Inf.*, **29** (2), 114–128.

5. Lee, E.S., Selkowitz, S.E., Clear, R.D., DiBartolomeo, D.L., Klems, J.H., Fernandes, L.L., Inkarojrit, V. and Yazdanian, M. (2006) Advancement of electrochromic windows. California Energy Commission, PIER. Publication number CEC-500-2006-052.
6. Reilly, S., Arasteh, D. and Selkowitz, S. (1991) Thermal and optical analysis of switchable window glazings. *Solar Energy Mater.*, **22** (1), 1–14.
7. Crawley, D.B., Lawrie, L.K., Winkelmann, F.C., Buhl, W.F., Huang, Y.J., Pedersen, C.O., Strand, R.K., Liesen, R.J., Fisher, D.E., Witte, M.J. and Glazer, J. (2001) EnergyPlus: creating a new-generation building energy simulation program. *Energy Build.*, **33** (4), 319–331.
8. Warner, J.L., Reillly, S.M., Selkowitz, S.E., Arasteh, D.K. and Ander, G.D. (1992) Utility and economic benefits of electrochromic smart windows. *Proceedings of the ACEEE Summer Study on Energy Efficient Buildings.*
9. Hopkinson, R.G. (1963) *Architectural Physics - Lighting*, Her Majesty's Stationery Office, London.
10. Mardaljevic, J. (2000) Simulation of annual daylighting profiles for internal illuminance. *Light. Res. Technol.*, **32** (3), 111–118.
11. Reinhart, C.F. and Herkel, S. (2000) The simulation of annual daylight illuminance distributions—a state-of-the-art comparison of six RADIANCE-based methods. *Energy Build.*, **32** (2), 167–187.
12. Reinhart, C.F., Mardaljevic, J. and Rogers, Z. (2006) Dynamic daylight performance metrics for sustainable building design. *Leukos*, **3** (1), 7–31.
13. Fernandes, L., Ward, G. and Lee, E.S. (2006) Radiance-Mathematica Optimization of Electrochromic Operations Foroccupant Comfort and Non-Energy Provisions. Paper LBNL-59821, Lawrence Berkeley National Laboratory.
14. Mardaljevic, J. and Nabil, A. (2008) Electrochromic glazing and facade photovoltaic panels: a strategic assessment of the potential energy benefits. *Light. Res. Technol.*, **40** (1), 55–76.
15. Lee, E.S., Selkowitz, S.E., Hughes, G.D., Clear, R.D., Ward, G., Mardaljevic, J., Lai, J., Inanici, M.N. and Inkarojrit, V. (2005) Daylighting the New York Times Headquarters Building. Final report LBNL-57602, Lawrence Berkeley National Laboratory.
16. Piccolo, A. and Simone, F. (2009) Effect of switchable glazing on discomfort glare from windows. *Build. Environ.*, **44** (6), 1171–1180.
17. Building Research & Information (2001) PROBE special issue: post-occupancy evaluation. *Build. Res. Inf.*, **29** (2), 158–163.
18. Clear, R.D., Inkarojrit, V. and Lee, E.S. (2006) Subject responses to electrochromic windows. *Energy Build.*, **38** (7), 758–779.
19. Lee, E.S., DiBartolomeo, D. and Selkowitz, S. (2000) Electrochromic Windows for Commercial Buildings - Monitored Results for a Full-Scale Testbed. Report LBNL-45415DA-414, Lawrence Berkeley National Laboratory.
20. Kelly, R., Painter, B., Mardaljevic, J. and Irvine, K. (2013) Capturing the user experience of electrochromic glazing in an open plan office. *CIE Midterm conference—Towards a new century of Light, Paris, France, April 12-19, 2013.*
21. Kelly, M.J. (2009) Retrofitting the existing UK building stock. *Build. Res. Inf.*, **37** (2), 196–200.
22. Arsenault, H., Hébert, M. and Dubois, M.-C. (2012) Effects of glazing colour type on perception of daylight quality, arousal, and switch-on patterns of electric light in office rooms. *Build. Environ.*, **56**, 223–231.
23. Mardaljevic, J., Kelly Waskett, R. and Painter, B. Neutral daylight illumination with variable transmission glass: theory and validation. *Light. Res. Technol.*, Accepted for publication.

20
Photoelectrochromic Materials and Devices

Kuo-Chuan Ho, Hsin-Wei Chen, and Chih-Yu Hsu

20.1
Introduction

Photoelectrochromic materials and devices have a wide range of applications such as architectural windows, displays and energy conversion. The primary focus of this chapter involves areas that are relevant to the R&D of the devices for renewable energy conversion and conservation. A photoelectrochromic device (PECD) can be designed to include an extra electrochromic (EC) layer in a light-driven self-powered battery, such as dye-sensitized solar cells (DSSCs), silicon solar cells or any other possible devices. Most of them are combined with DSSCs in the PECD field, which we mainly focus on in the following sections. In such a design, a PECD acts as both a DSSC to deliver power and a switchable electrochromic device (ECD) in a single configuration, without the need to supply an external voltage. Accordingly, PECDs offer dual application, since the generated electrons can either be transported out of the device as a power source or be charged within the device as an energy saving device. Based on this concept, some differently designed structures of the PECD were derived, which can be summarised as three types (Figure 20.1). Type I (the separated type) illustrates the separated layout for each light-sensitive and electrochromic layer. Types II (the combined type) and III (the non-symmetric type) are those with both materials constructed as the same layer, but sorted by full and partial coverage, respectively, of the light-sensitive material with respect to the electrochromic layer. Table 20.1 lists the PECDs based on the DSSCs' structure with different types and different EC materials summarised from the literature.

Electrochromic Materials and Devices, First Edition.
Edited by Roger J. Mortimer, David R. Rosseinsky and Paul M. S. Monk.

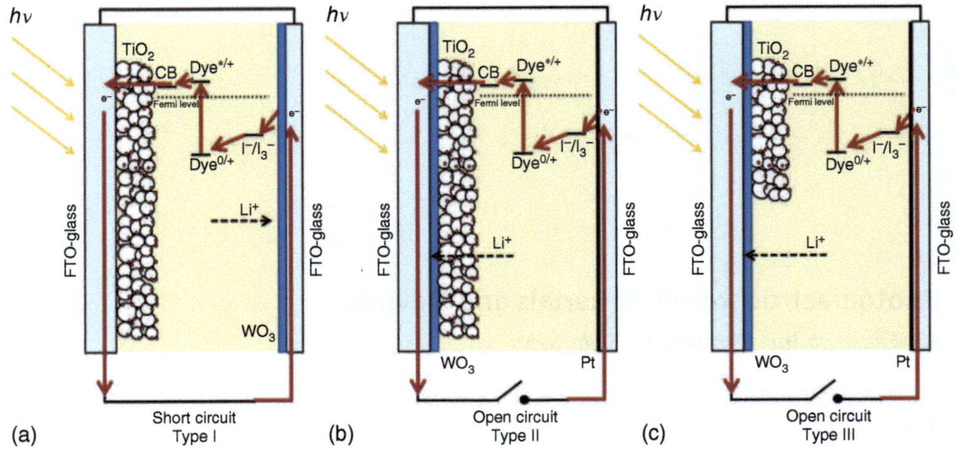

Figure 20.1 Sketch of the three types of photoelectrochromic devices including (a) the separated type, (b) the combined type, and (c) the non-symmetric type.

20.2
Structure Design of the PECDs

20.2.1
Separated-Type PECD (Type I): The Dye-Sensitized TiO_2 Layer is Separated from the Electrochromic Layer

Since 1996 Bechinger *et al.* [1] at NREL (National Renewable Energy Laboratory) published the first PECD in the Nature journal, researchers started to put vigorous efforts into the field of PECDs. This PECD combines a solar cell part and an EC layer as an integrated electrochemical device. The dye-sensitized TiO_2 mesoporous film will perform energy conversion to electricity, which harvests solar energy in order to provide the electric charge required for the reversible reaction of the optical properties of an EC layer, usually WO_3. As shown in Figure 20.1, there are three different types of PECD. Type I is the separated type of PECD, where the dye-sensitized TiO_2 and EC layer are deposited on different electrodes and separated by an electrolyte. The electrolyte contains counter ions, for example, Li^+, for the EC film to perform coloration and a redox couple (I^-/I_{3^-}) for internal charge transfer reaction of the sensitisers. These devices are coloured when light illuminates the PECD under short circuit, as displayed in Figures 20.2 and 20.3. The photoelectrons produced from the dye-sensitized TiO_2 film and injected into the EC film through the external circuit are recorded as the EC film darkens. In Figure 20.4, the transmittance difference (ΔT) between the (T_b) bleached state and (T_d) darkened state of the PECD is displayed. The working principle for the colour switch of the EC material, taking WO_3 for instance, is realised through the following equation:

$$WO_3(\text{transparent}) + xLi^+ + xe^- \rightleftharpoons Li_xWO_3(\text{coloured}) \qquad (20.1)$$

Table 20.1 Different types of photoelectrochromic devices.

PECD type/ EC layer		WE[a)]	Electrolyte	CE[b)]	ΔT[c)]	τ_d[d)] (s)	τ_b[e)] (s)	References
Type I	Inorganic	TiO_2/Ru dye	0.1 M LiI + 0.01 M 4-TBP[f)] in PC[g)]	WO_3	17% (788 nm)	50	250	[1, 2]
		TiO_2/Ru dye	0.5 M LiI + 0.005 M I_2 in PC	WO_3/Pt	44% (788 nm)	4	60	[3]
		TiO_2/Ru dye	0.1 M LiI + 0.01 M 4-TBP in γ-GBL[h)]	PB[i)]	70% (700 nm)	806	100	[4]
		TiO_2/Ru dye	0.5 M LiI + 1 mol% azobisisobutyronitrile (AIBN)[j)] + 900 mg PEGDMA[k)] (quasi-solid-state and solid-state electrolyte)	WO_3	20% (700 nm)	–	–	[5]
		TiO_2/Ru dye	Mixture of PEG[l)] (600) + TTIP[m)] + ethanol + LiI (quasi-solid-state and solid-state electrolyte)	WO_3	39% (620 nm)	20	18	[6]
	CP[n)]	TiO_2/Ru dye	0.5 M LiI + 0.04 M I_2 in PC:ACN[o)] (4 : 1 by volume)	PANI[p)]	5.4%	34	3	[7]
		TiO_2/Ru dye	0.5 M LiI + 0.04 M I_2 in PC:ACN (4 : 1 by volume)	PANI	–	7	6	[8]
		TiO_2/Ru dye	0.1 M LiI + 0.1 M $LiClO_4$ in PC	PEDOT[q)]	20% (630 nm)	50	60	[9]
		TiO_2/Ru dye	0.1 M LiI + 0.001 M I_2 in PC	PProDOT-Me_2[r)]	38% (637 nm)	2	4	[10]

(continued overleaf)

Table 20.1 (*Continued*)

PECD type/ EC layer	WE[a)]	Electrolyte	CE[b)]	ΔT[c)]	τ_d[d)] (s)	τ_b[e)] (s)	References
	TiO_2/organic dye	0.1 M $LiClO_4$ + 1.0 M TBABr[s)] in MPN[t)]	PProDOT-Et_2[u)]	33.7% (620 nm)	36	96	[11]
	TiO_2/organic dye	0.1 M $LiClO_4$ + 0.5 M TBABr in PC with PVDF-HFP[v)] : succinonitrile = 6 : 2 (quasi-solid-state and solid-state electrolyte)	PProDOT-Et_2	31% (620 nm)	3	20	[12]
Type II	WO_3/TiO_2/Ru dye	0.5 M LiI + 5 mM I_2 in PC	Pt	41% (620 nm)	120	120	[13]
	WO_3/TiO_2/Ru dye	0.5 M LiI + 5 mM I_2 in PC	Pt	46.2% (620 nm)	180	180	[14–16]
	WO_3/TiO_2/organic dye	0.5 M LiI in PC	Pt	36% (620 nm)	360	—	[17]
	WO_3/TiO_2/Ru dye	0.5 M LiI + 5 mM I_2 in mixture of PEG and PC (solid-state electrolyte)	Pt	60.4% (620 nm)	900	900	[18–21]
	WO_3/TiO_2/Ru dye	5 wt% PEDOT:poly(styrenesulfonate) (PSS)[w)] (solid-state electrolyte)	PEDOT	55% (800 nm)	10	5	[22]
Type III	WO_3/TiO_2/Ru dye	0.5 M LiI + 2.5 m M I_2 in PC	Pt	46% (620 nm)	180	–	[23]
	WO_3/TiO_2/Ru dye	0.7 M LiI + 5 m M I_2 + 0.5 M 4-TBP in ACN	Pt	46% (620 nm)	–	–	[24–26]
Other type	WO_3	–	a-Si[x)]	–	–	–	[27, 28]
	WO_3	0.6 M 1-propyl-3-methyl imidazolium iodide, 0.05 M I_2, 0.05 M LiI and 0.5 M 4-TBP in ACN	PB	36.7% (650 nm)	60	45	[29]

	WO_3	1 M H_2SO_4-PVA[y] gel electrolyte	Pt	25.9% (670 nm)	30	30	[30]
	–	0.05 M heptyl viologen ($HV(BF_4)_2$), 0.1 M $TBABF_4$ and 0.05 M TMPD[z] in PC	a-Si	70% (590 nm)	200	5	[31]

a) Working electrode.
b) Counter electrode.
c) Transmittance difference.
d) Darkening time.
e) Bleaching time.
f) 4-Tert-butylpyridine.
g) Propylene carbonate.
h) γ-Butyrolactone.
i) Prussian blue.
j) 2,2′-Azobis(2-methylpropionitrile).
k) Poly(ethylene glycol-400) di-methacrylate.
l) Polyethylene glycol.
m) Titanium(IV) isopropoxide.
n) Conducting polymer.
o) Acetonitrile.
p) Polyaniline.
q) Poly(3,4-ethylenedioxythiophene).
r) Poly[3,3-dimethyl-3,4-dihydro-2H-thieno[3,4- b][1,4]dioxepine].
s) Tetrabutylammonium bromide.
t) 3-Methoxypropionitrile.
u) Poly(3,3-diethyl-3,4-dihydro-2H-thieno[3,4-b][1,4]dioxepine).
v) Poly(vinylidene fluoride-hexafluoropropylene).
w) Poly(3,4-ethylenedioxythiophene):polystyrene sulfonate.
x) Amorphous silicon.
y) Polyvinyl alcohol.
z) *N,N,N′,N′*-tetramethyl-p-phenylenediamine.

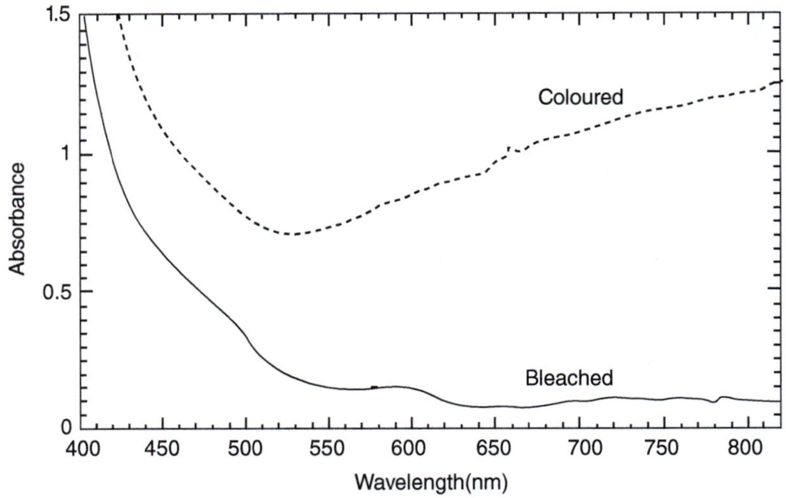

Figure 20.2 Absorption spectra of a PECD before (bleached) and after (coloured) illumination at an intensity of ~75 mW cm^{-2} for 1 min [1]. (From Ref. [1].)

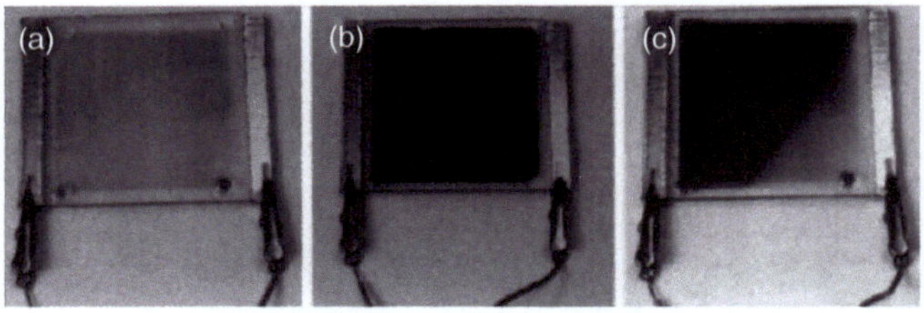

Figure 20.3 Photographs of a PECD with a 25 cm^2 active area. (a) The bleach state. (b) The coloured state is achieved by illumination at an intensity of ~75 mW cm^{-2} for 1 min. (c) The inhomogeneously coloured state is achieved by illumination with one corner of the cell masked [1]. (From Ref. [1].)

The counter ion, Li^+ for WO_3, penetrates into the EC film for charge balance as the electrons transported between the EC film and substrate, and consequently, the optical properties of the EC materials are changed. The darkening process is regarded as a charging process for a battery; the photovoltage and the photocurrent generated by the dye-sensitized TiO_2 film as a driving force and the electron source, respectively, enable the redox reaction of WO_3 film. Bleaching of the device is performed under open-circuit voltage (V_{OC}) when the illumination ceases, because the photovoltage returns to its original state as a reverse driving force for the bleaching. During bleaching, Li^+ is expelled from the WO_3 film and electrons are transferred back to the oxidized iodine species in the solution via the external circuit as shown in Figure 20.5 [2].

In the following sub-sections, the development of type I PECDs is divided into two categories based on the use of EC layers of inorganic materials and conjugated conducting polymer materials.

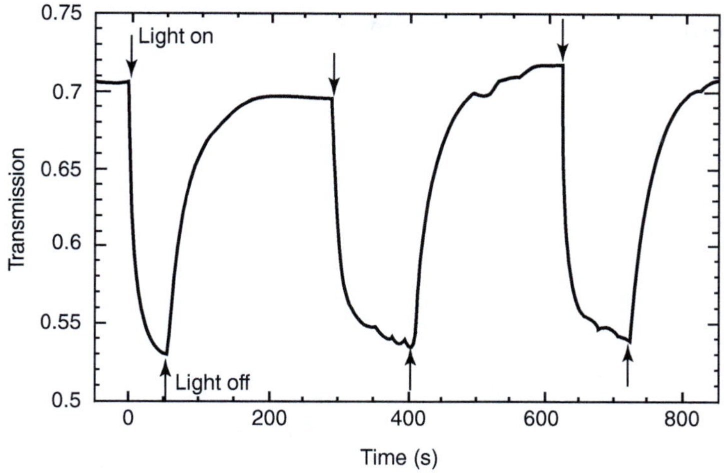

Figure 20.4 Change in transmission at 788 nm wavelength of a short-circuited PECD exposed to chopped white light [1]. (From Ref. [1].)

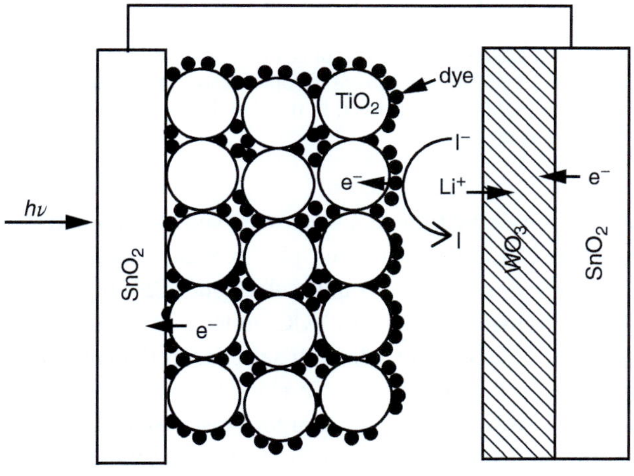

Figure 20.5 Schematic diagram of a PECD [2]. (From Ref. [2].)

20.2.1.1 **Inorganic Materials as EC Layers**

Inorganic materials are considered highly stable for application. In PECDs using inorganic materials as the EC layer, Wu *et al.* [3] demonstrated a photovoltachromic cell (PVCC) that utilised solar energy to stimulate electrochromic behaviour with the characteristic of tunable transmittance as shown in Figure 20.6. The cell was composed of a patterned WO_3/Pt electrochromic electrode and a dye-sensitized TiO_2 nanoparticle photoanode. Compared to the conventional PECD with non-patterned WO_3 electrochromic electrodes, the PVCC they proposed achieved a much faster bleaching time of only 60 s by blocking the light at short circuit. When the PVCC was bleached under illumination at open

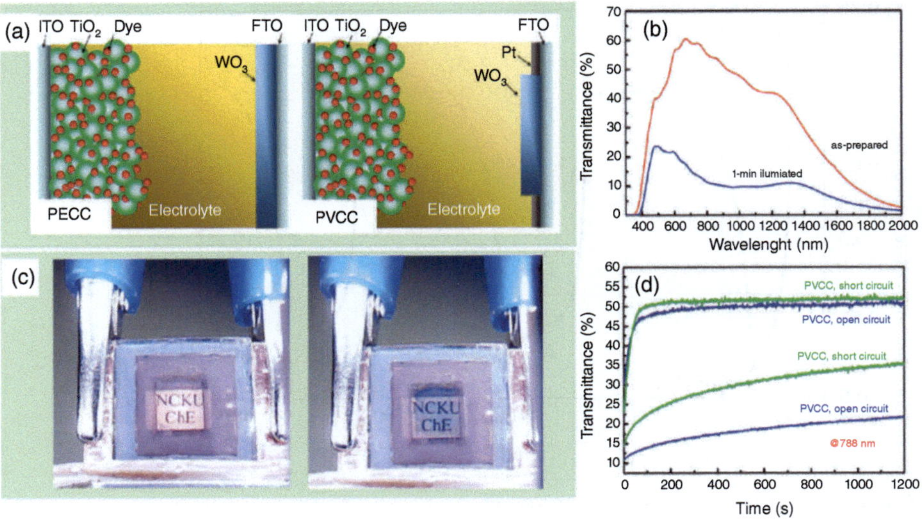

Figure 20.6 (a) Schematics of a photoelectrochromic cells (PECC) and a PVCC. (b) Transmittance spectra of a PECC in the as-prepared and the coloured states, measured at open circuit. (c) Photographs of a PVCC in the as-prepared (left) and coloured (right) states. The coloured photograph was taken at short circuit under illumination. The illumination makes the background letters rather clear even though the PVCC is coloured. (d) Evolution of cell bleaching at a wavelength of 788 nm for open- and short-circuited PECC and PVCC in the dark [3]. (From Ref. [3].)

circuit, an exceedingly short bleaching time of 4 s was achieved as demonstrated in Figure 20.7. Furthermore, PVCC still exhibited photovoltaic characteristics comparable to DSSCs, that is, using Pt as the counter electrode (CE). In contrast to conventional PECDs, the transmittance of a PVCC under a constant illumination can be adjusted by the resistance of a load in series with the cell. These characteristics attributed to the patterned WO_3/Pt electrode, which provided effective charge transfer pathways to facilitate the charging/discharging by Li ions and electrons via the photovoltaic potential and the Pt-electrolyte catalytic route, respectively.

Jiao *et al.* [4] reported a novel PECD composed of a dye-sensitized TiO_2 electrode and the Prussian blue (PB) CE sandwiched with an LiI electrolyte as illustrated in Figure 20.8. A maximum absorbance modulation of 0.44 recorded at 700 nm was obtained between the bleached and coloured states for the cell when the thickness of PB film was 452 nm, as shown in Figure 20.9. With illumination under different levels of light intensity or durations of time, the short-circuit cell presented an adjustable optical absorption from 470 to 840 nm. The *in situ* transmittance response depicted that the photobleaching response is 6.2 s for 70% transmittance change under 100 mW cm^{-2} illumination in short-circuit configuration, and the re-coloration time is about 600 ms under 2 V bias recorded at 780 nm, with an EC coloration efficiency of 103.4 cm^2 C^{-1}.

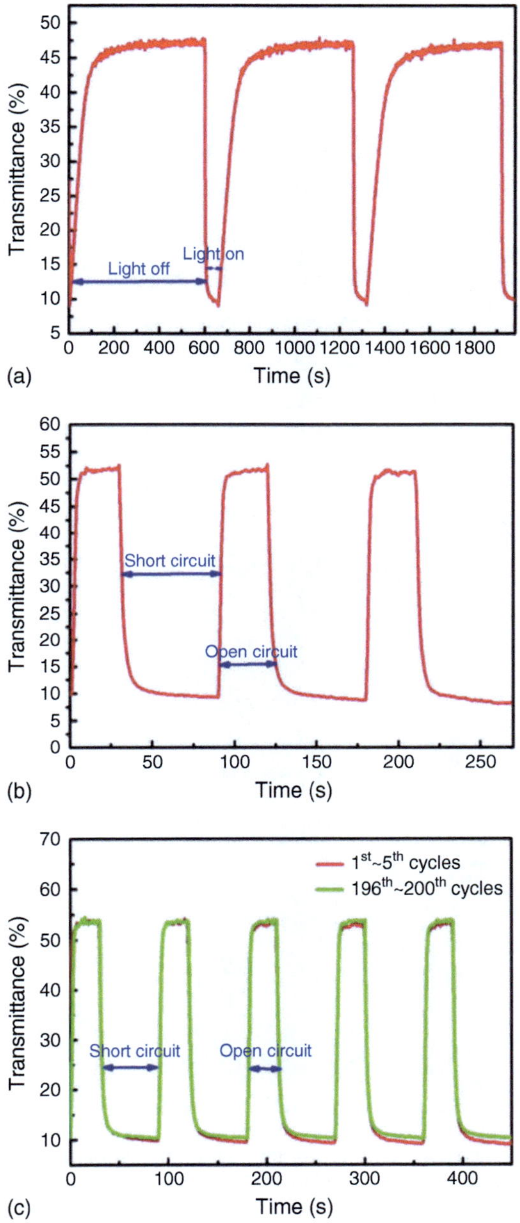

Figure 20.7 Change in transmittance of a PVCC measured at 788 nm: (a) exposed to chopped white light at short circuit; (b) subjected to open/short cycles under illumination and (c) stability of a PVCC performance under illumination after 200 open/short cycles. (Light source: white-light light emitting diode (LED) with an intensity of 4 mW cm^{-2}) [3]. (From Ref. [3].)

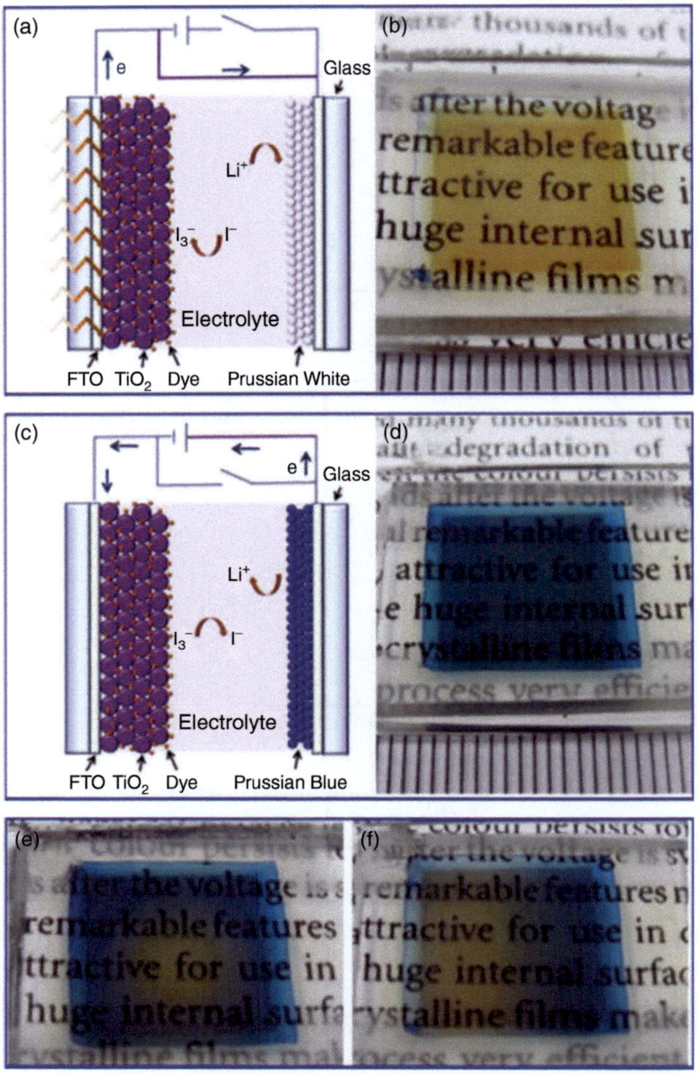

Figure 20.8 Schematics and photographs of a PECD with a 1.5 cm^2 active area. The duration for electrodeposition of PB film is 300 s. (a) Schematic of the short-circuit cell under illumination for bleaching. (b) Bleached cell achieved by illumination of 100 $mW\,cm^{-2}$ for 1 min in a short-circuit configuration. (c) Schematic of the cell under external voltage for colouring. (d) Photograph of the cell in the coloured state after applying a 2 V bias at the PB electrode for 30 s. (e,f) Spatially resolved bleaching states achieved by illuminating the cell with masks [4]. (From Ref. [4].)

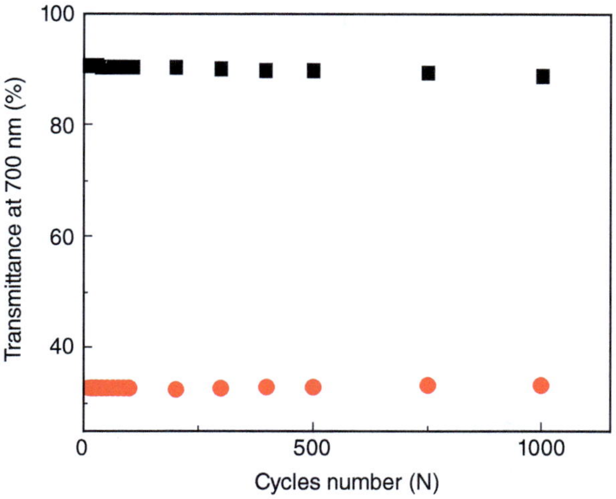

Figure 20.9 Transmittance change (ΔT) at 700 nm as a function of cycle number of the PECD, bleached under 100 mW cm^{-2} illumination for 60 s and coloured by adding a 2 V bias for 30 s [4]. (From Ref. [4].)

For practical applications, replacement of the traditional liquid-state electrolyte with quasi-solid-state or full-solid-state electrolytes was investigated. Pichot *et al.* [5] proposed a flexible PECD by using flexible conducting plastic substrates and solid-state electrolytes. The device was composed of a dye-sensitized TiO_2 electrode and a 500 nm thick WO_3 electrochromic layer, separated by a cross-linked polymer electrolyte containing LiI. A novel method for preparing conducting nanoporous TiO_2 films was described, which allowed for the fabrication of these devices on flexible plastic substrates. Colloidal solutions of TiO_2 free of surfactants were spin-coated onto indium tin oxide (ITO)-coated polyester substrates, resulting in highly transparent films ranging from 100 nm to 1 mm in thickness. Upon annealing at 100 °C, these films were strongly attached and displayed excellent photoconductivity as shown by their current–voltage characteristics. The devices typically transmitted 75% of visible light in the bleached state. After a few minutes of exposure to white light (75 mW cm^{-2}), the windows turned dark blue, transmitting only 30% of visible light. The transmittance difference (ΔT) achieved around 45% (wavelength at 650 nm) under illumination of 75 mW cm^{-2}. This is the first report of PECD being fabricated on a flexible plastic substrate.

Hechavarría *et al.* [6] reported a WO_3 EC film and dye-sensitized TiO_2 film based PECD with a composite polymeric electrolyte (PE). The PE was prepared by the sol–gel method using a viscous polyethylene glycol (PEG, molecular weight (MW) 600) and titanium isopropoxide (TTIP) as precursors in an acidic medium. LiI was added to the PEG–Ti complex as an electrolyte. The colour change of WO_3-based PEC devices under 100 mW cm^{-2} illumination under short-circuit conditions indicated that the introduction of titanium complex inside the PE accelerated the colouring speed from 26 to 9 s (Figure 20.10). Electrochemical impedance spectroscopy (EIS) analyses of the PECD showed that the presence of

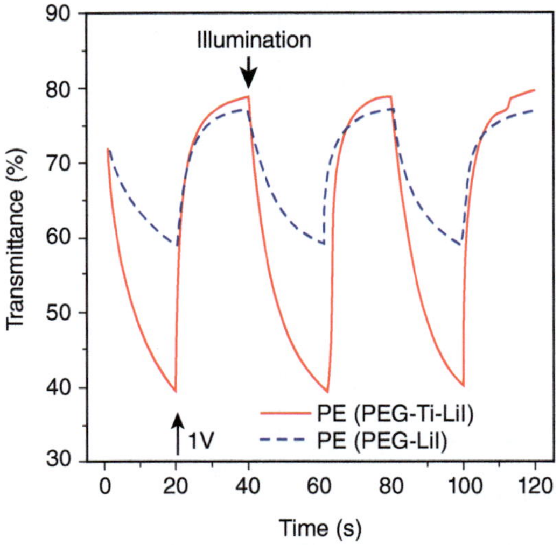

Figure 20.10 Transmittance transients of PECDs using PEG-LiI and PEG-Ti-LiI as electrolytes. Measurements were taken under white lamp illumination (100 mW cm^{-2}) and short-circuit condition (coloration process) and under 1 V polarisation without illumination (bleaching process) and a cycling time of 40 s [6]. (From Ref. [6].)

titanium complex in the PE reduced the charge transfer resistance and increased the charge storage capacity of the cells. This improvement came from the oxidation of iodides and the reduction of titanium ions inside the composite electrolytes that enhanced the mobility of Li ions.

20.2.1.2 **Conjugated Conducting Polymer Materials as EC Layers**

Although inorganic materials are considered stable, conjugated conducting polymers have attracted tremendous interest in the past several decades, owing to their tunable, designable and the variety of colours. Li *et al.* [7] presented a novel PECD containing a polyaniline (PANI, Figure 20.11) layer instead of the WO_3 film and a dye-sensitized nanocrystalline TiO_2 layer. The structure configuration is illustrated in Figure 20.12. PANI is one of the most interesting members of the group of conducting polymers that has been extensively investigated in recent decades. It is regarded as a promising electrochromic polymer material due to its high colour contrast, multiple colour possibilities and ease of processing compared to inorganic materials. It is generally accepted that PANI exhibits three main redox states: the fully reduced leucoemeraldine base form, the 50% oxidized emeraldine base form and the fully oxidized pernigraniline base form. Each redox state of PANI can exist in its base form or protonated (salt) form, which is produced by treating the base with an acid. After the protonation process, the emeraldine base form becomes the conducting emeraldine salt form of PANI. However, the EC properties of PANI depend not only on its redox state but also on its protonation state. Thus, PANI exhibits multiple colours, that is, transparent yellow, green, blue and black for the leucoemeraldine base, emeraldine salt, emeraldine base and the

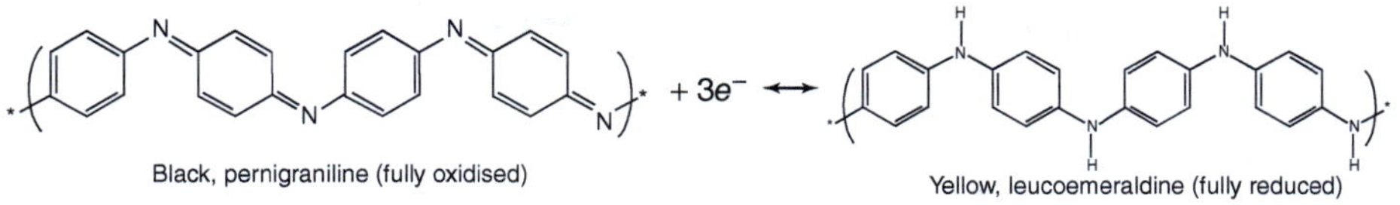

Figure 20.11 Electrochemical reaction of PANI.

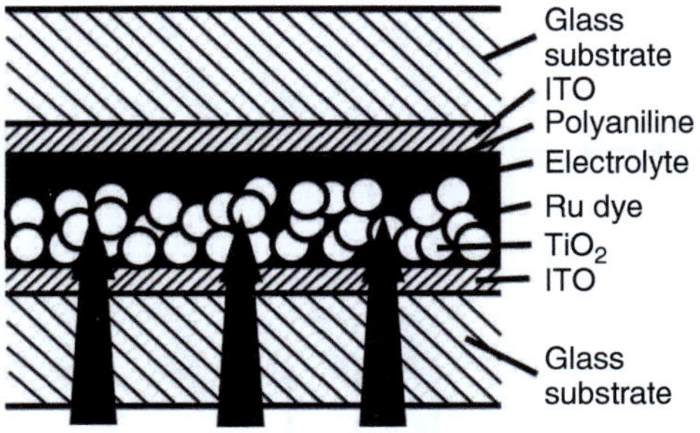

Figure 20.12 Schematic of the PECD with PANI-based EC layer [7]. (From Ref. [7].)

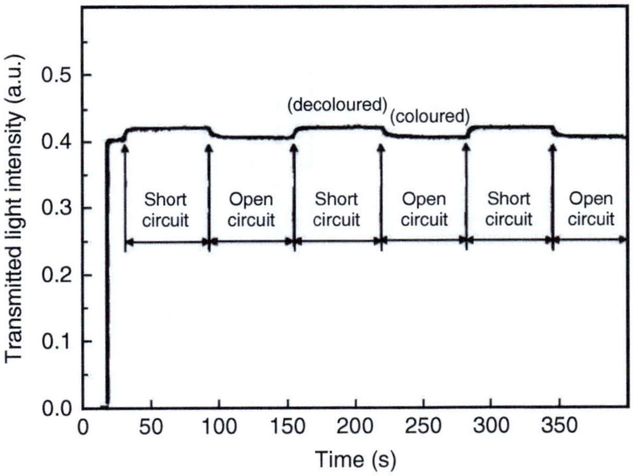

Figure 20.13 Change of the light intensity transmitted through the PECD [7]. (From Ref. [7].)

pernigraniline salt forms, respectively. In their PECDs, when the electrodes were short-circuited, electrons moved from the TiO_2 layer through the external circuit to the PANI layer, and the PANI layer was reduced and bleached. The redox reaction shown in figure 20.11 can be proposed.

PANI material EC layers have been prepared by electrochemical deposition and by spin coating a dispersion of a PANI solution. The influences of the preparation parameters on the micromorphology of the PANI layers were investigated. Although ΔT was only 5.4% (at 550 nm wavelength, Figure 20.13) under the illumination of 80.4 mW cm^{-2}, it was the very first attempt using a conjugated

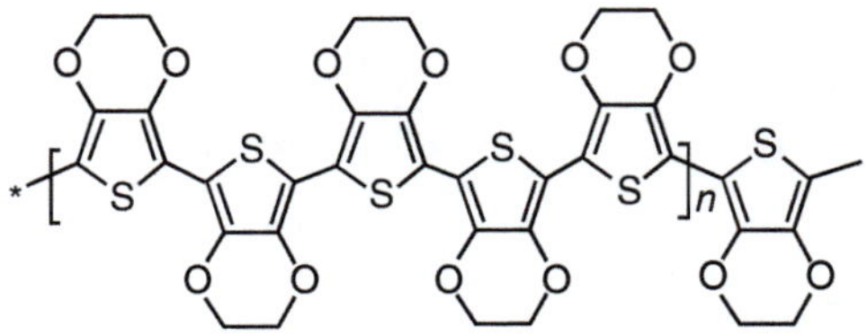

Figure 20.14 Polymer structure of PEDOT.

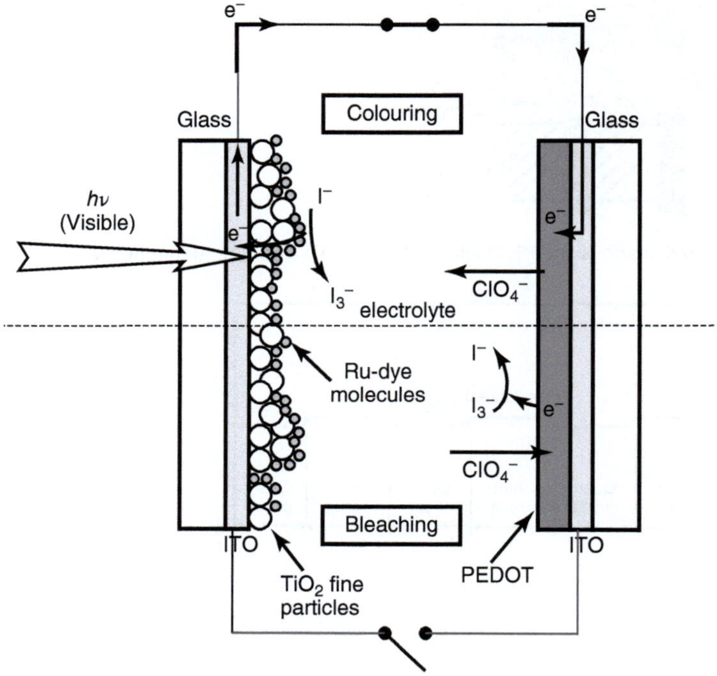

Figure 20.15 Schematic of the PECD with PEDOT-based EC layer [9]. (From Ref. [9].)

conducting polymer to replace the metal oxide EC layer. Subsequently, Yu *et al.* [8] fabricated PANI nanofibre films on an ITO glass substrate through a controllable electropolymerisation process. They found that the PANI nanofibre-based EC film benefited from a rapid response PECD compared to conventional thin films, due to the high aspect ratio of the nanostructured PANI layer.

Liao and Ho [9] replaced the PANI layers with another conducting polymer PEDOT (poly(3,4-ethylenedioxythiophene) structure shown in Figure 20.14) for the EC layer. The PECD structure is shown in Figure 20.15. Because of the excellent coloration efficiency of the PEDOT of 280 $cm^2 C^{-1}$, the required photoinduced charge is thus reduced. However, the insufficiently negative potential provided from the dye-sensitized TiO_2 resulted in a small transmittance

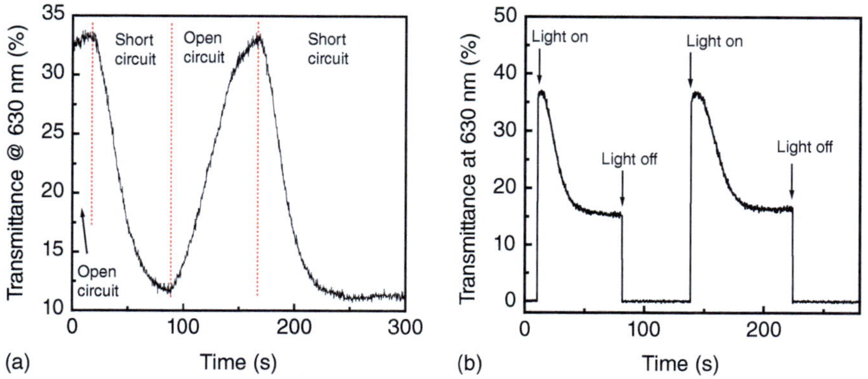

Figure 20.16 Transient transmittance for (a) a short/open-circuit control of an illuminated, reversible, separated PECD and (b) an on/off control of incident light of a short-circuit, reversible, separated PECD [9]. (From Ref. [9].)

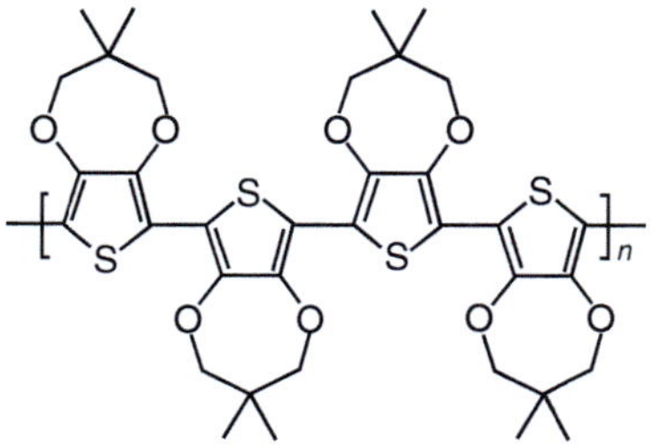

Figure 20.17 Polymer structure of PProDOT-Me_2.

attenuation of about 20% at 630 nm which can be observed from Figure 20.16. In addition, they discovered that the photoinduced, oxidized p-type PEDOT was responsible for the rising V_{OC} when a composite PECD was illuminated. A constant value of the V_{OC} can be regarded and analyzed as a 'potential step' in electrochemical modelling for a composite-type PECD system.

Yang *et al.* [10] reported another PECD based on an EC film of poly(3,4-(2,2-dimethyl-propylenedioxy)thiophene) (PProDOT-Me_2 structure shown in Figure 20.17) with a ruthenium complex dye. The addition of a thin layer of Pt under the EC layer to act as a catalyst allowed the assembled device to perform with fast response times (1 and 4 s for colouring and bleaching, respectively). The PECD demonstrated a 38% ΔT (at 637 nm wavelength) between the two states under 1 sun illumination as shown in Figure 20.18.

Later, Hsu *et al.* [11] replaced the PEDOT and the Ru-based dye with poly(3,3-diethyl-3,4-dihydro-2*H*-thieno-[3,4-*b*][1,4]dioxepine) (PProDOT-Et_2 structure shown in Figure 20.19) and an organic photosensitive dye, FL dye1. In the presence of the redox couple, the electron transfer reaction at the electrolyte/PProDOT-Et_2 conjugating conducting polymer interface was found to determine the electrochromism of the PECD. The rate constants of the electron transfer (k_0) for different redox couples at the PProDOT-Et_2 film were obtained by using EIS and were correlated with the coloration depth of the thin film. Since the I^-/I_3-

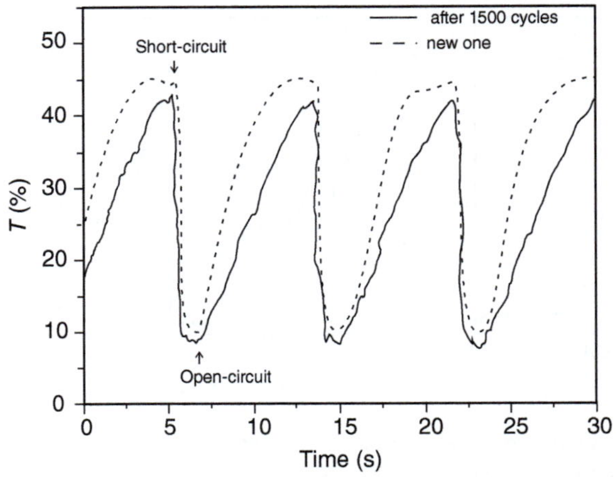

Figure 20.18 Transmittance switching repeatability of the PECD. The dotted line represents the initial cycles, and the solid line represents the cycles immediately after 1500 cycles. The switching period is 8 s [10]. (From Ref. [10].)

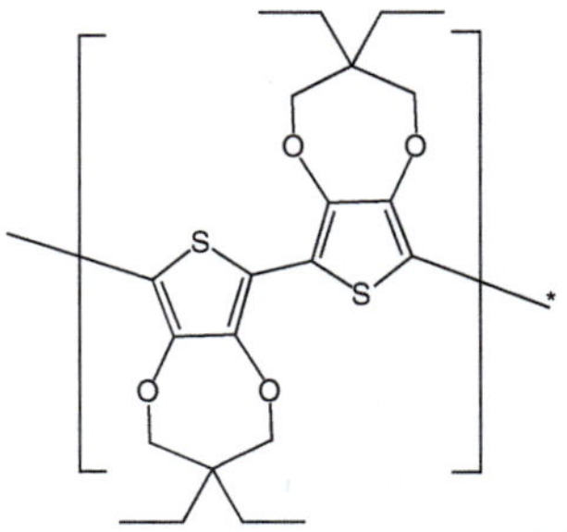

Figure 20.19 Polymer structure of PProDOT-Et_2.

and Br^-/Br_{3^-} couples possess the highest and the lowest k_0 value, respectively, the PECD was investigated using these two redox couples under constant light illumination of 50 mW cm^{-2}. The I^-/I_{3^-} couple provided a higher short-circuit current density (J_{SC}) when acting as a DSSC; thus, the switching response upon coloration (about 3 s) was much faster than that of the PECD using the Br^-/Br_{3^-} couple (about 2 min). On the contrary, the PECD using Br^-/Br_3^- exhibited a larger ΔT of about 33.7%, in comparison to that using I^-/I_3^- (about 14.1%).

Wu *et al.* [12] used the conjugated conductive polymer PProDOT-Et_2 layer as the EC film and the counter electrode for a DSSC as illustrated in Figure 20.20. An organic photosensitive dye, FL dye1, was adsorbed on a TiO_2-coated electrode for generating the photocurrent under light illumination. The electrolyte containing $LiClO_4$, Br^-/Br_3^- redox couple, succinonitrile and poly-(vinylidenefluoride-hexafluoropropylene) (PVDF-HFP) dissolved in propylene carbonate (PC) was proposed as the gel electrolyte for the PECD. When the volume ratio of PVDF-HFP to succinonitrile was 6 : 2, the DSSC part provided a photocurrent density of 0.164 mA cm^{-2}, a photovoltage of 0.75 V, fill factor (FF) of 0.48, with a conversion efficiency of 0.059% to drive the EC reaction. Figure 20.21 represents the optical

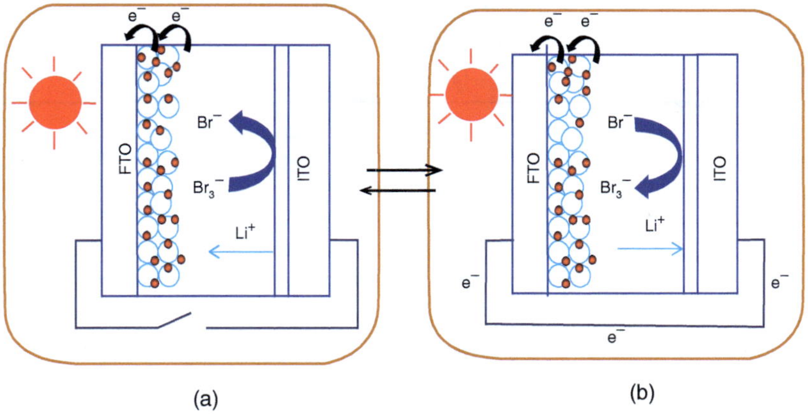

Figure 20.20 Working principle of the separated-type PECD. The reversible electrochromic switching can be repeated alternatively between the (a) open- and (b) short-circuit conditions [12]. (From Ref. [12].)

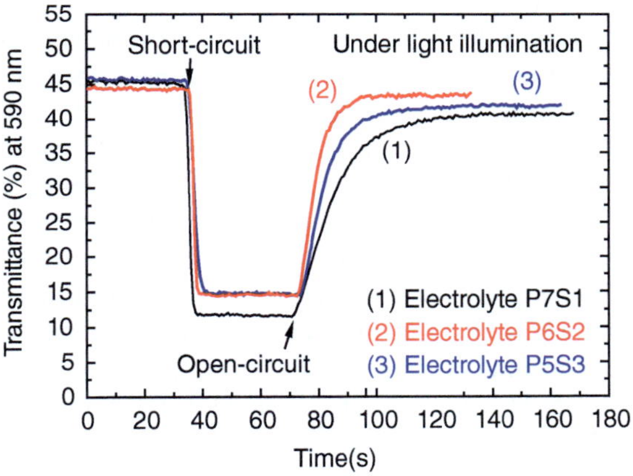

Figure 20.21 Transient transmittance responses at 590 nm of PECDs fabricated with electrolytes P7S1, P6S2 and P5S3 [12]. (From Ref. [12].)

response with summarised parameters in Table 20.2. This solar-to-electric power was sufficient for observing the electrochromism of PProDOT-Et_2, due to its high coloration efficiency. Moreover, when succinonitrile was added to the PVDF-HFP/PC electrolyte, the nucleus size of PVDF-HFP in the composite was reduced. The reason was that the effective space available for the primary nucleation in the composite was reduced, thus suppressing the crystallisation and increasing the ionic conductivity and the diffusivity of the electrolyte. Accordingly, a short bleaching time (20–35 s) of the PECD was achieved compared with the devices using succinonitrile (177 s) or PVDF–HFP (126 s) alone. The transmittance can be changed reversibly from 46% (bleached) to 15% (darkened) at 590 nm with ΔT of 31%.

Table 20.2 Optical performance of PECDs using electrolytes P, S and PxSy.

Electrolyte	T_b [a)] (%)	T_d [b)] (%)	ΔT	τ_d (s)	τ_b (s)
P	56	16	40	3	126
S	50	8	42	3	177
P7S1	46	12	34	3	35
P6S2	46	15	31	3	20
P5S3	45	15	30	3	21

a) Transmittance of bleached state.
b) Transmittance of darkened state.
The response time is based on the time it takes to reach 95% of the final transmittance level [12].
From Ref. [12].

20.2.2 Combined-Type PECD (Type II): The Dye-Sensitized TiO_2 Layer is Combined with the Electrochromic Layer

In type I PECDs, darkening and bleaching are regarded as competing processes that lead to either fast darkening and slow bleaching or slow coloration and fast bleaching [13, 14]. To overcome this problem, PECDs of type II have been proposed where both electrochromic and photovoltaic layers are deposited on the same electrode, with a Pt CE on the other side (as shown in Figure 20.1, type II). This device has the advantage that the bleaching process can be accelerated by the Pt catalyst independently from the coloration process. The electrolyte contains counter ions (e.g. Li^+) and a redox couple (I^-/I_3^-) in an organic solvent (e.g. PC). Under open circuit and under illumination, electrons are injected from the dye-sensitized TiO_2 film into the EC film performing coloration. Simultaneously, Li^+ is intercalated into the EC film and I^- is transformed into I_3^-. At short circuit, the electrons are transferred from EC to the Pt CE, causing the regeneration of I^- ions in the electrolyte, and Li^+ ions are expelling from the EC film. As a result, the EC layer is oxidized and becomes transparent again. Under the open circuit in the dark, the device bleaches by electron transfer from the TiO_2 or WO_3 to the electrolyte.

Hauch *et al.* [15] compared the PECD with different layer configurations. First, the TiO_2 layer situated above the WO_3 layer; second, the WO_3 layer situated above the TiO_2 layer; and third, the TiO_2 and WO_3 layers situated on opposite electrodes. For the first and second configurations, fast and intense colouring was possible together with fast bleaching, which is in principle not possible for the third case. The comparison between the first and second configurations demonstrated that it is possible to exchange the positions of the TiO_2 and WO_3 layers. The layers were prepared by dipping the samples in sols, and it was found that the material of the second layer penetrates the first layer, which influences the colouring and bleaching kinetics. For the first configuration, they showed that it is possible to switch the transmittance not only from the illuminated to the non-illuminated state but also during continuous illumination. For the first configuration and under 1 sun of illumination, the visual transmittance changed from 51% to 4.8%. Switching times were about 3 min for colouring and bleaching.

Through the variations of the light intensity, Georg *et al.* [16] found that the rate of the colouring process at open circuit was determined by the electron excitation at the dye by the incident light. The electron transfer from the WO_3 to the I_3^- in the electrolyte was the dominating loss reaction and the phase change of the WO_3 during the charge injection was also of importance. Transport processes in the TiO_2 or WO_3 and loss reactions at the TiO_2 were less significant. Leftheriotis *et al.* [17] presented an experimental study of a PECD for dynamic solar control in buildings. The fabricated devices consisted of a 400 nm thick WO_3 film, a 300 nm thick dye-sensitized TiO_2 layer, a 0.5 M LiI/PC electrolyte and an electrodeposited Pt counter electrode. All layers were developed aiming to increase the device transparency without reducing its functionality. The PECD exhibited fast coloration (about 12 min) under sunlight, considerable ΔT of 36% (varying from 50% to 14% and contrast ratio of 3.6 : 1) and a haze factor in the bleached state equal to 27.8% as shown in Figure 20.22. The PECD can withstand more than 100 continuous coloration-bleaching cycles. Krašovec *et al.* [18] proposed a type II PECD with solid electrolyte of a size up to 10 × 10 cm^2 employing a solid Li^+ ion conductor, based on organically modified silane (ormolyte) with a redox couple. The addition of the cationic surfactant to the electrolyte improved the stability, coloration and memory effect of the PECDs. The ΔT of the solid PECD was varied between 62% (bleached state) and 1.6% (coloured state). The corresponding colouring time under 1 sun illumination was 15 min, and 15 min was needed for bleaching [18–21]. Leftheriotis *et al.* [14] also discussed photocoloration efficiency and the stability of the PECDs for 70 days.

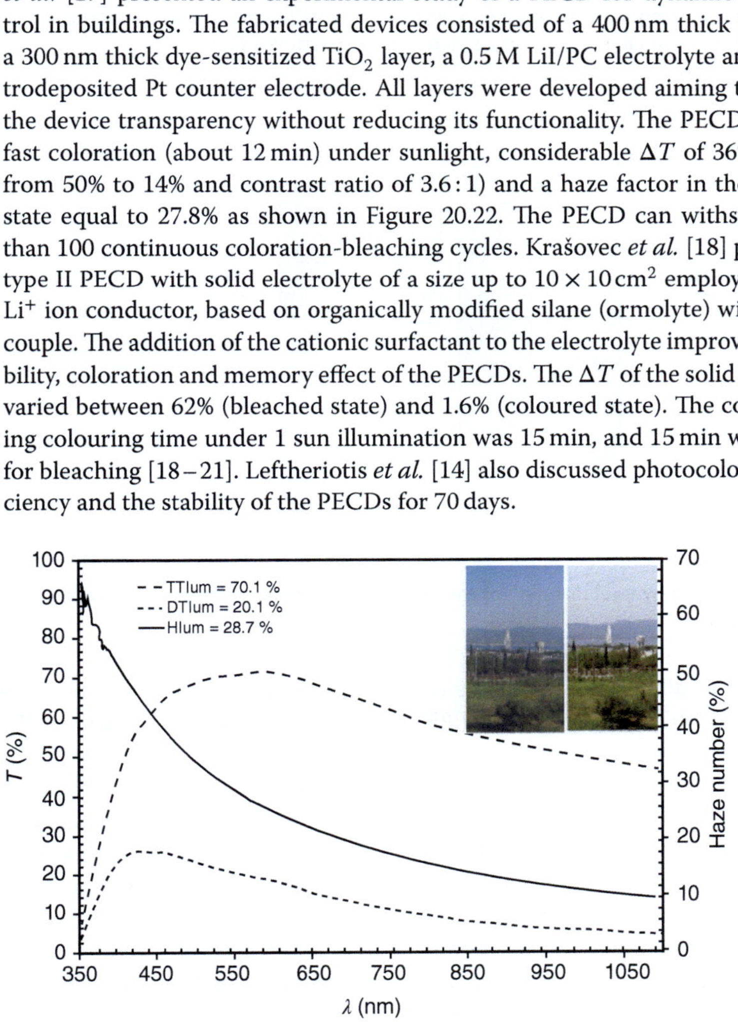

Figure 20.22 Total and diffuse transmittance spectra {total transmittance spectra (TT) (λ) and diffuse transmittance spectra (DT) (λ), respectively} and the resulting Haze number (H). Inset: view through the measured PECD (left) and view without the PECD (right) [17]. (From Ref. [17].)

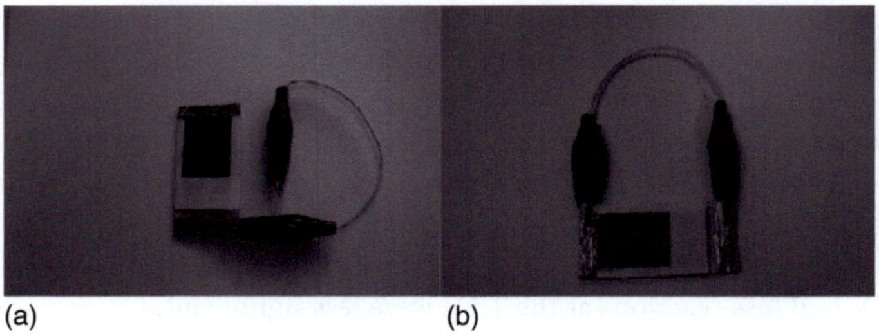

Figure 20.23 Photographs showing: (a) the coloured state in open-circuit conditions and (b) the bleached state in short-circuit conditions [22]. (From Ref. [22].)

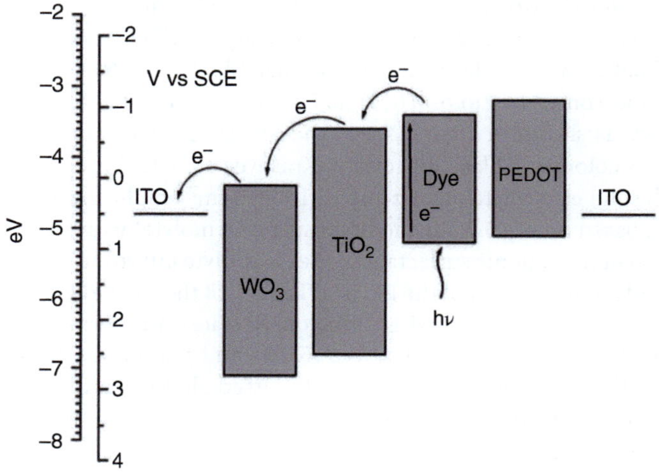

Figure 20.24 Energy level diagram for the all-solid-state photoelectrochromic device [22]. (From Ref. [22].)

De Filpo *et al.* [22] described the fabrication and the characterisation of an all-solid-state PECD using a PEDOT:PSS hole conducting layer. The image of the device is shown in Figure 20.23. In order to improve the electrical conductance, both the dye-TiO_2 and WO_3 layers were properly doped with single-wall carbon nanotubes (SWCNT) bearing COOH groups. A layer of PEDOT:PSS was cast between the dye-TiO_2 layer and the ITO/glass CE, without the use of any fluid component. The energy level of the device is adequate for the operation as shown in Figure 20.24. The coloration and bleaching times were 8 and 30 s, respectively, according to Figure 20.25. The responses were rather faster than other similar device (about 600 s). The reduction for coloration and bleaching time of about 2 orders of magnitude was achieved by using SWCNT. Besides, the carbon nanotube (CNT) functionalised with COOH groups allowed their easier and better dispersion in aqueous suspensions of TiO_2 and WO_3 nanoparticles. The improved miscibility with water and the good incorporation of COOH–SWCNTs in the TiO_2 and WO_3 films improved the uniformity of layers, their electrical

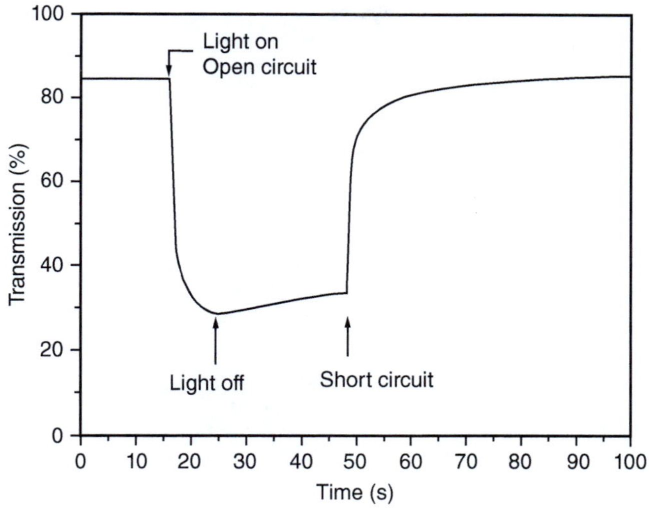

Figure 20.25 Variation of transmittance of the photoelectrochromic device. Under irradiation and open-circuit conditions, the devices become coloured. After switching off the light source and in short-circuit conditions, the devices bleach to the initial transmittance value in a few seconds [22]. (From Ref. [22].)

conductance and the electron transfer through the film and, consequently, remarkably reduced the response times.

20.2.3 Non-Symmetric-Type PECDs (Type III): The Active Area of the Dye-Sensitized TiO_2 Layer is Non-Symmetric to the Electrochromic Layer

In the combined-type PECD (type II), the active area of the dye-sensitized TiO_2 layer is fully covered, which means a thin, semi-transparent TiO_2 layer covers the entire area of the PECD. In type III PECDs, each layer has to be designed for maximising the device transparency without hampering its functionality. The requirement for high transmittance in the bleached state sets limitations on the device performance. The thin TiO_2 film (around 200 nm) cannot absorb neither sufficient amounts of dye nor sufficient photons when sensitised, resulting in a solar cell with a modest V_{OC} that limits the WO_3 speed and depth of coloration. Furthermore, the mesoporous TiO_2 film may cause a decrease of specular transmittance at short wavelengths (below 500 nm) due to scattering losses at the oxide grains. Accordingly, the non-symmetric-type (type III) PECDs are developed to overcome these issues [23]. Instead of a thin transparent TiO_2 film, the partly covered device of the type III PECDs employs a thick and opaque TiO_2 film, which covers only a fraction of the EC film on one end of the device as shown in Figure 20.1.

Leftheriotis *et al.* [23] proposed the non-symmetric-type PECDs with an opaque 5-μm-thick TiO_2 film covering only a fraction of WO_3 area as shown in

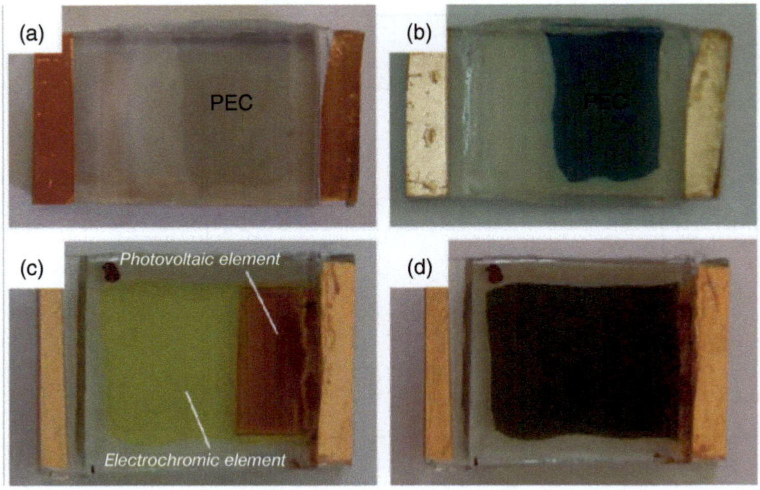

Figure 20.26 Photographs of the PECDs of the fully covered type (a) bleached, (b) coloured and the partly covered design with a 0.34 area ratio (c) bleached and (d) coloured [23]. (From Ref. [23].)

Figure 20.26. It was found that the partly covered devices were transparent in the visible light region (with luminous transmittance up to 51%). It also exhibited rapid coloration (around 3 min) upon solar irradiation in open-circuit conditions and achieved a considerable transmittance modulation (up to 10.2 : 1). Various ratios of TiO_2 to WO_3 area were tried, ranging from 0.23 to 0.48. Reduction of the area ratio decreased the speed of coloration but did not affect either the coloration homogeneity or the end coloration result. The use of N3 ruthenium dye has led to substantial photocoloration efficiencies (up to four times more than those achieved with Rhodamine B) and improved the durability of the devices.

Cannavale *et al.* [24] developed a specifically bifunctional counter electrode by depositing a C-shaped Pt frame which bounded a square region occupied by a WO_3 film onto a transparent conductive substrate. Figure 20.27 shows the PECD images. These two regions were electrically separated to make possible distinct operations on one or both of the available circuits. This novel concept made it possible to achieve a smart and fast-responsive control of the optical transmittance and relatively high photovoltaic conversion efficiency. In addition, they investigated the effect of the electrolyte composition on both photoelectrochromic and photovoltaic performances of such devices by systematically tuning the iodide content in the electrolyte. The best result was obtained by filling the cell with an iodine concentration of 0.005 M: a coloration efficiency of 61.10 $cm^2 C^{-1}$ at a wavelength of 780 nm and, at the same time, a photovoltaic conversion efficiency of 6.55% were achieved.

Cannavale *et al.* [25] fabricated devices that underwent a complete opto-electrical characterisation and the results obtained were employed as an input for

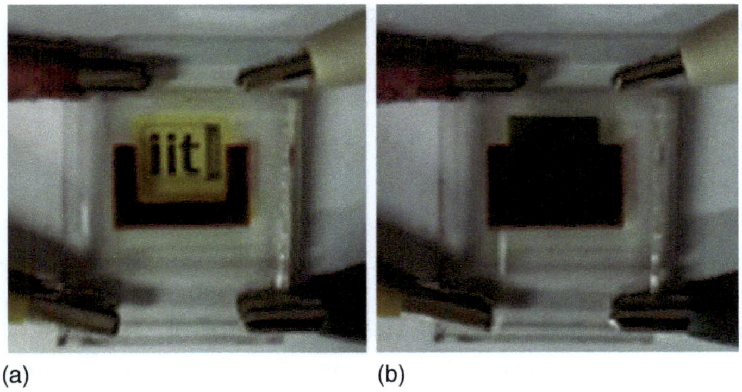

Figure 20.27 Images showing the C-shape partially covered PECD in the bleached (a) and coloured (b) state [24]. (From Ref. [24].)

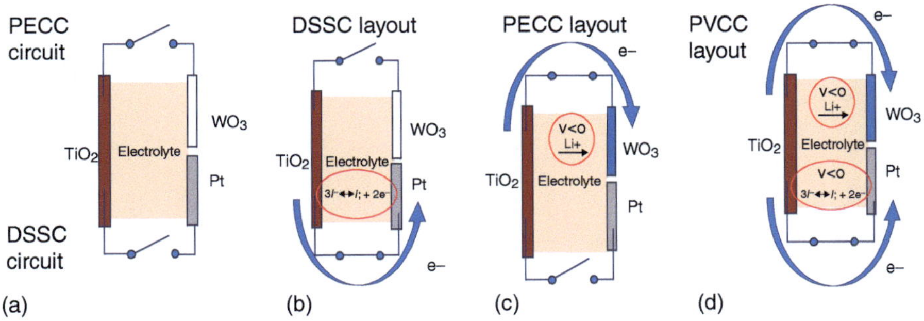

Figure 20.28 Schematic configuration of the PECD. (a) Fundamental configuration: DSSC and PECD open-circuit; (b) DSSC configuration; (c) PECD configuration and (d) PECD configuration [26]. (From Ref. [26].)

the simulation of building-integrated multifunctional windows. This multidisciplinary activity aimed at achieving relevant feedbacks from the simulation of real large area devices in order to adjust and even direct further experimental efforts, before reaching the production phase. Devices having different electrochromic capacitances were used and the optical measurements became useful inputs for the simulation task. The PECD showed a power peak of 4.22 mW cm^{-2} at the maximum power point and a smart modulation of optical transmittance of 50.16% (at 700 nm). Simulations of natural light penetration in office buildings showed that the integration of PECD in traditional windows could dramatically increase indoor visual comfort. Malara *et al.* [26] reported the results of a detailed electrochemical analysis aiming at investigating the electrochemical behaviour of these complex PECDs as shown in Figure 20.28. They investigated and provided a reasonable physical interpretation to the observed phenomena for the effect of Li^+ ions displacement during the coloration of the WO_3 on the performances of the photovoltaic unit.

20.2.4
Parallel-Type PECDs: Where the Dye-Sensitized TiO_2 Layer is Parallel and Separated with the Electrochromic Layer. The Electrolytes for Both Layers are Different for Their Optimal Performance

Apart from the aforementioned three different types of PECDs, there is another parallel-type PECD with an integrated photovoltaic device and an ECD. Shizukuishi *et al.* [27] proposed a solid-state ECD and a photoconductive amorphous silicon (a-Si) thin-film structure, that is, glass|ITO|a-Si|Cr_2O_3|$Cr_2O_3 \cdot V_2O_5$|WO_3|Au, as a parallel-type ECD, as shown in Figure 20.29. Deb *et al.* [28] developed a device employing a wide-band-gap a-$Si_{1-x}C_x$|H n−i−p (photovoltaic) PV cell as a semi-transparent power source and an Li_yWO_3|$LiAlF_4$|V_2O_5 EC device as an optical-transmittance modulator as shown in Figure 20.30. The EC device is deposited directly onto the top of a PV cell coating on a glass substrate. The a-$Si_{1-x}C_x$|H n−i−p PV cell has a gap of 2.5 eV and a transmittance

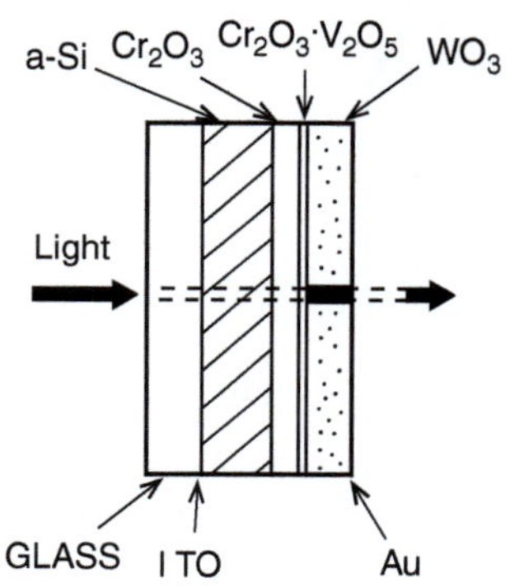

Figure 20.29 Illustration of a parallel-type PECD [27]. (From Ref. [27].)

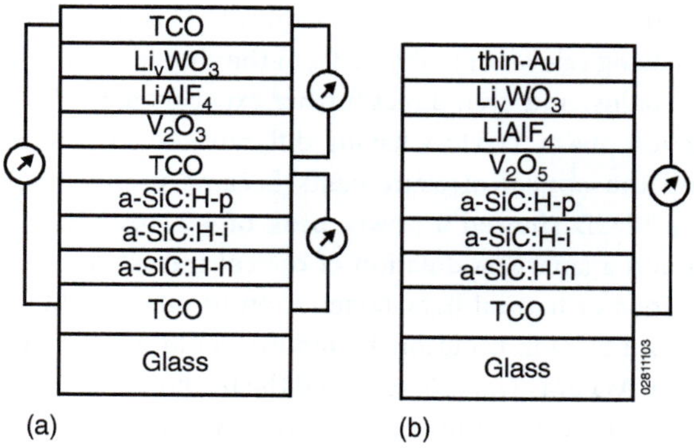

Figure 20.30 (a) Target structure of a PV-ECD device. The upper half is the ECD and the lower half is the PV. The middle contact permits battery and user control. (b) Structure of devices. When the top and the bottom conductors are connected, the device darkens [28]. (From Ref. [28].)

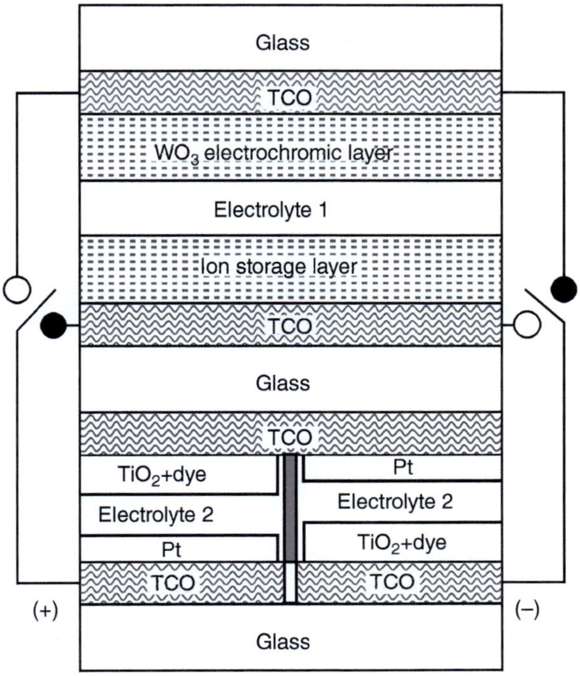

Figure 20.31 Schematic diagram of tandem parallel PECD device [29]. (From Ref. [29].)

of 60–80% over a large portion of the visible light spectrum. The prototype of 16 cm^2 PV-EC device modulated the transmittance by more than 60% over a large portion of the visible spectrum. The colouring and bleaching times of the EC device are approximately 1 min under normal operating conditions (±1 V).

Ahn *et al.* [29] proposed a tandem DSSC-EC device by using two-faced transparent conducting oxide-coated substrates, as shown in Figure 20.31. To supply sufficient voltage to drive the EC devices, two series-connected semi-transparent DSSCs were fabricated with 7-μm-thick, dye-adsorbed TiO_2 and 4-nm-thick Pt layers. The two series-connected, semi-transparent DSSCs had an V_{OC} and J_{SC} of about 1.35 V and 3.96 mA cm^{-2}, respectively, at 1 sun. The tandem DSSC-EC devices showed an optical density difference of 1.2 at 750 nm and reasonable response times of about 60 and 45 s during the colouring and bleaching processes, respectively. This was an indication that the two series-connected, semi-transparent DSSCs could be used as the power sources in the tandem photovoltaic-powered EC devices. Xie *et al.* [30] also proposed an integrated DSSC-EC device as the energy saving and storage application shown in Figure 20.32. Huang *et al.* [31] proposed an innovative concept of a solution-type photovoltaic electrochromic (PV-EC) device as shown in Figure 20.33. The device included a semi-transparent silicon thin-film solar cell (Si–TFSC) substrate, an electrochromic solution and a transparent non-conductive substrate, wherein the

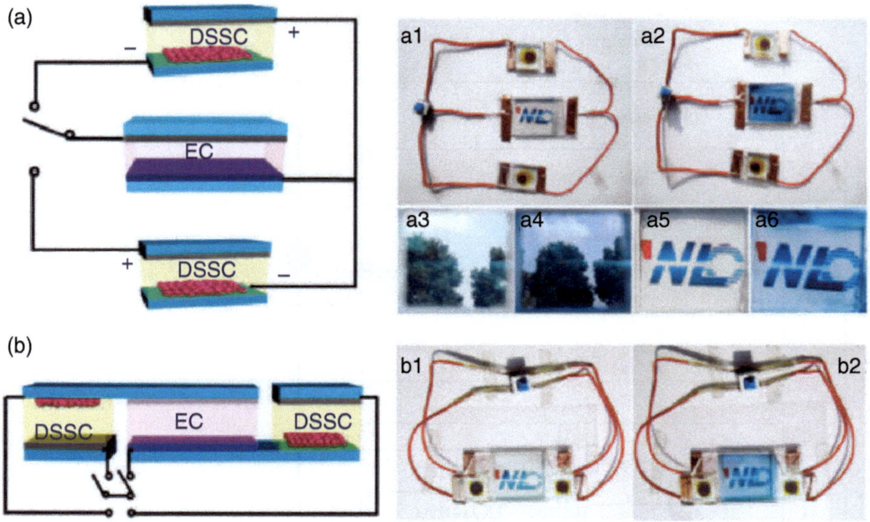

Figure 20.32 Photographs of a DSSC-driven EC smart window under illumination of sunlight: circuits of (a) the independent device (type I) and (b) the assembled device (type II) [30]. (From Ref. [30].)

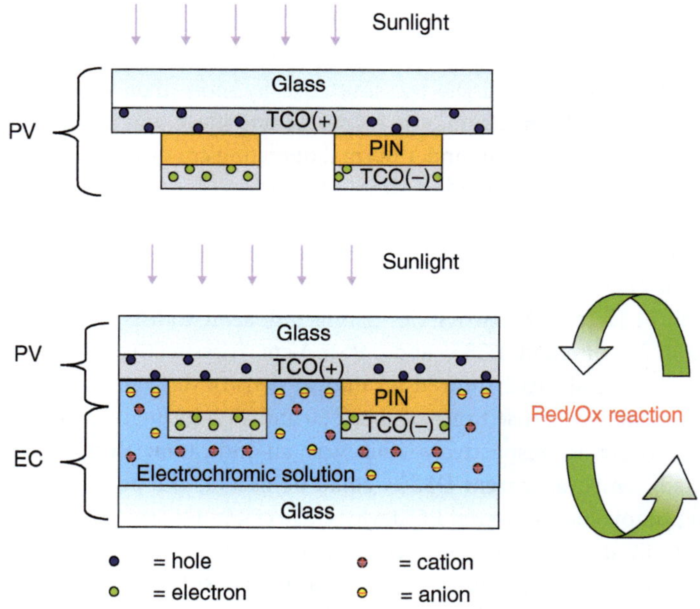

Figure 20.33 Working principle of solution-type PV-EC device. (From Ref. [31].)

electrochromic solution was located between the transparent non-conductive substrate and the Si–TFSC substrate. The solution–type PV–EC device had at least the following advantages: (i) the electrodes in the PV–EC device were planarly distributed in the whole semi-transparent Si–TFSC substrate to create a uniform electric field, which makes large-area PV–EC module application feasible; (ii) since the PV–EC device could be driven with low voltage and low current, the demand for photopotential and photocurrent produced by the Si–TFSCs was lowered and thus the thickness of an intrinsic layer in each of the Si–TFSCs could be thinned, which increases the transmittance of the device; and (iii) the electrical power generated by the PV–EC module could be controlled by an additional output switch layout coupled with the Si–TFSC. When illuminated by sunlight, a portion of electronic current produced by a Si–TFSC was converted into ionic current to cause colour changing of the PV–EC device, while the monolithically integrated Si–TFSC module generated electricity to a connected load. In view of photoelectric conversion and optical modulation properties, the PV–EC device could function both as solar cell module and as self-powered smart glass, which had great advantages in green energy applications.

20.2.5 Prospects

The advantages of photoelectrochromic windows, which combine energy saving and producing functions, seem superior to simple electrochromic or photochromic windows; however, more efforts are still required in this emerging field. The current status of the PECDs still focuses on the research and development stage, but its subsequent market and business opportunities have good potential in the near future. Developments and uses of new materials are undoubtedly one of the major ways to improve the performance of various components in the PECDs. The semiconductor metal oxide, dye molecules, EC materials and the electrolytes all play important key roles in the PECDs. For every type of PECD, the internal components require two transparent conductive oxide (TCO) substrates, which are ITO (tin-doped indium oxide) or FTO (fluorine-doped tin oxide)-coated substrates. However, indium is a rare element leading to a quite high cost compared to the FTO-coated substrate. Therefore, numerous studies attempt to look for alternative materials, such as zinc oxide (ZnO) and Aluminum Doped Zinc Oxide (Al:ZnO) (AZO) (Al: ZnO), which are considered promising new TCO materials. In addition, flexible PECDs with conductive plastic substrate are also important in practical usage, while the reported performances, that is, ΔT, T_d and T_b, are not as good as those of the PECDs with rigid conductive substrate. Hence, novel processes such as utilising low temperature conditions (<150 °C) to fabricate the PECDs provide possibilities for approaching ideal devices in future applications. Furthermore, the costly dye molecules are a very important component in PECDs. Most of the dyes employed in PECDs contain the rare ruthenium material; therefore, using an organic dye without rare materials is also an important way for designing cost-effective PECDs. In order to improve the durability of the PECD, the thermal stability, light stability and hydrophobic ability of the dye molecules should be verified. Sealing agents such as inert thermal glues and UV glues, which can stand high temperature and high humid conditions, should be carefully selected and tested. As for the electrolytes, the developments of the quasi-solid-state and solid-state electrolytes without leakage concerns have become a mainstream direction of lots of electrochemical devices. For the electrochromic materials, despite there being a variety of colours to choose, the manufacturing process and photoelectrochemical reactions compatibility should be taken into account. The vast majority of systems currently use WO_3 as the electrochromic layer for the PECDs. Conjugated conductive polymers may play a prevailing role in the near future, once the high cost, lower stability and poor lifetime of them are overcome with efforts from material chemists. As a summary, with the long-term usage and well-performed characteristics, PECDs could be an attractive and efficient energy conversion/saving architecture in the developments of human technologies.

References

1. Bechinger, C., Ferrere, S., Zaban, A., Sprague, J., and Gregg, B.A. (1996) Photoelectrochromic windows and displays. *Nature*, **383**, 608–610.
2. Gregg, B.A. (1997) Photoelectrochromic cells and their applications. *Endeavour*, **21**, 52–55.
3. Wu, J.J., Hsieh, M.D., Liao, W.P., Wu, W.T., and Chen, J.S. (2009) Fast-switching photovoltachromic cells with tunable transmittance. *ACS Nano*, **3**, 2297–2303.
4. Jiao, Z., Song, J.L., Sun, X.W., Liu, X.W., Wang, J.M., Ke, L., and Demir, H.V. (2012) A fast-switching light-writable and electric-erasable negative photo-electrochromic cell based on Prussian blue films. *Sol. Energy Mater. Sol. Cells*, **98**, 154–160.
5. Pichot, F., Ferrere, S., Pitts, R.J., and Gregg, B.A. (1999) Flexible solid state photoelectrochromic windows. *J. Electrochem. Soc.*, **146**, 4324–4326.
6. Hechavarría, L., Mendoza, N., Rincón, M.E., Campos, J., and Hu, H. (2012) Photoelectrochromic performance of tungsten oxide based devices with PEG-titanium complex as solvent-free electrolytes. *Sol. Energy Mater. Sol. Cells*, **100**, 27–32.
7. Li, Y., Hagen, J., and Haarer, D. (1998) Novel photoelectrochromic cells containing a polyaniline layer and a dye-sensitized nanocrystalline TiO_2 photovoltaic cell. *Synth. Met.*, **94**, 273–277.
8. Yu, X., Li, Y., Zhu, N., Yang, Q., and Kalantar-Zadeh, K. (2007) A polyaniline nanofibre electrode and its application in a self-powered photoelectrochromic cell. *Nanotechnology*, **18**, 015201.
9. Liao, J.Y. and Ho, K.C. (2005) A photo-electrochromic device using a PEDOT thin film. *J. New Mater. Electrochem. Syst.*, **8**, 37–47.
10. Yang, S., Zheng, J., Li, M., and Xu, C. (2012) A novel photoelectrochromic device based on poly(3,4-(2,2-dimethylpropylenedioxy)thiophene) thin film and dye-sensitized solar cell. *Sol. Energy Mater. Sol. Cells*, **97**, 186–190.
11. Hsu, C.Y., Lee, K.M., Huang, J.H., Justin Thomas, K.R., Lin, J.T., and Ho, K.C. (2008) A novel photo-electrochromic device with dual application based on poly(3,4-alkylenedioxythiophene) thin film and an organic dye. *J. Power Sources*, **185**, 1505–1508.
12. Wu, C.H., Hsu, C.Y., Huang, K.C., Nien, P.C., Lin, J.T., and Ho, K.C. (2012) A photoelectrochromic device based on gel electrolyte with a fast switching

rate. *Sol. Energy Mater. Sol. Cells*, **99**, 148–153.
13. Hauch, A., Georg, A., Baumgärtner, S., Opara Krašovec, U., and Orel, B. (2001) New photoelectrochromic device. *Electrochim. Acta*, **46**, 2131–2136.
14. Leftheriotis, G., Syrrokostas, G., and Yianoulis, P. (2013) Photocoloration efficiency and stability of photoelectrochromic devices. *Solid State Ionics*, **231**, 30–36.
15. Hauch, A., Georg, A., Opara Krašovec, U., and Orel, B. (2002) Comparison of photoelectrochromic devices with different layer configurations. *J. Electrochem. Soc.*, **149**, H159–H163.
16. Georg, A., Opara Krašovec, U., and Wittwer, V. (2005) Rate-determining processes in photoelectrochromic devices. *J. New Mater. Electrochem. Syst.*, **8**, 327–338.
17. Leftheriotis, G., Syrrokostas, G., and Yianoulis, P. (2010) Development of photoelectrochromic devices for dynamic solar control in buildings. *Sol. Energy Mater. Sol. Cells*, **94**, 2304–2313.
18. Krašovec, U.O., Georg, A., Wittwer, V., Luther, J., and Topič, M. (2004) Performance of a solid state photoelectrochromic device. *Sol. Energy Mater. Sol. Cells*, **84**, 369–380.
19. Krašovec, U.O., Topič, M., Georg, A., and Dražič, G. (2005) Preparation and characterisation of nano-structured WO_3-TiO_2 layers for photoelectrochromic devices. *J. Sol-Gel Sci. Technol.*, **36**, 45–52.
20. Georg, A. and Opara Krašovec, U. (2006) Photoelectrochromic window with Pt catalyst. *Thin Solid Films*, **502**, 246–251.
21. Georg, A., Graf, W., and Wittwer, V. (2008) Switchable windows with tungsten oxide. *Vacuum*, **82**, 730–735.
22. De Filpo, G., Mormile, S., Nicoletta, F.P., and Chidichimo, G. (2010) Fast, self-supplied, all-solid photoelectrochromic film. *J. Power Sources*, **195**, 4365–4369.
23. Leftheriotis, G., Syrrokostas, G., and Yianoulis, P. (2012) Partly covered photoelectrochromic devices with enhanced coloration speed and efficiency. *Sol. Energy Mater. Sol. Cells*, **96**, 86–92.
24. Cannavale, A., Manca, M., Malara, F., De Marco, L., Cingolani, R., and Gigli, G. (2011) Highly efficient smart photovoltachromic devices with tailored electrolyte composition. *Energy Environ. Sci.*, **4**, 2567–2574.
25. Cannavale, A., Fiorito, F., Resta, D., and Gigli, G. (2013) Visual comfort assessment of smart photovoltachromic windows. *Energy Build.*, **65**, 137–145.
26. Malara, F., Cannavale, A., and Gigli, G. (2013) Effect of lithium intercalation on the photovoltaic performances of photovoltachromic cells. *Prog. Photovoltaics Res. Appl.* doi: 10.1002/pip.2422
27. Shizukuishi, M., Shimizu, I., and Inoue, E. (1981) Application of amorphous silicon to WO_3 photoelectrochromic device. *Jpn. J. Appl. Phys.*, **20**, 2359–2363.
28. Deb, S.K., Lee, S.H., Edwin Tracy, C., Roland Pitts, J., Gregg, B.A., and Branz, H.M. (2001) Stand-alone photovoltaic-powered electrochromic smart window. *Electrochim. Acta*, **46**, 2125–2130.
29. Ahn, K.S., Yoo, S.J., Kang, M.S., Lee, J.W., and Sung, Y.E. (2007) Tandem dye-sensitized solar cell-powered electrochromic devices for the photovoltaic-powered smart window. *J. Power Sources*, **168**, 533–536.
30. Xie, Z., Jin, X., Chen, G., Xu, J., Chen, D., and Shen, G. (2014) Integrated smart electrochromic windows for energy saving and storage applications. *Chem. Commun.*, **50**, 608–610.
31. Huang, L.M., Hu, C.W., Liu, H.C., Hsu, C.Y., Chen, C.H., and Ho, K.C. (2012) Photovoltaic electrochromic device for solar cell module and self-powered smart glass applications. *Sol. Energy Mater. Sol. Cells*, **99**, 154–159.

Appendix
Definitions of Electrochromic Materials and Device Performance Parameters

Roger J. Mortimer, Paul M. S. Monk, and David R. Rosseinsky

The following definitions are invaluable when comparing the performance parameters of electrochromic materials – alone or incorporated within devices – between different laboratories. Some compilations of data are available in Refs. [1, 2].

A.1
Contrast Ratio *CR*

For electrochromes (studied singly or within devices), or for a system involving a colour change, a performance parameter is required that quantifies the intensity change between the bleached and coloured states, or between two coloured states. The 'contrast ratio' *CR* quantifies the ratio of the luminance of the brightest colour against that of the darkest colour. The *CR* is commonly employed to measure the intensity of colour formed electrochemically in an electrochromic material or electrochromic device (ECD), as seen by the human eye, Eq. (A.1),

$$CR = \left(\frac{R_o}{R_x}\right) \tag{A.1}$$

for a reflective system, where R_x is the intensity of light reflected diffusely through the coloured state of an electrochromic material or ECD, and R_o is the intensity reflected similarly but from a non-shiny white card [3]. *CR* values should best be quoted at a specific wavelength – usually at the wavelength of maximum absorbance, λ_{max}, of the coloured form. For electrochromic materials/ECDs operating in transmission mode, Eq. (A.2) applies,

$$CR = \left(\frac{T_b}{T_c}\right) \tag{A.2}$$

where T_b is the transmittance in the bleached form, and T_c is the transmittance in the coloured form. A high *CR* is desirable for display applications, so that images and text may be viewed easily, although lower values may be useful for depth and shadow effects.

Electrochromic Materials and Devices, First Edition.
Edited by Roger J. Mortimer, David R. Rosseinsky and Paul M. S. Monk.

When quoting *CR* values, it is important to explain the details of the experimental conditions. Many of the differences in published values can be attributed to such variant conditions during measurement.

Quantification of the colour contrast between electrochromic colour states may alternatively be expressed as the change in absorbance, ΔA, or the percentage change between the transmittance of the bleached and coloured forms, $\Delta\%T$ [4].

A.2 Response Time τ

The response time τ is the time required for an electrochromic material or ECD to change from its bleached to its coloured state (or vice versa) or between two distinct coloured states. It is generally unlikely that $\tau_{(\text{coloration})} = \tau_{(\text{bleach})}$. There are few reliable response times in the scientific literature since there is no consistency in the reporting and determination of cited data and especially in the way different kinetic criteria are involved when determining τ. For example, τ may represent the time required for some fraction of the colour (defined or arbitrary) to form or be bleached, or it may relate to the time required for an amount of charge (again defined or arbitrary) to be consumed in forming or bleaching colour at the electrode of interest. Not all ECD applications would require a rapid response time. Indeed, for electrochromic glazing, it is probably beneficial to have inherent 'inertia' in the switching rather than trying to follow rapidly changing conditions (Professor John Mardaljevic, Loughborough University, personal communication, 2015). For such 'smart' windows, there is also the matter of control rather than the glazing technology in isolation. There have, however, been instances where users were frustrated by the slow electrochromic response when a glare condition suddenly appears (Professor J. Mardaljevic, Loughborough University, personal communication, 2015).

A.3 Write–Erase Efficiency

The write–erase efficiency of an electrochromic material or ECD is the fraction (or percentage) of the originally formed coloration that can be subsequently electro-bleached or vice versa. The efficiency must approach 100% for a successful ECD, which is stringent test of design and construction.

A.4 Cycle Life

An adjunct to the write–erase efficiency is the ECDs cycle life, which represents the number of write–erase cycles that can be performed by the electrochromic material or ECD before any significant extent of degradation has occurred. The cycle life is therefore one of the principal experimental measures of electrochrome durability.

Since ECDs are usually intended for use in windows or data-display units, deterioration is best gauged by the human eye and with the same illumination, environment and cell driving conditions that would be employed during normal cell operation. While it may seem obvious that the cycle life should be cited this way, many tests of cell durability in the scientific literature involve cycles of much shorter duration than the ECD response time τ. Such partial tests are clearly of dubious value, but studies of cycle life are legion. Some researchers have attempted to address this problem of variation in severity of the cycle test by borrowing terminology [5] devised for the technology of secondary battery discharge and describing a write–erase cycle as 'deep' or 'shallow' (i.e. the cycle length being greater than τ or less than τ, respectively). As for a secondary battery, the cycle life would be highly dependent on charge and discharge rates, cut-off limits, depth of discharge, self-discharge rate and service temperatures. It also depends on whether the driving potential is constant or pulsed.

Maximising the cycle life is an obvious aim of all ECD fabrication. A working minimum of about 10^5 cycles is often stipulated.

A.5
Coloration Efficiency η

The amount of electrochromic colour formed or bleached by the charge consumed is characteristic of an electrochromic material or ECD. Its value depends on the wavelength chosen for study. The optimum value is the absorbance formed per unit charge density measured at the λ_{max} of the optical absorption band. The coloration efficiency η is defined according to Eq. (A.3),

$$\Delta A = \eta Q \quad \text{(A.3)}$$

where ΔA is the absorbance change on passing a charge density of Q. A graph of A against Q accurately gives η as the gradient. The ideal electrochromic material or ECD would exhibit a large absorbance change, giving rise to a high η. A general method for the consistent measurement of *composite* coloration efficiencies (CCEs) has been proposed [6], where chronocoulometric/chronoabsorptometric measurements are made in parallel, with CCEs being calculated at specific percentage transmittance changes, such as 90%, 95% and 98%.

References

1. Monk, P.M.S., Mortimer, R.J., and Rosseinsky, D.R. (1995) *Electrochromism – Fundamentals and Applications*, VCH, Weinheim.
2. Monk, P.M.S., Mortimer, R.J., and Rosseinsky, D.R. (2007) *Electrochromism and Electrochromic Devices*, Cambridge University Press, Cambridge.

3. Faughnan, B.W. and Crandall, R.S. (1980) in *Display Devices* (ed. J.L. Pankove), Springer-Verlag, Berlin, pp. 181–211.
4. Sialvi, M.Z., Mortimer, R.J., Wilcox, G.D., Teridi, A.M., Varley, T.S., Wijayantha, K.G.U., and Kirk, C.A. (2013) Electrochromic and colorimetric properties of nickel(II) oxide thin films prepared by aerosol-assisted chemical vapor deposition. *ACS Appl. Mater. Interfaces*, **5**, 5675–5682.
5. Vincent, C. A., with Bonino, F., Lazzari, M. and Scrosati, B., *Modern Batteries: An Introduction to Electrochemical Power Sources*, Edward Arnold, London, 1984, p. 57.
6. Gaupp, C.L., Welsh, D.M., Rauh, R.D., and Reynolds, J.R. (2002) Composite coloration efficiency measurements of electrochromic polymers based on 3,4-alkylenedioxythiophenes. *Chem. Mater.*, **14**, 3964–3970.

Index

Electrochromic Materials and Devices, First Edition.
Edited by Roger J. Mortimer, David R. Rosseinsky and Paul M. S. Monk.